MULTIDIMENSIONAL GLOBAL CHANGE

WILEY-PRAXIS SERIES IN REMOTE SENSING

Series Editor: David Sloggett, M.Sc., Ph.D.

Co-Director, Dundee Centre for Coastal Zone Research, UK

Deputy Chairman, Anite Systems, UK

This series aims to bring together some of the world's leading researchers working in the forefront of the analysis and application of remotely sensed data and the infrastructure required to utilize the data on an operational basis. A key theme of the series is monitoring the environment and the development of sustainable practices for its exploitation.

The series makes an important contribution to existing literature encompassing areas such as: theoretical research; data analysis; the infrastructure required to exploit the data; and the application of data derived from satellites, aircraft and *in situ* observations. The series specifically emphasizes research into the interaction of elements of the global ecosystem publishing high-quality material at the forefront of existing knowledge. It also provides unique insights into examples where remotely sensed data is combined with Geographic Information Systems and high-fidelity models of the physical, chemical and biological processes at the heart of our environment to provide operational applications of the remotely sensed data.

Aimed at a wide readership, the books will appeal to professional researchers working in the field of remote sensing, potential users of the information and data derived from the application of remote sensing techniques, and postgraduate and undergraduate students working in the field..

For further details of the books listed below and ordering information, why not visit the Praxis Web Site at http://www.praxis-publishing.co.uk

HANDBOOK OF SEAFLOOR SONAR IMAGERY
Philippe Blondel and Bramley J. Murton, Southampton Oceanography Centre, Southampton, UK

EARTHWATCH: The Climate from Space
John E. Harries, Professor of Earth Observation, Imperial College, London, UK

GLOBAL CHANGE AND REMOTE SENSING
Kirill Ya. Kondratyev, Research Centre for Ecological Safety, Russian Academy of Sciences, St Petersburg, Russia; A. A. Buznikov, Electrotechnical University of St Petersburg, Russia; O. M. Pokrovsky, Main Geophysical Observatory, St Petersburg, Russia

HIGH LATITUDE CLIMATE AND REMOTE SENSING
Kirill Ya. Kondratyev, Research Centre for Ecological Safety, Russian Academy of Sciences, St Petersburg, Russia; O.M. Johannessen, Nansen Environment and Remote Sensing Centre, Bergen, Norway; V.V. Melentyev, Nansen International Environmental Remote Sensing Centre, St Petersburg, Russia

MULTIDIMENSIONAL GLOBAL CHANGE
Kirill Ya. Kondratyev, Research Centre for Ecological Safety, Russian Academy of Sciences, St Petersburg, Russia

REMOTE SENSING AND GEOGRAPHIC INFORMATION SYSTEMS: Geological Mapping, Mineral Exploration and Mining
Christopher A. Legg, United Kingdom Overseas Development Administration, Forest and Land Use Mapping Project, Forest Department, Colombo, Sri Lanka

SATELLITE OCEANOGRAPHY: An Introduction for Oceanographers and Remote-sensing Scientists
Ian S. Robinson, Department of Oceanography, University of Southampton, UK

GEOGRAPHIC INFORMATION FROM SPACE: Processes and Applications of Geocoded Satellite Images
Jonathan Williams, Consultant, Space Division, Logica plc, Leatherhead, UK

Forthcoming titles

LIMNOLOGY AND REMOTE SENSING: A Contemporary Approach
Kirill Ya. Kondratyev, *et al.*, Research Centre for Ecological Safety, Russian Academy of Sciences, St Petersburg, Russia

PASSIVE MICROWAVE REMOTE SENSING OF OCEANS
Victor Yu Raizer and Igor V. Cherny, Centre for Program Studies, Russian Academy of Sciences, Moscow, Russia

REMOTE SENSING OF TROPICAL REGIONS
Eugene A. Sharkov, Head of Remote Sensing Laboratory, Space Research Institute, Russian Academy of Sciences, Moscow, Russia

MULTIDIMENSIONAL GLOBAL CHANGE

K. Ya. Kondratyev
Counsellor of the Russian Academy of Sciences
Research Centre for Ecological Safety
St Petersburg, Russia

JOHN WILEY & SONS
Chichester • New York • Weinheim • Brisbane • Singapore • Toronto

PRAXIS

Published in association with
PRAXIS PUBLISHING
Chichester

Published in 1998 by
John Wiley & Sons Ltd
in association with Praxis Publishing Ltd

Wiley Editorial Offices

John Wiley & Sons Ltd, Baffins Lane,
Chichester, West Sussex, PO19 1UD, England

John Wiley & Sons, Inc., 605 Third Avenue,
New York, NY 10158-0012, USA

Wiley-VCH Verlag GmbH, Pappelallee 3,
D-69469 Weinheim, Germany

Jacaranda Wiley Ltd, G.P.O. 33 Park Road, Milton,
Queensland 4001, Australia

John Wiley & Sons (Asia) Pte Ltd, 2 Clementi Loop #02-01,
Jin Xing Distripark, Singapore 12981

John Wiley & Sons (Canada) Ltd, 22 Worcester Road,
Rexdale, Ontario, M9W 1L1, Canada

Library of Congress Cataloguing-in-Publication Data

Kondratyev, K. Ya. (Kirill Yakovlevich)
Multidimensional global change / K. Ya. Kondratyev.
p. cm. -- (Wiley-Praxis series in remote sensing)
Includes bibliographical references.
ISBN 0-471-97179-0
1. Global environmental change. 2. Environmental policy--Social aspects. 3. Environmental policy--Economic aspects.
4. Environmental policy--Russia (Federation) I. Title. II. Series.
GE149.K66 1997
363.738'74--dc21 96-47601
CIP

A catalogue record for this book is available from the British Library

ISBN 0-471-97179-0

Printed and Bound in Great Britain by MPG Books Ltd, Bodmin

Table of contents

Preface

The Second UN Conference on Environment and Development (UNCED), which was held in Rio de Janeiro in June 1992, marked a new stage of understanding environmental problems as inseparable from societal dynamics. It has become much clearer than before that the multidimensional nature of global change should be the principal focus. The "Rio Declaration" adopted by the UNCED emphasized, in particular, the following two principles: "All States and all people shall cooperate in the essential task of eradicating poverty as an indispensable requirement for sustainable development in order to decrease the disparities in standards of living and better meet the needs of the majority of people of the world .... To achieve sustainable development and a higher quality of life for all people, States should reduce and eliminate the unsustainable pattern of production and consumption and promote appropriate demographic policies".

These basic principles have been further discussed during the UN General Assembly Special Session (UNGASS) in New York in June 1997 to analyse what happened after the UNCED. As is well known, the global ecological situation is far from being acceptable. Apart from what has been discussed earlier, most delegates of the UNGASS highlighted freshwater, energy and transport, forests and oceans as issues of new or priority concern. The importance of the cross-sectoral issues of poverty and changing consumption and production patterns has been emphasized because consumption and production patterns remain unsustainably high, official development assistance has actually declined, and developing countries lack essential "green technologies".

In connection with the assessment of progress made since UNCED, the UNGASS President, Ambassador Razali Ismail, pointed out that the meeting proved to be an honest attempt to try and make an appraisal of the results, and of how far we have gone from Rio. There was little attempt to try "to sweep things under the carpet or put a glass over something that's not there". On the other hand, the fact is that the state of global environment continued to deteriorate. Population growth rates have been declining globally, largely as a result of expanded basic education and health care. Overall, however, trends are worsening. Increasing levels of population threaten to exceed the capacity of the global environment to absorb them, increasing potential obstacles to economic and social development in developing countries. Despite a number of positive trends, deep concern has been expressed that the overall trends for sustainable development are worse today than they were in 1992. The issues that proved most difficult to resolve in 1992 are still problematic today. The accelerating pace of globalization, poverty and the growing gap between developed and developing countries are characteristic features of global ecodynamics. The debt situation remains a major constraint on achieving sustainable development. Technology transfer and technology-related investment from public and private sources has not been realized as it was earlier proposed. We are still very far from being able to reach interaction of economic, social and environmental objectives.

That is why the G-77/China made a statement that "while globalization is global, it is not universal in benefits. The world is crying for answers, and this session did not provide them". The UN Secretary-General, Kofi Annan, pointed out in his statement: "The main challenge of the twentieth century was whether the world's people could live together peacefully. The major challenge for the next century will be whether we can feed, clothe and provide for ourselves without destroying the natural resources base on which our existence depends".

In the preparatory process for Rio, the "Earth Charter" was supposed to be the main political statement to emerge from the Earth Summit. This effort has failed and the Rio Declaration on Environment and Development has been accepted instead, being a fairly vague document. A similar process took place in New York. Having failed to reach agreement on a political statement for the Heads of State and Government, the UNGASS Committee on the Whole adopted, in its place, a very general "Statement of Commitment".

The most important and still unsolved problem is an adequate substantiation of priorities in the field of integration of interactive social, economic and environmental objectives. A constructive approach to the solution of this problem has been discussed in the Introduction and Chapter 1 of the book on the basis of the analysis of the present-day global ecological situation and (methodologically) from the viewpoint of the principle of biotic regulation of the environment. A number of key aspects of contemporary global ecodynamics and ecopolitics have been considered in this context, including: global change and population dynamics; the notions of sustainable development and ecological safety (relevant criteria and indicators); social inequality: the "golden billion" and the rest of the world; market and non-market mechanisms of solving ecological problems; and others. An analysis has been made of basic international programmes such as the International Geosphere–Biosphere Programme (IGBP), the International Human Dimensions Programme (IHDP), and the World Climate Research Programme (WCRP).

It has so happened that during recent years the problem of global climate change has been the central theme of international discussions, with special emphasis on the anthropogenic contribution to climate change due to the increase of the greenhouse gases (GHGs) concentrations in the atmosphere. One of the basic documents approved by the UNCED was the Framework Convention on Climate Change (FCCC). Subsequent activities of the Intergovernmental Panel on Climate Change

(IPCC) contributed to relevant scientific assessments, including a controversial statement made in the IPCC Report "Climate Change 1995" that "the balance of evidence suggests a discernible human influence on global climate". Numerous discussions concerning the necessity to reduce GHGs emissions (including those held during the UNGASS) took place, as well as fundamental efforts to assess impacts, adaptation and mitigation of future climate change to analyse economic and social dimensions of climate change. All these and other problems relevant to global climate change have been discussed in detail in Chapter 2. The coupling between climate and ozone changes was also considered in Chapter 2, including a specific problem of the impact of total ozone variations on UV-B solar radiation changes, which are important from the viewpoint of their impacts on man and the biosphere. A survey of satellite remote sensing of minor gas and aerosol components of the atmosphere concluded Chapter 2.

European ecological problems in the context of global change formed the main subject of Chapter 3, based chiefly on the Dobriš Report and some other information sources (a recent monograph by S.G. Inge-Vechtomov, K. Ya. Kondratyev and A.K. Frolov (Eds) entitled "Ecological Problems of Northwestern Russia and Ways of Their Solution", Victoria Publ., St. Petersburg, 1997, 528 pp., is an important source of additional information). An important part of Chapter 3 is a general consideration of the problem of sustainable development indicators.

Problems of socio-economic development and environmental dynamics in China, a country with the largest (and still growing) population in the world which has valuable experience of solving ecological problems, have been discussed in Chapter 4. The concluding part of Chapter 4 is devoted to a survey of specific ideas of solving ecological problems developed by the International Institution of Industrial, Spiritual and Cultural Development under the leadership of Y.Y. Nakano. A brief summary of the ecomonitoring experience in Japan has also been given.

The environmental dynamics of Russia and adjacent territories has been considered in Chapter 5 written by K.S. Losev. Various problems have been discussed in this context, such as the general ecological situation and ecosystems dynamics; the environmental impact of industrial and agricultural activities; atmospheric, water and soil pollution (as well as land degradation); solid and toxic wastes, etc.

The purpose of the Conclusion is to substantiate the priorities of multi-dimensional global change studies and suggest a concept of the "Earth Charter" for further discussions in the course of the accomplishment of the IGBP, IHDP and WCRP.

Introduction

The aggravation of the ecological situation in many regions of the globe and over the entire planet requires greater and greater effort to be devoted to studying the regular features of regional and global ecodynamics. The central aspect of the problem is substantiation of priorities, and it has acquired great importance because the subject of discussion is an extremely complex interdisciplinary "multidimensional" problem (Morais et al., 1995). This has been convincingly demonstrated by the declarative nature of basic documents produced by the Second UN Conference on Environment and Development (UNCED-2) which was held in June 1992 in Rio de Janeiro (Danilov-Danilyan et al., 1994; Kondratyev et al., 1993; State..., 1996). It is the lack of precisely formulated priorities that resulted in the excessive attention on the part of UNCED-2 to the problem of reducing greenhouse gases (GHG) emissions to the atmosphere in order to prevent the potentially disastrous consequences of global climate warming. However, the discussion of the GHG problem proved to be useful because it revealed the socio-economic roots of environmental problems and the inseparability of ecology and policy. It has become clear that it is socio-economic development and environment that should be discussed, and not vice versa, as was the case with UNCED-2.

The conceptual basis of the book is the thesis, which has been soundly substantiated in a number of recent publications, that the main limiting factor of further quantitative development of humankind is biosphere resources, and the basic criterion of ecological well- (or ill-) being is the level of per anima consumption of biospheric resources. Such an assessment is based on the concept of biotic regulation of the environment (Gorshkov, 1990, 1995; Gorshkov et al., 1993, 1996). Hence the most acute social controversy of the contemporary world: the level of biospheric resources consumption in industrially developed countries exceeds by far that in the Third World, and it has already brought about a non-permissible perturbation of the biosphere. This disproportion can be eliminated through the advancement of current technologies (in particular, the "decarbonization" of energy production) and the development of new technologies; however, radical improvement of the situation can only be achieved on the basis of reducing the population number and establishing a new, fair economic order. So, global ecological perspectives can be promising if an optimal choice of acceptable ways of socio-economic development is made. This is what makes the problem multidimensional. When considering global change problems, it is very important to avoid exaggeration. Currently existing opinions vary within a broad range: from most gloomy predictions (e.g., the book by Carson (1994), the first edition of which played, however, a significant role because it stimulated public attention to environmental problems) to encouraging estimations (e.g., Bailey, 1995). Our purpose is to try and avoid the extremes.

1. THE ECOLOGICAL IMPERATIVE

The end of the twentieth century can be characterized as a time of drastic changes in stereotypes and paradigms in assessing the future of civilization. This is the result of historic developments which have stimulated human society as a whole and every individual human being on earth to work out a new mentality. The basic effect of this new mentality is the fact that, owing to the efforts of both the East and the West, the world has been brought to a considerable relaxation of its former ideological, political and military confrontation. This has made it possible to solve the most important ecological problem—a strong decrease in the probability, and hopefully the elimination of potential large-scale nuclear conflict and its catastrophic consequences for the biosphere and humanity. This opens up possibilities not only for the transition from a narrow-minded national or group mentality to global-scale human thinking, but also for stable peaceful development instead of the former military confrontation. Under the circumstances, the leading imperative becomes global ecological safety (Brown et al., 1996; Clayton and Radcliffe, 1996; Cohen, 1995; Glantz, 1994; Holdgate, 1996; Hunt, 1994a,b; Kondratyev, 1990, 1991a–b, 1992a, b, 1993a–b, 1994, 1995a–d, 1996; Leslie, 1996; McMichael, 1995; Pirages, 1996; Reid, 1995; Singh et al., 1994; Simon, 1996; Suzuki et al., 1996; Trzyna, 1995; Turekian, 1996, and many other publications). A radical mitigation of the global military threat makes it possible to better understand the basic problem of the survival of mankind under conditions of a progressing ecological crisis. Beginning with the 1960s, the scientific and, later, the political vocabulary came to be characterized by the frequent mention of such expressions as "ecological calamity", "ecological catastrophe", "ecological problem", "ecological safety". These have become part of the everyday language of mass media. Scientists and politicians concerned about the state of the biosphere use these expressions on a permanent basis. These are not just "fashionable" topics or a product of a propagandistic verbal drive; they are indicators of general anxiety of people who already suffer from the pressure of ecological problems. The subject of ecology often sets aside other problems, and occupies the leading position on both national and international scales. In some areas of the globe, ecological problems have become predominant.

Ecological problems have arisen as a result of natural development of our civilization. The patterns of human behaviour within society and in the interaction with the environment, which were worked out in earlier times when they were capable of ensuring environmental sustainability, have come into conflict with new conditions caused by the scientific and industrial revolution. Under these new conditions, in response to the challenge from environmental change, it has become necessary to form a new pattern of behaviour and new moral preferences, and to revise former ideas of socio-economic development (Bruce et al., 1996; Clark, 1991; Gore, 1993; Gorshkov, 1990, 1995; Gorshkov et al., 1993, 1994, 1996; Kondratyev et al., 1993, 1994, 1996b; McMichael, 1995; Pope, 1993; Reid, 1995; Simon, 1996; Suzuki et al., 1996; Viederman, 1996; Watson et al., 1996, etc.).

After World War II, the world seemed to have entered a phase of constant advancing economic progress. World production increased by a factor of 5 during the postwar period: the additional economic growth for each decade was equal to the production output achieved over the entire period from the start of our civilization to the 1950s. For the period mentioned, the production of crops increased by a factor of 2.6, and the world's population grew from 2.5 to 5.3 billion. The population growth rate increased from 50 to 84 million per year. During the same time period, the production of food per capita increased by a factor of 1.25, whereas the industrial output increased by a factor of 3–4 (State..., 1990, 1996).

A great number of new technologies have been developed. Such new technological systems as electronics and aerospace have been created; efficient medical treatment for many widespread diseases have been found; a new way of life has been formed in a society which can be called "an information society".

However, not every part of the world may be characterized as enjoying rapid economic development. In many developing countries, tragic social consequences have followed the course of industrial revolution. 750 million people on this planet are starving today, while another 1225 million suffer from real poverty. In developing countries, 95 per cent of children are born, live and die in poverty, being deprived of the benefits of scientific and technological achievements. About 2 million people suffer from water shortages and have to use contaminated water; the low quality of water is the cause of various diseases. Outbreaks of cholera and malaria, which were supposed to have been eliminated, have been registered again. New diseases have appeared, the most dangerous of which, so far, has been AIDS. The number of the poor and homeless is growing in developed countries as well. Drug addiction, violence and terrorism are spreading there. The number of technological accidents has increased sharply; the number of genetic and blood diseases grows, as does the incidence of malignant tumours.

Differences between the North and the South with regard to population growth have become more intense. In 1950, about 22 per cent of the global population lived in North America and Europe, and only 9 per cent in Africa. Now the proportion is reversed. The Southern developing countries comprise the major part of the world's population. The debt of this part of the world amounts to \$1.2 trillion, which constitutes one-half of the gross national product of the relevant countries. Poverty and starvation, diseases and general frustration are doubtless fraught with unpredictable conflicts and violence. There are numerous ecological refugees. The shortage of land and water resources is obvious, and this can cause conflicts and provoke wars. Dangerous features of development have appeared in the late 1980s: this has become the first period (since 1950) when the global crop yield failed to increase, when grain production output has dropped, and when the areas of arable and irrigated land per capita have increased. If, according to Adam Smith, "an invisible hand" appears to regulate the population growth problem, this hand is most likely to bring about the destruction of the biosphere followed by starvation and disease.

Forty years of technological progress have had an enormous impact on the environment, since this progress has been based on ruthless exploitation of natural resources. Economic estimations have practically never taken into account the depletion and damage to natural resources, or their direct consequence—a harmful effect on human health. Moreover, the results will be the deprivation of future generations of natural resources, and a long-term impact in the form of biospheric deformation.

Technologies involving the depletion of resources have proved most competitive, both under free market conditions, and within centrally planned economic systems.

The economic progress of the past forty years has proved to be catastrophic for the state of the biosphere: 20 per cent of the upper fertile soil of cultivated land has been lost completely; 20 per cent of tropical forests have been cleared; the process of desertification is continuing at a rate of 6 million ha annually, and deforestation amounts to 11 million ha annually (State..., 1990, 1996). The carbon dioxide concentration in the atmosphere has increased by 13 per cent during the last 40 years; the protective ozone layer is reduced annually by 2 per cent, and every spring an ozone hole appears over the Antarctic, signifying a sharp drop in the ozone level (Kondratyev, 1992a; Kondratyev et al., 1997). Dead lakes and damaged forests have now become inevitable components of the landscape of industrial countries. Most rivers and creeks in Europe and North America remain polluted throughout the entire year. In many cases, pollutants are mostly brought to waterbodies not so much from local sources, as from the atmosphere, owing to dry and wet deposition of industrial gases and aerosols, as well as from cropland runoff; 50 per cent of fertilizers and pesticides are washed out from croplands. Dangerous toxins—heavy metals, pesticide decomposition products, plastics, radionuclides, etc.—are accumulated in soil, in bottom sediments, and in plants and animals. Every day over a hundred species disappear altogether, the process being particularly intensive in tropical forests (Leakey and Lewin, 1995; Watson et al., 1996). On the whole, human beings consume up to 40 per cent of the primary land biological production and affect the environment in different ways, but always negatively. Less developed countries, because of their rapid population growth and technological inadequacy, stimulate the degradation of land, forest and non-renewable resources. Industrial countries possess the bulk of the world's industrial and energy potential; they are the main consumers of the resources (particularly, energy resources) and the main suppliers of pollutants. For example, in 1988 in North America, the carbon dioxide emissions to the atmosphere exceeded those in Africa by a factor of 6, in Latin America by a factor of 2.5, and in South-East Asia by a factor of 9 (State..., 1990, 1996).

The abovementioned tendencies require a revision of the outcome of socio-economic development over the last forty years; these decades should be reassessed as a period of considerable social degradation and serious aggravation of the global ecological crisis.

It is extremely important to realize that decisions in the field of ecological policy are made, as a rule, under the circumstances of considerable uncertainty (Bruce et al., 1996; Watson et al., 1996). This is relevant to links between the supposed policy and its potential ecological consequences, relationships between social priorities and ecodynamics. As noted by Hammitt (1995), two types of uncertainties, relevant to consequences and to value criteria, should be examined in this context.

Uncertainties of consequences appear as a result of the complexity of interactions between physical, chemical, biological, economic and other social processes which determine the efficiency of nature-protective and other measures of securing sustainable development. Some changes in environmental conditions may be considered to be direct consequences of an accepted ecological policy because they directly influence life (this relates, for example, to climate changes), whereas other changes can affect the economy and other aspects of socio-economic development determining the standard of living (for instance, climate change may entail a drop in crop productivity).

Value criteria uncertainties follow from the difficulty of assessing social priorities arising from the variability of environmental circumstances and economy. Many of the results (consequences) of the adopted ecological policy have nothing to do with market criteria, and therefore their economic expediency cannot be precisely characterized in quantitative terms.

In this connection, Hammitt (1995) examined the problem of uncertainty of value criteria in the substantiation of global ecological policy by making use of the following two cases as examples: (a) anthropogenically induced changes in the total ozone content (TOC) in the atmosphere, and (b) climate. Analysis of information obtained by the mid-1980s (the Montreal Protocol has been accepted during this period) and concerning potential TOC changes in the future under the influence of emissions of chlorofluorocarbons (CFC) has revealed such a high level of uncertainty that the Montreal Protocol and subsequent similar agreements should be considered as recommending (warning about potential danger), but not directive. The uncertainty of assessments is so great in this case that a situation cannot be excluded when unlimited emissions of CFC will not affect the global ecodynamics of TOC to any considerable extent. On the other hand, an opposite situation may arise when a catastrophic impact can take place under circumstances of considerably reduced CFC emissions. Still more important is the circumstance that probable TOC differences in the future, due to alternative solutions, will be smaller than possible variations relevant to the choice of a uniform ecological policy. Analysis of climate change problems has brought about a similar conclusion; in addition, the value criteria appear to be even more important in this case.

As has been noted above, all problems (from local-scale to continental-scale) should be considered in the context of global change (Kondratyev and Frolov, 1996), taking account of the interrelationship between socio-economic development and changes taking place in the environment (see References). It is for this reason that we shall first of all dwell on a brief characterization of the present state of global environment and natural resources (Brown et al., 1996; State..., 1990, 1996; Tolba et al., 1992).

Since public concern with regard to the state and evolution prospects of global environment and natural resources grows, regular analysis of the contemporary ecological situation becomes necessary. Relevant data are accumulated and publicized by various governmental and non-governmental, national and inter-national organizations. The UN Environment Programme (UNEP) summary reports hold the leading place (Grigoryev and Kondratyev, 1994; Tolba et al., 1992; Turner, 1993). Following the work by Grigoryev and Kondratyev (1994), we shall present a brief characterization of relevant UNEP documents summarizing the results submitted by numerous international organizations. A tremendous effort has been preparation for the Second UN Conference on Environment and Development (Rio de Janeiro, 3–14 June 1992) (Danilov-Danilyan et al., 1994; Kondratyev et al., 1992a; Kondratyev et al., 1993; Kondratyev et al., 1996b; Kondratyev and Cracknell, 1997).

2. POPULATION

Only 5–7 years ago, any discussion of the urgent problem of growing global overpopulation was carefully avoided. The Second UN Conference on Environment and Development (Rio de Janeiro, June 1992) became a turning point which broke the tradition and emphasized the necessity in a thorough and urgent analysis of global population dynamics and associated implications (Kondratyev, 1993a, b, 1995a–d; Kondratyev and Frolov, 1996; Kondratyev et al., 1992a, 1994). This is why the International Conference on Population and Development (ICPD) which followed UNCED-2 and was held in Cairo, 5–13 September 1994, and collected participants from 183 countries, was totally devoted to these problems. The World Summit for Social Development (Copenhagen, April–May 1995), the Fourth World Conference on Women's Problems (Beijing, September 1995) and the Second UN Conference on Human Settlements (Habitat II) (Istanbul, June 1996) were also focused on demographic aspects. A combined effort within the International Geosphere–Biosphere Programme (IGBP) and the International Human Dimensions Programme (IHDP) devoted to socio-economic aspects of the problem of global change has been an important step forward.

It should be mentioned that a complex approach to the solution of the problem of global change, including an adequate substantiation of the population aspect being a key facet of the entire problem, had already been elaborated in the early 1990s (see Gorshkov, 1990, 1995; Kondratyev, 1990, 1996) in the context of biotic regulation of the global environment.

In October 1993, a very true statement was made during the "Demographic Summit" in New Delhi: "mankind approaches a crucial point with respect to related problems of population, environment and development" (actually this point had been reached earlier (see Grigoryev and Kondratyev, 1996a)).

Polunin and Nazim (1994) emphasized that at present "there emerges the common recognition that the present pattern of resource consumption, and the models of economic growth and development, are the root causes of various manifestations which are threatening the stability of the planet. Rapid population growth is one of these manifestations, which is appearing as a symptom of maldevelopment under an unjust politico-economic system rather than the cause, as has so often been claimed in the past". Though the latter statement is questionable (of course, rapid population growth in developing countries is a result of unfavourable socio-economic development; nevertheless, the interactivity of cause and effect is doubtless), it reveals public concern about current demographic trends.

In this context, another quotation of interest can be presented (Brown, 1996): "The pace of change in our world is speeding up, accelerating to the point where it threatens to overwhelm the management capacity of political leaders. This acceleration of history comes not only from advancing technology, but also from unprecedented world population growth, even faster economic growth, and the increasingly frequent collisions between expanding human demands and the limits of the earth's natural systems".

Data characterizing demographic dynamics are well known; however, we shall present here some most essential figures. The present-day global population number is close to 6 billion people; by the year 2000, another billion people will be added to the figure, and by the year 2050, i.e. during the lifetime of a single generation, the world population will constitute about 8.5 billion. The African population grows at a maximum rate (2.9% per year); in East and West Africa the rate is even higher (3.1%), and in many African states the age of about one half of their population is below 16. The growth rates of the population of western and southern Asia (the Indian subcontinent) are 2.8% and 2.2%, respectively. Relevant figures for Asia and Latin America, taken as a whole, are 1.8% and about 2%, respectively. By the year 2050, the Asian population (approximately 3.2 billion in 1992) will reach the figure of 4.9 billion; the African population will be doubled (1.58 billion instead of 682 million). Relevant estimates for Latin America are 457 million and 701 million people.

Table 1 data (Brown et al., 1996) present information on the global population growth. As mentioned above, the growth of population has been very inhomogeneous geographically. For instance, in some countries, the process is slowing down. Because of the economic decline, a 0.6% population reduction was noted in Russia in 1995. This is the most rapid decline on record for a country not at war. As a result of an excess of deaths over births, the annual loss of population in Russia reached roughly 880 000 people.

The demographic situation in Japan is alarming. The statistics and official reports show that the country's population is expected to plummet by more than one half in the next 100 years from 125 million (as of the end of 1995) to a mere 55 million. In 1995, a Japanese woman was expected to give birth to only 1.43 children during her lifetime, whereas, if the current population is to be maintained, the rate has to be increased to 2.08.

The data of Table 2 (Brown et al., 1996) illustrate the growth of world urban population.

Forecasts of demographic dynamics in individual countries offer even more impressive figures. For instance, in Bangladesh, which is overpopulated as it is, the population number will practically be doubled during the period from 1990 to 2050. In many countries of Africa and the Middle East (e.g., Angola, Congo, Iraq, Jordan, Kenya, Libya, Madagascar, Malawi, Niger, Nigeria, Rwanda, Saudi Arabia, Syria,

Table 1. World population, 1950–95.

Year	Total (billion)	Annual addition (million)	Year	Total (billion)	Annual addition (million)
1950	2.555	37	1979	4.380	77
1955	2.779	53	1980	4.457	77
1960	3.038	41	1981	4.533	81
1965	3.345	70	1982	4.614	81
1966	3.414	69	1983	4.695	80
1967	3.484	71	1984	4.775	81
1968	3.555	74	1985	4.856	83
1969	3.629	75	1986	4.941	87
1970	3.704	78	1987	5.029	88
1971	3.782	77	1988	5.117	88
1972	3.859	77	1989	5.205	90
1973	3.936	76	1990	5.295	86
1974	4.012	74	1991	5.381	88
1975	4.086	73	1992	5.469	88
1976	4.159	73	1993	5.556	88
1977	4.231	73	1994	5.644	88
1978	4.304	76	1995 (preliminary)	5.732	87

Table 2. World urban population, 1950–95

Year	Population (billion)	Share of world population that is urban (per cent)
1950	0.738	29.3
1955	0.864	31.4
1960	1.033	34.2
1965	1.186	35.5
1970	1.353	36.6
1975	1.538	37.7
1980	1.752	39.4
1985	1.994	41.2
1990	2.277	43.1
1995 (preliminary)	2.584	45.2

Tanzania, Zaire) population growth rates markedly exceed 3%, and the total population number is doubled or trebled during the lifetime of a single generation. The severe reality is that some of these countries are already among the poorest in the world, and their population suffers from the deficit of vitally important resources, such as drinking water, etc.

Regrettably, consensus has not been reached thus far concerning this situation. The real truth is that this is not just a problem, but a threat. The opinion that the world is capable of feeding an unlimited number of its inhabitants is still shared by some scientists. But this is a myth, which has been revealed convincingly in the publications of Gorshkov (1995) and Gorshkov, Kondratyev and Losev (1992,1996).

Statistical data published in connection with the "City Summit" (Istanbul, June 1996) demonstrate a tremendous growth of urban population. Within the next few years, humanity, for the first time, will become an urban rather than a rural species. As recently as in 1975, only a third of the world's people lived in towns and cities; by the year 2025, nearly two-thirds of its greatly increased population will do so. Only in the Asia-Pacific region, the urban population will soar to 2.07 billion in 2025, compared with 923 million in 1993.

The most important circumstance is that the problems addressed by previous socio-ecological summits are at their highest level in cities. The contrast between wealth and poverty is at its most acute, population pressures and social dislocation are at their greatest (Lean, 1996). It is undoubtedly clear that runaway urban growth and the economic pull of cities will become the main demographic forces in coming decades.

A very important feature of population growth is the dramatic ageing of populations everywhere (Rousch, 1996). The industrial nations" investments in improved nutrition, medical technology, and public health have resulted in raising the average life expectancy in Europe and the USA from less than 47 years in 1980 to 75.5 in 1993. Japan has achieved even more, leading the world with an average life expectancy of 76.6 years for men and 83 for women. The proportion of people aged 65 and older in Japan is now about 14% (it is, however, no higher than in Western countries). During recent years, the developing countries have been showing similar gains. They now face massive increases in both the numbers and proportions of elderly people who, as a rule, will have no old-age pensions, who no longer can count on help from their families, and who will have access only to the most rudimentary medical care. Each month, the world sees a net gain of 800 000 people over 65, 70% of whom are in the developing world. Because fertility rates are declining in most countries, the growth of older populations is outpacing total population growth.

Brown (1996) points out that the world economy is growing even faster. It has expanded from $4 trillion in output in 1950 to more than $20 trillion in 1995. In just the 10 years from 1985 to 1995, it grew by $4 trillion more than from the beginning of civilization until 1950. As population has doubled since mid-century and the global economy has nearly quintupled in size, the demand for natural resources has grown at a phenomenal rate. Brown (1996) emphasizes in this context:

> "Since 1950, the need for grain has nearly tripled. Consumption of seafood has increased more than four times. Water use has tripled. Demand for the principal rangeland products, beef and mutton, has also tripled since 1950. Firewood demand has tripled, lumber has more than doubled, and paper has gone up sixfold. The burning of fossil fuels has increased nearly fourfold, and carbon emissions have risen accordingly.
>
> These spiraling human demands for resources are beginning to outgrow the capacity of the earth's natural systems. As this happens, the global economy is

Table 3. Refugees receiving UN assistance, 1961-95

Year	Total (million)	Year	Total (million)
1961	1.4	1979	4.6
1962	1.3	1980	5.7
1963	1.3	1981	8.2
1964	1.3	1982	9.8
1965	1.5	1983	10.4
1966	1.6	1984	10.9
1967	1.8	1985	10.5
1968	2.0	1986	11.6
1969	2.2	1987	12.4
1970	2.3	1988	13.3
1971	2.5	1989	14.8
1972	2.5	1990	14.9
1973	2.4	1991	17.2
1974	2.4	1992	17.0
1975	2.4	1993	19.0
1976	2.6	1994	23.0
1977	2.8	1995 (preliminary)	27.4
1978	3.3		

damaging the foundation on which it rests. Evidence of the damage to the earth's ecological infrastructure takes the form of collapsing fisheries, falling water tables, shrinking forests, eroding soils, dying lakes, crop-withering heat waves, and disappearing species."

An important feature of the contemporary demographic dynamics is a tremendous growth of migration shown in Table 3 (Brown et al., 1996). The main, and alarming tendency consists in the steady population number growth on the planet, in spite of the fact that the growth rate decelerates. According to the prediction of the United Nations Organization of 1984, there will be 10 billion inhabitants on the Earth by the year 2010; more recent forecasts (1990) yield an even greater figure—11 billion. In this connection, the problem of limiting the growth of population is vital for many countries. China has been the most successful state in this respect. It is generally known that China is the most populated country in the world. In 1990, its population was 1 135 495 000 people. Owing to a steady state policy directed towards restricting population growth, the mean birth rate in the country dropped from six children per woman in the 1950s to 2.2 in 1990. More precise global demographic information can be found in documents of the UN Conference on Population and Development which took place in September 1994 (Costanza, Norton and Haskell, 1993; Döös, 1994b; Kondratyev, Peña and Galindo 1997; Miller, 1994; Pearce and Barbier, 1994; Turner, 1993; Welt im Wandel..., 1994).

Urbanization is a most important problem of our times. During 20 years from

1970 to 1990, the population of cities increased very considerably, and its share in the Earth's population enlarged from 37% to 43%; by the end of the century, it may grow to 47%. In the year 1990, there were 69 cities with a population of over 3 million people (in 1970 there were 37), and 13 cities with a population of over 10 million people (in 1970 there were 4). The biggest city of the world became Tokyo (20 520 000 people).

Today the largest number of cities with a population of over 3 million people are situated in the developing countries (45 out of 69). It is in such states that the population grows particularly fast, sometimes in an explosive pattern; among them are Manila-Kason (from 3.6 to 8.6 million), Tehran (from 3.29 to 9.42 million), Greater Bombay (from 5.98 to 11.13 million), São Paulo (from 8.22 to 18.42 million), Mexico (from 9.12 to 19.37 million). The city of Medan in Indonesia grew by a factor of 5 (from 0.61 to 3.0 million people). This tendency to an abrupt increase in the share of urban population is first of all characteristic for the countries of Africa, as well as Asia, Central and South America. In many of them, the share of urban population tripled; for instance, in Mauritania it rose from 14% to 42%, in Yemen from 8% to 24%, in Guinea-Bisau from 18% to 47%, and in some states it increased by a factor of 4 (for example, in Thailand from 7% to 32%, in Mozambique from 6 to 24, and in Afghanistan from 5% to 21%).

The concentration of great numbers of people in cities is in many cases accompanied by certain negative consequences (a deficit of drinking water, antisanitary living conditions, etc., especially in the suburbs of big cities—the so-called "bidonvilles"). On the other hand, in the developed countries of Europe and North America, a far slower growth in the share of urban population has been observed during the last 20 years: in the most industrially developed European countries, the growth constituted 3–5%, in other states 10–18% at most. It seems symptomatic that the population growth has stabilized during 1970–90 in a number of the biggest cities of the world. In particular, the number of city dwellers increased in Philadelphia from 4.05 to 4.23 million people, in Chicago from 6.76 to 6.89, and in Paris from 8.34 to 8.75 million people. The number of Londoners and New Yorkers even decreased.

One of the most important factors of normal vital activity of population is its provision with clean water and sanitary services. On the whole in the world, the number of people provided with both is steadily increasing. The worst situation with drinking water is in Africa, and in particular, in Guinea-Bisau (only 22% of the population), the Central African Republic (12%), Mozambique (30%). The situation with drinking water is also very severe in Afghanistan (28% of the population is provided with it), Paraguay (35%) and Laos (39%). Out of 96 states, in 24 the water provision for the population is about 50% or less. In agricultural regions the situation is even worse: in 53 countries less than 51% of their population is provided with drinking water.

There are 25 countries where less than 51% of their population is provided with sanitary services. With regard to agricultural areas, this kind of situation is characteristic of 56 countries. Two densely populated countries, India and Pakistan, can only provide enough drinking water for 21% and 24%, respectively, of their population (among them, for the population of agricultural regions, for only 4% and 8%, respectively).

A characteristic feature of contemporary vital activity of the population of our planet is large-scale migration. Refugees cross state borders for various reasons: (1) political (including racial, religious or nationality aspects, political orientation, etc.) and wartime (people fleeing from zones of military actions); (2) economic (the deficit of job vacancies, low salaries, etc.); (3) environmental. The migration of the latter group may further be subdivided into the following categories: (a) due to natural calamities (earthquakes, volcanic eruptions, etc.); (b) caused by the deterioration of environmental quality as a result of man's economic activity; (c) provoked by periodic changes of natural conditions (for instance, expansion of desertification of climatic nature).

In 1985 the total number of refugees in the world was about 15 million people. 50% of the total number were refugees from three countries: Afghanistan (5 934 500), Mozambique (1 354 000) and Ethiopia (1 035 900). Palestinians constituted 2 340 500 refugees. The flows of refugees moved, as a rule, to neighbouring countries. For example, refugees from Afghanistan rushed mostly to Pakistan (3 579 000) and Iran (2 825 000).

Sometimes there are great flows of refugees witin the borders of their native state. This is characteristic, for instance, for Sudan (2–3 million people) or Afghanistan (2 million people). In the former USSR there were about 0.5 million internal refugees; in later years their number increased very considerably.

Human health is also one of the problems of vital importance. In this connection, many new negative factors affecting human health and due to environmental contamination should be mentioned (apart from polluted atmosphere and drinking water), such as: noise, radiation, badly disposed of industrial waste, and polluted air inside human dwellings. About 1 million people are poor, and this is a factor contributing to the number of sick people. Africa is notorious for the greatest number of human health problems (160 million Africans are underfed or ill). Poor nutrition and diseases cause the death of about 14 million children below 5 years of age per year in the developing countries.

Among the indicators of general state of health of population are: overall death rate, child death rate, food provision, health care, preventive vaccinations.

Every year about 50 million people die in this world, and 78% of them die in the developing countries. The structure of mortality (its causes) is different. Blood circulation troubles form the main cause of death all over the world. However, if in industrially developed countries blood circulation problems cause about 50% of lethal cases, in the developing states they cause only about 27% of deaths (the lowest figures for mortality caused by this group of diseases are observed in South-East Asia: 15.4% of men and 11.1% of women). Mortality from malignant tumours is also higher in the industrialized world (21%); in the developing countries, it constitutes 11% (and it is especially low in the countries of the eastern Mediterranean, 2.3% and 1.5%, and South-East Asia, 9.8% and 5.5%). In the former USSR the mortality from cancer (18.4% and 13.7%) was lower than in Sweden (20.9% and 21.2%), England (26.2% and 23.9%) or FRG (26.2% and 23.2%). At the same time, in the former

USSR mortality caused by infections and parasites of various kinds was much higher owing to a lower level of environmental cleanness.

Child mortality (the number of children who die aged from 1 to 5 years per 1000 1-year-old children) differs depending on the country. It is high in Africa, and particularly in Mali (147), Sierra Leone (147), and Gambia (139). In the developed European countries it is very low: in England (2%), Finland (1%), Italy (1%); it is also fairly low in the former USSR (5%). The child mortality value averaged over the planet during the period is 38 children per 1000.

Life duration increased from 50 years in 1960–65 to 63 years in 1990, and the increase is especially noticeable in the developing countries: from 46 to 61. The longest average lifetimes are in Iceland (78 years), Israel and the USA (76), the Netherlands and Norway (77); in the former USSR it was 70 years and later declined. Both these indices: child mortality and life duration (especially the first one) are representative indicators of the quality of living and environmental conditions.

Malnutrition is one of the main causes of numerous illnesses of both children and adults. Often it is the cause of death. The latter is specifically relevant to the developing countries. For instance, the deficit of vitamin A is the cause of illness of 40 million children; 300–500 million people suffer from anaemia which results from the deficit of iron.

The industrially developed countries are far ahead of the developing states with regard to the quantity of physicians (1 per 10 thousand people). In most African countries, this index hardly exceeds unity, whereas in Europe, as a rule, it is higher than 20. In the former USSR it was the highest in the world (42.1), but the level of medical service was nevertheless obviously much lower than in many industrialized countries.

A great number of children all over the world become invalids or die because they have not been vaccinated against infectious diseases. About 46 million children are not provided with such vaccinations; 2.8 million of them die and another 3 million become invalids.

Diseases related to the environment (infectious, cardiovascular and cancer) require special attention. In the developed countries, cardiovascular diseases cause about 40–50% of all deaths. However, mortality from cardiac ischaemia fell in a number of developed countries, such as the USA, Belgium, Japan, etc. At the same time it has been increasing in Eastern Europe and developing states.

Infections are causes of diseases and deaths of one-third of the inhabitants of our planet. Infectious diseases are certainly more widespread in the developing countries. This is due to malnutrition, the low quality of water, inadequate hygienic conditions, and overpopulation. In many regions of the world, cancer holds the second position (after cardiovascular diseases) as the cause of sickness and premature death. It has been proved that many forms of cancer are directly or indirectly connected with the poor quality of the environment. In 1986, 6.35 million people were cancer cases; 4.2 million out of this number died.

In most countries, accidents are one of the five main causes of death. The significance of this factor increases year in, year out. In the developed countries, the main cause of death from accidents are road accidents. A large number of people become victims of accidents caused by chemicals and radioactive substances (in industry or transport). About 200 serious accidents of this kind occur annually in the world. Among these, there also are bad (fatal) cases of food poisoning. The largest numbers of death cases in this category in the period from 1980 to 1988 were in Bhopal (India, 1984: 2988 people), Madrid (Spain, 1981: 340), Chernobyl (the former USSR): 29); the greatest numbers of sick people were: in Bhopal 100 000, in Madrid 20 000, in Yangtze (China, 1987) 15 400, in Chernobyl 300; the number of evacuated people were: in Bhopal 200 000, in Chernobyl 135 000. In the latter case, as is known, the long-term consequences of the accident have not been taken into account.

Local wars take the lives of great numbers of people.

Pesticides are most harmful to human health. Every year about 1 million people are badly poisoned by pesticides, and more than 220 thousand people die.

Goodland et al. (1994) are of the opinion that the only realistic way out is stabilization of population numbers by limiting the average family size in poor countries to 2.5–3.0 children, and in industrially developed states to 2.1 (to reach a balance between birth and mortality rates). These authors put forward the following priorities:

(1) Solution of the problem of family planning.
(2) Broader spreading of information about the interdependence between family size and economic well-being.
(3) Intensification of educational programmes.

Sadik (1994) has emphasized the simple truth that the realities of the contemporary world are such that humankind is not only facing the population growth problem, but is suffering from its numerous after-effects on a scale of separate countries, regions or whole continents. Today it is quite obvious that the exploitation of many resources, provoked by rapid population growth, has already reached a level which destroys, for example, the regenerative ability of the atmosphere. Another spectacular illustration of the above conclusion is the land-use problem. One of its most painful aspects is the continuing desertification which has by now encompassed 30% of the global land area and which reflects catastrophic losses of arable land.

The problem of per anima consumption becomes more acute every year. Analysis of data for the recent 40 years has revealed (Monitoring Environmental..., 1995) that the per anima consumption of copper, energy, meat, steel and timber has doubled; the number of automobiles and per anima consumption of cement increased by a factor of 4, that of plastic materials by a factor of 5, and of aluminium by a factor of 7. This has mostly taken place in industrially developed countries. It is interesting to compare data on per anima energy consumption in the USA and other countries. From this viewpoint a single American is equivalent to: 3 Japanese, 13 Brazilians, 14 Chinese, 38 Hindus, 168 inhabitants of Bangladesh, 280 of Nepal and 531 Ethiopians. While 260 million North Americans spend $5 billion annually to buy special, low-calory dietic food, approximately 400 million people (or more) suffer from starvation.

To maintain the lifestyle of the population of rich countries at the contemporary level, the per anima consumption of resources from 4–6 ha is required. If one proceeds from this norm, the total of 26.5 billion ha of land area would be required to meet the demands of the entire population of our planet (about 6 billion people). However, the total land area of the globe is 13 billion ha, out of which only 8.8 billion ha of land is suitable for agriculture or is occupied by forests.

Another essential aspect of the problem is food production, whose annual growth rate during the 1970s and 1980s reached 2.3% and exceeded the global mean population growth rate (1.8–1.9%). Relevant estimations have revealed, however, that during the next 15–20 years, a drop in food production growth rate will occur, and it will be faster than the decrease in the population growth rate. The growing population number causes more and more intense loadings on forest ecosystems and biodiversity, which, in its turn, is a serious threat for global ecological safety (Gorshkov et al., 1993; Gorshkov, Kondratyev, Danilov-Danilyan and Losev, 1993; Gorshkov, Kondratyev and Losev, 1996). Biological resources of seas and oceans are also subject to enormous loadings. The violation of admissible limits of fishing reduced the global catch from 86 to 80 million tons during three years (1989–92). This is due to the fact that half of the world population inhabits the 100 km coastal zone, and in South-East Asia two-thirds of the population lives in the coastal zone. 9 out of 10 biggest cities in the world are situated in the coastal zone, as well as two-thirds of the 50 biggest cities. This accounts for 20% of the world population being the inhabitants of seaside cities.

All this is strong evidence revealing not only the absence of environmentally sustainable development (ESD), but also the catastrophic nature of the present global ecological situation which is aggravated by the inadequacy of the socio-economic development characteristics because they do not take into account the damage to natural resources and the environment; very often the most profitable branches of industry prove to be ecologically harmful, and it is developing countries that suffer from this situation most of all. Therefore, it would only be fair if the rich countries of the North, pursuing their own interests, could help the poor states of the South in overcoming inequality and poverty, and in protecting their environment. Very important also are such priorities as the use of renewable sources of energy, more efficient utilization of fossil fuel, minimization and recycling of waste, development of ecologically beneficial methods in agriculture, etc.

An important aspect of the problem of overpopulation is the question of whether new technologies are capable of ensuring successful development in an over-populated world. The question has been discussed repeatedly, and one can only agree with Daly and Ehrlich (1992, Chapter 1) who state that it is hardly at all reasonable to rely on a technological miracle which will secure well-being for the Earth's population which will soon increase by billions of people.

In fact, hopes for new technologies are more than illusory, especially if one takes into account the following considerations:

(1) The negative impact of already existing technologies on the environment and man (e.g., the health of about 1 billion people in the world is subject to the destructive impact of various kinds of pollution);
(2) The expected doubling of global population over 42 years, and hence the

practically insoluble problem of levelling the per anima annual income in industrially developed countries (which constitutes \$18 330) and in the rest of the world. If one assumes an average per anima annual income of \$800, then, taking account of the expected population growth, the efficiency of technologies should be improved by $2 \times 23 = 46$ times. This unprecedented rate of technological progress can hardly be ever expected in the future (and it should be emphasized here that the basic restricting factor is the permissible level of biosphere resources consumption).

This most acute problem may also be regarded from the viewpoint of per anima energy consumption (Goldemberg, 1995; Monitoring Environmental..., 1995). In industrially developed countries, whose inhabitants constitute about 20% of the world population, in 1990 it was 7.5 KW on the average, with the total consumption of 9.0 TW (1 TW $= 10^{12}$ W). In the countries of the South (80% of the global population), the per anima annual energy consumption was only 1 KW (the total consumption was 4.1 TW). If one proceeds from an assumption that in the next century the world population will reach a stabilization level of about 12 billion people, and that the average per anima annual consumption will increase to 7.5 KW, it will imply a fantastic level of total energy consumption of 90 TW, and this figure exceeds the current one (already resulted in a situation of global ecologically unsustainable development) by a factor of 7.

Hemphill (1995, Chapter 1) has discussed six principles whose implementation might be beneficial for the energy problem solution in the conditions of the supposed global climate warming:

- the solution of relevant problems on the basis of the market approach (taking account, where possible, of non-market factors as well);
- a broader application of ecologically efficient "pure" technologies;
- coordination of power production development and environmental protection measures (taking account of the ecological "cost" of power production);
- creation of a system of adequate monitoring of the after-effects of power production development for the environment;
- support of the transfer of new "pure" technologies to other countries;
- ensuring the achievement of not only economic, but also ecological goals within the framework of ecological support projects.

One should emphasize, however, that all these measures can only be palliative.

3. ENERGY

Fig. 1 presents some examples of human needs in energy and energy conversion devices, in joules, on a logarithmic scale (Watson et al., 1996). Fig. 2 characterizes the energy system, including an energy supply sector and energy end-use (Watson et al., 1996). Fig. 3 illustrates the global energy balance and flows without anthropogenic interference (Watson et al., 1996). The production and consumption

Energy units and scales

Energy is defined as the capacity to do work and is measured in joules (J), where 1 joule is the work done when a force of 1 newton (1 N = 1 kg m/s^2) is applied through a distance of 1 metre. Power is the rate at which energy is transferred and is commonly measured in watts (W), where 1 watt is 1 joule per second. Newton, joule, and watt are defined as units in the International System of Units (SI). Other units used to measure energy are toe (ton of oil equivalent; 1 toe = 41.87×10^9 J), used by the oil industry; tce (ton of coal equivalent; 1 tce = 29.31×10^9 J), used by the coal industry; and kWh (kilowatt-hours; 1 kWh = 3.6×10^6 J), used to measure electricity. Some of the commonly used units of energy and a few examples of energy consumption levels, along with the Greek names and symbols for factors to power of ten (e.g., exa equals 10^{18} and is abbreviated as E; in 1990, the global primary energy consumption was 385 EJ).

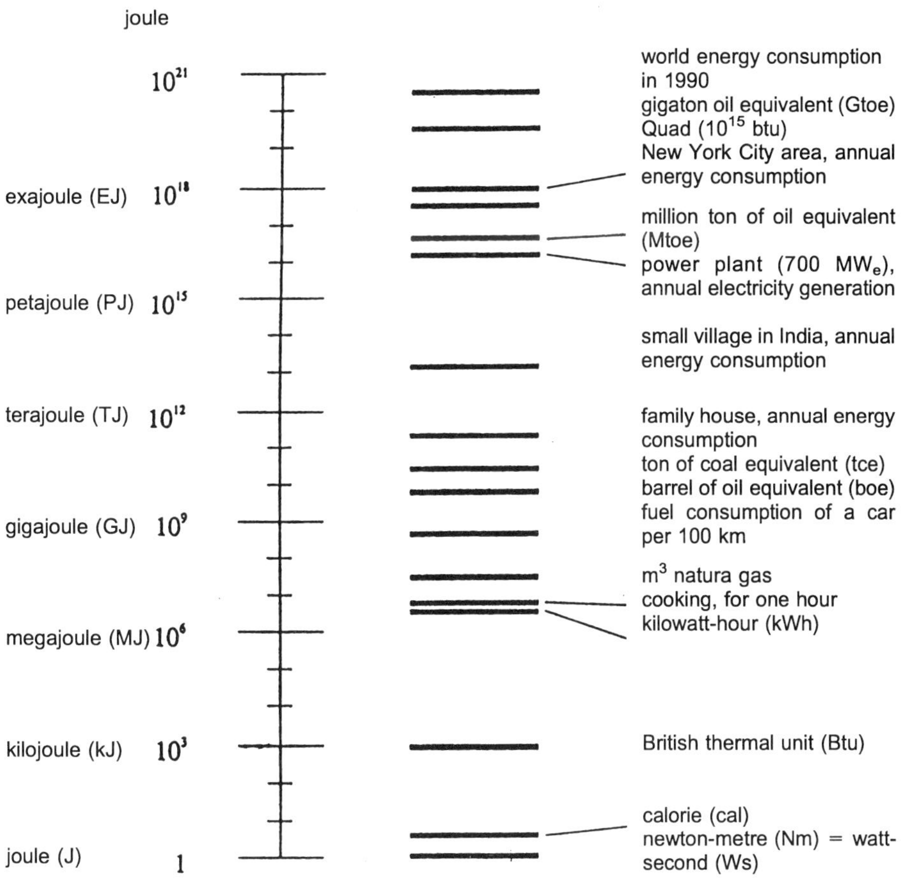

Fig. 1. Examples of some human energy needs and energy conversion devices, in joules, on a logarithmic scale.

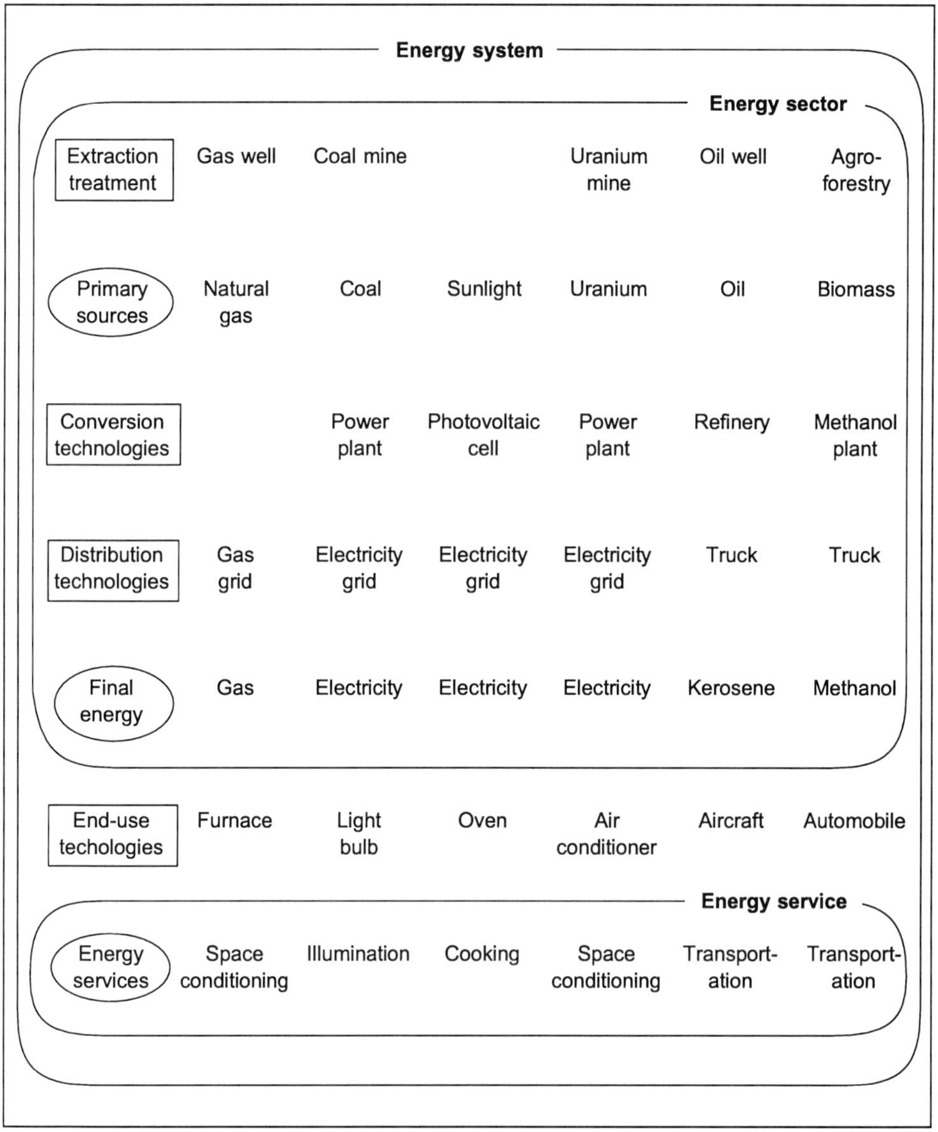

Fig. 2. The energy system: schematic diagram with some illustrative examples of the energy sector and energy end-use and services. The energy sector includes energy extraction, treatment, conversion, and distribution of final energy. The list is not exhaustive, and the links shown between stages are not "fixed" (e.g. natural gas is also used to generate electricity, and coal is not used exclusively for electricity generation).

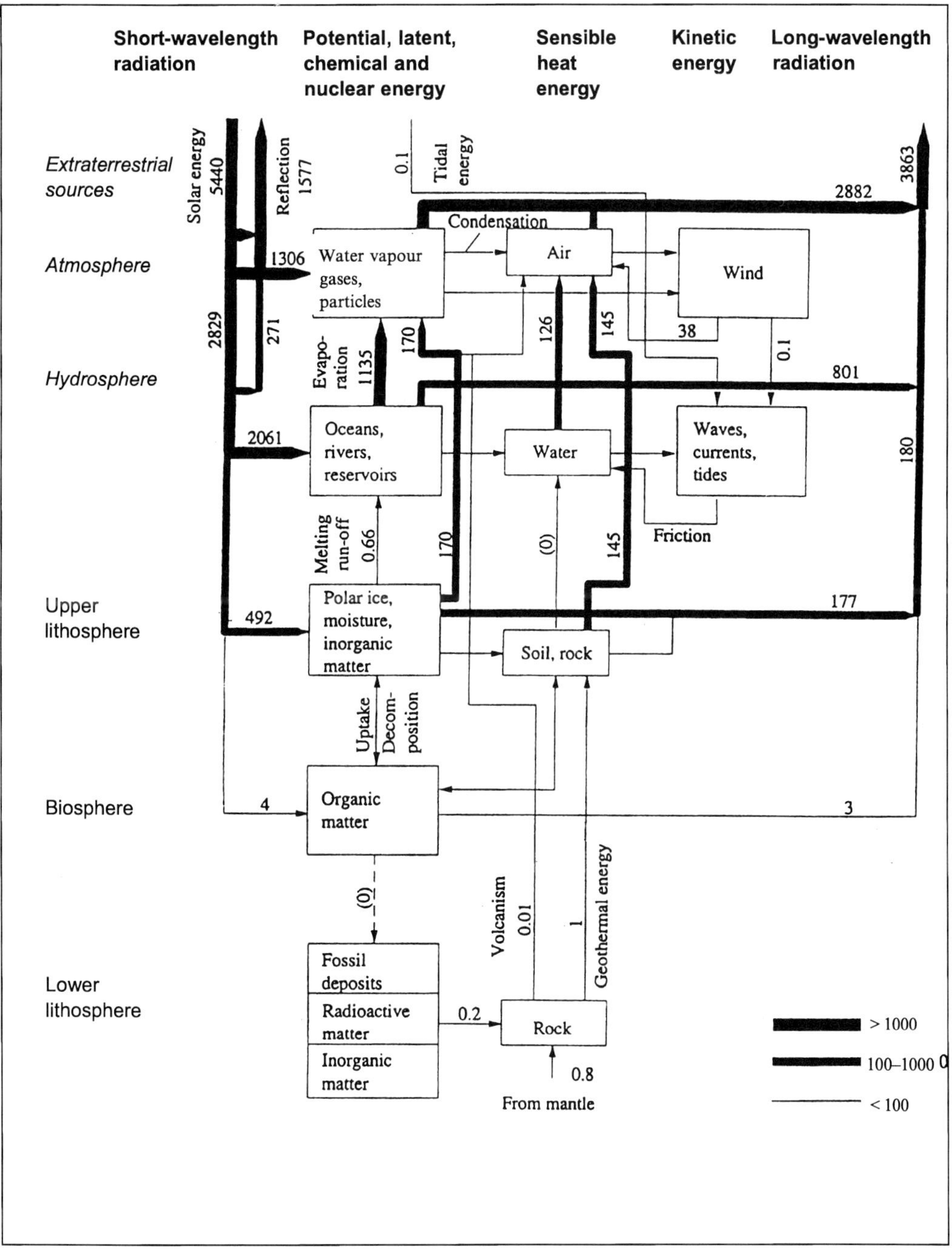

Fig. 3. Global energy balance and flows without anthropogenic interference. The energy flows are in units of 1000 EJ/yr. Numbers in parentheses are uncertain or rounded.

of some energy resources are causes of most dangerous ecological situations (Tolba et al., 1992). Countries richest in non-renewable energy resources are China (about 731×10^9 tons of coal), the former USSR (241×10^9) and the USA (215×10^9). The largest reserves of oil and gas have been found in the former USSR and in the Middle East. The former USSR possessed 38% of global natural gas reserves; Iran possessed 13%.

The data of Table 4 (Brown et al., 1996) characterize the world oil production during the time period 1950–95. Similar data for natural gas and coal are given in Tables 5 and 6, respectively.

Table 4. World oil production, 1950–95

Year	1950	1955	1960	1965	1966	1967	1968	1969	1970
Production (million tons)	518	767	1049	1509	1638	1743	1937	2050	2281
Year	1971	1972	1973	1974	1975	1976	1977	1978	1979
Production	2410	2547	2779	2803	2659	2901	2988	3023	3122
Year	1980	1981	1982	1983	1984	1985	1986	1987	1988
Production	2976	2779	2644	2619	2701	2659	2774	2754	2881
Year	1989	1990	1991	1992	1993	1994	1995	(preliminary)	
Production	2918	2956	2928	2999	2956	2986	3031		

Table 5. World natural gas production, 1950–95

Year	1950	1955	1960	1965	1966	1967	1968	1969	1970
Production (million tons of oil equivalent)	187	297	458	664	716	770	851	929	1026
Year	1971	1972	1973	1974	1975	1976	1977	1978	1979
Production	1106	1170	1241	1261	1267	1320	1354	1394	1495
Year	1980	1981	1982	1983	1984	1985	1986	1987	1988
Production	1455	1492	1482	1489	1625	1693	1732	1813	1900
Year	1989	1990	1991	1992	1993	1994	1995	(preliminary)	
Production	1962	2010	2045	2045	2091	2103	2114		

Table 6. World coal use, 1950–95

Year	1950	1955	1960	1965	1966	1967	1968	1969	1970
Coal use (million tons of oil equivalent)	884	1045	1271	1299	1321	1255	1317	1357	1359
Year	1971	1972	1973	1974	1975	1976	1977	1978	1979
Coal use	1355	1355	1413	1434	1450	1525	1581	1615	1681
Year	1980	1981	1982	1983	1984	1985	1986	1987	1988
Coal use	1708	1732	1751	1804	1877	1980	2001	2062	2183
Year	1989	1990	1991	1992	1993	1994	1995	(preliminary)	
Coal use	2189	2109	2074	2098	2085	2095	2109		

Table 7. Global energy consumption in 1990 by energy source and by sector, in EJ/yr

	Coal	Oil	Gas	Nuclear	Hydro[a]	Electricity	Heat	Biomass	Total
Primary	91	128	71	19	21	—	—	55	385
Final	36	106	41	—	—	35	8	53	279
Industry	25	15	22	—	—	17	4	3	86
Transport	1	59	0	—	—	1	0	0	61
Others	10	18	18	—	—	17	4	50	117
Feedstocks[b]	0	14	1	—	—	—	—	0	15

Notes. Primary energy is recovered or gathered directly from natural sources (e.g., mined coal, collected biomass, or harnessed hydroelectricity), then is converted into fuels and electricity (e.g., electricity, gasoline, and charcoal), resulting in final energy after distribution and delivery to the point of consumption.

[a] Nuclear and hydropower electricity have been converted into primary thermal equivalent, with an average factor of 38.5%.

[b] Feedstocks represent non-energy use of hydrocarbons.

In 1990, the world industrial sector consumed an estimated 98 EJ of end-use energy (including biomass) and 19 EJ of feedstocks to produce 6.7×10^{12} of value added (Watson et al., 1996). The available projections indicate that worldwide industrial energy use could rise from about 107 EJ/yr in 1990 to 140 and 190 EJ/yr in 2025 and 2100, respectively, if GHG emissions are stabilized.

The data of Table 7 illustrate the global energy consumption in 1990 by energy source and by sector, while the figures of Table 8 indicate global fossil energy reserves, resources, and occurrences (Watson et al., 1996).

The data of Table 8 show that the global fossil resource-base estimate is about 186 000 EJ, with additional occurrences of almost 1 million EJ. Included in the conventional resources are estimates of ultimately recoverable conventional oil and gas resources remaining to be discovered at 95%, 50%, and 5% probability levels, ranging between 1800 and 5500 EJ for oil and 2700 and 10 900 EJ for gas.

Table 8. Global fossil energy reserves, resources, and occurrences, in EJ

	Consumption 1860–1990	1990	Reserves identified	Conventional resources remaining to be discovered at probability			Unconventional resources		Resource base[a]	Additional occurrence
				95%	50%	5%	Currently recoverable	Recoverable with technological progress		
Oil										
Conventional	3 343	128	6 000	1 800	2 500	5 500			8 500	>10 000
Unconventional	—	—	7 100					9 000	16 100	>15 000
Gas										
Conventional	1 703	71	4 800	2 700	4 400	10 900			9 200	>10 000
Unconventional	—	—	6 900				2 200	17 800	26 900	>22 000
Hydrates	—	—								>800 000
Coal	5 203	91	25 200				13 900	86 400	125 500	>130 000
Total	10 249	290	50 000	>4500	>6 900	>16 400	>16 100	>113 200	>186 200	>987 000
Nuclear[b]	212	19	1 800		2 300		4 100	>6 000	>14 200	>1 000 000

Notes. All totals have been rounded; — = negligible amounts; blanks = data not available.

[a] Resource base is the sum of reserves and resources. Conventional resources remaining to be discovered at probability of 50% are included for oil and gas.

[b] Natural uranium reserves and resources are effectively 60 times larger if fast breeder reactors are used. Calculated from natural uranium reserves and resources (OECD/NEA and IAEA, 1993) into thermal equivalent for once-through fuel cycle with average factor of 1700 g per TJ thermal or 4440 g natural uranium per TJ of electricity (16 Mg natural uranium per TWh_e electricity).

The production and consumption of various kinds of energy resources differ very much from one country to another. The production (mining) of solid kinds of fuel (including coal, brown coal, peat and oil shale) constitutes 31% of all kinds of energy resources, and that of liquid types of fuel 41%. African countries are the main producers of energy; they also are (except the Republic of South Africa) its main exporters. These are Algeria (the production and consumption of energy, respectively: 3627×10^{15} and 976×10^{15} J), Nigeria (2756×10^{15} and 495×10^{15} J), Egypt (2132×10^{15} and 991×10^{15} J), Libya (2180×10^{15} and 339×10^{15} J). All these are countries exporting oil. Among the biggest producers of energy, the Republic of South Africa is perhaps a single country that consumes its energy (3956×10^{15} and 3154×10^{15} J).

In South America there are four big producers of energy, mainly of oil. However, two of the states, Argentina and especially Brazil, even suffer from a deficit of energy (compare figures for production and consumption: 1735×10^{15} and 1745×10^{15} J for Argentina; 2151×10^{15} and 3178×10^{15} J for Brazil). The other two, Colombia and Venezuela, produce sufficient amounts of energy (1480×10^{15} and 717×10^{15} J for Colombia; 5138×10^{15} and 1616×10^{15} J for Venezuela).

In Asia the biggest energy producers are China ($25\,930 \times 10^{15}$ J), India (6115×10^{15} J), Saudi Arabia (9697×10^{15} J), Iran (5463×10^{15} J) and Iraq ($4\,423 \times 10^{15}$ J). The latter three are the biggest producers of oil and oil products (8790×10^{15} J, 4783×10^{15} J and 4275×10^{15} J) and, at the same time, exporters of oil.

The majority of European countries suffer from a deficit of self-produced energy (among these, there are Spain, Switzerland, the Netherlands, Portugal, Austria, the Czech Republic and Slovakia, France, Belgium, Italy, FRG). England and Poland are among the biggest producers of energy (9732×10^{15} J and 5288×10^{15} J, respectively), and these two countries produce enough energy for their own consumption needs. Another big energy producer is Germany (4438×10^{15} J), but this country needs more energy than it produces.

In North America, the USA is the biggest producer of energy ($58\,242 \times 10^{15}$ J), but the amount is insufficient for consumption ($68\,079 \times 10^{15}$ J). Both Canada, and Mexico produce energy in excess (9739×10^{15} and 7316×10^{15} J, respectively). The biggest producer of energy in the world was the former USSR ($68\,457 \times 10^{15}$ J). This country consumed less energy than the USA ($54\,725 \times 10^{15}$ J).

The development of nuclear power production is strongly dependent on serious accidents which certainly provoke recessions in its progress. This is particularly relevant to the Chernobyl accident (1986). In 1989, 426 reactors were in operation. In 1988, energy produced by nuclear power stations (NPS) constituted 17% of the total amount of electric power. In 1989, the largest quantity of reactors was in the USA (110), then followed the former USSR (46), France (55), England and Japan (39 each).

In the total energy balance, an essential role belongs to sources of renewable energy, especially hydraulic. Its largest amounts were produced (for 1987) in the USA, the former USSR, and Brazil (249 695, 219 800, and 198 782 GW, respectively). The data of Table 9 indicate global renewable potentials by 2020–25, maximum technical potentials, and annual natural flows, in EJ thermal equivalent (Watson et al., 1996).

Table 9. Global renewable energy potentials by 2020-25, maximum technical potentials, and annual natural flows, in EJ thermal equivalent.[a]

	Consumption		Potential by 2020-2025[b]	Long-term technical potentials[c]	Annual flows
	1860–1990	1990			
Hydro	560	21	35–55	>130	>400
Geothermal	—	<1	4	>20	>800
Wind	—	—	7–10	>130	>200 000
Ocean	—	—	16–22	>2600	>3 000 000
Biomass	1150	55	72–137	>1300	
Total	1710	76	130–230	>4200	3 000 000

Notes. All totals have been rounded;— = negligible amounts; blanks = data not available.

[a] All estimates have been converted into thermal equivalent with an average factor of 38.5%.

[b] Range estimated from the literature. Survey represents renewable potentials by 2020–25, in scenarios with assumed policies for enhanced exploitation of renewable potentials.

[c] Long-term technical potentials are based on the IPCC Working Group II evaluation of the literature sources given in this Table. This evaluation is intended to correspond to the concept of fossil energy resources, conventional and unconventional.

Current use and practicable potentials of renewable energy technologies are presented in Table 10 (Watson et al., 1996).

Table 10. Current use and practicable potentials of renewable energy technologies (TWh/yr)

		Practicable potential estimate			
	Current use	Johansson et al. (1993) 2020	Swisher et al. (1993) 2030	WEC (1993) 2020	Read average to 2050
Solar	54		1395	489–1 592	
Wind	3.2		4931	20 148	
Hydro	228.2[a]	6000–9000	7077	8 295[b]	
Geothermal	37–57	>53	1499	178–405	
Ocean	0.6		247	48–240	
Traditional biomass	4170		8003	7031–7269	
Modern biomass[c]				about 35 000[d]	

[a] Includes 81.7 TWh/yr for small hydro.

[b] Includes 211–308 TWh/yr for small hydro.

[c] Modern biomass refers to the use of biomass (e.g., timber or sugar cane) for the production of electricity, liquid fuels, and heat using modern technology.

[d] Assumes 740 million hectares become available for biofuel production by 2050 (proportionately less according to technical progress with biofuel productivity per hectare) with a slow start and more rapid build-up after 2010. The 35 000 TWh would yield about 18 000 TWh of electricity given advanced generating technology expected to be in use next century.

Table 11 compares economic development and energy use in the main regions of the world as of 1990. The authors of the IPCC WG II Report (Watson et al., 1996) have emphasized that this comparison highlights the following striking differences between and within regions and may serve to clarify why the involvement of the developing countries in greenhouse gases policy formulation and implementation is imperative:

— GNP per capita varies from an average of \$440 for Asia to \$17 473 for the West, a ratio of 1:40. The differences can also be large within regions: sub-Saharan Africa has a per capita GDP of \$322 as against \$1507 for North and South Africa together (a ratio of 1:5).
— The relative use of traditional energy (woodfuel, crop residues, and animal dung) varies from 1% for the East to 37% for Africa. Within Africa traditional fuels account for 5% of energy use for all Africa and 61% for sub-Saharan Africa (1:12).
— The relative use of fossil fuels varies from 57% of total energy use for Africa to 98% for the Middle East; within Africa it amounts to 30% for sub-Saharan Africa and 92% for all Africa (1:3).
— The relative use of nuclear plus renewables varies from 1% for the Middle East to 22% for Latin America. Within Africa it amounts to 3% for all Africa and 9% for sub-Saharan Africa.
— The share of the main energy source as a percentage of total energy use varies from 36% for the East (natural gas) to 64% for the Middle East (oil). For Africa the main source is traditional fuels, for Asia coal, and for the West and Latin America oil.
— In terms of energy use per capita, the regions vary from 0.53 tonnes of oil equivalent (toe) for Asia to 5.06 toe for the West, or by a ratio of 1:10. Between Western subregions it still varies from 3.39 for Western Europe to 8.30 toe for North America.
— Energy intensity (in toe/\$1000 GNP) varies from 0.29 for the West to 1.19 for Asia (1:4). Between Western subregions, it varies from 0.17 for Japan/Australia/New Zealand to 0.4 for Northern America. However, energy intensity in toe/\$1000 GNP is an unreliable yardstick for comparisons between regions, especially between developed and developing countries, because of differences between nominal GNP and real GDP in purchasing power parities (PPP). If the energy intensity data in the table were expressed in toe/\$1000 PPP (correction based on UNDP Human Development Report, 1993 data), the energy intensity of the developing countries would become 0.35 but that of the developed countries would remain 0.34. However, this would still ignore the relatively higher energy content of the developing countries' imports and the lower energy contents of its exports. If these two factors are taken into account, energy intensity in the developing countries would be higher than that in toe/\$1000 PPP but probably lower than that in nominal terms.

A very important conclusion is that, due to most scenarios, during the next decades the growth in carbon emissions will increasingly take place in developing countries.

Of great interest is the possibility of producing power by making use of the energy

Table 11. Comparison of the development and energy situation in the main (sub-)regions

Region	Tr (%)	Fossil fuels (%) Co	Oi	NG	ΣF	Al (%)	ΣEner (Mtoe)	Pop. (million)	GNP (billion US$)	GNP ($ per capita)	Toe per capita	Toe per thousand US$
West	1	22	41	19	82	16	4296	849	14 840	17473	5.06	0.29
North America	2	22	38	24	83	15	2289	276	5 738	20809	8.30	0.40
Western Europe	1	22	43	16	80	19	1456	430	5 901	13726	3.39	0.25
JANZ	1	21	51	12	84	16	551	144	3 202	22279	3.83	0.17
East	1	25	29	36	91	8	1745	414	3 047	7367	4.22	0.57
Eastern Europe	1	46	25	19	90	9	352	125	388	3101	2.81	0.91
ex-USSR	1	20	31	40	91	8	1393	289	2 660	9215	4.83	0.52
North	1	23	38	24	85	14	6040	1263	17 888	14163	4.78	0.34
Africa	37	21	27	9	57	6	360	642	385	600	0.56	0.93
North and South Africa	5	na	na	na	92	3	154	150	227	1507	1.02	0.68
sub-Sahara	61	na	na	na	30	9	206	492	158	322	0.42	1.30
Asia	15	49	25	4	78	7	1475	2810	1 237	440	0.53	1.19
China	6	72	15	2	89	5	718	1139	393	345	0.63	1.83
India	25	41	23	4	68	7	253	853	287	337	0.30	0.88
Other	23	19	39	8	66	11	504	817	557	681	0.62	0.90
Latin America	15	4	45	14	63	22	556	448	842	1880	1.24	0.66
Brazil	30	5	34	2	41	29	185	150	375	2495	1.23	0.49
Mexico	5	5	49	22	75	20	142	89	170	1920	1.60	0.83
Other	9	4	51	20	74	17	228	209	297	1421	1.09	0.77
Middle East	1	1	64	32	98	1	233	129	370	2860	1.81	0.63
South	17	31	33	10	74	10	2624	4029	2 835	704	0.65	0.93
World	6	25	36	20	81	13	8664	5292	20 723	3916	1.64	0.42

Notes. Tr = traditional (woodfuel, crop residues and animal dung); Co = coal; Oi = oil; NG = natural gas; ΣF = total fossils; Al = alternatives (nuclear, hydro, wind, geothermal, etc.) (all % of total energy); ΣEner = total energy; (M)toe = (million) tons of oil equivalent; Pop. = population; GNP = gross national product in 1989; na = not available. Western Europe = OECD Europe (includes Turkey); Eastern Europe = non-OECD Europe (includes Cyprus, Gibraltar and Malta); JANZ = Japan, Australia and New Zealand.

Table 12. World wind energy generating capacity, 1980-95

Year	Capacity (MW)
1980	10
1981	25
1982	90
1983	210
1984	600
1985	1020
1986	1270
1987	1450
1988	1580
1989	1730
1990	1930
1991	2170
1992	2510
1993	2990
1994	3680
1995 (preliminary)	4880

Table 13. Net annual additions to world generating capacity, 1980-95

Year	Capacity (MW)
1980	5
1981	15
1982	65
1983	120
1984	390
1985	420
1986	250
1987	180
1988	130
1989	150
1990	200
1991	240
1992	340
1993	480
1994	720
1995 (preliminary)	1290

of sea tides. Tidal hydropower stations are in operation in seven countries only (1989): Canada (3), China, Korea, India, Russia, England and France. The biggest one producing more than one half of the whole (8000 GW) is located in England.

Considerable energy output gains came in wind electric generation capacity, which expanded by 33%, and in the sales of photovoltaic (solar) cells which climbed to 17%. This can be illustrated by Tables 12 and 13.

Thirty-two countries possess solar energy power stations. The biggest one is in the USA. The effect of such stations is very modest (in the USA 382 000, and in the former USSR 15 MW). Wind power stations are quite numerous (over 100 000 in 1989) but not very effective (a total of 1500 MW). Such stations are located in the USA, China, India, England, FRG, Denmark, etc. (16 countries). Still less important sources of power are geothermal stations and those operating on alcohol fuel. The expanding role of renewable energy resources, and the movement towards a solar/hydrogen economy constitute an optimistic aspect of the contemporary development (Brown et al., 1996).

India has become the first developing country with a real commercial market for wind power, while in the USA the wind industry was near the collapse in 1995 because of the uncertainty about the future structure of the electricity industry. Germany led the world in new capacity in 1995 with 505 MW, closely followed by India with 375 MW (Brown et al., 1996).

The global dynamics of nuclear power generation may be characterized by Table 14 data (Watson et al., 1996). In 1995, no government began construction on a new reactor project. Overall, just 34 reactors, representing 27 GW (gigawatts), remained under construction—the fewest in nearly 30 years. In Western Europe, nuclear

Table 14. World net installed electrical generating capacity of nuclear power plants, 1960–95

Year	1960	1961	1962	1963	1964	1965	1966	1967	1968	1969	1970
Capacity, GW	1	1	2	2	0	5	6	8	9	13	16
Year	1971	1972	1973	1974	1975	1976	1977	1978	1979	1980	1981
Capacity, GW	24	32	45		61 71	85	99	114	121	135	155
Year	1982	1983	1984	1985	1986	1987	1988	1989	1990	1991	1992
Capacity, GW	170	189	219	250	276	297	310	320	328	325	327
Year	1993		1994		1995		(preliminary)				
Capacity, GW	336		338		340						

expansion has stopped everywhere except in France, where four reactors remain under construction and are due to be completed by 1998. The two confirmed orders for new plants came from China and South Korea.

4. NATURAL RESOURCES

Abundant information on land and forest resources, timber production, food resources, fresh water reserves and fish resources can be found in the reports (Brown et al., 1994, 1996; State . . ., 1990, 1996; Tolba et al., 1992; United Nations . . ., 1991). The state of bioresources taken as a whole has also been discussed there, as well as the problem of diverse waste connected with the use of various resources.

It should be pointed out that there are controversial assessments of the global situation with regard to the availability of natural resources. Moore (Bailey, 1995) summarizes the situation as follows:

— The objective, scientific evidence available today shows that the prophets of doom were wrong in virtually every prediction they made in 1960s and 1970s when they forecast increasing natural resource scarcity and rising commodity prices.

— Every measurable trend of the past century suggests that humanity will soon be entering an age of increasing and unprecedented natural resources abundance.

— Technological improvements and advances in productivity have continually outpaced our consumption of natural resources and have led to the net creation of more resources available to future generations.

— Today, natural resources are about half as expensive relative to wages as they were in 1980, about three times less expensive than they were fifty years ago, and roughly eight times less costly than they were in 1900.

— In spite of Paul and Anne Ehrlich's projection that mineral supplies would be largely depleted by 1985, proven reserves of virtually all important minerals have skyrocketed since 1950.

— The continuing discovery of new mines, technological innovations in mining techniques, and the introduction of less expensive or superior substitutes for the use of some minerals have caused most minerals to become less scarce rather than more scarce over the past 100 years.

— Although Paul Ehrlich refers to the 1980s as a "catastrophic decade" in terms of consumption, the period witnessed tremendous increases in the supply of almost all raw materials.
— A well-publicized book, *The Energy Crisis*, projected in 1972 that the Earth held thirty years left of gas reserves and twenty years left of oil reserves. Since 1950, howcvcr, provcn rcscrvcs for oil and gas have climbed by over 700%.
— The very concept of "finite natural resources" embraced by geologists is a flawed way of thinking about the Earth and nature. "Natural resources" have value only when humanity invents a use for them.

Let us consider some concrete, presently available data.

4.1 Land resources

Global data on the dynamics of areas occupied by various kinds of land resources (arable land, pastures, meadows and forests) are very important. Comparative data for the years 1966–68 and 1986–88 have also been examined. During this time, considerable changes on a global scale have been noted only for forests, whose area decreased by 2%.

The area of "wild" land, not mastered by man, over the entire globe constituted by the mid-1980s the value $4\,806\,995 \times 10^3$ ha, i.e. 32% of the whole land area. Such wild lands are absolutely largest in the former USSR ($752\,022 \times 10^3$ ha), Canada ($640\,587 \times 10^3$ ha), Australia ($229\,431 \times 10^3$ ha), China ($210\,776 \times 10^3$ ha) and Brazil (202 061 $\times$x 10^3 ha), to say nothing, of course, about Antarctica and Greenland. In relative terms, however, these "wild" lands exist in countries rich in forest, such as Surinam (66%), Guyana (57%), French Guiana (45%) and in "desert" countries, such as Lesotho (70%), Mauritania (69%), Western Sahara (68%). In Europe (apart from the former USSR) these lands are practically absent, except a few "spots" in Scandinavian countries (especially in Norway where they occupy a territory of 5627×10^3 ha, i.e. 17%).

The world grain harvested area (1950–95) is characterized by Table 15 data, and the world irrigated area dynamics (1961–93) is presented in Table 16. As can be seen from Table 15, the world grain harvested area in 1995 totalled 666 million ha, down from 675 million ha the year before. In per capita terms, the grain harvested area per person has shrunk from 0.23 ha in 1950 to 0.12 ha in 1995. The data of Table 16 illustrate the dynamics of the world irrigated area.

4.2 Agricultural production

From the beginning of the current century, a stable growth of global agricultural production has been observed (see Table 17). However, in the 1980s, the tendency towards its further growth has markedly weakened in the developed countries in comparison with developing states. The world production of grain increased during the time period from 1966–68 to 1986–88. In some countries it increased very

Table 15. World grain harvested area, 1950–95

Year	Total (million ha)	Per capita (ha)	Year	Total (million ha)	Per capita (ha)
1950	587	0.23	1979	711	0.16
1955	639	0.23	1980	722	0.16
1960	639	0.21	1981	732	0.16
1965	653	0.20	1982	716	0.16
1966	655	0.19	1983	706	0.15
1967	665	0.19	1984	710	0.15
1968	670	0.19	1985	715	0.15
1969	672	0.19	1986	709	0.14
1970	663	0.18	1987	685	0.14
1971	672	0.18	1988	686	0.13
1972	661	0.17	1989	694	0.13
1973	688	0.17	1990	693	0.13
1974	691	0.17	1991	687	0.13
1975	708	0.17	1992	688	0.13
1976	717	0.17	1993	677	0.12
1977	714	0.17	1994	675	0.12
1978	713	0.17	1995 (preliminary)	666	0.12

Table 16. World irrigated area, 1961–93

Year	Total (million ha)	Per capita (ha per thousand population)	Year	Total (million ha)	Per capita (ha per thousand population)
1961	139	45.3	1978	204	47.5
1962	142	45.2	1979	209	47.8
1963	145	45.2	1980	211	47.4
1964	148	45.1	1981	214	47.3
1965	151	45.1	1982	217	47.1
1966	154	45.2	1983	216	46.2
1967	157	45.1	1984	222	46.6
1968	160	45.2	1985	226	46.5
1969	165	45.5	1986	227	46.0
1970	169	45.5	1987	232	46.2
1971	172	45.5	1988	231	45.3
1972	176	45.5	1989	235	45.1
1973	182	46.1	1990	239	45.1
1974	185	46.2	1991	243	45.1
1975	190	46.6	1992	249	45.5
1976	196	47.0	1993	248	44.4
1977	201	47.4			

Table 17. World grain production, 1950–95.

Year	Total (million tons)	Per capita (kg)	Year	Total (million tons)	Per capita (kg)
1950	631	247	1979	1428	326
1955	759	273	1980	1447	325
1960	847	279	1981	1499	331
1965	917	274	1982	1550	336
1966	1005	294	1983	1486	317
1967	1029	295	1984	1649	346
1968	1069	301	1985	1664	343
1969	1078	297	1986	1683	341
1970	1096	296	1987	1612	321
1971	1194	316	1988	1564	306
1972	1156	299	1989	1685	324
1973	1272	323	1990	1780	336
1974	1220	304	1991	1696	315
1975	1250	306	1992	1776	316
1976	1363	328	1993	1703	307
1977	1337	316	1994	1745	309
1978	1467	341	1995 (preliminary)	1680	293

considerably: in Brazil by 112%, Ghana by 76%, Cote d'Ivoire by 89%. At the same time, the grain production fell in several African states, such as Angola, Mozambique and Uganda. This has been caused by war. During the period from the late 1960s to the late 1970s, the per anima import of grain increased on all continents except Europe where it decreased. In Africa and North America the import of grain doubled.

The last few years have been characterized by a negative trend in world grain production (Table 17), which has led to a decrease in the world grain carryover stocks (Table 18), which reached an all-time minimum. Brown et al. (1996) emphasize that with the world grain production dropping in 1995, consumption exceeded production for the third consecutive year, further reducing grain stocks.

Avery (Bailey, 1995) made the following conclusions:

- Food is more abundant and cheaper today than at any other time before in history. Per capita grain supplies have increased 24% since 1950, while food prices have plummeted by 57% since 1980.
- The global trouble spots for soil erosion and habitat destruction—the Amazon rain forest, parts of Africa, and other scattered locations—are experiencing problems because local agriculture has not taken advantage of high-yield technologies and efficiencies. In the case of the rainforests, problems have been compounded by government subsidies for farming and logging.
- Global increases in average grain yields have lagged recently because affluent countries are trying to limit their surpluses and because of chaotic conditions in

Table 18. World grain carryover stocks, 1961–96.

Year	(million tons)	Stocks (days use)	Year	(million tons)	Stocks (days use)
1961	203	90	1979	328	84
1962	182	81	1980	315	81
1963	190	81	1981	288	71
1964	193	82	1982	309	77
1965	194	77	1983	357	87
1966	159	61	1984	304	72
1967	190	71	1985	366	85
1968	213	77	1986	434	100
1969	244	86	1987	465	104
1970	228	76	1988	409	90
1971	193	62	1989	316	70
1972	217	68	1990	301	65
1973	180	55	1991	342	73
1974	192	56	1992	317	69
1975	200	60	1993	351	76
1976	220	65	1994	313	65
1977	280	78	1995	296	61
1978	278	77	1996 (preliminary)	229	48

the former USSR. By contrast, grain yields in developing countries rose 32% between 1980 and 1992.

— Farmers are now close to reaching the point of diminishing returns in crop yields. In fact, the general trend for most crops is sustained growth, often at high and even accelerating rates of gain.
— Modern high-yield farming is responsible for preserving a great deal of the world's biodiversity. If crop yields had remained at levels typical of the 1950s, farmers would have had to plough up an additional 10 million square miles of wildlife habitat to raise enough food for the current world food supply.
— For 10 000 years soil erosion has been a major threat to humanity's sustainable existence. Traditional low-yield farming is responsible for the highest rates of soil erosion. Two modern soil-saving techniques, conservation tillage and no-till systems, cut erosion by 65% and 98%, respectively.
— The UN Food and Agriculture Organization estimates that, despite progress, the number of people with inadequate diets has increased in the past ten years, from 540 million to 580 million. Nonetheless, it should be relatively easy to boost the vast majority over the threshold of hunger, since 90% of those with inadequate diets are now within 10% of having enough calories for good health.
— The risk to humans and wildlife from agrochemicals is far less than the risk of losing wildlife to low yields and the consequent ploughing up of more habitat.

These conclusions are in general over-optimistic.

Of course, the worldwide growth of agricultural production was due to the

Table 19. World fertilizer use, 1950–95

Year	Total (million tons)	Per capita (kg)	Year	Total (million tons)	Per capita (kg)
1950	14	5.5	1979	111	25.3
1955	18	6.5	1980	112	25.1
1960	27	8.9	1981	117	25.8
1965	40	12.0	1982	115	25.0
1966	45	13.2	1983	115	24.5
1967	51	14.6	1984	126	26.4
1968	56	15.8	1985	131	27.0
1969	60	16.5	1986	129	26.1
1970	66	17.8	1987	132	26.3
1971	69	18.2	1988	140	27.4
1972	73	18.9	1989	146	28.0
1973	79	20.1	1990	143	27.0
1974	85	21.2	1991	138	25.7
1975	82	20.1	1992	134	24.5
1976	90	21.6	1993	126	22.7
1977	95	22.5	1994	121	21.6
1978	100	23.2	1995 (preliminary)	122	21.3

introduction of new, more advanced technologies of land use and fertilization. The data of Table 19 illustrate world fertilizer use in 1950–95 (Brown et al., 1996).

The use of fertilizers increased by 250% over 20 years (the data for 1966–68 and 1986–88). It increased most of all in China (by a factor of 7). For comparison, relevant figures for the former USSR and the USA increased by 3.5 times and by 26%, respectively. The application of pesticides also increased, which was helpful in eliminating pests. However, many negative aspects of their application have been revealed simultaneously (Carson, 1994). Probably this is why in many regions the application of pesticides became less frequent (for instance, in Africa and both Americas). At the same time, it increased in Oceania (including Australia), Europe and Asia.

The application of pesticides in 1975–84 fell in the USA, Brazil, Indonesia and Japan, was stable in India and China, and increased in the former USSR (by 1.5 times), South Korea (by more than 3.5 times), Argentina (by almost 2 times) and Canada (by more than 2 times).

Significant changes occur as a result of land irrigation (Table 16). During the time period from 1966–68 to 1986–88, the total area of irrigated land increased from $156\,738 \times 10^3$ to $226\,626 \times 10^3$ ha, i.e. by 45%. 15% (instead of 11%) of all arable land are watered areas (United Nations..., 1991). In comparison with these earlier figures, the data of Table 16 are less impressive. In the former USSR the irrigated land constituted only 9% of the whole arable land, whereas in South Korea it was 63%, in the Netherlands 59%, in Japan 62%, in Surinam and Oman 86%, and in Egypt 100%. The situation changed drastically in the former USSR during the last years: the decrease of over 0.5% in the world's irrigated area (Table 16) is mainly due to the decline in the irrigated area in eight of the former republics of the Soviet Union (more than 800 000 ha). Although land irrigation is most beneficial in principle, it is sometimes accompanied by a deterioration of land quality (mainly, its salinization) in regions where irrigation technologies are far from perfect. In South America 14% of the total land area became degraded to some extent as a result of man's economic activity, including irrigation development.

Quite interesting are recent trends (the data for 1967–69 and 1987–89) of the state of the total number of livestock (cattle, goats and sheep, pigs, horses, buffaloes and camels). On a global scale it increased by 20%. During these years in Asia, in contrast to other continents, the total number of pigs increased abruptly, especially in China (by 79%). At present the total number of pigs in China is $335\,641 \times 10^3$, which constitutes 81% of the total number of pigs in Asia and 4% in the world. The total number of pigs in the former USSR was $78\,210 \times 10^3$ and in India $10\,267 \times 10^3$ (United Nations..., 1991). Table 20 shows the world meat production dynamics which was slightly positive in the last few years. Brown (Brown et al., 1996) points out that the surge in the world meat production in the last two years reflects a robust growth in the world economy, with much of the increase concentrated in lower and middle-income countries in Asia.

Beginning in the 1950s, the world catch of fish steadily grows (Table 21). During the time period from 1976–78 to 1986–88, it increased by 37% (United Nations..., 1991). Japan, the former USSR and China were three leading states with regard to

Table 20. World meat production, 1950-95

Year	Total (million.tons)	Per capita (kg)	Year	Total (million.tons)	Per capita (kg)
1950	44	17.2	1979	126	28.8
1955	58	20.7	1980	130	29.1
1960	64	21.0	1981	132	29.2
1965	81	24.2	1982	134	29.0
1966	84	24.5	1983	138	29.4
1967	86	24.5	1984	142	29.7
1968	88	24.8	1985	146	30.1
1969	92	25.5	1986	152	30.8
1970	97	26.2	1987	157	31.2
1971	107	26.7	1988	164	32.1
1972	106	27.4	1989	166	31.9
1973	105	26.8	1990	171	32.4
1974	107	26.6	1991	173	32.1
1975	109	26.7	1992	175	31.9
1976	112	26.9	1993	177	31.9
1977	117	27.6	1994	184	32.7
1978	121	28.2	1995 (preliminary)	192	33.4

Table 21. World fish harvest, 1950-94

Year	Total (million. tons)	Per capita (kg)	Year	Total (million. tons)	Per capita (kg)
1950	21	8.6	1979	71	16.2
1955	29	10.4	1980	72	16.2
1960	40	12.5	1981	75	16.6
1965	54	16.1	1982	77	16.7
1966	57	16.7	1983	78	16.4
1967	61	17.2	1984	84	17.6
1968	64	18.0	1985	86	17.7
1969	63	17.4	1986	93	18.8
1970	66	17.8	1987	95	18.9
1971	66	17.5	1988	99	19.4
1972	62	16.1	1989	100	19.2
1973	63	16.0	1990	98	18.5
1974	67	16.7	1991	98	18.2
1975	66	16.2	1992	99	18.1
1976	69	16.6	1993	102	18.4
1977	68	16.1	1994	109	19.3
1978	70	16.3			

their fish catch (their annual catch values for 1986–88 were $11\,907 \times 10^3$; $11\,251 \times 10^3$ and 9235×10^3 tons, respectively). In the period mentioned, the world catch of freshwater species increased by 80%, and that of sea fish by only 32%. As it has been pointed out above, the situation has substantially changed recently. In particular, conditions for freshwater fish species have changed (see Table 22).

Platt (Brown et al., 1996) points out that the total fish harvest, which includes both the fish catch and fish farming output, climbed to 109 million tons in 1994, up from 102 million tons in 1993 (Table 21). The 7% increase in production came primarily from expanding aquaculture and larger catches in China, Peru, and Chile. These countries accounted for 37% of the world catch; Japan, the USA, India, and the former USSR contributed 20%. A very important circumstance is, however, that, despite the growth in total harvest, supplies cannot keep pace with increasing population.

Abramovitz (Brown et al., 1996; State . . . , 1996) emphasized that at least one-fifth of all freshwater fish have become extinct, threatened, or endangered in recent years, and whole freshwater faunas disappeared. In the northern third of the world, 77% of the water in the 139 largest river systems has been significantly affected by fragmentation of the rivers and their tributaries by dams, water flow regulation from reservoir operation, irrigation, and inter-basin transfers. Aquatic habitats are also degraded as their watersheds are logged, mined, or converted to agriculture or grazing; wetlands and coastal areas are converted into fish farms; and rain-fed or flood-dependent agriculture is replaced by irrigation schemes.

Serious concern over the worldwide problem of over-fishing has been expressed by Jeffreys (Bailey, 1996):

Table 22. Freshwater fish species: status and threats in selected areas

Area	Known freshwater fish species (number)	Extinct (%)	Imperilled (%)	Principal threats
Global	9000+	(20 combined)		
Amazon river	3000+	—	—	Habitat degradation; over-harvesting
Asia	1500+	—	—	Habitat degradation; competition for water; over-harvesting
North America	950	2	37	Habitat degradation; introduced species
Mexico (arid lands)	200	8	60	Competition for water; pollution
Europe	193	—	42	Habitat degradation; pollution
South Africa	94	—	63	Habitat degradation; pollution; competition for water
Lake Victoria	350	57	43	Introduced species
Costa Rica	127	—	9	Habitat degradation
Sri Lanka	65	—	28	Habitat degradation
Iran	159	—	22	Habitat degradation; competition for water
Australia	188	—	35	Habitat degradation; competition for water

— The commercial harvest of ocean fisheries has risen from an estimated 18 million metric tons in the late 1940s to well over 80 million tons today. An additional 24 million tons annually is estimated to be taken by local fishermen.

— After thousands of years of extensive human use, many of the ocean's resources have been seriously damaged or depleted by shortsighted practices.

— Out of 81 marine fishery stocks examined, 14 are still considered to be overexploited, and another 36 were being fished at the maximum capacity, according to the National Fish and Wildlife Foundation.

— The annual operating costs of the global marine fishing fleet in 1989 were US $22 billion greater than the total revenues obtained from the sale of fish.

— The worldwide problem of over-fishing has its source in the system of open access and free use, with its norm that only capture confers ownership of the fish.

— When oceans are part of the commons, users are more interested in extracting what they can than in stewarding what they cannot own.

- The government has contributed to the overfishing of marine fisheries by granting to all citizens the right to fish.
- The "open access" treatment of the oceans has allowed super-efficient commercial fishing fleets to deplete the fisheries, in spite of state action to limit technologies, impose catch quotas, and restrict access to the fisheries.
- Drift nets, which indiscriminately net and kill any species they entangle, are an example of the inappropriate measures taken in the absence of property rights.
- As a first step, individual transferable quotas—a personal right or license, which may be sold or exchanged, to catch a given quantity of fish—can encourage better management of fisheries.
- When water-based resources are privately owned, the same legal rights that enable shopkeepers and homeowners to prevent trespass or property damage can be used to ward off water resource abuse.
- Attaching property rights to coastal zones would encourage better management of the fisheries and create legal rights against polluters.
- If coastal property rights are to be effective, they must be clearly defined, defended against challengers, and transferable.
- The coastal fisheries cooperative associations (FCA) system in Japan is exemplary of a property rights-based fishing industry. FCA property rights are sufficient to block potentially harmful or polluting coastal developments.
- In Washington State, where oyster beds may be privately owned, the threat of a damages suit against terrestrial sources of pollution has protected the state coastal zones.
- The lobster industry in Maine has developed a fairly effective, if informal, system of property rights management. Along the coast, small boat fishermen strategically place traps within their exclusively owned areas. Self-enforcement by these fishermen is the common response to interlopers and trap thieves.
- Aquaculture has great potential to relieve the impact of over-harvesting fisheries while simultaneously increasing the food supply.

4.3 Forests

The destruction of forests—a most valuable resource—is aggravated every year. During the period from 1966–68 to 1986–88, the global forest area dropped by 6% and became equal to 130 691 $\times$ 10^3 km^2. The tropical forest is being destroyed most of all. As many as 5.9 million ha are logged annually in the tropics, 4.9 million of which are primary forests. If, on a global scale, the portion of land occupied by forest diminished by 2% (from 33% to 31%), it was reduced much more markedly in individual countries, such as Costa Rica (from 55% to 32%), Honduras (from 46% to 31%), Brunei (from 93% to 48%), and Côte d'Ivoire (from 56% to 20%).

The destruction of the famous Amazonian forest in Brazil is of special concern: its area diminished from 71% to 66%. According to satellite data, in 1987 and 1988 the areas of annihilated forest were 8 x 10^6 and 4.8 $\times$ 10^5 ha, respectively. It should be noted that in 1988 the destruction of selva was somewhat reduced due to special state measures taken as a result of public pressure in Brazil and other countries.

Table 23. Global forest status, by region

Region	Forest and other wooded land (million ha)	Annual change (thousand ha)	Forests per capita (ha)	Forests as share of land (%)
Industrial regions	2064	−79	1.1	27
Europe	195	+191	0.3	27
Former Soviet Union		942	+51	2.235
North America	749	−317	1.7	25
Developed Asia and Oceania	178	−4	0.5	9
Developing regions	3057	−9874	0.5	26
Africa	1137	−2828	0.9	8
Asia/Pacific	660	−999	0.2	19
Latin America and Caribbean	1260	−6047	2.2	48
All regions	5120	−9953	0.6	27

As far as temperate forests are concerned, an important fact is that, despite a rather stable global growth of temperate forests, there is a continuing degradation of their quality. In Western Europe, for instance, only about 1% of the region's forests can be classified as old-growth, and much of that now consists of small isolated remnants (Brown et al., 1996). Acid rains are increasingly playing a role in forest decline, especially in Europe.

Acharya (Brown et al., 1996) points out that between 1980 and 1990, the world lost an average of 9.95 million hectares of net forest area annually—roughly the size of South Korea. The data of Table 23 illustrate forest area dynamics in various parts of the globe.

According to Brown et al. (Watson et al., 1996), at the present time forests cover about 4.1 Gha on the Earth. Most of the forests are in the low latitudes (43%), followed by high latitudes (32%) and mid-latitudes (25%). Forest plantations are currently estimated to occupy about 0.1 Gha. The world's natural forests contain vast quantities of organic carbon, with an estimated 330 Gt carbon in vegetation (live and dead, above and below ground), 660 Gt carbon in soil (mineral soil plus organic horizon), and another 10 Gt carbon in plantations. Some components of the carbon pool are poorly known, however, including woody detritus and slash and dead roots. Most of the carbon pool is located in high-latitude forests (52%). The data of Table 24 demonstrate some features of the global carbon pool (Watson et al., 1996).

Brown et al. (Watson et al., 1996) have pointed out that three categories of promising forestry practices that promote sustainable management of forests and at the same time conserve and sequester carbon (C) are important: (1) management for

Table 24. Estimated carbon pools and flux in forest vegetation (above- and below-ground living and dead mass, including woody debris) and soils (oxygen horizon and mineral soil to 1 m depth) in forests of the world. Dates of estimate vary by country and region, but cover the decade of the 1980s. Estimates are based on complete carbon budgets in all latitudes, using data from original source or from adjustments for completeness.

Latitudinal belt	Carbon pools		Carbon flux (Gt/yr)
	Vegetation (Gt)	Soils (Gt)	
High			
FSU	46	123	+0.3 to +0.5
Canada	12	211	+0.08
Alaska	2	11	[a]
Subtotal	60	345	+0.48 0.2
Mid			
USA	15	21	+0.1 to +0.25
Europe	9	25	+0.09 to +0.12
China	17	16	−0.02
Australia	18	33	trace
Subtotal	59	95	+0.26 0.1
Low			
Asia	41–54	43	−0.50 to −0.90
Africa	52	63	−0.25 to −0.45
America	119	110	−0.50 to −0.70
Subtotal	212	216	−1.65 0.40
Total	331	656	−0.9 0.5

[a] Included with USA.

conservation of existing C pools in forests by slowing deforestation, changing harvesting regimes, and protecting forests from other anthropogenic disturbances; (2) management for expanding C storage by increasing the area and/or C density innative forests, plantations, and agroforestry and/or in wood products; and (3) management for substitution by increasing the transfer of forest biomass C into products such as biofuels and long-lived wood products that can be used instead of fossil-fuel-based products.

Brown et al. (Watson et al., 1996) have listed the following opportunities for forest management and conservation:

— The most effective long-term (> 50 years) ways in which to use forests to mitigate the increase in atmospheric CO_2 are to substitute fuelwood for fossil fuels and for energy-expensive materials. However, over the next 50 years or so, substantial opportunities exist to conserve and increase the C store in living trees and wood products (High Confidence).

— Under baseline conditions (today's climate and no change in the estimated available lands over the period of interest), the cumulative amount of C that could potentially be conserved and sequestered over the period 1995–2050 by slowing deforestation (138 Mha) and promoting natural forest regeneration (217 Mha) in the tropics, combined with the implementation of a global forestation programme (345 Mha of plantations and agroforestry), would be about 60 to 87 Gt—equivalent to 12–15% of the projected (IPCC 1992a scenario) cumulative fossil fuel C emissions over the same period (Medium Confidence).
— The annual C gain from the above programme would reach about 2.2 Gt/yr by 2050, or about four times the value of 0.5 Gt C/yr estimated in the 1992 assessment. The gradual increase over time occurs because of the time it takes for programmes to be implemented and the relatively slow rate at which C accumulates in forest systems (Medium Confidence).
— Uncertainty associated with the C conservation and sequestration estimates is caused mainly by high uncertainty in estimating land availability for forestation and regeneration programmes and the rate at which tropical deforestation can actually be reduced; estimates of the net amount of C per unit area conserved or sequestered under a particular management scheme are more certain (High Confidence).
— The tropics have the potential to conserve and sequester the largest quantity of C—45–72 Gt—more than half of which would be due to promoting natural forest regeneration and slowing deforestation. Tropical America has the largest potential for C conservation and sequestration (46% of the tropical total), followed by tropical Asia (34%) and tropical Africa (20%). The temperate and boreal zones could sequester about 13 Gt and 2.4 Gt, respectively—mainly in the USA, temperate Asia, the former USSR, China, and New Zealand (Medium Confidence).
— The cumulative cost—excluding land costs and other transaction costs—to conserve and sequester the above amounts of C range from US$247 billion to $302 billion at a unit cost of about $2–8/t C. These unit costs are considerably lower than those in the 1992 assessment, which ranged from $8/t C in tropical latitudes to $28 in non-USA Organization for Economic Cooperation and Development (OECD) countries. Transaction costs may significantly increase these estimated costs (Low Confidence).
— Costs per unit of C sequestered or conserved generally increase from low- to high-latitude nations (High Confidence) and from slowing deforestation and promoting regeneration to establishing plantations (Low Confidence). The latter trend may not hold if transaction costs of slowing deforestation are excessive.
— These cost estimates, although benefiting from improved data and methodology since the first assessment, generally represent only the cost of direct forest practices. These costs could be several times higher if land and opportunity costs and/or the costs of establishing infrastructure, protective fencing, training programmes, and tree nurseries were included; on the other hand, costs could be offset by revenues from timber and non-timber products. No complete cost estimates are available (Medium Confidence).

For comparison, here are conclusions made by Sedjo (Bailey, 1995):

- Though nearly 75% of the total industrial wood production comes from northern hemisphere industrial countries, the temperate forestlands of this region are expanding.
- The world's current industrial wood consumption requirements could be produced on only 5% of the world's total current forestland.
- About one-third of the Earth's terrestrial area remains wilderness.
- Tropical forests cover a land area that is almost exactly the size of South America.
- Tropical and temperate forests together cover a land area the size of the western hemisphere.
- The rate of conversion of the tropical forests increased from 0.6% in 1980 to about 0.8% in 1990.
- Fully two-thirds of the deforestation in the USA occurred in the 60 years prior to 1910 and most of the other third before 1850.
- Although the USA has been the world's number one timber producer since World War II, US forests have experienced an increase in volume in the past 50 years and have maintained roughly the same area over the past 75 years.
- The total area of forest and other wooded land in Europe increased by 2 million hectares annually between 1980 and 1990.
- The total forest area of the temperate region's industrialized countries increased between 1980 and 1990.
- During 1993, an estimated 4 million trees were planted in the USA each day.
- The forest biomass in the northern Rockies has increased by 30% or more since the middle of the eighteenth century.
- In recent years, private forestlands have accounted for 85% of total tree planting and seeding in the USA.
- The expansion of American forests has been made possible by improved tree-growing technology, the advent of tree plantations, improved control over wildfire, and the reversion of many agricultural lands, especially in the South and East, to forest.
- Deforestation in the Americas was probably greater before the Columbian encounter than it was for several centuries after.
- The recent proliferation of plantation forests in both the temperate and the tropical world has helped to deflect harvesting pressure from natural forests.
- Almost all of the timber harvested in the USA, Europe, and Nordic countries comes from second-growth/plantation forests.
- Commercial logging is not a major cause of deforestation; expanding agriculture is. In temperate countries, which provide over three-fourths of the world's industrial wood, reforestation is the rule, while in tropical countries forestland conversion to agriculture remains common.
- The developed countries in the temperate regions appear to have largely completed forestland conversion to agriculture and have achieved relative land-use stability. By contrast, the developing countries in the tropics are still in a

land conversion mode. This suggests that land conversion stability correlates strongly with successful economic development.

The role of forest dynamics in the global carbon cycle under conditions of increasing carbon dioxide concentration (including "CO_2 fertilization") will be discussed in Chapter 2.

4.4 Fresh water

Water resources are becoming a very serious threshold for further development in many countries (Kaczmarek et al. in: Watson et al., 1996). Brown (1996) points out that with water use exceeding the sustainable yield of aquifers in so much of the world, overpumping is now commonplace. Even at current levels of consumption, underground water tables are now falling in the southwestern USA, the US Great Plains, several states in India (including the Punjab, the country's breadbasket), in much of northern China, across northern Africa, in southern Europe, and throughout the Middle East. In some cases, the fall in water tables is measured in centimetres per year. In more extreme cases, it is gauged in metres per year.

Claims on rivers are also becoming excessive, draining some rivers dry before they reach the sea. China's Huang He (Yellow River) often runs dry before it gets to the Yellow Sea. At one point in early 1995, it dried up some 620 kilometres from the sea. Similarly, the Colorado River, the major river in the southwestern USA, rarely makes it to the Gulf of California; more often it disappears somewhere in the Arizona desert. And the Amu Dar'ya, which originates in the northern Himalayas and once fed the Aral Sea, is now drained dry by Uzbek and Turkmen cotton farmers along the way.

Conflicts over shared river systems are intensifying as well. Bangladesh is protesting India's excessive use of the Ganges, which leaves all too little for Bangladesh to irrigate its cropland. This kind of conflict has become typical for many countries all over the world.

The data of Table 25 characterize the dependence on imported surface water of various countries; this is an illustration of the scale of transboundary problems (State..., 1996). It is well known that the situation with fresh water reserves differs very much in various countries (United Nations..., 1991). The situation is tense in a number of states, particularly in those where a considerable part of renewable fresh water reserves is consumed, for instance, in Afghanistan 52% (1987), Tunisia 53% (1985), Egypt 97% (1985), Israel 88% (1986). For comparison, only 8% of renewable fresh water reserves was consumed in the former USSR (1980).

Fresh water is consumed in different ways. For example, in Asia and Africa during the 1980s, over 85% of fresh water was utilized for agricultural purposes. At the same time, in Europe only 30% of fresh water is used in agriculture, and its major part (55%) is spent for industrial needs.

About one-third (1000 km^3) of global fresh water consumption is by four countries: China, India, Pakistan and Japan, which spend it mostly for agricultural needs. This is understandable because 74% of the total irrigated land in Asia is situated within these four states.

Table 25. Dependence on imported surface water, selected countries

Country	Share of total flow originating outside of border (%)
Turkmenistan	98
Egypt	97
Hungary	95
Mauritania	95
Botswana	94
Bulgaria	91
Uzbekistan	91
Netherlands	89
Gambia	86
Cambodia	82
Syria	79
Sudan	77
Niger	68
Iraq	66
Bangladesh	42
Thailand	39
Jordan	36
Senegal	34
Israel[a]	21

[a] Includes only flows originating outside borders; a significant additional share of Israel's fresh water originates from occupied, disputed territories.

Postel (State . . . , 1996) has justly emphasized that without water, life and growth cease. At the present time, worldwide, agriculture uses about 65% of all the water removed from rivers, lakes, and aquifers for human activities, compared with 25% for industries and 10% for households and municipalities. It takes about 1000 tons of water to produce a ton of harvested grain. With the annual global population growth of about 90 million people, it requires an additional 27 billion cubic metres of water (about half of the annual flow of China's Huang He River). By 2025 (with the predicted population growth), additional 780 billion cubic metres of water will be necessary (more than nine times the annual flow of the Nile River). It is not at all clear where this water can come from.

Table 26 presents data on groundwater depletion in major regions of the world (State . . . , 1996). It is obvious that even now sources of groundwater in many regions of the globe are close to exhaustion. On the other hand, many of the major rivers are suffering from over-exploitation. A severe threat for water resources, depleted as it is, is rapid growth of urban population. In China, for instance, some 300 cities are at present experiencing a shortage of water.

A serious decline of the aquatic environment has taken place because of over-exploitation of water resources (excessive use of water runoff, large dams, etc.). In

Table 26. Groundwater depletion in major regions of the world, circa 1990

Region/Aquifer	Estimates of depletion
High Plains Aquifer System, USA	Net depletion to date of this aquifer that underlies nearly 20% of all US irrigated land total some 325 billion cubic metres, roughly 15 times the average annual flow of the Colorado River. More than two-thirds of this occurred in the Texas High Plains, where irrigated area dropped by 26% between 1979 and 1989. Current depletion is estimated at 12 billion cubic metres a year.
California, USA	Groundwater overdraft averages 1.6 billion cubic metres per year, amounting to 15% of the state's annual net groundwater use. Two-thirds of the depletion occurs in the Central Valley, the country's vegetable basket.
Southwestern, USA	Water tables have dropped more than 120 metres east of Phoenix, Arizona. Projections for Albuquerque, New Mexico, show that if groundwater withdrawals continue at current levels, water tables will drop an additional 20 metres by 2020.
Mexico City and Valley of Mexico	Pumping exceeds natural recharge by 50–80%, which has led to falling water tables, aquifer compaction, land subsidence, and damage to surface structures.
Arabian peninsula	Groundwater use is nearly three times greater than recharge. Saudi Arabia depends on non-renewable groundwater for roughly 75% of its water, which includes irrigation of 2-4 million tons of wheat per year. At the depletion rates projected for the 1990s, exploitable groundwater reserves would be exhausted within about 50 years.
African Sahara	Vast non-recharging aquifers underlie North Africa. Current depletion is estimated at 10 billion cubic metres a year.
India	Water tables are falling throughout much of Punjab and Haryana states, India's breadbasket. In Gujarat, groundwater levels declined in 90% of observation wells monitored during the 1980s. Large drops have also occurred in Tamil Nadu.
North China	The water table beneath portions of Beijing has dropped 37 metres over the last four decades. North China now has eight regions of overdraft, covering 1.5 million hectares, much of it productive irrigated farmland.
South-East Asia	Significant overdraft has occurred in and around Bangkok, Manila, and Jakarta. Over-pumping has caused land to subside beneath Bangkok at a rate of 5–10 centimetres a year for the past two decades.

California, for example, water development has decimated aquatic systems and the life dependent on them. The state has lost 95% of its wetlands. A real environmental tragedy is the drying out of the Aral Sea (Grigoryev and Kondratyev, 1996a).

Assessing the global water resources situation, Postel (State..., 1996) correctly emphasized:

> For all its impressive engineering, modern water development has been governed by a fairly simple calculus: estimate the demand for water and then build a new supply to meet it. It is an equation that ignores the complexities of the natural world, questions of human equity, concerns about other species, and the welfare of future generations. In a world of resource abundance, it may have served humanity adequately. But in a world of scarcity, it is a recipe for trouble.

Postel (State..., 1996) pointed out in connection with the water chapter of Agenda 21:

Although helpful as guiding principles, these documents lack concreteness. What is needed is a vision of a more sustainable water future, clear criteria and targets for measuring progress toward it, and a package of policy tools and practical steps to guide water management toward desired and realistic ends. At the core of such a strategy are the basic goals of sustainability—ecological integrity, efficiency, equity, and participatory decision-making. The setting of ecological criteria that ensure protection of aquatic systems is a critical first step.

Assessing the situation with fresh water resources, Kaczmarek et al. (Watson et al., 1996) concluded:

— Most of the regional water resources systems in the twenty-first century, particularly in developing countries, will become increasingly stressed owing to higher demand to meet the needs of a growing population and economy, as well as to protect ecosystems (High Confidence).
— Arid and semi-arid watersheds and river basins are inherently the most sensitive to changes in temperature and precipitation (High Confidence).
— Water demand for irrigated agriculture is very sensitive to climate change, especially in arid and semi-arid regions (High Confidence).
— The current generation of transient General Circulation Models (GCMs), though much improved, does not offer the degree of watershed-specific information or anticipated variability in future climate required to allow robust estimates to be made regarding changes in water availability (High Confidence).
— Water demand management and institutional adaptation are the primary components for increasing the flexibility of water resources systems to meet increasing uncertainties due to climate change (High Confidence).
— Increased streamflow regulation and water management regimes may be necessary to enable water systems to meet their goals (High Confidence).
— Isolated single-reservoir systems are less adaptable to climate change than integrated multiple-reservoir systems (High Confidence).

- Technological innovations and cost-effective technologies have already played a major role in water management; likely future technological changes can serve to mitigate many of the consequences of climate change (Medium Confidence).
- Changes in the mean and variability of water supply will require a systematic reexamination of engineering design criteria, operating rules, contingency plans, and water allocation policies (Medium Confidence).
- Temporal streamflow characteristics appear to be more variable under future climate scenarios, and amplification of extremes appears likely (Low Confidence).

Anderson (Bailey, 1995) summarized the situation in the following way:

- USA population has increased by 200% since 1900; per capita water usage grew 500% to 800% during the same period.
- Global water use has tripled since 1950.
- Water recycling increased more than 400% between 1954 and 1985 and is expected to double again between 1985 and 2000.
- Transferring just 5% of agricultural water to municipal uses would meet the needs of urban areas in the western USA for the next 25 years.
- Estimates show that a 10% increase in price would decrease agricultural water consumption in California by 6.5% and cut overall consumption by 3.7% in the 17 western states of the USA.
- Current government policies do not adequately reflect the costs of operation, maintenance, and capital for water projects and fail to account for the environmental damage associated with water storage and delivery.
- Special interest groups constitute a formidable political force to block the efficient allocation of water.
- Inefficient water use prevails in urban areas because consumers do not have to pay the full cost of the water they use.
- Water overuse adversely affects the environment by reducing water in streams. Subsidized agricultural drainage water loaded with leached salts and trace elements was dumped into California's Kesterson Wildlife Refuge, causing death and deformation of many animals, reptiles, and fish.
- Eliminating laws against water marketing and establishing private water rights would give consumers an incentive to use water more efficiently.
- Removing legal impediments to water markets would allow private firms to enter the water supply industry and take the burden off the public treasury. Market forces could pare demand, boost supply, reallocate water, and end the threat of water crises.

4.5 Protected territories

The increasing pressure on nature requires serious national and international efforts to protect it from destruction. One of the forms of such efforts is organization of protected territories, which can be subdivided into five categories: research reservations (natural reservations with a strict regime of their maintenance), national parks, natural monuments, managed natural reservations (biological preserves), protected landscapes.

Relevant state bodies do not take sufficient care of such territories whose total area constitutes only 3.2% of the whole land territory (United Nations..., 1991). On continents, the best protected is nature in North America where 9.6% of land is protected. In Europe protected territories occupy 7.6% of its total area, in South America 5.6%, in Africa 3.9%, and on the largest continent of the planet, in Asia, 33% (without the former USSR).

The former USSR took little care of its natural reserves. The absolute data show that the former USSR possessed the largest (in Europe) amount of territories that should be protected: 24 073 717 ha. However, protected territories constituted only 1.1% of the total area of the country. Great Britain, which is poor in vacant areas, protects 19% of its territory, and the USA 10.5% (98 341 751 ha).

Apart from the five abovementioned categories of protected territories, there are three more which should also be protected to some extent. These are: (1) biospheric preserves (differing from the abovementioned biological preserves); (2) territories of outstanding universal value (belonging to the rank of the world heritage); and (3) wetland (with abundant hydrophytes). The latter being one of the most biologically productive ecosystems, is under special protection in the USA. There are no global data available on the state and development of wetlands. With regard to the USA, 22 states of the country lost 50% or more of their wetlands during the last 200 years (1780–1980). About 91% of such land disappeared in California.

Urgent measures are necessary to prevent further loss of soil fertility and soil erosion. The data of Table 27 (Watson et al., 1996) illustrate the scale of soil erosion processes. In the context of the problem of preserving agricultural resources, Gardner (State..., 1996) has emphasized that

> indeed, if our planet were a bank, disbursing loans of natural resources, agriculture would be among its biggest debtors. Each year, agricultural practices erode mountains" worth of topsoil, drain continents of their

Table 27. Estimates of current global rates of soil erosion and likely future trends

Region	Area (10^6 ha)	Soil erosion by water			Soil erosion by wind	
		Denudation rate (mm/yr)	Dissolved load (10^6 t/yr)	Future trends	Area (10^6 ha)	Future trends
Africa	227	0.023	201	+	186	+
Asia	441	0.153	1592	+	222	+
South America	123	0.067	603	+	42	–
Central America	46	0.055 (Central and North America)	758 (Central and North America)	+	5	–
North America	60			+/–	35	–
Europe	114	0.032	425	+/–	42	+/–
Oceania	83	0.390	293	+	16	+
World	1094	0.079	3872	+	54.8	+

Notes: + = increased risks; – = decreased risks.

Table 28. Extent and severity of desertification

Region	Light		Moderate		Strong		Severe	
	Area	%	Area	%	Area	%	Area	%
Africa	1180	9	1272	10	707	5.0	35	0.2
Asia	1567	9	1701	10	430	3.0	5	0.1
Australasia	836	13	24	4	11	0.2	4	0.1
North America	134	2	588	8	73	0.1	0	0.0
South America	418	8	311	6	62	1.2	0	0.0
Total	4273	8	4703	9	1301	2.5	75	0.1

Notes. Area desertified in 10^3 km^2; % = area desertified as percentage of total drylands; drylands = arid + semi-arid + dry sub-humid.

> groundwater reserves, and eliminate plant genetic material in what is essentially colossal resource borrowing from future generations. Because today's farming uses many renewable resources at well beyond their rate of replenishment, and because little effort is made to replace them, agricultural resource debts continue to mount. The borrowing, however, cannot continue indefinitely. As this century draws to a close, clear signs of lender fatigue are evident: five decades of resource overexploitation have drained reserves of natural capital in many regions, and have limited agriculture's opportunities for future growth and resource borrowing.

A very important phenomenon is desertification, the extent and severity of which are characterized by the data of Table 28 (Watson et al., 1996).

5. BIODIVERSITY

Today one of the most acute problems causing general alarm is that of biodiversity on the Earth, which is more and more threatened by the consequences of man's economic activity (Graves and Reavey, 1996; Heywood and Watson, 1996; Leakey and Lewin, 1995; State of the World..., 1996).

It has been mentioned in *Global Biodiversity Assessment* (Watson et al., 1995) that

> the Earth is home to a rich and diverse array of living organisms, whose genetic diversity and relationships with each other and with their physical environment constitute our planet's biodiversity. *This biodiversity is the natural biological capital of the Earth, and presents important opportunities for all nations.* It provides goods and services essential to support human livelihoods and aspirations, and enables societies to adapt to changing needs and circumstances. The protection of these assets, and their continued exploration through science and technology, offer the only means by which the nations of the world can hope to develop sustainably. The ethical, aesthetic, spiritual, cultural and religious values of human societies are an integral part of this complex equation.

Biodiversity is defined by the Convention on Biological Diversity as "*the variability among living organisms from all sources, including terrestrial, marine and other aquatic ecosystems and the ecological complexes of which they are part; this includes diversity within species, between species and ecosystems*". More simply, biodiversity is the variety of the world's organisms, including their genetic make-up and the communities they form. Biodiversity is dynamic: the genetic composition of species changes over time in response to natural and human-induced selection pressures; the occurrence and relative abundance of species in ecological communities changes as a result of ecological and physical factors (Heywood and Watson, 1996).

Ecological diversity. Ecological systems do not exist as discrete units, but represent different parts of a natural continuum. Although terms such as forest, grassland, wetland and coral reef are commonly used to denote ecological systems, their delineation and spatial scale often depend on the intended purpose of the classification.

Some commonly used terms are:

Biome—a continental-scale region characterized by its distinctive vegetation and climate.
Ecosystem—the individuals, populations and species that occur in a defined area, including their interactions with each other and with their physical environment.
Ecological community—a group of species inhabiting a particular area.
Habitat—the biological and physical environment of a particular species.

Organismal diversity. The enormous diversity between species, ranging from common annual herbs to bacteria of the deep ocean trenches, their arrangement and classifications reflecting their phyletic relationships, and their complex patterns of variation and distribution that they show, provide the very substance of biodiversity. Groups of interbreeding individuals within a species form distinct populations. Groups such as plants, birds, mammals, fishes, reptiles and amphibians—the species with which we are most familiar—account for only 3% of the estimated total, while the majority of species belong to groups such as insects, arachnids, fungi, nematodes and microorganisms.

Genetic diversity. Genetic differences between the individuals of a species provide the basis for the diversity that is found between species. Molecular studies have revealed a wealth of genetic variability in most species: in fact, individuals of virtually all species are genetically unique. Genetic diversity can be described at multiple levels from single genes to visible multi-locus traits. It is expressed as genetic variability both within and between populations. The amount and distribution of genetic diversity vary extensively among species in ways which are incompletely understood. It is well established, however, that genetic diversity within species is necessary to allow them to adapt to changing environmental conditions.

An increasing danger for the biosphere and life on the planet is the loss of biodiversity. Heywood and Watson (1996) state that the primary causes underlying the loss of biodiversity are demographic, economic, institutional, and technological factors including:

— increasing demands for biological resources due to increasing population and economic development;
— failure of people to consider the long-term consequences of their actions, often due to a basic lack of knowledge;
— failure of people to appreciate the consequences of using inappropriate technology;
— failure of economic markets to recognize the true value of biodiversity;
— failure of economic markets to apply the global values of biodiversity at local level;
— institutional failure to regulate the use of biological resources resulting from the growth in urbanization, changes in property rights, and shifting cultural attitudes;
— failure of government policies to address the over-use of biological resources; and
— increasing human migration, travel, and international trade.

These underlying causes manifest themselves in the loss, fragmentation, and degradation of habitats; the conversion of natural habitats to other uses; over-exploitation of wild resources; the introduction of non-native species; the pollution of soil, water, and atmosphere; and, more recently, signs of long-term climate change.

What is missing in the *Global Biodiversity Assessment* (Heywood and Watson, 1996) is the recognition of the fact that the loss of biodiversity results in the violation of the closed nature of global biogeochemical cycles (Gorshkov, 1990, 1995; Leakey and Lewin, 1995; Lovelock, 1995), and consequently leads to the destruction of the most important life support system of the planet—the biosphere. Leakey (Leakey and Lewin, 1995) state clearly, "*Homo sapiens*, dominant species that it has become, may be in the process of causing a biological catastrophe of immense proportions, through eroding the diversity of life at an alarming rate", and further,

> The inputs and outputs are the interactions among species at all scales of life, from filaments of fungi nurturing the health of plant rootlets to the global chemical cycles of water, oxygen, and carbon dioxide. They are the ecosystem services to which I will also address the longer-term future, which puts our species in a larger geological context with the rest of the world's inhabitants. And I will suggest that the insights we have gained from the current intellectual revolution I formulated in the previous chapter demand that we adopt a certain ethical position on the impact of *Homo sapiens* on the biodiversity of which we are a part.
>
> Humans endanger the existence of species in three principal ways. The first is through direct exploitation, such as hunting. From butterflies, to songbirds, to elephants, the human appetite for collecting or eating parts of wild creatures puts many species at risk of extinction. Second is the biological havoc that is occasionally wreaked following the introduction of alien species to new ecosystems, whether deliberately or accidentally.... The third, and by far the most important, mode of human-driven extinction is the destruction and fragmentation of habitat, especially the inexorable cutting of tropical rainforests. The forests, which cover just 7 percent of the world's land surface, are a cauldron of evolutionary innovation and are home to half of the world's species. The

> continued growth of human populations in all parts of the world daily encroaches on wild habitats, whether through the expansion of agricultural land, the building of towns and cities, or the transport infrastructure that joins them. As the habitats shrink, so too does the Earth's capacity to sustain its biological heritage.

The monumental publication (Heywood and Watson, 1996) contains a very detailed consideration of the major issues relevant to biodiversity problems:

— characterization of biodiversity;
— magnitude and distribution of biodiversity;
— generation, maintenance and loss of biodiversity;
— biodiversity and ecosystem function—basic principles;
— biodiversity and ecosystem function—biome analyses;
— inventory and monitoring of biodiversity;
— the resource base for biodiversity assessments;
— data and information management and communication;
— biotechnology;
— human influences on biodiversity;
— economic values of biodiversity;
— measures for conservation of biodiversity and sustainable use of its components.

Therefore, we shall confine the discussion here to a few illustrations.

Representatives of the animate nature of Europe are subject to an especially great pressure. In a number of European countries, from 30% to 50% of their birds belong to the category of the risk of disappearance: in Germany, 98 species out of 305; in France, 136 out of 342; in the Netherlands, 85 out of 257; in Austria, 121 out of 201. For mammals the situation is even worse, and again it is the worst in Europe. In Germany 44 species of mammals out of 94 are on the threshold of dying out, in France 59 out of 113, in the Netherlands 39 out of 60 (United Nations..., 1991).

In comparison with the industrialized countries of Europe, and above all, Germany, France and Great Britain, the territory of the former USSR has suffered a far less damage in absolute terms, as far as biodiversity is concerned.

At first sight, it seems surprising that the situation with biodiversity is much less alarming in the USA and Japan, the two biggest industrial countries of the world, than in the former USSR. In these three regions (the former USSR, the USA, Japan) the threshold of the risk of disappearance has been surpassed, respectively, by mammals: 78 out of 357 species; 49 out of 466; 4 out of 186; by birds: 80 out of 765; 79 out of 1090; 35 out of 362; by reptiles: 37 out of 144; 26 out of 368; 3 out of 85.

The perilous situation with regard to African elephants should be mentioned specially. In East Africa over 10 years (1977–1987) their number dropped by 57, and in individual regions by 95. Simultaneously the export of ivory increased during the past decade. It reached 1 017 817 kg by 1983, after which, due to certain measures taken as a result of international efforts, the export of ivory started decreasing very considerably (to 182 212 kg in 1988).

Assessments of biodiversity changes are very much complicated by the poor data on this subject. Even reliable enough estimates of the number of species do not exist

Table 29. Approximate numbers of species in major groups. The groups of organisms estimated to contain more than 100 000 species, with vertebrates and others listed for comparison (all numbers in thousands).

Group	Number of described species[a]	Estimated total number of species in group
Viruses	4	400
Bacteria	4	1000
Fungi	721	500
'Protozoa'[b]	40	200
'Algae'[b]	40	400
Plants	270	320
Nematodes	25	400
Crustaceans	40	150
Arachnids	75	750
Insects	950	8000
Molluscs	c.70	200
Vertebrates	45	50
Others[c]	115	250
Totals	1750	13 620

[a] Estimated number of described species that are accepted today.
[b] The Protozoa and Algae are interpreted here in a broad sense.
[c] Mainly little-known animal groups.

as yet. It is known only that the total number of species on the Earth ranges from 7 to 20 million, and a good working estimate is between 13 and 14 million, of which only about 1.75 million species have been scientifically described; just under a fifth of them are plants and vertebrates. Table 29 characterizes approximate numbers of species in major groups (Watson et al., 1996).

It has been emphasized in (Heywood and Watson, 1996) that even for 1.75 million species that have been described, there is no comprehensive listing, and we have a highly incomplete and patchy understanding of their reproductive biology, their demography, the chemicals they contain, their ecological requirements and the roles they play in ecosystems. Genetic diversity within species is known well for only a very small number of species—primarily those that have direct importance for human health, scientific research and economic exploitation.

Leakey and Lewin (1995) have made such comments in this context:

> It is indeed a remarkable fact that we in this modern world, obsessed with measuring things, are so imprecise about the stuff of nature, to which we are intimately related and upon which we ultimately depend. We have a good estimate of how many stars there are in our galaxy, the Milky Way: some hundred billion. We know how many nucleotide bases constitute the human genetic blueprint: three billion. And we can calculate to within a few hours when a comet will collide with Jupiter, as it did at 4 p.m. (US Eastern Daylight

Table 30. Global human-induced conversions in selected forms of land cover

Cover	Date	Area ($\times 10^6$ km^2)	Date	Area ($\times 10^6$ km^2)	Percentage change
Cropland	1700	2.7	1990	14.4	+530
Irrigated cropland	1800	0.1	1989	2.0	+2000
Closed forest	pre-agricultural	46.3	1983	9.3	−15
Forest and woodland	pre-agricultural	61.5	1983	52.4	−15
Grassland/pasture	1700	68.6	1980	67.9	−1

> Time) on 16 July 1994. Yet we cannot put a secure number on current species diversity. It is not through lack of knowing how to obtain it, but through lack of commitment. Governments have spent hundreds of millions of dollars in the systematic study of the stars but only a tiny fraction of that sum on a systematic study of nature here on Earth.

The large-scale land cover transformation shown in Table 30 (Watson et al., 1995) is a very important factor of the loss of biodiversity.

Many species are subject to the danger of elimination. The data of Table 31 (Watson et al., 1995) show that the minimum estimates for numbers of globally threatened animal and plant species in 1994 were 5 400 and 26 000, respectively. About 13% of bird species, 18% of mammals, 18% of fish, and 11% of plants were catalogued as threatened.

It is important, however, to mention here what has been written by Leakey and Lewin (1995):

> To a paleontologist, death is a fact of life, and extinction is a fact of evolution. Some thirty billion species are estimated to have lived since multicellular creatures first evolved, in the Cambrian explosion. According to some estimates, thirty million species populate today's Earth. This means that 99.9 percent of all species

Table 31. Number of species considered 'threatened' by the World Conservation Monitoring Centre (WCMC classifies all species listed as 'endangered', 'vulnerable', 'rare' or 'indeterminate' as threatened species)

Threatened	Endangered	Vulnerable	Rare	Indeterminate	Total
Mammals	177	199	89	68	533
Birds	188	241	257	176	862
Reptiles	47	88	79	43	257
Amphibians	32	32	55	14	133
Fishes	158	226	246	304	934
Invertebrates	582	702	422	941	2647
Plants	3632	5687	11485	5302	26 06

that have ever lived are extinct. As one statistics wag put it, "to a first approximation, *all* species are extinct". Of course, it should be made clear that the above quotation is relevant to the natural (very slow) evolution and not the rapid loss of biodiversity due to man's impact, which is running at a rate some thousand to ten thousand higher than background extinction.

Serious dangers for biodiversity are invasions and introductions of genetically modified organisms, environmental pollution, climate change and other factors.

It has been emphasized (Heywood and Watson, 1996) that the three mutually reinforcing objectives of the Convention on Biological Diversity are conservation, sustainable use and equitable sharing of benefits from biodiversity. In the context of relevant national decision-making, it is important to recognize the social, cultural and economic aspects of various actions, including the importance of local knowledge and values. A special role of monitoring biodiversity has been outlined as follows (Heywood and Watson, 1996):

> Biological monitoring programmes can evaluate the trends in the status of species, communities and ecological systems over time. The conservation of biodiversity can be enhanced by monitoring programmes, providing a "feedback loop" that signals to managers and policy-makers the need to make appropriate changes. Monitoring programmes should be designed at a scale relevant to the ecological processes being monitored and the pertinent conservation or management questions being asked. Often, regional- or landscape-scale monitoring is necessary to evaluate the aggregate impact on species or ecosystems of local activities or events. Similarly, both a national and a global scale are needed to understand and effectively manage such species as migratory birds. Monitoring programmes need to be evaluated over time to refine techniques and to test new technologies.
>
> Remote sensing from satellites and aircraft can be used to monitor changes in the extent of land cover, the distribution of major vegetation types, and the functioning of some types of ecosystems. It can be used to stratify ecological sampling and to identify situations where more detailed, ground-based monitoring is warranted. Carefully chosen, indicator species provide an efficient means for monitoring the status and trends in the populations of some other species and certain ecosystem services. However, no small subset of indicator species can be used to monitor all aspects of biodiversity relevant to management.

Concerning sustainable use of biodiversity, it has been concluded:

> Sustainable use of biodiversity is considered today to be a prerequisite for sustainable social and economic development. Properly done, it ensures the continuing provision of goods and services from ecosystems and their components. In order to use biodiversity sustainably, it is essential to have a basic understanding of ecosystems, their components, and the social and economic pressures that affect them. Resource management practices that use our knowledge of the processes in intact ecosystems are often more effective and less

costly than those that do not. For example, forestry practices that mimic the frequencies of natural disturbances such as fires, appear to offer the best opportunities for maintaining the biodiversity associated with many forest ecosystems. In fisheries, better monitoring of fish stocks offers hope for increasing the sustainability of harvests and the conservation of marine biodiversity.

Social and economic measures may be more important than technical measures for ensuring sustainable use. For example, the assignment or recognition of clearly defined property rights and tenure systems is vital to sustainable forestry and fisheries. Of particular concern are equitable measures to transform open access regimes to private, community or other ownership systems to bring to a halt the overexploitation of forest, freshwater and marine ecosystems.

Edwards (Bailey, 1995) arrived at the following conclusions:

— An estimated 8 to 10 million different species live on the planet. Of these, only about 1.5 million have been named.
— Documented animal extinctions peaked in the 1930s, and the number of extinctions has been declining since then.
— Thirty-nine percent of the extinctions resulted from the introduction of non-native species, 36% from habitat destruction, and 23% from hunting (for sport and products for sale) and deliberate elimination of a species. The remaining 2% became extinct for a variety of causes, including pollution.
— More than 75% of the land on every continent except Europe is available for wildlife. It is this "undeveloped" land in developing countries that is of greatest importance to conserving biodiversity over the long term.
— Well over a billion hectares of the Earth's surface have been set aside as protected areas, representing over 10% of the land area.
— Since 1973, only 23 species have been removed from the US Endangered Species list: eight were listed in error, a court invalidated the listing of one, seven recovered, and seven became extinct. In other words, there is an equal probability that a species listed on the Endangered Species Act will recover or become extinct.
— Governments or conservation organizations need to begin to promote harvesting and using wild species rather than stopping such uses.
— Trade in wildlife products is of the same order of magnitude as trade in forestry and fish products. In 1986, international trade in primates, ivory, orchids, reptile skins, fur-bearers, and tropical fish was worth $5 billion.
— To enhance the probability of sustaining the wild harvest or use, the rural people living in developing countries with wildlife resources must have sufficient incentives to conserve the resources, free from excessive government interference. Environmentalists need to recognize that rural people can be allies in conservation, and not treat them as environmental criminals.
— Listing species to achieve legal protection is not working. More land and marine areas are being designated as protected; however, it is difficult to assess how effective protected areas are at conserving biodiversity.

The most important conclusion made by Leakey and Lewin (1995) is that: "we are but one of millions of species here on Earth, product of half a billion years of life's flow, lucky survivors of at least twenty biotic crises, including the catastrophic Big Five". The authors of this quotation define the "Big Five" as a succession of natural biotic crises, in which at least 65% of species became extinct in a brief geological instant: the end-Ordovician (440 million years ago), the Late Devonian (365 million years ago), the end-Permian (225 million years ago), the end-Triassic (210 million years ago), and the end-Cretaceous (65 million years ago). "The sixth extinction" has been defined as anthropogenically induced destruction of the biosphere through the loss of biodiversity. Leakey and Lewin (1995) conclude:

> We should be concerned because, special though we are in many ways, we are merely an accident of history. We did not arrive on Earth as if from outer space, set down amid a wondrous diversity of life, blessed with a right to do with it what we please. We, like every species with which we share the world, are a product of many chance events, leading back to that amazing explosion of life forms half a billion years ago, and beyond that to the origin of life itself. When we understand this intimate connection with the rest of nature in terms of our origins, an ethical imperative follows: it is our duty to protect, not harm, them. It is our duty, not because we are the one sentient creature on Earth, which bestows some kind of benevolent superiority on us, but because in a fundamental sense *Homo sapiens* is on an equal footing with each and every other species here on Earth. And when we understand the Earth's biota in holistic terms—that is, operating as an interactive whole that produces a healthy and stable living world—we come to see ourselves as part of that whole, not as a privileged species that can exploit it with impunity. The recognition that we are rooted in life itself and its well-being demands that we respect other species, not trample them in a blind pursuit of our own ends. And, by this same ethical principle, the fact that one day *Homo sapiens* will have disappeared from the face of the Earth does not give us license to do whatever we choose while we are here.

And this is the final quotation from Leakey and Lewin (1995) we have chosen:

> In the face of ignorance about how much of current biodiversity we need in order to sustain a healthy Earth's biota, is it more responsible to say (1) because we don't know if we need it all, we can safely assume we don't; or (2) we recognize the complexities of the system, and assume we do? The answer is obvious, because the costs of being wrong on the first count are enormous. In any case, many ecologists, extrapolating from the incomplete knowledge they have about the structure and dynamics of ecosystem services, believe that we do need all, or at least most, of what we currently have. Through continued destruction of biodiversity in the wake of economic development, we could push the natural world over a threshold beyond which it might be unable to sustain, first, itself and, ultimately, us. Unrestrained, *Homo sapiens* might not only be the agent of the sixth extinction, but also risks being one of its victims.

The Convention on Biological Diversity states that countries have rights over their biological resources, but they are also responsible for conserving their biological diversity and for using their biological resources in a sustainable manner (Keating, 1994). Nations that join the Convention shall:

— Identify the components of biological diversity important for conservation and sustainable use, and monitor activities which may have adverse impacts on this diversity.
— Develop national strategies, plans or programmes for the conservation and sustainable use of biological diversity.
— Make conservation and sustainable use of biological diversity part of planning and policy-making.
— Use the media and educational programmes to help people understand the importance of biological diversity and need for measures to conserve it.
— Establish laws to protect threatened species, develop systems of protected areas to conserve biological diversity, and promote environmentally sound development around these areas.
— Rehabilitate and restore degraded ecosystems and promote the recovery of threatened species, helping local people to develop and carry out these remedial plans.
— Establish means to control the risks from organisms modified by biotechnology.
— Use environmental impact assessment, with public participation, on projects that threaten biological diversity, in order to avoid or minimize damage.
— Prevent the introduction of, control or eradicate alien species which threaten ecosystems, habitats or species.

Many indigenous and local communities have a close dependence on biological resources, and nations should make use of this traditional knowledge of the conservation and sustainable use of biological diversity. Countries are to preserve and maintain such indigenous and local knowledge and promote its wider use. This is to be done with the approval and involvement of those who have such knowledge, and these people should benefit from the use of their practices.

The convention says that:

— Countries are to facilitate access to genetic materials within their borders for environmentally sound uses. Access will be allowed with the aim of sharing in a fair and equitable way the results of research and development and the benefits arising from the commercial and other uses of genetic resources.
— Developing countries are to have access to environmentally sound technologies that they need for the conservation and sustainable use of biodiversity. This access will be under fair and most favourable terms, and will recognize patent rights.
— Developing countries are to have access to technology that makes use of resources they provided. They are also to have a role in biotechnological research.
— Developing nations are to receive technical and scientific assistance, so that they

can develop their own institutions and expertise in sustainable use of biological diversity.

— Countries are to consider the need for an agreement on the safe handling and use of living organisms modified by biotechnology.

— Developed countries that sign the Convention shall provide new financial aid to developing countries to help them implement terms of the Convention. The initial funding will be handled by three United Nations organizations involved in environment and development.

6. ENVIRONMENTAL POLLUTION

This subject encompasses a number of vitally important problems which are reflected in many recent reports (Brown et al., 1996; Bruce et al., 1996; Climate Change..., 1995; State..., 1990, 1996; Tolba et al., 1992; United Nations..., 1991; Watson et al., 1996) considering key aspects of problems relevant to the state of the atmosphere: the greenhouse effect, ozone dynamics and urban pollution. It is well known (Climate Change..., 1995; Kondratyev, 1990, 1992a,b) that among gases responsible for the origination of the greenhouse effect of the atmosphere, the two basic ones are H_2O and CO_2. The data of the global observational network show that during the 1970s, 1980s and 1990s, a marked increase of the carbon dioxide concentration in the atmosphere occurred, and that by the year 1995 its concentration reached a new record level of 360.7 ppmv.

Global carbon emissions from fossil fuel burning and the CO_2 concentration dynamics data are shown in Table 32 (Brown et al., 1996).

The data of Table 33 demonstrate the distribution of carbon emissions among the 20 top emitters (State..., 1996).

As indicators of local carbon dioxide changes, the data of measurements performed in Alaska (327.86 ppm in 1974 and 354.88 ppm in 1989) and at Mauna Loa Observatory (329.72 ppm in 1974 and 352.25 ppm in 1989) are presented.

Data on CO_2 emissions of anthropogenic origin are doubtless very interesting (United Nations..., 1991). During the period from 1960 to 1988, a global increase of carbon dioxide emissions took place due to the use of fossil fuel and the production of cement factories. The increase was from $2\,490\,500 \times 10^3$ to $5\,669\,775 \times 10^3$ tons of carbon per year (t C/year). Data characterizing carbon dioxide emissions in individual countries are very interesting. In most European countries, the emissions increased by approximately a factor of 1.5 (here and in what follows, in tons per year). However, in some countries these emissions decreased, for instance, in Great Britain (from $160\,770 \times 10^3$ to $152\,520 \times 10^3$) and in Luxemburg (from 3175×10^3 to 2362×10^3). The biggest polluter of Europe outside Russia used to be Germany ($182\,704 \times$ x 10^3 for FRG within its previous borders, and $89\,262$ x 10^3 for the former DDR). The level of emissions in the former USSR ($1\,086\,000$ x 10^3) was equal to that of the entire remaining part of Europe ($1\,180\,202$ x 10^3). The situation has drastically changed, however, during recent years due to the economic decline.

Table 32. World carbon emissions from fossil fuel burning, 1950–95, and atmospheric concentrations of carbon dioxide, 1960–95

Year	Emissions (million tons of carbon)	Carbon dioxide (parts per million)	Year	Emissions (million tons of carbon)	Carbon dioxide (parts per million)
1950	1620	n.a.	1979	5239	336.7
1955	2020	n.a.	1980	5172	338.5
1960	2543	316.8	1981	5000	339.8
1965	3095	319.9	1982	4960	341.0
1966	3251	321.2	1983	4947	342.6
1967	3355	322.0	1984	5109	344.3
1968	3526	322.8	1985	5282	345.7
1969	3735	323.9	1986	5464	347.0
1970	4006	325.3	1987	5584	348.8
1971	4151	326.2	1988	5801	351.3
1972	4314	327.3	1989	5912	352.7
1973	4546	329.5	1990	5943	354.0
1974	4553	330.1	1991	6010	355.5
1975	4527	331.0	1992	5926	356.2
1976	4786	332.0	1993 (estimated)	5940	357.0
1977	4920	333.7	1994 (estimated)	5990	358.8
1978	4960	335.3	1995 (preliminary)	6056	360.7

The main source of global emissions of carbon dioxide is the USA (1.4 billion tons; 5.25 tons per capita annually). For Asian and African countries, a more intensive growth of emissions is characteristic, on the whole, than for other continents. In some such countries carbon dioxide emissions increased by several orders of magnitude: for example, in Cameroon (from 74×10^3 to 1532×10^3), Mauritania (from 10×10^3 to 853×10^3), or Qatar (from 40×10^3 to 4172×10^3, i.e. almost by a factor of 87). These data reveal the current trends of the transfer of "polluting industries" from industrially developed countries to developing states.

Since an important contributor of GHGs emissions is agriculture, Cole et al. (Brown et al., 1996) have analysed options for direct and indirect mitigation of GHGs emissions from agriculture (Table 34). More detailed information on carbon dioxide and other GHGs dynamics will be discussed in Chapter 2.

Among urban pollutants, special attention should be given to sulphur dioxide. Observational data for the period 1970–88 (United Nations..., 1991) demonstrate a decrease of its emissions in industrially developed countries of the world, such as the USA (from 28 200 to $20\,700 \times 10^3$ ton/year), FRG (from 3750×10^3 to 1300×10^3, i.e. by about a factor of 3), and the former USSR. This drop of SO_2 emissions in European or North American countries took place because relevant measures had been taken there. Corresponding data for developing countries are scarce; however, they are sufficient to show an opposite pattern.

Table 33. Carbon emissions from fossil fuel burning, top 20 emitters, 1994

Country	Total emissions (million tons)	Emissions per person (tons)	Emissions per dollar GNP[a] (tons per million dollars)	Emissions growth 1990–94 (%)
USA	1371	5.26	210	4.4
China	835	0.71	330	13.0
Russia	455	3.08	590	−24.1
Japan	299	2.39	110	0.1
Germany	234	2.89	140	-9.9
India	222	0.24	160	23.5
United Kingdom	153	2.62	150	−0.3
Ukraine	125	2.43	600	−43.5
Canada	116	3.97	200	5.3
Italy	104	1.81	110	0.8
France	90	1.56	80	−3.2
Poland	89	2.31	460	−4.5
South Korea	88	1.98	200	43.7
Mexico	88	0.96	140	7.1
South Africa	85	2.07	680	9.1
Kazakhstan[b]	81	4.71	1250	n.a.
Australia	75	4.19	230	4.2
North Korea[b]	67	2.90	960	n.a
Iran[b]	62	1.09	270	n.a.
Brazil	60	0.39	70	15.8

[a] GNP data are adjusted for purchasing power parity and are for 1993.
[b] Latest data available are for 1992.

Observations of another significant urban pollutant—nitrogen dioxide—are less systematic. On the whole, for the period 1970–88, a stable emission value was observed, or its minor increase (in the USA, Canada, Great Britain, and the former USSR).

The data of Table 35 characterize the world sulphur and nitrogen emissions from fossil fuel burning, and those of Table 36 illustrate the scale of global CFC production (Brown et al., 1996).

Global data on the concentrations of metal compounds in the urban atmosphere are too fragmentary. For the USA, a marked drop in the emissions of lead is characteristic (from 141 200 tons in 1977 to 8600 tons in 1986). A similar trend has been observed in a number of other countries.

Very interesting are data on the transboundary transport, as well as on the dry and wet deposition of atmospheric contaminants. This is one of the most important aspects of the relationship between man and nature; this problem more and more affects the interests of entire groups of states. A network of 102 stations is used in Europe for studying the transboundary transport of pollution. The amount of deposited sulphur in Europe (1988) varies from 10 kg/(ha × year) in Scandinavia to over 50 kg/(ha × year) in East European countries. The greatest amount of

Table 34. Options for direct and indirect mitigation of greenhouse gas emissions from agriculture

	CO_2	CH_4	N_2O
1. Land conversion and management			
— Reduced deforestation rate	H	M	M
— Pasture immediately after deforestation	M		L
— Conversion of marginal agricultural land to grassland, forest, or wetland	M		L
2. Agricultural land utilization and management			
— Restoring productivity of degraded soils	H	L	
— More intensive use of existing farmland	M	L	L
— Restricted use of organic soils	H		L
— Conservation tillage	M		L
— Reduction of dryland fallowing	M		
— Diversified rotations with forage crops	M		L
3. Biofuels			
— Energy crops for fossil fuel substitution	H		
— Agroforestry	L		
— Windbreaks and shelterbelts	L		
— Agroindustrial wastes for fossil fuel substitution	L		
4. Recycling of livestock and other wastes			
— Recycling of municipal organic wastes	L	L	M
— Biogas use from liquid manures		M	
5. Animal husbandry			
— Supplementing low-quality feed		M	
— Increasing feed digestibility		L	
— Production-enhancing agents		L	
6. Rice cropping systems			
— Irrigation management		M	L
— Nutrient management		M	L
— New cultivars and other		M	L
7. Plant nutrient management			
— Improved fertilizer use efficiency	L		H
— Nitrification inhibitors			M
— Legume cropping to bolster system productivity			M
— Integrating crop and animal farming			L
8. Minimizing overall N inputs			
— Reduced protein inputs in animal feed			M
— Reduced protein consumption by society		M	M

Notes: Mitigation potential: L = low, M = medium, H = high.

Table 35. World sulphur and nitrogen emissions from fossil fuel burning, 1950–93

Year	Sulphur (million tons)	Nitrogen (million tons)	Year	Sulphur (million tons)	Nitrogen (million tons)
1950	30.1	6.8	1981	61.9	22.1
1960	46.2	11.8	1982	62.1	22.2
1970	57.0	18.1	1983	63.0	22.5
1971	56.9	18.6	1984	64.5	23.3
1972	58.2	19.5	1985	64.2	23.4
1973	60.9	20.6	1986	65.2	23.6
1974	60.9	20.8	1987	66.5	24.3
1975	56.4	19.9	1988	68.4	25.3
1976	58.6	21.0	1989	70.8	26.5
1977	60.1	20.8	1990	68.7	26.3
1978	61.0	22.3	1991	69.9	26.7
1979	62.6	22.4	1992	70.0	26.5
1980	62.9	22.3	1993	69.5	26.7

Table 36. World CFC production, 1950–95

Year	Total (thousand tons)	Year	Total (thousand tons)
1950	42	1979	850
1955	86	1980	880
1960	150	1981	890
1965	330	1982	870
1966	390	1983	950
1967	440	1984	1050
1968	510	1985	1090
1969	580	1986	1130
1970	640	1987	1250
1971	690	1988	1260
1972	790	1989	1150
1973	900	1990	820
1974	970	1991	720
1975	860	1992	630
1976	920	1993	520
1977	880	1994	388
1978	880	1995 (preliminary)	300

deposited sulphur per unit area was observed in 1988 in the former DDR, and also in Czechoslovakia and Poland. Studies of sulphur dioxide concentration variations in precipitation for 6–8 years (1979–88, 1981–86) revealed their increase in the eastern part of the Gulf of Finland and southern Sweden.

A rather optimistic summary on atmospheric pollution has been made by Goklany (Bailey, 1996):

— Richer is cleaner. Anything that retards economic growth also retards ultimate environmental cleanup.
— Traditional pollutants like smoke and particulate matter, sulphur dioxide, and faecal coliform bacteria begin to decline when per capita incomes reach $3280; $3670, and $1375, respectively.
— Empirical evidence indicates that projections using the famous $I = PAT$ equation, widely publicized by biologist Paul Ehrlich, are dead wrong. I represents the cumulative impact on the environment; P, the population; A, affluence; and T, technology. Greater economic and technological progress are necessary preconditions for the solution, rather than the causes, of environmental degradation. They make cleaner environments more affordable, reduce the impacts per unit of production, and help moderate population growth.
— In the USA, sulphur dioxide emissions per capita are down 60% from the peak in 1920. Both particulates and carbon monoxide emissions per capita have declined more or less continuously since World War II—inhalable particulates down 79% since 1940 and carbon monoxide down 53% since 1945. Volatile organic compounds emissions per capita (contributors to smog) stayed roughly constant between 1925 and 1970, before they declined; in 1992 they were down 36% from the peak. Nitrogen oxide per capita peaked in 1980. It is now down about 10% from its peak.
— Ambient sulphur dioxide and particulate matter have been declining for at least the last two decades in most developed countries.
— Ambient sulphur dioxide levels in many world megacities have already declined significantly from the 1970s. Cities in the developed world have relatively low levels. Cities in the developing world have the highest levels.
— Ambient particulate matter is still a problem in most megacities. Among a sample of 20 world megacities, the three having low levels were all located in the developed world (New York, London, and Tokyo).
— The gravest air quality problem in the world is indoor air pollution due to heating and cooking with coal, wood, and dung or other biofuels in developing countries. Indoor particulate levels in poor countries rival those of the worst acute industrial episodes of the past.

Obviously, these conclusions are over-optimistic, but emphasis on economic growth is even dangerous. The principal purpose of this book is to demonstrate that unlimited economic growth is a sure way to global ecological catastrophe.

Data on the accumulation of soil pollution due to deposition from the atmosphere, summarized over the globe, are very fragmentary. In Europe (1988), over the territories of five states—Denmark, the Netherlands, FRG, Great Britain and France—about 330 thousand ha of land (4% of the entire area) were poisoned to a certain extent, 185 thousand ha of the area strongly poisoned. Pollution is an important cause of the decline in soil fertility (Carson, 1994).

Very fragmentary also are global data on various kinds of water pollution.

However, there are data available relevant to some rivers and regions (USA, Sweden, Norway). The rivers Rhine and Moselle have always been considered as the most polluted in Western Europe. Special measures have been taken, and, as a result, the concentrations of some heavy metals in both rivers decreased markedly, as revealed by measurement data. During the decade 1975–85, the Rhine has become clean due to the efforts of four countries: Switzerland, FRG, France and the Netherlands. The water quality of the Elbe has improved radically.

Data on the distribution of lakes with acid water in the USA are quite interesting. Lakes with the most acid waters are most frequent in the south-east (Florida, 22.1%) and the north-east of the country (4.6% of all lakes of the region). The eutrophication of lakes is a problem of general concern (Bukata et al., 1995; Kondratyev, 1990). It is regrettable that relevant data for the former USSR are lacking in the reports (Tolba et al., 1992; United Nations..., 1991). During the period from 1970 to 1987, the concentration of phosphorus decreased in the two largest lakes of Scandinavia, Melaren and Wettern, and the two largest North American lakes, Erie and Ontario; the concentration of nitrogen compounds in water dropped only in Melaren and Erie.

Information on the pollution of seas, gulfs and coastal waters are inadequate. There is no doubt, however, that in different regions of the world there are zones of gradually increasing eutrophication of sea waters, in particular, in the Mediterranean and Baltic seas. During the 1980s, the concentrations of nitrogen and phosphorus increased by factors of 4–6, and in the Kattegat Strait by more than 8.

Poisoning of the organic world, and in particular forests, by industrial emissions is most alarming. In 1989, among countries with greatest areas of degraded coniferous forests were Poland (34.5% of forests), Bulgaria (32.9%), Czechoslovakia (32%), Yugoslavia (34%), Denmark (24%). The least degraded coniferous forests were found in Spain (3.5%), Austria (4.1%), and, although to a lower degree, in Scandinavian countries: Sweden (12.9%), Norway (14.6%) and Finland (18.7%). Rapid increase of the amount of harmed coniferous trees in 1988–89 was observed in Bulgaria and Poland, and of degraded broad-leaved forests in Denmark, the Netherlands, Poland and Greece.

Data on the concentration of various substances in the organisms of water inhabitants are very convincing indicators of the state of the environment and the degree of chemical pressure it suffers. Unfortunately, such observations (in lakes, rivers, seas) are carried out most unevenly—both in time and space—in different countries of the world. Therefore, the data available are fragmentary and hard to compare. More representative are data on the content of mercury and chlororganic compounds in the organisms of the inhabitants of the Great Lakes. Annual concentrations of chlororganic compounds decreased in the late 1970s and early 1980s. This has been a result of the application of nature-protective technologies which made sinks less polluted. However, by the mid-1980s, the concentration of these substances in the organisms of fish increased again. This phenomenon has been attributed to the release of pollutants by the bottom deposition of the lakes.

Quite interesting are the results of investigations of the content of various harmful chemical compounds in food and in the human organism, as well as data on the radiation level in human dwellings. For example, human dwellings containing high concentrations of radon have been found in Great Britain, Norway and Sweden. The data on the radioactive pollution of territories in Europe followed from the Chernobyl nuclear power station accident have been discussed by Israel (1996).

Of great interest are data on the contamination of the natural environment in the historic past. Traces of this contamination can be found in sedimentary rocks, on the bottom of lakes, in the ice and snow covers (in the Arctic and Antarctic), in peat and tree rings. As might be expected, the concentrations of metals in the snow and ice covers of the Arctic and Greenland are higher than in the Antarctic. The concentration of lead in the ice of Antarctica was lower by a factor of 5 in the pre-industrial period than ever since. These and other similar data, obtained as a result of so-called historic monitoring of the state of various natural media, give a good perspective on the present state and trends of environmental pollution.

Waste is a real problem of today. Waste contributes to the deterioration of environmental quality, and consequently, to mass illnesses and even deaths of people. Judging by selective, rather limited data for different countries (no such data are available for the former USSR), the USA and Japan are the two main sources of solid waste in the world. The distribution of waste (in 10^3 ton/year) with regard to their types for 1985 in the USA and Japan, respectively, is as follows: municipal (178 000 and 41 530), industrial (628 000 and 312 000), agricultural (1 400 000 and 90 544), mining (1 300 000 and 26 017), the dry residue of sewage (8400 and 2003). The waste of the paper industry prevailed (34.7%) in 1985 in municipal waste in the USA.

The problem of harmful waste is extremely important (the waste of metal-working and chemical industries, as well as that of the production and processing of raw minerals, are especially harmful and poisonous). The "record" production of these types of waste in 1980 has been registered in the USA ($265\,000 \times 10^3$ ton/year) and India ($36\,000 \times 10^3$ ton/year). The greatest amounts of liquid radioactive waste have been produced in the USA and Canada, and in France (among all European countries). There are no relevant data available for the former USSR. From 1981 to 1984, an increase of such waste has been registered in most countries, and from 1985 an abrupt decrease began.

Waste disposal is a problem of primary importance because it is directly relevant to the improvement of the ecological situation. Municipal waste in the majority of West European countries, as well as in Canada and the USA is removed and stored in specially allotted areas. 70% (and in Great Britain as much as 90%) of waste is disposed of in this way. A certain portion of waste is recycled.

Dangerous industrial waste in the USA (70%) and the European Union countries (50%) is also stored in special areas, the remaining waste is recycled. In the developing countries, the recycling of waste is made difficult by the lack or deficit of relevant equipment. The release of waste to the sea is difficult to control. From 1946 to the 1980s, more than 850 regions have been discovered in North Atlantic and North Pacific where radioactive waste has been disposed of. The release of oil and oil products to the sea causes great damage to nature. The total amount of oil released to the sea from 1981 to 1989 was 945×10^3 tons; 56% of the total amount has been the result of 10 big catastrophes.

The waste amount may be reduced through its repeated (sometimes several times) utilization. Comparative data of this kind for the world are insufficient. For example, the share of the world production of glass by means of a secondary processing of the raw material increased from 21% in 1972 to 30% in 1988.

McMichael (1995) discusses in detail the impact of environmental pollution on human health.

7. CLIMATE CHANGE

Since this problem will be discussed in detail in Chapter 2, we shall only mention here very briefly the results of observations and numerical modelling of global average temperature changes which may be caused by the greenhouse effect, according to which the global mean temperature will increase at the rate of about 0.3°C per 10 years in the next century. As a result, by the year 2025, it may increase (in comparison with the pre-industrial time) by 2°C, and by the year 2100 by 4°C. Global warming must be accompanied by the intensification of precipitation which may constitute a several per cent increase by the year 2030. Besides, it may be expected that the world ocean level will increase by about 20 cm by the year 2030, and by 65 cm by the end of the next century (see Climate Change..., 1995; Kondratyev, 1990, 1992a,b). Assessments of these model data will be discussed in detail in Chapter 2.

According to observational data, during the 1980s and the 1990s, the global air temperature was higher than during any other decade throughout the past 140 years. Beginning with the present century, it increased by about 0.5°C.

The data of Table 37 (Tuneli, in Brown et al., 1996) illustrate the secular change of global average surface air temperature. It can be seen from Table 37 that 1995 was the warmest year since 130 years ago (the global average temperature was lower before 1950).

Very interesting are data on the correlation between the drop of precipitation in the sub-Saharan zone started from the early 1960s (i.e. during the period of drought) and anomalies in the surface temperatures of the Indian and Atlantic Oceans during the same period. According to the data of satellite observations, the annual mean extension of sea ice cover in the Arctic and Antarctic for the period 1979–93, was practically constant in the southern hemisphere, whereas in the northern hemisphere a weak trend of its decrease has been observed during the last few years.

There exist controversial judgements concerning increased anomalous weather events (typhoons, dust storms, etc.) under conditions of global warming (see Chapter 2). Observational data do not reveal substantial changes. On the other hand, it is a fact that the insurance industry payoffs for weather-related damage reached $48 billion in the 1990s, compared with $16 billion for the entire 1980s.

Table 37. Global average temperature, 1950–95

Year	Temperature (degrees Celsius)	Year	Temperature (degrees Celsius)
1950	14.86	1979	15.09
1955	14.92	1980	15.18
1960	14.98	1981	15.29
1965	14.88	1982	15.08
1966	14.95	1983	15.24
1967	14.99	1984	15.11
1968	14.93	1985	15.09
1969	15.05	1986	15.16
1970	15.02	1987	15.27
1971	14.93	1988	15.28
1972	15.00	1989	15.22
1973	15.11	1990	15.38
1974	14.92	1991	15.36
1975	14.92	1992	15.11
1976	14.84	1993	15.14
1977	15.11	1994	15.23
1978	15.06	1995 (preliminary)	15.39

There are, however, explanations of the statistics which are not relevant to weather changes (see Bruce et al., 1996).

8. NATURAL CALAMITIES

Natural calamities are a significant destabilizing factor in the global ecological situation. World statistics have revealed their growing frequency during the period from 1960 to 1989, with a peak in 1987 (United Nations . . . , 1991). The cause of this maximum is not clear. Possibly, the enhanced severity of calamity consequences have been due to the demographic situation (and in particular, to the total increase of the planet's population, especially in big cities, and to migration processes). It should also be pointed out that the available data on the frequency of dangerous meteorological events (tropical storms, torrential rains, floods, etc.) are still controversial. It has been shown, for example, that the global mean precipitation has not changed on decadal-to-multidecadal time scales, although fluctuations about the mean have increased significantly (Tsonis, 1996).

The probability of the occurrence of catastrophic (with a magnitude of 6–8 points on the Richter scale) earthquakes is the highest in approximately 35 countries. During the period from 1980 to 1990, the greatest earthquakes, which claimed numerous victims, took place mainly in Asia: in Iran (7.3 points, 50 000 dead, 1990), in the former USSR (7.6 points, 25 000 dead, 1988), in Turkey (7.6 points, 25 000 dead, 1988), and also in Mexico (8.1 points, 10 000 dead, 1985). The risk of a catastrophe is very big in any of the abovementioned countries; however, at present, great catastrophes are practically unpredictable and may occur in a completely different country (as was the case with Japan in December of 1994).

Underwater earthquakes and tsunamis also cause a considerable damage. At least 22 countries situated along the coasts of the Pacific Ocean are subject to tsunamis. However, only part of these states are prepared for such events. During a single decade from 1977 to 1986, about 6 600 people perished because of tsunamis. Numerous victims and substantial material damage are due to storm winds accompanying hurricanes and typhoons. During the period of 1980–85, 5078 individuals perished annually on the average because of storm winds; the material damage was \$3323 $\times 10^6$. Apart from the dead and wounded, hurricanes create large numbers of homeless people. For example, a typhoon in Sri Lanka left 200 million people homeless in 1985. In 1988, on Jamaica, a tropical cyclone made homeless 810 000 of the island's inhabitants. Great damage is caused to the population of our planet by floods which are due to both heavy precipitation, and anthropogenic impact. During the period from 1981 to 1988, five severe inundations took place on the Earth, each of which killed over 2000 people: in India (1981) 3000; in Peru (1982) 2520; in Bangladesh (1987) over 2500, (1988) over 2000; in China (1989) over 2500. Floods sometimes cause injury and damage to several hundred thousand and even millions of people. For example, 1985, Ecuador, 700 000 people; 1985, Brazil, 600 000; 1986, China, over 5 million; 1987, Bangladesh, twice, 3 million and 600 000, respectively; 1987, China, 860 000. The record number of people harmed by a flood is relevant to the inundation of 1988 in Bangladesh, 45 million injured plus 28 million homeless people. The material damage caused by floods is enormous. Thus, for instance, the material damage caused by the flood of 1983 in France constituted \$604 million.

Landslides entail less harm or material damage. During the period from 1981 to 1989, only one out of a hundred serious landslides killed about 1 000 people (Ecuador, 1987). In this period, the greatest number of people affected by a landslide (including the wounded and sick) was registered in Nepal in 1987, 351 000 people. The maximum number of homeless people was in Sri Lanka in 1989 (about 250 000), and the largest material damage caused by a landslide was registered in Switzerland in 1988 (\$7 million). Of course, the consequences of landslides are not comparable with those of floods or earthquakes. However, they are frequent phenomena and should be carefully monitored. Dust storms are also very dangerous (Grigoryev, 1991).

In the Reports discussed (Tolba et al., 1992; United Nations . . . , 1991) there are gaps with regard to statistical data for individual countries. Some are due to the lack of observational data (especially where developing countries are concerned). Other gaps are due to military or political considerations (such is mostly the case with data lacking for the former USSR). At the same time, the Reports contain unique information which has been summarized with regard to various aspects: countries, regions, continents, kinds of production, etc. The data presented make it possible to follow global trends in the propagation and distribution of various phenomena.

9. TRANSPORT AND TOURISM

Automobile roads and traffic are most important indicators of the level of development of a country. At the same time, their development entails negative consequences, such as road accidents and environmental pollution. The greatest extension of roads (1988) is in the USA, Japan, Brazil and the former USSR. The thickest networks of roads (1988) are in Japan (2.92 km/km^2), FRG (1.98 km/km^2) and the former USSR (0.07 km/km^2). During the period from 1978 to 1988, in 58 countries (out of 77 under control) the road network expanded. At the present time, the transport sector—including passanger travel and freight movements by road, rail, air and water—is responsible for about 25% of 1990 world primary energy use and 22% of carbon dioxide emissions from fossil fuel use. Energy use was estimated to be 61–65 EJ. Without new measures, this might grow to 90–140 EJ in 2050 (Watson et al., 1996). In 20 countries for which detailed data are available, road traffic uses roughly 80% of total transport final energy consumption, air traffic 13%, rail 4.4%, and inland water transport 2.6%.

The greatest number of motor vehicles (four-wheeled, excluding tractors) is in the USA (732 units per 1000 individuals). In Japan, France and the former USSR in 1988 there were 427 481 and 51, respectively. In 1988 the greatest number of deaths in road accidents were in the former USSR (47 197 people) and the USA (46 385 people). All in all, there are about 500 million cars in the world (1988), and most of them are in the developed countries (United Nations..., 1991). Out of the total amount of energy in the world consumed by automobile transport, 2% are spent in the developing countries, and 35% in the USA. During the next two or three decades, motor transport is expected to double. However, in spite of this general progress, there still are many countries whose population is deprived of the benefit of transportation. Thus, for instance, in the agricultural regions of Kenya 90% of travel is made on foot, and only 31% of journeys involve automobiles or buses (1989).

Road haulage becomes more and more substantial. During the 10 years from 1978 to 1988, the transportation of cargo increased by almost a factor of 2 (from 113 540 to 211 295 ton/km). North America has the greatest volume of this kind of transportation (83 165 ton/km). Cargo transportation grows most of all in the Asian-Pacific region. The most important cargo of sea carriage is oil and its products. During 20 years (from 1970 to 1990), the world number of tankers somewhat decreased (from 3235 to 3015). At the same time, their carrying capacity increased by 80%.

Table 38 shows the matrix of transport system improvement options which might lead to the decrease of environmental pollution due to transport (Watson et al., 1996). The role of tourism increases permanently, and the transportation of tourists (in volume) is only inferior to that of oil. The greatest amount of tourists falls on Europe (250.3×10^6 people in 1988 and 184.4×10^6 in 1978). The tourist flow to North America is much less (49.4×10^6 people).

Table 38. Matrix of transport system improvement options

Options	Situation		
	High level of existing infrastructure and strong influence on technology	Rapidly developing infrastructure and relatively little influence on technology	Small towns, relatively simple infrastructure, little influence on technology
Reducing vehicle emissions	Impose vehicle standards or emission-related taxes/rebates Set up inspection and maintenance programmes Driver and police training for efficient driving and traffic control Public-awareness campaigns Research, development, and demonstration (RD&D)	Work with other cities and regions to encourage use of improved vehicles Set up inspection and maintenance programmes Driver and police training for efficient driving and traffic control Public-awareness campaigns	Work with other cities and regions to encourage use of vehicles Set up inspection and maintenance programmes Driver and police training for efficient driving and traffic control Public-awareness campaigns
Shifting to cleaner or alternative fuels	Fuel-quality standards, mandates, and taxes Evaluate options for alternative fuels, especially in large fleets including public transport RD&D	Work with other cities and regions to encourage use of cleaner fuels Evaluate options for alternative fuels, especially in large fleets including public transport	Work with other cities and regions to encourage use of cleaner fuels Evaluate options for alternative fuels, especially in large fleets including public transport
Shifting to modes with lower emissions	Provide facilities for non-motorized modes Consider non-motorized streets and zones Exclusive bus lanes and bus/tram priority Consider subsidies/fare controls for public transport controls Consider investment in public transport	Exclusive bus lanes and bus priority Segregate non-motorized modes Consider possibility of rail conversion to exclusive bus lanes Consider subsidies/fare for public transport Consider investiment in public transport	Begin to give priority to mass transity in urban transport policy Ensure access to public transport Exclusive bus lanes and bus priority in new development Segregate non-motorized modes
Transport demand management	Regulate or charge for parking Consider vehicle-free zones or restrictions Consider road-user fees, and vehicle and fuel taxes Investigate telecom muting, etc., as alternatives to transport	Restrict on-street parking Consider vehicle-free zones or restrictions Consider road-user fees, and vehicle and fuel taxes Investigate improved communication as alternative to transport Car-pool and high-	Formal parking controls or fees Consider vehicle-free zones or restrictions Investigate long-term role of road-user fees to manage traffic Investigate improved communication as

Table 38. *(continued)*

Options	Situation		
	High level of existing infrastructure and strong influence on technology	Rapidly developing infrastructure and relatively little influence on technology	Small towns, relatively simple infrastructure, little influence on technology
	Car-pool and high-occupancy vehicle incentives	occupancy vehicle incentives	alternative to transport Car-pool and high-occupancy vehicle incentives
Transport infra-structure development	Computerized traffic control Maintain and provide infrastructure for non-motorized and public transport Segregated non-motorized and public transport corridors	Improved traffic control Maintain and provide infrastructure for non-motorized and public transport Segregated non-motorized and public transport corridors Consider bypasses	Improved traffic control Give priority to public transport and non-motorized transport in new infrastructure
Land-use planning	Consider de-emphasizing central zone as major focus of activity Encourage mixed use, non-motorized/public transport-oriented suburban development Promote mixed use and non-motorized/public transport access in urban redevelopment Discourage parking provision in development and redevelopment	Desegregate land use; allow mixed use in central business district Promote mixed use and nonmotorized/public transport access in new development Encourage mixed use, non-motorized/public transport-oriented suburban development	Ensure transport development that enhances rather than hampers economic health of central business district Take transport trends into account and plan transport services in coordination with other development

10. ECONOMIC AND SOCIAL TRENDS

To avoid repetitions and what has already been considered in the Introduction and will be discussed in Chapter 1, we shall only dwell upon a few aspects of economic and social trends of global development. It is important to mention in this context that relevant problems have been recently discussed in a number of fundamental publications (see the References), especially in Brown et al. (1996), Bruce et al. (1996), Kennedy (1993), McMichael (1995), Munn and la Riviere (1996), Polunin and Nazim (1994), Pope (1993), Reid (1995), State... (1996), Suzuki et al. (1996), Theodore and Theodore (1996), Tolba (1992), United Nations... (1991).

The most important feature of contemporary global development is the continuing growth of population, and the economy which, on the one hand, is natural, and, on the other, is creating serious annoyance with regard to the future of humankind. Pope (1993) writes in his book:

> The 90s will birth about as much change in the way people think and act and as much consciousness growth as the last four decades put together. Given the last four decades, this is bound to be difficult to conceive. The pace of the 20th century is still increasing, pushing people and their social systems to grow quickly and then be ready to move on again. Political systems will be panting to catch up. The first decade of the 21st century will see plenty of action and growth too, but many of the changes of those years will be consciously created out of the slam-bam education endured in the 90s.
>
> Consciousness, whether individual or collective, tends to have a life of its own. For instance, the hyper-materialism of the 80s will likely cause those years to remain known as the "Greed Decade". Crave, buy, charge and steal material goods, show them off and enjoy them; these were the values that permeated the air. Apparently, the US and a good part of the Western world needed one last gagging look at me-first materialism before something new could truly emerge. The 90s are not about greed; probably you realize that already. Awareness is growing that present systems—political, environmental, educational, economic and social—are out-of-date, not handling increasing levels of problems. Many are already breaking down of their own accord or by their own corruption. The times are ripe for a primary rearrangement of human priorities. Getting rid of so much that is so much that is old may mean "Detox Decade" will be the label that eventually sticks on the 90s. Already that label makes sense for what most of Eastern Europe and the former USSR are going through. The five mid-decade years especially will see incredibly fast movement, a monumental shift really, in the way humans live, act and interact. Shifts of this magnitude don't happen often or at random on a planet; there is a predictable pattern, a sequence to epochal change of this nature (though having a sequence does not necessarily prevent chaos in-process).

And further:

> From our perspective, it is clear that nearly six billion people exceeds the carrying capacity of the Earth—if you want a healthy planet and humans with healthy futures. Even if the world were sensibly and fairly run, a human population in excess of four billion severely pushes the limits of systems and resources. Not even at the height of Atlantean times was the planet's population so high, and never again is it likely to be.

Let us now have a look at factual data discussed by Brown et al. (1996). The data of Table 39 illustrate the dynamics of gross world product, while those of Tables 40 and 41 characterize the development of steel and automobile production.

Table 39. Gross world product, 1950–95

Year	Total (trillion 1987 $)	Per capita (1987 $)	Year	Total (trillion 1987 $)	Per capita (1987 $)
1950	3.8	1487	1979	14.0	3197
1955	4.9	1763	1980	14.1	3165
1960	6.1	2008	1981	14.3	3156
1965	7.9	2362	1982	14.4	3124
1966	8.3	2430	1983	14.8	3156
1967	8.6	2468	1984	15.4	3228
1968	9.1	2560	1985	16.0	3297
1969	9.7	2673	1986	16.4	3321
1970	10.1	2727	1987	17.0	3382
1971	10.5	2776	1988	17.8	3480
1972	11.0	2850	1989	18.4	3534
1973	11.7	2973	1990	18.8	3549
1974	11.8	2941	1991	18.7	3471
1975	11.9	2912	1992	19.0	3448
1976	12.5	3006	1993	19.5	3464
1977	13.0	3073	1994	20.1	3577
1978	13.5	3137	1995 (preliminary)	20.8	3629

Table 40. World steel production, 1950–95

Year	Total (million tons)	Per person (kg)	Year	Total (million tons)	Per person (kg)
1950	190	74	1979	747	171
1955	271	97	1980	716	161
1960	347	114	1981	707	156
1965	451	135	1982	645	140
1966	469	137	1983	664	141
1967	493	142	1984	710	149
1968	524	147	1985	719	148
1969	571	157	1986	714	144
1970	595	161	1987	736	146
1971	582	154	1988	780	152
1972	631	163	1989	786	151
1973	696	177	1990	770	145
1974	703	175	1991	736	137
1975	644	157	1992	724	132
1976	675	162	1993	731	132
1977	675	160	1994	730	129
1978	717	166	1995 (preliminary)	748	131

Table 41. World automobile production and fleet, 1950–95

Year	Total (million)	Fleet (million)	Year	Total (million)	Fleet (million)
1950	8	53	1979	31	308
1955	11	73	1980	29	320
1960	13	98	1981	28	331
1965	19	140	1982	27	340
1966	19	148	1983	30	352
1967	19	158	1984	30	365
1968	22	170	1985	32	374
1969	23	181	1986	33	386
1970	22	194	1987	33	394
1971	26	207	1988	34	413
1972	28	220	1989	36	424
1973	30	236	1990	36	445
1974	26	249	1991	35	456
1975	25	26	1992	35	470
1976	29	269	1993	34	470
1977	30	285	1994	35	480
1978	31	297	1995 (preliminary)	36	486

Everything reveals growth. The only encouraging fact is that the world bicycle production increased considerably during the last few years, as can be seen from the data of Table 42 (Brown et al., 1996).

Table 42. World bicycle production, 1950–95

Year	Production (million)	Year	Production (million)
1950	11	1979	54
1955	15	1980	62
1960	20	1981	65
1965	21	1982	69
1966	22	1983	74
1967	23	1984	76
1968	24	1985	79
1969	25	1986	84
1970	36	1987	98
1971	39	1988	105
1972	46	1989	95
1973	52	1990	90
1974	52	1991	96
1975	43	1992	103
1976	47	1993	108
1977	49	1994	111
1978	51	1995 (preliminary)	114

Table 43. World exports, 1950–95

Year	Exports (billion 1990 $)	Year	Exports (billion 1990 $)
1950	330	1979	2132
1955	451	1980	2207
1960	586	1981	2222
1965	804	1982	2147
1966	860	1983	2200
1967	905	1984	2393
1968	1031	1985	2483
1969	1150	1986	2502
1970	1270	1987	2659
1971	1336	1988	2933
1972	1460	1989	3142
1973	1643	1990	3334
1974	1732	1991	3482
1975	1640	1992	3658
1976	1828	1993	3857
1977	1897	1994	4235
1978	1994	1995 (preliminary)	4317

Global economic development made countries all over the world much more mutually dependent than before. This is clearly demonstrated by the data of Table 43 on world exports (Brown et al., 1996). Sometimes the real situation is frightening. For instance, under the conditions of economic decline in contemporary Russia, 70% of the goods sold in its two biggest cities, Moscow and St Petersburg, appear to be imported ones.

Since the contemporary development of civilization is obviously unsustainable, there appears a problem of crucial importance: how to make it sustainable. We shall discuss it in subsequent chapters.

Reid (1995) has pointed out that three principal issues stand out in this context:

(1) The increase of human-related activities and their impacts, and an accompanying decrease of the resources of the planet.
(2) The growing inequity between the rich and the poor—between rich and poor nations, and between the rich and the poor within some countries.
(3) Population growth.

We have already discussed these issues earlier. Relevant information will also be considered in subsequent chapters. Meanwhile, let us look at some specific aspects of the problem of economic and social development. The data of Table 44 (Reid, 1995) illustrate the scale of North/South inequality. The data of Table 45 are even more striking in this respect.

Table 44. Consumption of selected items, North/South

	Share (%) North	Share (%) South	ADR North/South	EDR North/South
Meat	64	36	6	52
Cereals	48	52	3	6
Round wood	46	54	1	6
Paper, etc.	81	19	14	115
Fertilizer	60	40	5	6
Iron and steel	80	20	13	22
Cars	92	8	24	320
Electricity	81	19	13	46
Carbon dioxide	70	30	8	27

Notes. The average disparity ratio (ADR) is the ratio of per capita consumption levels in North and South. For example, on average a person in the North eats six times as much meat as a person in the South. The extreme disparity ratio (EDR) is the ratio between the richest and poorest countries, using the USA and India. Thus on average an American eats 52 times as much meat as an Indian. The eastern European countries are included in the North, and the newly industrialized countries of Asia are included in the South. The ADR is therefore smaller than the real disparities between many European countries and the low income countries of Asia and Africa.

Table 45. Ratio of income of richest 20% to poorest 20% of world population

1960	30 : 1
1970	32 : 1
1980	45 : 1
1989	59 : 1
1991	61 : 1

The widening economic gaps between regions (percentage of global economic activity) are clearly demonstrated by the UNDP 1992 data in Table 46, referenced by Keating (1994).

A problem of utmost importance is substantiation of an optimal interrelationship between market and non-market approaches to solving economic problems (obviously, this interrelationship should be different for various countries). McMichael (1995) has been quite right to point out in this context that:

> Orthodox economists anticipate the future by linear extrapolation of the past. They presume that the system is "open"; besides, limits can always be pushed back with new technology (innovation, substitution and discovery). Theirs is the dominant paradigm in political and financial circles. Ecologists, aware of the complex, changeable and non-linear aspects of natural systems, perceive "development" as being subject to the confines of the biosphere. Where the economist views ecological concerns as a subject of economics, to be dealt with market-place corrections, the ecologist views the economy as a subject of the global ecosystem. (Meanwhile, an ecologically-attuned minority of economists

Table 46. Widening economic gaps between regions (percentage of global economic activity)

	Global GNP		Global trade		Global commercial bank lending		Foreign private investment	
	1960	1989	1970	1989	1970	1989	1970	1989
Sub-Saharan Africa	1.9	1.2	3.8	1.0	0.3	0.3	24.8	15.0
South Asia	3.1	2.8	1.3	0.9	0.1	0.1	1.5	3.1
East and South-East Asia (excluding China)	1.7	2.9	4.1	8.1	5.7	12.8	11.7	33.3
China	3.0	2.0	0.8	1.9	0.4	0.2	11.8	7.0
Arab states	1.5	2.5	3.3	4.1	1.5	2.4	9.8	13.2
Latin America and the Caribbean	4.7	4.4	5.6	3.3	0.8	6.4	40.4	28.4
Developing countries	15.9	15.8	18.9	19.3	8.8	22.2	100.0	100.0
Least-developed countries	1.0	0.5	0.8	0.4	0.2	0.1	1.7	2.2
Industrial countries	84.1	84.2	81.1	80.7	91.2	77.8		

> argues that economic practice need not be inimical to the environment, and that good ecological practice would be good long-term economics.)
>
> Some of today's environment-damaging trends will take decades to slow and reverse. Decisions to desist from use of ozone-damaging chemicals or to eliminate the emissions that cause acid rain are *relatively* easy. In contrast, strategies for capping the growth of world population or for curbing the release of greenhouse gases are unprecedentedly difficult; and they strike at the value base of our society. Lasting solutions will therefore not come from the fine-tuning of markets nor from ingenious technical "fixes". The problems go deeper than that. The realisation that much of the environmental health impact will, eventually, bear on *everyone*—not just on the world's poor—should stiffen our resolve to act.

Leemans et al. (Watson et al., 1996) have discussed important issues related to land use and land cover from the viewpoint of the systems approach, and pointed out in this respect that:

> A wide range of future mitigation options and strategies involving various uses of land areas has been suggested. The primary issue in evaluating these options is whether the world can continue to support an increasing population with its growing needs for food and fiber and, at the same time, expand the amount of land used for production of biomass for energy. Analyses of land-use trends and patterns make clear that substantial land can be made available for biomass energy resources only if high rates of improvement in agricultural

> productivity continue throughout the world. Alternatively, moderate increases in agricultural productivity, reduced population growth rates, and reduced emphasis on meat consumption in the future could provide an adequate global food supply with reduced requirements for agricultural land. Even making optimistic assumptions, however, there will be regional dislocations and imbalances as demand for food, limitations in agricultural productivity, and increasing populations come into conflict. Important stresses on land availability for agriculture, biomass plantations, and other land uses will emerge if population growth continues unabated, high-meat dietary preferences remain stable or increase, the rate of increase in agricultural productivity slows or stops, land degradation continues, or the development and penetration of new technologies cannot be extended to developing countries.

Leemans et al. (Watson et al., 1996) have considered integrated assessment models (IAM) which can be used to evaluate the implications of extensive biomass plantations and other measures on land use and availability. They conclude:

> A review of conceptual approaches for counterbalancing anthropogenic climate change through geoengineering indicates that many options entail important adverse environmental consequences. Thus, these approaches do not provide an alternative that would readily permit the continued and expanded use of carbon-based fuels. Proposed concepts for geoengineering that have been examined include, for example, the deployment of solar radiation reflectors in space and the injection of sulfate aerosols into the atmosphere to mimic the cooling influence of volcanic eruptions. Most of these approaches are likely to be expensive to sustain and/or to have serious, but poorly understood, side effects, making them unattractive as possible mitigation options. Projections of the chemical and climatic effects of implementing such approaches are, for most options, at least as uncertain as those for future development of technologies and agricultural productivity. Geoengineering options also generally impose added costs on society and become essentially permanent commitments. Although it is perhaps appropriate to keep geoengineering approaches in reserve in case of unexpectedly rapid climatic change, investments in nontraditional approaches would seem much more effective if directed toward developing fossil fuel-free energy sources rather than at efforts to counterbalance the consequences of fossil fuel use.

To assess future trends of civilization, it is important to answer the following questions which have been raised by Reid (1995):

— Is it meaningful to talk of sustainable development when we may not be able to tell with certainty what the needs of future generations may be, which particular ecological, social and economic conditions represent the "sustainable state", and how close to such conditions we may be?
— Is it possible to devise a set of operational guidelines which would, if implemented, satisfy the twin criteria of "meeting the needs of the present" and "not compromising the ability of future generations to meet theirs"?

— Is any national or international programme for sustainable development likely to win political approval? Or is sustainable development merely an idea, which provides a basis for many theoretical studies but which cannot survive in the world of economic "realities", short-term political priorities, North–South suspicions, growing inequities, political oppression, and the prevalence of corruption, which in more or less blatant forms afflicts both North and South?
— Is the concept of sustainable development so open to misinterpretation that established interests can ignore its radical implications and be content with minor adjustments to their practices to allow them to claim they are following a sustainable path?
— Is it possible that the more radical implications of any set of operational guidelines can survive their mediation through the institutional machinery of mainstream development, staffed by an international desk-bound elite who have allowed bureaucracies to proliferate, but who have failed to arrest either the continuing degradation of the ecological systems of the planet or the deterioration in the welfare of many people throughout the world?

We shall address these questions in Chapter 1.

BIBLIOGRAPHY AND REFERENCES

Agafonov N.T., Isliayev R.A. 1994. Regional ecological policy and territorial organization of society. Relationships, problems. Regional Policy, 6, p.24–38.

Arski Yu.M., Danilov-Danilyan V.I., Zalikhanov M.Ch., Kondratyev K.Ya., Kotlyakov V.B., Losev K.S. 1997. Ecological Problems: What Is Going on, Who Is Responsible, and What to Do? MNEPU, Moscow, 331 pp. (in Russian).

Asrar G., Dokken D.J. (Eds.). 1995. The State of the Earth Science from Space. Past Progress. Future Prospects. AIP Press, Woodbury, New York, 159 pp.

Azuela A., Carabias J., Provencio E., Quadri G., 1993. Desarrollo Sustentable. Hacia una Politica Ambiental. Universidad Nacional Autonoma de Mexico (UNAM), Mexico, 176 pp.

Ausbel J.H., Meyer P.S. 1994. Graphic representations of world population growth. Human Dimensions Quart., 1, No.2, p.17–19.

Bailey R. (Ed.). 1995. The True State of the Planet. The Free Press, New York, 472 pp.

Bailey R.G. 1996. Ecosystem Geography. Springer Verlag, Berlin, 204 pp.

Bakkes J.A., van der Born G.J., Helder J.C., Swart R.J., Hope C.W., Parker J.D.E. 1994. Overview of Environmental Indicators: State of the Art and Perspectives. UNEP/EATR, 94-1. RIVM/402000 1001, Nairobi, 72 pp.

Balick M.J., Cox P.A. 1996. Plants, People and Culture: The Science of Ethnobotany. Freeman/Sci. Amer. Library, 228 pp.

Banuri T. 1993. The landscape of diplomatic conflicts. In: W. Sechs (Ed.). Global Ecology. Zed Books, London.

Basiago A.D. 1995. Methods of defining "sustainable development". Sustain. Development, 3, p.109–119.

Berner E.K., Berner R.A. 1996. Global Environment: Water, Air, and Geochemical Cycles. Prentice Hall, Upper Saddle River, NJ, 376 pp.

Bongaarts J. 1994. Population policy opinions in the developing world. Science, 263, p.771–776.

Boyden S., Dovers S. 1992. Natural resource consumption and its environmental impacts in the western world: impacts of increasing per capita consumption. Ambio, 21, No.1, p.63–69.

Bretherton F. 1994. Perspectives on policy. Ambio, 23, No.1, p.96–97.

Broecker W.S. 1993. Labor Erde. Bausteine für einen lebensfreundlichen Planeten. Springer Verlag, Berlin, 300 pp.

Brown L.R., Kane H. 1994. Full House. Reassessing the Earth's Population Carrying Capacity. Earthscan, London, 224 pp.

Brown L.R. et al. 1994. State of the World 1994. A World-Watch Institute Report on Progress toward Sustainable Society. Earthscan, London, 288 pp.

Brown L.R. 1996. The acceleration of history. In: State of the World 1996 (Ed. by L.R. Brown et al.). W.W. Norton & Co., New York/London, p.3–20.

Brown L.R., Flavin C., Kane H. 1996. Vital Signs 1996. The Trends that Are Shaping Our Future. W.W. Norton & Co., New York/London, 166 pp.

Bruce J.P. 1994. Natural disaster reduction and global change. Bull. Amer. Meteorol. Soc., 75, No.10, p.1831–1835.

Bruce J.P., Lee H., Haites E.F. 1996. Climate Change 1995. Economic and Social Dimensions of Climate Change. Cambridge Univ. Press, 448 pp.

Bukata R.P., Jerome J.H., Kondratyev K.Ya., Pozdnyakov D.V. 1995. Optical Properties and Remote Sensing of Inland and Coastal Waters. CRC Press, Boca Raton, FL, USA, 384 pp.

Busch L., Lacy W.B., Burkhardt J., Hemken D., Moraga-Rujel J., de Silva J. 1966. Making Nature, Shaping Culture: Plant Biodiversity in Global Context. Univ. of Nebraska Press, 261 pp.

Calabrese E.J., Baldwin L.A. 1993. Performing Ecological Risk Assessments. CRC Press, 288 pp.

Campbell B. 1995. Human Ecology. 2dn edn. Aldine de Groyer, New York, 213 pp.

Carson R. 1994. Silent Spring. (With an Introduction by Vice President Al Gore.) Houghton Mifflin, Boston, 368 pp.

Chapman J.L., Reiss M.J. 1992. Ecology, Principles and Applications. Cambridge Univ. Press, 294 pp.

Clark W.C. 1991. Managing Planet Earth. In: Global Climate Change (Eds. J.C. White, W. Wagner, C.N. Beal). Elsevier, New York, p.1–11.

Clarke J. (Ed.) 1993. Nature in Question. An Anthology of Ideas and Arguments. Earthscan, London, 192 pp.

Clayton A.M.H., Radcliffe N.J. 1996. Sustainability. A System Approach. Earthscan, London, 258 pp.

Climate Change 1995. 1996. The Science of Climate Change (Ed. by J.T. Houghton, L.G. Meira Filho, B.A. Callander, N. Harris, A. Kattenberg and K. Maskell). Cambridge Univ. Press, 572 pp.

Cohen J.E. 1995. How Many People Can the Earth Support? W.W. Norton & Co., London/New York, 532 pp.
Cohen P., Polunin N. 1990. The visible culture. Environm. Conservation, 17, No.1, p.3–6.
Colvin M. 1994. Bringing economics down to earth. The Center for Values and Social Policy Newsletter, Boulder, CO, USA, XIII, No.2, p.1–3.
Contzen J.-P., Ghazi A. 1994. The role of the European Union in global change research. Ambio, 23, No.1, p.101–103.
Corson W.H. 1996. Measuring sustainability: indicators, trends and performance. In: Building Sustainable Societies (Ed. by D.C. Pirages). M.E. Sharpe, Inc., Armonk, NY, p.325–352.
Costanza R., Norton B.G., Haskell B. (Eds.). 1993. Ecosystem Health. New Goals for Environmental Management. Island Press, Earthscan, London, 160 pp.
Cothern C.R. (Ed.). 1993. Comparative Environmental Risk Assessment. CRC Press, Boca Raton, FL, 336 pp.
Cropper M.L., Griffiths Ch. 1994. Interaction of population growth and environmental quality. Amer. Econom. Rev., Papers and Proc., 84, No.2, p.250–265.
Daly H.E. 1996. Physical growth versus technological development: odds in the sustainability race. In: Building Sustainable Societies (Ed. by D.C. Pirages). M.E. Sharpe, Inc., Armonk, NY, p.167–173.
Danilov-Danilyan V.I., Gorshkov V.G., Arsky Yu.M., Losev K.S. 1994. Environment between the Past and Future: The World and Russia (an Experience of Ecologo-Economic Analysis). VINITI Press, Moscow, 133 pp.
Döös B.R. 1994a. Why is environmental protection so slow? Global Environm. Change, 4, No.3, p.179–184.
Döös B.R. 1994b. Global food production and risk. Tiempo, No.12, p.17–22.
Dovers S.R., Handmer J.W. 1992. Uncertainty, sustainability and change. Global Environm. Change, 2, No.4, p.262–276.
Dovers S.R., Handmer J.W. 1993. Contradictions in sustainability. Environm. Conservation, 20, No.3, p.217–222.
Dovers S., Norton T. 1994. Population, environment and sustainability: Reconstructing the debate. Sustain. Development, 2, No.1, p.1–7.
Eddy J.A. 1994. The present population of the world. Human Dimensions Focus, 1, No.1, 3 pp.
Eden S.E. 1994. Using sustainable development: the business case. Global Environm. Change, 4, No.2, p.160–167.
Ehrlich A.H. 1996. Toward a sustainable global population. In: Building Sustainable Societies (Ed. by D.C. Pirages). M.E. Sharpe, Inc., Armonk, NY, p.151–165.
Eldridge D., Nisar T.M. 1995. Institutional development and sustainability: approaches for public sector project organizations. Sustain. Development, 3, No.3, p.130–139.
Ellsaesser H.W. (Ed.). 1992. Global 2000 Revisited. Mankind's Impact on Spaceship Earth. Paragon House, New York, 436 pp.
Encyclopedia of Earth System Science. 1991. Volumes 1-4. W.A. Nierenberg (Ed.), Academic Press, New York, 2845 pp.

Faber M., Proops J.L.R. 1993. Evolution, Time, Production and the Environment. Springer, Environm. Sci., XX, 288 pp.

Factor 10. 1996. Developing Ideas, 1, p.4.

Frey W.H. 1994. Global aging. Human Dimensions, 1, No.1, p.13–19.

Gallopin G. et al. 1974. The Limits to Poverty: The Bariloche Report.

German Advisory Council on Global Change (WBGU). Alfred-Wegener-Inst., Bremerhafen, Germany (Ed.). 1996. World in Transition: Ways towards Global Environmental Solutions. Annual Report 1995. Springer Verlag, Berlin, 255 pp.

Ghimire K.B. 1993. Linkages between Population, Environment and Development. Case Studies from Costa Rica, Pakistan and Uganda. UNRISD Report 93.1, Geneva, 199 pp.

Girardet H. 1995. Sustainability in the age of the city: Countdown to Istanbul. No.5, p.5.

Glantz M.H. (Ed.). 1994. The Role of Regional Organizations in the Context of Climate Change. Springer Verlag, Berlin, 208 pp.

Goldemberg J. 1994. Energy for a sustainable world population. In: Population and Global Security. The Foundation for Environm. Conservation, Geneva, p.205–219.

Goldemberg J. 1995. Energy needs in developing countries and sustainability. Science, 269, No.5227, p.1058–1059.

Goodland R., Herman D. 1993. Ten Reasons Why Northern Income Growth Is Not the Solution to Southern Poverty. World Bank Publ., Environment Dept., Washington, DC.

Goodland R.J.A., Daly H.E., Kellenberg J. 1994. Imperatives for environmental sustainability: decrease overconsumption and stabilized population. In: Population and Global Security. The Foundation for Environm. Conservation, Geneva, p.85–99.

Goodman R., Daly H., El Serafy S. 1993. Population, Technology and Lifestyle. Island Press, Earthscan, London, 140 pp.

Gore A. 1993. Earth in the Balance. Ecology and the Human Spirit. Plume, New York, 408 pp.

Gorshkov V.G. 1990. Biospheric energetics and sustainability of the environment. Itogi Nauki I Tekhniki, Theoretical and General Aspects of Geography, 7, VINITI Press, Moscow, 238 pp. (in Russian).

Gorshkov V.G., Kondratyev K.Ya., Danilov-Danilyan V.I., Losev K.S. 1993. The Environment: From New Technologies to a New Thinking. VINITI Press, Moscow, 22 pp. (in Russian).

Gorshkov V.G., Kondratyev K.Ya., Losev K.S. 1994. The natural biological regulation of the environment. J. Ecol. Chem., 3 (2), p.85–90.

Gorshkov V.G. 1995. Physical and Biological Bases of Life Stability. Man, Biota, Environment. Springer Verlag, Berlin, 340 pp.

Gorshkov V.G., Kondratyev K.Ya., Losev K.S. 1996. Ecological problems of contemporary civilization. Vestnik of Russian Acad. Sci., No.2, p.119–128 (in Russian).

Goulet D. 1996. Authentic development: is it sustainable? In: Building Sustainable Societies (Ed. by D.C. Pirages). M.E. Sharpe, Inc., Armonk, NY, p.189–205.

Gowdy J.M. 1992. Economic growth versus the environment. Environm. Conservation, 19, No.2, p.102–104.

Gowdy J.M. 1994. Natural capital and the growth economy. Sustain. Development, 1, p.12–16.

Graves J., Reavey D. 1996. Global Environmental Change. Plants, Animals & Communities. Longman, Harlow, UK, 226 pp.

Grigoryev A.A. 1991. Ecological Lessons of the Historic Past and Modern Times, "Nauka" Publ., Leningrad, 252 pp.

Grigoryev A.A., Kondratyev K.Ya. 1994. The present state of global environment and natural resources. Izvestiya Russian Geograph. Soc., 126, issue 3, p.12–26.

Grigoryev A.A., Kondratyev K.Ya. 1996a. Satellite monitoring of natural and anthropogenic calamities. Studying the Earth from Space. No.4, p.3–14 (in Russian).

Grigoryev A.A., Kondratyev K.Ya. 1996b. Global change: a problem of ecologically sustainable development indicators. Izvestiya of Russian Geograph. Soc., 128, No.4, p.15–27.

Hammitt J.K. 1995. Outcome and value uncertainties in global-change policy. Clim. Change, 30, No.2, p.125–145.

Hammond A., Adrianse A., Rodenburg E., Bryant D., Woodward R. 1995. A systematic approach to measuring and reporting on environmental policy performance. In the Context of Sustainable Development, World Resources Inst., New York, N.Y., 43 pp.

Heywood K.H. (Exec. Ed.), Watson R.T. (Chair). 1996. Global Biodiversity Assessment. Cambridge Univ. Press., 1040 pp.

Holdgate M.V. 1994. Hopes for the future. In: Population and Global Security. The Foundation for Environm. Conservation, Geneva, p.241–246.

Holdgate M. 1996. From Care to Action. Earthscan Publ. Ltd., London, 346 pp.

Holmberg J., Thomson K., Timberlake L. 1993. Facing the Future. Earthscan, London, 48 pp.

Hunt C. 1994a. The sustainable development debate revisited. Current Affairs Bull., 70, No.10, p.4–11.

Hunt C.A.G. 1994b. Ecologically sustainable development: spatial, economic and institutional issues in its implementation. Ph.D. Thesis, Flinders Univ. of South Australia, 484 pp.

Hunt C.A.G. 1994c. Sustainable development in a region: a practical approach to targeting environmental impacts. Sustain. Development, 2, No.1, p.23–30.

International Institute for Sustainable Development Annual Report 1993–1994. 1994. Winnipeg, Manitoba, Canada, 16 pp.

Israel Yu.A. 1996. Radioactive Fallout after Nuclear Explosions and Accidents. St.Petersburg: Progress-Pogoda Publ., 356 pp. (in Russian).

Jänicke M., Bolle H.-J., Carius A. (Eds.). Umwelt Global. Veränderungen, Probleme, Lösungen. Springer, VIII, 256 pp.

Jørgensen S.E. 1994. Fundamentals of Ecological Modelling. 2nd edn., Elsevier, Amsterdam, 632 pp.

Karl T.R., Knight R.W., Easterling D.R., Quayle R.G. 1996. Indices of climate change

for the United States. Bull. Amer. Meteorol. Soc., 77, No.2, p.279–292.
Katz L.S., Orrick S., Honig R. (Eds.). 1993. Environmental Profiles. A Global Guide to Projects and People. Garland Publ., Hamden, 1400 pp.
Kearney A.R. 1994. Understanding global change: a cognitive perspective on communicating through stories. Clim. Change, 27, No.4, p.419–442.
Keating M. 1994. The Earth Summit's Agenda for Change. Centre for Our Common Future, Geneva, 70 pp.
Kemf E. (Ed.). 1993. Indigenous Peoples and Protected Areas. The Law of Mother Earth. Earthscan, London, 480 pp.
Kennedy P. 1993. Preparing for the Twenty-First Century. Vintage Books, New York, xvi + 428 pp.
Kirkby J., O'Keefe P., Timberlake L. 1995. The Earthscan Reader in Sustainable Development. Earthscan, London, 371 pp.
Kondratyev K.Ya. 1990. Key Problems of Global Ecology. Itogi Nauki I Tekhn. Theoretical and General Problems of Geography, Vol.9. VINITI Press, Moscow, 454 pp. (in Russian).
Kondratyev K.Ya. 1991a. Priorities of global ecology. Izvestiya of USSR Acad. Sci., Ser. Geography, No.6, p.21–30 (in Russian).
Kondratyev K.Ya. 1991b. Global security and ecology. Geofis. Intern., 29, No.3, p.123–138.
Kondratyev K.Ya. 1992a. Global Climate. "Nauka" Publ., St. Petersburg, 359 pp. (in Russian).
Kondratyev K.Ya. 1992b. Global Ecology and Requirements to Observational Data. "Nauka" Publ., St. Petersburg, 93 pp. (in Russian).
Kondratyev K.Ya., Buznikov A.A., Pokrovsky O.M. 1992a. Global Ecology: Remote Sensing. Itogi Nauki I Tekhniki. Atmosphere, Ocean, Space. The 'Cross-Sections' Programme, 14, VINITI Press, Moscow, 310 pp. (in Russian).
Kondratyev K.Ya., Ortner J., Preining O. 1992b. Priorities of global ecology now and in the next century. Space Policy, 8, No.1, p.39–48.
Kondratyev K.Ya. 1993a. Ecology and policy. Part I. Izvestiya Russian Geograph. Soc., 125, issue 1, p.3–11 (in Russian).
Kondratyev K.Ya. 1993b. Ecology and policy. Part II. Izvestiya Russian Geograph. Soc., 125, issue 2, p.1–10 (in Russian).
Kondratyev K.Ya., Danilov-Danilyan V.I., Donchenko V.K., Losev K.S. 1993. Ecology and Policy. Russian Acad. Sci. Publ., St. Petersburg, 286 pp. (in Russian).
Kondratyev K.Ya., Moreno-Peña F., Galindo I. 1994. Global Change: Environment and Safety. Univ. de Colima Publ., 47 pp.
Kondratyev K.Ya. 1995a. Key aspects of ecological policy. 1. Ecodynamics. Izvestiya of Russian Geograph. Soc., 127, No.3, p.36–43 (in Russian).
Kondratyev K.Ya. 1995b. Key aspects of ecological policy. 2. Socio-economic problems. Izvestiya of Russian Geograph. Soc., 127, No.4, p.1–10 (in Russian).
Kondratyev K.Ya. 1995c. Earth in the Balance: Ecology and the Human Spirit (a review of A. Gore's book). Sustain. Development, 3, No.2, p.104–105.
Kondratyev K.Ya. 1995d. Studies on the problems of global change, state of the art: the U.S. program. Studying the Earth from Space, No.2, p.98–105 (in Russian).

Kondratyev K.Ya. 1996. Global change and demographic dynamics. Izvestiya of Russian Geograph. Soc., 128, No.3, p.1-12 (in Russian).
Kondratyev K.Ya., Buznikov A.A., Pokrovsky O.M. 1996. Global Change and Remote Sensing. Wiley/Praxis, Chichester, 380 pp.
Kondratyev K.Ya., Frolov A.K. (Eds.). 1996. Ecodynamics and Ecological Monitoring of the St.Petersburg Region in the Context of Global Change. Nauka Press, St.Petersburg, 443 pp. (in Russian).
Kondratyev K.Ya., Johannessen O.M., Melentyev V.V. 1996. High Latitude Climate and Remote Sensing, Chichester, Wiley/Praxis, 208 pp.
Kondratyev K.Ya., Pozdnyakov D.V. 1996. Land-ocean interactions in the coastal zone: the LOICZ project. Il Nuovo Cimento C, Vol. 19C, No.3, pp.339–354.
Kondratyev K.Ya., Romanyuk L.P. 1996a. Environmental problems and the World Bank. Izvestiya of Russian Geograph. Soc., 128, No.2, p.1–11 (in Russian).
Kondratyev K.Ya., Romanyuk L.P. 1996b. Sustainable development: conceptual aspects. Izvestiya of Russian Geograph. Soc., 128, No.6, p.1–12 (in Russian).
Kondratyev K.Ya., Peña, F.M., Galindo, I. 1997. Sustainable Development and Population Dynamics. Universidad de Colima, Colima, Mexico, 128 pp.
Kormondy E.J. 1996. Concepts of Ecology. 4th edn. Prentice Hall, Upper Saddle River, NJ, 559 pp.
Kramer H.J. 1994. Observation of the Earth and Its Environment. Springer, XXI, 580 pp.
Lavrov S.B., Sdasyuk G.V. 1985. This World of Contrasts: Geographical Aspects of Some Global Problems. "Mysl" Publ., Moscow, 207 pp.
Lavrov S.B. 1994. Geopolicy of Russia: Eurasian Chance. Regionalnaya Politika, 6, p.86–98 (in Russian).
Leakey R., Lewin R. 1995. The Sixth Extinction. Biodiversity and Its Survival. Weidenfeld and Nicholson, London, 271 pp.
Lean G. 1996. Eyes down for last great UN gathering of the century. Independent, June 16.
Leslie J. 1996. The End of the World. The Science and Ethics of Human Extinctions. Routledge, 305 pp.
Livshitz A.Ya., Novikov A.V., Smirnyagin L.V. 1994. Regional strategy. Regionalnaya Politika,6, p.3–24 (in Russian).
Long C.S., Miller A.J., Lee H.-T., Wild J.D., Przywarty R.C., Hufford D. 1996. Ultraviolet index forecasts issued by the National Weather Service. Bull. Amer. Meteorol. Soc., 77, No.4, p.729–748.
Lovelock J. 1995. The Ages of Gaia: A Biography of Our Living Earth. W.W. Norton & Co., New York/London, 255 pp.
Lutz W. (Ed.). 1994. The Future Population of the World. What Can We Assume Today? Earthscan, London, 288 pp.
Mainguet M. 1994. Desertification. Natural Background and Human Mismanagement. 2nd edn. Springer Verlag, Berlin, 314 pp.
Maitra A.K. 1995. Small cities: the key of sustainability. Countdown to Istanbul, No.5, p.5–17.
Making Development Sustainable. The World Bank Group and the Environment. Fiscal 1994. 1994. World Bank Publ., Washington, DC, 270 pp.

Marien M. 1996. Infoglut and competing problems: key barriers suggesting a new strategy for sustainability. In: Building Sustainable Societies (Ed. by D.C. Pirages), M.E. Sharpe, Armonk, NY, p.299–311.

Masood E. 1996. Industrial world is unlikely to meet promised carbon cuts. Nature, 382, No.6587, p.103.

Maybank J., Bonsal B., Jones K., Lawford R., O'Brien E.G., Ripley E.A., Wheaton E. 1995. Drought as a natural disaster. Atmosphere-Ocean, 33, No.2, p.195–222.

McMichael A.J. 1995. Planet Overload. Global Environmental Change and the Health of the Human Species. Cambridge Univ. Press, xviii + 352 pp.

Meadows D.H., Meadows D.L., Rangers J., Behrens W.W., III. 1972. The Limits to Growth. Report to the Club of Rome. A Signet Book, New York, 207 pp.

Meadows D.L., Behrens W.W., III, Meadows D.H. et al. 1974. Dynamics of Growth in a Finite World. Wright Allen Press, New York, 511 pp.

Menza V., Lupien J.R. 1994. World population and nutritional well-being. In: Population and Global Security. The Foundation for Conservation, Geneva, 125 pp.

Mesarovic M., Pestel E. 1974. Mankind at the Turning Point. Dutton, New York, 139 pp.

Mészáros E. 1993. Global and Regional Changes in Atmospheric Composition. CRC Press, Boca Raton, FL, 192 pp.

Milbrath L.M. 1996. Becoming sustainable: changing the way we think. In: Building Sustainable Societies (Ed. by D.C. Pirages), M.E. Sharpe, Inc., Armonk, NY, p.275–297.

Miller R.B. 1994. Interactions and collaboration in global change across the social and natural sciences. Ambio, 23, No.1, p.19–24.

Millington A.C., Pye K. (Eds.). 1994. Environmental Change in Drylands. Biogeographical and Geomorphological Perspectives. John Wiley & Sons, Chichester, 456 pp.

Monitoring Environmental Progress (a report on work in progress). 1995. The World Bank, Washington, DC, 72 pp.

Mooney H.A., Fuentes E.R., Kronberg B.L. (Eds.). 1993. Earth System Responses to Global Changes. Contrasts between North and South America. Academic Press, London, 365 pp.

Morais J., Rapley C., Grassl H. 1995. Why the International Global Change Research Programmes need the Human Dimensions of Global Environmental Change Programme. IGBP Newsletter, No.24, p.23–25.

Munn R.E., la Riviere J.M.M. (Eds.). 1996. Policy Making in an Era of Global Environmental Change. Kluwer Acad. Publ., Dordrecht.

Myers N. 1994. Global Population and Emergent Pressures. In: Population and Global Security. The Foundation for Environmental Conservation, Geneva, p.17–48.

Myerson G., Rydin Y. 1996. The Language of Environment. A New Rhetoric. UCL Press, London, 264 pp.

Nakano Y.Y. 1994. Miraculous Life Chain. The Essence of Evolution from the Universe to Mankind. Helix Editions, Ltd., Oxon, UK, 107 pp.

N'Dow W. 1995. Editorial. Countdown to Istanbul, No.5, p.2.

Nihoul J.C.J. 1994. Modelling sustainable development. Model., Identif. and Contr., 156, No.3, p.165–169.
O'Connor J.C. 1994a. Towards Environmentally Sustainable Development. Draft. The World Bank, Washington, DC, 18 pp.
O'Connor J.C. 1994b. Environmental Performance Monitoring Indicators. Draft for Discussion. The World Bank, Washington, DC, 15 pp.
Our Changing Planet. The FY 1995. U.S. Global Change Research Program. 1994. A Supplement to the President's Fiscal Year 1995 Budget. Coordination Office of the US GCRP, Washington, DC, 133 pp.
Pearce D., Barbier E. (Eds.). 1994. Blueprint 4: Sustaining the Earth. Capturing Global Value. Earthscan, London, 192 pp.
Peet J. 1993. Energy and the Ecological Economics of Sustainability. Island Press, Earthscan, London, 300 pp.
Pielke R.A., Jr. 1994. Scientific information and global change policy making. Editorial essay. Clim. Change, 28, No.4, p.315–320.
Pirages D.C. (Ed.). 1996. Building Sustainable Societies. A Blueprint for a Post-Industrial World. M.E. Sharpe Inc., Armonk, NY, 361 pp.
Polunin N., Nazim M. (Eds.). 1994. Population and Global Security (Environmental Changes II). The Foundation for Environmental Conservation, Geneva, 285 pp.
Pope J. 1993. Upcoming Changes. The Next 20 Years. Emerald Wave Publ. Co., Fayetteville, AR, 225 pp.
Population stability within reach? Nature, 370, No.6491, p.583–589.
Ramphal Sh. 1994. Where is the time-bomb ticking? In: Population and Global Security. The Foundation for Environmental Conservation, Geneva, p.49–52.
Rapport D.J., Gaudet C.L., Calow P. (Eds.). 1995. Evaluating and Monitoring the Health of Large-Scale Ecosystems. NATO ASI Series I: Global Environmental Change. Vol.28. Springer Verlag, Berlin, xi + 454 pp.
Reid D. 1995. Sustainable Development. An Introductory Guide. Earthscan, London, 261 pp.
Roe D., Dalal-Clayton B., Hughes R. 1995. A Directory of Environmental Assessment Guidelines. IIED Publ., London.
Rousch W. 1994. Population: the view from Cairo. Science, 265, No.5176, p.1164–1167.
Rousch W. 1996. Live long and prosper? Science, 273, No.5271, p.41–46.
Sadik N. 1994. Population expansion and global stability. In: Population and Global Security. The Foundation for Environmental Conservation, Geneva, p.1–15.
Schaefer N., Kreisel W. 1994. Health of planet. In: Population and Global Security. The Foundation for Environmental Conservation, Geneva, p.101–124.
Schoon N. 1996. EU 'failing' to meet pollution target. The Daily Yomiuri, 26 May.
Serageldin I., Steer A. (Eds.), with Cernea M.M., Dixon S.A., Lutz E., Margulis S., Munasinghe M. and Rees C. 1994. Making Development Sustainable: From Concepts to Action. Environmentally Sustainable Development. Occasional Paper Series No.2. The World Bank, Washington, DC, 340 pp.
Serageldin I., Goodland R., Daly H. 1995. The concept of sustainability. In: Wouter van Dieren (Ed.): Taking Nature into Account—A Report to the Club of Rome. Springer Verlag, New York, pp.99–123.

Shaw R.V. 1992. Indicators of ecologically sustainable development. Izvestiya of Russian Geograph. Soc., 124, No.2, p.105–115 (in Russian).
Simon J.L. (Ed.). 1996. The State of Humanity. Blackwell, Oxford, 694 pp.
Simpson J.C. 1948. The Tempo and Forms of Evolution. Inostr. Literat. Publ., Moscow, 358 pp.
Singh N., Amaratunga C., Shea C.P., Romanov J. 1994. Designing Work for Sustainability. Int. Inst. for Sustain. Development, Winnipeg, Canada, 61 pp.
Snape W.J., III, Houck O.A. (Eds.). 1996. Biodiversity and the Law. Island Press, Washington, DC, 259 pp.
Sokolov V.E. 1994. Fundamental biological and ecological studies. Vestnik Russian Acad. Sci., 64, No.9, p.797–803 (in Russian).
State of the World 1990. A Worldwatch Institute Report on Progress toward a Sustainable Society. (Ed. by Linda Starke). W.W. Norton & Co., London/New York, xi + 263 pp.
State of the World 1996. A Worldwatch Institute Report on Progress toward a Sustainable Society. (Project Director: Lester R. Brown; Ed. by Linda Starke). W.W. Norton & Co., London/New York, xiii + 249 pp.
Summary of the Programme of Action of the International Conference on Population and Development (ICDP-94). 1995. New York, 27 pp.
Suter G.W., II (Ed.). 1993. Ecological Risk Assessment. CRC Press, Boca Raton, FL, 560 pp.
Suzuki Y., Ueta K., Mori S. (Eds.). 1996. Global Environmental Security. From Protection to Prevention. Springer Verlag, Berlin, xii + 262 pp.
The Centre for Our Common Future, 1992: Our Common Future. 1992. Reconvened World Commission on Environment and Development, April 22–24, Geneva, 32 pp.
The Ecologist. 1994. Too Many for Whom. Population Control and the Environment. Earthscan, London, 224 pp.
The International Conference on Population and Development Report, 1994. Cairo, Sept. 5-13, 152 pp. (in Russian).
Theodore M.K., Theodore L. 1996. Major Environmental Issues Facing the 21st Century. Prentice Hall, Upper Saddle River, NJ, 502 pp.
Tisdell C.A. 1994. Population, economics, development and environmental security. In: Population and Global Security. The Foundation for Environmental Conservation, Geneva, p.63–84.
Tolba M.K. 1992. Saving Our Planet. Challenges and Hopes. Chapman & Hall, London, 287 pp.
Tolba M., El-Kholy O.A., El Hinnawi E., Holdgate M.W., McMichael D.F., Munn R.E. (Eds.). 1992. The World Environment. 1972–1992. Two Decades of Challenge. Chapman & Hall, London, 884 pp.
Tomas C.D. 1990. Fewer species. Nature, 347, p.237.
Trzyna T.C. (Ed.). 1995. A Sustainable World. Defining and Measuring Sustainable Development. IUCN—The World Conservation Union. Earthscan, London, 272 pp.
Tsonis A.A. 1996. Widespread increases in low-frequency variability of precipitation over the past century. Nature, 382, No.6593, p.700–702.

Turekian K.K. 1996. Global Environmental Change: Past, Present, and Future. Prentice Hall, Upper Saddle River, NJ, 200 pp.

Turner B.L., II. 1993. The Earth as Transformed by Human Action. Cambridge Univ. Press, 736 pp.

United Nations Environment Programme. Environmental Data Report, 3rd edn. 1991. London, Basil Blackwell, 408 pp.

United Nations Environment Programme. 1995. Programme of Work on Indicators of Sustainable Development, New York, E/CN. 17/1995/18.

United Nations Work Programme on Indicators of Sustainable Development. 1995. U.N. Department for Policy Coordination and Sustainable Development, New York, 4 pp., Appendices.

Vernadsky V.I. 1994. Living Substance in the Biosphere. "Nauka" Publ., Moscow, 672 pp.

Viederman S. 1996. Sustainability's five capitals and three pillars. In: Building Sustainable Societies (Ed. by D.C. Pirages), M.E. Sharpe Inc., Armonk, NY, p.45–53.

Vitousek P.M., Ehrlich P.R., Ehrlich A.H.E., Matson P.A. 1986. Human appropriation of the products of photosynthesis. Bioscience, 36, p.368–375.

Watson R.T., Heywood V.H., Baste I., Dias B., Gamez R., Janetos T., Reid W., Ruark G. 1995. Global Biodiversity Assessment. Summary for Policy Maker. Cambridge Univ. Press, 46 pp.

Watson R.T., Zynyowera M.C., Moss R.H., Dokken D.J. (Eds.). 1996. Climate Change 1995. Impacts, Adaptations and Mitigation of Climate Change: Scientific-Technical Analysis. Cambridge Univ. Press, 878 pp.

Weizsaecker E.U., von. 1994. Earth Politics. Zed Books, London/New York, 234 pp.

Welt im Wandel: Die Gefärdung der Böden. 1994. Wissenschaftlicher Beirat der Bundesregierung. Globale Umweltveränderungen. Jahresgutachten 1994, Bremerhaven, 325 pp.

Wichmann R. 1995. The link between poverty, environment and development. Countdown to Istanbul, No.5, pp.1,3,4.

Wilson E. 1992. The Diversity of Life. Harvard Univ. Press—Belknap Press, Cambridge, MA (USA).

Woods A., Bernasconi M.P. 1995. Choosing a space age or a stone age. Space News, Oct. 2–8, 19 pp.

World Commission on Environment and Development (The Brundtland Commission). 1987. Our Common Future. Oxford Univ. Press, New York.

World in Transition: Basic Structure of Global People–Environment Interactions. 1993. Annual Report. German Advisory Council on Global Change. 1994. Economica Verlag GmbH, Bonn, 214 pp.

World Resources 1987. A Report by the International Institute for Environment and Development. The World Resources Inst. 1987. Basic Books, New York, 369 pp.

World Resources, 1989–1990. 1989. Basic Books, New York, XII, 372 pp.

1

Global ecodynamics and sustainable development

By the end of the twentieth century, humankind has entered a stage of its development when the words of V. I. Vernadsky become realistic. Vernadsky, a great Russian natural scientist and philosopher, predicted that man's economic activity would become a geological force capable of changing the world by putting it on the edge of a global ecological catastrophe (*V. I. Vernadsky and Modern World Times*, 1990). This is precisely what determines the exceptional urgency of global ecology problems, and the formulation of the political concept of the priority of human values.

The contemporary epoch is characterized by a great controversy between the socio-economic development of human society, on the one hand, and the limited capacity of the biosphere to reproduce natural resources and make provisions for stable life on the planet, on the other hand. The natural resources of land and ocean are being exhausted, various kinds of plants and animals are being annihilated; a technogenic breaking of biogeochemical cycles of matter is observed accompanied by the pollution of all components of natural environment; the simplification and degradation of ecosystems is taking place (see the References at the end of this chapter).

1.1 KEY ASPECTS OF BIOSPHERIC AND ECOLOGICAL STUDIES

All kinds of usage of the natural environment—industrial, agricultural, recreationary, etc.—are accompanied not only by obtaining desirable and expected results, but also by ecological, ecologo-economic and ecologo-social consequences which have created a critical ecological situation in many regions of the world and on a global scale. Changing the strategy of using nature—its continuous improvement to overcome negative effects in different regions of the Earth and over the entire planet—this is the basic scientific problem of optimal use of nature. To resolve the problem, research in the ecological, technological and economico-legal (humanistic) directions is necessary, which includes (Kondratyev, 1990a–f; Marchuk and Kondratyev, 1992):

(1) Study of regular features of the functioning of the biosphere as a unity of ecosystems forming it and natural-economic systems of different scales, from local to global. Study of biological, geochemical, geophysical and anthropogenic factors and processes, governing and changing the reproduction of natural resources and conditions in the biosphere. Revealing characteristic forms of the transformation of ecosystems and natural-economic systems. Development of mathematical models of ecosystems and natural-economic systems of various scales. Long-term forecasting of their variation for different "scenarios" of nature use. The employment of models for ecological normalization of economic impacts on the natural environment, for ecological expertise of industrial objects, both operational and planned, for mutual correlation of measures planned by different branches of the economy to protect the natural environment.

The most important directions of research include the evolution of the biosphere, energy- and mass exchange in the biosphere, biological aspects of nature protection, biospheric chemistry, the methodology of systems research and mathematical modelling of biospheric processes, geoecoinformatics.

(2) Development of an optimal strategy of nature use in power production, industry, agriculture and forestry, city planning and building and other most important branches of the economy, which might eliminate the anthropogenic impact of physical, chemical or biological nature. Finding promising ways of reducing energy-, material- and water-consumption of production processes; complex use of natural raw material resources; utilization of waste through recycling within the same branch of industry or for further application to other industries; reproduction of soil organics on the basis of its recycling and further rotation between plant growing and cattle production, and the recirculation of matter in landscapes; prevention from chemical or bacterial pollution of the natural environment and food production in agriculture, etc. Substantiation of the methodology of a regional low-waste system of technological processes.

The leading directions of research are: the ecology of industrial production, the ecology of agricultural production, human ecology, regional ecological problems.

(3) Development of methods of ecological assessment of natural resources and the consequences of the impact on the biosphere. Development of methods ensuring economic and social efficiency of production, taking into account its full-scale ecological, economic and social consequences. Creation of an economic mechanism stimulating the meeting of requirements of environmental protection and rational use of natural resources in industry, branches of economy, administrative regions and the entire country. Improvement of the scientific basis of legal acts regulating the use of nature by private persons, enterprises, branches of economy, administrative authorities, etc. Development of strategic directions of ecological education of the entire population of the country.

A number of problems of global ecology (i.e. climate and ozone layer changes, dynamics of vegetation cover and the biosphere on the whole) point to possible catastrophic consequences which are a threat to the very existence of modern civilization. However, the solution of not only global, but also regional ecological problems is difficult for the following two reasons: (1) the lack of sufficiently representative observational data characterizing real ecological situations; (2) the schematic nature of mathematical simulation models intended for the reproduction (and forecast) of processes taking place in nature.

From the viewpoint of the range of problems of global ecology, more and more important and promising becomes a combined use of conventional and satellite observation means for the assessment of the state of the natural environment and its potential changes in the future. The task of studying the Earth as an integrated natural sustem, formulated in the International Geosphere–Biosphere Programme (IGBP), can only be coped with on the basis of broad application of satellite observation means (Kondratyev, 1987, 1990a–d, 1991a–b, 1994; Kondratyev, Adamenko, Henderson-Sellers et. al., 1990; Kondratyev, Buznikov and Pokrovsky, 1992, 1996; Kondratyev and Cracknell, 1997; Kondratyev, Cracknell and Vaughan, 1994; Kondratyev, Johannessen and Melentyev, 1996; Kondratyev, Kozoderov, Fedchenko et al., 1994; Kondratyev, Ortner and Preining, 1992; Kondratyev and Pozdnyakov, 1996; Kondratyev and Romanyuk, 1996a,b). Therefore, the implementation of the programme of the International Space Year (ISY) has been so important (Kondratyev, 1990d,f).

It should be noted that the representatives of the Russian Academy of Sciences and the National Academy of Sciences of the USA signed an agreement in December 1988 concerning the formation of an Inter-Academic Committee on Global Ecology, and formulated the following key problems (Kondratyev, 1990c):

(1) Global energy and biogeochemical cycles—industrial and biospheric processes:
  — the perspectives of power production development and its potential impact on ecological systems;
  — monitoring of the "health" of global soils and vegetation, including controlled systems (at the initial stage, attention should be focused on the data for Russia and the USA);
  — determination of the amount of industrial pollution emissions in the USA and Russia in the past and present; exchange of relevant data and situation assessments.
(2) Substantiation of observing systems for estimation of global change:
  — development of new observing systems based on the ground and satellites;
  — programmes of exchange of satellite data;
  — analysis of the adequacy of the existing and planned observing systems for estimation of global change.
(3) Global change in biodiversity:
  — description and observation of biodiversity in Russia and the USA, and, as far as possible, over the entire world;
  — development of effective observing systems for approximate assessment of biological extinction;
  — ways of mitigating the process of dying out of species and loss of biodiversity;

— discussion of links between biogeochemical cycles and biodiversity.

(4) Theoretical and methodological basis for understanding ecological changes:
- — finding approaches based on the methods of nonlinear physics and contemporary mathematics for studying the unsustainablity of ecological systems;
- — analysis of the methodological basis for forecasting global ecological changes.

(5) Discussion and support of relevant international efforts:
- — International Geosphere–Biosphere Programme (IGBP), etc.;
- — bilateral and regional intergovernmental programmes;
- — study and, if necessary, support of international agreements in the field of environmental research.

The putting forward to the foreground of human values following from the enhancing interdependence of states and the increasing urgency of ecological problems, the improvement of political situation in the world and the weakening of the threat of a nuclear war, taken together, require a broader understanding of the problem of global security. Of course, the central point of the problem is global ecology, and particularly its socio-economic aspects. One of the important efforts planned along these lines was a three-year Global Security and Risk Management Project (GSRMP) developed at the International Institute of Applied Systems Analysis (IIASA).The key role in the Project belongs to the substantiation of priorities (Kondratyev, 1990a).

The main goal of the GSRMP is analysis of the necessity and perspectives of new approaches to the solution of the problems of global security and risk management.

(1) Investigation of the most significant indications of risk which can affect the survival of the world community, including ecological, demographic, economic, political, military, humane and social aspects. The aim of such investigations should be establishing admissible risk thresholds ("boundary conditions") whose violation may result in a global threat to human life and civilization. Very important in principle is the development of an interactive approach to the solution of problems discussed and substantiating (on this basis) of priorities.
(2) Development of approaches for early recognition and assessment of the danger involved in each of the risk components, as well as the role of their interaction.
(3) Finding ways for more efficient use of the existing agreements, programmes and institutions for the solution of global security problems, and for a broader and deeper international cooperation to achieve a more comprehensive and reliable monitoring of anthropogenic impacts creating risky situations.
(4) Expansion of the possibilities of multilateral organizations, and especially the system of the UN agencies and organizations, from the standpoint of their role in ensuring global security in accordance with the principles of the UN Charter.
(5) Development of practical proposals substantiating actions required for ensuring a global security.
(6) Providing access to the results obtained and the UN recommendations to governments and private persons all over the world.

(7) Functioning of a consulting body for reporting to the UN, including its Secretariat and the Secretary-General.

The following directions of research on problems of management, global security and risk seem to be most important: (1) global environment, the impact on it and problems of risk; (2) political and military security, factors affecting it and problems of risk; (3) demographic changes on a global scale; developed and developing countries; (4) development of economics and technologies.

Potential strategies of the reaction to situations that may occur involve: (1) a new definition of the notion of security; (2) analysis of possibilities of environmental management; (3) problems of military/political security; (4) possible forms of international cooperation and joint actions.

The notions of global (international) security and risk have not been defined precisely and uniformly enough thus far because they involve threats that have not been fully realized. Up to quite recently, international security was considered to be identical to military security, although it must be clear enough that this is an inadequate understanding. Security in a broad sense should imply the capability to withstand threats to human life, health, well-being and basic rights; to sources of life support and resources necessary to provide the above; to social order; to the ability of people to adapt themselves to changing circumstances. There are three main categories of threats to security:

(a) *Military threats:* a global nuclear war; the development and spreading of weapons of mass destruction, as well as the increasing accuracy of their systems of delivery; international transportation of weapons; large-scale wars with the application of conventional weapons (especially in regions of possible production of nuclear weapons); local conflicts.
(b) *Economic and social threats:* mass poverty originating starvation; economic collapses; destabilization of processes of capital transfer; excessive population growth and urbanization; international mass migration; gene manipulations; pandemics.
(c) *Ecological threats:* changes in the chemical composition of the atmosphere and their consequences (the "greenhouse gases" (GHG) concentration growth and global warming; the acidification of the environment, etc.); the pollution of natural fresh waters, oceans and coastal waters; deforestation and desertification, soil erosion, loss of soil fertility; risk relevant to the biotechnology; dangerous emissions of pollutants, as well as production, transportation and application of toxic substances; differing (over countries) opinions concerning ensuring human health and safety; the transfer of dangerous technologies to developing countries creating threats to their population.

Prevention of these threats requires preventive actions whose strategy must envisage a substantiation and priorities of targets; determination of the scale of risk, monitoring and assessment of the role of relevant processes; development of ways of resolving problems, taking account of scientific and technological possibilities, socio-economic aspects and the existing international agreements; estimation of the efficiency of efforts.

It is reasonable to examine three aspects of the problem of global security:

(1) perturbing impacts which are an object of regulation, limitation and control;
(2) consequences of perturbing impacts, the elimination of which requires applying various strategies of adaptation to ensure sustainable development;
(3) response to variations taking place in order to change the perturbing impacts and ensure global security.

One of the most important "perturbing impacts" is population dynamics which reveals itself in a drop of growth rate (except for the year 1988 when a slight rise (from 1.7% to 1.8%) was observed instead) but the increasing absolute population number (it looks like a maximum of the order of 8 billion people or more will be reached by the middle of the next century), and this is extremely significant. On the one hand, poverty, illiteracy and starvation decrease on a global scale; on the other hand, the absolute number of people suffering from these evil things becomes greater. Poverty and illiteracy increase the birth rate (and, consequently, child mortality). Population growth in developing countries is accompanied by the concentration of population in cities, which causes unemployment growth resulting, in its turn, in a stronger socio-economic tension.

The impact of the developing industry and agriculture on the environment and biosphere is of key importance. It provokes grave problems which are generally acknowledged, such as global climate and ozone layer changes, deforestation and desertification, acid rains and their influence on the biosphere, the pollution of the atmosphere, soils and oceans. Most dangerous is the continuing loss of biodiversity. This is revealed in biospheric dynamics determining the closedness of global cycles of matter with an extremely high accuracy, owing to biodiversity (Gorshkov and Kondratyev, 1990; Gorshkov, Kondratyev and Losev, 1992, 1994, 1996; Gorshkov, 1990, 1993, 1995a,b).

1.2 ECOLOGICAL PROBLEMS OF MODERN CIVILIZATION

(V.G. Gorshkov, K.Ya. Kondratyev, K.S. Losev)

Today there hardly is any doubt that the human environment deteriorates and that the natural biota (fauna and flora) is destroyed under the pressure of the growing impact of man's economic activity. However, the observed deterioration of the human environment and the destroyal of the biota have so far not been considered as interrelated processes. Hence the idea that the problems of the reconstruction of the environment and biota can be resolved independently, one by reducing the level of industrial pollution, and the other by preventing further reduction of biodiversity.

It is within this kind of problem formulation that discussions are held about how to feed the growing world population, stop global changes and put an end to the dying out of biological species. Those who suppose that these problems can be solved are considered to be "optimists", whereas others, who are of the opinion that these problems are extremely difficult or can hardly be resolved at all, are labelled "pessimists", "alarmists" or "extremists".

It will be shown in what follows that this is an incorrect formulation of the problem. One need not save the environment alone or the biota alone. A task like that would be insoluble indeed. It would be sufficient, however, to weaken the anthropogenic pressure on the natural biota, thus reducing the extent of its destruction. Then the biota would be quite capable of restoring the quality of the environment that might be appropriate for human life (Gorshkov, 1995a,b; Gorshkov, Kondratyev and Losev, 1996).

To substantiate the problem formulation we propose, we shall first compare quantitative information characteristics of the natural biota and human activity (civilization).

1.2.1 Information volumes in the biota and civilization

Of special interest is genetic information which is encoded in a DNA molecule with the help of four "letters"—nucleotide pairs (np), which are molecular memory cells—and each of these can memorize two pairs of different "letters", i.e. two bits of information. A human genome contains $M = 3 \times 10^9$ np (Lewin, 1987). Therefore, the value (volume) of human genetic information is $2M = 6 \times 10^9$ bits. The information storage in a man's memory is approximately equal to his genetic information volume. In fact, a human brain is capable of assimilating not more than 100 bits/second. Out of this amount, about 10 bits/second apparently incomes a man's long-term memory, on the average (Dudkin, 1985). Active assimilation of fresh information by man continues during his first 20 years of life, i.e. during 6×10^8 s. After that, the memory information increases insignificantly, and one may assume that its order of magnitude remains the same. As a result, the information volume of human memory may be estimated as 6×10^8 s $\times$ 10 bits/second $= 6 \times 10^9$ bits.

Information of the civilization is composed of non-overlapping portions of information contained in the memories of all the people on the Earth. Information introduced into books or computers is not dead and lost only if it is contained in the memories of some living members of human society. The upper limit of the information volume of the civilization can be assessed as the product of the world population number, 5×10^9 persons by the information content of each person's memory, 6×10^9 bits/person, i.e. a value of the order of 10^{19} bits. Obviously, this upper limit is overestimated considerably because the major part of memory information is the same for all contemporaries. Unique, non-coincident portions of information are contained in the memories of scientists-experts in various fields, as well as of authors, musicians, artists. Such scientists constitute about 1% of the population (factor 1/100). Every field of knowledge can be considered stable if no less than 100 leading experts in the field are simultaneously working there (one more factor 1/100). Thus, a realistic estimate of the information volume of the civilization can be obtained by multiplying the upper limit calculated above by the factor 10^{-4}, which yields a figure of 10^{15} bits.

A specific feature of human society is the presence of cultural heritage of the civilization (the notion of culture is implied here in its broad sense and includes science, arts and sports). A unique heritage is concentrated most of all in the brains of professionals. An important feature of human society is that it includes scientists, authors, musicians and artists—professionals of science and culture (and our contemporary politicians and businessmen should bear this fact in mind and stop the decay of science and culture).

The volume of genetic information of the natural biota of the entire biosphere equals the product of the average information of one kind of genome (which may be considered to be coinciding with the genome information of the most numerous group of species—insects—which is of the order of 10^8 bits (Lewin, 1987)) and the number of species in the biosphere—of the order of 10^7 (Thomas, 1990), i.e. 10^{15} bits. The intra-species genetic diversity includes mostly the "decay" genome variations (Gorshkov, 1995a,b) and therefore does not increase the estimate obtained. So, the volume of genetic information of the biosphere and the volume of cultural information of the civilization coincide in order of magnitude.

The memory volume of a modern personal computer (PC) reaches 10^9 bits and approximately corresponds to the volume of genetic and cultural information of a single human. Thus, the memory volume of the entire world modern computer technology is capable of encompassing the volumes of both the cultural information of the civilization, and the genetic information of the biosphere. So we see that, as far as the memory volume and information content are concerned, our contemporary civilization is equal to the natural biota.

However, life is a dynamic process which can be characterized by information fluxes and the amount of work that can be fulfilled per unit time by living organisms, in their interaction with the environment. With regard to these characteristics, there is an insurmountable quantitative precipice between the civilization and biota.

1.2.2 Information fluxes in the biota and civilization

We shall now consider the most important information fluxes. The mean "information" rate of evolution (the rate of changing of the volume of genetic information over the entire Earth's biota) can be estimated, proceeding from paleontological data about the average time of changing of the species composition of the biota, equal to 3×10^6 years (Van Valen, 1973) and the obtained above volume of genetic information in the global biota—10^{15} bits. The changing of the species composition of the biota takes place in the transition from some species to their closest-related species. Only 1% of genetic information changes in such a transition. A complete replacement of genetic information takes place during a time period which is greater by a factor of 100, i.e. during 3×10^8 years, or approximately 10^{16} s. As a result, we obtain an "information" evolution rate of 10^{15} bits/10^{16} seconds, or approximately 0.1 bits/second. This rate has been sufficient to ensure a stable development of the biosphere, i.e. to support the evolution of a biota capable of compensating directional changes in the environment caused by cosmic and geophysical factors.

The "information" rate of progress of the civilization, i.e. the rate of changing of cultural information, is determined by the human capability of assimilating and generating new information. At present, the collective brain of the humankind

(5×10^9 individuals) is capable of assimilating an information volume not exceeding 5×10^9 by 10 bits/second $= 5 \times 10^{10}$ bits/second. The current rate of progress is determined by the present-day time of changing of technologies, which is of the order of 10 years $= 3 \times 10^8$ s, and by the above-presented estimate for the cultural information volume of the civilization of about 10^{15} bits, i.e. the information rate of progress is of the order of 10^{15} bits/3×10^8 seconds, i.e. about 3×10^6 bits/second. Therefore, the ratio of the newly generated information to the assimilated information, on the average, does not exceed 10^{-4}. The civilization progress rate is more than seven orders of magnitude greater than the biota evolution rate; that is why man is *hors concours* in comparison with other species as far as his capacity for destroying the biosphere is concerned.

Information fluxes in the civilization's computers can be estimated as the product of information fluxes in a personal computer (about 10^8 bits/second) by the number of individuals possessing such computers. Assuming that at present there is one computer per every 100 persons, we obtain a figure of the order of 10^{16} bits/second for information fluxes in all computers of the civilization. This information flux can hardly increase by more than six orders of magnitude even if each individual on the planet is provided with a PC and if fluxes in each PC are increased by four orders of magnitude, i.e. the ultimate figure cannot exceed 10^{22} bits/second.

Today information fluxes in computers exceed by a million times the possibilities of humans to assimilate the information (with the human assimilation capacity of about 10^{10} bits/second). The use of computers is worthwhile insofar as they accelerate data processing and fulfil automatic management in accordance with computer programs (software) compiled by humans. However, the output information fluxes of computers can be controlled by humans alone. All information yielded by computers together with other information technology media (TV and radio, cinema and video production, theatres, the music industry, etc.) above this value is actually information pollution which affects all the five senses of man (mostly, light and noise contamination) and is a serious threat to human psychic health.

Information fluxes connected with the human metabolism, where all molecules of the cells of the human body take part, correspond to an equivalent food consumption power which is $q = 140$ W (about 3000 Kcal/day) divided by the energy of molecule excitation over thermal noise, which is of the order of $C_B T$ ($C_B = 1.4 \times 10^{-23}$ J K^{-1} molecules $^{-1}$ is the Boltzmann constant, $T = 310$ K is the absolute temperature of the human body). Assuming that one molecule represents one memory cell with two states—basic and excited (one molecule is approximately equal to one bit)—we obtain that $q/C_B T \approx 3 \times 10^{22}$ bits/second. This value exceeds the asymptotic computer information power of the entire civilization and is almost 12 orders of magnitude greater than the collective assimilation capability of all humans populating the Earth.

A human body contains over 10^{14} cells. Each cell processes about 10^8 bits/second of information, which is in perfect agreement with the information processing rate of a modern personal computer. Finally, the biosphere contains about 10^{28} of living cells which, consequently, process an information flux of the order of 10^{36} bits/

second, which value exceeds by 20 orders of magnitude the volume of information fluxes in all modern civilization's computers taken together. Unlike the interaction between man and computer, in the natural biota (as well as in the human organism), molecular storage registers in a cell are combined with elements of interaction with the medium surrounding the cell. Therefore, the entire information flux processed by the biota is used directly in the course of its interaction with the environment. It follows from the above estimates that no computer can be created which might be capable of modelling the activity of a living human organism, to say nothing of modelling the functioning of the biospheric biota on the whole.

1.2.3 Efficiency of energy consumption by the biota and civilization

The energy consumption of humankind today constitutes 10^{13} W (Starke, 1990), which is only an order of magnitude less than the photosynthesis power of the biota equal to 10^{14} W (Whittaker and Likens, 1975). One can imagine that if mankind increases its energy consumption by a factor of 10, it will be able to regulate the environment exactly in the same way as the biota. However, since the gap between the information volume of the biota and that of the civilization constitutes about 10^{20} (as shown above), the energy consumption of humankind can be used, and actually is used with an extremely low efficiency. In the course of its development, mankind has become capable of creating high-efficiency means for moving people and mechanisms, and for transporting energy. In the process, the external energy is transformed, with a high efficiency, into the kinetic energy of a moving or a stationary but operating mechanism. The motion of a macroscopic body as a whole is encoded by a small amount of information in a "macroscopic memory cell" relevant to the body (the motion is determined by four values only—the mass and velocity vector). In principle, kinetic energy can be transformed into gravitational or electrical energy with an efficiency of the order of 1; later on this energy might be used for the excitation of molecular processes of complex correlation, such as those taking place in a living cell. However, in the process of transportation of macroscopic bodies, kinetic energy is spent to overcome friction, does not generate any ordered molecular processes and is subject to dissipation when it is transformed into heat.

The biota also uses motion when it transforms the metabolic power of an animal into the mechanical power of its movement, with subsequent dissipation of the power. However, the biota spends less than 1% of its entire energy on this most extravagant channel of energy utilization (Gorshkov, 1995a,b). In the meantime, mankind spends more than half of its energy consumption on the production of transportation means and their functioning. The remaining energy is spent in a way which is no less extravagant. The expense of large energy amounts for overcoming and liquidation of consequences of natural calamities, such as fires, hurricanes, sand-storms, earth-quakes, big meteorite falls, etc. is equally wasteful. Since man is unable to govern the energy of natural calamities, this energy becomes a destructive force which annihilates the ordered structures of the biota, civilization and their environment. On the other hand, the increased power of civilization also becomes a destructive force and a "natural disaster" for the natural biota and global environment.

In the case of unicellular organisms, almost all processes taking place in the cell are realized in direct contact with the environment. Multicellular plants preserve this property of the unicellular ones owing to the extremely high efficiency of their organs—leaves, branches, roots. In the case of a big moving animal, its metabolism takes place mostly in its internal organs situated in an internal environment and deprived of contacts with the external environment. The larger the organism, the less the diversity of possibilities of its interaction with the environment. The most complicated metabolic transformations inside its organism are mainly used for the conversion of metabolic power into the mechanical power of motion, biotic biomass consumption and excretion. These simple functions do not require the complicated biochemical organization which has been inherited by big animals from unicellular organisms. This kind of work can be fulfilled quite successfully by technical means created by man, which are much more primitive than the organisms of big animals. Contrary to the generally adopted but erroneous idea, big animals represent the most primitive part of the biota, whose function, in particular, is periodic perturbation of the environment, which is a kind of "training" for the biota supporting its ability to compensate external perturbations. Small animals are "programmed" for extremely complex and specific types of interaction with vegetation which cannot be fulfilled by any technical means. It will suffice to refer to pollination of flower plants by insects.

Man has preserved a strategic program of behaviour based on the genetic programme of positive and negative emotions inherited from big animals, which can be reduced to the consumption of natural biotic resources: the destruction of the biota, transportation of consumption products and removal of waste—excretion. All goals of humans in their interaction with nature are determined by this program. The intellectual heritage accumulated by humans allows man to increase his impact on the environment without any limitations. However, the nature and diversity of this impact remain within man's inherited strategic program.

1.2.4 Possibilities of managing the environment and biota

According to above-presented estimates, humankind is apparently capable of reducing the gap between the information fluxes of the civilization and natural biota by six orders of magnitude (from 10^{20} to 10^{14}). Nevertheless, even if this gap could be eliminated altogether and the civilization could develop information fluxes identical with biotic ones, humans would still be unable to create a technical management system equivalent to the system of biotic regulation of the environment. This becomes obvious from the analysis of the well-known problem of automatic and manual control.

Manual control is implemented on the basis of man's genetic and cultural information. It takes place due to feedbacks which appear together with information incoming from the environment, and is limited by the rate of assimilation of this information by the human brain. Automatic control is fulfilled on the basis of a program introduced in the computer. Its rate is limited by the speed of information processing by the computer, and this speed may exceed the capacity of the human brain by a factor of 1 million.

Automatic control can only be used if the program serving its basis is supported by rigid theories verified in numerous independent experiments, which allows them to be considered completely reliable and eliminates the necessity to check management results. For instance, well-established laws of physics make it possible to fulfil automatic control of space flights. However, any malfunctions or extreme situations during the flight require passing over to manual control which is inevitably accompanied by a reduction of operational speed by millions of times.

Biotic regulation of the environment is equivalent to the kind of management where fluxes of assimilated information exceed by 34 orders of magnitude the intellectual abilities of the human brain, and by 20 orders the capability for automatic management of the modern civilization. This regulation is based on the genetic program of the global biota and the extragenetic memory programs of moving animals, and represents a system of natural automatic management of the environment which has been working according to the program of an "experiment" taking place for billions of years. Mankind is unable to build adequate mathematical models of the functioning of the natural biota. It cannot create a system of elements affecting the environment that would be equivalent to the natural biota, where every micrometre of the Earth's surface is controlled by dozens of uni- and multicellular living organisms functioning independently of each other. Humankind is unable to replace the genetic program of the biota by any technological program, even if its energy and information outputs could be of the same order of magnitude as biotic ones, because the correctness of any technological program requires experimental verification. The search, verification and correction of such programs can only be performed at the rate of human manual management, which means that billions of years may be necessary to fulfil a task of this kind.

The natural biota is subject to evolution, and its complete change takes place every 3×10^8 years. It follows that throughout the entire life history lasting for 3×10^9 years no more than ten different programs of environmental management have taken place. A verified and working program of biotic regulation of the environment apparently is a single one for each epoch. Evolution takes place but the principle of succession (continuity) is preserved, and in the process of evolution, the universal nature of biochemical organization of life and its fundamental properties are preserved. The "construction" of new programs by life is subjected to a detailed testing. The replacement of biotic programs occurs as a result of the impacts of geophysical and cosmic changes on the biosphere. Every acting program is maintained by the biota in a stationary state as long as possible.

Thus, although the volume of cultural information of the civilization has already approached that of genetic information of the biota, the biotic information flux exceeds by approximately 20 orders of magnitude the information flux of modern civilization. The major part of the genetic programme determines processes taking place in living organisms, i.e. the structure of their information fluxes. With regard to these factors, the presence of a gap between the biota and civilization shows that the information volume accumulated by the civilization is absolutely insufficient for managing the environment. Civilization is quite helpless in comparison with biota.

The natural biota of the Earth, including people, alone may claim to be called life on this planet.

Mankind must learn to respect the natural biota which is far closer to perfection than man-made civilization; however, this civilization is capable today of fully assimilating or destroying the biosphere. There is no sense in replacing the presently existing, smoothly performing global biota by a different one. Humankind should desist from such attempts because it cannot help the natural biota to control environmental quality. The best it can do is not to prevent the biota from controlling it. Therefore at the moment, the only goal of human community should be preservation of the presently functioning biota and restoration of its capability of regulating the environment on a global scale.

1.2.5 Evolution and progress

Humankind as a biological species possesses a unique feature which makes it different from all the other species. This is its cultural heritage continuously accumulated by every generation and transferred to subsequent generations. The ability and necessity for accumulating cultural heritage are genetic characteristics of normal individuals of the *Homo sapiens* species. The development of the cultural heritage cannot be stopped. The ceasing of cultural development would be equivalent to the annihilation of humanity as a species.

As has been noted by Gorshkov, Kondratyev and Losev (1996), the present-day speed of the development of science and technology can be quantitatively characterized by the mean time of radical technological changes, and nowadays this time is of the order of 10 years. The natural "biotic change of technologies" equivalent to this process can be quantitatively characterized by the mean evolutionary time of the changing of species, whose genetic program can be completely replaced over a hundred million years. Thus, the rate of progress exceeds that of evolution by seven orders of magnitude. No "sustainable development of the biosphere" is possible that can be compared with the speed of human development. Illusions concerning the possibility to include "biosphere development" in the development of civilization as its composite part are based on erroneous attempts to identify rapid processes of decay and relaxation of natural biota (flora and fauna), involving artificial selection of cultivated plants and animal species, with the processes of the evolution of species and the biosphere on the whole. A consequence of this mix-up of evolution and decay processes is negation of the observed biotic regulation of the environment. Such illusions may prove to be very dangerous for the future of humankind.

So, the human race is now facing phenomena that are strictly conflicting: the inevitability of rapid development of civilization, on the one hand, and the impossibility of sustainable development of the biosphere at the same rate, on the other hand (Gorshkov, 1995a,b; Gorshkov et al., 1994; Beder, 1995). To resolve this conflict, the development of civilization should be restricted in such a way that the biosphere (biota and its environment) will not be destroyed. For this purpose, one should establish quantitatively the thresholds of permissible impacts on the biosphere because exceeding these thresholds may result in a breakdown of biotic regulation of the environment and the origination of its global changes. Anthropogenic impact on the biosphere is completely determined by the value of global mean population density (Gorshkov, 1995a,b). Therefore, there actually is a single limitation for the development of civilization: the average population density value should not exceed the threshold marking the starting point of biosphere destruction.

The threshold value for global mean population density is determined by the level at which global environmental changes begin to occur. Cases of local exceeding of such a threshold can easily be found in history. Information about the ecologically admissible population density must doubtless be contained in the normal human genome (just as with other species). This population density will probably be close to the number of collectors and traditional fishermen. It is possible that the mass bloody wars that have broken out many times in the history of mankind (an unprecedented phenomenon in the animal world, at least within the same species) were indicators of sustainable local exceedings of ecologically admissible population density.

For thousands of years various countries struggled and competed with one another, striving to capture and possess territorial and material resources of the biosphere through wars resulting in mass losses of people. Warrriors were the more respected and admired, the larger the number of enemies they killed. Empires which possessed numerous colonies were considered the greatest powers. Countries at war were weakened, but wars slowed the rate of biosphere destruction by civilization only very slightly. That is why Malthus was wrong to suppose that wars may be a solution to the controversy between the growing consumption of the civilization and the stationary state of the biosphere.

Nowadays colonies are a feature of the past. It is probable that wars as a means of competing for possession of biospheric resources will gradually die out. But no less cruel economic wars have now replaced them, and their goal is the same: possession of biospheric resources. Individuals who have managed to use the resources most efficiently and thus obtain the greatest profit have acquired a prestigious position in contemporary society. As a rule, economic wars under conditions of a free market are not accompanied by human losses, which makes them most effective with regard to the utilization of resources, and this leads to an extremely rapid destruction of the biosphere. Traditional plunder of the biosphere can stop only when people realize that the biosphere is not an unlimited source of resources that may be mastered, used and annihilated. Civilization can and must develop, but not at the expense of unrestrained utilization of the biosphere as a resource of development. In this respect, the development of civilization should be restricted.

1.2.6 Restrictions on civilization development

Restrictions on civilization development are not violence to a human personality or a limitation of human rights. Humankind has long been accustomed to physical limitations based on the laws of nature. Nobody feels depressed by the well-established facts that the speed of motion cannot be greater than the velocity of light, that a *perpetuum mobile* of the first or second kind cannot be built, or that human life is limited in time. A good number of restrictions, that are legally secured but are at variance with human nature, have accumulated in human society, as the population density has grown. These restrictions are necessary to maintain the sustainability of the contemporary overpopulated society.

Ecological restrictions on development, if substantiated and defined precisely enough, will not be a threat to human personality either. Reliable knowledge about ecological development thresholds and corresponding rules of human behaviour should by all means be included in the educational and cultural systems of mankind. By adhering to these natural restrictions, people might be able to cancel the major part of strict legal limitations oppressing the life of a contemporary human individual.

The present ecological difficulties of mankind are due to the fact that the natural-scientific investigation of ecological restrictions to development has come so late. The human race has been too long in a state of euphoria with regard to the "unlimited" opportunities of opening up and managing the resources of the Earth and space, and sticking to the idea of non-standard development of civilization in principle, by analogy with the development of the Universe as a whole. There is no doubt (Gorshkov, 1995a,b) that anthropogenic perturbations of the biota and environment have already considerably exceeded all admissible ecological thresholds, and the presently observed global changes are direct consequences of this situation. Therefore, the most important problem today is how to reduce the degree of biosphere perturbation to an admissible threshold value. The complexity of the problem and the lack of traditional ways to resolve it often provoke unwillingness to even admit the existence of ecological thresholds for economic development and population growth.

Beginning with the middle of the past millennium, humankind has been existing under conditions of continuous population growth. The resulting demographic explosion has become possible owing to economic growth based on progress in science and technology. Many generations of people have already been existing in conditions of exponential economic growth; therefore, the situation itself is considered the natural state of humanity. The growth in population results in an annual increase of production, bank deposits, shares and their costs. And this is regarded as routine development.

Production fulfilled by mankind is based on the use of raw materials and energy, i.e. natural resources. Some of these are non-renewable: fossil minerals and fuel which are not utilized by the biota. Renewable resources are the production of the biota itself which is utilized by man.

The limits to exponential economic growth following from the exhaustion of non-renewable natural resources and approaching the level of consumption when all renewable natural resources are consumed, have been persuasively demonstrated to the world public over twenty years ago in a series of publications of the Club of Rome (Meadows et al., 1972, 1974; Mesarovic and Pestel, 1974). However, all these publications treated the biosphere-biota and its environment from the standpoint of modern economics, i.e. as a resource, along with fossil minerals. Environmental pollution by civilization has been assumed to be compensated by relevant civilization activity based on the consumption of natural resources. The question whether humankind is capable of making this compensation in principle has never been put forward. The problem of natural sustainability of the biosphere and the existence of biotic regulation of the environment has been ignored completely.

Even today many publications treat the problem of population growth only from the point of view of feeding the growing population by means of exploitation of renewable natural resources. The so-called optimistic forecasts are based on the fact that the present-day direct consumption of biospheric produce by humans does not exceed 10% of the global biota production, and that consequently, the presence of this renewable resource makes it possible (taking account of the law of conservation of matter) to increase consumption (and hence, the population number) by an order of magnitude. However, resources are renewable until their consumed share exceeds the threshold of biotic sustainability. Beyond this threshold, resources cease to be renewable. Renewable resources can be restored if the level of their consumption is reduced to below the threshold value. However, then the notion of a resource becomes senseless (see below).

1.2.7 Resources and economic growth

Economic growth on the basis of a free market takes place owing to the possibility of determining the capital and the production of goods which are in demand and bring profit. The profit is partly used for the enrichment of the owner of the business and the executives involved, and partly for further expansion of production. If the business is a success, the result is the exponential growth of the production and its owner's profit taking place during a period which is much less than an average human lifetime. A businessman's wealth determines his competitiveness and social position. Every normal human individual is genetically programmed to strive to increase his prestige. This is certainly characteristic for a businessman. Socialism and centralized economic management have lost the battle with capitalism, first of all because this genetic feature of human beings has been ignored.

Any kind of production is an ordered process supported by an external energy flux. Production is made out of raw materials which are considered to be natural resources. Finally, the product is in demand either if it is biospheric production necessary for man, or if it enables man to increase the consumption of biospheric production. A free market enables supply to equal demand, and introduces a uniform monetary system to measure the cost of any produced goods. Modern economics as a science is based on the accounting for possibilities of economic growth. Its goal makes no sense if this growth stops. Economic growth increases the production of any goods, and consequently it enlarges their monetary equivalent. The latter value is accepted to be the main index of the standard of living. Therefore, politicians do their best to stimulate economic growth and fear economic stagnation most of all.

Every living organism in the biosphere is adapted to its ecological niche. Therefore, the best *modus vivendi* for a species is living in conditions as close as possible to the natural ecological niche. A human individual normally enjoys this kind of life during a vacation. Most people have a 2–4 week-long vacation per year, which constitutes less than 10% of their lifetime. It follows that by the age of 70 a person has spent no more than 7 years under conditions which are natural for a human. However, such conditions are gradually reduced in the transformed biosphere. From the standpoint of contemporary economics, these conditions are more and more evaluated as a very costly resource. This resource can be bought by very wealthy people alone. The great majority of people have to spend their lives under conditions distorted by civilization.

In a completely distorted biosphere, no money can buy terms of existence that are natural for man. The intensive work of practically every individual in the conditions of modern civilization is directed towards the conversion of the environment into a comfortable "prison cell" with the highest level of per anima production of any kind of goods. This cannot be an adequate substitution for the natural conditions of an ecological niche; this cannot either be considered an increase in the standard of living. And this cannot certainly be the meaning of a human's life.

It is usually assumed that the question of the meaning of one's life is extremely complex, is possibly non-scientific and cannot have an unambiguous answer. Nevertheless, man returns to this question again and again, and this in itself is a fact of natural science which requires an explanation. Obviously, the meaning of life of every human individual involves the fulfilment of work required for the reproduction and genetic stabilization of mankind as a biological species. This has always been general knowledge. In recent millennia, it has become obvious that the meaning of life of man also includes necessary actions oriented towards the stabilization of culture and civilization. In the recent decades, it has become evident that the basic meaning of a human's life must be his work directed towards the stabilization of the biosphere—the human environment, and the biota as a regulator of this environment. Human culture cannot develop independently of the above-said and must be completely governed by this latter requirement, as well as genetic information.

The meaning of life of every individual of each natural biotic species is not simply reduced to its/his existence in its/his environment, adjustment to it and using its resources. Out of all possible species capable of living under terrestrial conditions, the biota has selected only those which are able to fulfil the necessary actions within their communities, i.e. able to fulfil a certain work directed towards the stabilization of their environment. The human being's meaning of life cannot (in this context) differ from those of other living creatures of the natural biota.

Just like an ant in an ant-hill, a bird in a nest or a bear in its lair, a human can and must spend part of his time in the conditions of civilization he created. But the region of the biosphere occupied by civilization (similarly to the ant-hill, bird's nest or bear's lair) should occupy a small portion of the area occupied by natural biota. The main work fulfilled by man should be directed not to extensive development of civilization, but to conservation of natural biota in an unperturbed state in such a way that the threshold of biospheric sustainability is not reached.

The functioning of the natural biota is not based on the consumption of environmental resources. The environment itself has been formed by the biota and is maintained by it in a state which is optimal for life. The notion of resources is alien to the biosphere because any kind of biotic production is compensated by its destruction, and both processes are based on the processing of materials created by the biota itself. Exactly in the same way, man does not consider food products he produces as resources. Any ordered processes fulfilled by the biota form a closed cycle of substances. Free competitive interaction of individuals in the natural biota serves for the removal from the biosphere of "decaying", non-competitive individuals destroying the existing environmental condition (Gorshkov, 1995a,b). In a natural community, the "gangster" species, ruining correlations within the community and making it incapable of regulating the environment, cannot be permanently present. Individuals of such species would use the biota and environment as their own resource, not caring about consequences. And these consequences would be the destruction of the community, the cessation of biotic regulation and the distortion of the environment. The loss of competitiveness by the community, accompanying these processes, would have resulted in its ousting (together with the gangster-species) by normal communities having no such species.

Thus, the notion of "resource" is senseless in a sustainable self-regulating system "community + environment". Since there is no notion of "resource", competitive interaction aimed at using it efficiently becomes impossible. The stratification of any kind of population to produce "the rich" and "the poor" is not possible either. In the natural conditions of an ecological niche, decaying non-competitive individuals form a negligibly small portion of the whole population. The majority of the population is formed by normal individuals which are equally competitive, capable of fulfilling equal portions of work to stabilize the environment and getting equal amounts of the common wealth necessary for life.

The natural conditions of an ecological niche are the conditions of a true "democracy" for most individuals of the population, who themselves determine the state of the environment. However, in the distorted conditions, the "decaying" individuals are accumulating who may finally form the majority. The decaying individuals have no program of fulfilling the necessary work directed to the stabilization of the environment. To bring the population back to its normal state, the functioning of the population must be determined by the behaviour of the few remaining normal individuals. Under such terms, "democracy", which makes normal and decaying individuals equal, would only increase the number of decaying individuals and thus accelerate the drop of the organization level and increase the probability of the dying out of the species.

The notion of "resource" makes sense only during a finite time period of the destruction of the whole system after the gangster-species has been introduced into it. There are numerous well-known examples of the destruction of natural communities and their environments when foreign gangster-species have been artificially introduced into them: the islands of Oceania, New Zealand, Australia. The gangster-species do not possess information on the necessity to preserve the community and its environment. If several gangster-species are introduced, they may compete for the same resources. These gangster fights reveal the most competitive among the gangsters, i.e. those who are quickest in using the resource and destroying the system. This kind of competition is not relevant to the competitive interaction of individuals within the populations of natural species which results in the removal of decaying individuals, breaking the self-regulation of the system.

Modern economics, which considers the biosphere to be a resource of humanity, brings about the destruction of the biota and the environment. The actions of businessmen permitted by legal norms with respect to natural biota do not differ in many cases from the actions of criminal structures prohibited by the same legal norms with respect to economics itself. Strictly speaking, the competition of businessmen for effective use of natural resources and the increase of production per anima are no different from the competitive fighting of criminal structures for their "spheres of influence". Gangsters destroy the economy. The destruction of the biosphere deprives man of his natural ecological niche, which cannot be compensated by the per anima production of any kind of products.

The present economic crisis in Russia is a kind of respite for the natural biota because it decelerates the development of an ecological crisis. Regrettably, many politicians see the only way out of the current economic crisis in stimulating business people to use biospheric resources, i.e. provoking a more profound ecological crisis. However, there is no doubt that to preserve a healthy biosphere—the environment and biota regulating it—the activity of civilization should be maintained below the relevant threshold level.

1.2.8 Sustainability thresholds of life

To conserve biospheric sustainability, the consumption of its resources should not exceed 1% of pure primary production of the global biota. This figure has been obtained on the basis of several quantitative evaluations performed independently of each other (Gorshkov, 1995a,b; Gorshkov, Kondratyev and Losev, 1996). The present-day direct consumption of biospheric land production by civilization constitutes about 10%, i.e. one order of magnitude more than the permissible threshold value. The proportion of land economically managed and transformed by civilization is about 60% (World Resources, 1988).

Balanced reduction of biospheric production consumption to its threshold level may only be achieved by reduction of global population number to about its level at the beginning of the eighteenth century. This can take place during a time period of less than one century if an average family on our planet becomes a one-child family. This kind of reduction can be accompanied by a continuous increase of the standard of living of people with no offence to human rights or moral norms. To make this possible, human society should overcome its non-scientific conservatism revealing itself in the negative attitude to depopulation on the part of professional politicians, demographers, economists, and the mass media which are at their service. People must learn, at last, to respect future generations and their interests. All these problems have been discussed already (and will be further discussed) at the most important international conferences, including the Cairo Conference on Population (September 1994) and the Copenhagen Conference on Social Development Problems (March–April 1995). An international society (Negative Population, Inc.), headed by its President, Donald V. Munn, has been organized. The world-renowned scientist, Linus Pauling, was one of its active members to his last day. These are indicators of the broadening recognition of the fact that there exists an acute problem of global demography.

The culture of modern civilized society is quite prepared for the possibility of depopulation. Demographic processes which are at present taking place in Russia provide ample evidence in favour of the above statement. The economic recession and the drop in the gross national product (GNP) of the country are partly compensated by a decreased birth rate, i.e. by negative feedbacks. This must lead to a rise in the average standard of living and contribute to a more humane attitude to children already born. Of course, the current economic situation in Russia cannot be considered normal.

A considerable decrease of the level of anthropogenic perturbation of the biosphere, even stopping or reversal of currently existing tendencies of global change, can also be achieved with the present-day population number of the planet by way of unbalanced reduction of certain kinds of consumption of biospheric production. For example, even if today's rate of consumption of fossil fuel is conserved, the carbon cycle global changes and the accumulation of CO_2 in the atmosphere can probably be stopped if people cease to exploit forests and reduce the economically managed portion of land from 60% to 20%, i.e. to a value which is twice as large as the area of arable land. It is obvious, however, that, if population growth continues, mankind will not be able to exist for long without consuming wood and with an abruptly reduced developed territory. Therefore, in the future, in order to restore balanced consumption of biospheric production along with preserving its sustainability (and it is such a situation that may be defined as "ecologically sustainable development"), a reduction of population to the above-mentioned threshold value will inevitably be required. This process, however, can be extended in time and fulfilled without any undesirable pressure upon the currently existent population. So, the present ecological crisis has a way out, and the preservation of human environment has a realistic perspective, but their implementation certainly requires a consensus with regard to goals and means of resolving the problem on a global scale.

Humankind will have to maintain a permanent population number which must be an order of magnitude lower than today. In such conditions, there will be no need to take care of environmental protection on a global scale. Similarly to what was the case during the previous billions of years, the natural biota of the biosphere will fulfil this function. In a minor biospheric region occupied by civilization and serving as a "home" for humans, there will be industry and agriculture, as well as technological means for reducing man-made pollution to a level determined by maximum admissible concentrations (from the viewpoint of human health). This kind of industrial and agricultural sanitation is an analogue of sewage disposal, which is familiar to everyone. (Regrettably, the main goal of ecology is too often reduced to this kind of industrial "sewage disposal".)

True ecologically sustainable development will be achieved if mankind ceases to regard the biosphere as a totality of its own resources. Similarly to what can be observed in the natural biota, society will not be divided into two categories: the rich and the poor. Extensive economic growth will stop. The words "mastering" and "war" will be excluded from the human vocabulary for good, and the notion of "resource" will only be used to characterize the long lifetimes of instruments and mechanisms created by man.

The progress of culture and civilization will continue. The process of changing of technologies on the basis of new fundamental scientific discoveries will gradually reach saturation, and possibly the speed of progress in science and technology will become equal to the rate of evolution. The accumulation of knowledge pertaining to the humanities, the development of pure mathematics, the advancement of sporting games, etc. will probably take place at the same rate. This kind of progress is similar to the "organic growth" discussed more than 20 years ago by Mesarovic and Pestel (1974) but taking place in the conditions of preserved natural biota and human environment. The perspective described here is by no means a utopia because all biospheric species have existed in a similar state for billions of years.

It is possible to draw the following two conclusions from the foregoing:

(1) a critical condition for ensuring a global ecological security and sustainable socio-economic development is the preservation of the biosphere;
(2) an urgent task is the development and implementation of a global system of ecological observations, involving conventional and satellite observation means, to guarantee adequate monitoring of processes in the biosphere and environment in order to prevent further perturbation of the biosphere and control the efficiency of measures necessary for its restoration.

1.3 SOCIO-ECONOMIC ASPECTS OF ECOLOGICALLY SUSTAINABLE DEVELOPMENT

Special attention should be paid to two aspects mentioned above: global dynamics of population and modern trends in the development of technologies which take place under conditions when the level of prices for energy carriers does not stimulate the working out of energy- and resource-saving (low-waste) technologies. Demographic contrasts between developed countries (the falling population number and increasing average age) and developing ones (the opposite pattern) have been intensified. The process of industrialization, which is of key importance for developing countries, originates their greater contribution to the exploitation of biospheric resources (which is first of all relevant to deforestation) and environmental pollution, which reveals itself, for instance, in the growing concentration of "greenhouse gases" (GHG) in the atmosphere.

Four categories of threats caused by environmental changes should be taken into consideration, viz. (1) direct threats to human existence (hunger, diseases, radiation, etc.); (2) threats to territories (desertification, ocean level rise, transboundary transport of pollution, misuse of river runoff by some countries, etc.); (3) threats to life support systems (agriculture and fishery, fresh water resources, forests, etc.); (4) threats to economic development (deficit of natural resources, instability of the financial system and markets, etc.).

The predominance of cheap raw materials as an item of export from developing countries results in their increased debts and aggravates global socio-economic tension. A tight knot of global security problems has finally been formed, whose ecological, socio-economic, political and other aspects have proved to be inseparable. There are numerous examples that may illustrate this conclusion: global climate change and its impact on agriculture, forestry and water resource management; acid rains and forest degradation; the reducing reserves and deteriorating quality of drinking water; the loss of biodiversity and its unpredictable consequences, etc. All these processes stimulate instability on the regional and the global scale, and determine a situation when problems of national security can only be solved in the context of global safety on the basis of efficient international cooperation. Problems such as the broadening of possibilities of possessing nuclear arms, or access to chemical and biological weapons, also require collaboration on a global scale.

To ensure global security, interdisciplinary research is required to substantiate recommendations to governments and to the United Nations Organization that can be both effective and feasible.

At the moment, the following most important recommendations can be formulated (Kondratyev, 1990a):

(1) An international agreement at the UN General Assembly is necessary concerning the development of an optimal global system of ecological monitoring, based on a combined use of conventional and satellite observation techniques. The use of such observation data as input information for numerical simulation models of global ecodynamics (biosphere dynamics, changes in climate and ozone layer, dynamics of water and forest resources, etc.), which require further advancement, will make it possible to operate a global early warning service about dangerous trends in the development of the ecological situation, as well as to substantiate potential development scenarios and measures necessary to prevent catastrophic consequences of anthropogenic impact on the environment and biosphere.
(2) Of course, the data of global ecological monitoring and assessment of trends in the development of the ecological situation should be accessible to all states. This problem can be solved, and a relevant agreement on the creation of an optimal global monitoring system reached only if a special body within the UN system is organized which will be capable of taking up the responsibility of handling the problem of global ecological security.
(3) Since there are sound grounds to believe that the Earth is already overpopulated, urgent efforts at the UN level are necessary to stimulate further studies of ecologo-demographic problems and work out recommendations whose implementation can ensure a sustainable socio-economic development over the entire planet.

The Second UN Conference on Environment and Development (UNCED-2) and the UN Conference on Population and Social Development Problems have been two landmarks on the road to the solution of the problems discussed above.

Prior to summing up the results of UNCED-2, we shall note that the participants of the "Social Summit" in March 1995 in Copenhagen (including 192 state and government leaders) have discussed in detail the problems of poverty, unemployment and social integration (*The Bulletin*, 1995), and formulated the following vitally important goals:

(1) Creation of environmental conditions necessary to ensure an ecologically sustainable development (ESD) (of course, the very notion of ESD requires a solid substantiation, as well as qualitative and quantitative indicators of ESD).
(2) Elimination of poverty.
(3) Elimination of unemployment.
(4) Support of social integration.
(5) Overcoming the inequality of sexes and support of the leading role of women at all levels.
(6) Support of socio-economic development of Africa and other least-developed countries.
(7) Support of structural reconstruction programmes, including reaching the aims of social development.
(8) Increase and/or improvement of the efficiency of the utilization of resources intended for the solution of problems of socio-economic development.
(9) Intensification of international, regional and subregional cooperation in the solution of social development problems.
(10) Worldwide support of educational and cultural development.

Many aspects of the practical realization of these goals are not quite clear today. The idea that further economic growth is the best means of reaching these goals is rather doubtful. One can hardly answer directly the question of how this growth can be implemented: by intensifying production and open economy within the scope of the worldwide trade system? What level should be taken as a reference standard? Should it be that of rich developed countries (the USA, Western Europe, Japan), with their average annual per anima income of \$20 000 and CO_2 per person emissions to the atmosphere of between 15 and 20 tons per year? One can see that the problem is much more complicated and cannot be reduced to economic growth as it is. The poor should have better living conditions in the direct sense, i.e. their dwelling and its security, a healthy environment, etc.

Let us now consider in detail the problem of sustainable development (SD), especially in the context of sustainable development (SD) indicators and demographic dynamics.

1.4 SUSTAINABLE DEVELOPMENT: CONCEPTUAL ASPECTS

1.4.1 Introduction

Since antiquity, nature has been regarded as a single whole. It is only now that the contemporary world begins to realize this simple truth, after having faced the results of technogenic intrusion of society into the biosphere on a scale comparable with geospheric processes. In the second half of the twentieth century, mankind has finally realized the importance of problems of interaction between nature and society in their global dimension.

It is already some years ago now that a number of Russian scientists substantiated the necessity of a complex approach to global change problems, taking into account their demographic aspect as the key one in the context of biotic regulation of global environment (see Gorshkov, 1995a,b; Kondratyev, 1987–1996; Kondratyev et al., 1975–1996).

1.4.2 Formulation of the problem

The question of whether the limited resources of our planet can continue meeting the demands of growth and development indefinitely goes back a long way at least to the middle of the eighteenth century, and to Thomas Malthus.

In recent years, gloomy neo-malthusian prophecies of a global decay have been revived (Homer-Dixon, Boutwell and Rathjens, 1993; Kaplan, 1994; Kennedy, 1993). According to the conclusion drawn in 1993 by the Council for a Parliament of the World's Religions, "the world is in agony" (Gladwin, Krause and Kennelly, 1995). This is confirmed by summarized data of global characteristics of socio-economic development (see Introduction). They reveal that efforts aimed at preserving the biospheric potential alone, without ensuring social equity and balanced economic development, may merely amount to a "rearrangement of deck chairs on a global *Titanic*" (Union of Concerned Scientists, 1992).

The understanding of the interdependence of the abovementioned processes determined the necessity for working out a strategy of sustainable, ecologically safe development of human society on a planetary scale. The problem of global demographic dynamics has acquired a key importance in this regard (Kondratyev and Frolov, 1996).

The UN experts forecast that by the year 2050, the world's population (now approaching 6 billion, with African, Asian and Latin-American countries responsible for over 95% of the population growth (Gorshkov, Kondratyev and Losev, 1992)) will reach 7.9 billion (low growth rates) or even 11.9 billion (high growth rates). This will aggravate social inequity in the world. Ramfal (see Polunin and Nazim, 1994) notes that the level of resource consumption typical for a native of the northern hemisphere during his lifetime exceeds by a factor of 20–30 that typical for the developing countries, and their levels of waste production are in the same ratio. From the viewpoint of resource consumption, 1.2 billion people living in the industrially developed countries are equivalent to 24 billion of those living in the South. These and other circumstances require an urgent change in socio-economic development, and a global-scale action plan to save humanity and the environment.This initiative must come from OECD nations; however, the developing countries should also contribute to the implementation of such an action plan.

The British Overseas Development Administration has worked out a "Strategy for Studying Renewable Natural Resources (RNR)" which is aimed at the improvement of natural resources productivity in developing countries along with ensuring an ecologically sustainable development. The "Strategy" emphasizes that ecological aspects of planned projects should by all means be taken into account, viz. climate change, biodiversity, ozone layer depletion, environmental pollution, urbanization after-effects. In what follows, we shall discuss to what extent the solution of a problem like this is practically feasible in the conditions of the present ecological and socio-economic situation.

Dr N. Sadick, Secretary-General of the International Conference on Population and Development (ICPD) (September 5–13, 1994, Cairo), emphasized the importance of a clear definition of the term *global sustainability* (*sustainable development* or *ecologically sustainable development)* which had by then become highly popular. She was quite right to state that our world has hardly ever been sustainable and is certainly not at present. The same problems have been discussed in detail by Gorshkov (1990, 1993, 1995a,b). Goodland et al. (1994) have been correct to say that as humankind is approaching the twenty-first century, it is moving away progressively from ecologically sustainable development (ESD) at an unprecedented rate. In this context, we shall now discuss the ESD problem in more detail, considering its two aspects: population growth and increasing consumption of natural resources.

1.4.3 A concept of sustainable development

The notion of global ecologically sustainable development (ESD), in the context of the concept of biotic regulation of the environment, can be not only clearly defined but also made concrete quantitatively with the help of such indices as the level of consumption of biospheric resources and, respectively, the indicator of closedness of global biogeochemical cycles (Gorshkov, 1990, 1993, 1995a,b; Kondratyev, 1990b; Marchuk and Kondratyev, 1992; Gorshkov et al., 1994). The general concept of sustainable development is a much more complicated matter.

The term *sustainable development* was first popularized by the International Union for the Conservation of Nature (IUCN). In a publication "Our Common Future", which is generally known as the Brundtland Commission Report, sustainable development has been defined as "development that meets the needs of the present without compromising the ability of future generations to meet their own needs" (World Commission..., 1987). This simple but effective definition provoked numerous interpretations; as a result, by the year 1989, there were over 60 definitions of the term *sustainable development* (Polunin and Nazim, 1994) which interpreted in one or another way the idea behind the concept.

The world community started working out various theoretical bases to substantiate the concept of sustainable development, later documented by the Second UN Conference on Environment and Development (UNCED-2) (Kondratyev, 1990–1996; Kondratyev and Romanyuk, 1996a,b). However, some key aspects of the problem are still open to discussion. For instance, von Weizsäcker (1994) was absolutely right to emphasize the fact that the concept of *sustainable development* is by no means helpful in solving problems humankind is facing. To make this concept more realistic, "a new hierarchy of values is needed, together with a new culture and a new totality of incentives...".

It is clear that different interpretations of a comprehensive notion, such as *sustainable development*, may result, especially at an early stage of comprehension of the notion and its introduction to public mentality, in great variance of opinions where national programmes and plans aimed at achieving sustainable development come in. The concept should be theoretically substantiated; otherwise, the road from its basic principles to their practical implementation may be very long.

It should be mentioned that initially the idea of sustainable development was perceived very often as a concept of ecological efficiency, since social welfare is impossible without sustained reproduction of resources and the absorption of human-made wastes by natural systems. The ecological component, however, is a necessary but not a single prerequisite of sustainable development.

Goodland et al. (1994) who justly emphasized the interrelationship between ESD, sustainable social and economic development, offered the following definitions of these notions.

- Sustainable social development (SSD) implies such a regime of socio-economic development which ensures self-control in the solution of problems of environmental and natural resources management. The use of resources should be oriented towards achieving social equality and justice. The goals of SSD must be the priority of qualitative improvement over quantitative growth, and a price policy which provides complete covering of the expenses of production, including social expenses. These goals can only be achieved under condition of social partnership. Although not subject to quantitative estimation, still important are such forms of "social capital" as social well-being, cultural development, discipline, honesty, etc. Such forms of moral or political capital require support and renewal on the basis of the cultural life of the community. The main objectives of SSD are equal rights for women, demographic sustainability and elimination of poverty.
- Economically sustainable development (ECSD) is usually defined as "support of capital" (the term was used as far back as in the middle ages by merchants). However, out of all forms of capital (manmade, natural and human), economists have so far preferred taking into account natural capital alone (healthy forests, clean air, etc.). Expressing the value of goods in terms of money, the followers of classical economics preferred "externalizing" expenses for environmental protection. At present, a tendency has appeared towards the internationalization of such expenditures, and the consideration of such principles as non-damaging, uncertainty and risk.
- Ecologically sustainable development (ESD) is determined by human needs and social concern, and its main objective is ensuring human well-being by protecting the sources of raw materials used for the needs of people and the environment (by means of adequate sinks for anthropogenic emissions). Mankind must learn to observe limitations arising from the necessity of preserving a proper state of the environment, which is simultaneously a source of resources and a sink for wastes. The latter implies that the level of emissions should not violate the assimilative capacity of nature. With regard to non-renewable resources, the application of the condition of achieving quasi-ESD, when the rate of their spending is equivalent to

that of their compensation by replacement with renewable components.

A schematic estimation of the ESD conditions characterized by value I is possible through taking account of the anthropogenic impact according to the ratio (Kondratyev et al., 1996b):

$$I = P \times A \times T \qquad (1)$$

where P is population number, A is per anima consumption of resources, T is the level of technological development (the latter can be characterized as environmental protection expenses per unit of production).

However, since quantitative estimations by means of this ratio are complicated, data on per anima energy consumption are often used as a defective substitute. If one proceeds, for instance, from a per anima GNP level of $4000 (as marking the boundary between rich and poor nations), one will find out that the damage to the environment from a single inhabitant of the developed part of the world is 7.5 times greater than that from a representative of the Third World. It follows from expression (1) that the three conditions for ESD are:

(1) Limitation of population growth, especially in developing countries.
(2) Limitation of per anima resources consumption (above all, in developed countries).
(3) Advancement of technologies.

In his book Dr von Weizsäcker (1994) has very justly put an emphasis on the "North–South" problem. He writes how necessary it is to formulate "a new set of values, a new culture, a new set of incentives to make millions of actors both in the North and the South act differently. We need to enter and to create a new era of human history". Of crucial importance becomes "the cruel reality of ecological devastation which will confront us in our everyday lives and which will inevitably shape our civilization if present trends of destruction continue for just another decade or two".

Serageldin et al. (1994) proposed the following classification of the sutainable development objectives:

(1) Ecological objectives, including the range of problems relevant to the integrity of ecosystems, the carrying capacity of our planet (more precisely, the biosphere), biodiversity and the state of global environment (changes in climate, ozone layer, etc.).
(2) Economic objectives, including problems of economic growth and principles of equality and efficiency.
(3) Social objectives, including living conditions, social equality, mobility and compatibility, cultural identity and institutional development.

Emphasizing the acuteness of the social "dimension" of sustainable development, Cernea (see Serageldin et al., 1994) noted the urgency of its two aspects:

(1) Substantiation of concepts which make it possible to explain the nature of social actions; the interrelationships between people; complex forms of social

organization of the society and institutional structures; the specific features of cultures, motives, incentives and values which determine human relationships and man's attitude to natural resources.

(2) Development of methods of choosing and substantiating social actions, as well as of social infrastructure. Analysis of the results of the implementation of 25 projects supported by the World Bank (WB) in Africa, Asia and Latin America has revealed (Making Development..., 1994) that ignoring the social "dimension" has largely contributed to the failure in 13 cases to improve the level of sustainable development 6–10 years after the completion of relevant projects. The role of creating effective organizational structures has been clearly revealed in the implementation of irrigation projects in several developing countries.

Khan (1995) (as well as Goodland et al., 1994) emphasizes that "the paradigm of sustainable development which seeks to pursue growth and equity within the context of intergenerational resource stability sees development as *achieving the interlinking objectives of social, economic, and environmental sustainability both in the short and in the long term*".

There are a great number of urgent problems requiring the creation of an integrated (interactive) model of sustainable development. Khan identifies the following three main tasks which imply the development of:

(1) appropriate methodological tools to appraise projects by weighing equally the requirements of social, economic, and environmental sustainability;
(2) new methods of analysis of man's economic activity which take into account the cost of natural resources;
(3) adequate indicators within the integrated framework of social, economic, and ecological sustainability, to replace the classical method based on the notion of "gross national product" (GNP).

Serageldin et al. (1994) recognize at least four categories of the global "capital" essential for sustainable development; these complement each other but are not interchangeable:

— Man-made capital (factories, machines, buildings, etc., including the infrastructure). It is this kind of capital that has conventionally been regarded as the most essential (partly because it can be subject to a simple quantitative evaluation), and the primary goal has been considered the growth of this kind of capital.
— Natural capital (soils, minerals, forests, waters, the atmosphere, etc.), which is a source of renewable and non-renewable natural resources possessing the commercial value or not. The depletion of natural capital has originated a critically acute situation and determined the necessity of threshold limitations of development.
— "Human" capital, whose importance has been very clearly realized in the course of the latest three decades, as well as the profitableness of "investment" in people, especially in developing countries. Investments in health, education and

nutrition are increasingly becoming the first priority of national financing strategies, although there still are numerous unsolved problems, particularly in view of high unemployment levels and increasing number of senior citizens in industrially developed countries. The latter circumstances determine the essential role of the fourth category of capital.

— Social capital, whose low level makes ecologically sustainable development absolutely impossible (Somali and Rwanda provide spectacular evidence of this situation). An unambiguous definition of social capital has not been worked out thus far; however, it is certainly characteristic of a powerful civic community governed by a system of free-will horizontal associations (as opposed to hierarchic vertical associations). A good example can be provided by the contrast between rapidly developing industrialized Northern Italy and the less efficiently developing south of the country, where autocratic tendencies prevail.

So, the notion of sustainable development should be defined more precisely; it should imply the necessity to preserve (or better further augment) the capital of all four types for future generations.

Thus, it can be concluded that the notion of *sustainable development* differs from the conventional notion of *development* in several aspects. Mikesell (1992) distinguishes the following aspects:

— Sustainable development is based on taking account of natural resources as the main restraining factor, whereas conventional development puts special emphasis on man-made capital.
— Sustainable development includes an additional element of intergenerational responsibility for the use of natural resources.
— Sustainable development assumes that the assimilation of waste is the main function of the environment and an essential limiting factor of economic growth.
— Sustainable development takes into account the priority of all social benefits and costs, including the depletion of natural resources.

Bruff and Wood (1995) offer four key aspects of the notion of *sustainable development* which briefly summarize the abovementioned notions and definitions:

- *futurity*—concern about the well-being of future generations;
- *environment*—recognition of the integrity and viability of the environment;
- *quality of life*—the multi-dimensional nature of well-being;
- *equity*—fair distribution of costs and benefits.

As has been mentioned above, Russian scientists also contributed to the theoretical formulation of the concept of sustainable development. As early as 1990, two monographs of conceptual value were published (Gorshkov, 1990; Kondratyev, 1990b); these were followed by a series of works treating global development priorities (Marchuk and Kondratyev, 1992), the problem of ecologically sustainable development indicators (Grigoryev and Kondratyev, 1996a), the principles and methods of the implementation of integration processes oriented towards the achievement of sustainable, ecologically secure development of the society (Report of the International Conference..., 1994; Donchenko, 1995; Zubov, 1993; Zubakov, 1995; Kondratyev, 1995a–c, 1996a; Kondratyev, Asimakopoulos and Varotsos, 1995; Kondratyev, Buznikov and Pokrovsky, 1992; Kondratyev, Danilov-Danilyan and Donchenko, 1993; Kondratyev and Fedchenko, 1992; Kondratyev, Grigoryev and Pokrovsky, 1975; Kondratyev and Pokrovsky, 1977, 1989; Kondratyev and Romanyuk, 1996a,b).

It should be noted that the publications by Donchenko (1995) and Kondratyev, Danilov-Danilyan and Donchenko (1993) discuss a problem which is not often treated in scientific publications, viz. that of international relations in the context of sustainable development. For example, obviously, in the absence of an international norm base and due to the inadequate mechanism of control (the result of the absence of a unified system of environmental monitoring), the transboundary impact of some countries upon others remains unchecked and not regulated economically. Under these circumstances, Russia, whose natural environment accounts for a considerable part of the global biospheric potential, compensates, often gratis, the lost biospheric potential of many neighbouring countries.

Culture and ethics form an important aspect of the concept of sustainable, ecologically safe socio-economic development. Voracious consumption habits have threatened nature for a long time. Nature was treated as our common and unlimited benefit. The question was how to distribute this benefit. This erroneous concept has been replaced by comprehension of the fact that the "ecologization" of mentality is necessary as an additional component in the system of the formation of social development goals and strategies of their achievement in a global context (V. I. Vernadsky . . . , 1986).

The demand side of sustainable development is a deep and painful process of moral transformation of the society which must reject the stereotype of consumerism. The paradigm of sustainable development requires a fundamental change in the nature and goals of human progress. It shifts human values, views and social rules "from economic efficiency to social equity, from luxury to necessity, from today to tomorrow, and from growth that benefits a few to genuine human development that benefits all" (Gladwin, Krause and Kennelly, 1993).

1.4.4 Global demographic dynamics and sustainable development

Data characterizing demographic dynamics are well known; however, we shall present here some most essential figures. The data will be treated more in detail in section 1.5. The present-day global population number is about 5.6 billion people; by the year 2000, another billion people will be added to the figure, and by the year 2050, i.e. during the lifetime of a single generation, the world population will constitute about 8.5 billion. The African population grows at the highest rate (2.9% per year); in East and West Africa the rate is even higher (3.1%), and in many African states the age of about one-half of their population is below 16. The growth rates of the population of western and southern Asia (the Indian subcontinent) are 2.8% and 2.2%, respectively. Relevant figures for Asia and Latin America, taken as a whole, are 1.8% and about 2%, respectively. By the year 2050, the Asian population (approximately 3.2 billion in 1992) will reach the figure of 4.9 billion; the

African population will be doubled (1.58 billion instead of 682 million). Relevant estimates for Latin America are 457 million and 701 million people.

Forecasts of demographic dynamics in individual countries offer even more impressive figures. For instance, in Bangladesh, which is overpopulated as it is, the population number will practically be doubled during the period from 1990 to 2050. In many countries of Africa and the Middle East (e.g., Angola, Congo, Iraq, Jordan, Kenya, Libya, Madagascar, Malawi, Niger, Nigeria, Rwanda, Saudi Arabia, Syria, Tanzania, Zaire) population growth rates markedly exceed 3%, and the total population number is doubled or trebled during the lifetime of a single generation. The severe reality is that some of these countries are already among the poorest in the world, and their population suffers from the deficit of vitally important resources, such as drinking water, etc.

Regrettably, consensus has not been reached thus far concerning this situation. The real truth is that this is not just a problem, but a threat. The opinion that the world is capable of feeding an unlimited number of its inhabitants is still shared by some scientists. But this is a myth, which has been revealed convincingly in the publications of Gorshkov (1995a,b) and Gorshkov, Kondratyev and Losev (1996).

Goodland et al. (1994) are of the opinion that the only realistic way out is stabilization of population numbers by limiting the average family size in poor countries to 2.5–3.0 children, and in industrially developed states to 2.1 (to reach a balance between birth and mortality rates). These authors put forward the following priorities:

(1) Solution of the problem of family planning.
(2) Broader spreading of information about the interdependence between family size and economic well-being.
(3) Intensification of educational programmes.

Sadik (1994) has emphasized the simple truth that the realities of the contemporary world are such that humankind is not only facing the population growth problem, but is suffering from its numerous after-effects on a scale of separate countries, regions or whole continents. Today it is quite obvious that the exploitation of many resources, provoked by rapid population growth, has already reached a level which destroys, for example, the regenerative ability of the atmosphere. Another spectacular illustration of the above conclusion is the land-use problem. One of its most painful aspects is the continuing desertification which has by now encompassed 30% of global land area and which reflects catastrophic losses of arable land.

The problem of per anima consumption becomes more acute every year. Analysis of data for the recent 40 years has revealed (Monitoring Environmental..., 1995) that the per anima consumption of copper, energy, meat, steel and timber has doubled; the number of automobiles and per anima consumption of cement increased by a factor of 4, that of plastic materials by a factor of 5, and of aluminium by a factor of 7. This has mostly happened in industrially developed countries. It is interesting to compare data on per anima energy consumption in the USA and other countries. From this viewpoint a single American is equivalent to: 3 Japanese, 13 Brazilians, 14 Chinese, 38 Hindus, 168 inhabitants of Bangladesh, 280 of Nepal and 531 Ethiopians. While 260 million North Americans spend $5 billion annually to buy special, low-calorie dietic food, approximately 400 million people (or more) suffer from starvation.

To maintain the lifestyle of the population of rich countries at the contemporary level, the per anima consumption of resources from 4 to 6 ha is required. If one proceeds from this norm, the total of 26.5 billion ha of land area would be required to meet the demands of the entire population of our planet (about 5.5 billion people). However, the total land area of the globe is 13 billion ha, out of which only 8.8 billion ha of land is suitable for agriculture or is occupied by forests.

Another essential aspect of the problem is food production, whose annual growth rate during the 1970s and 1980s reached 2.3% and exceeded the global mean population growth rate (1.8–1.9%). Relevant estimations have revealed, however, that during the next 15–20 years, a drop in the food production growth rate will occur, and it will be faster than the decrease in the population growth rate. The growing population number causes more and more intense loadings on forest ecosystems and biodiversity, which, in its turn, is a serious threat for global ecological safety (Gorshkov, 1993, 1995; Gorshkov, Kondratyev, Danilov-Danilyan and Losev, 1994; Gorshkov, Kondratyev and Losev, 1996). Biological resources of seas and oceans are also subject to enormous loadings. The violation of admissible limits of fishing reduced the global catch from 86 to 80 million tons in three years (1989–92). This is due to the fact that half of the world population inhabits the 100-km coastal zone, and in South-East Asia two-thirds of the population lives in the coastal zone. Nine out of the ten biggest cities in the world are situated in the coastal zone, as well as two-thirds of the 50 biggest cities. This accounts for 20% of the world population being the inhabitants of seaside cities.

All this is strong evidence revealing not only the absence of ESD, but also the catastrophic nature of the present global ecological situation which is aggravated by the inadequacy of the socio-economic development characteristics because they do not take into account the damage to natural resources and the environment; very often the most profitable branches of industry prove to be ecologically harmful, and it is developing countries that suffer from this situation most of all. Therefore, it would only be fair if the rich countries of the North, pursuing their own interests, could help the poor states of the South in overcoming inequality and poverty, and in protecting their environment. Very important also are such priorities as the use of renewable sources of energy, more efficient utilization of fossil fuel, minimization and recycling of waste, development of ecologically beneficial methods in agriculture, etc.

An important aspect of the problem of overpopulation is the question of whether new technologies are capable of ensuring successful development in an over-populated world. The question has been discussed repeatedly, and one can only agree with Daily and Ehrlich (1992) who state that it is hardly at all reasonable to rely on a technological miracle which will secure well-being for the Earth's population which will soon increase by billions of people.

In fact, hopes for new technologies are more than illusory, especially if one takes into account the following considerations:

(1) The negative impact of already existing technologies on the environment and man (e.g., the health of about 1 billion people in the world is subject to the destructive impact of various kinds of pollution);
(2) The expected doubling of global population in 42 years, and hence the practically insoluble problem of levelling the per anima annual income in industrially developed countries (which constitutes \$18 330) and in the rest of the world. If one assumes an average per anima annual income of \$800, then, taking into account the expected population growth, the efficiency of technologies should be improved by $2 \times 23 = 46$ times. This unprecedented rate of technological progress can hardly be ever expected in the future (and it should be emphasized here that the basic restricting factor is the permissible level of biosphere resources consumption).

This most acute problem may also be regarded from the viewpoint of per anima energy consumption (Monitoring Environmental..., 1995). In industrially developed countries, whose inhabitants constitute about 20% of the world population, in 1990 it was 7.5 kW on the average, with the total consumption of 9.0 TW (1 TW = 10^{12} W). In the countries of the South (80% of the global population), the per anima annual energy consumption was only 1 KW (the total consumption was 4.1 TW). If one proceeds from an assumption that in the next century the world population will reach a stabilization level of about 12 billion people, and that the average per anima aanual consumption will increase to 7.5 kW, it will imply a fantastic level of total energy consumption of 90 TW, and this figure exceeds the current one (already resulting in a situation of global ecologically unsustainable development) by a factor of 7.

Hemphill (1995) has discussed six principles whose implementation might be beneficial for the energy problem solution in the conditions of the supposed global climate warming:

— the solution of relevant problems on the basis of the market approach (taking account, where possible, of non-market factors as well);
— a broader application of ecologically efficient "pure" technologies;
— coordination of power production development and environmental protection measures (taking account of the ecological "cost" of power production development);
— creation of a system of adequate monitoring of the after-effects of power production development for the environment;
— support of the transfer of new "pure" technologies to other countries;
— ensuring the achievement of not only economic, but also ecological goals within the framework of ecological support projects.

One should emphasize, however, that all these measures can only be palliative.

1.4.5 Socio-economic factors of development in the context of ESD

Goodland et al. (1994) formulated the following priorities of socio-economic development for industrially developed countries (the North):

(1) Changing the predominant "consumeristic" lifestyle for a different way of life, oriented towards the implementation of ESD (Gorshkov, 1995a,b; Kondratyev, 1995a-c, 1996a; Kondratyev et al., 1975, 1977, 1989, 1992, 1993, 1995, 1996a,b).
(2) Compensation of expenses for environmental protection by increasing prices for energy and accelerating the transition to renewable energy sources (Goodland et al., 1994).
(3) Reimbursement of expenses for the neutralization and burial of toxic and non-toxic waste by their producers and prevention of the export of such waste to developing countries (such measures are very efficient for reducing the production of toxic waste in industrially developed countries).
(4) Rapid advancement of technologies and their transfer to developing countries, which is beneficial for economic development without damaging the environment; writing off debts, especially in cases when their payment (by selling raw materials) is seriously dangerous for the environment; increasing the number of grants allotted to the South countries specially intended for providing the generally adopted standards of environmental quality.
(5) Making broader and deeper the analysis of economic efficiency of environmental protection costs.
(6) Measures to support biophysical infrastructures on which economic development is based.
(7) Focusing on direct help to the poorest part of the population of developing countries instead of relying upon general economic development which will eliminate poverty one day in the future.

Goodland et al. (1994) made the following recommendations with regard to priorities of achieving ESD in countries with a low consumption level:

(1) Rapid transition to stable standards of living of the population; support of family planning.
(2) Rapid transition to using renewable energy sources.
(3) Support of progress in the improvement of education, qualification and employment, especially for young women.
(4) Elimination of poverty: increasing the role of mothers and children; social protection measures.
(5) Intensification of measures for protection of the environment and natural resources (forests, waters, agricultural land, etc.).

Goodland et al. (1994) emphasized that it would not be natural or ethical to expect from low-income countries any restrictions in their socio-economic development implemented on the basis of their economic growth. However, it would only be just to expect this from industrially developed countries characteristic of high standards of living; these countries might limit their economic growth (the GNP level) for the sake of the transition to ESD. This problem can only be solved on certain terms, viz. stabilization of the population number, serious measures to save energy and materials, minimization (and recycling) of waste, etc. However, as noted by Ramphal (see Polunin and Nazim, 1994), so far not a single country of the North has declared readiness (or developed relevant programmes) to reduce per anima consumption and the population number.

There is no doubt that the problems of socio-economic development, demographic dynamics and the environment are interrelated. However, concrete analysis of various aspects of this kind of interactivity is a complicated problem. In this context, a discussion has been symptomatic which was held within the framework of the forum "The Present State of the World" on October 1, 1995, in San Francisco, with the participation of M. Thatcher, G. Bush and M. Gorbachev, and broadcast by CNN. The views of the first two participants of the forum on the present state of the world can be summarized like this: the well-being of the world will take place if the system of free enterprise with market economy and free international trade becomes predominant everywhere. Gorbachev mentioned the problem of consumption, although he has not defined it clearly. Not a single participant of the forum mentioned the Second UN Conference on Environment and Development (UNCED-2) or subsequent forums (including the Fourth World Women's Congress which was held in Beijing in September 1995). All this reveals the necessity of clear formulation of socio-economic factors of ESD. Since the range of problems involved is very broad, we shall only make a few comments here in addition to some aspects discussed above.

One of the most important aspects of the entire problem is free trade (the absence of trade barriers) which does not exist in reality. Ramphal (see Polunin and Nazim, 1994) mentions, for example, that the economic relations between the European Community and the countries of Africa, the Caribbean and Pacific basins are regulated by a convention which defines the rules of selective access to the market and a whole system of preferences (trade barriers), especially with regard to such agricultural products as sugar, beef, vegetables, for which a complex system of quotas and market limitations has been established. The existence of broadly practised protectionism is a significant feature which predetermines the economy of both developed (like Japan) and developing countries.

In spite of all the talk about free trade and the persistent demand made of developing countries to consider the opening of their markets, the USA and the European Community countries pursue the policy of double standards with respect to developing countries; the policy is based on such labels as "governed economy", "negotiations oriented towards the results", "European preferences", etc. One can just wonder how, under these conditions, the Third World countries can develop trade relations beneficial for the required demographic transition. As a result of this situation, the foreign debt of African states surpassed 100% (in Latin American countries it is 37%). Every Zambian has the doubtful privilege of receiving an annual credit of $766, which is twice as large as the average annual per anima income. In Mozambique the foreign debt exceeds the DNP by a factor of four.

These and other numerous facts show that the recommendations of the UNCED-2 ("Agenda 21") can hardly be carried out in the present conditions of economic inequality without relevant financial support (Kondratyev, 1992b; Losev et al., 1993). As has been mentioned above, at present a quarter of the world's population uses about three-quarters of the global resources of various kinds. The share of industrially developed countries in the world consumption constitutes: 75% of globally produced energy, 90% of timber, 81% of paper, 80% of iron and steel, 70% of milk and meat, 60% of fertilizers. The rest of the planet's population must be satisfied with the remaining 25% of the resource "pie". Under current conditions, a British family with two children appears to be equivalent (from the point of view of global mean consumption) to a "globally averaged" family having 10–15 children. Although only 5% of the global population growth is the North's share, this minor value proves to be a much greater load for the planet (from the viewpoint of consumption) than the remaining 95% (the South's share). Hence the main conclusion mentioned above: to achieve ESD, the level of consumption should be reduced in the North, and the population number should be reduced in the South.

It seems beyond doubt that the beginning of the next century will be marked by gradual shifting of the economic "centre of gravity" from North America and Europe to east Asia, with a particularly important role for Japan. In the course of further development, the importance of Latin America, south Asia, and finally Africa will increase. Here the crucial point is the goals of this development. It would certainly be disastrous if the population of these countries were to strive to reproduce the consumerism of the Western lifestyle.

1.4.6. Conclusion

Substantial progress in the assessment of the present state and perspectives of sustainable socio-economic development can only result from proper understanding of the key role of demographic dynamics and the necessity of accounting for (and protecting) the natural biotic regulation of the environment (Gorshkov, 1993, 1995; Kondratyev, 1995a–c, 1996a; Kondratyev et al., 1975, 1977, 1989, 1992, 1993, 1995, 1996a,b), although the current assessments are rather contradictory. The Programme of Action of the International Conference on Population and Development (ICPD), developed for the next 20 years, approves a new strategy based on the interrelationship between the problems of population and development (Summary of the Programme..., 1995). The Programme is focused on the priority of satisfying the demands of individuals, and not on the achievement of some common demographic goals. This is a justified approach; however, one cannot feel completely satisfied because not a single document produced by the ICPD contains a clear estimation of the severe reality which reveals the instability of global ecodynamics. The Programme of Action (Report of the International..., 1994) puts forward only one new objective: achieving slower population growth rates than the average ones supposed in the UN forecasts. It is with regret that one has to state that, just as in the case of preparations for the UNCED-2 (Kondratyev, 1992b), important conceptual Russian publications have not been given proper consideration (Gorshkov, 1995a,b; Kondratyev, 1988, 1990e,f, 1992a–c, 1993a,b, 1994, 1995a–c, 1996a,b; Kondratyev et al., 1975, 1977, 1989, 1990, 1992, 1993, 1994, 1995, 1996a,b; Kotlyakov, 1995; Losev et al., 1993; Gorshkov, Kondratyev and Sherman, 1990; Gorshkov, Kondratyev and Losev, 1994).

In this connection, an extremely important declaration should be mentioned which was made by 1680 scientific leaders from 70 countries (including 104 Nobel Prize winners) in 1992. These scientists formed the Union of Concerned Scientists. In their Warning to Humanity (1992) the Union members state that humankind and nature are on the way to collision, that the Earth's capacity to meet the demands of its growing population is limited, and that we are rapidly approaching many of the limits. Pressure on the natural world, which is a result of unrestricted population growth, puts such requirements to the natural world which can overweigh any attempts to achieve sustainable development. Mankind has only one or a few decades to make amends; otherwise there will be no chance to eliminate the forthcoming threats, and perspectives for the human race will be drastically reduced. Argumentation used in the abovementioned Russian conceptual works confirm completely that this is a well-founded conclusion. Attempts of more optimistic assessment of the situation are wide open to criticism. Gorshkov (1995b) is absolutely right when he concludes his book as follows: "Modern civilization does not ensure either normal living conditions for man, or sustainable existence of life on the Earth".

1.5 GLOBAL CHANGE AND DEMOGRAPHIC DYNAMICS

Just a few (5–7) years ago the characteristic general tendency was to get away from the most acute problems relevant to the continuing growth of the global population. The Second UN Conference on Environment and Development (Rio de Janeiro, June 1992) has to a certain extent become a turning point in the very approach to the problem, revealing the necessity of an urgent analysis of global demographic dynamics and related after-effects (Kondratyev, 1992b; Kondratyev, Danilov-Danilyan and Donchenko, 1993; Losev et al., 1993). Therefore, the International Conference on Population and Development (IPCD) which was held in Cairo in September of 1994, with the participation of 183 countries, was devoted entirely to this range of problems (Report of the International Conference..., 1994; Summary of the Programme..., 1995). The discussion of demographic aspects was an important part of the World Summit on Social Development (Copenhagen, April–May 1995), the Fourth World Women's Conference (Beijing, September 1995) and the Second UN Conference on Human Settlements (Habitat II). An important step forward in 1995 has become an integration of efforts within the International Geosphere–Biosphere Programme (IGBP) and the International Human Dimensions Programme (IHDP) concerning socio-economic aspects of global change problems.

It should be noted here that a complex approach to the solution of the problem of global change, including a substantiation of its demographic aspect as a key one, has been long enough developed by Russian scientists in the context of biospheric regulation of the global environment (Gorshkov, 1995a,b; Gorshkov, Kondratyev and Losev, 1994; Kondratyev, 1990b, 1992b, 1995a–d; Kondratyev, Danilov-Danilyan and Donchenko, 1993; Kondratyev, Ortner and Preining, 1992;

Kondratyev and Pozdnyakov, 1996). The purpose of this section is to summarize the results of the Cairo Conference. Back in 1993, during the "Demographic Summit" in New Delhi, it has been stated that humankind is approaching a crucial point where the interrelated problems of population number, environment and development are concerned (and it would only be more precise to declare that this point has already been reached (Gorshkov, 1995a,b; Gorshkov, Kondratyev and Losev, 1994).

Polunin and Nazim (1994) have been absolutely right to emphasize that at present the fact has been generally recognized that the structure of the consumption of natural resources, as well as the existing models of economic growth and development are the basic reasons for various manifestations of danger threatening the sustainability of development on the planet. Rapid population growth as one of such manifestations seems to be a *symptom*, and not the *cause* (as was often claimed to be in the past) of unfavourable development in the conditions of an unfair politico-economic system. Although the latter statement cannot be considered indisputable (of course, the continuing rapid population growth in developing countries is a consequence of unfavourable socio-economic development, but at the same time, the interactivity of the nature of causes and effects is beyond doubt), it is convincing evidence of general concern about contemporary demographic trends.

According to the UN forecasts, by the year 2050 the world population number may be between 7.9 billion (low growth rates) and 11.9 billion people (high growth rates). Today the figure is rapidly approaching 6 billion; 95% of the population growth is the share of African, Asian and Latin American states (Report of the International..., 1994). Such conditions provoke an intensification of social inequality in the world. As noted by Ramphal (see Polunin and Nazim, 1994), each inhabitant of the North consumes 20–30 times more resources and produces 2-30 times more waste during his lifetime than his counterpart in the South. From the viewpoint of resource consumption, 1.2 billion people living in industrially developed countries are equivalent to 24 billion of the inhabitants of developing states. These and other circumstances require different ways of socio-economic development and an action plan whose implementation would make it possible to save both humanity and the environment. The initiative should come from developed countries, but developing states should also contribute to it substantially.

The Secretary-General of the ICPD, Sadik (1994) attracted general attention to the principal importance of a clear definition of the notion of global sustainability (sustainable development or ecologically sustainable development), having emphasized the fact that the contemporary world has hardly ever been sustainable and certainly is not such now (relevant problems have been earlier discussed by Gorshkov, 1995a,b; Gorshkov, Kondratyev and Losev, 1994). Goodland et al. (1994) have been quite right to say that the nearer mankind approaches the twenty-first century, the faster it moves away from the state of ecologically sustainable development.

Since the demographic situation all over the world has reached a critical level, urgent measures should be taken now. Family planning must be the basis for successful solution of this painful problem. Tisdell (1994) analysed those factors (mostly economic) which must be conducive to family size reduction (see Table 1.1).

Table 1.1. Factors conducive to family size reduction

Factors weakening high birth rate and reducing family size	Factors determining growth of expenses to support large families
Social approval of small families	Improvement of social standards relevant to child upbringing and growth of relevant expenses (for education, health care, etc.)
Lower employment opportunities for children under conditions of a growing family	Compulsory (mostly paid) education for children
Growth of labour efficiency of surviving children	Growth of children maintenance costs (for example, in broken families when parents have jobs outside their homes)
Improvement of medical care	Increase of prices for children's goods (clothes, food, medical care, etc.)

Other factors conducive to family size reduction are:

— Reduced prices for contraceptives, their higher reliability and availability.
— Education of population oriented towards problems of family planning and use of contraceptives.
— Increase of general education level.
— Increase of socio-economic status of women.
— Growth of urban population share (urbanization).

Of course, the real situation is much more complicated than the scheme presented above. The economic and political instability in Russia, whose consequences have become falls in birth rate and population number, can serve as an example. It is especially important to analyse the effect of population growth on the resource base of socio-economic development (situations of this kind are particularly frequent in Africa). Since demographic factors strongly affect the environment, the most important aspect of the problem is human health, especially in cities (Sadik, 1994), and in this regard, the problem of nutrition (Menza and Lupien, 1994).

The sad fact is that the very first victims of ill-nutrition and hunger are children and women. The number of people in developing countries who suffer from hunger is growing. There are data revealing that the number of starving people reached 1.2–1.3 billion in 1991–92. The number of people in Africa suffering from inadequate nutrition increased from 101 million in 1961–71 to 168 million in 1988–90. However, one should note the great success of China and India, the two most populated countries in the world, which have fully overcome the problem of starvation.

An alarming symptom is the growing dependence on the import of food products for the increasing urban population of many countries (it is assumed that by the year 2015, the urban population will constitute 56% of the whole, in comparison with 45% in 1994 (Report of the International..., 1994)). This often happens because the imported goods are subsidized, which suppresses the possibility of producing and selling local food products (such is, for instance, the current situation in Russia).

The problem of migration, resulting from various causes, has recently been acquiring a threatening scale (Myers, 1994).

There have been attempts of a more optimistic assessment of the current socio-economic situation in the world (see, for example, Holdgate, 1994), but the argument used does not hold water.

1.6 INDICATORS OF ECOLOGICALLY SUSTAINABLE DEVELOPMENT

1.6.1 Introduction

The main perspectives of achieving a more profound understanding and forecasting of global change (including both their natural-science aspects and "human dimensions") are doubtless dependent on further working out and application of various methods of numerical simulation modelling of processes governing the interaction between socio-economic development and ecodynamics. It is clear enough that the substantiation of adequate models is impossible in the absence of sufficiently reliable representative observation data (Grigoryev, 1991; Grigoryev and Kondratyev, 1994; Kondratyev, 1990b, 1992a; Kondratyev, Buznikov and Pokrovsky, 1992, 1996a; Kondratyev, Johannessen and Melentyev, 1996). In recent years, the notion of *sustainable development* has become very popular not only in the scientific literature, but also with mass media. It has not been clearly defined, however, thus far, and is not correct when translated into Russian.

Not going into the details of the problem (for which see, e.g., Kondratyev, Danilov-Danilyan and Donchenko, 1993; Kondratyev and Romanyuk, 1996a,b; Losev et al., 1993; Hunt, 1994a,b) and limiting ourselves to intuitive conceptions, we shall note here that from the standpoint of the concrete content of the notion of *sustainable development*, progress with the development of global change models should be of decisive value. The problem is, however, that models available today are far from being capable of explaining (to say nothing of predicting) global change. Even the theory of climate, which has been developed over many years and is fairly advanced, cannot offer a satisfactory explanation, for instance, of the extent to which the global climate warming of the recent 100 years is due to natural or anthropogenic factors (Kondratyev, 1995c; Kondratyev, Buznikov and Pokrovsky, 1992; Kondratyev, Johannessen and Melentyev, 1996; Ellsaesser, 1992).

The development of a concept of ecologically sustainable development (ESD) of the society in its interaction with nature presents the most important target of contemporary science. The understanding of means for achieving ESD can be reached only on the basis of further development of various, sometimes very remote from one another, branches of science and technology, such as agriculture and nuclear power production. And this is quite natural because ESD is relevant at once to mankind as a whole, and to each human being in particular. It is a well-known fact that the world community is not at present developing in an optimal way: the equilibrium of nature is more and more often violated, and this influences the vital activity of man to an increasing extent.

For a long time, the basic attention of experts has been paid to the study of the consequences of anthropogenically induced disturbances, their overcoming and the search for more advanced technologies. Such, doubtless useful, studies have yielded positive results: partial removal of the natural equilibrium disturbances and improved living conditions for the population. However, these results have been partial and temporary, as might have been expected.

The development of a concept of ESD (Kondratyev, Danilov-Danilyan and Donchenko, 1993; Making Development..., 1994; Pykh and Maklina-Pykh, 1994; Serageldin and Cohen, 1995) has been aimed at reaching optimal relations between man and his environment on a global scale. Such purpose-oriented investigations were started in the second half of the 1980s and were initiated, first of all, by the Commission on Environment and Development (the Brundtland Commission). In 1987 this Commission formulated the notion of ecologically sustainable development as a situation which meets the demands of the present generation without threatening the needs of future generations; subsequently this definition has been reformulated and made more precise (Bakkes et al., 1994; Dovers and Norton, 1994; Making Development..., 1994).

The concept of ESD was the focus of a number of international forums, including the Second UN Conference on Environment and Development (Kondratyev, Danilov-Danilyan and Donchenko, 1993; Losev et al., 1993; Hunt, 1994a). In subsequent years, this problem has been adequately treated in a number of UNEP and World Bank publications (Kondratyev and Romanyuk, 1996a; Bakkes et al., 1994; Making Development..., 1994; Monitoring Environmental..., 1995; Novek, 1995; O'Connor, 1994a,b).

In this connection, the development of environmental dynamics indicators acquired a special importance, taking account of the fact that the specific features of socio-economic development have not been stipulated, and that the substantiation and application of such indicators makes it possible to assess the ESD perspectives. There are five most significant aspects of the problem (Bakkes et al., 1994):

(1) Population number and human health.
(2) Production, consumption and technologies.
(3) Renewable and non-renewable natural resources.
(4) Environmental dynamics.
(5) The unity and interdependence of ecosystems.

The problem is to substantiate representative indicators of the state and dynamics of the above mentioned subsystems, taking account of their interrelationship on different scales (from local to global). Of course, the most consistent and reliable solution of the problem should be based on the analysis of relevant numerical modelling results; however, as has already been noted, this kind of approach is not yet feasible.

Naturally, the very notion of an *indicator* must be clearly defined. According to Bakkes et al. (1994), an indicator, as an information element, (a) is a characteristic used in the interests of the process of management and can be applied for planning such a process; (b) plays a role which is beyond the limits of its direct significance.

Examples of indicators of the state and dynamics of the global environment and natural resources can be the characteristics of the global cycles of various substances (water, carbon, nitrogen, sulphur, etc.). The totality of socio-economic development indicators may include a broad range of indices of usual or aggregated indicators, taking account of the expected life duration, child mortality, education level and per anima GNP. In the context of matter cycles, indicators can be developed which characterize the efficiency of economy from the viewpoint of power and material consumption (including waste) or the efficiency of using environmental resources as a factor of production (agriculture, forestry, etc.). Areas of studies requiring a most urgent development of such indicators are: atmospheric and natural water pollution; natural resources and biodiversity; climatic change; ozone content depletion; human health; demography; production, consumption and technologies; waste; the cost of environmental protection measures.

What is needed is a complex analysis of environmental dynamics and factors of socio-economic development affecting it. Under the present circumstances, it is, however, reasonable to focus on the problem of the quality of various natural media: soils, water, air, etc. The solution of relevant problems (particularly from the standpoint of interaction between media) requires a considerable improvement of national and international systems of environmental monitoring. As in other cases mentioned above, further development of numerical modelling of ecodynamics is necessary for the purpose. It is for this reason that interactions between biophysical, chemical and socio-economic aspects of ecodynamics have not been adequately studied thus far (Kondratyev, 1995a,b).

One of the difficult aspects of the development of SD indicators (SDI) is due to the fact that this kind of work is mostly fulfilled by the suppliers and not users of ecological information. No less important is the circumstance that experts in the field, as a rule, represent industrially developed countries, and their results may sometimes be inadequate in the context of the global ecological situation.

1.6.2 Conceptual aspects of the problem of indicators

Environmental state indicators (ESI) are first of all needed to substantiate and stimulate environmental protection measures. It soon becomes obvious that the indicators are specific for every problem, and that there is no universal set of indicators. Therefore, the choice of adequate indicators depends on the specific nature of problems that should be solved. However, there are two common requirements concerning the choice of indicators: (1) an indicator must have a broader value than its direct significance as a quantitative characteristic (thus, for instance, the level of sulphur dioxide emissions has been broadly used as a general indicator of atmospheric pollution); (2) an indicator must describe deviations from the norm. There are several purposes requiring the introduction of environmental indicatiors, viz. (1) assessment of environmental conditions and relevant trends over national, regional and global scales; (2) comparison of conditions in various countries and regions; (3) forecasting of possible trends; (4) providing early warning about potential danger; (5) assessment of environmental conditions from the viewpoint of achieving certain objectives. Criteria determining the choice of ESI can be classified into three categories:

(2.1) Indicators which are of practical value for users and may be taken into account in policy making should:

- provide a representative description of the environment, the pressure on the environment or the response of society to changes in environmental state;
- be simple enough, easy to interpret and capable of characterizing temporal trends;
- be sensitive to environmental changes and anthropogenic impacts;
- be appropriate for international comparisons;
- be of either national nature, or suitable for the solution of regional environmental problems which are important for a given state;
- be suitable for comparison with threshold values and for supplying the user with information about environmental conditions.

(2.2) From the standpoint of the analysis of environmental conditions, indicators should:

- be theoretically well-substantiated;
- rely upon international standards and be internationally coordinated;
- be suitable for the use in economic models and information systems.

(2.3) With regard to input information for calculating ESI, the following is required:

- the presence of economically justified (the cost–benefit relationship) input data;
- the presence of adequate documentation and information about the input data quality;
- providing the possibility of regular input data renewal.

With regard to the possibility of classifying ESI themselves, it can be realistic if three interdependent aspects are taken into account (Bakkes et al., 1994):

(a) directions of ESI applications (to early warning systems, in policy making, etc.);
(b) objects described by ESI: the particular kind of natural environment (atmosphere, natural waters, etc.) or process (acidification, salinization, desertification, etc.).
(c) the position of ESI in the cause-and-effect characteristic (ecological stresses, environmental dynamics, the influence of environmental protection measures, etc.).

The socio-economic development of the world community is governed by a group of factors, among which those of the first priority are:

(1) ecological factors (including natural processes and phenomena and anthropogenically-induced violations of the environment;

(2) economic factors;
(3) social factors.

In this connection, of interest are the investigations of the staff of the World Bank (WB) in the field of so-called primary indicators for the ESD matrix (O'Connor, 1994a,b).

The list of ESD indicators proposed by the WB contains three groups of phenomena: (1) economic; (2) social; (3) ecological. The first group involves poverty, productivity, expenses, labour. The second group includes urbanization (the housing problem, water and air quality), health, nutrition, transport, the role of a woman in the society. The largest is the third group. Its indicators are subdivided into five subgroups: (a) global scale (climate change, stratospheric ozone, sea resources); (b) national interests (biodiversity, water); (c) market objects (gas, oil and coal, metals and minerals, forest resources, plough land–soil quality); (d) the carrying capacity (eutrophication, acidification, substances, waste) and e) indicators of general nature.

In their turn, the phenomena-indicators of all the abovementioned groups have been subdivided into three subgroups: (a) indicators of anthropogenic impact; (b) indicators of state; (c) indicators of the response of society. Indicators of anthropogenic impact reflect its various kinds, such as the impact of industry, agriculture, transport and power production upon the environment. Environmental components whose state should be monitored are air, water, soils, natural resources. Finally, the third type of indicators (societal response) must include indicators of the response of administrative bodies and private persons, as well as different initiatives and international efforts.

The system of indicators proposed makes it doubtless possible not only to evaluate the state of the environment, but also to plan necessary concrete measures for its optimization, and to substantiate an optimal socio-economic policy. The great number of ESD indicators (about a hundred) makes it possible to work out a sufficiently detailed assessment of the present state of the environment and socio-economic development. It is obvious, however, that for operational use of the indicators, the difficult task of substantiating the priorities should be fulfilled.

In ESD studies, the notion of sector indicators has become widely spread; these characterize some single facet of a natural (economic or social) phenomenon. For example, forest can be a sector phenomenon-indicator; its concrete indicators can be the forest area, the wood volume, the distribution of forest massives and their value (O'Connor, 1994b).

Another example of a sector indicator is water. In this case, individual indicators may be the intensity of the utilization of water resources, their availability to the population, as well as measures for their protection, regulation and consumption.

It is clear that, although informative and inevitable, the sector approach is somewhat one-sided. Therefore, the idea of the main role of integrative approaches to ESD studies seems to be quite reasonable (O'Connor, 1994a).

1.6.3 Some aspects of ecodynamics

Let us now analyse examples of possible substantiation of choosing ESD indicators to characterize various aspects of environmental dynamics, including relevant socio-economic aspects. Here it should be emphasized that the results discussed below are only preliminary.

1.6.3.1 Atmospheric pollution

It can be said that the most substantial experience of using ESI (although not comprehensive enough) has been accumulated for this particular case. Such indicators as the maximum admissible concentration (MAC) or maximum admissible emissions (MAE) have long been used as threshold values; the level of atmospheric pollution is determined by comparison with these threshold values. In the context of the Framework Climate Convention, the inventory making of CO_2 and other "greenhouse" gas emissions has been broadly implemented in many countries (Kondratyev, 1992a; Kondratyev, Danilov-Danilyan and Donchenko, 1993; Ellsaesser, 1992).

In this case, just emission levels serve as ESI, but another widely spread indicator is the global warming potential (GWP), whose adequacy, however, is a subject of debate (Kondratyev, 1992a; Kondratyev, Danilov-Danilyan and Donchenko, 1993). Very urgent is the problem of substantiating a set of indicators of climate change and factors responsible for it. Thus far, the most frequently used indicators have been air and ocean surface temperatures. The continuing accumulation of satellite observation data makes it possible to begin analysing the trends of such indicators as snow and ice cover extension, the Earth's radiation budget components, etc. The functioning of the Global Climate Observing System (GCOS), the Global Terrestrial Observing System (GTOS) and the Global Ocean Observing System (GOOS) opens up perspectives for using a much broader range of climate change indicators. The signing of the Convention on far transboundary pollution transport (whose European analogue is the EMEP Programme) stimulated a similar effort of inventory of sulphur and nitrogen oxide emissions. Within the framework of EMEP, summarized data on the emissions and deposition of such gases as SO_2, NO_x, NH_3 and VOC (volatile organic compounds) are regularly published. Data on atmospheric contamination with heavy metals and organic compounds are subject to archiving and analysis within the framework of the Paris Convention on preventing sea pollution by ground-based sources (PARCOM) and the Helsinki Convention on protecting the Baltic Sea from pollution (HELCOM). It is regrettable that in many developing countries, where soil acidification has become a serious problem, there is no adequate emission monitoring as yet. The existing networks of stations for observing atmospheric background pollution are mostly concentrated in industrially developed countries (this refers, for example, to the BAPMoN network supervised by the World Meteorological Organization (WMO)). Observation networks for pollutants (including carbon monoxide, tropospheric ozone and aerosol) in cities are comparatively well developed. Attempts have been made at substantiating an integral (complex) indicator of atmospheric pollution but they have not been successful.

Analysis of the impacts of atmospheric pollution on human health, agricultural crops, forests and other ecosystems is a difficult task because the results of such impacts, as a rule, depend on numerous factors. However, MAC and MAE can still be established, as well as threshold equations of acid depositions from the viewpoint of their influence on the ecosystems of land and water basins.

Among problems that are still waiting to be solved there are:

— expansion of observation networks in developing countries;
— data quality improvement and comparability;
— extension of the range of pollution kinds which are subject to monitoring (this refers, in particular, to hydrocarbons, especially to stable organic pollution);
— improvement of means of data processing and analysis;
— more reliable substantiation of environmental quality standards.

Considering the problem of atmospheric pollution, it should be emphasized that its most important aspects are anthropogenic impacts on the stratospheric ozone layer, related trends of total ozone content (TOC) drop and solar ultraviolet radiation enhancement at the Earth's surface level which originate various dangers for man and ecosystems (Kondratyev, Asimakopoulos and Varotsos, 1995). Natural ESIs in this case are TOC and UV radiance. If the global TOC field can be retrieved from the data of satellite and ground-based observations, measurements of solar UV radiation are very fragmentary and irregular. Within the framework of the programme of TOC monitoring, very important are observations of components responsible for ozone depletion (above all, chlorine compounds), and the use of such an indicator as the ozone depletion potential (ODP). Biological indicators requiring substantiation can doubtless also be very significant. The Montreal Protocol and subsequent Supplements agreed upon in London and Copenhagen determine the necessity for regular monitoring of numerous chlorofluorocarbons (CFCs) and their substitutes (both categories are greenhouse gases). In this connection, serious efforts are carried out within the scope of the global emission inventory programme (GEIA) supported by the International Geosphere-Biosphere Programme (IGBP).

1.6.3.2 Pollution of seas and freshwater basins

The table that follows characterizes the possibilities of using indicators of the state of natural waters and resource parameters of great socio-economic value. It can be seen that some indicators can be determined in a simple way; others require the use of rather complicated equipment (the cases of the chemical composition of natural waters, phytoplankton concentration, etc.). A method of aggregation of water medium state indicators has been developed in the Netherlands and designated as AMOEBA. The method is based on the use of a set of fundamental characteristics (biological, physical, chemical) describing the undisturbed state of ecosystems, which forms a certain standard for comparisons. The choice of characteristics is determined by the following factors: the availability of relevant initial data and easy access to such data; the sensitivity of the chosen characteristic to anthropogenic impacts; the information content of the characteristics from the viewpoint of environmental state

Table 1.2. Indicators of various types of natural water

Indicator	Freshwater lakes	Saline lakes	Rivers	Seas and oceans
Variations of hydrological conditions:				
Water plane area	*	*	*	
Water temperature	*	*	*	*
Dry season duration	*	*	*	
Minimum/maximum runoff			*	
Suspended matter concentration			*	
Salinity	*	*	*	*
Eutrophication:				
Total nitrogen and phosphorus	*	*	*	*
Chlorophyll-a	*	*	*	*
Contamination with waste:				
O_2, BOD, COD	*	*	*	*
Organic and inorganic pollution:				
Heavy metals	*	*	*	*
Micropollution	*	*	*	*
Ecological state:				
Bioproductivity	*	*	*	*
Species composition/biodiversity	*	*	*	*
Fish diseases	*	*	*	*

description; the representativity of the entire totality of the characteristics, their social and political attractiveness.

Indicators which characterize socio-economic factors (directly non-related to water type) are as follows: fish catches, drinking water resources, bacterial contamination (drinking water, bathing, water organisms used as food), salinized and non-suitable for agriculture water basins, wetlands suitable for agriculture and hunting.

1.6.4 Socio-economic factors of ESD

1.6.4.1 Conceptual aspects

To provide ESD indicators, it is important to formulate the notions of the above mentioned four kinds of natural capital which characterizes the resource potential of the world community (and individual states). Serageldin et al. (1994) single out: (1) economic, (2) social, (3) "human" and (4) natural capital. It should be noted here that up to very recently, when different states were compared, basic attention was paid to economic indices (economic capital), whereas the information about social development and the "human" capital was obviously underestimated. The latter notion encompasses various aspects of social development, such as cultural, biological, psychological and physical (Monitoring Environmental..., 1995).

An essential component of the capital (resource potential) of any state, or of the world community as a whole is natural capital. Natural capital (natural resources expressed in their market values) determines to a great extent socio-economic development. By the way, the notion of natural capital is comparatively new and requires more precise definition (for example, some kinds of natural capital, such as flora or fauna, are extremely difficult to evaluate) (see Kondratyev, 1995a,b; Gore, 1993; von Weizsäcker, 1994).

The progress of humankind, which is inevitably accompanied by the destruction of the natural environment, requires (from the viewpoint of achieving ESD) defining the limits (boundaries) of possible exploitation of the natural capital and working out of indicators of the kind of land use that would guarantee sustainable socio-economic development. At present, in many countries over-exploitation of natural resources is obvious (in the absence of due inventory of their reserves). This can certainly result in rapid exhaustion of vitally important natural resources, in an economic and social collapse.

An essential aspect of substantiating a matrix of ESD indicators is detection of interrelated factors of socio-economic development and ecodynamics. For example, the information content of an indicator such as population number (or its growth rate) increases considerably if another indicator—child mortality—is taken into account simultaneously.

ESD indicators describe phenomena which are characteristic of different time scales. It is important in this connection that the characteristic time of economic and political decisions does not usually exceed several years (which is to a great extent determined by election dates). Changes in natural systems, on the contrary, take place, as a rule, during larger time periods. For example, forests reach their ecological maturity in about a hundred years or more. Significant changes in fish populations occur with intervals from one to ten years. The phenomenon named El Niño which causes considerable changes in the global climatic system happens every several years. These circumstances show that ESD indicators can be subdivided into two categories: short-term and long-term ones. The first group comprises phenomena which take place during a short time and have quick after-effects (positive or negative). The second group encompasses indicators of long-term environmental and socio-economic changes. For example, the second group indicators includes the level of carbon dioxide emissions to the atmosphere, the total ozone content in the atmosphere, the world ocean level, etc.

Estimating various ESD indicator listings, for instance, those suggested by the Brundtland Commission and the World Bank (Kondratyev, Danilov-Danilyan and Donchenko, 1993; Kondratyev and Romanyuk, 1996a,b; O'Connor, 1994a,b), it should be mentioned that they are not at all comprehensive; in particular, indicators of a number of catastrophic natural phenomena are lacking there, such as earthquakes, volcanic eruptions, river inundations and coastal area floods, periodic droughts (Grigoryev, 1991). Any one of these phenomena is of greater practical importance than, for instance, the potential increase of the world ocean level which is necessarily mentioned in all listings of ESD indicators. Catastrophic after-effects of natural disasters do great damage to the socio-economic development of numerous states. The centres of destabilization of ESD in individual states impact the entire world economy. It is enough to say that large-scale droughts cause crop failure, especially poor wheat harvests. Forecasting of natural disasters is regrettably inadequate (particularly, this refers to earthquakes and volcanic eruptions).

It is also reasonable to classify ESD indicators as global, national and subnational or regional. Among global ESD indicators there are such that reflect phenomena which are extremely important for the world community (their consequences or symptoms) and which may lead to anomalies in the global eco-situation, the world economic and social development. Here are some examples.

A key ESD indicator is population number. It is a well-known fact that the planet's population growth is ahead of food production. And this ends in mass undernourishment and starvation, sickness and untimely death for millions of people. Under current global socio-economic conditions, biospheric resources are insufficient to meet the needs of the ever-growing population (Gorshkov, 1990; Gorshkov and Kondratyev, 1990; Losev et al., 1993). The world population may reach the figure of 10–12 billion by the middle of the twenty-first century (Gladky, Grigoryev and Yagya, 1990; Kondratyev, Grigoryev and Pokrovsky, 1975). The information content of the population number (and its dynamics) as an ESD indicator increases if it is complemented by certain demographic indices, such as child mortality, the number of population which is beneath the limit of poverty, average lifetime, etc. According to some estimates, the global ESD is possible if the planet's population number does not exceed 0.5–1.0 billion people (Kondratyev, Danilov-Danilyan and Donchenko, 1993; Gorshkov, Kondratyev and Losev, 1994).

Another key indicator of global ecological security is the degree of closedness of global biogeochemical cycles (Gorshkov, 1990; Gorshkov and Kondratyev, 1990; Losev et al., 1993; Gorshkov, Kondratyev and Losev, 1994), which is determined by the level of utilization of biospheric resources. The relative threshold value of this level in conditions of a sustainable biosphere constitutes about 1% of the entire biotic production. Today this threshold level has been substantially exceeded, which reveals an early stage of the process of global biospheric destruction (Gorshkov, 1990; Gorshkov and Kondratyev, 1990; Gorshkov, Kondratyev and Losev, 1994).

It seems worthwhile to consider two more global indicators which are specific but closely connected with the abovementioned key indicator. These are the characteristics of biodiversity and forest cover.

1.6.4.2 Biodiversity

A reduction of the number of many organisms of the flora and fauna of the Earth, an impoverishment of the species composition and a decrease in the number of species take place every year. The main cause is the continuously growing anthropogenic stress on nature.

It is not quite clear yet what negative consequences may follow from the reduction of biodiversity on our planet or what permissible limits of this reduction may be. It is obvious, however, that in the long run the closedness of global biogeochemical cycles will be broken and therefore a global economic catastrophe will become inevitable.

As an indicator of biodiversity on a global scale, the area ratio (in per cent) can be used of large complexes that are protected, unprotected or subject to slight anthropogenic pressure, such as biomes (landscape zones) (Monitoring Environmental..., 1995). It can be mentioned, for example, that in the steppes of the mid-latitude belt only a small percentage of land is protected, and about 25% of land is in a disturbed state. At the same time, in such a biome as sclerophyll evergreen forests, only 2.8% of the territories are slightly disturbed, whereas the proportion of protected land is greater (4.7%). Comparative data of this kind show that the biome of evergreen forests is subject to a much stronger anthropogenic pressure, and that more efforts should be made for its protection.

Another potential biodiversity indicator on a global scale can also be landscapes forming natural (landscape) zones of the Earth. This is possible because every landscape with its specific ecological conditions is also characteristic of special combinations of biocenoses. The destruction of landscapes is the destruction of habitats of living organisms and it thus leads to the reduction of biodiversity.

Of course, landscapes (both natural and those disturbed by anthropogenic impacts or by natural causes, such as fires, desertification, etc.) used as indicators can only yield approximate information about the state of biodiversity. This kind of information may be a signal of initial or critical changes in the habitats of living organisms on a global scale. Certain difficulties in using landscapes as global biodiversity indicators are due to the absence of a world standard in the definition of the notion of a landscape.

1.6.4.3 Forests

One of the important global ESD indicators can be forest cover characteristics. The significance of this indicator is determined, in particular, by the role of forest in hydrological processes (including its influence on groundwater) and in gas exchange with the atmosphere, as well as as a habitat of various living organisms. The importance of forest for man's vital activity is very well known.

The simplest indicator of the state of forest ecosystems can be the area occupied by forest (apart from lakes, swamps, rivers, clearings). A more useful indicator is the changes in the forest cover (as a percentage) fixing the reduction or increase of the area of forest massives. The global tendency towards deforestation is characterized by the data on the reduction of forest areas over 20 years (from 1966-68 to 1986–88). The reduction constituted 6%, or 130 691 $\times$ 10^3 km^2. Regions where the forest cover situation is most stable (both indicators are close to global average) are the territories of the USA, India and Argentina. Global assessments of forest cover have revealed regions of the greatest forest destruction, especially in the tropics. If on the planet as a whole the forest area reduced by 2% relative to other land types, in some countries the reduction was disastrous (by 23% in Costa Rica, 15% in Honduras, and 45% in Brunei) (Monitoring Environmental..., 1995).

Apart from the abovementioned indicators, other first priority global ESD indicators are the characteristics of land desertification and provision of food for population (or the number of starving people).

When choosing first priority ESD indicators for individual regions and countries, one should take into account the specific nature of relevant territories. On a regional (subnational) level, the complexes of ESD indicators oriented towards practical purposes can be most specific. This is determined by their dependence on the natural and economic features peculiar to the region, its social conditions and the human capital (human resources). Hunt (1994a,b) offered an example of developing ESD indicators for the northern part of Queensland, Australia. The development of this region is closely connected with tourism; hence the substantial anthropogenic impact on nature in the area. As indicators determining the basic features of this impact the following ones have been chosen: population number, per anima resource consumption, and the intensity of impact on the environment (per anima) determined by currently existing technologies.

Similarly, the sets of ESD indicators for such areas in Russia as the Evenk or Kuzbass regions should be very specific. The development of ESD indicators for such regions should be based on accounting for particular natural and economic features, such as the state of pastures on meadows in river valleys and the state of forests in watersheds, and deer-breeding and hunting in the first case, and, in the second case, the state of partially-wooded steppe, foothills and mountainous landscapes, the situation in the mining and chemical industries, and pollution emissions (above all, to the atmosphere).

ESD indicators on a national scale vary from country to country, since states differ very much with regard to their natural, economic, social and human capital, and their peculiar features of anthropogenic impact. At the same time, a whole number of indicators can be common, viz. the extent of emissions of greenhouse and ozone-destroying gases, the eutrophication of waterbodies, the acidification of natural waters, toxic waste, biodiversity, urban environmental quality parameters, soil degradation, the characteristics of the state and utilization of water, forest and fish resources, as well as a number of indicators reflecting population growth, the development of industry, agriculture, transport and power production (O'Connor, 1994a,b).

1.6.5 Conclusion

The substantiation (prioritization, optimization) of ecologically sustainable development indicators is at the initial stage of its development. It is doubtless, however, that search for ESD indicators should become the key direction of investigations relevant to the study of global change. The primary target here must be substantiation of priority indicators on the global, national and regional levels. It is important to choose a limited number of priority global indicators. Table 1.3 contains a tentative list of SD indicators in accordance with *Agenda 21.*

Although national and subnational ESD indicators may differ from country to country, it is necessary to try and choose a set of relatively universal indicators which make it possible to compare data for different countries.

ESD indicators should not only reflect the most essential features of environmental dynamics and socio-economic development; they should also be informative, comparable and accessible for determination. The latter circumstance is important from the viewpoint of the organization of systems of monitoring ESD indicators and their input parameters.

The currently existent system of ESD indicators should be appropriate to be capable of answering today the question: what countries or regions on this planet are close to the optimal state from the ESD standpoint, or on the contrary, reveal dangerous tendencies. ESD models are urgently needed for individual countries and regions. They could be very helpful in revealing weak spots of individual states and on the planet as a whole.

Table 1.3. Sustainable development indicators

Category (1)	Chapter of *Agenda 21* (2)	Impact factors (3)	Indicators of state (4)	Response (5)
Social problems	Chapter 3: Extreme poverty	— Unemployment growth rate, % — man's ratio to woman's	— Number of people living in absolute poverty — Income coefficient (according to G.)	
	Chapter 5: Demographic dynamics and sustainable development	— Population growth rate, % — Migration rate	— Population density (person/km^2)	— Birth rate
	Chapter 36: Support of education and vocational training (including women's problems)	— Schoolchildren growth rate — Share of children finishing elementary school, % — Share of children getting full-scale school education, %	— Share of literate adults, % — Share of population finishing 5 grades, % — Mean duration of school education	— GDP per cent expended for — Number of women per 100 men in the sphere of working activity, %
	Chapter 6: Protection and support of human health	— % of people having no access to enough good-quality water — % of people not provided with adequate sanitation — % of people pollution harmful for human health — % of people not provided with food products	— Child mortality (per 100 children born) — Expected life duration at the moment of birth (years) — Maternal mortality (per 1000 births)	— % of GDP expended for health care — % of population provided with primary medical care — % of population provided with vaccinations — % of women capable of childbearing participating in family planning

Table 1.3. *(continued)*

Category (1)	Chapter of *Agenda 21* (2)	Impact factors (3)	Indicators of state (4)	Response (5)
	Chapter 2: Human settlements (including transport problems)	— Urban population growth rate, %	— % of urban population — Area and population number in unfavourable zones (km^2, persons) — Human victims and economic losses due to natural catastrophes (persons, $) — Living space per person, m^2 — Dwelling cost to income ratio	— Expenses for cheap houses ($) — Expenses for public transportation ($) — Per anima expenses for infrastructure ($) — Credits for housing
Economy	Chapter 2: International co-operation	— Per anima real GDP growth rate, % — Exports of goods and services ($) — Imports of goods and services ($)	— Per anima GDP ($) — Share of production costs in GDP, % — Share of exports, %	— Share of investments to GDP, % — Participation in regional trade agreements (yes/no)
	Chapter 4: Structure of production and consumption	— Depletion of mineral resource reserves (% of prospected ones) — Per anima annual energy consumption, J	— Prospected reserves of mineral raw materials (tons) — Prospected energy resources (oil equivalent) — Lifetime of prospected energy resources (years) — Share of industry with intensive use of natural resources, % — Share of producers in trade export, %	— Relation between consumption of renewable and non-renewable energy resources, %

Table 1.3. *(continued)*

Category (1)	Chapter of *Agenda 21* (2)	Impact factors (3)	Indicators of state (4)	Response (5)
	Chapter 33: Financial resources	— Resources to GDP ratio, %	— Debt/GDP, % — Debt/export servicing, %	— Share of expenses for environmental protection in GDP, % — Share of ecological taxes and subsidies in public revenues, % — Amount of new or additional financing of sustainable development projects, beginning with 1992, $
	Chapter 34: Transfer of technologies			— Programme of economic and ecological assessments (yes/no) — Debt reduction
Environment	Chapter 18: Freshwater resources	— Share of use of under ground and surface waters (m^3) in relation to their resources — Per anima domestic water consumption (m^3)	— Resources of underground waters (%) — Concentration of fecal impurities in fresh water bodies (per 100ml) — BOD and COD concentrations in water basins (mg/l)	— Released water treatment (% of population provided with purified water)
	Chapter 17: Protection of oceans, all kinds of seas and coastal regions	— Marine species yield (tons) — Fish population numbers growth in coastal waters, % — Oil emissions to coastal waters (tons)	— Deviation of marine yields (MSY) from maximum but sustainably possible, % — Relation between MSY and average indices, % — Algae index	— Participation in marine agreements (yes/no)

Table 1.3. *(continued)*

Category (1)	Chapter of *Agenda 21* (2)	Impact factors (3)	Indicators of state (4)	Response (5)
Land	Chapter 10: Planning and management of land resources	— Land use	— Areas, confirmed soil erosions (km^2)/erosion index	— Reform policy in land use) (yes/no) — Decentrilization of land resource management
	Chapter 12: Fighting of desertification and droughts	— Per anima consumption timber as fuel (m^3) — Cattle number per km^2 of arid land (or intensity) — Share of arid land inhabitants living below the level of poverty	— Land which is subject to desertification (km^2)/desertification index — Frequency of droughts	
	Chapter 13: Sustainable development of mountain regions			
	Chapter 14: Support of sustainable agriculture	— Use of pesticides (tons/km^2) — Use of fertilizers (tons/km^2) — Per anima amount of arable land (ha/person) — Share (%) of irrigated arable land	— Areas subject to salinization (km^2)	— Investments to agricultural studies and projects ($) — Area of restored land (km^2)
Other natural resources	Chapter 11: Fighting deforestation	— Deforestation rate (km^2/year) — Annual timber production (m^3)	— Timber reserves (m^3) — Forest area (km^2) — Share of timber in energy consumption (%)	— Afforestation rate (km^2/year) — Share of protected forests, %

Table 1.3. *(continued)*

Category (1)	Chapter of *Agenda 21* (2)	Impact factors (3)	Indicators of state (4)	Response (5)
	Chapter 15: Preserving biodiversity		— Share of threatened species (%)	— Share of protected land, %
	Chapter 16: Biotechnology			
Atmo-sphere	Chapter 9: Protection of the atmosphere	— Emissions of greenhouse gases (tons) — Emissions of SO_x and NO_x (tons) — Ozone depletion scale (tons)	— Mixing ratios of SO_2, CO, No_x, O_3, etc. in urban regions (mln^{-1})	— Expenses for prevention of atmospheric pollution ($) — Reduction of emissions of greenhouse gases So_x and No_x (% per year)
Environ-ment pollution	Chapter 21: Solid emissions and the problem of waste	— Industrial and municipal emissions (tons)	— Per anima emissions (tons)	— Costs of collection and recycling of waste ($) — Waste recycling rate (%) — Per anima municipal emissions (tons/person) — Emission reduction rate per unit GDP (tons/year)
	Chapter 19: Toxic components			
	Chapter 20: Dangerous emissions	— Formation of dangerous emissions (tons)	— Land area polluted by dangerous emissions	— Costs of dangerous emission treatment (tons)

Table 1.3. *(continued)*

Category (1)	Chapter of *Agenda 21* (2)	Impact factors (3)	Indicators of state (4)	Response (5)
Institutional problems	Chapter 35: Science			
	Chapter 37: Creation of ecological institutions			
	Chapters 6,38,39,40: Decision-making structures		— Possessing mandate for assessment of impacts on the environment (yes/no) — Programmes of national ecological statistics (yes/no) — Sustainable development strategies (yes/no) — National councils on sustainable development (yes/no) — Number of telephone lines per 100 inhabitants — Representatives of local minor nationalities in national development councils (yes/no) — The presence of databases of conventional information (yes/no)	
	Chapters 23–32: The role of major population groups		— Representation of major population groups in national councils on sustainable development	

1.7 THE SECOND UN CONFERENCE ON ENVIRONMENT AND DEVELOPMENT: SOME RESULTS AND PERSPECTIVES

The Second United Nations Conference on Environment and Development (UNCED-2) which took place on 3–14 June 1992 in Rio de Janeiro has certainly become an important step in the evolution of the global ecological situation. M. Strong, Secretary-General of the UNCED-2 has formulated this as follows (The Spirit of Rio . . . , 1992): "we have carried out successfully the largest intergovernmental conference ever held on this planet. The right people have been here and the people of this planet have watched and wondered. The world will not be the same after this conference. Diplomacy will not be the same. The United Nations will not be the same. And prospects for our Earth cannot . . . must not . . . be the same". Making this general assessment of the Conference more concrete, N. Desai, its Deputy Secretary-General, has formulated the main results as follows (Desai, 1992): "First, it has secured a set of arguments between governments which mark a significant advance in international cooperation on development and environmental issues. Second, it has secured political commitment to these agreements at the highest level and placed the issue of sustainable development at the heart of the international diplomatic agenda. Third, it has opened new pathways for communication and cooperation between official and non-official organizations working towards developmental and environmental ends. Fourth, it has led to an enormous increase in public awareness of the issues that were tackled in the process".

Touching upon the perspectives, M. Strong (The Spirit of Rio . . . , 1992) emphasized: "We must form a new global partnership, built on the conventions, move quickly to a desertification convention and a forestry regime acceptable to all. We must move from a *Declaration* to an *Earth Charter* and translate goodwill into financial commitments. And the participatory process must be expanded in implementation of *Agenda 21* and the United Nations itself. Finally, this decade must be a time of transition that finally moves us onto a new pathway and a new economy, one that is truly an economic system".

1.7.1 The Rio Declaration on Environment and Development

When stating that the UNCED-2 has not been a great success, we proceed, first of all, from the analysis of the contents of the *Rio Declaration*, its basic but, regrettably, highly rhetorical document. It is important, however, as noted by Desai (1992) that this document "places human beings at the centre of things" (where else can human beings be placed?), "enunciates the right to development and includes clear statements of the "polluter-pays principle" and the precautionary principle, as well as the principle of shared but differentiated responsibility".

Two of the principles presented in the Declaration seem to be of special significance, viz.:

— "All States and all people shall cooperate in the essential task of eradicating poverty as an indispensable requirement for sustainable development in order to decrease the disparities in standards of living and better meet the needs of the majority of the people of the world".

— "To achieve sustainable development and a higher quality of life for all people, States should reduce and eliminate unsustainable pattern of production and consumption and promote appropriate demographic policies".

Essential are those statements of the Declaration which dwell with problems of environmental legislation, economics and international trade. However, separate parts of the document, as well as the Declaration as a whole are very declarative (which is even reflected in its title). Initially a completely different document was supposed to be worked out and entitled the Earth Charter. The idea was cancelled because agreement had not been reached with regard to priority issues. The failure must be due to the fact that the most qualified experts were not invited to participate at the preparatory stage. To show that this is a non-biased judgement, the reader should refer to the scientific literature devoted to the entire range of problems of the environment and development.

It will hardly be an exaggeration to state that Russian scientists made a major contribution to the conceptual preparation of UNCED-2. Two monographs of principal value have been published in 1990 (Gorshkov, 1990; Kondratyev, 1990b); later a whole series of relevant publications appeared (Gorshkov, Kondratyev and Sherman, 1990; Gorshkov, Kondratyev and Losev, 1992, 1994; Gorshkov, 1993, 1995a,b; Gorshkov, Kondratyev, Danilov-Danilyan and Losev, 1994; Kondratyev, 1993a,b; 1995a,b,c; 1996a,b; Kondratyev, Asimakopoulos and Varotsos, 1995; Kondratyev, Buznikov and Pokrovsky, 1992, 1996a; Kondratyev, Danilov-Danilyan and Donchenko, 1993; Kondratyev and Fedchenko, 1992; Kondratyev and Frolov, 1996; Kondratyev and Romanyuk, 1996a,b; Kotlyakov, 1995; Losev et al., 1993; Marchuk and Kondratyev, 1992, etc.). Ample material presented in these publications shows that the problem of priority issues, if applied to global ecology, does not offer great mysteries, although it does have its "surprise aspects", and these may be illustrated by a publication by Kondratyev and Fedchenko (1992).

A somewhat schematic conceptual consideration of the range of problems relevant to global ecology may be reduced to the following two key aspects (see Sections 1.1 and 1.2):

1.7.1.1

The level of consumption of biospheric resources which determines the ecological well-being (or ill-being at present) of the planet. Hence follows the necessity to make biospheric dynamics the focal point of natural studies at large. At the same time, one should take into account that the impact on the biosphere is anthropogenic, and therefore socio-economic aspects of relevant problems are of primary importance. It should be emphasized here that this conclusion is extremely important: another conclusion immediately follows from it, namely that *the problem of socio-economic development and environment should be the centre of discussion* (the subject has been analysed in detail by Kondratyev (1990a,b; 1995b).

With regard to biospheric dynamics, we shall limit ourselves to a brief commentary because the subject has been discussed in literature in great detail (Gorshkov, Kondratyev, Sherman, 1990; Gorshkov, Kondratyev, Losev, 1994; Gorshkov, 1995a,b; Gorshkov, Kondratyev, Danilov-Danilyan and Losev, 1994; Kondratyev, 1993a,b; Kondratyev, Belchansky and Petrosyan, 1992; Marchuk and Kondratyev, 1992). It should first of all be emphasized that our understanding of the regular features of biospheric dynamics is still fragmentary and approximate (it should suffice to mention that we still do not have reliable information about the number of species populating our planet).

The inadequacy of the present-day knowledge about the biosphere accounts for the rough and approximate nature of current estimates of the threshold value of permissible impact on the biosphere from the viewpoint of consuming its resources (of course, it would be extremely important to be able to establish the value). An essential aspect of the problem is the lack of adequate observational information about the biosphere. If not the already adopted Declaration, then perhaps the future Earth Charter must contain an important section characterizing the level of knowledge about the biosphere and attracting attention towards the problem of the optimization of a global system of ecological observations, whose major component should be satellites (Kondratyev, 1987, 1991a, 1995 c,d; Kondratyev, Adamenko, Henderson-Sellers, et al., 1990; Kondratyev, Buznikov and Pokrovsky, 1992, 1996a; Kondratyev and Cracknell, 1997; Kondratyev, Ortner and Preining, 1992; Kondratyev and Pozdnyakov, 1996; Kondratyev and Romanyuk, 1996b).

1.7.1.2

The relationship between the level of biospheric resources consumption and population growth. The mention of the necessity of an adequate demographic policy which can be found in the Declaration is most timely. The statement, however, should be expanded and have a clearer meaning, as is the case in quite a number of publications (see, for instance, Gorshkov, 1990, 1993, 1995aa,b; Gorshkov and Kondratyev, 1990; Gorshkov, Kondratyev and Losev, 1992, 1994; Gorshkov, Kondratyev and Sherman, 1990; Gorshkov, Kondratyev, Danilov-Danilyan and Losev, 1994; Kondratyev, 1990c; Raven, 1988, etc.). Perhaps, it is this aspect of the whole problem of sustainable development and environment that shows the significance of priorities in the most impressive way. The intention to achieve a standard of living on the entire planet that would meet the needs of all its inhabitants may be very fair (as expressed in the Declaration). However, a question immediately arises whether this is a realistic approach. Is this possible in principle? The abovementioned publications answer this question in the negative, and the answer is based on assessments, according to which the threshold of admissible impact on the biosphere was surpassed by mankind as far back as in the previous century. This conclusion certainly needs further confirmation and can be challenged. But at the moment it can hardly be ignored, since there have not been any publications refuting the conclusion. It is important in principle for the concept (and from the standpoint of priorities) of the Declaration.

Mata et al. (1994) proposed to use a demotechnical index D to characterize the relationship between consumption and population number. Index D represents the

Table 1.4. Real population number and population number reduced to the level of consumption, from data for 1990

Country	Population number (million)	Reduced population number (million)
China	1 139	9 329
India	853	3 907
USSR	289	16 828
USA	249	22 993
Canada	27	3 159

ratio between technological energy consumption (as an indicator of total resource consumption) and physiological energy consumption, expressed in identical D units which are determined by the per anima annual average physiological energy consumption, viz. 852 249 kcal/year (2333 kcal/day). Alongside this, the notion is introduced of the population number of a certain country, reduced to the level of consumption, which equals the sum of population number multiplied by (1 + D). It is clear that the reduced population number characterizes the relationship (for a given country) between the physiological energy consumption (in the form of food) and total energy consumption (including technological).

In the case of Canada, for example, the population number in 1990 was 26.521 million, and the technological energy consumption in 1989 was equal to 421.76 GJ. For Canada, the unit D constitutes 3.5709 GJ, and therefore D is equal to 421.76/3.5709, i.e. to 118.11. So, the reduced population number (in D units) is: 26.521 (1 + 118.11) = 3158.916 million.

Calculations have revealed that index D varies within broad limits, from 198.49 (Qatar) to 0.39 (Comoro Islands). Naturally, developing countries can be found at the end of a list compiled in the order of the diminishing of index D. The data of Table 1.4 illustrate the relation between the real and reduced population number for several big countries.

Experts working within the United Nations Programme on Socio-Economic Development (UNDP) proposed a Human Development Index (HDI) as an indicator of meeting human needs in different countries, expressed in relative units (from 0 to 1). HDI values are calculated from the data about four parameters (their minimum and maximum values are given in brackets): expected life duration (25 and 85 years), level of illiteracy of the adult population (0% and 100%), mean duration of school tuition (0 and 15 years) and per anima annual income ($200 and $40 000). It has been assumed that countries for which HDI 0.5 belong to the category of low-level development. Accordingly, HDI constituting 0.5–0.8, or exceeding 0.8, correspond to the medium and high levels of development. In 1994 maximum HDI values were characteristic for Canada, Switzerland and Japan, and developed countries having minimum HDI values comprised Poland, Romania, Bulgaria and former Republics of the USSR. The Third World countries were situated within a wide range of HDI values: from Barbados and Hong Kong (maximum HDI) to Burkina Faso and Guinea (minimum HDI).

It is interesting to note that in many cases there is no correlation between material wealth and the well-being of people, i.e. between HDI and GNP. For example, in countries such as Angola, Gabon and United Arab Emirates, the income rank exceeds by far the HDI rank, which shows the presence of a considerable potential for making use of the income in the interests of the well-being of the people. At the same time, in countries such as China, Colombia, Costa Rica, Cuba, Guinea, Madagascar, Sri Lanka, the HDI rank is far ahead of the income rank. From the viewpoint of differences between HDI and GNP, situations are completely different in China (+49 points) and Gabon (−72 points).

Countries with the same GNP may differ very much in levels of social development. For example, the per anima income in Sri Lanka and Pakistan is about $400–500, whereas HDI figures are over 0.6 and 0.39, respectively. The latter is determined by the priority of military expenditures.

The circumstances described above show very distinctly that economic growth alone is by no means a guarantee of improved standards of living. It should be emphasized that it is impossible to substantiate a unique index characterizing sustainable socio-economic development. The problem of "indexing" is extremely complicated and has not been resolved so far (*Monitoring Environmental*..., 1995). We shall come back to the problem later.

A thorough analysis of the *status quo* reveals the sad fact that our modern world is a kind of a "time bomb": contradictions between the prospering North and the largely backward (sometimes starving) South are being accumulated and can hardly be resolved in the present conditions. Since there are no realistic perspectives for a cardinal solution of the demographic problem (reduction of the world population number), analysis is urgently needed of possible ways of changing lifestyles and social priorities, and overcoming the prevailing stereotype of the "consumer society". It is certainly most disappointing that relevant assessments are lacking in the text of the Declaration.

1.7.2 *Agenda 21*

As has been mentioned above, this non-binding (from the standpoint of the States' responsibilities) document must be an action plan on global ecologically harmonious and sustainable socio-economic development. This comprehensive document covers more than 100 programme areas ranging from poverty alleviation to strengthening the role of the society in the solution of problems of nature protection. Regrettably, some most important issues, such as population, consumption patterns, the Third World debt, the role of military expenditures, the export of hazardous wastes, and deforestation, have either been omitted or inadequately treated.

Agenda 21 contains a very significant chapter concerning financial resources and mechanisms. It commits developed countries to reach the accepted UN target of 0.7% of GNP for official development assistance by the year 2000, or as soon as possible thereafter. Naturally, parties concerned responded differently to the call made in the document. For example, the European Union intends to increase its environmental aid by \$4 billion (the precise time frame has not been mentioned), Germany plans to enlarge its GNP share allotted for the same purpose from 0.42% to 0.7% by the year 2000. Japan will allocate about \$7 billion during the next 5 years for the implementation of environmental projects in developing countries (in addition to the \$3.1 billion of its present contribution). Mr John Major, the UK Prime Minister, recommended that the World Bank should increase its assignations for ecological assistance from \$2 to 3 billion.

To ensure an adequate follow-up, a high level Commission on Sustainable Development has been created within the United Nations. This Commission is responsible for coordination within the UN system, and will report to the General Assembly through the Economic and Social Council of the United Nations on the implementation of the programme of *Agenda 21*.

Briefly summing up the above, one is obliged to state that *Agenda 21*, this highly declarative document, ignores the most important thing—the responsibility of the States involved—and thus it really remains a non-binding document.

1.7.3 The Framework Convention on Climate Change

As has been mentioned earlier, the Framework Convention on Climate Change (FC), which formulates certain responsibilities taken and signed by 154 States (except Malaysia which characterized the FC as "meaningless" because it lacked binding targets for greenhouse gas reductions), has been a product of long evolution which was finally completed (on a working level) at the Fifth Session of the Intergovernmental Negotiating Committee on FC (Kondratyev, 1991c, 1992a–c, 1994; Kondratyev, Adamenko, Henderson-Sellers et al., 1990; Kondratyev, Cracknell and Vaughan, 1994; Kondratyev, Kozoderov, Fedchenko et al., 1994; Kondratyev, Ortner and Preining, 1992).

Regrettably, the final version of the FC, approved by the UNCED-2, seems to be the worst possible variant: it is vague and ambiguous even within the "greenhouse" concept of the document. Berreen and Meyer (1992) have been quite right to note that the basic proposals formulated in the FC are nothing else but vague declarations about how undesirable excessive greenhouse emissions are. The FC was criticized before (Kondratyev, 1991c, 1992c); today the criticism is no less valid. Therefore, to avoid repetitions, two circumstances of primary importance should be considered:

1.7.3.1

The two Conventions (on climate and biodiversity) and the Statement on Forests were prepared by different Working Groups as independent, non-linked documents. The preparation of the FC is a good example of the inconsistency of such an approach. The thing is that the solution of environmental problems is impossible without due account of the interactivity of its components.

The ecological well-being of the Earth is determined by the presence of a slightly disturbed (and capable of recovery) biosphere. The biosphere, which functions on the basis of the Le Chatelier principle, secures the compensation of the perturbing forces and regulation of environmental properties owing, to a great extent, to biodiversity existing in nature (Gorshkov, 1990, 1993, 1995a,b; Gorshkov and

Kondratyev, 1990; Gorshkov, Kondratyev and Losev, 1992, 1994; Gorshkov, Kondratyev, Danilov-Danilyan and Losev, 1994; Gorshkov, Kondratyev and Sherman, 1990). We observe a close interrelationship between problems relevant to climate, biodiversity and forest dynamics.

Nature is a single whole. This is a fundamental principle that cannot be ignored; otherwise one will find himself on a wrong track. The obvious conclusion is this: further efforts in working out similar documents should be based on close and efficient coordination, which may be constructive only if a preliminary agreement is reached concerning the key aspects of the interactivity of problems under consideration.

1.7.3.2

The Supplement Report to the Intergovernmental Panel on Climate Change (IPCC, 1992) contains the following statement with regard to global climate warming (within 0.3–0.6°C) during the past century: "the size of this warming is broadly consistent with predictions of climate models, but it is also of the same magnitude as natural climate variability. Thus the observed increase could be largely due to natural variability; alternatively, this variation could be due to human factors that offset a still larger induced greenhouse warming; the unequivocal detection of the enhanced greenhouse effect from observations is not likely for a decade or more".

Thus, it clearly follows from the new IPCC document that there are no grounds for a one-sided support of the stereotype of greenhouse warming and for putting forward the gas emission reduction as the central problem if we are preoccupied with the task of preventing undesirable after-effects of global climate change. It is most surprising to see that the greenhouse concept has been made the basis of the FC. Since the subject has been treated at length in numerous monographs (Kondratyev, 1990b, 1992a; Kondratyev and Cracknell, 1997) it will suffice to mention here that a vital factor of anthropogenic impact on global climate is the destruction of the biosphere which leads to increased emissions of carbon dioxide and other greenhouse gases (GHG) into the atmosphere by the continental biosphere. This is a consideration of primary importance; hence, our concern should be focused on the preservation of the biosphere as the main factor of global ecological safety. Forest dynamics is a problem that should be considered in this context.

Summarizing the above, one should emphasize that the ambiguous nature of the FC which caused a painful response on the part of Malaysia, its inadequate conceptual basis require further serious efforts for improving the FC. In this connection, the decision to keep the IPCC seems reasonable. It is equally obvious, however, that mechanical continuation of previous activities will not end in a great success. The circle of experts involved in relevant efforts should be broadened to bring about more valuable ideas. A considerable increase of the efficiency of international cooperation in the field is absolutely necessary.

1.7.4 The Convention on Biodiversity

As is the case with other documents, the Convention on Biodiversity (CB) is the result of long-term efforts of coordinating contradictory interests of various States. 153 States signed the Convention; however, the United States of America initially refused to join them. This fact shows that the situation was really tense.

The CB commits the participating States, in particular: to develop national plans, strategies or programmes for conservation and sustainable use of biodiversity; to secure inventory and monitoring of biodiversity components and processes which affect it adversely; to develop and enhance *in situ* mechanisms for conservation of biodiversity, both within and outside of protected areas; to intensify restoration of degraded ecosystems and recovery of endangered species; to secure regulation of the income of genetically modified organisms; to preserve and maintain indigenous and local systems of biological resource management and equitable sharing of benefits with local communities; to carry out assessments of the impact on biodiversity of the implementation of various projects, programmes and policies.

Because the CB contains numerous ambiguous terms and expressions, such as "as far as possible and as appropriate", "by mutual agreement", etc., it is not clear to what extent it will lead to concrete action. What is clear is that no less than 2–3 years will be required to make preparations for its practical realization.

Miller and Barber (1992) are of the opinion that the CB contains quite a number of unconvincing and/or contradictory statements which can be reduced to the following:

1.7.4.1

The provisions of intellectual property rights and the transfer of technology are exceptionally confusing, and several industrial countries have expressed their discomfort with these sections, while one—the United States—initially refused to sign the convention on the grounds that the intellectual property rights of biotechnology companies (and their patented technological processes) would be compromised and the growth of that industry stunted. This may require a separately negotiated Protocol to elaborate more precise statements in the CB.

1.7.4.2

An important financing mechanism is the Global Environmental Facility (GEF) established as an interim structure. Its status requires revision (with special emphasis on the governance of the GEF).

1.7.4.3

Policy and institutional issues that the Global Biodiversity Strategy (launched by the World Conservation Union, the World Resources Institute and UNEP in February 1992) identifies as key ones for conserving biodiversity and reducing the Third World debt, were treated in a weak manner or not at all in the convention.

1.7.4.4

The CB does not completely take into account the possibilities for non-governmental organizations.

In addition, perhaps the most important issue should be emphasized: the CB does not reflect the fact that biodiversity is the basis of global ecological safety because it is a factor securing the sustainability of ecosystems and the biosphere as a whole (Gorshkov, 1990, 1993, 1995b; Gorshkov and Kondratyev, 1990; Gorshkov, Kondratyev and Losev, 1992, 1994; Gorshkov, Kondratyev, Danilov-Danilyan and Losev, 1994; Gorshkov, Kondratyev and Sherman, 1990). In this connection, an issue of serious concern is the low level of biodiversity studies; these must include the documenting of its structure, composition, global distribution, and the analysis of the entire hierarchy of biosystems from molecules to the biosphere, as well as the elaboration on this basis of rational measures for biodiversity protection on the local, regional and global scales, with scrupulous accounting for the unique nature of every form of life on the Earth. The solution of these problems requires multidisciplinary studies on the basis of adequate information and modern systems of monitoring, taking proper account of the hierarchy, dynamics and spatial distribution of biosystems. Kondratyev, Belchansky and Petrosian (1992) offer a discussion of some conceptual aspects of monitoring biodiversity from space which have not been reflected thus far in the existing programmes of global monitoring of biospheric dynamics.

The Intergovernmental Committee on the Convention on Biological Diversity (ICBC) adopted a recommendation concerning the need for regular (beginning with 1993) ICBC conferences in order to coordinate efforts aimed mostly at promotion of necessary studies and projects. One may hope that in the future a much broader (than up to now) interdisciplinary base will be created to work out relevant recommendations.

1.7.5 Land resources: deforestation

Although a binding convention on forests has not been initially accepted because of opposition from developing countries (it has been done later), the adoption of the Statement of Principles on the Management, Conservation and Sustainable Development of All Types of Forests is extremely important, since it paves the way for future efforts. The Statement (*The Spirit of Rio*..., 1992) quite appropriately focuses attention on the consideration that "programmes for agricultural and rural development, and for water resource development seek to combine a basic development concern with sound environmental management".

Further, the Statement declares that it is necessary to recognize the key role of all types of forests in the maintenance of ecological processes and balance on local, national, regional and global scales through protection of unstable ecosystems, water catchment areas and freshwater resources, and also as accumulators of biodiversity and biological resources and sources for genetic material for biotechnological production, as well as for carbon.

Finally, it is important to attract general attention to the necessity of further "interactive" efforts in the development and implementation of the conventions of climate, biodiversity and forests.

1.7.6 Conclusion

There is no doubt that the most important, but surely most difficult thing in securing a sustainable socio-economic development is the necessity of adaptation to the resources of the global biosphere. The Statement adopted by the reconvened World Commission on Environment and Development (WCED, or the Brundtland Commission) focused general attention on the following (*Our Common Future*. . ., 1992):

1.7.6.1

Broadening of discussions on population problems to take into account the economic and political causes of poverty. Poverty, environment and population can no longer be dealt with as separate issues.

1.7.6.2

The necessity of documenting the commercial bank debt of the Third World which has reached the figure of $1.3 thousand billion. The economic policy, standards of production and consumption of industrialized countries put a limit to the capability of developing states to achieve a sustainable development and eliminate poverty.

1.7.6.3

Substantiation of the polluter-pays principle and working out of measures for the application of this principle, including prices that must reflect full environmental costs.

1.7.6.4

Implementation of the Official Development Assistance target of 0.7% of GNP of developed countries by the year 2000.

The recommendation of the Brundtland Commission concerning the creation of a new, independent international body outside the United Nations system to monitor, analyse and assess the state of the environment has been very timely.

UNCED-2 has doubtless played a most important role by attracting general attention to crucial environmental issues. Stronger and broader efforts are needed to accomplish the programme of action outlined because the scale and complexity of problems involved are unprecedented. Raven (1987) had every right to state: "The large-scale problems of overpopulation and overdevelopment are eradicating the land and organisms that sustain life on this planet. If we can solve these problems, we can lay the foundation for peace and prosperity in the future. By ignoring these issues, drifting passively while attending to what seem more urgent, personal priorities, we are courting disaster".

1.8 PRESENT-TIME ECOLOGICAL PROBLEMS: POSSIBLE DEVELOPMENT CONCEPTS AND SOLUTIONS

(V. G. Gorshkov, K. Ya. Kondratyev, K. S. Losev)

Analysis of the perspectives of global ecological development is at present the focus of general attention (see Sections 1.1 and 1.2). Of course, it presents a broad spectrum of varying judgement (see the References). However, the existing ideas can be summarized and presented as two main concepts of development.

1.8.1 Two concepts of development

The Second UN Conference on Environment and Development (Rio de Janeiro, 1992), The International Conference on Population (Cairo, 1994) and the World Summit on Social Development (Copenhagen, 1995) have once again confirmed the fact that socio-economic development, demography and environment are closely interrelated. However, not one of these conferences discussed an alternative to the existing resource concept of the development of the world system—the concept which regards the Earth as just (or mainly) a source of resources. There are various realizations of global resource models of different generations predicting future development up to the middle and the end of the twenty-first century. They study the behaviour of such components of development as economics, natural resources, population, food resources and the state of the environment. These models mostly substantiate catastrophe scenarios which must happen by the middle or the end of the next century due to the exhaustion of resources caused by population and economic growth. According to different scenarios of development, the stabilization of population growth must occur at levels from 7 to 12 billion people. These scenarios assume that today the environment has not yet reached a catastrophic state. An alternative concept of development may be called "biospheric". This concept assumes natural biological regulation of the environment; therefore, unlike the case of the resource concept, no *a priori* assumptions concerning future development are required here.

The essence of the theory of natural biological regulation of the environment may be formulated like this: the biosphere, which comprises the biota and its environment, possesses powerful mechanisms for stabilization of environmental parameters to provide such conditions for the existence of living organisms which can be considered close to optimal (Gorshkov, 1990, 1993, 1995a,b; Gorshkov and Kondratyev, 1990; Gorshkov, Kondratyev, Danilov-Danilyan and Losev, 1994; Gorshkov, Kondratyev and Losev, 1992, 1994; Gorshkov, Kondratyev and Sherman, 1990). On the basis of competing interaction between highly correlated communities of organisms forming the biota, a high degree of closedness of the cycles of substances (nutrients) is achieved, which guarantees environmental stability. The dynamically closed cycles of substances secured by the biota exceed by many orders of magnitude the environment perturbation levels; this is what makes it possible for the biota to compensate practically immediately any unfavourable environmental changes which are revealed as breaks in the cycle closedness. Over time scales of the order of 10^3–10^4 years, the environment returns to its former state; during larger time periods, the transition to new levels of sustainability of the environment takes place.

The existence of biotic regulation of the environment can be confirmed by the fact that geophysical processes, by causing purposeful environmental changes, may change environment characteristics by 100% during a time period of the order of 100 thousand years, and may render the environment completely unsuitable for life during about 1 million years. However, this has not happened during the 4 billion years of the existence of life.

The theory of biotic regulation of the environment has been developed in Russia during the 1980s (see the References cited earlier in this section). The essence of this theory consists in the definition of the sustainability threshold of the global biosphere, since it is obvious that there exists a threshold value of environmental disturbance, and if this value is exceeded, the sustainability of the biota and its habitat is violated. Using two methods which are independent of each other and based on observational data, it has been demonstrated that the biosphere is sustainable, i.e. is capable of compensating for any perturbations caused by man's economic activity until the level of biotic production consumption by humans does not exceed 1%. The remaining 99% is required for the biota to stabilize the environment. So, only one kind of human activity is critical: the consumption of biotic production (photosynthesis production), or, in other words, the interrelationship between man and nature when man is in the capacity of a consumer and thus fulfils his role as a species.

1.8.2 Choosing the concept of world development

Out of these two concepts, preference has so far been shown to the resource one. This can be seen from numerous international documents, such as the materials of the Rio Conference and the Population Conference in Cairo, as well as from numerous publications and international agreements concerning environmental protection. However, comparison of the two concepts provides evidence rather in favour of the biospheric concept.

In fact, the resource concept has no theoretical substantiation. It is actually based on man's practical experience; the most complex subject is simplified to the level of presentation insisting that mankind's economic activity is wrong, and that the only two things needed are the renovation of technologies and the resource-saving policy. However, any technologies require the consumption of resources, which means that there cannot be any technologies that would not change and destroy the biosphere. Man produces nothing but waste: direct (in the process of production) and delayed (in the process of consumption). The history of science and technology progress shows that man destroys natural ecological systems in a growing volume and on a growing time scale; in the process of this destruction, first local, and later global ecological crises were originated.

The resource concept has no "turning point" for many characteristics of the Earth's system, which might determine the possibility of sustainable development. For example, with regard to the environment, it is supposed that it has not yet reached a critical point, and therefore additional economic loadings are permissible. There are assumptions concerning some mineral resources that these resources can be exhausted during the nearest 10, 50 or 100 years; various assumptions lead to different output model results.

As has been mentioned above (see the References cited earlier in this section), the biotic concept (and the corresponding model) is based on a sufficiently developed theory. This theory has an essential "turning point" for the environment—the value of permissible perturbation of the environment. The theory of biotic regulation makes it possible to not only determine the threshold value of permissible impact and its measurement units, but also to establish, on the basis of observational data, the time moment of exceeding this threshold: global changes that have already been observed indicate unambiguously that this threshold has already been exceeded. According to the biospheric concept, the ecological threshold of the development of humankind has been reached earlier than any other growth limits which are regarded as hypothetical within the framework of the resource concept; the biotic concept reveals the fact that the catastrophe has already happened. The above shows that preference should be made in favour of the biotic concept of the world development.

1.8.3 Global ecological catastrophe

The threshold of permissible perturbation of the environment has been exceeded as far back as in the early twentieth century, which is indicated by the estimates of the closedness of the global carbon cycle and the level of consumption of biotic production at the beginning of the present century which reached about 1%. Today the biota is no longer capable of securing global environmental sustainability. There are some remaining biospheric regions which tend to the return to stability, but they are not powerful enough (Gorshkov, 1993, 1995b). The consumption of 1% of primary biotic production is connected with energy expenditure of the order of 1 TW. The greater the consumption of primary production, and therefore the stronger the energy power of a country, the more it contributes to biospheric destruction and environmental destabilization. Therefore, an integral index of anthropogenic destabilization of the environment may be the value of power production of a given country per unit of its area. If such values are related to global average power production per unit area, all countries can be put in order in accordance with their share of anthropogenic destabilization of the environment (the anthropogenic stress coefficient). Such data are presented in Table 1.5 and are relevant to the year 1986 (Gorshkov, Kondratyev and Losev, 1992; Gorshkov, Kondratyev, Danilov-Danilyan and Losev, 1994). In later years, relevant indices increased considerably in some countries (e.g., China (to 1.5), India, etc.).

The data of Table 1.5 show that it is the developed countries that have mainly contributed and continue to contribute to the situation of an ecological crisis. However, today there also are some Third World states (known as "little dragons") which play an identical role in provoking an ecological crisis. Big, heavily populated developing countries certainly contribute markedly to the dangerous tendency.

Table 1.5. Coefficient of anthropogenic stress and share (%) of preserved natural territories

Country	Netherlands	FRG	Japan	USA	South Korea	India	China	Mexico	Russia	Global
Coefficient	42	19	16	3.4	4	1	1.1	1.2	0.7	1
Natural territories	0	0	0	4	0	1	20	2	45	39

The "ecologically unfavourable" countries , first of all, have destroyed their own nature (see Table 1.5), which can be illustrated by the practical absence of territories occupied by natural communities. In China the comparatively high percentage of such territories has been preserved only due to the low-productivity deserts and high mountains in Tibet. Countries of this kind have built and developed their well-being by exhausting their resources and destroying their nature, or restructuring it in such a way that the closedness of biogeochemical cycles has been violated. It is important to mention here that developed countries often use the resources of less developed states. Parts of the biosphere in some countries which have been preserved intact have accepted the load of stabilizing the environment, together with the world ocean. In the northern hemisphere, there remained just a few such countries. First of all, such are Russia and Canada (the coefficient of anthropogenic stress for Canada is 0.4), where the biggest areas of highly productive natural forest communities and wetlands have been preserved (800 and 642 million ha, respectively), as well as Algeria (152 million ha, coefficient of anthropogenic stress = 0.16), where low-productivity deserts have mainly remained intact. The remaining countries of the northern hemisphere have at least one order of magnitude smaller territories which have not been disturbed by economic activity.

In the southern hemisphere, there are two states which mainly contribute to the stabilization of the environment due to territories conserved in their natural state: Brazil (273 million ha, coefficient of anthropogenic stress = 0.2) and Australia (251 million ha, with the same coefficient of anthropogenic stress). The contribution of Australia is much less because low-productive deserts prevail in the country, whereas in Brazil it is mostly highly productive humid tropical forests. On the whole, about 39% of land territories have been conserved in their natural state (exclusive of Antarctica).

Mankind at present consumes about 6–8% of biotic primary production directly in the anthropogenic "channel", and another 30–32% indirectly, by making use of 10 TW of power obtained due to fossil fuel (Gorshkov, 1990, 1995b). Thus, the permissible level of consuming biospheric resources has been exceeded by almost an order of magnitude. The production of the biota is withdrawn from the nutrition of other organisms and does not participate in the regulation and securing the sustainability of the environment. In spite of the fact that a gigantic amount of power is used to obtain annually about 20 tonnes of raw material per anima on a global scale, which are then processed into finite products by using 800 tonnes of water, there are about 1.5 billion absolutely pauperized people in the world and about 0.5 billion people suffer from starvation (Gorshkov, 1993; Gorshkov, Kondratyev and Losev, 1992; Losev et al., 1993). Therefore, our modern civilization cannot ensure a bearable standard of living for a considerable part of the inhabitants of our planet, but, at the same time, it leaves only 60% of photosynthetic production at the disposal of many millions of species of living organisms on this land which, inhabiting highly correlated natural communities, have been securing environmental sustainability during billions of years.

If the current rates of economic development and population growth continue, 30–40 years from now mankind will use, directly or indirectly, about 80% of primary biological production. It means that, starting from the very beginning of the twentieth century, the world system is in a state of ecological catastrophe, and the last decade of the current century is critical; hence the necessity of working out a long-term strategy and urgent practical measures of ensuring sustainable development.

1.8.4 Choosing a strategy

The notion of "sustainable development" which appeared during the recent decade, has not been helpful so far for understanding the ways of liquidation (or at least mitigation) of the developing ecological catastrophe. Its definition (development is sustainable if it meets the demands of the current generation without threatening the capability of future generations to meet their demands) (see World Commission . . . , 1987) has been borrowed from the resource concept. The biospheric concept offers a different definition: "Sustainable development is the improvement of the well-being of people in the conditions of a sustainable biosphere, i.e. in the conditions when economic activity does not lead to the exceeding of the permissible threshold of biosphere perturbation, or when such a volume of natural environment is preserved which can secure biospheric sustainability, including man's economic activity".

If humankind could come back to the limits of the economic capacity of the biosphere, the ecological problems would disappear automatically, and anthropogenic changes of the environment would stop. However, this requires specific measures, especially in the stabilization, and later reduction of the planet's population. The world is hardly ready to take these measures right now. The UN Conference on Population and Development (Cairo, 1994) has revealed the fact that group, national and confessional interests dominate over those common to the whole of mankind. These interests sometimes are above the natural fear of facing a catastrophe. Besides, many political figures and scientists do not admit the existence of an ecological catastrophe. Views like this have been formed mostly under the influence of resource models, according to which the resource limit must be reached earlier than the ecological one.

For various reasons, there is no possibility at present to adopt a global strategy of population reduction. However, the biotic concept indicates a different solution: mankind can still have some time to be prepared to pass over to a new thinking, which assumes joint efforts of all nations and states in the solution of the problem of eliminating an ecological catastrophe and survival of humankind (Gorshkov, Kondratyev and Losev, 1992).

The thing is that those ecosystems that have been preserved intact can become centres of the restoration of the volume of natural communities of organisms which is required for a sustainable environment. Such natural regions should be preserved by all means, and later expanded. The restoration of natural communities is a natural process which requires hundreds of years, but their regulatory potentials can be restored much faster—over several decades (Gorshkov, 1993, 1995b). Only 39% of intact territories have been conserved on land; they need preservation and extension.

To implement the strategy of preservation and expansion of natural territories, relevant efforts of all states, and individual political leaders should be united. A tax for using the biosphere should also be introduced because, in accordance with the biotic concept, the biosphere is the fundamental basis of life (it is interesting to mention E. U. von Weizsäcker's (1994) judgement on the subject). Such a tax would be similar to that for the use of a dwelling or for public utilities. It must be proportional to the coefficient of anthropogenic stress, and it must take into account the natural productivity of a territory used for economic activity with respect to global mean productivity: for instance, the productivity of a tropical rain forest is four times greater than that of a temperate zone forest, and the productivity of the latter is an order of magnitude higher than that of the tundra. The money collected should be redistributed in favour of those countries where considerable areas of highly productive land have been preserved and where the growth of such land would be most efficient. A tax like this might essentially change the relationship between man and nature.

If the strategy of preserving and expanding the part of nature that remained intact could be adopted, the reduction of population and economic activity might be postponed to a later time. It has been established within the framework of the biotic concept (see the References cited earlier in this section) that if the present scale of economic activity is conserved (10 TW), in order to stabilize global changes (first of all, carbon emission by the land biota), the area of land disturbed by man should be reduced from 61% (today) to 38%. It means that man must reduce the cultivated land area by 32 million km^2 provided that the primary biological production over this territory will be equal to global average production. For example, it may be the zone of temperate latitudes, with forests and wetlands. If natural land regions are restored in the zone of tropical forest, the territory can be smaller. Intact territories should be used only for recreation, tourism and sport hunting; all mechanic transport means should be forbidden.

Parallel with efforts for reducing land regions disturbed by economic activity, measures should be taken towards the stabilization of population number and gradual decrease of the power capacity of human economic activity due to energy- and resource-saving technologies. Only after that, another stage will follow—that of population number reduction. To stabilize the population level, the strategy "one family – two children" can be adopted; to reduce it, a different strategy "one family – one child" should be accepted.

In the northern hemisphere, where the majority of industrially developed countries is concentrated, there are only two states which are ready for such a transition: Russia and Canada, and especially Russia. Russia possesses the largest area of land not disturbed by economic activity, and this land is mainly represented by forests and wetlands. The country is in a transitional state both ideologically and spiritually, and the vacuum could be filled with the ideas of sustainable development on the basis of the biotic regulation concept. Russia is not facing the problem of stabilization because its population number has been reducing.

1.8.5 The world of myths

We are at present living in a period of a rapidly progressing global ecological catastrophe caused by man's economic activity which, in the course of just a few decades, has broken the balance maintained by the biosphere for hundreds or billions of years. As a rule, this situation is ignored or reduced to a regional scale, as well as to its individual manifestations, such as pollution or the danger of radioactivity. The latter is extremely important; however, the roots of the entire catastrophic situation should be realized. Numerous scientific publications on the subject ignore the disastrous nature of the situation, to say nothing of political leaders who, as a rule, reject the presence of an ecological catastrophe and the violation by man of the very mechanism of environmental regulation and stabilization. Such a position is exacerbated by various causes.

The well-being of developed countries gets in the way of a non-biased approach. Such countries are capable of cleaning the environment within their borders or in individual regions needing such measures. Developing countries cannot realize the catastrophe because they are poorly developed economically; they are facing the problem of starvation and actually are in a state of a social catastrophe. A minor part of developing countries have achieved a spurt in their economic progress and are in a state of euphoria which prevents them from understanding the catastrophic ecological situation. Finally, countries whose economy is transitional, are facing a great deal of urgent socio-economic problems which shield all others, including that of an ecological disaster.

Just a few humans can think of their coming death. Ecological catastrophe is a good example to show that humankind as a whole follows the same pattern, trying to avoid any thoughts about being doomed, instead of starting the programme of survival. However, this postponement of real actions directed towards overcoming the ecological catastrophe only accelerates our approaching a collapse.

The history of the biosphere shows that throughout almost 4 billion years it has been facing and overcoming all kinds of shocks: levelling the consequences of most powerful volcanic eruptions and the falls of big celestial bodies, passing over to new sustainability levels on a glaciated or ice-free Earth, in the presence of the junction or destruction of entire continents. In all cases, the biosphere has always been a stabilizer which kept the environment on a level acceptable for life. It is probable that certain destructive species did originate in the past, but nature apparently cut them off; as demonstrated by palaeontology, a vanished species never comes back to life. There are no grounds to suppose that the laws of the existence and development of the biosphere are not valid for man. Therefore, if *Homo sapiens* continues living in the world of myths instead of taking measures for his salvation in the conditions of an ecological catastrophe he has created, nature may find a way to get rid of this species as well. This sad fate can, however, be avoided if mankind takes the road of reasonable balanced actions within the framework of the biotic regulation concept of development which is based on the laws of physics, chemistry and biology.

1.9 ECOLOGY–POLICY–ECONOMICS: POSSIBLE WAYS FOR SUSTAINABLE DEVELOPMENT

There can hardly be any doubt that socio-economic and political aspects (and the "human measurement" in general) of changes taking place in the environment cause more and more concern. Population growth, economic level and other factors of socio-economic development are essentially the causes of ecological ill-being which is observed on local, regional and global scales (see the References cited earlier in the previous section).

1.9.1 On the threshold of the century of the environment

Comprehensive discussions of the complex set of problems relevant to the interrelationship between ecology, politics and economy can be found in a monograph by von Weizsäcker (1994).

In the Preface to his book "Earth Politics" Professor von Weizsäcker states that, by making use of the notion "earth politics", he emphasizes the necessity of an effective global environmental policy, and thus of political measures that will be inevitably associated with it. His main objective is to propose a pragmatic strategy for "earth politics", whose conceptual basis is presented in Part I of his book. Part II outlines five main aspects of the global environmental crisis; Part III offers possible solutions of the entire problem and, above all, its economic aspect (from this viewpoint, the results of the Second UN Conference on Environment and Development—UNCED-2—held in 1992 have been assessed as unsatisfactory). The closing Part IV of the monograph is devoted to subjects which are beyond today's pragmatism.

Estimating the outcome of the UNCED-2 and considering the difference between the "South" and the "North" in their approaches to such an estimation (the subject is also discussed in subsequent sections), von Weizsäcker states that for the developing world (the South), the main problem is to overcome poverty and global economic inequality. This is fair and understandable; however, this is no way to solve the problem of poverty because the conventional way, that of expanding material production, means a further increase of the anthropogenic stress on the environment (because it means higher per capita consumption of water, energy, biomass, mineral resources, and a stronger environmental pollution). Eliminating economic inequality has no perspective, for the same reason: the limited amount of resources.

Today a single German consumes roughly ten times as much resources as one Argentinian, Egyptian or Filipino (to say nothing of Somalians or Ethiopians). If the 5.5 billion population of the planet gets access to the "German" level of consumption, the global biosphere and the environment will be totally destroyed.

The myopia of the North is no less fallacious. The North puts forward the problem of environmental protection (and that of birth control, in developing countries, of course!). Important as this objective may be, it is obvious that the South, which must solve difficult problems of economic development to improve the standard of living of its population, does not possess the necessary resources to afford the costly pollution-control technologies.

In the context of problems discussed, it is clear that the rather inexact concept of *sustainable development* does not in the least open up ways for the solution of problems mankind is at present facing. To escape the "North–South" dilemma and make the concept of sustainable development a reality, we must formulate "a new set of values, a new culture, a new set of incentives to make millions of actors both in the North and the South act differently... We need to enter and to create a new era of human history" (von Weizsäcker, 1994).

Our century could bear the name the "Economic Century": the world is divided into parts according to economic criteria (developed countries, developing countries, etc.). It was not so before 1900. In the seventeenth century, at least in central Europe, religious wars determined cultural development. The eighteenth century was that of monarchs' rule. In the nineteenth century, bourgeois revolutions originated the appearance of national states, provoked industrialization and the emergence of the working class. The system of values of our century—a century of economics—is based on the anti-ethics of private profit.

The author was quite right to note: "the real danger, as I see it, is that the supremacy of economics in its present form will cause irreparable damage to the Earth and to the people who live on it", and further, "the wonderful days of naive economic consensus are numbered", since the entire population of our planet cannot be provided with natural resources on the per capita level which is characteristic for industrially developed countries (inhabited by a minor portion of the world's population). Besides, the threat of global ecological catastrophes increases due to the ozone layer depletion, the enhancement of the biologically dangerous UV-B solar radiation, climate changes and loss of biodiversity.

The main thesis of von Weizsäcker's monograph (1994) has been formulated as follows: "we are now entering a "Century of the Environment", when short-term economic goals (important as they may be) will not be achieved if they are not subordinated to the ecological imperative. Of crucial value becomes "the cruel reality of ecological devastation which will confront us in our everyday lives and which will inevitably shape our civilization if present trends of destruction continue for just another decade or two". Hence it follows that, from its very beginning, the twenty-first century will bear the mark of a massively endangered natural environment, and this will more and more determine decision-making in all sectors of politics: from foreign policy and perspectives of socio-economic development to various fields of science, technology and education. It is the ecological imperative that will become the dominant determinant for law and administration, city planning and agriculture, the arts and religion, technology, economic and social development. All relevant political goals should thus be defined by the notion "earth politics", which must be very pragmatic (and at the same time, taking account of long-term perspectives) and internationalistic (we must get out of the habit of thinking in terms of the nation, which is a nineteenth century concept; on the other hand, we must respect people's material and cultural needs on the basis of the principle "think globally – act locally").

1.9.2 Classical environmental policy: pollution control

It is only in the twentieth century that mankind started to think systematically about environmental protection. Quite a number of various measures have been taken that have resulted in a better quality of the environment in some countries. For example, the notorious English smog disappeared as a result of certain measures (which made the British believe that their environmental problems were over, but not for long). In the USA and some other countries, the use of agricultural pesticides and DDT has been prohibited; environmental legislation aimed at the protection of nature has been developed intensively. A major event of global-scale significance has become the first UN Conference on the Human Environment (Stockholm, 1972). However, at that time, most developing countries took the position that economic development was their top priority task; they left environmental problems as a responsibility of the rich industrialized North.

The environmental thinking of Western Europe of that time was mainly reduced to two targets: nature protection and lowering environmental pollution. France was the first European Community (EC) country to create a Ministry of Environment. In West Germany modern environmental protection came into being in North-Rhein-Westphalia in the 1960s, and Willy Brandt was the first Federal Chancellor to give high priority to environmental protection in a government programme. The Federal Environment Programme which was adopted in 1971 established three principles of environmental policy: the polluter-pays, precautionary and co-operation principles. The Federal Environment Ministry was created only in 1986 (in the wake of the Chernobyl disaster).

The classical environmental policy remains on the agenda in all major industrial countries. The defeat of communism in the former USSR and eastern Europe, the severe ecological and economic consequences of the many years of functioning of a monopolistic, central-planning economy, on the one hand, and certain achievements in environmental protection in industrialized Western countries, on the other, have brought about a moral that seems clear: become rich and you will be able to shoulder the cost of protecting the environment.

Von Weizsäcker is absolutely correct in pointing out, however, that a formula like this leaves unsolved a problem of extreme importance: becoming rich (in the conventional Western sense) means a very high level of consumption. The per capita and material consumption level in rich (Northern) states is 10–20 times as great as that in the poor (Southern) countries. It is this circumstance that creates practically insoluble socio-economic controversy between the North and the South. The circumstance has become a fact of fundamental significance. It is essential to emphasize here that neither the system of command and control, nor the market economy have proved effective from the ecological standpoint. Gore (1993) presents interesting ideas in this respect.

In the context of the subject discussed and from the viewpoint of the global perspective, the ecological legislation and special environmental protection measures taken in Western Europe, North America and Japan are fundamentally inadequate because: (1) they are geographically limited to the rich countries of the North; (2) they are thematically restricted, since they do not provide effective resource-saving measures relevant to energy, minerals, land and water; (3) they are mostly limited to specific environmental media (air, water, soil); (4) they are expensive (in fact, unacceptable for the poor countries of the South). Therefore, it looks like "pollution control and bureaucratic regulation appear to be incapable of solving the major environmental problems of our time".

1.9.3 Europe's ecodynamics in the context of global change

Since this subject is discussed in detail below (see Chapter 3), we shall dwell very briefly on the most essential issues. Von Weizsäcker (1994) notes that "in the early years of the European Economic Community (EEC) there was virtually no interest in the environmental questions, yet rapid economic development on the American model led to an enormous increase in environmental damage. Energy consumption, traffic, urban sprawl, the use of agrochemicals, air and water pollution and waste (including toxic) all reached hitherto unknown levels, and this at a time which is now recorded in the history books as the happy post-war period of peaceful reconstruction and reconciliation between old enemies". The crude motto "grow or make way" became the EEC doctrine. The victim was the environment. In spite of an intensive development of the environmental legislation, it was only the Maastricht Union Treaty (entered into force in November 1993) that specified economic growth as "sustainable" and "respecting the environment". By the end of 1992, within the framework of the EEC, decisions have been made that determined the creation and functioning of the European Environment Agency (EEA). Today our common goal is integration of the efforts of all European countries for the solution of European environmental problems in the context of global change.

1.9.4 Global change trends

In this connection, very important have become the initiatives of the Club of Rome, and in particular, a monograph by Meadows et al. *The Limits to Growth* (1972), a response to which was a book by Gallopin et al. entitled *The Limits to Poverty* (1974) (see Introduction). The attention of the scientific community and public in general has thus been attracted to the Third World and its two central problems: development and equality. Global ecological problems relevant to climate change, stratospheric and tropospheric ozone dynamics, biodiversity, etc. require urgent consideration. To avoid repetition, we shall refer our reader to numerous abovementioned publications discussing these problems in every detail. The set of problems relevant to the global ecological crisis in the context of its five most important aspects are treated in Part II of von Weizsäcker's (1994) book.

1.9.5 Energy and material resources

During the current century, an economy based on science and technology developed intensively, and a close correlation established between prosperity and energy consumption. After World War II, when the horror of the atomic bomb was tempered by the promise that swords would be turned into ploughshares and that nuclear power would only be used for peaceful purposes, it seemed that energy was enthroned as the queen of the economy and progress. Then, in October 1973, a petroleum crisis broke out when the OPEC countries increased fourfold the world oil prices. This stimulated such a rapid and effective progress in the development of energy-saving technologies in industry that, after only a few years of the "energy crisis", it became clear that "economic growth could be elegantly decoupled from energy consumption". However, in spite of this, the goal of energy consumption reduction remains most urgent, since the use of not only non-renewable, but also renewable energy sources entails a negative impact on the environment (even wind power, generally considered environmentally harmless, turned out to be hazardous to birds). Assessments of the perspectives of nuclear energy production are very controversial. Of key importance becomes the increasing efficiency of energy use (it has been mentioned above that certain opportunities exist to improve the situation) and relevant economic stimulation.

According to the classical theory of economics and under the circumstances of the traditional free-market economy, energy prices should reflect the meeting point between the demand and supply. With the command and control system of economic management in the former USSR and Eastern Europe, energy prices were made too low; this (and other factors) resulted in an economic collapse. However, under the circumstances of market economies, energy prices are also much too low because they do not tell "the ecological truth", i.e. they do not reflect all material losses due to long-term consequences of the stress on the environment. Hence an issue of key importance arises—that of the necessity for introducing an additional "environmental tax" (the problem will be discussed at length in what follows). An essential (and, at first sight, paradoxical) fact is that the analysis of the level of economic development for twelve industrially developed countries of the world revealed a correlation between this level and energy prices.

The situation is very similar with material resources, first of all because there is a limit to such resources. Although enormous progress in such spheres as geology, mining, transport, processing, etc. was perhaps overshadowed by developments in nuclear power production, space technology, microelectronics and information technology, biotechnology and medicine, that progress was much more important for everyday life, including its impact on the environment. For example, the world has actually gone through a long period of exponential growth of human-driven material flows, and among them the production of metals. Obviously, the exponential rise in the use of raw materials is disastrous for nature. Hence numerous questions which are so difficult to answer. Where is nature still intact? When and by whom will the ever-growing mountains of waste be cleared up? What is the ultimate fate of manufactured goods produced today and, moreover, during the coming decades? (Gorshkov (1995b) provides convincing evidence that the entire amount of material production is nothing else but delayed waste.) What shall we do with tons of plutonium, thousands of tons of cadmium and mercury, hundreds of thousand tons of lead and copper, millions of tons of waste generated in the mining industry? In the past, the environmental policy was concentrated on the problem of industrial pollution; in the future the problem of waste disposal will outweigh all others (the chain "from the cradle to the grave").

Just as for energy, the nub of a rational environmental policy should be the adjustment of prices in accordance with the long-term cost materials cause throughout their life. Material prices at present reflect only mining and transport, refining, synthesizing and processing costs—and all of these at energy prices which are far too low. All other kinds of ecological expenses are usually latent. In this connection, von Weizsäcker suggests four principles that must determine the strategy of utilizing raw materials and waste: (1) adoption of overestimated prices for secondary raw materials; (2) imposition of levies for waste; (3) adoption of increased prices for energy (which will certainly make production and transportation more expensive); (4) establishment of rules whereby the producer of durable goods (such as cars, refrigerators, etc.) will carry ecological responsibility throughout the entire period of their use, while the customer pays a leasing price (which will stimulate the re-use of the raw materials).

1.9.6 Transport

Like energy, transport plays a central role in economic development. Passenger and freight miles can be used like megawatt hours to indicate economic activity. At the same time, transport is one of the main causes of damage to nature and strain on the environment. For instance, in Germany transport accounts for almost a quarter of total energy consumption, 60% of nitric oxides pollution and 70% of carbon monoxide pollution; transport communications embrace a substantial part of land. Automobile and air transportation go on developing (the EEC's Transport 2000 Strategy favours this development); hence compensating these costs, including ecological consequences, is an urgent task.

In this connection, an important transport dilemma arises: rail or road? There are some estimates which show that, in terms of primary energy consumption per kilometre, the car is less efficient than the train by a factor of 3 or 4; railroads occupy far less land surface than automobile roads. Car-induced emissions of carbon dioxide and other atmospheric pollutants are incomparably greater than those due to railroad transport. However, railways and buses cannot win the battle against the car. (This is partly due also to petrol being relatively inexpensive.) An obvious way out is increasing taxes for petrol and diesel fuel. Of course, such measures are extremely unpopular, and new solutions must be looked for. Among these, there is "obligatory leasing" of cars. In different countries, certain experience has been accumulated concerning stimulation of the use of public transportation in cities. This is done in various forms, including, for instance, prohibiting (or at least limiting) the use of private cars in central regions of cities. In Athens there is a rule allowing the use of private cars every other day only.

1.9.7 Agriculture

It is a well-known fact that arable and livestock farming changed the face of the Earth. Modern agriculture is characteristic of high levels of mechanization, chemical application and specialization. This has resulted in a tremendous growth of agricultural productivity, but, on the other hand, in a destructive impact on the environment. For a long time, this negative side of modern agriculture was denied by the "farmer lobby". It was only in 1983 that Professor Wolfgang Haber published a comprehensive report which summarized the results obtained by the German Council of Experts on Environmental Affairs. The report entitled "Environmental Problems in Agriculture" contained a detailed analysis of ecological aspects of contemporary agricultural practice. In 1993 Professor Haber was awarded the first German Environment Prize (the "Green Nobel Prize").

Top of the list of environmental damage resulting from agriculture is the loss of biodiversity. In Germany agriculture became a major threat to the loss of species and varieties, in fauna and flora alike, for to several reasons (excessive application of fertilizers, pesticides and herbicides, and of agricultural machinery, the effect of drainage schemes, etc.). In less industrially developed countries, the loss of biodiversity was even greater (and it was mostly due to deforestation).

The second major environmental problem in Europe caused by agriculture is pollution of underground and surface waters. The runoff of waste nutrients into rivers and lakes can lead to excessive algae growth and ultimately to more or less fatal eutrophication. The excessive influx of nutrients becomes a real danger to the North, Baltic, Black and Mediterranean Seas. Other negative ecological consequences are soil erosion, loss of fertility, acid rains (it is possible that agriculture is mainly responsible for acid rains in the Netherlands). Finally, agriculture is one of the main energy consumers.

The abovementioned and other negative effects of agriculture on the environment require the development and implementation of a system of measures oriented towards the "ecologization" of agricultural production. For example, the use of chemical fertilizers must be made more expensive. Such measures require additional expenditure and are therefore not easy to undertake. One of possible perspectives is stimulating the use of organic fertilizers alone ("organic farming"). Additional sources of income for farmers should be actively promoted in any modern policy for the rural areas.

1.9.8 The Third World

The situation in the Third World requires special attention because it is this part of the world that mostly contributes to environmental destruction on a global scale. Perhaps 90% of species extinction, soil erosion, forest and wilderness destruction and desertification are taking place in developing countries (with Russia and other CIS states rapidly joining the Third World). In the solution of nature protection tasks in countries of the South, it is necessary to take into account their peculiar features, such as: (1) location mainly in climatic zones where processes of wind and water erosion of the soil are particularly intensive; (2) predominance of natural species of plants and animals, as well as rich biodiversity; (3) a severe imbalance in trade between industrialized and developing countries, and predominance of raw material resources in the export of the South (at present, the export of biomass to industrial countries exceeds by about an order of magnitude that of the colonial period); (4) the ecological inaccessibility of new, ecologically acceptable technologies for developing countries; (5) a strong dependence of the rural population on the direct use of local natural resources (forest, water, soil); (6) the growing population and increasing need in the intensive exploitation of natural resources; doubtless, the rich countries of the North must share the responsibility for these problems (especially, in the context of the trade imbalance): "To a considerable degree, poverty and other reasons for environmental destruction have their roots in unequal North–South relations" (von Weizsäcker, 1994).

Of course, population growth is a very serious problem in the Third World, although there are statistical data available which show that during the period from 1965 to 1991, the number of children per family reduced from 6 to 3.9 in developing countries (the issue has been the subject of detailed discussion during the UN Conference on Population and Development, Cairo, September 1994; see Introduction). On the other hand, it should be emphasized here that the 0.6% annual population growth in OECD countries creates a stronger ecological stress on the global environment than the countries of the South.

A great danger for the global ecological situation is the traditional international division of labour between the North, the producer and supplier of high-technology production, and the South, the main source and producer of raw materials. Low prices for raw materials have greatly contributed to the so-called "debt crisis" of the Third World, where numerous countries have debts increasing on a gigantic scale (sometimes astronomical). Since these countries can pay their debts in one way only—by over-exploitation of their natural resources—it follows that this debt crisis is actually a global ecological crisis which is more and more aggravated. Unfortunately, no common concept of overcoming this critical situation has been worked out thus far. The Second UN Conference on Environment and Development has not produced a concept of this kind either. Von Weizsäcker (1994) is absolutely right to comment that *Agenda 21*, the basic document produced by UNCED-2, "is a fairly conventional bag of growth wishes, although with some relation to environmental protection, but in no way does it address the paramount global problem, the unsustainable lifestyles prevailing in the North".

The Western (and Japanese) model of development fascinates people in developing countries. However, as has been mentioned above, it is impossible to use this model all over the world. If one just imagines the possibility of a worldwide distribution of the Western per capita consumption level for such resources as water, soil, minerals, a global ecological catastrophe will be unavoidable. Moreover, in Professor von Weizsäcker's opinion, if this level of consumption is not decreased, "the Earth will suffer irreversible and disastrous damage within five to ten years", and further: "In order to bring the developing countries back to health, not only is it necessary to correct unfair trade structures but also to wean them from their vision of the North's throwaway society". A first-priority task for the North is to find as quickly as possible and substantiate for itself different models of ecologically beneficiary development. From the ecological standpoint, new models of development in the North must radically limit both pollution, and the consumption of non-renewable resources. With regard to similar perspectives for the South, which entail a more intense exploitation of their natural resources, including transformation of primary forest into farmland, etc., it is high time to stop defining such activity as "development aid" and supporting these efforts financially using Northern funds.

1.9.9 Biodiversity and genetic engineering

Wilson (1992) declared that the destruction of biological diversity was the environmental sin for which future generations would forgive us least, and he compared the biological community faced with present ecological destruction with art lovers seeing the Louvre and other museums in flames but unable to stop the fire. Only about 1.5 million plant and animal species have been scientifically identified up to now. The majority of the yet unknown species is believed to dwell in tropical forests, often under circumstances creating the danger of anthropogenic impact. Approximate estimations have revealed the fact that some 50 species face extinction every day (possibly 100 species because of the increasing rate of tropical forest destruction since the mid-1970s). The present rate of species extinction may exceed by a factor of 1000 the "natural" one, especially in the tropical rainforest areas due to forest fires and intensive logging. The situation is particularly desperate in Madagascar with its unique fauna, where over 90% of the original vegetation has been destroyed, in the monsoon rainforests at the foot of the Himalayas, and in New Caledonia which has 83% of endemic plant species, etc. It is not just forest habitats that are threatened. River systems, swamps, mountain and desert ecosystems are equally under threat, and their biodiversity is continuously reduced. The situation in Europe is somewhat better, but there also are large numbers of species which have been brought to the verge of extinction during the last thirty years.

Very important has been the development and adoption at the UNCED-2 in 1992 of the Convention on Biological Diversity, signed by some 150 states (the USA initially refused to sign the Convention, but then joined the others in April 1993). As far as the ecological significance of biodiversity is concerned, it consists, first of all, in the creation of "free market" mechanisms in the biosphere which are capable of securing the high accuracy of closedness of global biochemical cycles (Gorshkov, 1990, 1995b; Gorshkov and Kondratyev, 1990; Gorshkov, Kondratyev and Losev, 1992, 1994; Gorshkov, Kondratyev, Danilov-Danilyan and Losev, 1994; Gorshkov, Kondratyev and Sherman, 1990). Regrettably, the interdependence of the Conventions on Biodiversity and Climate has not been generally recognized (Kondratyev, 1995a,b; Kondratyev, Belchansky and Petrosian, 1992).

In scientific literature (and particularly in science fiction), genetic engineering is sometimes presented as a hope for the future, especially in relation to the restoration of biodiversity, but in reality it is just an illusion. Contemporary genetic engineering mostly consists in the implantation of the existing genes into other organisms. New genes can also be created (within a narrow range), but without any prospect for assistance in the conservation of the biodiversity. In practice, it is far more significant that the application of genetic engineering is accompanied by the disturbing tendency towards uniformity, monoculture and depletion of genetic variance. Conservation in gene banks is not equivalent to *in situ* preservation under real-life circumstances.

Von Weizsäcker (1994) formulated the following criteria and requirements which should be fulfilled when genetically engineered organisms are released into the environment: (1) the licensing procedures should be open to the public; (2) the granting of permission for a release should be dependent on an environmental impact assessment; 93) the liability for risks involved in releasing new organisms should remain with those responsible for the release; (4) the practice of gene technology should lessen the dependence on chemicals; it must include detoxification of waste and cultivation of sites and areas which are not hospitable to ordinary crops (e.g., salinized areas); (5) certain ethical criteria should be properly defined, such as "no manipulation of human heredity", etc. The most important thing is to understand that biodiversity is not just a commodity (like oil) but a unique gift to humanity from nature.

Having examined five most important issues of the global ecological crisis, which reveal the lack of effective approaches to policies of overcoming or preventing relevant dangerous tendencies of ecodynamics, Professor von Weizsäcker (1994) made an attempt in Part III of his book (Chapters 10–14) to substantiate a uniform strategy based on recommendations offered in Part II, and on fundamental economic measures. The latter, first of all, refers to a new system of price formation ("prices should tell the ecological truth as closely as possible"), a system that must ensure an "efficiency revolution" in the use of scarce natural resources. The objective is a substantiation of "economy-friendly" environmental policies. In this context, of special importance is the problem of cities and adjoining regions.

1.9.10 Prices should tell the "ecological truth"

From the time of Adam Smith, the market has been viewed as the generator of wealth growth. In the early days of the environmental movement, however, it looked as if environmental degradation went strictly hand-in-hand with the increasing wealth. This, in particular, stimulated the development of environmental legislation which is intended to protect the environment from the market "monster". However, even the most industrially and economically developed states have not been very successful in pursuing these policies. Now the task is to substantiate a set of simple rules and mechanisms which can make prices and costs tell the true ecological cost, since the existing market prices are a long way from telling the ecological truth. Under the circumstances, an "externalization" of costs takes place, i.e. the transfer of a substantial part of the real cost beyond the limits of the relationship between the producer and the consumer. The ecological truth, of course, will never be precisely quantifiable in any scientific manner. There are estimates of the external costs arising from environmental degradation in West Germany: about DM 100–200 billion based on the prices of 1985, i.e. 5–10% of that year's GDP. On a global scale, the external effects from burning fossil fuels alone may be about 14% of the global GDP, an incredibly high amount. Such estimates of the externalities can be compared with the annual expenditures for environmental protection measures. For Germany they consist only 1.1–1.2% of the GDP.

One is able to state, therefore, that today's prices represent from only a quarter to one-tenth of the "ecological truth". In other words, if the costs of the use of the resources and environmental pollution were increased by a total amount of approximately 5% to 10% of the GDP, then the market mechanisms could be assumed to work reasonably well to protect the environment.

One of possible solutions is to raise the level of requirements to standards of admissible impacts on the environment. This is a reasonable enough method; however, much more efficient is the strategy of fixing prices reflecting the "ecological truth", and, in this connection, the implementation of the polluter-pays principle. This principle can function, however, only if the following three conditions are fulfilled: (1) the polluter-pays principle should be made increasingly valid in energy production, distribution, transport, agriculture and other sectors of economy; (2) polluters should be subject to financial burdens (e.g. special charges) even when they remain below the permitted levels; (3) market-based economic instruments must be given much more weight in environmental protection.

Beginning with the mid-1980s, the EC and OECD declared that they would give more prominence to economic instruments of the environmental policy. Analysis of over 150 instances of the use of economic instruments has revealed that the main instruments included the following: (1) emissions charges (for waste water, aircraft noise, etc.) to both producers and consumers; (2) product charges (non-returnable bottles in Scandinavia, mercury and cadmium batteries in the Netherlands and Norway, etc.); (3) administrative charges (e.g., for the registration of new chemicals); (4) tradeable permits for emissions into the air or water; (5) deposits (for bottles, drinks cans used, etc.); (6) tax breaks for environmentally friendly products, etc. Additional measures have also been put into practice, such as charges on hazardous substances and subsidies for preventive technologies; tax exemptions for environmental investments, fines for standards infringement; user advantages, e.g. permits to use cars provided with catalytic converters only when smog is forecast; various kinds of environmental taxes, etc. Of particular significance has been Japan's high charge on sulphur dioxide emissions, which was abolished (because it was no longer necessary) in 1989. The money went to a fund of assistance to victims of pulmonary diseases.

Efficient as these "classical" measures have proved to be, they are still insufficient: they cannot solve all environmental problems outlined in Part II of von Weizsäcker's book because they are incapable of considering all long-term ecological consequences. A more radical environmental tax reform is therefore urgently needed.

1.9.11 Ecological tax reform

It is highly unlikely that any government will ever get away with special charges at anything approaching 5–10% of the GDP. The assumption that the total revenue from special charges can attain five times its present size, i.e. 0.5% of the GDP, is perhaps over-optimistic as well. Even if this assumption is adopted, a redistribution of taxes is required in favour of an increased (by a factor of 50–100) ecological tax. The rate and level of taxes should be politically determined and not tied to calculations of the costs associated with the environmental damage or its restitution. Proposals concerning ecological taxes made during the recent years are very different and rather controversial. However, the ecological crisis is becoming more and more acute, and therefore the problem of ecological taxes should be solved; it can also be solved in such a way that it will be economically and politically attractive. However, certain criteria and conditions require justification in this connection.

Environmental taxes should be raised on goods and services which are generally agreed to place the greatest burden on the environment. An attractive way of starting a tax reform might be a systematic cutting of unreasonable tax privileges and subsidies. (Of course, such a reform must be socially just.) Environmental taxes should be simple, "transparent" and easy to administer. They should be introduced gradually, e.g. during 40 years, and they should be internationally harmonized. Following this "gradualist" approach, prices for the ecologically most essential issues, such as energy and natural resources, may be increased annually by 5%, first by cutting all subsidies and tax exemptions, and later raising taxes on these factors. This annual increase may be agreed for a minimum period of 20 years, preferably 40 years; a doubling of relevant prices will then be reached after 14 years, a quadrupling after 28 years, an octupling after 42 years. Other taxes should be reduced by the same total amount.

If such a reform is effected, gradual reduction of the number of energy-consuming and polluting manufacturing units will take place in favour of relatively clean production units. This will secure stable income and employment levels. Candidates for taxation are: fossil fuels based on carbon content (originating carbon dioxide emissions); nuclear energy, possibly differentiated according to safety criteria; water, possibly only above a certain (small) minimum consumption; commercial fertilizers; land used for building purposes; metals and other raw materials; chlorine and other halogens; toxic compounds, etc. If certain substances or factors need to be phased out quickly, increases higher than 5% may be considered.

Of course, the environmental tax reform discussed here is not perfect. For example, its social aspects require a more detailed analysis, since environmental taxes hit the poor harder than the rich. It is feared that ecological taxes may prove to be a too strong load for the economy. But opinion polls carried out in Germany have revealed that the majority of the population supports the necessity of an ecological tax reform. This agrees fairly well with the expectation that ultimately the ecological tax reform is likely to produce macro-economic gains, not losses. The past experience of some countries shows that high energy prices have not prevented these countries from prospering and becoming more competitive on world markets.

1.9.12 Cooperation not harassment

To ensure ecologically favourable sustainable socio-economic development, co-operation between the state, business circles and the broad public is required. This is clear from the experience of many countries. In Professor von Weizsäcker's opinion, five basic criteria may be useful for the assessment of the prospects of sustainable development:

(1) All important changes to the terms of reference should be discussed beforehand with industry.
(2) The state should apply strict self-restraint and limit itself to the setting of the

terms of reference rather than intervene in many details. Where things go wrong, it is a matter for the courts and not the government.

(3) Environment policy should as far as is possible be harmonized internationally. (Being the pioneer in environmental protection has in the past strengthened rather than weakened the international competitiveness of different countries.)

(4) Environmental policy should be planned in a long-term and reliable manner. The timescale of environmental policy should be roughly in line with the timescale of investment decisions, i.e. ten or more years. The planning must be flexible enough and take account of situations demanding immediate reaction (e.g. the "ozone hole" in the Antarctic, etc.).

(5) Environmental policy should accept the criterion of economic efficiency.

On the whole, an economy-friendly environmental policy should be based on three favourite terms in today's business language: voluntary agreements, compensation arrangements and tradeable permits.

Voluntary agreements may be useful in phasing out certain toxic substances and in achieving waste reduction. The experience of German environmental policy is that major voluntary agreements take place only if the state has shown determination to solve the problem rapidly. Compensation arrangements mean that certain obligations can be met by a "trading" agreement at a geographically distant place. For instance, the USA and the Netherlands have committed themselves to afforesting areas in developing countries to make good for the carbon combustion at home. Tradeable permits mean that companies thus obtain the right to fixed levels of raw material consumption or emissions. Tradeable carbon dioxide emission permits would be an instrument for providing international equity, especially if granted on a per capita basis. However, the solution of the problem of monitoring aimed at the control over the use of emission quotas will be extremely complicated.

Sustainable and ecologically safe development of the economy is only possible if a radical structural change is put into effect. In this context, the tendencies of economic development are most controversial. On the one hand, in the years since 1970s, the more prosperous countries experienced a rapid increase in wealth, and simultaneously, their per capita consumption of energy, cement, steel and freight transport—perhaps the most easily measured indicators of environmentally damaging industries—declined. In many other countries, on the other hand, the increase in wealth over the same period was matched by the usual increase in environmental degradation. It would certainly be a mistake to assume that, in itself, progress leads to greater environmental friendliness. The goal can be achieved only under condition of an active ecological policy which can make effective the necessary tax reform and structural change. It should be emphasized that, thus far, no economic means have been found to solve such problems as the extinction of species and varieties, the threat to the climate or the soil, the chemical and genetic risks, and the negative environmental effects that wealthy countries still manage to "export" to the less fortunate states. If these factors were taken into account, the favourable downward trend shown for Germany, Japan and some other prosperous countries would have changed for its opposite.

Negative ecological changes cannot be stopped just by making more efficient the production of energy, steel, etc., or by lowering the level of pollutant emissions to the atmosphere. The goal can only be achieved on the basis of a total restructuring of the entire infrastructure of modern civilization, which, of course, is a long-term task requiring at least 20–40 years. What is important, however, is that this is a realistic perspective. For over a century, progress implied a kind of technical progress, which meant putting nature to the use and service of humankind, under the circumstances of a "classical" market economy. But Professor von Weizsäcker (1994) emphasizes that "in five years, at the very most, the euphoria at the victory of the market economy will be gone. The market economy will have to stand the test of social acceptability and environmental sustainability". The world of business has no other solution than that based on ecologically favourable structural changes which can only rely on a new form of consensus.

1.9.13 Town and country

It is a well-known fact that the "town-environment" problem becomes more acute as the environmental quality continuously deteriorates in big industrial centres. This is an urgent problem, especially if one takes into account the existing forecasts that the world's population is to redouble in the coming century, and that most of the additional 5000 million come to live in towns. Then we will need at least a further hundred cities the size of today's Mexico City (20 million inhabitants) or a thousand cities the size of Naples (2 million). It is unlikely that these giant cities will be able to free themselves from mass poverty, street gangs and drug epidemics, from traffic chaos or the chronic threats to health due to poor hygiene. That is why the problem of population growth and the standard of living in the countries of the South is so painful.

Special attention should be given to the fact that big cities prosper and develop in a parasitic way. Water, air, food, space for dumping rubbish and for recreation, etc. are all things from which cities benefit, and for which they pay practically nothing. Cities pay nothing for fresh air, which they pollute, and declare themselves generous if they reduce emissions a little, which they do mostly for themselves. Since cities have a strong negative effect on rural areas surrounding them, the problem of big cities should be expanded to that of "urban–rural complexes". Naturally, from the standpoint of economy, to solve the problem of the "urban–rural crisis", the principle "the parasite should pay" must be put into effect. Gradual realization of the ecological tax reform considered above will make it possible to solve part of the problem. However, from the viewpoint of the long-term perspective, what is most important here is a drastic change of our lifestyles which are presently characteristic of excessive consumption, and hence—an enormous amount of waste.

1.9.14 Environmental policy after UNCED-2

Since the results of the Second UN Conference on Environment and Development (UNCED-2) have been broadly discussed in the scientific literature (see also Section 1.7), we shall touch upon the subject only very briefly, and first of all, in the context of the ecological conflict between the North and the South. This conflict is reflected

Table 1.6

	Diagnosis	Therapy
South	The environmental crisis results from poverty in the South, wastefulness in the North, external debts and destructive trade relations.	Sustainable growth in the South; debt relief; fair trade relations; modest consumption in the North.
North	The environmental crisis results from over-population, poor management and lack of technology.	Birth control in the South; environmental impact assessment in development projects; transfer of technology and management skills; debt relief with green conditionality.

in the different understanding of the "diagnosis" and "therapy" as applied to the solution of the problem of the socio-ecological crisis (see Kondratyev, (1995b) and Table 1.6). The documents (decisions) of the UNCED-2 (to a considerable extent, non-binding for the parties involved), open up certain perspectives for eliminating these conflicts; however, they cannot be very effective because they are declarative. For example, the Framework Convention on Climate recommends stabilization of carbon dioxide emissions at 1990 levels until the year 2000 (Kondratyev, 1991c, 1992c). But this is hardly a realistic recommendation for countries which urgently need a rapid economic development. That is why China, whose contribution to global carbon dioxide emissions has already reached the figure of 13%, plans to increase the number of electric power stations operating on coal. The actual achievement of sustainability, equity and fairness in the North–South relations is a goal for a distant future. One of the most painful aspects of this perspective is the solution of the problem of the debts of developing countries which reached astronomical figures. The problem of the "ecological debt" of the North is no less serious.

From the standpoint of searching for possible ways of solving problems discussed above, Professor von Weizsäcker's ideas concerning remote perspectives of the Century of the Environment outlined in Part IV of his book are extremely important.

1.9.15 The problem of green technologies

As a result of rapidly progressing developments in the field of ecologically friendly technologies during the recent decades, technology has become a successful sector of the economy. Thus, the stereotype of technology being "enemy number one" of the environmental movement ceased to exist (at least, in principle). There are amazing practical results: for example, rivers the Rhine, the Elbe and the Thames have practically become clean; the chemical giant Bayer-Leverkusen succeeded in its waste water treatment; great progress has been achieved with regard to the desulphurization and denitrification ("de-nox") technologies which are now established air pollution control technologies, etc. There also is a considerable progress with energy- and material-saving technologies. Between 1973 and 1985, for instance, Japan achieved a reduction in its energy use of 31% (from 18.9 to 13.1 megajoules per constant dollar GNP), while in Canada, with its low energy prices, there was a reduction from 38.3 to 36 only, and three times as much energy is used per unit GNP. Of course, Japan's success has to a considerable extent been achieved at the expense of developing countries by transferring many of its polluting and energy-intensive industries elsewhere. This shows that, instead of solving environmental problems, prosperous countries sometimes aggravate the situation.

A "green revolution" should become the long-term perspective for technological development. Professor von Weizsäcker offers seven criteria to define this green revolution:

(1) Cleanliness (little or no emissions).
(2) Energy productivity and solarization.
(3) Minerals and water productivity (in particular, separate use of drinking and non-drinking water).
(4) Ecological land use.
(5) Information intensity (computerization, communications, etc.).
(6) "Error friendliness" (the capability to withstand erroneous decisions).
(7) Suitability for the informal (e.g., domestic) economy.

Possible ways of implementing the above-enumerated criteria are diverse. The first four criteria can be made effective only by state commitment to the ecological tax reform or through a system of tradeable permits. International regulation can also be very efficient.

1.9.16 The impact of science

A cardinal task of science consists in the development of interdisciplinary aspects of research dealing with the dynamics of modern world: "Science should not close its mind to the destruction of the Earth's biological diversity, to population growth, climatic risks or any of the thousands of local environmental problems, and it should also fight for better funding of the related research". Professor von Weizsäcker has been quite right to note: "It is quite possible that we are today on the threshold of a new revolution which may encompass all spheres of knowledge". Of decisive value become complex developments based on interdisciplinary efforts and close cooperation of experts in natural sciences and the humanities.

1.9.17 Freedom of work

Since nowadays people's spare time increases is increasing as a share of the whole (for instance, in Germany its average value is 90% with respect to the whole, including the sleeping hours), a perspective becomes more and more significant (even within the framework of "realistic politics"—"Realpolitik" in Professor von Weizsäcker's terms) when well-being (including its material portion) will be much less dependent on the fact of having a job or on consumption of the products of the "formal" economy. Hence follows the necessity of realizing the importance of combining the formal and the informal economies. If such a combination is achieved, the GNP will decrease in parallel with the benefits for the environment, health and happiness. Building in insulation to diminish heat loss or installing solar panels on roofs to obtain hot water are examples demonstrating possibilities of saving energy (and consequently, of reducing the GNP). Another example could be the replacement of 30% of the present-day professional medical care by various forms of self-help (including herb medication). Still another example is domestic repairs. In this context, *Eigenarbeit* (work for oneself) will make its way in the future world.

1.9.18 New models of wealth

It has been repeatedly emphasized above that the basis of the present world development is consumption growth, and hence the increased exploitation of natural resources and the increasing stress on the environment. New models of wealth, sustainable and copyable ways of life are diametrically opposed to present reality, and yet inescapable. One of the dangers of practical implementation of these new models may be ecotyranny in the Century of the Environment. Professor von Weizsäcker (1994) suggests three policies to be pursued to prevent this ecotyranny: (1) ecological transformation should start in good time, while there is still plenty of time and room for manoeuvre; (2) environmental policy instruments should be promoted which, besides being effective, leave much room for individual response (such measures should not be repressive); (3) the liberties should be specially protected (human rights must not be eroded; on the contrary, they should be further developed to include the rights of future generations).

Education and culture should play their proper roles in the implementation of the new tasks the world is facing. Proper and full-scale understanding of the arising (and already existing) ecological threats on this planet, which is our common home, must be a real stimulus for the realization of measures that are so necessary but, at first sight, may appear idealistic.

1.9.19 On paradigms of global ecology

A fundamental contribution to the discussion of key aspects of global ecology has become the book *Earth in the Balance. Ecology and the Human Spirit* publishd by Albert Gore, the US Vice-President (see Gore, 1993), although this book has been subject to fair criticism (Kondratyev, 1993a,b; Baden, 1994). The Foreword to the book is conceptual in its nature. It offers an assessment of the Second UN Conference on Environment and Development (the Earth Summit), which was seriously criticized for producing several watered-down documents, but "nevertheless marked a historic turning point in the long struggle to increase international awareness of the true nature of the global environmental crisis". Discussing various consequences of the "collision" between the world civilization and the global ecosystem, Gore offers a very precise characteristic of the current situation: "This collision is occurring principally because of three large changes in the nature of our relationship to the earth: first, the population explosion is now adding the equivalent of China's population every ten years; secondly, the scientific and technological revolution has increased our power to manipulate nature and has vastly magnified our ability to have an impact on the world around us; and third, for reasons discussed in Part II of the book, one way of thinking about our relationship to the environment has changed (not for the better, unfortunately) as we have yielded to the intense modern pressures to deny responsibility for the future consequences of our present actions".

But there are many reasons for renewed hope. UNCED-2 laid the groundwork for important shifts in policies throughout the world designed to stop destruction of the global ecological system. Even though the formal documents signed in Rio were short in substance, the Summit spawned widespread recognition of a powerful change in thinking that is now under way in both industrial and developing countries. Moreover, most participants and observers agreed that this conference would be only the first in a continuing series of worldwide meetings to discuss and implement a new generation of global treaties aimed at promoting sustainable economic progress and healing the relationship between civilization and the fragile ecological system of the Earth.

After formulating an adequate concept of the global ecological situation (it would, however, be worthwhile to analyse its socio-economc roots in a clearer and more comprehensive manner (see Kondratyev, 1990b, 1993a,b, 1995a,b)), Gore mentioned that regrettably President George Bush's administration had not been quite up to the mark during the Second World Conference on Environment and Development because it "not only failed to lead but actively fought against needed progress on the environment", whereas "people are determined to see that our nation does offer leadership on the environment—with vision, courage and a real commitment to the future". And this is taking place when Japan is demonstrating the great effort of its people within the framework of an ambitious "100-Year Plan" aimed at saving the global environment on the basis of a strategy that makes it possible to ensure simultaneously the preservation of the environment and a substantial increase in quality, productivity and economic efficiency. "Many Japanese businesses... are searching for ways to redesign the entire production process, this time with an eye in eliminating unnecessary pollution at every step along the way. What they are finding is that waste in the form of pollution is also economic waste".

In the Introduction to the book, Gore is quite right to emphasize that the "ecological perspective begins with a view of the whole, an understanding of how the various parts of nature interact in patterns that tend toward balance and persist over time. But this perspective cannot treat the earth as something separate from human civilization; we are part of the whole too... And if we do not see that the human part of nature has increasingly powerful influence over the whole of nature—that we are, in effect, a natural force just like the winds and the tides—then we will not be able to see how dangerously we are threatening to push the earth out of balance". One can only agree completely with this kind of assessment which fully coincides with conclusions made in relevant scientific literature (see Gorshkov, 1990, 1993, 1995a,b; Gorshkov and Kondratyev, 1990; Gorshkov, Kondratyev, Danilov-Danilyan and

Losev, 1994; Gorshkov, Kondratyev and Losev, 1992, 1994; Gorshkov, Kondratyev and Sherman, 1990; Kondratyev, 1990b; Kondratyev, Buznikov and Pokrovsky, 1996a; Kondratyev and Cracknell, 1997; Kondratyev, Johannessen and Melentyev, 1996; Kondratyev, Ortner and Preining, 1992; Kondratyev and Pozdnyakov, 1996; Kondratyev and Romanyuk, 1996a,b) and which has been discussed above.

According to Gore's definition, "ecology is the study of balance, and some of the same principles that govern the healthy balance of elements in the global environment also apply to the healthy balance of forces making up our political system. In my view, however, our system is on the verge of losing its essential equilibrium". These fair judgements might be supplemented by the following considerations: (1) in the interaction between the socio-economic development and the environment, the role of the first factor is decisive; therefore, the primary source of global ecological safety should, above all, be international actions of political nature (Kondratyev, 1990a,b, 1991b); (2) the global ecosystem is already beyond the threshold of possible influence on it, and its destruction, started right after the industrial revolution two centuries ago, has already originated the state of a global ecological crisis (Gorshkov, 1990, 1993, 1995a,b; Gorshkov and Kondratyev, 1990; Gorshkov, Kondratyev, Danilov-Danilyan and Losev, 1994; Gorshkov, Kondratyev and Losev, 1992, 1994; Gorshkov, Kondratyev and Sherman, 1990). Consumption per capita is a decisive factor now, and the enormous difference in consumption between the industrial and developing countries determine a situation comparable with the presence of an ecological "time bomb" on a global scale.

In this context, very important is the following conclusion the author has come to in the Introduction: "I have therefore come to believe that the world's ecological balance depends on more than just our ability to restore a balance between civilization's ravenous appetite for resources and the fragile equilibrium of the earth's environment; it depends on more, even, than our ability to restore a balance within ourselves as individuals and the civilization we aspire to create and sustain. In the end we must restore a balance within ourselves between who we are and what we are doing. Each of us must take a greater personal responsibility for this deteriorating global environment; each of us must take a hard look at the habits of mind and action that reflect—and have led to—this grave crisis".

The author continues: "The need for personal equilibrium can be described in an even simpler way. The more deeply I search for the roots of the global environmental crisis, the more I am convinced that it is an outer manifestation of an inner crisis that is, for lack of a better word, spiritual. As a politician, I know full well the special hazards of using 'spiritual' to describe a problem like this one. For many, it is like one of those signs that warns a motorist: Steep Slope—Truckers Use Brakes. But what other word describes the collection of values and assumptions that determine our basic understanding of how we fit into the universe?", and further: "Now no one can afford to assume that the world will somehow solve its problems. We must all become partners in a bold effort to change the very foundation of our civilization. But I believe deeply that true change is possible only when it begins inside the person who is advocating it. Mahatma Gandhi said it well: 'We must be the change we wish to see in the world' ".

The words presented above are not in the least the rhetoric of a politician: they are a reflection of the real state of the matter. The problem actually is how to clearly formulate the obligations of every individual and his role in maintaining environmental stability, and consequently, socio-economic development. The outcome of UNCED-2 demonstrates in a most spectacular way how difficult it is to solve this kind of a problem. The contents of the book considered reflect, to a great extent, the kind of difficulties one confronts because these difficulties are originated by the controversies in the understanding of the nature of the global ecological crisis.

Part I of the book (*Balance at Risk*) is devoted, in particular, to the discussion of a number of regional ecological problems. The programme of Gore's field investigations included a study of the effect of radioactive waste on the fauna of the White Sea, an expedition to the Trans-Antarctic Mountains, participation in acoustic soundings of polar ice at high latitudes of the northern hemisphere, etc. In connection with the cutting down of tropical forests in Amazonia, the author attracts the reader's attention to a most important problem—that of loss of biodiversity (one of its most convincing examples is the "bleaching" of coral reefs in the Caribbean Sea which Gore observed during his underwater investigations). The basic conclusion that can be drawn from the results of numerous observations in different regions of the globe is that indicators of the distress of the global environment are commonly seen anywhere. One can therefore speak about new threats to the global environment which can be classified as strategic ones. Among these, there is a problem of global climate warming. This problem is treated in detail in Chapters 2–4 of the book.

Regrettably, in the context of global climate change, the author doubtless overestimates the role of the greenhouse effect. Since the subject will be treated in detail in Chapter 2 of this book, we shall only mention here that the main problem which has not yet been solved consists in finding a relationship between natural and anthropogenically induced climatic changes. The most important aspect of the problem is the key role of the biosphere in climate formation. Today we hardly have any grounds to state in a general and categorical form that "human civilization is now the dominant cause of change in the global environment"; however, it is a fact that the anthropogenic impact is not only substantial, but can be felt on a global scale (which is especially true where the destruction of the biosphere is concerned (see Gorshkov, 1990, 1993; Gorshkov and Kondratyev, 1990; Gorshkov, Kondratyev and Losev, 1992)). In this context, of conceptual value is analysis of the interactivity between climate change and biospheric dynamics. It has so happened that the UNCED-2 participants have failed to solve the problem; therefore, for instance, the Conventions on Climate and Biodiversity, approved by the Second UN Conference, are documents lacking any interrelationship (moreover, the Bush administration refused to support the Convention on Biodiversity).

While stating very truly that "a choice to 'do nothing' in response to the mounting evidence is actually a choice to continue and even accelerate the reckless environmental destruction that is creating the catastrophe at hand", Gore, obviously contradicting the common truth, and at variance with the documents of the Second

World Climate Conference (Kondratyev, 1992b), declares: "it is clear that doubling CO_2 will in fact increase global temperatures". On the one hand, it is true that "we need to act now on the basis of what we know", and very true is the old Tennessee saying: "When you are in a hole, stop digging", but, on the other hand, the situation is far from being clear, including the actual possibility to "stop digging" (Kondratyev, 1992b; Kondratyev et al., 1996b). In fact, Professor Richard Lindzen of the Massachusetts Institute of Technology is quite right when he says (Lindzen, 1990) that the discussion of the global warming problem has proved to be "a largely political issue", and that the role of the mass media has rather been negative due to propaganda of the stereotype of "global warming"—a largely one-sided point of view.

The idea of a "thermostat" mechanism in climate regulation (Ramanathan and Collins, 1992) seems to be quite reasonable. This is relevant to the fact that an intensive development of the penetrating convection over the tropical ocean begins when the threshold value of sea surface temperature (SST) exceeds 300 K. In this connection, an important question arises: why is the difference between the maximum observed and the threshold SST values as little as just a few degrees, in regions of convection such as, for instance, the western warm basin of the Pacific Ocean? According to Ramanathan and Collins (1992), an answer to this question can be found on the basis of the thermostat hypothesis, viz. that, when the process of the penetrating convection is developing, a thick cloud cover of "anvils" is formed, which is responsible for the increase of the earth's albedo, and consequently, for limiting the SST increase (some critical remarks concerning the hypothesis will be discussed in Chapter 2).

According to the Intergovernmental Panel of Climate Change (IPCC) Report (1992 IPCC Supplement, 1992), basic conclusions relevant to global climate change can be summarized as follows:

(1) A considerable increase of anthropogenically induced emissions of greenhouse gases (GHGs) to the atmosphere takes place: such as carbon dioxide, methane, chlorofluorocarbons, nitrogen dioxide, tropospheric ozone.
(2) Assessments based on the analysis of data of observations and numerical climate modelling, make it possible to consider as the most probable, the following limits of increase in annual global surface air temperature (SAT), with a doubled CO_2 concentration: 1.5–4.5°C.
(3) Inadequate understanding of the regular features of climate determines the presence of many uncertainties in forecasting its possible future changes from the point of view of the value, time and regional spatial distribution of climate changes.
(4) During the present century, there has been an increase of global average SAT within 0.3-0.6°C.
(5) The dimension of climate warming is, on the whole, in fair agreement with the forecasts made on the basis of climate models, but its value is the same as that of the natural climatic change. So, the observed increase of SAT can be attributed to natural variability; on the other hand, this kind of variability and other anthropogenic impacts might have compensated even a more substantial (than anticipated) anthropogenically induced "greenhouse" warming.

(6) Unquestionable detection of the enhancement of the greenhouse effect, judging by observational data, is hardly probable sooner than in 10 or more years.

Michaels and Stocksbury (1992) had every reason to note (although there exists presently a widely spread notion of the possibility of global climate warming due to the increasing greenhouse gas concentration in the atmosphere which contributes to the enhanced greenhouse effect) that more and more arguments are accumulated which indicate a different conclusion: the threat of an ecologically and economically dangerous global warming must be more remote than previously anticipated. This, of course, means that a revision is required of the reliability of those prognostic assessments of climate changes and their consequences that are contained in the Framework Convention on Climate Change (Kondratyev, 1991c).

The necessity for revision of the "greenhouse" stereotype of global warming is first of all determined by two circumstances: (1) the inadequacy of the contemporary theory of climate changes and insufficient reliability of the results of numerical modelling of anthropogenically induced climate change which formed the basis of the "greenhouse stereotype"; (2) the discrepancy between the results of numerical modelling and the surface air temperature (SAT) observation data.

It is a known fact that the SAT increase observed in the northern hemisphere from 1860, took place mainly prior to 1940, and not after World War II when an intensive growth of the GHGs concentration began. The linear trend of SAT in the northern hemisphere after 1935 was close to zero. On the other hand, data for the "water portion" of the southern hemisphere, where the warming should have been slower and weaker, revealed a more pronounced presence of the "greenhouse" signal in climate changes. Statistical analysis of global mean SAT values has revealed that warming had taken place mainly before 1945.

It is also important that the analysis of data on the trends in SAT and SST has revealed the presence of just a little time shift between data relevant to the land and to the oceans. This shows how insignificant is the thermal inertia of the ocean which delays the display of climate warming. Taking into account the influence of urban "heat islands" on the data of some meteorological stations (estimations of the global mean level yielded values less than 0.1°C), it may be concluded that the SAT increase, which has occurred during the last 100 years, constituted 0.35–0.40°C.

When assessing the reliability of the results of numerical modelling of the effect of the GHGs concentration growth on climate, one should take into account, first of all, the continuing inadequacy of the consideration of the two fundamental climate-forming factors: (1) the interaction between the atmosphere and ocean (including its deep layers); (2) cloud cover dynamics, as well as the interaction between cloudiness and radiation. Also important is the "fertilization" effect, the presence of which has been established from the data of observations. For example, an increase in the biomass of European forests during two decades (the 1970s and the 1980s) has been discovered, in spite of acid rains which are harmful for the forest. It follows that a better assimilation of CO_2 by vegetation, as its concentration grows, must hamper noticeably this increase.

Attention has recently been attracted to estimations of climate cooling which must follow the increasing content of anthropogenic sulphate aerosol in the atmosphere due to growing emissions of sulphur dioxide in the atmosphere and subsequent gas-phase reactions of the formation of aerosol (Kondratyev, 1992a; Kondratyev et al., 1996b). There are some indirect observational data that show, for instance, that now the concentration of sulphates in the atmosphere of the northern hemisphere is comparable with that observed after the Tambora volcano eruption, when during a short period there was a drop in SAT of about 1–2°C.

Theoretical assessments indicate that the global mean forcing (cooling) impact due to aerosol must be from -1.5 to -2.0 W/m^2, i.e. it is comparable with the greenhouse effect which constitutes 1.5–2.5 W/m^2. Therefore, it is the influence of aerosol-induced cooling that can account for the fact that, in spite of a more intensive growth of the CO_2 concentration during the second half of the twentieth century, during this period the SAT growth has almost stopped in the northern hemisphere, where the basic mass of anthropogenic sulphate is formed; however, SAT continued growing in the southern hemisphere, although somewhat slower.

The enhancement of the gas-phase formation of anthropogenic sulphate aerosol must have resulted in the growing cloudiness of the lower layer. If this really taken place, it must have been accompanied by the following consequences: (1) increased heating at night (due to the enhancement of both the greenhouse effect and cloudiness); (2) lower heating in the daytime (as a result of the increasing albedo of clouds) and a decreased amplitude of the daily temperature variation; (3) the cloud-induced effect of night-time warming must be stronger (weaker) during longer (shorter) nights in winter (summer); (4) the strongest (weakest) cooling (the daytime albedo effect) must be expected in summer (winter) when the day duration is maximum (minimum); (5) the most considerable intensification of cloudiness and of the abovementioned climatic effects should occur in regions (North America, Eurasia) where sources of anthropogenic aerosol (condensation nuclei) are concentrated. The consequences of the impact of anthropogenic aerosol on climate, which have been enumerated above, are mostly confirmed by the data of observations.

Today the situation in the field of global climate studies can be called paradoxical because: (1) following numerous long-period efforts of many experts in substantiating the global warming stereotype, the IPCC developments have brought about conclusions about the lack of convincing proofs of the "greenhouse" effect being the cause of global climate warming in the previous century, and about the inadequacy of palaeo-analogues as climate predictors (Kondratyev, 1992a; Kondratyev et al., 1996b); (2) the well-grounded understanding is now commonly shared that studies of the internal variability of the climatic system is a first priority item; hence we find in the foreground a study of the climate-forming role of the dynamics of cloud cover (especially, the interaction between cloudiness and radiation), the interaction between the atmosphere and ocean, and the role of the biosphere as an interactive component of the climatic system (Kondratyev and Cracknell, 1998; Kondratyev et al., 1996b); (3) in recent years, a number of interesting investigations in the field of climate diagnostics have appeared, which have introduced an essential change into the earlier notions and assessments of the secular variation of the surface air temperature (SAT) during the past century. A characteristic aspect of the abovementioned paradoxical situation is that the extent of uncertainty of climate forecasting has increased, and the most important factor of uncertainty is the inadequately known sensitivity of climate to external impacts.

With regard to the present-day understanding of the regular features of the global carbon cycle, which determine the carbon dioxide concentration growth, information remains highly controversial on the relationship between such components of the global balance and carbon cycle as emissions to the atmosphere due to fossil fuel burning, changes in land use and cutting down of forests (i.e., the destruction of the continental biosphere), the bioproductivity of the marine biosphere (the "ocean pump"), the cement industry, etc.

Gore (1993) very justly emphasizes: "Global warming, ozone depletion, the loss of living species, deforestation—they all have a common cause: the new relationship between human civilization and the earth's natural balance", having noted somewhat earlier that "this century has witnessed dramatic changes in two key factors that define the physical reality of our relationship to the earth: a sudden and startling surge in human population, with the addition of one China's worth of people every ten years, and a sudden acceleration, which has allowed an almost unimaginable magnification of our power to affect the world around us by burning, cutting, digging, moving and transforming the physical matter that makes up the earth". All this has created a situation where "the problem is not our effect on the environment so much as our relationship with the environment... the real solution will be found in reinventing and finally healing the relationship between civilization and the earth... the key changes will involve new ways of thinking about the relationship itself". One cannot but appreciate this kind of a conclusion.

Summing up the discussion on the problem of global climate change, one is ready to agree with Gore in that "throughout the world, we are witnessing the first stirrings of a new political will to slow down our momentum toward environmental catastrophe. The challenge now is to accelerate the widespread recognition of this pattern and organize to change our current direction—before our momentum carries us past the point beyond which an ecological collapse is inevitable". It should be added that many of the measures required to prevent a global ecological catastrophe are envisaged in *Agenda 21*, although the problem of reaching a consensus on adequate priorities is regrettably far from being solved. This is specifically relevant to the human dimension, and it is only very true that "it is the human factor in all of these feedback loops that is critical to saving the global environment".

1.10 ENVIRONMENTAL PROBLEMS AND THE WORLD BANK

The special role of the World Bank (WB) and Global Environmental Facility (GEF) in supporting efforts aimed at achieving ecologically sustainable development makes it necessary to discuss the range of problems the WB is involved in.

1.10.1 Introduction

Beginning with 1987, when the World Bank took a decision to organize its Environment Department, the WB efforts were started in support of developments in the field of environmental problems based on the substantiation of relevant priorities. In 1992 a fundamental report entitled "Development and Environment: World Development Report 1992" was published, in which the range of problems of socio-economic development and environment has been discussed in detail, proceeding from the realization of the fact that development and environment are closely interrelated. In January 1993 a new position—that of Vice-President for Ecologically Sustainable Development—has been created within the WB structure. This means that the leading role of ecological problems in the WB activity has been recognized. A collective work by Serageldin et al. (1994), as well as an earlier WB publication (*Making Development*..., 1994), offer substantiations of conceptual ideas forming the basis of relevant activity. The basic peculiar feature of this kind of ideas is taking into account the necessity of an interdisciplinary (systems) approach to the solution of sustainable development problems by joint efforts of sociologists, ecologists, economists and other experts in the fields of natural and social sciences.

In the context of documents adopted by the Second UN Conference on Environment and Development (UNCED-2), Serageldin et al. (1994) have been quite right to emphasize that in the absence of more concern about the environment, the socio-economic development will be made difficult, and that if the development of poor countries is not accelerated, political efforts aimed at the solution of environmental problems will fail.

There are well-known sad facts, such as that, for example, the standard of living of approximately 1 billion people is beneath the threshold of profound poverty (and another 2 billion are close to that), that 1 billion people have no access to clean drinking water, and 1.7 billion people to sanitary means. 2–3 million children die annually due to the lack of water or sanitation. And, in spite of this situation, the annual growth rate of the world population is 90 million people. During the present century, the global population number has tripled, the scale of economy has grown by a factor of 20, fossil fuel consumption has increased by a factor of 30, and industry has expanded by about 50 times. About 6 million km^2 of forests have been cut down, beginning in the eighteenth century, i.e. over an area exceeding that of Europe (Kondratyev, Asimakopoulos and Varotsos, 1995).

Serageldin et al. (1994) have emphasized that the Rio Conference has become a signal to the world that, after decades of economic growth at the expense of the quality of the environment, politicians have finally realized the crucial importance of the interrelationship between the environment and development. Mankind must learn to live, taking account of restrictions imposed by the environment, which fulfils two functions: a source of resources and a sink for waste. We must realize that, even if environmental degradation does not reach a level threatening life itself, it may lead to a considerable decline in our world's quality.

If recently the environment and natural resources could be regarded as "free goods" and their unlimited use was growing, today, as noted by Rees (Serageldin et al., 1994), three fundamental ecological principles should be observed:

(1) Man's ecological activity is a subsystem which exists within a larger-scale (but limited) global ecosystem, whose violation leads to the damage of life-support systems important for economic development.
(2) As a result of the increasing economic development and population growth, more and more intensive use of natural resources takes place alongside growing emissions of pollutants, and the process sometimes exceeds the limits determined by the carrying capacities of ecosystems.
(3) Some impacts on the environment relevant to economic development originate its long-term (sometimes irreversible) changes (cutting down of forests is one of the most spectacular examples of such impacts).

This is evidence in favour of the critical importance of a systems approach to the problem of environment and development. In this connection, a most timely effort on the part of the WB has been the support of national action plans of environmental protection (by March 1994, the elaboration of such action plans has been completed in 33 developing countries).

According to Dixon and Steer (Serageldin et al., 1994), the WB efforts relevant to environmental problems should be carried out in the following four directions:

(1) Support to countries in achieving positive relationships between socio-economic development and environment. The target of primary importance here is the solution of the interrelated problems of population growth, poverty and the efficiency of using natural resources.
(2) Support to countries in eliminating negative relationships between socio-economic development and environment (monitoring of pollution and urban environment, environmental protection in agricultural regions, improvement of constitutional structures, etc.).
(3) Assessment and prevention of potential ecologically negative consequences of projects implemented with the WB support.
(4) Solution of global environmental problems (mainly via the Global Environmental Facility—GEF). In this context, projects relevant to the solution of the following problems have been supported: global warming (greenhouse gas emissions to the atmosphere), loss of biodiversity, pollution of international waters, global ozone layer depletion. In March 1994, representatives of over 70 countries supported proposals concerning the restructuring and reinforcement of the GEF; besides, the sponsors agreed to allocate over $3 billion to the Gore fund for the next three years.

1.10.2 Priorities

The UNCED-2 documents considered above, and especially *Agenda 21*, contain ample information about various and numerous environmental problems on scales from local to global. The non-binding conventions on climate, biodiversity and desertification (Kondratyev, 1992b; Desai, 1992) offer substantiations of different concrete measures required to solve the most urgent environmental problems. The Cairo Conference on Population and Development (1994) and the "Social Summit" in Copenhagen (1995) attracted general attention to the urgency of the demography problem. Under present circumstances, which are characteristic of a great number and variety of environmental problems, the substantiation of priorities becomes essential. The problem has been discussed in many recent publications, for example, by Gorshkov (1990, 1993, 1995a,b), Gorshkov and Kondratyev (1990), Gorshkov, Kondratyev and Losev (1992, 1994, 1996), Gorshkov, Kondratyev, Danilov-Danilyan and Losev (1994) and Gorshkov, Kondratyev and Sherman (1990). However, no consensus has been reached on the subject within the international scientific community.

1.10.2.1 Food production

Proceeding from the assumption that in the course of the next 40 years the global population will reach 9 billion people, one can easily see that the production of food must then be doubled. Since many of the sources of the food production growth rate per year (about 2% for the latest 30 years) have by now been exhausted, the task of maintaining this growth rate appears to be rather difficult; to provide the doubling of food production in about 40 years from now, a 1.6% annual growth rate is required. Under such circumstances, there arises an alternative: (1) to intensify production on the land area already used (this kind of approach ensured 90% of the agricultural production growth during the last three decades); (2) to hope for another "green revolution". The latter possibility is naturally difficult to predict; therefore, it is not very reliable; the practical implementation of the first possibility has already resulted in serious (often unacceptable) stress upon the environment (contamination of waterbodies accompanied by their eutrophication, salinization of soils, etc.).

1.10.2.2 Urbanization and pollution

The fact that about 90% of the global population lives in cities or their suburbs originates extremely complicated problems of overcoming or preventing negative social and ecological consequences of this situation. The necessity to provide water supplies for additional 3.7 billion city dwellers during the forthcoming 40 years is only one example illustrating future difficulties. In order to reduce the level of environmental pollution, it is necessary to reduce it by 90% per unit of industrial production before the year 2030.

1.10.2.3 Population growth

Continuing population growth and the increasing number and variety of negative impacts on the environment it creates (cutting down of tropical and temperate forests, reduction of wetland areas, pollution of drinking water sources, etc.) require urgent political decisions capable of preventing the approaching global ecological catastrophe. It should be noted here that conceptual results which are important in this context can be found in the works of Gorshkov (1990, 1993), Gorshkov and Kondratyev (1990), Gorshkov, Kondratyev and Losev (1992) and Kondratyev (1990b). These results, discussed in detail earlier, can be briefly formulated as follows:

(1) Global ecological well-being is determined by the state of the biosphere due to biotic regulation of the environment. The degree of closedness of global biogeochemical cycles is an indicator of this well-being, which is dependent on the level of utilization of biospheric resources, whose relative threshold value is about 1% of the entire biotic production.
(2) Today the threshold level has been considerably exceeded, i.e. the process of destruction of the global biosphere has begun.
(3) Important as it may be to adopt low-waste, resource- and energy-saving technologies, the only radical way of preserving the biosphere at the pre-threshold level of anthropogenic impact on it is the reduction of global population number down to approximately 1 billion people, which can be achieved if mankind follows the principle: "one child per one family".

1.10.3 The notion of "sustainable development"

In continuation of what has been discussed above, one can repeat that in the publication, "Our Common Future", which is generally known as the Brundtland Commission Report, sustainable development has been defined as "development that meets the needs of the present without compromising the ability of future generations to meet their own needs" (World Commission..., 1987). A more concrete interpretation of this notion is required, and it can be performed on a multidisciplinary basis, with the participation of experts in ecology, economics, sociology and other social sciences. Von Weizsäcker (1994) has been absolutely right to emphasize that, in order to make the concept of sustainable development more realistic, it is necessary to formulate "a new set of values, a new culture, a new set of incentives to make millions of actors both in the North and the South act differently... We need to enter and to create a new era of human history". Of crucial importance becomes "the cruel reality of ecological devastation which will confront us in our everyday lives and which will inevitably shape our civilization if present trends of destruction continue for just another decade or two".

Investigating the problem of political decision-making, one can see that some key aspects relevant to four broad categories remain unsolved, viz. assessment of the "cost" of the environment; decision-making in the presence of threshold limitations and uncertainties; politics and constitutional structures; social stability. We shall now discuss these subjects more in detail, following Serageldin et al. (1994).

1.10.4 The "cost" criteria for the environment

For a long time, the development of industrial countries was characteristic of their striving for production growth. Therefore, the model of development adopted by the Third World after World War II was also determined by the priority of economic growth. However, in 1960 more reasonable models were elaborated which took into account such social aspects as redistribution of of profit and mitigation of pauperization. During the 1980s, these models were integrated on the basis of the concept of sustainable development which reflected the growing concern about the state of the environment. Nevertheless, the priority of economic growth is still valid in developing countries; the understanding prevails that environmental problems can only be solved after the main economic goals have been achieved. At the same time, politicians are preoccupied with the problem of securing an ecologically sustainable development. This uneasy problem can obviously be solved in a systematic and multi-level manner, and the process must include different-scale actions from individual projects and branches of economy to macroeconomics and international efforts.

The problem of "cost" criteria is the focal point of the whole consideration. Although it is quite clear in principle that the environmental cost criteria should be substantiated and taken into account in the process of decision-making, many facets of the problem remain vague. In particular, there is a most difficult question of working out methods for quantitative assessment of cost indices because not only the direct value should be considered (e.g., soil fertility or drinking water quality), but also the qualitative consequences of negative impacts upon the environment both today and (what is especially difficult) in the future. Another difficulty consists in the circumstance that in many cases economic estimations are very complicated in principle (this relates, for instance, to the multi-aspect problem of biodiversity). It should also be taken into account that the world of nature has an "internal" value, and it is beyond human capability to estimate its valuableness.

For the substantiation of individual cost indices, various methods were proposed, which could take into account, for instance, expenses for the restoration of natural resources (afforestation, removal of environmental pollution, etc.). It should be emphasized, however, that relevant efforts must be assessed only as the initial stage of the solution of an extremely complicated problem which cannot be reduced to the substantiation of "ecological" costs (von Weizsäcker, 1994). Serageldin et al. (1994) offer a table demonstrating the complexity of the problem discussed (see Table 1.7). Even a brief analysis of the scheme presented by Serageldin shows that it must be more precise. To achieve this, time- and labour-consuming efforts are required. For example, the significance of the problems of habitation zones or threatened species is determined by not only their moral or religious aspects, but also their important role in securing biospheric sustainability (Gorshkov, 1990, 1993, 1995a,b; Gorshkov and Kondratyev, 1990; Gorshkov, Kondratyev and Losev, 1992, 1994; Gorshkov, Kondratyev, Danilov-Danilyan and Losev, 1994; Gorshkov, Kondratyev and Sherman, 1990; Kondratyev, 1990b). The accounting for the "ecological" functions of nature in terms of cost parameters is also a difficult task. For example, it is always hard to assess the consequences of forest cuttings or soil fertility loss.

As is well-known, an integral index of the level of economic development of a country is the gross national product (GNP), which, however, is a very imperfect criterion of a nation's well-being. Steer and Lutz (Serageldin et al., 1994) propose to take into account at least four additional information sources to be able to judge about the well-being of people, viz.:

(1) Inequality and poverty.
(2) Standards of living (child mortality, level of literacy, school education, access to medical aid, etc.).
(3) Environment.

Table 1.7. Parameters of the economic value of the environment

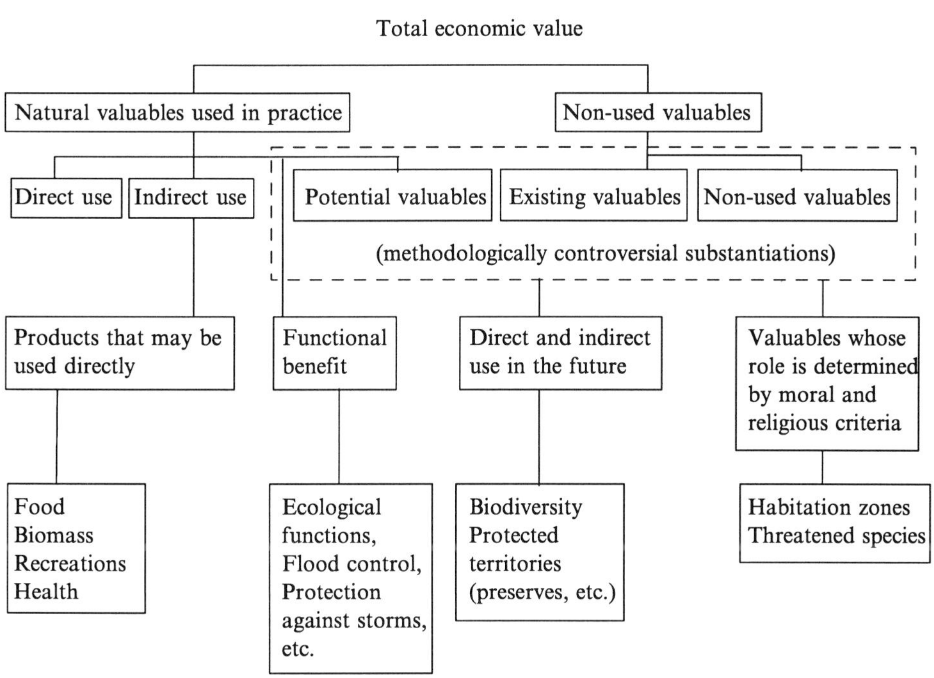

(4) Social, political and spiritual aspects of well-being (the "economic" well-being indicators are incapable of taking account of social, political, spiritual aspects of human well-being, and often it is these factors that are most important).

Steer and Lutz (Serageldin et al., 1994) proposed a scheme of ecologization ("greening") of national expenses (Table 1.8); its main objective is adaptation of of conventional national expenditures taking account of the following three factors:

(1) Impacts on the environment and natural resources whose monetary equivalent can be evaluated over the scale of the entire country.
(2) Impacts which do not have monetary equivalents (health, global warming, ozone layer depletion, etc.).
(3) Impacts which can be evaluated in economic terms.

Of course, practical implementation of such a scheme is an extremely complicated task. Approximate estimates obtained for Mexico (a country with serious ecological problems) revealed that, according to the data for 1985, EDP1 (EDP2) constituted about 94% (87%) with respect to NDP. It is obvious that, to obtain sufficiently reliable estimates in this or other cases, far more comprehensive information about ecodynamics and socio-economic development parameters is required.

Table 1.8. A scheme of ecologization of national expenses (EDP is ecologically adapted national product, NDP is gross national product)

Conventional national expenses	A Physical impacts (without a monetary equivalent) Decrease of natural resources reserves	B Impacts on health, etc. (without a monetary equivalent) Decrease of natural resources reserves	C Impacts determined in terms of a monetary equivalent Decrease of natural resources reserves
	Deterioration of environmental quality	Deterioration of environmental quality	Deterioration of environmental quality
	Dependence "impact-response" Adapted national budgets		Economic assessment
	EDP1 = NDP—decrease of resources reserves EDP2 = EDP1—deterioration of environmental quality		

Estimations of the cost of future ecological losses are certainly most difficult. One may consider that the well-being of children not yet born is no less important than our own; however, economists point out that the interests of future generations will be protected if, for instance a 1 dollar investment today is paid back as a few dollars in the future. Such an approach can be justified from the point of view of short-term perspectives, but it is more than doubtful in a long-term context, especially if ecologically sustainable development is to be secured.

Munasinghe (Serageldin et al., 1994) proposed a classification of environmental value estimates presented in Table 1.9, based on the market approach. The most useful estimates are based on the consideration of the direct response of the market to environmental changes: the influence on production (from the standpoint of the production quantity, quality and cost), the effect on human health (loss of production due to illness and death), protective or preventive measures costs for removal of the consequences of environmental impacts. Both compensation for losses, and "shady" projects imply future expenditures for restoration of the damaged environment. Situations are considered "shady" when projects are implemented to restore damaged elements caused by other projects (for example, expenses for afforestation after building electric power stations). If direct market estimations are impossible, "implicit" market estimates can be used (additional expenses for tourism, increased salaries in polluted regions of zones of natural calamities, the growing cost of real estate property in ecologically clean regions, the market cost of replacing such components of the environment which do not figure in the market because they represent "indirect" goods, etc.). If the market approach is not feasible, it can be imitated in the form of a "designed" market (readiness to compensate for ecological expenses determined by an artificial market, or for unforeseen expenses connected, for example, with drinking water supplies, etc.).

Table 1.7. Classification of environmental value estimates

Regime of market "behaviour"	Market type		
	Conventional market	"Implicit" market	"Designed" market
Based on real behaviour	Effect on production and health. Protective or preventive expenses	Tourism expenses Differentiated salaries Cost of real estate property	Artificial market
Based on supposed behaviour	Cost of compensation for damaged environment. "Shady" projects		Unforeseen expenses

An important circumstance is that the approach to the substantiation of ecological cost estimates must be a systems (multicriterial) approach. One of the most convincing illustrations of this consideration is the problem of drinking water. Although the ecological value of water can be measured, the substantiation of relevant ecological and social consequences in terms of their cost is a task of great complexity; ultimately, the problem should be solved taking account of the optimization of three factors: economic efficiency (in monetary terms), social equity, environmental pollution (water quality).

1.10.5 Threshold limitations and uncertainties

This extremely complicated problem has been subject of broad discussion in the scientific literature (see, for instance, Gorshkov, 1990, 1993, 1995a,b; Gorshkov and Kondratyev, 1990; Gorshkov, Kondratyev and Losev, 1992, 1994; Gorshkov, Kondratyev, Danilov-Danilyan and Losev, 1994; Gorshkov, Kondratyev and Sherman, 1990; Kondratyev, 1990b). However, the relationship between man's economic and other activity and ecological processes has been studied inadequately. In this case, the difficulty consists, in particular, in the possibility of critical ("ruptured") relationships of this kind, when unpredicted and irreversible collapse of ecosystems occurs caused by anthropogenic impacts. Situations unpredictable in time happen even if the relationships discussed are not interrupted. This happened, for example, in the case of the use of chlorofluorocarbons for most humane purposes (for fighting fires or in refrigerators) which resulted, however, in a destructive effect on the ozone layer and the potential threat of a global ecological catastrophe (Kondratyev, 1990b, 1992a). Strictly speaking, anthropogenically induced climate changes ("greenhouse" warming) are still unpredictable because the relationship between natural and anthropogenic climate changes has not been reliably established (or predicted) thus far (Kondratyev, 1992a; Kondratyev et al., 1996b).

One of the most important examples of irreversible and continuous changes is the loss of biodiversity. It should be emphasized here that it is the question of not only irrevocable loss of species and the threat of the loss of genetic diversity; it is the matter of possible violation of the mechanism which ensures the closedness of global biogeochemical cycles, and this may be fraught with a global ecological catastrophe (Gorshkov, 1990, 1993, 1995a,b; Gorshkov and Kondratyev, 1990; Gorshkov, Kondratyev and Losev, 1992, 1994; Gorshkov, Kondratyev, Danilov-Danilyan and Losev, 1994; Gorshkov, Kondratyev and Sherman, 1990). Regrettably, this aspect of the problem has not been reflected in the Convention on Biodiversity (Losev et al., 1993); hence it follows that further efforts concerning the biodiversity problem should be undertaken.

In the presence of a great number of uncertainties, it is most important to stress that such a situation is not a basis for refusing to make decisions; under the circumstances of the contemporary global socio-economic dynamics (population growth, poverty enhancement, etc.), such a refusal may be more expensive than insufficiently justified preventive actions (though certainly, under the circumstances, it becomes very difficult to make decisions, and the problem of choosing priorities becomes of primary importance).

1.10.6 Policy and institutional structures

The main purpose of correct ecological policy is decision-making following the principle of a reliably substantiated system of priorities under the circumstances of limited financial resources. One is obliged to state that, in this connection, the situation in industrially developed countries is as far from being adequate as is the financial support to developing states. For example, the broad support of measures for the reduction of "greenhouse" gas (GHG) emissions to the atmosphere to prevent "greenhouse" warming is a result of vulgar exaggeration. Non-biased data obtained from present-day scientific investigations (Kondratyev, 1992a; Kondratyev and Cracknell, 1997; Kondratyev et al., 1996b) do not offer convincing arguments in favour of such measures; at the same time, such measures are useful because GHGs are also atmospheric pollutants, and freons are not only GHGs but also ozone-destroying gases (see the next chapter). One should not forget about another circumstance: the requirement of reducing the emissions of CO_2 (the main agent of "greenhouse" warming) is hardly compatible with the needs of socio-economic development of countries with a transitional economy or the Third World states. Even the reduction of CO_2 emissions by the year 2000 to their level of 1990 recommended by the Framework Convention on Climate (Kondratyev, 1991c, 1992a) is difficult to fulfil on a global scale (whereas an alternative decision would be inadequate).

A most significant aspect of ecological policy is a foothold upon a system of indicators taking into account human health, various ecological functions and other factors. As has been noted above, however, only the very first steps have been made in the development of such indicators. The nearest perspective consists in the testing of the existing techniques with respect to individual countries. The complicated facet of the matter is the necessity to substantiate a reasonable (optimized) approach to the utilization of market mechanisms and state control, which is certainly very much dependent upon the specific situations in different countries. In this context, very timely could be the search for and support of the most effective institutional structures (the WB has already supported over fifty projects of this kind (see Serageldin et al., 1994)).

Since the human being is the focus of the matter discussed, because man is simultaneously the instrument, the user and the victim, it is extremely important to attract broad population masses to participation in resolving the problems of the environment in the interests of all, and, above all, the poorest part of the population and women.

1.10.7 Conclusion

Efforts undertaken by the World Bank (and the Global Environmental Facility) during recent years to support states with transitional economies and developing countries (in the form of loans and permanent loans) in projects oriented towards achieving ecologically sustainable development have certainly been fruitful. Financial assistance on a large scale has already brought about essential results. However, there exists a problem of further increasing the effectiveness of investments, and its key aspect is substantiation of priorities. In this connection, more representative scientific discussions on an international level would be very useful. It is somewhat disappointing that Russia's participation in such discussions and in the WB programmes has been very limited: paradoxical as it may sound, a single case of supporting a Russian initiative by the WB has been a project on monitoring gas pipeline leaks implemented by "Gasprom". This cannot be considered an adequate solution reflecting either Russia's problems of ecologically sustainable development (see Losev et al., 1993) or the potential of the Russian science and technology in the solution of national and global ecological problems.

A serious drawback to support programmes realized via the World Bank is the underestimation of the urgency of outlining the entire totality of global ecological monitoring problems. The problem of substantiating an optimized system of global monitoring, making use of both conventional, and satellite observation means has not been solved thus far (Kondratyev, 1990b; Kondratyev, Grigoryev and Pokrovsky, 1975; Kondratyev and Pokrovsky, 1977, 1989; Kondratyev and Fedchenko, 1992; Kondratyev, Buznikov and Pokrovsky, 1992). Of course, the leading role in providing the satellite component of the system of monitoring belongs to industrially developed countries (although the contributions of such countries as Russia, China, India, Brazil, etc. must not be underestimated); however, as far as routine (ground-based) observations are concerned, the contributions of developing countries are essential and should be supported.

Of special importance are the development and implementation of the Global Climate Observing System (GCOS) and two programmes complementing it: the Global Ocean Observing System (GOOS) and the Global Terrestrial Observing System (GTOS) (see Kondratyev, 1992a; Kondratyev and Cracknell, 1997; Kondratyev et al., 1996b). As noted by Karl et al. (1995), the documents of the Intergovernmental Panel on Climate Change (IPCC) attract the readers' attention to the unsatisfactory situation with respect to present-time observing means and methods of observational data control; therefore, it is impossible to reliably characterize the climate of the present or the past, or to answer questions about current climatic changes (global warming), water and energy cycles, the circulation of the atmosphere and ocean. Without an answer to these questions, one cannot understand the causes of climate change, still less make climate forecasts. In particular, it has not been possible thus far to establish the proportion of natural and anthropogenic contributions to climate change. Requirements to observational data need an unbiased substantiation; otherwise, adequate planning (optimization) of observing systems is impossible (Kondratyev et al., 1996a).

1.11 NATIONAL GLOBAL CHANGE RESEARCH PROGRAMMES DEVELOPED IN THE USA AND FRANCE

In the context of global ecology priorities, it is of interest to mention Global Change Research Programmes developed in the USA and France.

1.11.1 The United States Global Change Research Program (USGCRP)

Supervised by the Committee on Earth and Environmental Sciences (CEE), intensive research is under way in the USA within the framework of the national Global Change Research Programme (USGCRP). It is a component part of the World Climate Research Programme (WCRP) and the International Geosphere-Biosphere Programme (IGBP) (see Kondratyev, 1990 b–d; Kondratyev and Pokrovsky, 1989), and its main objectives are as follows:

1.11.1.1

Creation of an integrated, comprehensive and long-term programme of documenting the Earth system dynamics on a global scale;

1.11.1.2

Implementation of a programme of purposive (problem-oriented) studies to better understand physical, geological, chemical, biological and social processes governing the Earth system dynamics, including its trends on a global and on regional scales;

1.11.1.3

Development of integrated conceptual and prognostic models of the Earth system.

An essential feature of the USGCRP is the interactivity of its all sections in combination with the central strategic component: a programme of adequately combined conventional and satellite observation means. Conventional (*in situ*) observations, combined with theoretical studies of physical, chemical, biological and geological processes, should be supplemented with a broad programme of space research capable of providing observations of key environmental parameters on a global scale (according to information available, a full-scale functioning of a satellite Earth Observing System (EOS) is planned to be started in 1998).

Proceeding from the recommendations of the International Panel on Climate Change (IPCC) (*Our Changing Planet*, 1990, 1996), the Committee on Earth and Environmental Sciences (CEES) formulated the following main goals of the USGCRP:

1. Stimulate studies aimed at restricting the limits of uncertainties relevant to conclusions on causes and scales of climate change (as marked by the IPCC), especially from the viewpoint of assessing:
1.1. The sources and sinks of greenhouse gases (GHG), knowledge about which determines the level of reliability of predictions of GHG concentrations in the future.
1.2. The dynamics of cloud cover and the radiation budget, which determines, to a great extent, climate changes on a global-to-regional scale.
1.3. Processes in oceans, as well as the interaction between the atmosphere and ocean, which are key climate-forming factors.
1.4. Processes on the land surface which influence the formation of regional climates and water resources.
1.5. Polar ice dynamics which determines potential changes in the world ocean level.
1.6. The dynamics of ecosystems subject to the impact of climate change.

2. Speed up complex studies of greenhouse gases focusing on the following priorities:
2.1. Analysis of the sources and sinks of GHG (not separate, but on the basis of complex studies).
2.2. Development of quantitative indicators of radiative forcing;.
2.3. Substantiation and application of economic models aimed at predictions, taking account of the total set of GHGs (as climate change factors) and various sectors of the economy.

Taking account of the IPCC recommendations (Houghton, Jenkins and Ephraums, 1990; *1992 IPCC Supplement*, 1992), four key directions of interdisciplinary research have been formulated:

I. Numerical modelling and climate forecasting, both of which will be a basis for a broader modelling of the Earth system. In this connection, of decisive value are the improvement of four-dimensional assimilation methods, the parameterization of climate-forming processes, the development of interactive climate system models and their realization making use of new generations of computers. Quite important remain palaeoclimatic studies as a source of information for verification of the reliability of climate models; however, palaeo-analogues should not be used for future climate forecasting (see Kondratyev, 1990b; Kondratyev et al., 1996b).

II. Global cycles of water and energy. In this connection, the main objectives are: a better understanding of water cycles (precipitation, evaporation, evapotranspiration, soil moisture, ice and snow cover dynamics) and energy cycles (climate warming/cooling, radiation budget, latent heat, solar activity). Basic attention should be paid to studying:

II.1. Cloud cover dynamics (phase transformation processes, radiative properties of clouds, etc.).
II.2. Energy- and mass-exchange between the ocean (taking account of ice cover) and the atmosphere; between upper deep ocean layers; transport in the ocean thickness.
II.3. Energy- and mass-exchange between the ecosystems of land and the atmosphere, taking account of anthropogenic impacts on the ecosystems.
II.4. World ocean level changes as indicators of the impact of climate variations on the Earth system (taking account of the thermal expansion of the ocean, the melting of glaciers and the dynamics of polar ice sheets).
III. Global carbon cycle. Relevant research should be focused on the quantitative assessments of natural and anthropogenic sources and sinks of basic carbon compounds on land and in the ocean (e.g., carbon dioxide and such precursors in the processes of formation of methane and ozone as carbon monoxide and non-methane hydrocarbons), taking account of chemical reactions in the atmosphere and ocean involving the abovementioned carbon compounds. Very important is studying fluxes of these compounds due to physical, chemical and biological processes, as well as changes in these fluxes caused by climate variations and the GHG dynamics.
IV. Ecosystems and population dynamics. The main goal in this case is development of methods for more reliable estimations of the impacts of global change over regional scales. First priority is given to the analysis of the response to global change on the part of such ecosystems on land and in the ocean which are subject to strong anthropogenic impacts; the efforts should be focused on the following directions of research:
IV.1. The species composition of ecosystems.
IV.2. The spatial extension and distribution of ecosystems.
IV.3. Productivity of ecosystems.

However, the US Global Change Research Programme is broader than its above-enumerated key directions of research. For instance, studies in the fields of nature-use economics, solar–terrestrial relationships or global ozone dynamics are to be duly supported. As a rule, the research efforts planned are part of relevant international studies conducted within the framework of various international programmes, and above all, the WCRP and IGBP. In this context, essential also are such WCRP subprogrammes as TOGA (Tropical Oceans/Global Atmosphere) and GEWEX (Global Energy and Water Experiment), as well as a number of the IGBP core projects, such as: JGOFS (Joint Global Ocean Flux Study), IGAC (International Global Atmospheric Chemistry Project), BAHC (Biospheric Aspects of the Hydrological Cycle), GCTE (Global Change and Terrestrial Ecosystems), PAGES (Past Global Changes). It should be mentioned that these projects have been discussed in detail by Kondratyev (1990b).

In a recent document entitled *Our Changing Planet* (1996), the following key global change environmental issues have been chosen: (1) seasonal-to-interannual climate variability; (2) climate change over decades-to-centuries; (3) changes in ozone, UV radiation, and atmospheric chemistry; (4) changes in land cover and in terrestrial and marine ecosystems.

It is justly emphasized in the document *Research Strategies for the US Global Change Research Program* (1990) that for the political leaders on national or international scales, those aspects of global change are of primary interest which are relevant to favourable living conditions on the Earth. Nevertheless, the contents of the document hardly reflects adequately problems arising in this connection or possible ways of resolving such problems. Because the programmed goals of global change studies are presented in detail in a monograph by Kondratyev (1990b), it will be sufficient here to attract the reader's attention to just two problems of utmost importance with regard to the USGCRP, consisting in:

(1) The lack of a well-formulated thesis that global security (favourable living conditions on the planet) is determined by the closedness of global biogeochemical cycles (Gorshkov, 1990, 1993, 1995a,b; Gorshkov and Kondratyev, 1990; Gorshkov, Kondratyev, Danilov-Danilyan and Losev, 1994; Gorshkov, Kondratyev and Losev, 1992, 1994; Gorshkov, Kondratyev and Sherman, 1990). It is very good, of course, that the problem of global carbon cycle is among the most essential tasks of the USGCRP; in fact, however, the problem is much more comprehensive. It seems strange that such central problems as water resources and desertification are almost neglected in the USGCRP.
(2) A certain underestimation of the role of socio-economic "roots" of global ecology problems, and especially their most painful issue—population growth on Earth (see Gorshkov, 1990; Gorshkov and Kondratyev, 1990; Gorshkov, Kondratyev and Losev, 1992). It is fully apparent that one cannot analyse the long-term global ecological perspective without proper consideration of current demography problems or taking account of the gigantic contrasts in development levels between the industrialized countries and those of the Third World. Nevertheless, the problem of population growth dynamics is just vaguely mentioned in the US Global Change Research Program. It should be mentioned, however, that in a recent version (*Our Changing Planet*, 1996), much more attention has been paid to "human dimensions".

In this context, it is interesting to follow the evolution of the US Global Change Research Program (USGCRP). Its initial stage (*Our Changing Planet*, 1990) was later shaped into a more comprehensive programme (*Our Changing Planet*, 1996), and certain measures (above all, organizational) were taken for its more efficient implementation.

The Committee on the Earth and Environment (CEE) was replaced by a new Subcommittee on Global Change Research (SGCR) of the Committee on Environment and Natural Resources (CENR). This happened after the US President decided on November 23, 1993, to create the National Science and Technology Council (NSTC). The SGCR prepared the reports *Our Changing Planet* and similar subsequent publications. The creation of the CENR was the turning point in the organization of studies of the environment and natural resources on a federal level.

There are numerous CENR subcommittees which are responsible for various aspects of the totality of ecological problems: global change; biodiversity and ecosystems; land use and management of natural resources; toxic pollution and dangerous emissions; air quality; water resources; coastal and marine environment; mitigation of the consequences of natural catastrophes. The nearest goal of the CENR is substantiation of a coordinated federal strategy of studies and projects in the field of the environment and natural resources. Great importance is attached to interdisciplinary studies which combine the natural scientific aspects of the entire problem and its "human dimension" (i.e. projects in the fields of socio-economic and political sciences). First priority support is stipulated to the implementation of the Earth Observing System (EOS). The funding allotted for the abovementioned projects in the fiscal year 1995 and subsequent budget years constituted about $1.8 billion. Apart from this, additional money was allocated for some other related efforts, for instance, the development of methods and mechanisms to reduce GHG emissions to the atmosphere, etc.

Priorities in global change research. According to the Act on Global Change Research adopted in the USA in 1990, global change is defined as changes in the global environment (including changes in climate, the productivity of land, ocean and other water resources, atmospheric chemistry and ecological systems) which may modify the Earth's capability to maintain life. The following is implied: the understanding and forecasting of causes and impacts, as well as of potential response to long-term climate change and greenhouse warming; stratospheric ozone depletion and increased ultraviolet (UV) radiation at the Earth's surface level; natural seasonal to interannual climate fluctuations. Highly important aspects of global change certainly are such phenomena as desertification, deforestation, land use, preservation of ecosystems and biodiversity.

Although no special emphasis on the "human dimension" of global change has been made at the beginning, sufficient attention has been given to the subject later, and the demand for focusing on the interactive nature of the system "nature–society" has been declared. Regrettably, the well-known overestimation of the role of greenhouse effects on climate in the context of global change is present in the report. As has been noted above (Kondratyev, 1993a,b), the "greenhouse" stereotype can also be traced in the US Vice-President A. Gore's book *Earth in the Balance. Ecology and the Human Spirit* (1993).

The main goals of the USGCRP encompass: monitoring of the global environment, understanding of its changes and development of methods for their quantitative assessment and prediction, analysis of global change consequences. The following efforts are planned to pursue these purposes:

(1) Development and implementation of ground-based, marine and satellite observation systems, making use of the Earth Observing System (EOS) and other space-based means, as well as conventional network and problem-oriented observations.
(2) Development of techniques for global data set management (including archiving, distribution, etc.).

(3) Support of problem-oriented studies of physical, chemical and biological processes in the Earth's system determining global change. The following directions of research are envisaged:
  (a) climate dynamics and hydrological systems;
  (b) dynamics of chemical and biogeochemical processes;
  (c) dynamics of ecosystems;
  (d) anthropogenic impacts, socio-economic aspects;
  (e) past ecological changes;
  (f) interaction with the Earth's crust;
  (g) influence of solar activity;
(4) Development of interactive models of the Earth system intended for predicting the volume, time and spatial coverage of global change, viz.:
  (a) modelling of the Earth system and long-term forecasting;
  (b) seasonal to interannual prediction;
  (c) assessment of the reliability of prediction based on observational data.
(5) Analysis of global change impacts on the environment and society; elaboration of strategies for prevention of undesirable changes, viz.:
  (a) impacts on the environment and potential adaptation mechanisms;
  (b) impacts on socio-economic development and adaptation mechanisms;
  (c) the strategy of preventing undesirable changes.
(6) Working out of patterns of the implementation of national and international policy of responding to changes taking place:
  (a) study of socio-economic and political aspects of global change;
  (b) potentialities of a systems approach to study the system "nature–society";
  (c) decision making and information distribution;
  (d) use of natural resources and management possibilities.

According to the document *Our Changing Planet* (1996), the key aspects of global change include the following:

- *Climate change and greenhouse warming*, which is necessary to estimate the effects of anthropogenic emissions of greenhouse gases and aerosols on the global climate, environment as a whole, and economic activity.

Since recent monographs (Kondratyev, 1992a; Kondratyev et al., 1996b) offer a detailed analysis of the range of problems relevant to global climate change (see also Chapter 2), we shall limit ourselves to mentioning estimates of possible impact of anthropogenic aerosols on climate; such estimates reveal that the value of "greenhouse" warming is close to that of "aerosol" cooling (it should be emphasized here that the reliability of such estimates is doubtful). Observational data have revealed unexpected (and not yet accounted for) variations in concentrations of some greenhouse gases (GHGs) during the recent few years: a slowing down in the carbon dioxide concentration growth (with a subsequent increase); a slowing down in the concentration growth of chlorofluorocarbons (CFCs) which is certainly the result of special measures intended to prevent CFC emissions; a deceleration in the methane concentration growth whose cause is not clear; an increase in the tropospheric ozone concentration (relevant data are not completely reliable); understanding of the necessity of studying the role of fully fluorinated ("perfluorinated") chlorofluorocarbons (PFCs, such as CF_4, C_2F_6 and C_6F_4) whose lifetimes may exceed thousands of years. The decrease of TOC observed during two years after the Mt Pinatubo volcanic eruption evoked great interest (a decrease in the ozone content in the lower stratosphere may cause a temperature drop near the Earth's surface, and hence a certain compensation for the "greenhouse" warming).

- *Total ozone content depletion and solar ultraviolet (UV) radiation enhancement* as a consequence of anthropogenic impacts (emissions of ozone-destructing gases) on the ozone layer and subsequent enhancement of UV-radiation at the Earth's surface level which have a pathologic effect on man and the biosphere. In this connection, a global network for monitoring spectral UV-radiation is urgently needed.
- *Considerable seasonal to interannual climate fluctuations* affecting man's economic activity (above all, agriculture). Special attention should be given to changes in the precipitation regime (originating periods of intensive precipitation or drought) caused by the phenomenon known as El Niño/Southern Oscillation (ENSO) occurring in the tropical Pacific. One of the nearest goals of the USGCRP is ENSO forecasting one year in advance. In this context, the Seasonal-to-Interannual Climate Prediction Program (SCPP) has been worked out whose two main purposes are consideration of the interactive system "atmosphere–ocean" and taking account of socio-economic consequences of climate change. One of the key components of the SCPP is creation of an international network of research centres to carry out relevant research and applied efforts. The abovementioned priority directions of research do not exhaust all projects of primary importance.

Socio-economic and political aspects. The new stage of developments within the USGCRP makes provisions for focusing on its socio-economic and political aspects. In particular, stronger support is planned to research in the area of global change impacts on human health, economy, urban settlements and social structures, as well as to efforts aimed at the substantiation of adaptation strategies and preventive measures for mitigation of global change consequences. The ultimate aim is achieving a closer link between the results of research and political decisions. In this context, an important role belongs to the development of integrative models (IAM) for the assessment of socio-economic consequences of global change; these models can be both global- and regional-scale, and they must be able to reproduce adequately the interacting physical, chemical, biological and geological processes taking account of the dynamics of socio-economic development. Of great interest are, for instance, the estimates of admissible threshold impacts on the environment. At the initial stage of regional modelling, the regions of the Great Lakes, Great Plains and the south-east of the USA are supposed to be considered. Here are some priority complex regional- and global-scale projects.

- *International trends of population number dynamics and the standard of living:*
  - interaction between population growth, migration and environmental changes;
  - demographic and social factors which are particularly important from the standpoint of societal vulnerability in response to the impact of environmental change;
  - the influence of environmental change impacts on human health from the viewpoint of the purchasing power and living conditions of the population.
- *The structure of trade and global economic activity:*
  - the influence of global economic growth and the structure of international trade on the use and valuableness of the environment resources such as wetlands, coastal zones and forests;
  - the influence of economic relations on global environmental resources;
  - the effects of the development and implementation of new technologies on global change;
  - the role of changes such as political and economic liberalization in the use of environmental resources.
- *Adaptation and attenuation of anthropogenic impacts, including the utilization of environmental resources and their management:*
  - the experience of the past and the solution of problems of adaptation to ecological stresses;
  - expenses and useful results relevant to the application of various approaches to the policy of using such natural resources as soil, energy, water and coastal zones;
  - the influence of institutional and legal measures on the use of resources which are public property;
  - international mechanisms and processes that can be used to form an international coalition which is necessary to ensure a stable level of greenhouse gas concentrations.

Unlike the economy which is regulated by market prices, the environment allows no "simple" mechanisms of the calibration of relative value of its "products". Therefore, alternative paradigms are required for economic assessments in the field of environmental economics which must be substantiated taking account of social and behavioural factors reflecting the psychological, cultural and ethical "dimensions" of environmental value characteristics. Within the framework of the USGCRP, studies are planned of global change impacts on the environment and society, including the solution of problems of agriculture, protection of human health, accessibility to water and its quality, etc. The analysis of measures for the prevention of undesirable impacts is concentrated on the necessity of reducing emissions of greenhouse and ozone-depleting gases and intensifying their sinks. Among the central adaptation goals, there are developments of new technologies ensuring the easing of anthropogenic stress on the environment. Independently of the USGCRP, and action plan has been adopted relevant to the global climate problem (the 1995 national budget provides its funding totalling to $280 million) which makes provisions for 50 projects in the fields of industry, transport, forestry and agriculture, electric power consumption, etc., all of which are aimed at the reduction of GHG emissions.

Important problems are being solved within the framework of the International Human Dimensions of Global Environmental Change Programme (IHDP), including: (1) land surface changes during the last 300 years due to land use; (2) analysis of individual anthropogenic impacts in the process of land use, and of their geographic specific features; (3) assessments of possible influence of peculiar features of land use in the past which may be felt during the next 50 years; (4) the influence of global climate change and biogeochemical processes on land use.

Information obtained as a result of these and other investigations should be concentrated at the Consortium of the International Earth Science Information Network (CIESIN) to be distributed through the Socio-Economic Data Center (SEDAC) network.

In this context, it is most important to develop complex ecological models suitable for prognostic assessments of the Earth system dynamics, including socio-economic aspects. The following efforts are at present under way:

— A pilot project has been undertaken to train a group of experts from different countries in the field of seasonal-to-interannual climate variability prediction on regional scales for the purpose of decision making on the management of natural resources which are subject to the effect of ENSO.
— Databases of ENSO observation data have been considerably broadened and supplemented.
— To advance numerical modelling of global change (especially, the general atmospheric and oceanic circulation), parallel supercomputers have been used.
— A preliminary analysis of the results of the implementation of the Atmospheric Model Intercomparison Project (AMIP) has been conducted which revealed a close agreement in predictions of the latitudinal distribution of surface pressure and wind in the upper troposphere, a fair enough conformity between the main latitudinal zones of heavy and slight precipitation, and a fairly good correspondence between the results of numerical modelling and observations of atmospheric processes for the period 1979–88.

The plans for 1995 envisage, in particular, an intensification of efforts aimed at a better understanding of the physiological and ecological response of plants and animals to global climate change, atmospheric chemical composition and UV-B radiation enhancement. The goal of the Global Marine Ecosystem Dynamics Program (GLOBEC) will be analysis of the relative role of physical and biological processes in the ocean responsible for the dynamics of fish populations and zooplankton under the influence of global climate change. The study of the effect of UV radiation on the carbon cycle in the ocean in connection with global warming is planned. Great attention will be paid to the analysis of climate change impacts on regional levels, including the problem of biodiversity and other relevant subjects.

Of course, the implementation of the USGCRP can be effective only if efficient international cooperation is achieved. Education and dissemination of information must certainly be given serious attention.

- *Global change monitoring*. Further development of programmes of global change observations is one of the focal points of the USGCRP. This is especially relevant to the use of space-based observation means (the corresponding debit item of the budget envisages the biggest financial support), among which the central place belongs to the Earth Observing System (EOS).

— *The Earth Observing System (EOS)*. To avoid repetition, we shall dwell on the most significant components of the EOS, whose purpose is providing information required for the solution of the following problems:
  (i) energy and water cycles (in the context of atmospheric and oceanic dynamics) which control the temperature field, as well as processes of the formation, development and dissipation of cloud cover (taking account of the interaction between cloudiness and radiation);
  (ii) biogeochemical processes contributing to the formation, transport and transformation of optically active minor gaseous components (MGC) of the atmosphere and aerosols;
  (iii) climatological processes which govern the interactions of land and ocean surfaces with the atmosphere through the transport of water, heat, mass, and momentum;
  (iv) ecological processes as an interactive component of global change;
  (v) geophysical processes which have shaped and continue to modify the Earth's surface (volcanism, the melting of glaciers and sea ice, etc.).

The data of Table 1.10 characterize the main objectives and implementation stages of the EOS, a joint effort of the USA and some other countries and institutions, such as Canada, Japan, the European Space Agency (ESA), etc. The

Table 1.10. EOS implementation stages and main mission objectives

Satellite	Key observations
EOS-AM, "morning" crossing	Clouds, radiation, aerosols, land and ocean surface temperature, vegetation cover, ocean phytoplankton, global bioproductivity, tropospheric chemistry
EOS-Color	Ocean phytoplankton and bioproductivity (in the afternoon)
EOS-Aerosol	Aerosols, stratospheric chemistry
EOS-PM, "afternoon" crossing	The same as for EOS-AM
EOS-Altimetry	Ocean circulation and level, topography of glaciers and land surface, stratified structure of cloud cover
EOS-Chemistry	Tropospheric and stratospheric chemistry, aerosols, solar radiation

first satellite EOS-AM is to be launched in 1998. The EOS-AM observational data will be concentrated at the EOS Data Center (EOSDIS) and will be easily accessible.

— *Other satellite observation means.* Apart from the EOS system satellites, an important source of global environmental information is a number of satellites which are in operation and other space vehicles whose launching is planned for the nearest years.

For example, a large volume of data on the total ozone content (TOC) has been obtained with the Total Ozone Mapping Spectrometer (TOMS) installed on the satellites "Nimbus-7" and "Meteor-3" (launched in August 1991). The latter's data reveal that, up to the summer of 1993, the lowest TOC values took place, after which they decreased to values registered before 1992. Similar observations were continued in 1994 from the Earth Probes Satellite and in 1996 from the Japanese satellite ADEOS.

Ample valuable information concerning the upper atmosphere chemistry has been obtained from the data of the UARS satellite launched in 1991 (in particular, very high concentrations of ClO have been discovered in the Arctic and the Antarctic in winter-time in the stratosphere, within the circumpolar vortex; global mapping of polar stratospheric clouds has been conducted for the first time).

Very important information on the concentrations of various MGC in the stratosphere has been obtained from the data of instruments launched on the Shuttle within the framework of the ATLAS Programme (in particular, this relates to the UV backscatter spectrometer SSBUV for measuring TOC). Of great value are developments of active radar sounding means, such as the radioaltimetry instrumentation TOPEX/Poseidon, intended for high-precision measurements of the ocean surface level height, studies of tides and oceanic circulation, as well as the video radar SIR-C (the Shuttle) for hydrological, geological and oceanographic investigations, the satellite lidar LITE (the Shuttle) for remote sensing of clouds and aerosols, the SeaWiFS instrument for measuring the ocean colour necessary to retrieve its optical parameters. The abovementioned Japanese satellite ADEOS will carry the NSCAT scatterometer intended for the retrieval of the wind speed field at the ocean surface. According to the American–Japanese joint programme TRMM (Tropical Rainfall Measurement Mission), a satellite is planned to be launched in 1997 for studying the processes of energy- and water-exchange, precipitation formation in the tropical zone. Observations of the solar constant (in particular, from the Russian satellite "Meteor-3") and the Earth's radiation balance (also from "Meteor-3") will be continued. The NOAA and Landsat satellites are most valuable sources of comprehensive information for determining various characteristics of land surface and snow-ice cover. The creation of the Global Climate Observing System (GCOS) is a project of extreme importance (Kondratyev, 1992a; Kondratyev, Buznikov, Pokrovsky, 1992).

An important component of the global system of ecological monitoring are various goal-oriented programmes of field observations. Among these there are: the

Atmospheric Radiation Measurement Programme (ARM), the Tropical Oceans/Global Atmosphere Programme (TOGA), the World Ocean Circulation Experiment (WOCE), the Global Energy and Water Experiment (GEWEX), the Global Ocean–Atmosphere–Land System Programme (GOALS), the Climate Variability and Predictability Programme (CLIVAR) which is very significant as a component of the WCRP, etc. The International Geosphere–Biosphere Programme (IGBP) and its core projects must greatly contribute to the solution of numerous problems relevant to global change studies. A detailed discussion of the abovementioned and other programmes can be found, for instance, in monographs by Kondratyev (1990b, 1992a) and Kondratyev and Cracknell (1997).

Conclusion

The US National Global Change Research Program (USGCRP) which is under way at present, obviously reveals the fundamental contribution of the USA to the solution of a broad range of problems involved, and particularly in connection with today's better understanding of the importance of the "human dimension". What is regrettable, however, is a certain narrowness of methodological sources for substantiation of the programme, which are limited by the national scope. So, for instance, the authors of the report *Our Changing Planet* (1996) have not been aware of numerous Russian publications devoted to the substantiation of priorities of the entire problem of global change (see Kondratyev, 1992a; Marchuk and Kondratyev, 1992; Gorshkov, 1995), and publications summarizing relevant results (such as Kondratyev, 1995a,b). Here are basic conclusions drawn in these publications (their consideration would be certainly beneficial for the USGCRP):

(1) A methodology for the optimization of systems of global and regional change observations has been proposed; making use of the methodology can make such systems much more adequate and economical (see Kondratyev, Buznikov and Pokrovsky, 1992).

(2) The key aspect of the global change problem (the level of consumption of biospheric resources) has been quantitatively substantiated, and the admissible upper limit of impact on the global biosphere has been assessed, which, in its turn, resulted in estimating the "ecological carrying capacity of the Earth" from the standpoint of optimal population number (see Gorshkov, Kondratyev and Losev, 1992; Gorshkov, 1995). Of course, these results should be further developed and made more precise. Ignoring them, however, may result in a gap in global change research because some of its aspects of crucial importance will be missing.

1.11.2 The Global Change Research Programme of France

Of great interest are similar research efforts developed in France, a brief review of which can be found in the report (*Scientific Orientation*..., 1990). The national programme of environmental studies developed by the French National Scientific Research Centre (CNRS), which was started in 1991, is based on the idea of the necessity of complex efforts integrating research in natural and social sciences. These will include the following directions of research:

(1) Law and economics (where relevant to environmental problems).
(2) Environmental problems in their historical development.
(3) Analysis of the most important natural systems and their evolution (land hydrology, coastal environment, urbanization, etc.).
(4) Understanding of the main mechanisms and evidence of anthropogenic impacts on the environment (the response of populations and ecosystems to the impacts, biological intrusions, etc.).
(5) Key aspects of global change (the atmospheric component of global biogeochemical cycles, global-change-induced dynamics of ecosystems, etc.).

All these and other efforts should be primarily oriented towards the solution of human ecology problems, since human health must be the focal point.

One of the most difficult problems is substantiation of the priorities of individual projects. With regard to ecosystem studies, special attention should be paid to the following: tropical and moderate latitude forests; steppe and savannah; land hydrological systems (rivers and lakes), seas and oceans; transitional zones (ecotones); urbanized systems; agroecosystems; "stress" ecosystems (deserts, the Arctic and the Antarctic, high-mountain regions, etc.); synthetic ecosystems (e.g., orbital stations).

The CNRS Programme puts special emphasis on the significance of "human dimensions" of ecological problems, and on the urgent necessity of more intensive efforts in this respect. It stresses the need in the analysis of the compatibility of demographic and economic growth under the circumstances which can be characterized by increasing loadings upon the environment.

In conclusion, one is obliged to state that "programme provisions" for the range of global ecology problems are inadequate, and efforts are insufficient both in individual countries and on the part of international organizations. This provokes the suggestion to discuss global change problems (above all, from the viewpoint of providing favourable living conditions on the Earth, and not only in the near, but also in the remote future) at a conference called by the United Nations Secretary-General. A conference of this rank might be able to define the priorities and then stimulate the development and implementation (on the international basis) of an adequate programme directed towards the investigation and solution of the global change problem. It is regrettable that these goals cannot be fully achieved through the International Geosphere–Biosphere Programme (the subject has been discussed earlier by Kondratyev, 1987; 1990c). One is also obliged to state that adequate progress has not been made in the documents adopted by the Second World Climate Conference either, in particular, in its decision to prepare a Framework Convention on Climate Change (FC) (see Kondratyev, 1991c). The Second UN Conference on Environment and Development has not been very successful either. For instance, its decisions relevant to climate reveal that the stereotype of global warming due to growing greenhouse gas concentrations prevailed again. Hence the call for reducing GHG emissions prevailed over the unbiased (and justified by the IPCC) accounting for the numerous and diverse factors determining global climate change. That is why further relevant discussions are urgently required.

1.12 THE INTERNATIONAL GEOSPHERE–BIOSPHERE PROGRAMME (IGBP)

1.12.1 Key issues

The evolution of the geosphere and biosphere can be characterized by ever-increasing interaction between their components, which is due, above all, to the growing anthropogenic impact on the environment. This impact is first of all revealed in the form of the transformation of nature carried out by man, and in the more and more enhanced anthropogenic loading on the environment, which can be traced in practically all components of the geosphere and biosphere (in some cases, on a global scale). Hence the need in a systematic approach to the study of the processes under consideration (Gorshkov, 1990; Kondratyev, 1990c; Kondratyev, Asimakopoulos and Varotsos, 1995). To reveal certain laws of the impact of geospheric and biospheric components, interdisciplinary studies of unprecedented complexity are required; so the whole range of relevant problems is brought to the foreground of contemporary natural science.

The scientific community has been aware of the necessity for complex studies of the geosphere and biosphere for a long time. During recent decades, this awareness has revealed itself in a most spectacular way through the implementation of the following scientific programmes: the International Geophysical Year (IGY), the International Quiet Sun Year (IQSY), the World Weather Watch (WWW), and the Global Atmospheric Research Programme (GARP). Recently, the large-scale Middle Atmosphere Programme (MAP) has been launched, as well as the International Lithosphere Programme, the World Climate Programme (with the World Climate Research Programme (WCRP) as a component), specialized International Satellite Cloud Climatology Project (ISCCP) and the International Satellite Land Surface Climatology Project (ISLSCP) and a large-scale Russian programme "Sections" aimed at studying the role of oceans in short-term climate changes. There are two new international programmes with the same objective: The World Ocean Circulation Experiment (WOCE) and The Tropical Ocean and Global Atmosphere (TOGA) programme. Some important international projects are under way which are accomplished under the aegis of SCOPE and UNEP, and are aimed at studying the biogeochemical cycles of various atmospheric components. The American Programme on Global Tropospheric Chemistry, the International Hydrobiological Programme and the Man and Biosphere Programme (within the framework of UNESCO) are certainly very promising.

The following directions of research within SCOPE illustrate the variety of the IGBP problems:

(1) The cycles of carbon and nutrients in lakes and estuaries.
(2) The sulphur cycle in the continental and water ecosystems.
(3) The processes of acidification in tropical countries.
(4) The phosphorus cycle in the continental and water ecosystems.
(5) The budgets of organic substances.
(6) The effect of pesticides.

(7) Organisms generated by genetic engineering in the environment.
(8) The lowering of coastal zones and rising of the sea level.
(9) Ecotones in the varying environment.
(10) The exchange of minor gaseous components between the biosphere and atmosphere.
(11) The ecosystem experiments.
(12) The pollution of subsurface waters.

The following projects were also under consideration:

(1) Flow of particles into the ocean.
(2) Studies of the problem of genetic engineering and the environment.
(3) Long-term ecological investigations.
(4) Biogeochemical metabolism of small water catchment areas.
(5) Use of scientific information for resolving problems relevant to sustainable development of the biosphere.
(6) Biogeochemical "trajectories" of radionuclides.
(7) Effect of climate change on human health.

The Russian programme of biospheric and ecological studies is broader still.

Each of the programmes mentioned above (and many that have not been mentioned) are characteristic of a complex approach to the solution of relevant problems. However, analysis of the set of problems relevant to global ecology shows that a closer coordination of efforts is required and that those efforts should be oriented towards achieving the particular goals of various programmes, and the substantiation of a kind of "superprogramme" encompassing the key aspects of geospheric studies.

It is these circumstances that have stimulated the development of the long-term International Geosphere–Biosphere Programme (IGBP) (the full-scale development of observational systems was started in the 1990s). Its main goal is the study of global changes in the geosphere, especially from the viewpoint of the anthropogenic impact on the biogeochemical cycles of carbon, nitrogen, sulphur, phosphorus, water, and the dynamics of such life-supporting factors as radiation, the quality of air and natural waters, and soil fertility, as well as interactions between biospheric and geospheric phenomena.

The current situation requires, first of all, a proper analysis of key problems of the geosphere–biosphere studies. And, in this connection, the following considerations are of primary importance.

The first one is that, on a time scale of the order of a decade or more, it becomes quite evident that geospheric components compose a single, coupled system. The interaction between individual components of the system is characteristic of the presence of synergism (feedbacks which sometimes lead to a mutual intensification of various processes). It is this circumstance that makes so important the systems approach, the interdisciplinary studies of physical, chemical and biological processes based on the realization of an integrated programme.

The second important fact is that the ever-increasing level of anthropogenic impact on the geosphere–biosphere indicates even more clearly (than consideration of naturally induced processes alone) the necessity of studying processes and changes on a global scale. From the viewpoint of planning an observing system, this indicates a leading role of satellite observation means (although conventional observing systems must play their positive role now, as well as in the future). Of specific interest for the totality of problems involved are palaeo- and comparative-planetary analogies, which means that geospheric research should be closely associated with the study of the origin and evolution of the solar system. One of the aspects here which is especially significant is that the geosphere–biosphere system is not closed in the sense that it is subject to the influence of various extraterrestrial factors (solar activity, galactic cosmis rays, etc.). Thus, there is no doubt that in geosphere–biosphere studies, the key role belongs to such objects of these studies as the atmosphere, ocean, lithosphere and solar/cosmic factors (Kondratyev, 1987, 1991a).

An important aspect of the problem is that some anthropogenic impacts on the global geosphere are already appreciable (for instance, through the growth of carbon dioxide and nitrogen oxide concentrations—evidence of impact on the global cycles of carbon and nitrogen), and may become very considerable in a few decades' time, whereas our capability of taking relevant countermeasures requires an equal (or larger) characteristic time scale; hence, the urgency for studying the set of problems under examination.

It is only natural that, when an interdisciplinary problem of exceptional complexity is developed, of primary importance becomes the establishment of its priority aspects. One of these, in particular, is the problem of climate and its change. Climate is formed as a result of the interaction of the components of the climatic system (the "atmosphere–hydrosphere–lithosphere–cryosphere–biosphere"), for which the existence of numerous feedbacks is typical. At present, there is hardly any other problem in the natural sciences of this degree of complexity on a global scale, and requiring consideration of the most important and complicated anthropogenic impacts.

Directly relevant to the climate problem is the role of the interaction between the geosphere and biosphere. This role is very pronounced, for example, in the consideration of the problem of the influence of the anthropogenically induced carbon dioxide concentration growth on climate. In recent years, this problem has given rise to contradictory statements concerning climate change assessments (Abakumov, 1993; Kondratyev, 1990b; Kondratyev, 1992a–c). This is mainly due to two circumstances: (1) the lack of sufficiently reliable models of the global carbon cycle (the contribution of marine and continental biota has not been clearly established so far); (2) the inadequacy of the theory of climate from the viewpoint of interactive accounting of processes in the geosphere-biosphere. Although the climate problem may serve as an illustration of the importance of considering the effect of extraterrestrial factors, established from observational data, the physical mechanisms determining this influence are not fully known today.

The problem of climate is also so important because of its exceptional practical significance. It is a well-known fact that man's economic activity (and above all, agricultural activity) is strictly dependent on climate conditions. Recently, a new aspect of the problem has become evident–that of the possible effect of a nuclear war on climate and the biosphere (Kondratyev, 1988). The assessments made reveal the danger of a global ecological catastrophe which may be caused, in particular, by a strong climate cooling (the "nuclear winter") resulting from a decreased income of solar radiation to the Earth's surface owing to its marked attenuation by smoke—the after-effects of fires in forests and cities—the consequences of nuclear explosions. Still more important may be climate instability resulting from a strong disturbance of the climate system.

Considering the importance of the problem of climate, a respective interdisciplinary section within the IGBP must be formulated, taking into account the WCRP and other specialized programmes (WOCE, TOGA, ISCCP, etc.). Examination of specific features of programmes of this kind shows that studies of closely interrelated problems of atmospheric physics and chemistry require special attention within the IGBP, including investigation of biogeochemical cycles of such components as carbon, nitrogen, phosphorus, and sulphur, and of solar–atmospheric relationships.

Various aspects of atmospheric physics and chemistry are highly important from the viewpoint of estimating the prospects of the existence of the biosphere (its productivity) under the circumstances of an ever-growing anthropogenic stress (Gorshkov, 1990, 1993; Gorshkov and Kondratyev, 1990; Gorshkov, Kondratyev and Losev, 1992; Mohnen, 1984). Special attention should be given to the problem of the impact on the ozone layer in the stratosphere, which protects the biosphere from harmful effects of hard ultraviolet radiation.

Although assessments of the anthropogenic impact on ozone vary within a broad range of judgement (Kondratyev, 1990b), there is no doubt that there still is deep general concern about the influence on the total ozone content and the vertical ozone concentration profile of the emissions of chlorofluorocarbons (CFCs), carbon dioxide, nitrogen oxides and methane. It is clear now that the effects of these components are closely interrelated, and therefore estimates for individual (isolated) components cannot be considered correct.

There is convincing evidence (obtained from observational data) of increased concentrations of CFC, methylchloroform, carbon dioxide, nitrogen monoxide and methane. However, the lack of reliable data on the sources and sinks of the abovementioned gases makes it impossible to obtain realistic forecasts of their trends in the future (this is especially true for methane).

The solution of the ozone problem (considering that ozone is an important atmospheric component) requires the implementation of a broad programme of laboratory, field and theoretical studies of the interrelated physical, chemical and photochemical processes in the troposphere and stratosphere, and monitoring changes in the atmospheric chemical composition (for a large number of components) on a global scale.

With regard to tropospheric chemistry, most urgent must be studies of: 1) the water cycle; (2) processes determining the dynamics of oxidants, especially of O_3, SO_2 and NO_2; (3) the mechanisms of dry and wet deposition of pollutants at the land surface and in water basins; (4) processes determining the concentrations of aerosols and such gases as CO_2, N_2O, CH_4, NH_3 and H_2O in lower atmospheric layers; (5) the effect of optically active atmospheric components on the radiative regime and climate. Just as in all other cases, the interaction between chemical processes in the troposphere and stratosphere plays a most important role.

For a better understanding of biogeochemical cycles, it is necessary to study such factors of their control as: (1) natural and anthropogenic emissions of minor gaseous components to the troposphere; (2) their transport over large distances; 3) chemical transformations (including gas-to-particle conversions); (4) removal from the stratosphere of minor gaseous and aerosol components (in this connection, of special interest is the role of cloud cover). Of primary interest are studies of such biospheric components as forests in the tropics and at temperate latitudes, steppes and savannahs, tundra, and agricultural land, and also of coastal upwellings and processes in the open ocean.

Climatic changes are determined by the important role of minor optically active atmospheric components (including ozone) in the formation of the greenhouse effect of the atmosphere, whose variations are the primary energy source of anthropogenic impact on climate. This may be illustrated by the increased content of ozone in the troposphere detected during recent years. Assessments have revealed that the contribution of this increase to the enhancement of the greenhouse effect is comparable with the effect of increased concentration of carbon dioxide.

In the field of simulation modelling, most urgent is the task of developing interactive models, taking account of dynamical and photochemical processes, for the simulation of global budgets of O_3, CO_2 and other components of crucial importance. Sufficient attention must be paid, as before, to approximate 1-D and 2-D models in the studies of regional transport, geochemical balances, and in the search for new important chemical reactions. Long-term tasks must include examination of the role of complex heterogeneous processes in the formation of the global cycles of various components; here special attention should be given to poorly studied processes in remote (background) regions of the atmosphere.

In many respects, the chemical composition of the atmosphere has been formed and is at present subject to changes as a result of the interaction with the biosphere. The biosphere is a major source of hydrocarbons, and of such a volatile halocarbon as methylchloroform. Interactions between respective processes are not always direct and unequivocal. For instance, chemical reactions involving nitrogen oxides in the troposphere may lead to both an increase and a decrease in the ozone content, and this, in turn, influences the vegetation cover. Variations in the N_2O, CH_3Cl and CH_4 concentrations in the troposphere affect the ozone content in the stratosphere, which influences the biosphere.

A problem of major importance is the study of physical, chemical and biological processes responsible for the release and assimilation by the biosphere of various components (within different ecosystems), and also, on the other hand, of the reaction of the biosphere to the variability of physical characteristics and the chemical composition of the atmosphere. A first priority task, in this connection, appears to be the monitoring of such gases whose concentrations are already subject to changes as a result of man's economic activity. Such gases are CO_2, CH_4, N_2O, NO_x and SO_x, as well as carbon monoxide.

There are sound grounds to believe that chemical and photochemical processes in the stratosphere and mesosphere which have been changing under the influence of cosmic factors (solar proton events, galactic cosmic rays) are responsible for mechanisms determining the effect of cosmic factors on climate.

A key factor in the formation of the atmospheric chemical composition in its interaction with the ocean, which is a sink area for CO_2 and nitrogen oxides, but a source (in regions of coastal lowlands and productive zones) of CH_4, H_2S and $(CH_3)_2S$, as well as of halogens, nitrous oxide and water vapour for the atmosphere. An urgent problem is that of examining the gas exchange between the ocean and the atmosphere as a consequence of respective chemical, physical and biological processes. The cardinal role of the ocean in the sulphur cycle formation requires special attention. The interaction between processes in the ocean and on the land through rivers, estuaries and coastal zones also deserves close consideration. In this case, assessments of the input to the ocean of fresh waters, carbon, phosphorus and suspended matter are needed above all.

Atmospheric aerosol plays an important role because it is an optically active atmospheric component affecting climate formation (Kondratyev, 1990b; Kondratyev et al., 1996b). Volcanic stratospheric aerosols causing strong and long-term disturbances in the radiative regime, and climatic changes accompanying them, produce an essential factor of the impact of natural aerosols on radiation and global climate (dust storms are an important regional factor). The presence of anthropogenic tropospheric aerosols is responsible for such consequences as the trend of growing atmospheric turbidity and the formation of an aerosol haze at high latitudes of the northern hemisphere. The mechanism of gas-to-particle conversion is highly important in all cases (it plays a leading part in the case of volcanic stratospheric aerosols), which determines the interaction between the biogeochemical cycles of sulphur and nitrogen and the processes forming atmospheric aerosols. The effect of aerosols on the processes of formation and destruction of cloud cover has hardly been studied at all, and this is a serious problem. The crucial importance of smoke aerosols produced by fires originated by nuclear explosions has already been mentioned above.

Another first-priority component of the IGBP is doubtless the problem of biological productivity on the land and in the ocean. Conditions required for biological productivity on the land surface are determined by the resources of water, nutrients and light; for the ocean, the latter two factors are significant. Hence it follows that to assess land surface conditions, one needs information about the spatial–temporal distribution and intensity of precipitation. Flows of nutrients carried by rivers and brought to oceanic coastal zones are of decisive value as factors of oceanic bioproductivity. They are mixed and eventually brought to deep ocean layers. Equally important is the transport of matter from the land surface to the ocean through the atmosphere.

Estimations of the second factor are complicated because contemporary climate models are unreliable from the point of view of describing the global distribution of rainfall, and also because it is necessary to take into account the increasing anthropogenic emissions of optically active minor gaseous components. This leads to changes in the temperature regime, affects the water cycle, causes the formation of acid rains and changes the concentrations of toxic gases, soil fertility, the content of nutrients in rivers and the shelf zone. Essential for the impact on the bioproductivity can also be the variation of the intensity of ultraviolet solar radiation at the Earth's surface level caused by atmospheric contamination. An important task is, firstly, to reveal the naturally induced variability in bioproductivity and, secondly, to analyse the anthropogenic impact on the biosphere. To solve this very complicated problem, a whole number of projects along the lines of basic directions of research must be accomplished.

One of these is the global cycle of water, whose formation is determined by the interaction of the thermal regime and circulation of the atmosphere, the processes of evaporation (including evapotranspiration), transport and phase transformations of water in the atmosphere; here atmospheric circulation is interactively coupled with the radiative regime, which is subject to the effect of anthropogenically induced variations in the atmospheric composition (chemically and biologically active components included). Among difficulties arising in the course of resolving the water cycle problem, there are: insufficient reliability of the estimates of evapotranspiration on the land surface, evaporation and precipitation in the oceans; inadequacy of the description of the processes of cloud formation and rainfall. In particular, the role of vegetation cover in the transport of water from the soil to the atmosphere should be further studied. Data on the global climatology of cloud cover, the distribution of precipitation over oceans and evaporation from the ocean surface remain quite fragmentary.

The leading role of the ocean in the formation of the global water cycle and climate determines the key importance of the problem of the interaction between the atmosphere and ocean, and in this connection, the key part to be played by programme "Razrezy" ("Sections") (Kondratyev, Buznikov and Pokrovsky, 1992). The interaction between the atmosphere and land surface has obviously been studied insufficiently; that is why the International Conference on the Parameterization of Land Surface Processes was held in 1982, and later the International Satellite Land Surface Climatology Project (ISLSCP) was initiated.

Snow and ice covers are of fundamental climate-forming significance, as well as their variability. The high albedo of snow and ice determines a strong (and interactive) dependence of climate on variations in the extent and properties of snow and ice covers. Sea ice plays an important role as a thermal insulation layer (which accounts for the great contribution of heat exchange through polynyas and leads at high latitudes), and as a thermal–inertial component which transforms the annual temperature variation by shifting the extrema due to the latent heat release during freezing in autumn and the heat expenditure for melting in spring. As a sensitive component of the climate system, ice cover is a good indicator of climatic changes.

A fact of principal importance is that the continental biomass has been assessed to an accuracy of about a factor of 2; changes on a scale of decades are known only very approximately. The situation is aggravated by the huge extent of the impact of man's economic activity on natural ecosystems during the last hundred years (deforestation, broad application of monocultures in agriculture, etc.). Far more precise estimations are required for the total biomass of individual ecosystems, and the analysis of the effect of changes in land use, climate, and the income of nutrients, on the global primary bioproductivity. Here are first priority directions of research: (1) the effect of the income of biogenic components (in particular, of nitrogen, phosphorus, sulphur), the variation of oxidant concentrations, acid rains and heavy metal deposition on carbon reservoirs; (2) the biota's counter-response to changes in the environment; (3) factors regulating the carbon ratio between living and dead components of ecosystems, and their dependence on anthropogenic impacts.

The International Geosphere–Biosphere Programme (IGBP), whose implementation started on 10 December 1990, comprises seven key directions of research, viz. (*IGBP* ..., 1994):

1. Regular features of chemical processes in minor gaseous component cycles. Research within this subject includes two core projects (CP).
1.1. The International Global Atmospheric Chemistry Project (IGAC) is a core project stipulating:
1.1.1. Study of processes responsible for the variation of the chemical composition of the atmosphere.
1.1.2. Examination of the interactions between the atmospheric chemical composition, biospheric and climatic processes;
1.1.3. Forecast of the effect of natural and anthropogenic factors on the atmospheric chemical composition.
1.2. The Stratosphere–Troposphere Interactions and the Biosphere (STIB) is a core project planned to:
1.2.1. Analyse the consequences of ozone concentration variations in the stratosphere from the viewpoint of the penetration of biologically dangerous UV radiation to the Earth's surface.
1.2.2. Obtain quantitative estimates of the processes of interaction between the stratosphere and troposphere.
1.2.3. Assess the natural variability of the stratosphere and the impact on it from anthropogenic factors.
1.2.4. Obtain quantitative estimates of the effect of aerosols on climate.
1.2.5. Assess the influence of processes in the stratosphere on climate. A project of this kind has been approved as part of the WCRP.
2. The effect of biogeochemical processes in the ocean on climate and the counter-effect of climate changes on the ocean. Two projects are envisaged:
2.1. The Joint Global Ocean Flux Study (JGOFS).
2.1.1. Study on a global scale of processes responsible for the dynamics of the income of carbon fluxes and other accompanying nutrients to the ocean, as well as assessment of gas exchange between the atmosphere, the sea bottom, and continental boundaries.
2.1.2. Development of methods of forecasting on a global scale of the response of biogeochemical processes in the ocean to anthropogenic disturbances, especially those relevant to climate change.
The first field investigations within the project were accomplished in 1989 in the Atlantic Ocean.

2.2. The Global Ocean Euphotic Zone Study (GOEZS), which takes account of the interaction between physical, chemical and biological processes.
3. The effect of changes in land use on the resources of coastal zones of seas and oceans, as well as the effect of changes in the ocean level and climate on coastal ecosystems. This direction of research will be realized through the LOICZ project.
3.1. The Land–Ocean Interactions in the Coastal Zone (LOICZ) is aimed at:
3.1.1. The development of methods of forecasting the effects of changes in climate, land use and ocean level on the functioning and stability of coastal ecosystems on a global scale, taking account of the first-priority significance of interactions between the varying processes on the land surface and in the ocean, and possible feedbacks of the effects of physical processes in the environment.
4. Interaction between the vegetation cover and physical processes responsible for the formation of the global water cycle.
4.1. The Biological Aspects of the Hydrological Cycle (BAHC). Research within this project is planned to be supplementary to the Global Energy and Water Cycle Experiment (GEWEX), which is a component part of the World Climate Research Programme (WCRP).
4.1.1. Study of the impact of the biosphere on the water cycle on the basis of field observations aimed at the development of models of energy and moisture exchange in the system "soil–vegetation cover–atmosphere", with a spatial–temporal scale from the elements of vegetation cover to the cell of general atmospheric circulation models.
4.1.2. Development of databases which may be used to characterize interactions between the biosphere and physical processes in the environment, as well as to verify the reliability of simulation models of such interactions.
5. The effect of global change on continental ecosystems.
The Global Change and Terrestrial Ecosystems (GCTE)—a core project.
Development of methods for forecasting the effects of climate change, carbon dioxide concentration and land use on ecosystems, as well as assessment of consequences of such effects on the formation of feedbacks between the biosphere and the environment. The project will focus on the following three basic directions of research: (1) the physiology of ecosystems; (2) the dynamics of ecosystems; (3) the impact on agriculture and forestry.
Development of methods for forecasting the relationships between global change and biodiversity (ecological complexity).
6. Palaeoecological and palaeoclimatic changes and their consequences.
6.1. Past Global Changes (PAGES)—a core project.
6.1.1. Reconstruction of a detailed history of changes in climate and the environment on a global scale over a period from 2000 BC to nowadays, with a temporal resolution of no less than 10 years (better, one year or one season);
6.1.2. Reconstruction of the history of climatic and environmental changes during the entire glaciation cycle, in order to achieve a more profound understanding of natural processes responsible for climate change.
7. A systematic approach to numerical simulation modelling of the Earth's system to enable the forecasting of its evolution in the future.

Global Analysis, Interpretation and Modelling (GAIM)—a core project.
Quantitative estimations of the interaction between global physical, chemical and biological interactive processes in the Earth's system during the past 100 000 years, with special emphasis on the analysis of potential future changes.

Five out of the six abovementioned projects (IGAC, JGOFS, BAHC, GCTE, PAGES), which are in the category of approved core projects, have been formulated clearly enough. The implementation of two of them (JGOFS and IGAC) has already been started, whereas the other three (BAHC, GCTE and PAGES) are still in the stage of detailed development. The same relates to the projects STIB, LOICZ and GAIM.

Two important directions of research to be carried out independently, each of which encompasses the needs of all the CP, here: (1) the development of the Data and Information System (DIS) to ensure the archiving and broad access to the IGBP data for long periods of time; (2) the creation of regional research centres (RRCs) oriented towards stimulation and support of regional ecological research, especially in developing countries. The latter programme has later been transformed into a system of analysis, research and education in the area of global change (START) which is under way.

Within the framework of cooperation between the IGBP and the IHDP (International Human Dimensions of Global Environmental Change Programme), a new core project has been approved which was named the Land-Use and Land-Cover Change (LUCC; *Land-Use*..., 1995); in collaboration with the Scientific Committee on Ocean Research (SCOR), the Global Marine Ecosystem Dynamics Program (GLOBEC) has been adopted. The Fourth Session of the Scientific Advisory Council for the IGBP (SAC IV) (Resolution and Recommendations..., 1995) approved the following new directions of research:

(1) the problem of food supply;
(2) freshwater resources;
(3) interaction between hydrological and biological processes;
(4) the effect of geological processes on biotechnical cycles.

The participants of the First IGBP Congress (18–22 April, 1996) have discussed the following "hyper-projects" which have various regional implications:

— LBA (Large-scale Biosphere–Atmosphere Experiment in Amazonia);
— Weather Generator;
— Mountain Hydrology;
— ISLSCP Global Data Sets;
— Arctic Drainage Basin and Climate Change;
— PAGES-START (PEP III, Africa);
— LUCC as a Driver of Global Change;
— Joint JGOFS-LOICZ CMTT (Continental Margins Task Team) Meeting;
— PAGES and its relation to other projects;
— Net Primary Productivity;
— Palaeo Trace-Gas Evolution over the last 150 kyr.

1.12.2 Ecological monitoring

The exceptional complexity of modelling, and more so that of forecasting global ecodynamics require creating an adequate system of ecological monitoring to be able to obtain sufficiently comprehensive observational data. The most promising research directly relevant to the above is being at present implemented in the USA within the programme "Mission to Planet Earth" (MTPE), whose observational component is the Earth Observing System (EOS). In what follows, we shall briefly characterize various efforts which are being undertaken in this respect. Naturally, the substantiation of the EOS must proceed from the analysis of the priorities of global and regional ecology. Although the subject has been treated in quite a few publications during recent years (see Kondratyev, 1990c; Marchuk and Kondratyev, 1992; *Research Strategies...*, 1990), we shall examine some of its key aspects concerning the substantiation of priorities and requirements to observation data.

Moore (1984) has been absolutely right to say that, from the viewpoint of a researcher of the geosphere–biosphere, of special interest is the study of biogeophysical cycles of carbon, nitrogen, phosphorus and sulphur, which are governed by both natural, and anthropogenic factors. The latter are of particular significance for the carbon cycle, which has not been sufficiently investigated so far. The main sources of uncertainties are processes relevant to the contribution of the continental biomass dynamics (deforestation, etc.), and also to changes in the total productivity of ecosystems which are due to variations of other components' cycles (Kondratyev, 1990c). This is what determines the need for satellite monitoring of the biomass dynamics on a global scale with "telescopy" to obtain assessments of detailed spatial-temporal variability of the biomass in individual regions, which can be reliable only if there are adequate control data of direct ground-based measurements.

The key aspect of the carbon cycle relates to estimates of the contribution of the marine biota and the carbon exchange between the atmosphere and the ocean. Of principal significance is the accounting of the interaction between the cycles of carbon and nitrogen. The nitrogen cycle dynamics is determined by three processes with different time constants: long-term processes of the transport to and out of the atmosphere; a faster nitrogen exchange within the biosphere; the growing income of anthropogenically induced nitrogen (owing to high-temperature combustion processes alone, about 20 Mt of nitrogen per year is emitted into the atmosphere, and the application of fertilizers adds about 40 Mt of bound nitrogen per year to this amount). However, the global anthropogenic impact on the nitrogen cycle still remains negligible in comparison with nitrogen reservoirs in the soil and in the oceans. It may become quite substantial in about several decades (or one century from now).

Unlike the cycles of carbon, nitrogen and sulphur, the phosphorus cycle is independent of processes in the atmosphere. In this case, the main processes are connected with dissolved and particle-shaped phosphorus that can be found in river waters, with soil weathering and biogenesis processes in sedimentary rocks. The important problems of the contribution of river-carried phosphorus, involved in the biological cycle, and the characteristic time scales of the transport of phosphorus to the ocean, have not been resolved so far. Among the basic aspects of studying the global phosphorus cycle are: (1) mechanisms controlling the presence of phosphorus in soils, including the effect of anthropogenic factors (for instance, acid rains); (2) the transport of phosphorus to the ocean by river waters; (3) the relationship between the ocean's bioproductivity and the phosphorus concentration (in particular, the influence of phosphorus income on the fixation of nitrogen in coastal zones).

The problem of anthropogenic impact on the sulphur cycle is of great interest. There are indirect assessments which indicate that by now the emissions of gas-phase sulphur to the atmosphere (due to fuel burning processes) have reached the order of magnitude of the contribution of natural factors. Of special interest are: (1) estimations of biologically induced gaseous compounds of reduced sulphur, studies of processes and anthropogenic impacts determining them; and (2) study of the transformation of anthropogenic emissions of sulphur dioxide and transformations of gaseous sulphur compounds in the atmosphere in general, including gas-to-particle conversion reactions involved in the formation of sulphate aerosol, which may affect climate to a great extent (Kondratyev, 1990b). Since the processes of anthropogenic impact on biogeochemical cycles are slow and sluggish, it would be very important to be able to trace the dangerous tendencies well beforehand.

Delwiche (1984) has demonstrated that satellite remote sensing techniques form a most reliable basis for studying the peculiar features of the geographical distribution of ecosystems and boundaries between them, the scales and rates of the anthropogenic impact on ecosystems, and the effect of climate and other factors on ecosystems. These techniques are especially important from the viewpoint of monitoring the variability of biogeochemical cycles, controlling the primary biomass dynamics and chlorophyll concentration. Table 1.11 presents a brief characterization of requirements to observational data. Lidar sounding (mainly, employing the principle of differential absorption) appears to be a most promising method.

The World Ocean (see Chase and McGoldrick, 1984) plays the key role in the formation of global cycles of biogenic components. With regard to biogeochemical cycles, the ocean functions as a huge reservoir of biogenic components. Component parts of crucial importance of the functioning of this reservoir are: the exchange of matter between the upper and deep oceanic layers, between ocean and land, between ocean and atmosphere. In many cases, the rates of this exchange are not reliably known. The ocean bioproductivity, low as it is per unit of area, totals to a considerable share of the biospheric productivity owing to the great area of the oceans, and it is closely connected with biogeochemical cycles. Thus, for instance, the rate of transformation of nitrate into organic compounds of nitrogen is dependent on the productivity of phytoplankton, which in turn is determined by the presence of nutrients.

In the sphere of studies of global biogeochemical cycles and ocean productivity, of special interest are: (1) mapping of the spatial–temporal variability of the biomass and phytoplankton productivity in the World Ocean; (2) interaction between biological and physical processes and its effect on the distribution and growth rate of the phytoplankton biomass. In this connection, of key importance are the following aspects: (1) detection of regions with marked mesoscale variability of productivity;

Table 1.11. Requirements put to data of remote observations of biogeochemical sources, sinks and fluxes

Observed values	Observation target	Spectral region
Ecosystem boundary	Indicators of variations caused by environmental changes	Visible or near-IR
Turbidity plumes in rivers	Indicators of erosion and eutrophication	Ditto
Chlorophyll content in the ocean	Indicators of primary productivity and its variations	Visible (including observations of forced fluorescence
Atmospheric parameters: CH_4—total content in the troposphere CH_4—source intensity (and gradient) in individual regions	Intensity of biological and anthropogenic sources, lifetime	IR-absorption or lidar sounding
N_2O—total content in the atmosphere and (desirably) gradients over individual regions	Biological and anthropogenic (agricultural) sources	IR-absorption or lidar sounding
CH_3Br—total content and (desirably) gradients $(CH_3)_2S$, $(CH_3)SH$, H_2S	Recognition of assumed biological sources sounding	IR-absorption or lidar
Total content and (desirably) gradients	Recognition of assumed biological sources and the duration of their functioning	IR-absorption or lidar sounding
NH_3—gradients in the troposphere over individual regions (forests, emission regions, etc.)	Intensity of biological sources	?
Non-methane hydrocarbons —gradients in the troposphere over individual regions	Recognition of assumed biological sources and the duration of their functioning	?

(2) analysis of the influence of this kind of variability on global productivity and biogeochemical cycles; (3) "climatology", the seasonal and interannual variability of the biomass productivity; (4) contrasts between the oceans of the southern and northern hemispheres, the coastal and oceanic waters; (5) the vertical structure of the biomass and productivity, its temporal variations, the effect on the processes of horizontal redistribution.

In the investigation of the interaction between biological and physical processes, the priority issues are the following: (1) the relationship between mesoscale phenomena (vortices, meandering fronts, coastal jet stream currents, etc.) and mesoscale distributions of biomass and productivity; (2) the effect of the bottom and coast topography on biological processes; (3) the connection between meso-scale and large-scale processes; (4) processes controlling the exchange of matter between coastal and oceanic waters; (5) the influence of the general oceanic circulation (including surface currents, return currents and long waves) on remote geographical regions of the World Ocean; 6) the effect of climate on biological processes; (7) the distribution of basic nutrients; (8) the income of nutrients from deep waters to surface waters; (9) the parameterization of mixing in the surface layer.

From the viewpoint of requirements of observational data, the main role belongs to information about the spatial–temporal variability of the phytoplankton biomass and primary productivity, which can be obtained with instrumentation of the type of the ocean colour scanner for the 350–800 nm region (with a 10-nm resolution), which ensures that global information may be provided for 24 hours, with a spatial resolution of above 1 km in coastal zones, and 4×4 km in the open sea. To observe the chlorophyll fluorescence, data in the 600–750 nm interval with a 5-km resolution are required, and global mapping of bioluminescence will provide data in the 450–550 nm interval (with a 10-nm resolution) with the spatial resolution of 5×5 km. Satellite observations should be accompanied by control ship and buoy observations.

It is a well-known fact that the World Ocean plays the key role in mitigating the Earth's climate, owing to its gigantic inertia, the transport of heat to the poles, and the formation of cloud cover. Of equal importance is the contribution of the ocean to the global cycle of carbon, as well as the processes of transport and transformation of biogenic components governing bioproductivity conditions (Chase and McGoldrick, 1984).

A classical picture of the general oceanic circulation built on the basis of the analysis of ship observational data cannot be absolutely reliable for the following two reasons: (1) the fields of oceanographic characteristics compiled from fragmentary (and time-differing) data of ship observations cannot be considered representative (although a relatively large amount of energy constitutes the share of low-frequency components of the transport, an essential role belongs to the dynamic and kinematic interaction between smaller-scale and larger-scale motions); and (2) the contribution of variations in the ocean surface level to the formation of the pressure field in the oceanic thickness remains unknown. Therefore, one of the most urgent problems is that of the combined application of conventional and satellite observation means to achieve more comprehensive investigations of the dynamics and thermal regime of the World Ocean.

Table 1.12. Requirements put to observational data relevant to physical oceanography

Observed values	Observation target	Observational data
Ocean surface topography	Oceanic circulation: geostrophic velocity near the surface, localization of mesoscale phenomena (fronts, vortices, rings); interaction between physical and biological processes	Radioaltimeter on satellite with orbit inclination angle 65° (non-helio-synchronous orbit with repetition of orbits 10–20 days): IR and microwave radiometer for atmospheric correction.
Significant wave height ($H_{1/3}$)	Forecast of wind, waves and swelling sea	Radioaltimeter, synthetic aperture radar (SAR)
Sea surface temperature	Interaction between atmosphere and ocean, and between physical and biological processes	IR radiometers of the type used on meteorological satellites
Wind stress near the ocean surface	Oceanic circulation (wind-induced, ageostrophysical component, upwellings), interaction between ocean and atmosphere	Scatterometer in Ku band, measurements every 50 km in band $\pm 70°$ of latitude every two days
Spectrum of surface water directions	Forecast of waves and wind, monitoring of storm movement	SAR, scatterometer, scanning short-pulse radioaltimeter.

Table 1.12 reflects the order of priority for the physical characteristics of the ocean and requirements put to relevant observational data. It should also be mentioned that there is an additional possibility of finding the velocity vector of sea currents by following drifting buoys and acoustic generators from satellites. What is important, in principle, is the combination of the analysis of observational data and numerical simulation experiments.

Melack (1984) states that the central aspect of biospheric studies is reduced to finding to what extent the Earth's surface, atmosphere and hydrosphere are subject to the impact of biological, and not abiotic, processes. Considerable progress may be expected in this connection from the use of satellite observation means for monitoring the biological productivity and that part of biogeochemical cycles that is relevant to water systems.

So far we have been unable to obtain reliable quantitative estimates of the influence of living organisms on the atmospheric or hydrospheric composition, or sedimented rocks. Hence the necessity for estimating the role of diatomic algae and swamps in global biogeochemical cycles, which requires: (1) evaluation of the areals of these organisms or ecosystems; and (2) measurements of the fixation of carbon, the rates of accumulation of deposits and gas formation.

In the light of large-scale anthropogenic impact on the carbon, nitrogen, sulphur and phosphorus cycles, there is an urgent need for measurements of natural and total fluxes of these elements on a global scale. This requires development of remote sensing of the biological productivity, as well as the fluxes of various components between inland waterbodies, ocean, land and atmosphere, taking account of a number of specific features of inland waterbodies: (1) a broad range of sizes (from 100 km to 1 km in length); (2) the wide territorial spread; (3) the extended boundary between water and land; (4) an extremely broad range of optical conditions (e.g., the chlorophyll concentration may vary from 0.01 to 100 mg/m^3); (5) rapid changes in biological, chemical and physical conditions with time (with characteristic time variations from seconds to weeks, but, for changes of special ecological significance, of the order of several days); (6) the importance of episodic and periodic fluctuations; (7) the presence of strong horizontal and vertical gradients of biological and physicochemical characteristics.

Table 1.13 contains schematic requirements put to observational data for inland waterbodies. The biological productivity can be found from the data of measurements of photosynthetic activity, which requires measuring the amount of photosynthetizing substance and the carbon fixation rate, and determining the portion of fixed carbon consumed by other organisms or accumulated in precipitation. Fluxes of various components from land to inland waterbodies are mostly dependent on the manner of land use in the water catchment basin and on hydrological conditions. For instance, estimations of the income of nutrients to lakes from remote sensing data and information on the water catchment area, including land use and snow cover, have already become feasible.

Table 1.13. Requirements to observational data for inland waterbodies

Object of observation	Observation target	Instrumentation characterization
Regions of swamps and floods	Biological productivity, nitrogen fixation. Sources reduced gases	Multispectral images (including of SAR), with resolution 30 m, once per day / twice per week.
Chlorophyll of phytoplankton and biomass	Spatial–temporal inhomogeneity of biological productivity	Multispectral images (30 m), including spectrometric of measurements of fluorescence
Land use over water catchment area	Content of nutrients and suspended matter (as related to water quality and productivity	Multispectral images (including SAR), 30 m, once per month/ year
Runoff and relevant hydrologic parameters	Content of nutrients and suspended matter	Visible, IR, and microwave regions, once per day/week
Surface temperature evaporation	Hydrography, and productivity	IR images, 30 m, once per day/ week.

Considering the important role of forests in the formation of biogeochemical cycles of carbon and nitrogen, and also the impact of forests on the energy and water budgets, Zinke (1984) mentions the necessity of solving the following problems: (1) precipitation on a large spatial scale; (2) the influence of forests on the local and regional energy budgets, and hence on climate; (3) the role of forests as a source or sink of carbon dioxide; (4) the effect of such parameters of inland water quality as the temperature and turbidity of water in rivers and lakes, etc., on the practical aspects of land use resulting in variations in vegetation and forests in the water catchment basin.

Table 1.14 illustrates approximate requirements put to observational data which can be obtained by remote sensing methods and are aimed at estimating the components of the hydrologic and energy budgets, and also biogeochemical cycles. One such example is the study of the contribution of tropical forests in the Amazon river basin as a source of water, which requires an analysis of the water balance of forests taking account of precipitation and evapotranspiration.

Table 1.14. Requirements put to data of observations of forests and natural vegetation

Observed values	Observation target	Observational data
Mass and energy fluxes incident and reflected shortwave radiation, relationship between fluxes of radiation, latent and sensible heat	Components of energy budget of forests and natural vegetation; effect of deforestation on microclimate and spread of diseases	Visible and IR spectral regions spatial resolution 50–60 m, at least once a month; surface temperature, albedo, water vapour content, 50–60 m, once per month
Hydrologic cycle components: precipitation, evapotranspiration, river runoff, moisture content of soil, snow content	Local and regional water budget for main rivers, and effect on it of land use and deforestation	Microwave measurements of moisture content in soil (1 km^2) and in atmosphere; measurements of runoff and stagnant water surface; once per month
Density, structure and biomass of vegetation cover	Interrelationship between vegetation characteristics and energy and water budgets	Multispectral images (brightness fields), 60 m
Soil characteristics	Carbon and nitrogen content in soils	Regression relationship between multispectral satellite and ground-based data
State of vegetation	Assessment of state of vegetation, its classification and monitoring of seasonal dynamics	Multispectral images

Table 1.15. Requirements put to observations of biological processes on land

Observed values	Observation target	Observational data
Leaf area index, phytomass	Latent heat and carbon dioxide flux, phytomass of major ecosystems	Visible, IR and microwave regions, resolution 30 m, observations of individual regions at 13–15 h, every 3–5 days
Morphology of vegetation	Species recognition, light absorption, withering determination; spatial variability of vegetation characteristics	Ditto. Observations at three angles of vision Ditto. Resolution 10 m
Soil moisture	Latent heat, productivity, energy balance	Microwave radiometry (0–5 cm layer)
Surface temperature	Energy balance, drought, desertification	IR radiometry, 30 m, once per day, around noon

Data presented in Table 1.15, compiled by Kanemasu (1984), illustrate requirements which are put to observations of parameters characterizing biological processes on land. In this connection, one of the most important problems is the presence of water and the capability of plants to extract water from the soil, both of which strongly influence the processes of transpiration and photosynthesis. The key position belongs to studying the specific features of the energy balance of various ecosystems: deserts, forests, savannahs, agricultural areas, etc. The quantity and quality of vegetation in these systems are in a considerable interdependence with the character of processes taking place in neighbouring regions. For instance, it is obvious that deforestation and desertification heavily affect areas which are outside the limits of their direct influence, and lead to noticeable changes in carbon dioxide and water vapour fluxes, as well as nutrient fluxes over vast regions.

River runoff and the motion of suspended matter introduce radical changes into the water medium properties for downstream river currents. The surface albedo variation affects the thermal balance and climate. The interaction between the land surface, atmosphere and ocean, which takes place under such conditions, is of primary interest, and for complete understanding of this interaction, more comprehensive observational data are required. Data on the heat balance of the Earth's surface which must be obtained from combined conventional and satellite observation means are certainly first-priority data. It can be seen from Table 1.15 that quite a number of biophysical characteristics (leaf area index, phytomass, morphology and temperature) may be found by making use of remote sensing techniques.

On a time scale of a decade or more, the atmosphere, land and ocean function as an interdependent system with regard not only to physical, but also to chemical and biological processes. In this connection, Mohnen (1984) analysed the role of the troposphere from the viewpoint of basic processes taking place there, which govern the chemical composition and cycles of various tropospheric components and determine its response to external impacts.

The following problems are of key importance: (1) the contribution of biological sources to the formation of the cycles of the compounds of carbon, sulphur, nitrogen and halogens (the most significant areas are forests at middle and tropical latitudes, zones of grass cover, intensively cultivated regions, coastal waters and salted swamps, continental shelf zones, as well as those of upwelling and tundra; the open ocean; regions of biomass burning); (2) the intensity of aerosol sources (dust, fuel combustion, etc.) on regional and continental scales (the desert aerosol is of special interest); (3) the global distribution of basic gaseous and aerosol components taking part in chemical cycles, and the hydrological cycle components (water vapour, clouds, precipitation), with special reference to studying the photochemical processes, and dry and wet deposition of gases and particles (here, numerical simulation modelling becomes very important). Table 1.16 illustrates approximate requirements put to remote sensing data concerning the chemical composition of the troposphere.

Table 1.16. Requirements to observational data relevant to the chemical composition of the troposphere

Observed values	Observation target	Observational data
CO_2, CO, CH_4	Understanding of biogeochemical cycles	Retrieval of gas content in three layers (0-15 km) with horizontal resolution 10 km and errors; $\pm$ 0.3 ppm (CO_2), 10 ppb $\div$ 3 ppm (CO), 100 ppb $\div$ 3 ppm (CH_4)
OH	Lifetime in the troposphere of such gases as CO and CH_4	Minimum detection limit 0.5×10^4 mol/cm^3
NO_2, NO, NH_3, N_2O	Nitrogen cycle	Detection limit 0.1 ppb
HNO_3, NO_3	Nitrogen cycle	Detection limit 0.05 ppb
SO_2, H_2S, COS and other sulphur compounds	Sulphur cycle	Detection limit 0.05 ppb
H_2, H_2O	Hydrogen cycle	Range to 0.02 ppm (H_2) and 1 ppm $\div$ 0.5×10^3 ppm (H_2O)
O_3	Oxygen and oxidant cycle	Range 2 ppb $\div$ 2000 ppb
Aerosol	Aerosol cycle (including sulphur and nitrogen)	Range 0.1 μg/m^3 $\div$ 100 μg/m^3
Temperature, wind velocity, cloudiness, precipitation intensity, lightning	For interpretation of all cycles	Vertical resolution 1 km, wind velocity to 1 m/s

Table 1.17. Requirements of observational data for the troposphere

Observed values	Observation target	Requirements of observational data
Air humidity	Relationship between large-scale circulation with soil moisture and evapotranspiration	Microwave, IR-radiometry, near IR region, lidar, not less than three levels below 500 with spatial averaging 2° latitude × 2° longitude, error ± 20%, twice per day, global information
Moisture of upper soil layer (1 m)	—	Active and passive microwave instruments, IR-radiometry; resolution not worse than 2° by 2°, twice per day over continents
Evapotranspiration	—	The same instrumentation, estimates of moisture content of vegetation cover
Precipitation over oceans	Annual and interannual variations of tropical oceanic moisture source	The same instrumentation, precipitation observations four times per day (ground-based control data are of crucial importance). To assess evaporation, data on humidity and wind profiles are required, as well as SST and cloudiness
Clouds (temperature and upper boundary height, thickness, albedo, moisture and water content), wind in tropics	Relationship between annual variation and interannual variability of climate (especially effect of clouds on thermal budget)	Visible, near-IR and IR spectral regions. Cloud amount (0–10) points; upper boundary height ± 0.5 km, four times per day, 2° × 2° grid
	Study of circulation in the tropics	Doppler lidar not less than four times per day, 100 × 100 km^2 grid, error ± 2 m/s, at least at 3 levels

The perspectives of further development of satellite remote sounding of the troposphere are mainly relevant to the solution of the problems of the global water cycle, cloud cover dynamics, interaction between the atmosphere and ocean, atmospheric circulation in the tropics and the changing gaseous composition of the troposphere. Relevant requirements of observational data are characterized in Table 1.17 (information on the gaseous composition is treated in a different section).

Although the most important aspects of the middle atmosphere have been fairly well investigated during recent years, quite a number of problems remain unsolved so far and, among these, the problem of interaction between the troposphere and the middle atmosphere (MA), expressed as an exchange of mass, energy and chemical components. As shown by Gille (1984), particularly important is the long-term evolution of the MA composition, which is due, to a great extent, to anthropogenic impact. To understand this evolution, one must have data on all long-living basic components of the MA (including those that are sources of radicals, radicals themselves and molecules "sinks" causing the removal of active components of the MA) during at least a single 11-year solar activity cycle. Minimum requirements of observational data are presented in Table 1.18. These requirements have mostly been met by making use of instruments installed on the UARS satellite for the purpose of studying the upper atmosphere.

Two most important problems arise in connection with the investigation of the role of polar glaciers in the formation of climate (see Kondratyev, 1994): (1)

Table 1.18. Requirements to observational data relevant to the middle atmosphere

Observed values	Observation target	Requirements of observational data
Temperature	Chemistry, dynamics, transport, energetics of MA	Layer 0–150 km, with resolution 0.5 of height scale over vertical and errors not over $\pm$ 2 K (0–80 km), $\pm$ 5 K (80–120 km), $\pm$ 10 (120–150 km); errors or relative values—less by factor of 2
Wind	Dynamics, transport	Layer 0-50 km (error $\pm$ 3 m/s); 50–150 km ($\pm$ 10 m/s)
Composition: molecules —sources of radicals O_3, N_2O, CH_4, $CFCl_3$, CF_2Cl_2 —"reservoir" molecules HCl, HNO_3, H_2O_2 HNO_4, $ClONO_3$ —radicals: ClO, NO, NO_2, OH, HO_2	Chemistry, transport	At various heights with error 10% (realistic) and 5% (desirable), vertical distribution 0.5 of height scale
Emissions: O_2, ($^1\Delta g$), OH bands, 1–4 μm, NO (2.8; 5.3 m); CO_2 (4.8; 10.4 μm)	Energetics of upper mesosphere–lower thermosphere (non-	Vertical resolution 0.5 of height scale, $\pm$ 25% (error of LTE) relative values $\pm$ 10%)

assessment of the contribution of glaciers to the hydrological cycle in the "ocean–atmosphere–glaciers" system; (2) analysis of ice cores aimed at the retrieval of palaeoclimatic characteristics, as well as of variations of the concentration in the atmosphere of such components as CO_2, No_x, SO_2, aerosols, etc., for the past 200 000 years. Approximate estimates have revealed that the variability of glaciers caused by internal factors and the temperature and precipitation regime can heavily affect climate and the World Ocean level on a time scale of one to several centuries. This shows that a study of the dynamics of glaciers based on numerical modelling and observational data is a task of primary importance. Table 1.19 presents a summary of approximate requirements to observational data.

In considering the problem of the influence of sea ice on climate, of special interest is investigation of the interaction between ice cover and the atmosphere, particularly in the presence of polynyas and leads when this interaction is intensified in a radical way. This kind of investigation should be based on the employment of interactive simulation models of the "ocean–ice cover–atmosphere" system and the utilization of complex observational data for verification of the adequacy of models. Table 1.20 shows requirements fo data of observations of the most essential values. Since the observations discussed are performed in remote regions, it is reasonable to use a special communication satellite for data transmission.

Analysis of the characteristics of some satellite instrumentation, functioning at present (see Table 1.21), reveals a considerable "intersection" of various parameters. Attention is drawn to the similarity of choosing spectral sounding channels used in different instruments which are often installed on board the same satellite. A multispectral scanner MSS is an instrument whose data are used by a broad range of experts in natural resources research. Therefore, the boundaries of spectral intervals of relevant sounding channels have been selected very approimately. Relevant characteristics of the thematic mapper TM have been selected, proceeding from the interests of a narrower group of problems. The spectral characteristics of the shelf zone colour scanner, SZCS, reflect the requirements of a single group of problems ("Oceanology"). The HRV instrumentation is intended for surveying aquatic areas.

The correspondence between the demands of individual information users and their choice of the spectral characteristics of satellite instrumentation can be followed by making use of the data of Kondratyev, Buznikov and Pokrovsky (1992), Kondratyev, Grigoryev and Pokrovsky (1975) Kondratyev and Pokrovsky (1977, 1989). This kind of analysis stimulated the authors to formulate a problem of planning satellite multipurpose remote sensing systems.

Table 1.19. Requirements to observational data relevant to polar glaciers

Observed values	Observation target	Requirements of observational data
Surface level height	Dynamics of glaciers	Profile measurements using a laser altimeter, with error 2 m on glacier and 0.2 m in shelf zone. Distance between profiles is 1 km
Level height variation	Dynamics and balance of glacier mass	Radioaltimeter observations for routes every 20 km, with interval 1–2 years. Error 0.1–1.0 m
Glacier boundary (from sea-side)	Speed of boundary displacement	Observations on shelf using SAR every 2 years (error 30 m) and inside glacier every 5 years (error 10 m)
Accumulation rate and surface temperature	Dynamics of glaciers	Area data about accumulation every 10–100 km (error 5 cm/year). Temperature every 100 km (data of microwave and IR radiometry with error 1°C/year)
Lateral stress on the surface	Modelling of ice dynamics	Localization of individual points from satellites with error 1 m and repetition rate 5-10 times per year (according to laser reflector)
Melting on the surface	Balance of mass and energy	Microwave radiometry in summer, 1–3 times around noon (interrogation of groundbased stations from satellite)
Breaking off of icebergs	Variations depending on hydrological conditions	Analysis of breaking off of icebergs from icebergs satellite SAR images. Estimation of regional production with error 10%

Table 1.20. Requirements of observational data relevant to ice cover

Observed values	Observation target	Requirements of observational data
Position of ice cover boundaries	Study of ice cover characteristics, heat exchange between ocean and atmosphere	Microwave radiometers or SAR, radioaltimeter, daily, error 5–20 km
Ice concentration		The same instrumentation, daily for 10% of whole area, error about 10%
Ice albedo		Averaged over area, with step 25–100 km and error 0.02
Ice motion		Displacement of fixed points with error 0.5 km/day, with spatial resolution 5–100 km, daily
Type and thickness of ice		Microwave radiometers, SAR, scatterometer, radioaltimeter, once per week
Ice hummocks		SAR, Landsat data, daily
Polynyas	Effect of thermal perturbations	The same instrumentation, daily
Ice field distribution		SAR, Landsat, etc., error in floe size about 10 m, once every 1–7 days
Surface melting		Microwave radiometry, daily in summer; in pack ice zone—throughout the year
Ice surface temperature		Microwave and IR radiometry, control data from buoys
Wind speed		Scatterometer, buoys, SAR
Sea surface temperature		Microwave and IR radiometry, error 1°C

Table 1.21. Characteristics of space-borne instruments in the visible and IR spectral ranges

Instrument	Spectral channels (μm)	Spatial resolution (nadir)	Viewing bandwidth (km)	Remarks
MSS	0.50–0.60	80 m	185	Multispectral scanner
	0.60–0.70			
	0.70–0.80			
	0.80–1.10			
TM	0.45–0.52	30 m	185	Landsat-4.5 thematic mapper
	0.52–0.60			
	0.63–0.69			
	0.70–0.90			
SZCS	0.43–0.45	825 m	1636	SZCS
	0.51–0.53			
	0.54–0.56			
	0.66–0.68			
	0.70–0.80			
	1.05–1.25			
HRV	0.50–0.59	20 m	600	Water basin photography
	0.61–0.68			
	0.79–0.89			
	0.51–0.73			

Optimal planning of observing systems

The urgent need in optimal planning of space-based remote sensing systems is determined by a number of reasons, including the costliness of satellite experiments and limitations on the weight of the instrumentation, on the power of energy sources and on the volume of in-flight memory. Besides, special attention should be paid to the scientific aspects of the problem of satellite data interpretation, such as the mathematical non-correctness of inverse problems (parameters of a relevant medium must be retrieved from the characteristics of emitted, reflected or absorbed radiation), uncontrollable fluctuations of atmospheric optical characteristics, etc.

In the solution of the problem of planning remote sensing experiments, two extreme cases should be excluded: the excess or the deficit of information obtained through inadequate choice of conditions selected for the experiment. In each of these cases, the value of the solution of an inverse problem may be decreased because of its incorrectness.

Two approaches have been developed to the solution of the problem of planning multipurpose remote sensing systems: the informational and the economic approaches. In both cases it is assumed that the users of information, representing the interests of different branches of economy, are capable of formulating precisely

Table 22. Choice of spectral intervals (μm) using the factor analysis technique

Groups of problems	Factors 1	2	3	4	5	6	7
I. Oceanology	0.45–0.53	0.64–0.69	0.70–0.90	0.57–0.64	0.82–1.30	0.50–0.58	0.40–0.44
	0.45–0.53	0.64–0.69	0.70–0.75	0.62–0.64	0.96–1.10	0.50–0.57	0.42–0.44
II. Hydrology	0.70–1.30	0.51–0.59	0.44–0.52	0.67–0.74	0.57–0.67	0.30–0.40	0.38–0.49
	0.96–1.10	0.51–0.57	0.44–0.52	0.67–0.74	0.62–0.67	0.70–0.75	0.42–0.49
III. Geology	0.51–0.63	0.62–0.70	0.70–0.89	0.40–0.51	0.80-1.30		
	0.51–0.57	0.62–0.70	0.70–0.75	0.42–0.51	0.96–1.10		
IV. Forestry	0.51–0.57	0.59–0.70	0.75–0.78	0.70–0.74	0.40–0.51	0.55–0.58	0.88–1.30
and agriculture	0.51–0.57	0.62–0.70	0.76–0.78	0.70–0.74	0.42–0.5	0.55–0.57	0.96–1.10

their requirements to basic characteristics of such information (spectral channels, the experiment geometry, the spatial–temporal resolution, etc.). In what follows, we use a summary of such requirements presented by Kondratyev, Grigoryev and Pokrovsky (1975) relevant to four groups of problems corresponding to four different branches of economy: I, oceanology (102 requirements); II, hydrology (33); III, geology (36); IV, forestry and agriculture (32).

The informational approach

The idea of this approach (Kondratyev et al., 1996a) consists in the selection of the most important factors and determination of relevant vectors of factor loadings in a formalized space of requirements. A technique of building a space of requirements and a modification of the factor analysis scheme used has been suggested. The results of finding optimal sets of spectral intervals are presented in Table 1.22 for groups of problems I to IV. The results of calculations for various complexes of economy problems have been presented. It should be emphasized that the data of Table 22 (column 1) have been obtained exclusively on the basis of the consideration of the demands of information users; these data do not take into account either the selectivity of information distortion by the atmosphere in various regions of the electromagnetic spectrum, or the economic characteristics of satellite experiments. Table 1.22 (column 2) presents precise boundaries of spectral sounding channels owing to elimination of absorption bands of H_2O, O_3 and NO_2 in the following intervals: 0.38–0.42; 0.57–0.62; 0.755–0.760; 0.785–0.790; 0.810–0.845; 0.910–0.960; 1.10–1.16 μm. A substantial contribution of molecular scattering by the atmosphere accounts for the low information content of the 0.30–0.40 μm channel. This is why this channel is not used in the instrumentation of remote sounding systems (see Table 1.21).

Analysis of instrumentation capabilities taking account of information users' needs

If we arrange the spectral sounding channels selected in a sequence, in the order of increasing wavelengths, a relationship may be established between similar channels corresponding to different groups of problems (see Table 1.23). It can be seen that

Table 1.23. Optimal spectral sounding channels corresponding to main groups of problems

Groups of problems	Spectral sounding channels								
	1	2	3	4	5	6	7	8	9
Oceanology	0.42–0.44	0.45–0.53	0.50–0.57	—	0.62–0.64	0.64–0.69	0.70–0.75	—	0.96–1.10
Hydrology	—	0.44–0.52	0.51–0.57	—	0.62–0.67	0.67–0.74	0.70–0.75	—	0.96–1.10
Geolology	—	0.42–0.51	0.51–0.57	—	0.62-0.70	—	0.70–0.75	—	0.96–1.10
Forestry and agriculture	—	0.42–0.51	0.51–0.57	0.55–0.57	0.62–0.70	—	0.70–0.74	0.76–0.78	0.96-1.10

the shortwave channel 0.42–0.44 μm is necessary only in the case of oceanology problems. Considering the data of Table 1.21, we find that the corresponding channel 0.43–0.45 μm is also used in a specialized oceanographic instrument SZCS only. The set of channels presented in Table 1.23 for the "Oceanology" group has much in common with that for the SZCS scanner (see Table 1.23). Differences in the boundaries of relevant spectral intervals are due to the circumstance that the scheme of the oceanographic scanner does not duly take into account the users' needs (see Kondratyev, Grigoryev and Pokrovsky, 1975) or the optical characteristics of the atmosphere. For example, the spectral channel 0.62–0.64 μm, which is present in the requirements put by every problem of the "Oceanology" group, is absent in the scheme altogether.

The data of Table 1.23 also show that measurements in two narrow channels, 0.55–0.57 and 0.76–0.78 μm, are required only for problems of the group "Forestry and agriculture". However, experts interested in this group of problems cannot obtain information in these spectral intervals with the help of any currently existing instruments. Narrow channel 6 is necessary for information users interested in "Oceanology" and "Hydrology" problems. Experts in "Geology" and "Forestry and agriculture" problems need information obtained in a broad spectral interval from 0.62 to 0.70 μm, integrating channels 5 and 6 for problems of groups I and II (Table 1.22). In the instruments available today, this spectral region is presented as a single spectral interval. This means that the instruments available do not satisfy information users interested in "Oceanology" and "Hydrology" problems as far as observation data from channels 5 and 6 are concerned.

Comparing the data of Tables 1.21 and 1.23, one can see that information obtained with the TM thematic mapper is in fair agreement with problems of the "Geology" group, and not so well with the "Hydrology" problems. The multispectral scanner MSS is incapable of satisfying the requirements of information users. This happens because it does not provide any information for channel 2 (Table 1.23). Channels 5 and 6 of the optimal set are actually transformed into a single channel in the MSS scheme. Information for channels 0.70–0.80 and 0.80–1.10 μm is largely subject to distortion due to reflected radiation absorption by atmospheric gases.

These data indicate that satellite instrumentation available is not adequate enough to the needs of satellite information users who present various branches of economy.

The economic approach

Consideration of the fact that future development of space-based remote sensing systems should be financed by those branches of economy which are interested in obtaining satellite information, provokes the question of the cost-effectiveness of planned instrumentation complexes intended for long-term operation on board space platforms. The multizonal nature of satellite information (SI) and its multipurpose application make it possible to analyse the problem of optimal planning of relevant observing systems from the standpoint of maximization of the integral economic efficiency index of a multipurpose observing system V, provided that individual efficiencies for separate problems V_i ($i = 1, \ldots, N$) are regulated by special restrictions of the type of inequalities from above and from below.

Let us formulate the problem more in detail. We define: c_{ij} is economic efficiency of using the jth measurement channel ($j = 1, \ldots, M$) for the ith partial problem; d_j is full expenses for the functioning and processing of the information of the jth measurement channel of the multipurpose measurement system; $b^+{}_i$, $b^-{}_i$ are maximum and minimum values of economic efficiency from the application of observation data to the ith problem. Then, for economic efficiency of using the jth measurement channel, k_j, we obtain the expression:

$$k_j = \sum_j c_{ij}\, d_j \; (j = 1, \ldots, M) \qquad (1)$$

By introducing the variable y_j, which is an indicator of including the jth channel in the observing system $y_j = \{{}_1{}^0$, we shall formulate the basic problem of optimization planning in the form of a standard problem of the Boolean integer programming (see Kondratyev et al., 1996a):

$$y^{\min}\,(y_1, \ldots, y_M)\,\{-\sum \mathrm{k_j y_j}\}; \qquad (2)$$

$$y_j = \{{}^0{}_1, (\mathrm{j} = 1, \ldots, \mathrm{M}); \qquad (3)$$

$$b_i \leqslant \sum_{j=1}^{\mathrm{M}} c^{ij}\, y^j \leqslant b^+{}_i \; (i = 1, \ldots, N) \qquad (4)$$

One of the most efficient algorithms for the solution of problem (2)–(4) is a method proposed by Balash. This method, realized in a standard package of EC computer programs, was used in calculations made for planning multipurpose systems of remote sensing of natural objects from space. As mentioned above, problems discussed by Kondratyev, Grigoryev and Pokrovsky (1975) have been distributed between four comprehensive groups: "Oceanology", "Hydrology", "Geology" and "Forestry and agriculture". Each of these groups, in its turn, is subdivided into subgroups. The number of individual problems within each of the main groups varies from 9 to 14. For illustrative purposes, we shall consider the results of choosing an optimal set of spectral sounding channels for multipurpose surveying in the interests of the "Oceanology" group problems (Kondratyev, 1990b).

Here is a list of particular problems relevant to the "Oceanology" group: (1) sea surface state; (2) water turbidity; (3) sea ice; (4) general-purpose near-coastal surveys; (5) mapping of coastal currents and tides; (6) global mapping of currents, ocean surface surveying; (7) mapping of the coastline and shallow water; (8) bathymetry and topography of ice; (9) propagation and migration of sea organisms; (10) coastal water pollution; (11) impact of pollution on marine environment; (12) study of bars and reefs.

The total number of elementary spectral sounding channels for 12 problems ($N = 12$) relevant to the "Oceanology" group is 37 ($M = 37$). Here the number of elementary spectral intervals m_j, included in the requirements for the ith particular problem, are within: $18 \leqslant m_j \leqslant 33$. This means that the information volume for each particular problem is sufficiently large. It constitutes at least one half of the total information volume for the entire "Oceanology" group. When developing multichannel measurement instrumentation, one may proceed from the consideration that the expenses involved will be in inverse proportion to the spectral interval width, and in direct proportion to the spectral resolution. Let β be the cost of a single measurement channel with a resolution 1 nm. Then the cost $d_{\Delta\lambda}$ of a channel with the resolution $\Delta\lambda$ (nm) will be $d_{\Delta\lambda} = b/\Delta\lambda$. Let b denote the total efficiency of using satellite information in the "Oceanology" group. Let us introduce b_i to denote the efficiency of using satellite information in the ith problem of the "Oceanology" group. Then $\Sigma\, b_i = b$. Let all b_i be identical. Then $b_i = b/N$ will be average efficiency relevant to a single particular problem. By introducing a threshold value α $(0 < \alpha < 1)$ of permissible deviations from average efficiency values, we shall determine the constraints

$$b_i^+ = (1 + \alpha)\, b_i\, ; \qquad b_i^- = (1 - \alpha)\, b_i \qquad (i = 1, \ldots, M)$$

in (3). Let us denote as I_i the totality of channel numbers included in the requirements to the observing system for the ith particular problem. We introduce an indicator function of the jth measurement channel, according to the equality:

$$f_j\,(I_i) \quad = \quad \frac{1, \text{ if } j \in I_i}{0, \text{ if } j \notin I_i}$$

Then the set of coefficients c_{ij}, which are used in the formulation of the optimization problem (1)–(3), can be determined from the equality:

$$c_{ij} = b_i/m_i\, f_j\,(I_i), \qquad (i = 1, \ldots, N)$$

Making use of concrete numerical values: $\beta = 5$ and $\alpha = 31$ (Marchuk and Kondratyev, 1992), all parameters included in (1)–(3) were calculated. Calculations of cost-efficiency indices of measurement channels, k_j, show that some spectral regions are potentially non-profitable (negative k_j values from the viewpoint of the multipurpose functioning of the observing system). For the "Oceanology" group, these are channels in the 0.40–0.45 and 0.69–0.73 μm regions.

Measurements are more justified from the economic standpoint in the following spectral channels: (a) 0.53–0.55 μm; (b) 0.66-0.68 μm; (c) 0.59–0.61 μm; (d) 0.61–0.64 μm; (e) 0.55–0.57 μm. Here maximum values of $\{k_j\}$ are reached. So, even rather simple calculations make it possible to single out the most economically efficient sounding channels among those listed above.

One can therefore see that channels 1 and 7 are the least suitable for problems relevant to the "Oceanology" group out of those presented in Table 1.23. However, this information should only be taken into account in the case of developing specialized oceanographic instruments. There are similar "non-profitable" spectral regions for other groups of problems. For instance, for problems of the "Forestry and agriculture" group, the most cost-effective channel is 1.5–1.8 μm, and the least profitable is channel 9 (0.96–1.10 μm). Consideration of channels suitable for the "Hydrology" problems revealed that all channels were "profitable".

Current developments in the field of satellite remote sensing of the environment have been discussed in detail in monographs by Kondratyev, 1992a, 1994; Kondratyev, Buznikov and Pokrovsky, 1992, 1996a; Kondratyev and Cracknell, 1997; Kondratyev, Johannessen and Melentyev, 1996. Therefore, we shall confine ourselves here to a brief characterization of the most recent developments relevant to the implementation of the Earth Observing System (EOS) (see Bakkes et al., 1994; Asrar and Greenstone, 1995).

1.12.2.1 Processes on the land surface

Numerous agencies take part in developments in this field, but the greatest achievements during the recent time are connected with the creation and application by NASA experts of such active remote sensing (RS) means as a multifrequency multipolarization satellite radar and a satellite lidar (Asrar and Greenstone, 1995).

The radar, developed in collaboration with experts from Italy and Germany, has been tested when launched on two shuttles (Satellite Radar Laboratories (SRL)). The first one, *Endeavor* (STS-59; apogee and perigee heights: 218.0 and 204.0 km, respectively; period of rotation 88.7 min; orbit inclination angle relative to the equator 56.0°) was launched into orbit on April 9, 1994, and its flight continued for 11 days 5 h and 50 min. The second shuttle, also *Endeavor* (STS-68; 226.0 and 213 km; 88.9 min; 57.0°) was launched on September 30, 1994, and landed on October 11, 1994.

The goals of both SRLs were investigations in the fields of ecology, meteorology, oceanography, geology, volcanology, etc. Tentative analysis of data obtained made it possible to monitor the dynamics of many natural phenomena, and in particular, the shift in boundaries between high- and moderate-latitude forests. The US Geological Survey is involved in the mapping of the regions of North and South Africa, Asia and the north-west of the United States. Because the radar signal registered is formed in the upper ground layer, whose thickness is very considerable, especially for dry sand, analysis of radar images opens up opportunities for detection (particularly in desert conditions) of various subsurface structures (including the purpose of studying geological indicators of climate change). The data of SRL-1 obtained during April–May 1994 have been used in the interests of the American–Canadian experiment BOREAS (study of the interaction between the atmosphere and boreal ecosystems to determine the role of the biosphere in climate formation). For instance, a complex analysis of ground-based observation and remote sensing data made it possible to reveal a very low level of water table in the boreal forests of central Canada. Such data show that a correction is necessary of models overestimating the atmospheric moisture content. In reply to requests on the part of the Ukraine and Georgia, these countries were provided with the SRL-2 data for relevant orbits.

A satellite lidar intended for remote sensing of clouds and atmospheric aerosol was launched (the LITE experiment) on the shuttle *Discovery* (STS-64; 269.0 and 259.0 km; 89.5 min; 56.9°) on September 9, 1994 (the mission's duration was 10 days, 22 h, and 50 min).

Because the launch of the natural resources satellite Landsat-6 was not successful, special attention was paid to the processing of the data of the still functioning satellites Landsat-4 and -5, as well as preparations to the launch of Landsat-7 planned for December of 1998. It has been discovered, in particular, that deforestation in the Amazon river basin (Brazil) appeared to be not so catastrophic as assessed previously, even though the area of cut-down forests increased by a factor of 3 during the past 10 years. The fragmentation of humid tropical forests and the "edge effect" along the perimeter of the fragments, which appear as a result of forest cutting, may prove to be very dangerous for biodiversity (about half of the existing species of plants and animals is concentrated in the tropics). Combined analysis of the results of satellite, aircraft and ground-based remote sensing has been very important for the restoration of Californian vineyards which are seriously injured by pests (the cost of an annual vineyard production is about $10 billion).

Remote sensing developments are carried out in many US agencies. For example, the EROS Data Center, which belongs to the US Geological Survey, is responsible for archiving and distribution of the data of a moderate resolution videospectrometer and an advanced satellite radiometer for measuring the outgoing longwave and reflected solar radiation. Both instruments will be installed on the first "polar platform" of the EOS system, whose launch is planned for 1998, and the Landsat-7 satellite data. The National Geophysical Data Center (NGDC), belonging to the National Oceanic and Atmospheric Administration (NOAA), is fully responsible for processing, archiving and distribution of the data of meteorological satellites of the Department of Defense (DMSP). Managing information fluxes and providing information to its users constitute one of the main goals of the NOAA programme on climate and global change (C&GC). From the viewpoint of data interpretation, of primary importance is analysis of the variability of the global environment, the relationship between the contributions of natural and anthropogenic factors of this variability, and assessment of possible socio-economic consequences of global climate change. In this regard, of special significance is the implementation of the *Pathfinder* programme, whose main objective is making the huge database of satellite observations accumulated during several decades really available for its various and numerous users. Within the framework of this programme, the US Geological Survey prepared standardized data of a multichannel scanner, used on Landsat satellites three times: in 1973, 1986 and 1992. The data are intended for studying the landscape dynamics in North America, and humid tropical forests in the Amazon basin, Africa and South-East Asia for a period of the order of 20 years. To be able to compile global maps of vegetation cover dynamics in 1992 (from the vegetation index data), 93 000 images with a 1 km resolution have been accumulated; the images have been obtained from advanced very high resolution radiometers (AVHRR) installed on NOAA satellites. A group of experts representing different agencies made an analysis of the results of monitoring of big floods in the valleys of the

Mississippi and Missouri rivers which occurred in 1993; both satellite and conventional observation data have been utilized.

The Bureau of Indian Affairs (BIA) used Landsat (Thematic Mapper) and SPOT (panchromatic images) satellite data to monitor the land use dynamics and forest fire consequences over the territories of a number of Indian reservations in the States Arizona and New Mexico. The Bureau of Reclamation (BOR) used similar information (as well as AVHRR and aerial photography data) to analyse the variability of natural conditions and efficiency of nature protective actions in a number of regions in the USA; the recently organized National Biological Service (NBS) started detection of such land areas where no measures have been taken to protect biodiversity (it is planned to complete the investigation in the entire country by the year 1999). The interests of the Fish and Wildlife Service (FWS) have been concentrated on ecosystem studies of habitation conditions of the animal world (migration of birds, threatened species, wetland mapping, etc.). The US Bureau of Mines (USBM) used the data of remote sensing mainly to investigate the territories of derelict mines.

The efforts of the Foreign Agriculture Service were aimed at the evaluation of harvest and basic agricultural crops production on a global scale. Relevant prognostic models were developed for the purpose; the objective was providing information to the Department of State concerning food supply, first of all, in the former Republics of the USSR, especially in the Ukraine which is often subject to droughts. Special attention was also given to the dynamics of the summer monsoon in India, the damage to coffee plantations in Brazil caused by frosts, the droughts of Eastern Australia where wheat is grown. The functioning of the satellite service of early warning concerning various threats to agricultural production made it possible to perform a correction of prices and thus ensure maximum profit for American farmers. The Service of National Agricultural Statistics used remote sensing data to evaluate the agricultural production conditions in the USA, and the Agricultural Research Service (ARS) utilized the data to analyze the dynamics of weeds in the National Park of Northern Dakota, as well as for other purposes (mapping of soil salinization, the biomass, harvest, all of which measures have been aimed at the improvement of the harvest of the main agricultural crops).

The Forest Service (FS) used remote sensing data to solve its main problem: that of fighting forest fires (estimation of fire danger extent, damage from fires, etc.), whose propagation is very considerable. Over a controlled area of 191 million acres, the number of fires reached a record figure in 1994: as a result of 66 000 fires, 3.8 million acres of forest was burnt out. After a forest fire begins, the basic means of its monitoring is aerial photography from aircraft. An important source of information about soils for the Soil Protection Service are the data of digital orthography. The Environmental Protection Agency (EPA) is implementing a broad programme of ecological monitoring, making use of remote sensing data in the interests of global change problems, as well as studies of environmental pollution, ecosystem dynamics, etc.

1.12.2.2 The atmosphere

Great attention has been drawn to the monitoring of the total ozone content (TOC) depletion in the atmosphere as a consequence of the Pinatubo volcano eruption in June 1991 (see Kondratyev and Cracknell, 1997); the study was conducted from the TOMS spectrometer data, intended for TOC mapping and installed on the NOAA-9 and Meteor-3M satellites. The volcanic eruption also caused a considerable change of the Earth's radiation budget and the effect of global climate cooling which revealed itself in the course of a few subsequent years. A combined analysis of the data of TOMS and microwave sounding instruments installed on the VARS satellite, the information obtained from the ER 2 stratospheric aircraft and balloon soundings made it possible for the first time to assess the interdependence, on a global scale, of the fields of ozone content, ClO and other trace gases, volcanic aerosols and temperature.

The programme of upper atmospheric studies (UARS) was focused in 1994 on the investigation of the impact of stratospheric aviation upon the ozone layer in the southern hemisphere. Studies of the ozone layer dynamics are an important part of the *Mission to Planet Earth* (MTPE) programme. In this connection, over twenty field investigations are planned for the period to 1998 (the first stage of MTPE) making use of the data of satellite, aircraft, balloon and ground-based observations. An important event was the launch on April 13, 1994, of a geostationary satellite (GOES-8) to be later followed by GOES J-M satellites. As has already been mentioned above, the second MTPE stage will be opened by the launch of the first "polar platform" in 1998. However, due to budget reduction of the EOS programme, these initial plans may be modified (Kondratyev, 1994, 1995c, 1996b). Simultaneously, some additional efforts are being planned now, among which there is possible launching of the SAGE satellite intended for occultation sounding of the stratosphere in the year 2000, as well as the installation of instruments similar to those used on Landsat satellites on the EOS AM-2 polar platform. Very serious efforts are being made to create the EOSDIS system of processing, archiving and distribution of EOS data. The marked progress of the *Pathfinder* programme will be most helpful in this respect.

Supported by the National Science Foundation, studies are conducted of the role of extra-atmospheric insolation variations and anthropogenic factors in the formation of global change. The implementation of the SunRISE programme will make it possible to obtain detailed data of observation of various solar activity indicators determined by the dynamics of sunspots, faculae, the Sun's diameter and other characteristics. Very timely are efforts in numerical modelling of global climate taking account of the concentration growth of both "greenhouse" gases (including methane) and anthropogenic sulphate aerosols (see Kondratyev, 1992a).

Although investigations relevant to satellite meteorology have not been adequately treated in the report by Bakkes, van der Born, Helder et al. (1994) (possibly, the agency to which the authors belong is to some extent responsible for this inadequacy), still some efforts have been mentioned there in the context of the need for aeronautics in meteorological information. This, for instance, refers to efforts within the framework of a programme of observing wind shift (turbulence zones) in front of an aircraft with the help of an on-board radar system, and the turbulent "wake" behind the aircraft, making use of lidar sounding. The development of a complex ground-based system was continued for meteorological service of aviation; the system was tested in the airports of Memphis and Orlando; it includes a new meteorological radar. Studies were conducted to improve flight security in the presence of icing and lightning.

An event of great significance has been the decision made by the US President on May 5, 1994, concerning integration of meteorological satellite systems of the Department of Defense (DMSP) and civilian space-based systems (POES), which must be effected by the beginning of the next century.

1.12.2.3 Oceanographic studies

During 1994, the accumulation, processing and analysis of the data of the radioaltimeter installed on the French–American satellite TOPEX/Poseidon, launched in August 1992, were continued. High-precision data on ocean level height have revealed that the amplitude of the annual variation of ocean level in the northern hemisphere is approximately 50% greater than that in the southern hemisphere. This asymmetry reflects the presence of a more intensive heat exchange between atmosphere and ocean in the northern as compared with the southern hemisphere. Data obtained for the period of the longest El-Niño phenomenon in the last 40 years in the Pacific Ocean of the northern hemisphere (1991–93) have revealed the presence of accompanying changes, such as a displacement northwards of the Kuroshio current and the appearance of a fast current south-eastwards from Japan, which may seriously influence the weather in North America. A considerable contribution to the understanding of processes of the formation of precipitation, clouds and radiation field, of the interaction between atmosphere and of ocean, and oceanic processes has been the implementation in 1993 of a joint project TOGA–COARE (studies of the tropical ocean and global atmosphere with studies of the interaction between atmosphere and ocean), making use of aircraft laboratories ER2 and DC8.

Preparations were made for the processing and analysis of the data of the ocean colour scanner SeaWiFS for the purpose of mapping the global phytoplankton dynamics, which is important, for instance, for the assessment of the carbon dioxide exchange between the atmosphere and ocean. Experts from NOAA have completed the development and in-flight testing of a scanning low-frequency microwave radiometer for mapping the salinity of sea water in coastal zones, as well as preparations for above-water control observations necessary for the interpretation of the data of the scannerometer and ocean colour scanner, which must be launched in 1996 on the Japanese satellite ADEOS.

1.12.2.4 Conclusion

Even a brief and non-comprehensive display of ecological space-based studies in the USA reflects their large scale and depth of penetration (which is even more characteristic for space-based studies on the whole). This is also indicated by generous (although somewhat reduced recently) financing, which amounted, during the fiscal year considered (for summarized satellite studies of all agencies), to $26.653 billion (the share of NASA is 14.568 billion and that of the Department of Defense, coming second, is 13.166 billion).

What is not satisfactory is a certain fragmentary nature of the report discussed (the lack of a systematic approach), which is by no means occasional. As has been mentioned earlier (Kondratyev, 1995a–d, 1996a,b), the EOS programme cannot be called an optimal one; it requires considerable revision and polishing from the viewpoint of establishing priorities with regard to choosing the most urgent problems and relevant observing systems. It is regrettable that even the already existing developments and publications, including the monograph by Mr A. Gore, the US Vice-President (Gore, 1993) and comments on this work (Kondratyev, 1993a,b) have not been used to improve the situation.

The main difficulty in establishing priorities with regard to global change and, consequently, in establishing an optimized global observing system, consists in the number and diversity of environmental problems, which cannot be considered separately from the "human dimension" of the problem of global change.

1.12.3 Satellite monitoring of natural and anthropogenic catastrophes

The problem of using the data of satellite remote sensing aimed at monitoring natural and anthropogenic catastrophes occupies a special place. In this context, we shall consider such disastrous phenomena as floods, volcanic eruptions, earthquakes, deforestation and desertification, forest fires, dust and sand storms, and technogenic catastrophes in cities.

1.12.3.1 Introduction

In the twentieth century catastrophes have become an integral part of the development of modern civilization. In spite of all achievements of technological progress, the broadening of our knowledge about their causes, and all prevention measures taken, catastrophes become more and more appreciable, and the scale of their consequences increases. This is, first of all, due to the planet's population growth, the concentration of masses of people in places which are very vulnerable to dangerous natural disasters, and also to the violation of the rules of technological security when dealing with complicated systems (which is specifically relevant to nuclear power stations and ships).

As far back as the very beginning of the space era, very promising prospects were noted of using satellite observation data for monitoring and studying various natural disasters (above all, typhoons, hurricanes and floods). These promising prospects were connected with the promptness and efficiency of satellite observations, the possibility of great spatial coverage of catastrophe zones and of following the development of disasters. As the number of specialized satellites, including geostationary ones, increased and technical observation means became more abundant, their broad application to monitoring dangerous phenomena was started (Grigoryev and Kondratyev, 1994, 1996b; Kondratyev, Buznikov and Pokrovsky, 1992).

By now, humankind possesses a great experience of using the data of the Earth imagery from space for monitoring various catastrophes. This experience should be thoroughly analysed to enable the elaboration of a complex system of monitoring natural and anthropogenic catastrophes. The purpose of this survey to analyse of the best results of satellite observations. Attention is mainly paid to the most dangerous catastrophes, on the one hand, and prevalent (although perhaps less dangerous) ones, on the other hand.

At present, the particular attention of scientists, governments and the public is attracted to potential global catastrophes of the near future, such as global climate warming, due to the growing concentration of greenhouse gases in the atmosphere, and the depletion of the ozone layer.

1.12.3.2 Floods

Floods are among the most devastating of phenomena, causing (along with earthquakes) the greatest number of victims. Floods can be classified into two large groups: river and lake inundations, and coastal zone floods. Sometimes, in river deltas, both types of floods are joined into a single inundation. A number of disastrous floods have taken place during the twentieth century; each caused over 100 000 human victims (1911, the Yangtze river, China: about 100 000 people; 1931, ibid.: 140 000; 1970, the Ganges river, Bangladesh: 265 700). The material damage caused by floods is enormous (1951, Kansas and Missouri, USA: $1 billion; 1972, eastern USA: $2 billion) (Grigoryev, 1991).

Observations from space make it possible to promptly detect the flooded regions and follow the dynamics of flood development. For example, such observations were effectively used in 1973 in the USA during a large flooding of the Mississippi and its tributaries, caused by heavy rains in the upper Mississippi. In 1993 a huge flood happened there again (according to assessments, a flood of this scale is possible only once every 500 years). Satellite images made it possible to compile a map of the flooded area along the banks extending for many hundreds of kilometres over a few days.

During the 1973 inundation (floods of this scale occur once every 100 years), the mapping of high water made from satellites images made it possible to detect an inundated zone which was not known before—a high water-meadow which had seldom been flooded previously.

A combination of satellite and aircraft pictures (and, above all, those obtained with synthetic aperture radar (SAR) providing all-weather information) yields the best results. It is in this way (among others, from NOAA and Landsat satellites) that in 1991 in China a successful monitoring of the dynamics of floods during an inundation on the Yangtze and its tributaries was accomplished. From that time on, satellite and aircraft observations became a key part of the system of flood monitoring in the country.

In India operational hydrological maps on the scale 1:50 000 are compiled weekly from satellite images. These maps make it possible to assess the scale of a flood, the tendencies of its development, and form a basis for taking short- and long-term measures. Such satellite observations, including those from the Indian satellite IRS-1A, were performed, for instance, to monitor a flood in the basin of the Brahmaputra river.

Radar satellite imagery is especially promising for monitoring floods. For example, images from the Cosmos-1500 satellite were used to follow the flood of the Amur river. The basic parameters of flooded regions were put under control.

To estimate the development of inundations and to forecast them attempts (especially for mountain rivers in Siberia) have been made to model these phenomena taking account of information (including that obtained from satellites) about cloudiness, snow cover, soil moisture.

In some countries (the USA, Bangladesh, Pakistan), the floods of coastal zones cause great devastation, especially in river deltas. To predict such floods, one must be able to forecast the development of cyclones. As far back as in 1970, by making use of satellite information, the flooding of a low coastal area in Pakistan caused by sea waves driven by a tropical cyclone was predicted in advance. Attempts to use satellite data on cyclone movements for modelling inundations in coastal zones of India and Bangladesh also proved to be successful. The mapping of flooded areas, in particular, in the Ganges delta, from high-resolution satellite images was performed to determine the risk zone.

1.12.3.3 Earthquakes

Earthquakes cause the greatest numbers of human victims and huge material damage. During the period 1900–1989, about 1.2 million people perished on the planet as a result of earthquakes. The former USSR was among the six states whose victims constitute more than 82% of the total number. Russia is among 35 countries where the probability of catastrophic earthquakes (of the magnitude of 6–8 points on the Richter scale) is very high. The tragic consequences of the 1995 earthquake in Neftegorsk (Sakhalin island) are widely known.

Earthquakes are extremely difficult to predict. Therefore, some useful information from space is very valuable. Above all, such information is important for detecting and mapping active tectonic perturbations. Investigations of this kind (started in the early 1970s) made it possible to establish a relationship between the centres of earthquakes and active disruptions of rocks in various seismically dangerous regions of the world, for instance, in California, Alaska and the Caucasus. Very useful satellite observations are performed in China, where every year underground shocks kill more people (545) than other natural disasters. This figure exceeds the global average (51%), which is the result of the overpopulation of this highly seismic zone. From satellite pictures of a number of regions in China, it has been possible to detect numerous active tectonic perturbations and, what is particularly important, to establish a high correlation with the orientation of groups of earthquake epicentres reflected on the newest geological map of 1991 (Grigoryev, 1991).

Of great interest is investigation of the phenomenon of precursors of underground shocks. Among these, for instance, are gas emanations which rise to the surface via dicontinuities resulting from tectonic perturbations. Sometimes these emanations are activized and then they contribute to the formation of clouds extended along rock cracks, which is reflected in satellite pictures. Of special interest are specific clouds appearing above active fractures in the regions of earthquakes. For example, it has been noticed that sometimes in seismically dangerous areas, rings of clouds (of the type of smoke rings) are formed several days prior to underground shocks. Such formations have been reflected in satellite images for the Gusli region in central Asia one day before each of two strong underground shocks occurred in April and May 1976.

An important earthquake indicator may sometimes be thermal anomalies of the surface, which can be seen in infrared satellite pictures. For instance, in China thermal anomalies have been established in the areas of the development of big earthquakes: one, which happened in November 1986 in the Yun-Nan province (the Lanshend region), and two others, which occurred in October 1989 in the Shengshi province (the Datong region). These anomalies in the Earth's surface temperature, formed just before the underground shocks, have been detected both from satellite and ground-based observations.

Forerunners of powerful earthquakes may sometimes be two simultaneous phenomena: thermal anomalies (on a scale of 2–10 days) and accumulations of frontal cloud fields (appearing 1–2 days prior to an earthquake). Phenomena of this kind have been repeatedly registered in a number of regions of central Asia directly before powerful underground shocks, from pictures obtained from Meteor and NOAA satellites.

Surveying from satellites is extremely useful for the assessment of various consequences of destructive underground shocks, and for the determination of areas and the extent of perturbations in natural landscapes and technogenic formations. The perspectiveness of using satellite radar images and interferograms should be mentioned, especially for studying tectonic movements (with an accuracy up to a few centimetres). For example, a 10–12 cm displacement of rocks was detected from the ERS-1 satellite after an earthquake in California on May 17, 1991.

1.12.3.4 Volcanic eruptions

With regard to their scale and depth of impact on ecosystems and human society, volcanic eruptions are far behind earthquakes and floods because they are not so frequent; however, they are probably the most menacing and destructive natural catastrophes. If a volcanic eruption takes place in a densely populated area, the number of victims and the extent of material damage may be huge. In the twentieth century, only two volcanic eruptions (1902, Mount Pelée, Martinique, and 1985, Ruiz, Colombia) killed 28 000 and 25 000, respectively.

Satellite pictures make it possible to register very promptly the starting point of an eruption from distinctly visible "smoke" emissions (Grigoryev, 1991). This is how volcanic eruptions were detected in very remote and hardly accessible regions of the planet, such as the Arctic and the Antarctic. In particular, the development of an extended volcanic plume was repeatedly detected in the vicinity of Bennet island (the East-Siberian Sea), whose origination must be due to the activization of an underwater volcano.

The beginning of an explosive eruption of the Mount St Helens volcano (18 May 1980) was registered from the spectrometer data in the IR spectral region from a US Air Force satellite. Observations of an aerosol-gaseous cloud which appeared one minute after the underground shock, which caused the eruption, were conducted simultaneously from two satellites; this enabled triangulation measurements. Thus, consecutive stages of the development of the volcanic eruption could be followed. Satellite data on the extension, direction and temperature of volcanic plumes can be used for evaluating the eruption energy.

Satellite surveying of volcanoes in the infrared spectral region is especially interesting. Satellite pictures reveal thermal anomalies around many active volcanoes all over the globe. They make it possible to assess the functioning of volcanoes even through a thin cloud cover (for instance, this kind of monitoring was performed with regard to the revived volcano Beeren Berg on Jan-Majen island in the Atlantic as far back as in 1970). Such thermal anomalies detected, for example, with a Thematic Mapper installed on a Landsat satellite, made it possible to follow the activities of the volcanoes Erebus in the Antarctic and Lascara in Chile. The appearance of a thermal anomaly around the latter was detected from a satellite a year before the volcano actually erupted in September 1986.

Of course, for predicting volcanic eruptions, as well as earthquakes (often interrelated phenomena), it is very important to collect information in the operational regime; such information is obtained from and transmitted via satellites by means of special ground-based automatic sensors, which register, in particular, variations in the composition of gas emanations. A number of the Earth's volcanoes are already monitored by an observational network, which collects information from satellites. Analysis of satellite information makes it possible to assess the consequences of volcanic eruptions, which is important, for example, for the evaluation of damage caused to population (e.g., devastation of agricultural areas). Space-derived images reveal distinctly not only contemporary but also ancient lava fields which were formed in the historical time, in particular, on the slopes of Etna (Italy) and Mauna-Loa (Hawaii). This is important to assess the risk for human settlements located on the slopes of both volcanoes.

Observations from space proved very useful for the estimation of the risk of avalanching caused by hydrothermal volcanic activity. For instance, from Landsat pictures, the temperature activity has been assessed of fumaroles on the Mt Baker volcano (Washington state), the activity of which is connected with the movement of mud masses along the slopes (Kondratyev, 1996b).

1.12.3.5 Forest and other fires

Fires also cause a lot of damage. They result in considerable material damage, land devastation and destruction of human settlements. Big fires, which are mostly of natural–anthropogenic, or simply anthropogenic origin, occur in various regions and natural zones of the Earth. They happen frequently in forests and swamps, as well as in steppes and savannahs.

During very big forest fires in Indonesia in 1983, the result of a drought and careless behaviour of people in eastern Kalimantan and northern Borneo, forest was damaged over an area of 3.5 and 1 million ha, respectively (Grigoryev, 1991).

Data of satellite observations are used very efficiently for the detection of forest fires; they are especially important for prompt detection of fires in thinly populated or inaccessible areas, for instance, in the Siberian taiga or forest-tundra. In the daytime, fires are recognized in the visible spectral region from smoke plumes. The entire zone of forest fires can be mapped very quickly. This was the case, for instance, with forest and swamp fires in densely populated central Russia which happened as a result of an over-dry summer, during the summer and autumn of 1972. The entire extensive area of forest fires was detected and mapped along the left-hand bank of the Ob' river, according to the pictures from meteorological satellites. The whole area, marked by several dozens of smoke plumes, reached 200 000 square kilometres.

Permanent monitoring of the entire zone of development of large-scale fires, which often cover territories of hundreds of square kilometres in Siberia or the Far East and continue for many weeks, is possible from satellites only. The data from geostationary satellites can also be very helpful.

To determine the precise boundaries of a fire centre, infrared images are used most efficiently; spectral intervals 3.5–4.2 and 4.4–5.0 μm, where the radiation contrast between the fire centre and its background is revealed most distinctly, have proved to be optimal for the purpose. IR images are also helpful for monitoring fires at night and, what is even more important, for detecting underground, soil-hidden and deep fires in swamps.

During the large-scale forest and swamp burning of 1983 in Indonesia, the fire development was monitored from IR images from the NOAA-7 satellite. A smoke barrage did not hinder the monitoring. It appeared possible to detect from the pictures the centres of underground fires (in swamps) preserved after rainfall.

In China, where the number of annual burnings amounts to nearly 4000, NOAA satellite pictures have been successfully used for monitoring forest fires, beginning in 1985. In particular, they have been used to obtain information about catastrophic forest fires in May–June 1987, in a mountain region Xinan. Space-derived images with different territory coverage and different resolution from the Landsat, NOAA and SPOT satellites have made it possible to detect new sources of burning and to monitor zones of fire propagation.

Observations from space also yield good results in the assessment of the propagation of smoke from fires and burns, in the evaluation of the danger of fires (from the data about soil moisture, snow cover melting, etc.), in territory classification according to the conditions of fire development, and also in the search for "resource" cloudiness (powerful cumulus and cumulus-rainy clouds) to provoke precipitation. To improve the effectiveness of image decoding, various processing techniques are used. For instance, to map on a scale of 1: 250 000 the consequences of forest fires in Provence (France) in the early 1980s, the following operations were performed on the basis of Landsat pictures: coding at 16 levels, colour coding, compilation of histograms, the variation of the contrasts and scale of images in the interactive regime (Losev, Gorshkov, Kondratyev et al., 1993).

1.12.3.6 Deforestation of territories

One of the present-day global-scale problems is massed and progressing annihilation of forests. The human onslaught makes forests retreat on all continents and in almost all countries. During the last 20 years (according to the data for the period from 1966–68 to 1986–88), the global forest area decreased by 6% and constituted $130\,691 \times 10^3$ km^2.

The annihilation of forests leads not just to deforestation of huge territories, but to the transformation of natural conditions of entire geographical regions and even the mix-up of the boundaries of natural zones, and, ultimately, to the violation of ecological equilibrium. The latter is fraught with the danger of a large-scale ecological catastrophe. The value of satellite observations, in particular, consists in the circumstance that these observations yield an integrated pattern of the actual propagation of forests preserved in one or another region. Forest cuttings are a major cause of the destruction of forest ecosystems. Fresh cuttings, which are absent from the maps, can be very clearly seen even in small-scale low-resolution satellite pictures (the first hundreds of metres); the woods of the Komi Republic or the Kirov and Arkhangelsk regions may serve as examples.

An issue of special concern is the cutting of tropical rain forest, which, as is well known, is accompanied by a considerable loss of biodiversity, the disturbance of the hydrologic cycle, and impacts on climate. It is in the tropics, and, as established by satellite observations, particularly in the forests of South America, that rapid deforestation of the territory takes place. Within just seven countries of the continent, forests are destroyed annually over an area of 6880×10^3 ha. During the period from 1981 to 1990, the forest area was reduced from $797\,100 \times 10^3$ ha to $729\,300 \times 10^3$ ha. According to surveying from space data, in 1988 selva in Amazonia (Brazil) was annihilated over an area of 4.8×10^5 ha. Selva is destroyed most of all alongside roads, and in particular, along the Trans-American Highway.

Forest degradation over a huge territory is due not only to wood-cutting, but also to fires and to damage caused by pests. To assess such damage, maps on scales from 1:100 000 to 1:500 000 are compiled from satellite pictures. In China images from Landsat and the Chinese satellite FY1 are used for the purpose. On the basis of spectral differences, for example, forests were classified into two groups: those slightly injured and those severely damaged by pests.

In India the assessment of the state of great forest massives was conducted from images obtained from the IRS-1A satellite with a resolution of about 36 m. The accuracy of maps made on their basis was sometimes 80%, which was verified by ground-based investigations.

Surveying forests from space is most helpful also for the evaluation of deforestation consequences. The utilization of satellite microwave survey data in studying the drying-up of Amazonia (Gorshkov, 1990), connected with the destruction of selva, is very interesting. An inventory (from satellite data) of the forest biomass might be a promising study; its results would be useful for the assessment of carbon dynamics in the "forest–atmosphere" system. An attempt at such an inventory has been made in India (the Rajari national park); it has been found that approximately 0.188×10^9 of carbon is annually absorbed over an area of 1 km^2. Global biomass estimation is already possible on the basis of mapping the vegetation index (which takes account of brightness characteristic differences at various wavelengths), in particular, from NOAA satellite data. Thus, realistic opportunities open up for the assessment of gas exchange disturbances in large regions and on the planet as a whole (Kondratyev, Asimakopoulos and Varotsos, 1995).

1.12.3.7 Land desertification

Land desertification is a phenomenon which is usually defined as ecosystem degradation accompanied by a decrease in bioproductivity; it takes place both in desert areas, and (especially) in so-called marginal (neighbouring) zones—semideserts, dry steppes, etc. All these changes are registered, even in savannahs and steppes. The landscape degradation (of the ecosystems of marginal areas encircling deserts if these are sufficiently salinized) originates ecological catastrophes. The two greatest catastrophes of this kind in the twentieth century occurred during the 1930s in the Midwest of the United States, and in Sahel (Africa) during 1968–73. Such phenomena have become common over the entire planet, including in Russia (mainly Kalmykia) and Central Asian countries (around the Aral Sea).

In the Aral region, as has been established from satellite observation data, there are marked signs of land degradation in the ecosystems of the Amu-Darya and Syr-Darya river deltas. For instance, the following phenomena have been registered there: the drying-up of forests growing along river arms, degradation of the delta vegetation between the river arms, formation of soil retirement regions, defilation. Many of the lakes are also dried-up and therefore become smaller or disappear altogether; river arms and swamps become dried-up too, and sometimes the land becomes saline (Gorshkov, Kondratyev and Losev, 1992). All these changes are registered from the data of remote sensing in the visible and infrared spectral regions. IR images are especially helpful for detection of changes in the moisture condition of river deltas. Since the Aral Sea level continues to decrease, the tendency of landscape degradation in river deltas, which was registered in 1961, also remains.

Similar processes have been developing during the recent decades in Kalmykia, the semideserts of the Black Lands and the surrounding region of the Salsk steppes. Analysis of space-derived pictures has revealed the progressing propagation of desertification "nuclei", which are usually formed around wells (Grigoryev, 1991).

Study of satellite images of Sahel (the transitional zone of semideserts and desert-savannahs to savannahs in Africa) made it possible to detect the propagation of the Sahara desert southwards. But, what is even more important, it has also been discovered that the desertification of Sahali takes place not only due to the broadening of Sahara proper, but also inside Sahali itself owing to the formation and broadening of desertification nuclei around settlements and wells. These desertification nuclei become wider, merge and form big centres of desertification, which reach the boundary of the Saharan desert zone.

Important information concerning large-scale desertification tendencies on the Earth, including Sahel, is provided by data about the vegetation index, which is calculated from satellite information and reflects seasonal and intraseasonal changes and productivity variations.

1.12.3.8 Dust storms

The natural, short-term phenomenon of a dust storm is rarely accompanied by human victims, but it causes great damage to the soil cover owing to the blowing-out of the fertile soil layer. Dust and salt blow-aways are harmful to vegetation cover. Because it is a large-scale phenomenon bringing about negative consequences, it may be considered catastrophic. The largest-scale catastrophe of this kind in the twentieth century may be considered the dust storms in the so-called "dust bowl" which was formed in the USA during the 1930s. The development of such dust storms was regarded as a national catastrophe. Nowadays, special attention is attracted to dust storms, in so-called marginal zones surrounding deserts, which are anthropogenically induced.

Observations from space make it possible to follow the propagation of dust fluxes, to determine their dynamics and temporal variability, as well as to detect the centres of their origination and monitor the peculiar features of their development and functioning. The Aral region, formerly the territory of the USSR, and now that of Turkmenistan, Uzbekistan and Kazakhstan, is one of the areas subject to dust storms. This region has been acknowledged to be a zone of ecological disaster on an international level. Satellite surveying made it possible to determine the starting point of the development of powerful dust blow-aways in the region (1975), the main directions of their movement (south-westwards, to the densely populated oasis in the Amu-Darya delta), their expansion (about 200 000 km^2), the average (250 km) and maximum (500 km) extension of dust fluxes (Grigoryev, 1991).

The data of surveying from space (from meteorological and Earth resources satellites) for the period from 1975 to 1990 have revealed an annual increase of the number of days with powerful dust blow-aways from 4.8 (1975–82) to 5.5 (1985-1990), which shows the deepening of the ecological crisis in the Aral region. The area of dust storm centres has been found to increase. New dust storm centres continue to be formed along the southern coast and along the periphery of the Amu-Darya delta (in addition to earlier centres along the eastern coast). Finally, the data of space imagery (in combination with other sources) makes it possible to approximately estimate the amount of dust lifted into the air: on the average, 45 million tons per year in 1975–82 and about 90 million tons per year in 1982-90. The mass of salts, which are harmful for the ecosystems, blown away by dust storms has been estimated at 0.9–1.5 million tonnes.

Dust storms monitored from satellites have become a reliable indicator of desertification, or an unfavourable ecological situation, in other regions of the world as well, for example, in Kalmykia, and, especially, in Sahel. In the late 1980s, threatening signs of ecosystem degradation were discovered in the Midwest of the USA: a large-scale dust cloud was detected from satellites. It grew out of the area previously sombrely known as the "dust pot", and extended for 2.4 thousand kilometres along the Atlantic shore.

1.12.3.9 Catastrophes in cities

Cities, which are centres of concentration of large numbers of people and dangerous industry, should be under constant surveillance. Observations of cities and their conurbations from space are very useful. These observations are most effective for the assessment of natural catastrophes threatening cities. For example, it can be seen now that many cities are located most unfavourably in zones of impacts from volcanic eruptions, earthquakes and other natural phenomena.

Photography from space makes it possible to detect active fractures, through which a discharge of the Earth's bowels takes place. One of the first maps of this kind was compiled from images obtained on board the Apollo-7 manned spacecraft for the region of Los Angeles located in a seismically dangerous zone. The map has proved very helpful for the assessment of more or less probable areas of underground shocks (Grigoryev, 1991; Grigoryev and Kondratyev, 1996b).

Satellite photography supplies useful information about large-scale floods, in particular river inundations, which are a threat to the economic activity of cities. During a catastrophic inundation on the Mississippi river in 1973, acknowledged as a national disaster, a broad belt of land along the banks extending for about 2000 km was flooded, including the suburbs of numerous cities and towns. The flood dynamics was followed owing to satellite pictures, which facilitated taking operational measures.

Satellite imagery supplies essential information about various urban phenomena that are due to anthropogenic (technogenic) causes. This has proved to be the only possible way of detecting large-scale smoke pollution in Western Europe and the eastern part of North America; the overall area in the first case amounted to 200 000 km^2, and in the second one, above 1 million km^2. Many cities, which are not themselves active sources of pollution, appeared to be in the zone of developing atmospheric contamination.

Observations from space proved to be useful for the assessment of the development of large-scale fires in cities and their suburbs, and in particular in Los Angeles in 1992. Thermal IR images (at the 3.7 m wavelength) from NOAA satellites made it possible to detect considerable thermal anomalies corresponding to the regions of substantial fire development, and to follow their dynamics. This kind of information is important for judgement about the situation in cities during big fires.

Explosions, which sometimes happen at industrial sites in cities, can be registered from space. Often enough such explosions are dangerous for neighbouring cities and towns and even for large regions. Information of this sort has been obtained from satellite observations on the Chernobyl nuclear power station: an explosion and the beginning of a fire were registered by satellites. It is this information which served a warning about potential radioactive pollution of the European continent (which was extremely important because, at the first stage of developments, the information was kept a secret).

1.12.3.10 Conclusion

The experience of satellite observations of natural and anthropogenic (technogenic) catastrophes is sufficiently rich. It indicates that the efficiency of such observations has considerably increased during the last 25 years, mainly due to a broader range of technical means for obtaining the information and advanced techniques of its processing. However, this experience might and should be used on a broader scale in practice, for operational purposes. Observations from space are most often used as an operational aid for monitoring the development of hurricanes and, to some extent, of sea coastal or river floods. There have been cases of their application to monitoring the development of fires and dust storms. Regrettably, the majority of catastrophic phenomena, such as land-slides and snow-slides, and particularly the most dangerous ones, volcanic eruptions and earthquakes, are still outside the system of permanent monitoring from space. This is partially due to objective (organizational) difficulties (especially with regard to earthquakes), and, to a certain extent, to underestimation the value of satellite observations.

Taking account of considerations presented by Clark and Munn (1986), the basic goals relevant to satellite monitoring of catastrophes are: (1) working out requirements of remote sounding data; (2) elaborating a uniform system of observation of catastrophic phenomena and regions of their possible development; (3) creating a uniform information structure (database) for the analysis of phenomena observed; (4) combined use of the data of satellite and conventional observations; (5) training experts for remote sensing data processing.

A prototype system of continuous operational monitoring of the environment, intended for detecting and following the consequences of natural disasters and catastrophes caused by man's reckless activities, is being developed in Japan (Kuroda and Koizumis, 1992). The launching of 26 polar-orbit and six geostationary satellites is proposed for the purpose. As a result, a system like this will provide satellite information with a resolution of 20 m for any region of the globe. Observations of the entire planet will be performed at least once every day. If necessary, images with a resolution of 2 m will be obtained. It is supposed that all countries will have access to information obtained in Japan.

REFERENCES

Abakumov V.A. 1993. On the peculiar features of biospheric thermodynamics. Problems of ecological monitoring and modelling of ecosystems. **15**, p.21–36.

Aeronautics and Space Report of the President. Fiscal Year 1994 Activities. 1995. Washington, DC, 143 pp.

Alekseyev B.V., Kryshev I.I., Sazykina T.P. 1992. Physical and Mathematical Modelling of Ecosystems. Gidrometeoizdat Publ., St. Petersburg, 368 pp.

Anderson V. 1991. Alternative Economic Indicators. Routledge, London, 106 pp.

Asrar G., Greenstone R. (Eds.). 1995. MTPE EOS Reference Book. NASA GSFC Publ., Greenbelt, MD, USA, 277 pp.

Baden J.A. (Ed.). 1994. Environmental Gore. Pacific Res. Inst. for Public Policy, San Francisco, 270 pp.

Bakkes J.A., van der Born G.J., Helder J.C., Swart R.J., Hope C.W., Parker J.D.E. 1994. An Overview of Environmental Indicators: State of the Art and Perspectives. UNEP/EATR.94.-01. RIVM/40200 1001, Nairobi, 72 pp.

Barbier E., Burgess J., Folke C. 1994. Paradise Lost? The Ecological Economics of Biodiversity. Earthscan, London, 224 pp.

Beder Sh. 1995. The role of technology in sustainable development. Technol. Society Magaz., 13, No.4, p.14–19.
Belchansky G.I., Kozlenko N.N., Ovchinnikov G.K., Shevchenko V.I. 1994. Documenting of tundra ecosystem parameters using data of spacecraft "Almaz", ERS-1 and "Landsat-TM". Studying the Earth from Space, No.4, p.81–87.
Belniak E., Górka K. 1994. Sources of financing of the environmental protection in Poland. In: Industrial and Environmental Policy in Poland, Cracow, p.87–95.
Beran M. (Ed.). 1995. Carbon Sequestration in the Biosphere. Springer Verlag, Berlin, 298 pp.
Berreen J., Meyer A. 1992. A package marked "Return to sender". Some problems with the Climate Convention. Network'92, No.18, p.6–7.
Beruchashvili N.L. 1994. Seasonal dynamics of global landscapes (results of computer experiments). Izvestiya Russian Acad. Sci. Ser. Geography, No.6, p.25–40.
Bierbaum R., Watson R. 1995. Recasting U.S. federal environmental R&D priorities. An editorial comment. Climatic Change, 29, No.2, p.123–130.
Boden T.A., Kaiser D.P., Sepanski R.J., Stoss F.W. 1994. Trends'93: A compendium of data on global change. CDIAC Communications, issue No.20, p.8–10.
Bolin B. 1994. Science and policy making. Ambio, **23**, No.1, p.25–29.
Bongaarts L. 1994. Population policy options in the developing world. Science, 263, p.771–776.
Booth D.E. (Ed.). 1994. Valuing Nature. The Decline and Preservation of Old-Growth Forests. Roseman and Littlefield Publ., London, MD, USA.
Borisenkov E.P., Kondratyev K.Ya. 1988. Carbon Cycle and Climate. Gidrometeoizdat Publ., Leningrad, 320 pp.
Boyle T.J.B., Boyle C.E.B. (Eds.). 1994. Biodiversity, Temperate Ecosystems and Global Change. Springer, Environm. Sci., **VI**, 456 pp.
Bromley D.W., Segerson K. 1992. The Social Response to Environmental Risk. Policy Formulation in an Age of Uncertainty. Kluwer, Dordrecht, 240 pp.
Bruff G., Wood A. 1995. Sustainable development in English metropolitan district authorities: an investigation using unitary development plans. Sustain. Development, 3, p.9–19.
Bruntland G.H. 1989. Global change and our common future. Environment, 31, No.5, p.16–20; 40–44.
Buiten H.J., Clevens J.G.P.W. 1994. Land Observation by Remote Sensing. Theory and Observations. Gordon & Breach, Reading, UK, XVI, 642 pp.
Bukata R.P., Jerome J.H., Kondratyev K.Ya., Pozdnyakov D.V. 1995. Optical Properties and Remote Sensing of Inland and Coastal Waters. CRC Press, Boca Raton, FL, 362 pp.
Calvert J. (Ed.). 1994. The Chemistry of the Atmosphere: Its Impact on Global Change. Blackwell, London, 394 pp.
Carley M., Christie I. 1993. Managing Sustainable Development. Univ. of Minnesota Press, Minneapolis, USA, 303 pp.
Chaning M.-L. (Ed.). 1993. The Role of the Stratosphere in Global Change. NATO ASI Series, Subseries I "Global Environmental Change", 8, 557 pp.
Chase R., McGoldrick L. 1984. Oceanic transport. NASA Techn. Memo. 86129, Goddard Space Flight Center, Greenbelt, MD, USA, p.34–36.
Clark W.C., Munn R.E. (Eds.). 1986. Sustainable Development of the Biosphere. Cambridge Univ. Press, Cambridge, 491 pp.
Cobb C.W., Cobb J.R. 1994. The Green National Product: A Proposed Index of

Sustainable Economic Welfare. Univ. Press of America, Inc., Lanham, MD, USA, 343 pp.
Cohen J.E. 1995. Population growth and Earth's human carrying capacity. Science, 269, No.5222, p.341–346.
Daily G.C., Ehrlich P.R. 1992. Population, sustainability, and earth's carrying capacity: a framework for estimating pollution sizes and lifestyles that could be sustained without undermining future generations. Bio Science, 42, p.761–771.
Dalal-Clayton D.B., Bass S., Sadler B., Thomson K., Sardbrook R., Robins N., Hughes R. 1994. National Sustainable Development Strategies: Experience and Dilemmas. IIED Environm. Planning Issues, No.6, London, 56 pp.
Daly H. 1992. Allocation, distribution, and scale: towards an economics that is efficient, just, and sustainable. Ecological Economics, 6, p.185–193.
Delmas R.J. (Ed.). 1995. Ice Core Studies of Global Biochemical Cycles. Springer Verlag, Berlin, 475 pp.
Delwiche C.C. 1984. Biogeochemical considerations. NASA Techn. Memo. 86129, 1 (Pt.2). Goddard Space Flight Center, Greenbelt, MD, USA, p.8–10.
Dentener F.J., Crutzen P.J. 1996, A three-dimensional model of the global ammonia cycle. J. Atmos. Chem., 19, No.4, p.331–370.
Desai N. 1992. The outcome of Rio. Network'92, No.18, p.1, 18, 19.
Development and Environment: World Development Report 1992. 1992. Oxford Univ. Press, New York.
Dieren W. van. 1995. Taking Nature into Account: A Report to the Club of Rome. Copernicus (Springer Verlag), New York, 332 pp.
Donchenko V.K. 1995. Ecological Integration. Part 1. Socio-Economic Aspects of the Ecological Integration of Russia into the World Community. Research Centre for Ecol. Safety Publ., St. Petersburg, 163 pp.
Dooge J.C.I. (Ed.) et al. 1992. An Agenda of Science for Environment and Development into the 21st Century. Cambridge Univ. Press, 352 pp.
Dovers S., Norton T. 1994, Population, environment and sustainability: reconstructing the debate. Sustain. Development, 2, Pt.1, p.1–7.
Dudkin K.N. 1985. Visual Perception and Memory. "Nauka" Publ., Leningrad, 205 pp.
Dworak T.Z. Fizyka Srodowiska Atmosferycznego. Wygwnictwa AGH, Krakov, 111 pp.
Ecological Economic Aspects of Nature and Society. 1994. Course papers from the Ecological Economic Graduate Course. Spring 1994. Dept. of Systems Ecology, Stockholm Univ. Techn. Report No.17, Stockholm.
Ehrlich P.R., Ehrlich A.H. 1988. Population, plenty and poverty. Nat. Geograph., 174, No.6, p.914–945.
Ellsaesser H.W. (Ed.). 1992. Global 2000 Revisited. Mankind's Impact on Spaceship Earth. Paragon House, New York, 436 pp.
Factor 10 Club. Carnonles Declaration. 1995. Wuppertal Institute Publ., Wuppertal, Germany, 6 pp.
Famielec J., Górka K. 1994. Economic losses due to environmental pollution in Poland. In: Industrial and Environmental Policy in Poland, Cracow, p.57–61.
Fleagle R.G. 1994. Global Environmental Change: Interactions of Science, Policy, and Politics in the United States. Greenwood Publ. Group, Westport, CT, 264 pp.
Folke C., Holling C.S., Perrings Ch. 1994. Biodiversity, Ecosystems and Human Welfare. BEIJER Discussion Paper Ser. No.49, Stockholm, 13 pp.
Foody G., Curran P. (Eds.). 1994. Environmental Remote Sensing from Regional to Global Scales. Wiley, Chichester, 250 pp.

Friedmann J., Rangan H. 1993. In Defense of Livelihood. Comparative Studies on Environmental Action. UNRISD. Kumarian Press, West Hartford, CT, 220 pp.
General Assembly Resolution 42/186 of 11 Dec. 1987 on the Environmental Perspective to the Year 2000 and Beyond. 1988. UNEP, Nairobi, 34 pp.
Ghai D. Conservation, Livelihood and Democracy: Social Dynamics of Environmental Changes in Africa. 1992. UNRISD Discussion Paper DP-33, Geneva, Switzerland, 22 pp.
Ghai D. (Ed.). 1994. Development and Environment. Sustaining People and Nature. UNRISD, Blackwell, Oxford, 263 pp.
Gille J.C. 1984. Middle Atmosphere Science. NASA Techn. Memo. 86129, **1**, Pt.2, Goddard Space Flight Center, Greenbelt, MD, USA, p.49–52.
Gillroy J.M. (Ed.). 1993. Environmental Risk, Environmental Values and Political Choices: Beyond Efficiency Trade-Offs in Public Policy Analysis. Westview Press, Boulder, CO, USA, 189 pp.
Gladky Yu.N., Grigoryev A.A., Yagya V.S. 1990. The Horizons of Oykumena. Lenizdat, Leningrad, 272 pp.
Gladwin T.N., Krause T.-S., Kennelly J.J. 1995. Beyond eco-efficiency: towards socially sustainable business. Sustain. Development, **3**, p.35–43.
Global Change and Arctic Terrestrial Ecosystems. 1995. Ecosystems Research Report-10. European Commission, EVR 15519 EN, Luxembourg, 329 pp.
Goklany I.M. 1995. Strategies to enhance adaptability: technological change, sustainable growth and free trade. Climatic Change, 30, No.4, p.427–449.
Goldenberg J. 1994. Energy for a sustainable world population. In: Population and Global Security. The Foundation for Environm. Conservation, Geneva, p.205–219.
Golubev G.N., Danilov-Danilyan V.I., Kondratyev K.Ya., Kotlyakov V.M., Losev K.S. 1995. Once again about the basis of sustainable development of nature and society. Vestnik Russian Acad. Sci., No.6, p.516–519.
Goodland R., Daly H., Kellenberg J. 1994. Burden sharing in the transition to environmental sustainability. Futures, 26(2), p.146–155.
Gore A. 1993. Earth in the Balance. Ecology and the Human Spirit. Plume, New York, 408 pp.
Górka K. 1994. The role of the state in industrial reconstructuring in Poland. In: Industrial and Environmental Policy in Poland, Cracow, p.8–26.
Gorshkov V.G. 1990. Biospheric energetics and the sustainability of environmental state. Itogi Nauki I Tekhn. Theoretical and General Aspects of Geography, **7**, Moscow, 239 pp.
Gorshkov V.G. 1993. Present environmental changes and possibilities for preventing them. Doklady Russian Acad. Sci., 332, No.2, p.802–805.
Gorshkov V.G. 1995a. Physical and Biological Bases of Life Stability. VINITI Press, Moscow, 470 pp.
Gorshkov V.G. 1995b. Physical and Biological Bases of Life Stability. Man, Biota, Environment. Springer, Heidelberg, 330 pp.
Gorshkov V.G., Kondratyev K.Ya. 1990. The Le Chatelier principle as applied to the biosphere. Ecology, No.1, p.7–16.
Gorshkov V.G., Kondratyev K.Ya., Sherman S.G. 1990. The global carbon cycle change: Le Chatelier principle in the response of biota to changing CO_2 concentration in the atmosphere. Il Nuovo Cimento C, 13, Ser.1, No.5, p.801–816.
Gorshkov V.G., Kondratyev K.Ya., Losev K.S. 1992. Global ecological perspectives. Vestnik Russian Acad. Sci., No.5, p.70–81.

Gorshkov V.G., Kondratyev K.Ya., Danilov-Danilyan V.I., Losev K.S. 1994. Environment: From New Technologies to New Thinking. VINITI Press, Moscow, 24 pp.
Gorshkov V.G., Kondratyev K.Ya., Losev K.S. 1994. The natural biological regulation of the environment. J. Ecol. Chem., **3(2)**, p.85–90.
Gorshkov V.G., Kondratyev K.Ya., Losev K.S. 1996. Ecological problems of modern civilization. Vestnik Russian Acad. Sci., **66,** No.2, p.119–128.
Gowdy J. 1994. Coevolutionary Economics: The Economy, Society and the Environment. Kluwer, Dordrecht, 264 pp.
Gowdy J.M., Olsen P.R. 1994. Further problems with neoclassical environmental economics. Ethics, 16, p.161–171.
Grace J., Lloyd J., McIntyre J., Miranda A., Meir P., Miranda H., Moncrieff J., Massheder J., Wright I., Gash S. 1995. Fluxes of carbon dioxide and water vapour over an undisturbed tropical forest in south-west Amazonia. Global Change Biology, **1**, No.1, p.1–12.
Graedel T.E., Crutzen P.J. 1993. Atmospheric Change: An Earth System Perspective. W. H. Freeman & Co., 446 pp.
Grigoryev A.A. 1991. Ecological Lessons of the Historical Past and Present Times. "Nauka" Publ., Leningrad, 252 pp.
Grigoryev A.A., Kondratyev K.Ya. 1994. The present state of global environment and natural resources (a review of the UNEP Report for 1991–1992). Izvestiya Russian Geograph. Soc., **126**, issue 3, p.12–26.
Grigoryev A.A., Kondratyev K.Ya. 1996a. Global change: the problem of sustainable development indicators. Izvestiya Russian Geograph. Soc., 128, issue 1.
Grigoryev A.A., Kondratyev K.Ya. 1996b. Satellite monitoring of natural and anthropogenic catastrophes. Studying the Earth from Space, No.3.
Gritsay O.V., Kotlyakov V.M., Preobrazhensky V.S. 1994. The changing world and the evolution of geographic thought. Izvestiya Russian Acad. Sci., Ser. Geography, No.6, p.5–24.
Grubb M., Koch M., Thomson K., Munson A., Sullivan F. 1993. The Earth Summit Agreements. A Guide and Assessment. Earthscan, London, 208 pp.
Haub C. 1995. Global and US national population trends. Consequences, 1, No.2, p.2–11.
Hemond H.F., Fechner E.J. Chemical Fate and Transport in the Environment. Academic Press, New York, 338 pp.
Hemphill J.G. 1995. Six principles for a sustainable energy future. Climate Change Bull., issue 8, p.7–8.
Henderson-Sellers A., McGuffie K. 1995. Global climate models and "dynamic" vegetation changes. Global Change Biology, 1, No.1, p.63–76.
Hinsman D.E. (Ed.). 1995. Environmental Programmes and Agencies. Preprint, WMO, Geneva, 17 pp.
Holdgate M.W. 1994. Hopes for the future. In: Population and Global Security. The Foundation for Environmental Conservation, Geneva, p.221–246.
Holmberg J., Thomson K., Timberlake L. 1993. Facing the Future. Beyond the Earth Summit. IIED, Earthscan, London, 48 pp.
Homer-Dixon T.F., Boutwell J.H., Rathjens G.W. 1993. Environmental change and violent conflict: growing scarcities of renewable resources can contribute to social instability and civil strife. Sci. American, 268, No.2, p.38–45.
Houghton J.T., Jenkins G.J., Ephraums J.J. (Eds.). 1990. Climate Change. The IPCC Scientific Assessment. Cambridge Univ. Press, Cambridge, 364 pp.

Hull R. 1993. European Community Environment Policy. Post-UNCED. Environm. Policy & Practice, 3, No.2, p.151–160.
Hunt C.A.G. 1994a. Ecologically Sustainable Development: Spatial, Economic and Institutional Issues in Its Implementation. PhD Thesis. The Faculty of Social Sciences, Flinders Univ. of South Australia, 484 pp.
Hunt C. 1994b. Sustainable development in a region: a practical approach to targeting environmental impacts. Sustain. Development, 2, Part 1, p.23–30.
IGBP in Action: Work Plan 1994–1998. 1994. Global Change Report No.28, 151 pp.
IGBP Global Modelling and Data Activities 1994–1999. 1994. Global Change Report No.30, 87 pp.
Investing in People. The World Bank in Action. 1995. The World Bank Publ., Washington, DC, 83 pp.
1992 IPCC Supplement. Intergovernmental Panel on Climate Change. 1992. WMO/UNEP, Geneva, 7 pp.
Ireland E.C., van (Ed.). 1994. International Environmental Economics: Theories, Models and Applications to Climate Change, International Trade and Acidification. Elsevier, Amsterdam.
Israel Yu.A., Semenov S.M., Khachaturov M.A. 1993. Bioclimatological aspects of complex global monitoring. Problems of Ecological Monitoring and Modelling of Ecosystems. Gidrometeoizdat, St. Petersburg, **15**, p.8–20.
James D.E. 1994. The Application of Economic Techniques in Environmental Impact Assessment. Kluwer, Dordrecht, 320 pp.
Jänicke M., Bolle H.-J., Carius A. (Eds.). 1995. Umwelt Global. Veränderungen, Probleme, Lösungssätze. Springer Verlag, Heidelberg, **XI**, 252 pp.
Jørgensen S.E. 1992. Integration of Ecosystem Theories: A Pattern. Kluwer, Dordrecht, 389 pp.
Kanemasu E.T. 1984. Land biology. NASA Techn. Memo. 86129, 1, Part 2, Goddard Space Flight Center, Greenbelt, MD, USA, p.21–23.
Kaplan R.D. 1994. The coming anarchy. The Atlantic Monthly, 273, No.2, p.43–76.
Karl T., Bretherton F., Easterling W., Trenberth K. 1995. Long-Term Climate Monitoring of the Global Climate Observing System (GCOS). Draft Report, preprint. Asheville, NC, USA, 16 pp.
Kassas M. 1989. The biosphere and the threat of global industrialization: limits of the biosphere. The Environmentalist, 9, No.4, p.261–268.
Kelly M., Granich S. 1992. A step in the right direction? Tiempo, issue 2, p.1–3.
Kennedy P. 1993. Preparing for the Twenty-First Century. Random House, New York.
Khan A.M. 1995. Sustainable development: the key concepts, issues and implications. Sustain. Development, **3**, No.32, p.63-69.
Kondratyev K.Ya. 1987. The International Geosphere–Biosphere Programme: the role of satellite observations means. Studying the Earth from Space, No.1, p.104–118.
Kondratyev K.Ya. 1988. Climate Shocks: Natural and Anthropogenic. Wiley, Chichester, 296 pp.
Kondratyev K.Ya. 1990a. Global security and its ecological component. Izvestiya USSR Geograph. Soc., 122, issue 3, p.212–220.
Kondratyev K.Ya. 1990b. Key Problems of Global Ecology. Itogi Nauki I Tekhn. Theoret. and General Aspects of Geography, 9, Moscow, 454 pp.
Kondratyev K.Ya. 1990c. The International Geosphere-Biosphere Programme: the status quo and perspectives. Izvestiya USSR Geograph. Soc., 122, issue 5, p.393–403.

Kondratyev K.Ya. 1990d. The International Space Year: priorities and perspectives. Studying the Earth from Space, No.1, p.3-13.
Kondratyev K.Ya. 1990e. Global security and ecology. Geofis. Intern., 29, No.3, p.123–128.
Kondratyev K.Ya. 1990f. The International Space Year: Priorities and Perspectives. Il Nuovo Cimento, **13**, Ser.1, No.5, p.835–846.
Kondratyev K.Ya. 1991a. The International Geosphere-Biosphere Programme: the role of observations from space. Izvestiya USSR Geograph. Soc., 123, issue 3, p.2–7.
Kondratyev K.Ya. 1991b. Priorities of global ecology. Izvestiya USSR Acad. Sci., Ser. Geography, No.6, p.21–30.
Kondratyev K.Ya. 1991c. On the International Framework Convention on Climate Change. Izvestiya USSR Acad. Sci., Ser. Geography, No.5, p.117–122.
Kondratyev K.Ya. 1992a. Global Climate. "Nauka" Publ., St. Petersburg, 359 pp.
Kondratyev K.Ya. 1992b. The Second World Climate Conference. II. Assessments of global climate change and measures for prevention of global change. Izvestiya Russian Geograph. Soc., 123, issue 6, p.473–480.
Kondratyev K.Ya. 1992c. On the Framework Convention on Climate Change. Il Nuovo Cimento C, 156, No.1, p.87–98.
Kondratyev K.Ya. 1993a. Paradigms of global ecology. I. Environment. Studying the Earth from Space, No.4, p.109–122.
Kondratyev K.Ya. 1993b. Paradigms of global ecology. II. Socio-economic aspects. Studying the Earth from Space, No.5, p.118–126.
Kondratyev K.Ya. 1994. Satellite observations in global ecology. In: Remote Sensing and Global Climate Change. Eds. R.A. Vaughan, A.P. Cracknell. Springer Verlag, Berlin, p.67–94.
Kondratyev K.Ya. 1995a. Key aspects of ecological policy. I. Ecodynamics. Izvestiya Russian Geograph. Soc., 127, issue 3.
Kondratyev K.Ya. 1995b. Key aspects of ecological policy. II. Socio-economic problems. Izvestiya Russian Geograph. Soc., 127, issue 4.
Kondratyev K.Ya. 1995c. The current stage of development of global change problem studies: the USA National Programme. Studying the Earth from Space, No.2, p.98–105.
Kondratyev K.Ya. 1995d. "Earth in the Balance: Ecology and the Human Spirit" (a review of the book by A. Gore). Sustain. Development, 3, No.2, p.104–105.
Kondratyev K.Ya. 1996a. The mysterious life chain. Izvestiya Russian Geograph. Soc., 128, issue 2.
Kondratyev K.Ya. 1996b. Development of satellite remote sensing in the USA: Summing up progress for 1994 fiscal year. Studying the Earth from Space, No.6, p.118–126.
Kondratyev K.Ya., Adamenko V.N., Henderson-Sellers B. et al. 1990. Using large lakes as analogues for oceanographic studies. In: Modelling Marine Ecosystems. Ed. A.M. Davies, 2, CRC Press, Boca Raton, FL, p.299-344.
Kondratyev K.Ya., Asimakopoulos D.N., Varotsos C.A. 1995. European ecodynamics in the context of global change. Studying the Earth from Space, issue 3, p.104–118.
Kondratyev K.Ya., Belchansky G.I., Petrosian V.G. 1992. Conceptual aspects of satellite monitoring of biodiversity. Doklady Russian Acad. Sci., 322, Moscow, p.200.
Kondratyev K.Ya., Buznikov A.A., Pokrovsky O.M. 1992. Global ecology: remote sensing. Itogi Nauki I Tekhn. Atmosphere, Ocean, Space—"Sections" Programme, 14, Moscow, 310 pp.

Kondratyev K.Ya., Buznikov A.A., Pokrovsky O.M. 1996a. Global Change and Remote Sensing, Wiley–Praxis, Chichester, 370 pp.

Kondratyev K.Ya., Cracknell A.P. 1998. Observing Global Climate Change. Taylor & Francis, London (in press).

Kondratyev K.Ya., Cracknell A.P., Vaughan R.A. 1994. Politics and climate change. In: Remote Sensing and Global Climate Change. Eds. R.A. Vaughan and A.P. Cracknell. Springer Verlag, Berlin, p.465–478.

Kondratyev K.Ya., Danilov-Danilyan V.I., Donchenko V.K. 1993. Ecology and Policy. Res. Centre for Ecol. Safety Publ., St. Petersburg, 286 pp.

Kondratyev K.Ya., Fedchenko P.P. 1992. Fine Structure of the Solar Spectrum and Its Role in Biosphere Evolution. "PROPO" Publ., St. Petersburg, 40 pp.

Kondratyev K.Ya., Frolov A.K. (Eds.). 1996. Ecodynamics and Ecological Monitoring of the St. Petersburg Region in the Context of Global Change. "Nauka" Publ., St. Petersburg.

Kondratyev K.Ya., Donchenko V.K., Losev K.S., Frolov A.K. 1996b. Ecology–Economics–Policy. St. Petersburg Scientific Centre Publ., 828 pp.

Kondratyev K.Ya., Grigoryev A.A., Pokrovsky O.M. 1975. Information Content of Satellite Remote Sensing Data on Environmental and Natural Resources Parameters. Leningrad Univ. Publ., Leningrad, 146 pp.

Kondratyev K.Ya., Johannessen O.M., Melentyev V.V. 1996. High Latitude Climate and Remote Sensing. Wiley–Praxis, Chichester, 200 pp.

Kondratyev K.Ya., Kozoderov V.V., Fedchenko P.P., Bozo P., Matrai M. 1994. Determination of vegetation biomass by satellite sensor measurements. Proc. Int. Symp. on the State of the Art of Remote Sensing in Biosphere Studies, 18–22 Sept. 1989, Moscow. UNEP Publ., Nairobi, p.6.1–6.20.

Kondratyev K.Ya., Ortner J., Preining O. 1992. Priorities of global ecology now and in the next century. *Space Policy*, **8**, No.1, p.39–48.

Kondratyev K.Ya., Pokrovsky O.M. 1977. Planning multipurpose experiments in remote sensing of environmental and natural resources parameters. Izvestiya USSR Acad. Sci., Ser. Geography, No.3, p.83–89.

Kondratyev K.Ya., Pokrovsky O.M. 1989. The International Geosphere-Biosphere Programme. Key aspects of requirements to data of Earth observation from space. Izvestiya USSR Acad. Sci., Ser. Geography, No.1, p.20–28.

Kondratyev K.Ya., Pozdnyakov D.V. 1996. Land-ocean interaction in the coastal zone: the LOICZ project. Il Nuovo Cimento C, 19C, No.3, p.339–354.

Kondratyev K.Ya., Romanyuk L.P. 1996a. Environmental problems and the World Bank. Izvestiya Russian Geograph. Soc., **128**, issue 2, p.1–10.

Kondratyev K.Ya., Romanyuk L.P. 1996b. Making development sustainable. IEEE Technol. and Society Magazine, **15**, No.2, p.9–11.

Kotlyakov V.M. 1995. A forum of defenders of the environment. Vestnik Russian Acad. Sci., 65, No.12, p.1139–1142.

Kuroda T., Koizumis S. 1992. A plan for the world environment and disaster observation system. World Space Congress Book Abstr., Washington, DC.

Land-Use and Land-Cover Change. Science/Research Plan. 1995. IGBP Report No.35/HDP Report No.7, Stockholm, 132 pp.

Leovy C.B. 1984. Tropospheric science. NASA Techn. Memo. 86129, **1**, Part 2, Goddard Space Flight Center, Greenbelt, MD, USA, p.44–48.

Lewin R. 1987. *The Genes*. "Mir" Publ., Moscow, 480 pp.

Lindzen R.S. 1990. Some coolness concerning global warming. Bull. Amer. Meteorol. Soc., **71**, No.3, p.288-299.

Losev K.S., Gorshkov V.G., Kondratyev K.Ya., Kotlyakov V.M., Zalikhanov M.Ch., Danilov-Danilyan V.I., Golubev G.N., Gavrilov I.T., Reviakin V.S., Grakovich V.F. 1993. Ecological Problems of Russia. VINITI Publ., Moscow, 350 pp.

Lovelock J.E. 1988. The Ages of Gaia. A Biography of Our Living Planet. W.W. Norton & Co., New York, 252 pp.

Lovelock J.E. 1989. Geophysiology, the science of Gaia. Revs. Geophys., **27**, No.2, p.215–222.

Madeley J. 1994. Cracks in the World Bank's structural adjustment programme. IIED Perspectives, No.13, p.6–7.

Making Development Sustainable. The World Bank Group and the Environment, Fiscal 1994. 1994. The World Bank Publ., Washington, DC, 270 pp.

Marchuk G.I., Kondratyev K.Ya., Kozoderov V.V. 1990. Earth Radiation Budget: Key Aspects. "Nauka" Publ., Moscow, 231 pp.

Marchuk G.I., Kondratyev K.Ya. 1992. Priorities of Global Ecology. "Nauka" Publ., Moscow, 264 pp.

Marou P., Daly H., Goodland R. 1994. Environmental sustainability through project appraisals. Sustain. Development, **2**, Part 3.

Massom R.A. 1994. Satellite Remote Sensing of Polar Regions. Wiley, Chichester, UK, 288 pp.

Mata F.J., Onisto L.J., Vallentyne J.R. 1994. Consumption: the other side of population and development. Paper for Int. Conf. on Population and Development, Cairo, 3014 Sept. 1994, 17 pp.

Matschullat J., Müller G. (Eds.). 1994. Geowissenschaften und Umwelt. Springer Verlag, Heidelberg, xxiii + 3645.

Mathews J.T. 1990. Environment, development and international security. Bull. Amer. Acad. Arts & Sci., **XLIII**, No.7, p.10–26.

Meadows D.H., Meadows D.L., Randers J., Behrens W.W., III. 1972. The Limits to Growth. Potomac, New York, 207 pp.

Meadows D.L., Behrens W.W., III, Meadows D.H. et al. 1974. Dynamics of Growth in a Finite World. Wright Allen Press, New York, 511 pp.

Melack J.M. 1984. Inland aquatic resources and biogeochemical cycles. NASA Techn. Memo. 86129, **1**, Part 2, Goddard Space Flight Center, Greenbelt, MD, USA, p.15–17.

Menza V., Lupien J.R. 1994. World population and nutritional well-being. In: Population and Global Security. The Foundation for Environmental Conservation, Geneva, p.125–139.

Mesarovic M., Pestel E. 1974. Mankind at the Turning Point. Dutton, New York, 10 pp.

Michaels P.J., Stocksbury D.E. 1992. Global warming: a reduced threat? Bull. Amer. Meteorol. Soc., **73**, No.10, p.1563-1577.

Miller K., Barber Ch. 1992. Biodiversity after the Earth Summit: Prospects for the Convention on Biodiversity. Network'92, No.18, p.5.

Mikesell R.F. 1992. Economic Development and the Environment. A Comparison of Sustainable Development with Conventional Development Economics, Mansell, New York.

Mitlin D., Satterthwaite D. 1994. Sustainable Development and Cities. Global Forum'94 and IIED, 82 pp.

Mohnen V. 1984. Tropospheric chemistry. NASA Techn. Memo. 86129, **1**, Part 2, Goddard Space Flight Center, Greenbelt, MD, USA, p.24–27.

Moiseyev N.N. 1995. Strategy of the transitional period. Vestnik Russian Acad. Sci., No.4, p.291–295.
Monitoring Environmental Progress. A Report on Work in Progress. 1995. The World Bank, Environment Dept., Washington, DC, 72 pp.
Moore B. 1984. Global biogeochemistry: an overview. NASA Techn. Memo. 86129, **1**, Part 2, Goddard Space Flight Center, Greenbelt, MD, USA, p.5–7.
Morais J., Rapley C., Grassl H. 1995. Why the international global change research programmes need the Human Dimensions of Global Environmental Change Programme. Global Change Newsletter, No.24, p.23–25.
Munn R.E. 1988. Towards sustainable development: an environmental perspective. Preprint for Int. Conf. on Environment and Development, Milano, Italy, March 1988, 22 p.
Munn R.E. (Ed.). 1996. Looking Ahead: The Inclusion of Long-Term Futures in Cumulative Environmental Assessments, IES Environm. Monograph No.11, Institute for Environmental Studies, Univ. of Toronto, Toronto, Canada, 282 pp.
Myers N. 1994. Global population and emergent pressures. In: Population and Global Security, The Foundation for Environm. Conservation, Geneva, p.17–48.
Nasem S., Thompson L.J., Lawler S.P., Lawton J.H., Woodfin R.M. 1995. Declining biodiversity can alter the performance of ecosystems. Nature, 368, p.734–737.
Nature Management and Sustainable Development. 1989. Ed. W.D. Verwey. IOS, Amsterdam, 531 pp.
Novek J. 1995. Environmental impact assessment and sustainable development: case studies of environmental conflict. Society and National Resources, 8, No.2, p.145–160.
O'Connor J.C. 1994a. Towards environmentally sustainable development. Draft. The World Bank Publ. 01.16, 18 pp.
O'Connor J.C. 1994b. Environmental performance monitoring indicators. Draft for discussion. The World Bank Publ. 5.9.
Opschoor J.B., Turner R.K. 1994. Economic Incentives and Environmental Policies: Principles and Practice. Kluwer, Dordrecht, 312 pp.
O'Riordan T., Cameron J. (Eds.). 1994. Interpreting and Precautionary Principle. Earthscan, London.
Osipov V.I. 1995. Natural catastrophes are the focal point for scientists. Vestnik Russian Acad. Sci., No.6, p.487–495.
Our Changing Planet: the FY 1991 Research Plan. The US Global Change Research Program. 1990. A Report by Committee on Earth and Environm. Sciences, Reston, VA, USA, 169 pp., Appendices.
Our Changing Planet. The FY 1997. 1996. US Global Change Research Program, Washington, DC, 162 pp.
Our Common Future. Reconvened World Commission on Environment and Development, 22-24 April 1992. 1992. The Centre for Our Common Future, Geneva, 32 pp.
Our Common Future. 1989. Report of Intern. Commission on Environment and Development. "Progress" Publ., Moscow, 374 pp.
Overseas Development Administration. Environmental Research for Sustainable Development. 1995. The Globe, issue 28, p.2–4.
Perry J.S. 1995. A Sustainable Future: Catastrophe or Cornucopia; Crusade or Challenge? Editorial Review Essay. Climatic Change, 29, No.3, p.259–263.
Pethig R. 1994. Valuing the Environment: Methodological and Measurement Issues. Kluwer, Dordrecht, 376 pp.

Pimm S.L. (Ed.). 1994. The Balance of Nature: Ecological Issues in the Conservation of Species and Communities. The Univ. of Chicago Press, Chicago, IL, USA.
Polunin N., Nazim M. (Eds.). 1994. Population and Global Security (Environmental Challenges II). The Foundation for Environm. Conservation Publ., Geneva, 285 pp.
Prinn R.G. 1994. The interactive atmosphere: Global atmospheric–biospheric chemistry. Ambio, **23**, No.1, p.50–61.
Pykh J., Maklina-Pykh I.C. 1994. Environmental indicators and their applications (trends of activity and development). Working Pap. of Intern. Institute for Appl. Syst. Analysis, 169 pp.
Ramanathan V., Collins W. 1992. Thermostat and global warming. Nature (London), **357**, No.6380, p.649.
Raven P.H. 1987. The global ecosystem in crisis. A MacArthur Foundation Occasional Pap., Chicago, Il, USA, 24 pp.
Raven P.H. 1988. State of the world: 2000—what we should do to affect it. Preprint. Wingspread Conf., Apr. 1988, Missouri Botanical Garden Publ., 29 pp.
Report from the START Regional Meeting for Southeast Asia. 1992. Global Change Report, No.22, Stockholm, 114 pp.
Report of the International Conference on Population and Development, Cairo, 5–13 Sept. 1994. 152 pp.
Research Strategies for the US Global Change Research Program. 1990. National Res. Council, Washington, DC, 291 pp.
Resolution and Recommendations of the Fourth Scientific Advisory Council for the IGBP (SAC IV). 1995. Global Change Newsletter, No.24, p.10.
Robin G. de J. 1984. Polar glaciology. NASA Techn. Memo. 86129, **1**, Part 2, Goddard Space Flight Center, Greenbelt, MD, USA, p.37-40.
Roe D., Dalal-Clayton B., Hughes R. 1995. A Directory of Environmental Assessment Guidelines. IIED Publ., London.
Ruth M. 1993. Integrating Economics, Ecology and Thermodynamics. Kluwer, Dordrecht, 264 pp.
Sadik N. 1994. Population expansion and global stability. In: Population and Global Security. The Foundation for Environm. Conservation, Geneva, p.1–15.
Scaefer M., Kreisel W. 1994. Health of people, health of planet. In: Population and Global Security. The Foundation for Environm. Conservation, Geneva, p.101–124.
Schimel D.S. 1995. Terrestrial ecosystems and carbon cycle. Global Change Biology, **1**, No.1, p.77–91.
Schnaiberg A., Gould K.A. (Eds.). 1993. Environment and Society, the Enduring Conflict. St. Martin's Press, New York.
Scientific Orientation and Programming. Interdisciplinary Research Environment Program. 1990. CNRS Publ., Paris, 20 pp.
Securing the Future. 1990. A Report on a Joint WFUNA–IIASA Initiative on Global Risk Management, IIASA, Laxenburg, 28 pp.
Seitz F. 1994. The Human Use of Science. The George C. Marshall Inst., Washington, DC, 16 pp.
Senanayke P. 1994. Women and the family planning imperative. In: Population and Global Security. The Foundation for Environm. Conservation, Geneva, p.155–174.
Serageldin I., Steer A. (Eds.) with Cernea M.M., Dixon S.A., Lutz E., Margulis S., Munasinghe M., Rees C. 1994. Making Development Sustainable. From Concepts to Action. Environm. Sustain. Development Occasional Paper Series No.2, The World Bank, Washington, DC, 340 pp.

Serageldin I., Cohen M.A. (Eds.). 1995. The Human Face of the Urban Environment. The World Bank, Environmentally Sustain. Development Proc. Ser. No.5, 55 pp.
Serrano I.R. 1995. After the Social Summit—what? Bull. Centre for Our Common Future, issue 26, p.53–56.
Singh N., Titi V. (Eds.). 1995. Empowerment. Toward Sustainable Development. Ferwood Publ., Winnipeg, Canada, 198 pp.
Slezynski J. 1996. Carbon tax as an environmental policy instrument. In: Industrial and Environmental Policy, Cracow, p.125–147.
Space Technology and Remote Sensing Techniques for Sustainable Development in and around the Mediterranean. 1995. EURISY Coll. Inst. Geograf. Nacional, Madrid, Spain, May 31–June 1, 1995.
Starke L. (Ed.). 1990. State of the World, 1987. Norton, New York, 53 pp.
Sterner T. (Ed.). 1994. Economic Policies for Sustainable Development. Kluwer, Dordrecht, 344 pp.
Summary of the Programme of Action of the International Conference on Population and Development. 1995. ICPD'94, New York, 27 pp.
The Bulletin. 1995. A Publication of the Centre for Our Common Future, issue 26, Geneva.
The Nairobi Declaration on Climate Change. 1990. Intern. Conf. on Global Warming and Climatic Change: African Perspectives (2–4 May, 1990). Acta Press, Nairobi, Kenya, 33 pp.
The National Science Foundation Global Change Research Program. Research Opportunities 1995-1996. 1995. NSF 94–95, Washington, DC, 42 pp.
The Noordwijk Declaration on Climate Change. Atmospheric Pollution and Climate Change. 1989. Ministerial Conf. (Noordwijk, the Netherlands, 6–7 Nov. 1989), 12 pp.
The Spirit of Rio: Earth Summit 1992. 1992. Brundtland Bull., issue 16, p.3–5.
Thomas C.D. 1990. Fewer species. Nature, 347, p.237.
Thomas G.B., Lester J.P., Sadeh W.Z. 1995. International cooperation in remote sensing for global change research: political and economic considerations. Space Policy, 11, No.2, p.131–141.
Tisdell C.A. 1994. Population, economics, development and environmental security. In: Population and Global Security, Foundation for Environm. Conservation, Geneva, p.63–84.
Tolba M.K. 1992. First Plenary Lecture. The Second Princess Chulabhorn Science Congress. Environment, Science and Technology: Challenges for the Twenty-First Century, Bangkok, 3 Nov. 1992, 8 pp.
Transportation and Global Climate Change. 1993. Amer. Council for Energy, Efficient Economy, Washington, DC, 357 pp.
Trends'93: A Compendium of Data on Global Change (Eds. T.A. Boden, D.P. Kaiser, R.J. Stepanski, F.S. Stoss). 1994. Carbon Dioxide Information Analysis Center, Oak Ridge Natl. Lab., Oak Ridge, 984 pp., Appendices.
Trofimov A.M., Kotlyakov V.M., Seliverstov Yu.P., Khuzeyev R.T. 1994. The theoretical aspect of geoecological studies. I. The problem of interests and conflicts in geoecological systems. Izvestiya Russian Geograph. Soc., 126, issue 5, p.1–11.
UN General Assembly Document A/CONF.151/PC/WG III/L.33/Rev.1, 2 Apr. 1992, 1992. 6 pp.
Union of Concerned Scientists. World Scientists' Warning to Humanity. 1992. Union of Concerned Scientists, Cambridge.
United Nations Framework Convention on Climate Change. 1992. UN General

Assembly Document A/AC.237/18 (Part II), Add.1, 15 May 1992, 29 pp.
United Nations Environment Programme. Environm. 1991. Data Report, 3rd edn, Basil Blackwell, London, 408 pp.
Van Valen. 1973. A new evolutionary law. Evol. Theory, 1, p.1–30.
Vari A., Tamas P. (Eds.). 1963. Environmental and Democratic Transition: Policy and Politics in Central and Eastern Europe. Kluwer, Dordrecht, 296 pp.
V. I. Vernadsky and Modern Times. 1986. Eds. V.S. Sokolov and A.L. Yanshin. "Nauka" Publ., Moscow, 390 pp.
Vitousek P.M., Ehrlich P.R., Ehrlich A.H., Matson P.A. 1986. Human appropriation of the products of photosynthesis. Bio Science, 36, p.368–373.
Wackernagel M., Rees W. 1996. Our Ecological Footprint. Reducing Human Impact on the Earth. New Society Publ., Gabriola Island, BC, Canada, 160 pp.
Weber G.R. 1996. Global Warming. The Rest of the Story. Dr. Boettiger Verlag, Wiesbaden, 188 pp.
Weizsäcker E.U., von. 1994. Earth Politics. Zed Books, London, 234 pp.
West G.A., Milchand R.W. (Eds.). 1995. Principle of Environmental Health and Safety Management. Government Inst. Inc., Rockville, MD, USA, 362 pp.
Whittaker R.H., Likens G.E. 1975. The biosphere and man. In: Primary Productivity of the Biosphere. H. Lieth and R. Whittaker (Eds.), Springer Verlag, Berlin, p.305–328.
Wilson E. 1992. The Diversity of Life. Belknap Press, Cambridge (USA).
Woods A., Bernasconi M.P. 1995. Choosing a space age or a stone age. Space News, Oct. 2–8, p.19.
World Commission on Environment and Development (The Brundtland Commission). 1987. Our Common Future, Oxford Univ. Press, New York.
World Resources 1988–1989. Land Cover Settlements. New York, p.264.
Yaffee S.L. 1994. The Wisdom of the Spotted Owl: Policy Lessons for a New Century. Island Press, Washington, DC.
Young J. 1995a. Towards a new culture of consumption. People and the Planet, **4**, No.1, p.10–12.
Young J. 1995b. The material revolution. People and the Planet, 4, No.1, p.13.
Zamolodchikov D.G. 1993. Estimates for ecologically admissible levels of anthropogenic impact on fresh water ecosystems. Problems of Ecological Monitoring and Ecosystem Modelling, **15**, p.214–233.
Zecca A., Brusa R.S. 1992. SO_2 cooling and detection of the greenhouse effect. Il Nuovo Cimento, **150**, No.4, p.481–483.
Zhirmunsky A.V., Kuz'min V.I. 1994. Critical levels of the world population number growth. Izvestiya Russian Acad. Sci., Ser. Biology, No.5, p.839–842.
Zinke P. 1984. Forest environments. NASA Techn. Memo. 86129, **1**, Part 2, Goddard Space Flight Center, Greenbelt, MD, USA, p.18–20.
Zubakov V.A. 1990. Ecological crisis and the future of mankind. Izvestiya USSR Geol. Soc., 122, No.2, p.143–153.
Zubakov V.A. 1993. Evolution and mankind. In: Evolution of Geological Processes in the Earth's History, "Nauka" Publ., p.226–236.
Zubakov V.A. 1995. Our Future—Technosphere or Ecogeism? St. Petersburg, 19 pp.
Zubov V.I. 1993. Apocalypse. The Behests of Gone Generations. "Mobilnost' Plius" Publ., St. Petersburg, 56 pp.
Zylicz T. 1994. National environmental policy outline of economic instruments after 3 years. In: Industrial and Environmental Policy in Poland, Cracow, p.63–73.

2

Global climate change and ozone dynamics: key issues

2.1 INTRODUCTION

Global climate change studies have become a focal point of the present-day environmental research of great practical importance. The most persuasive illustration of such a conclusion is the Second UN Conference on Environment and Development (UNCED) and the Framework Convention on Climate Change (FCCC) approved of by the UNCED and then by the UN General Assembly (Kondratyev, 1992i,j, 1993a, 1995b,c; Kondratyev et al., 1996a,b).

Certain features of the development of the global climate studies may seem controversial: the deeper level of understanding results in increased uncertainties. McBean (McBean et al., 1996) has justly pointed out: "It should be noted that each increment of better understanding may not necessarily reduce uncertainties, at least, initially. Eventually, however, such increased understanding should help to narrow uncertainties, i.e., to move the high and low range limits closer together. Further, we pay attention to areas where currently imaginable but highly uncertain "surprise" outcomes may exist.... Complex systems, such as the climate system, can respond in non-linear ways and produce surprises. The nature of the system forces us to take a flexible approach in research planning and societal response and makes exact predictions of research advances impossible".

There is, in fact, no controversy at all: we simply have to deal with the most complex system the dynamics of which may be understood only on the basis of an adequate consideration of all significant interactions of the system's components, including what is called human dimensions. Global climate change is a problem of interaction between nature and society. Even more: its roots originate in socio-economic dynamics. One more complication is that there is still no adequate global climate observing system and, hence, reliable enough data to judge what is going on with climate. It is very difficult, of course, to study climate change in conditions of the deficit of observational data.

The most important task of an interdisciplinary research is to reach good enough levels of coordination and cooperation. This is why "The Climate Agenda" (1995) and its accomplishment acquires a decisive role. The Climate Agenda, A proposal for an integrating framework for international climate-related programmes, has been prepared in response to the request by the international organizations involved in climate-related activities: the UN Food and Agriculture Organization (FAO), the non-governmental International Council of Scientific Unions (ICSU), the UN Environment Programme (UNEP), the UN Educational, Scientific and Cultural Organization (Unesco) and its Intergovernmental Oceanographic Commission (IOC), the World Meteorological Organization (WMO); the UN Development Programme (UNDP) which acts primarily as a funding agency; and others, such as the World Health Organization (WHO) and the non-governmental International Social Council (ISSC) which operates the climate-related programmes. Below we quote certain parts of the Climate Agenda, then make a few comments concerning this document.

The following programmes and activities have been considered within the scope of the Climate Agenda:

- World Climate Programme (WCP):
  - World Climate Applications and Services Programme (WCASP);
  - World Climate Data and Monitoring Programme (WCDMP);
  - World Climate Impact Assessment and Response Strategies Programme (WCIRP);
  - World Climate Research Programme (WCRP);
- The climate-relevant parts of:
  - International Geosphere–Biosphere Programme (IGBP);
  - International Human Dimensions of Environmental Change Programme (IHDP);
  - FAO food security, environment and natural resources programmes;
  - Hydrology and Water Resources Programme (HWR), including the World Climate Programme—Water (WCP—Water);
  - ICSU World Data Centre System;
  - Unesco International Hydrology Programme (IHP);
  - Unesco Man and Biosphere Programme (MAB);
  - WMO Agricultural Meteorology Programme (AgMP);
  - World Health Organization (WHO) climate-related programmes;
  - Global Atmosphere Watch (GAW);
  - World Weather Watch (WWW);
  - Global Environment Monitoring System (GEMS) and the Global Resource information Database (GRID) of the UNEP;
- Observing systems:
  - Global Climate Observing System(GCOS), including the climate-relevant part of both the Global Ocean Observing System (GOOS) and the Global Terrestrial Observing System (GTOS);
  - the climate-relevant part of the IOC ocean observing systems such as the Integrated Global Ocean Services System (IGOSS) and Global Sea-Level Observing System (GLOSS).

Cooperation within the WCP, and between the WCP and other climate-related activities is maintained through a number of advisory bodies. In addition to the sponsoring international organizations, a large number of national organizations play key roles in the successful implementation of these programmes.

(1) Achievements to date

Governments reviewed the ability of the WCP and its associated activities to meet the obligations arising from international agreements and the needs for national socio-economic development. Governments recognized that programmes are at different stages of development and that further work remains to be done but concluded that achievements to date have been significant:

- Research into the basic processes of the climate system and predictions of climate has resulted in improved understanding of the climate system culminating in an ability to reproduce, within models, some key features of global change over the past 150 years. In addition, experimental seasonal forecasts are now available which have allowed governments in some tropical regions to anticipate and alleviate adverse climatic effects on their economies and populations, for example with respect to agricultural productivity, the viability of commercial fisheries and the frequency, severity, and paths of storms.
- Observations of the climate system, made by operational and research systems, provide basic data of fundamental importance. As a result of international planning and coordination of these nationally funded observation systems, a number of high-quality, global datasets are now available for use in understanding the global climate system and these are beginning to allow the evaluation of strategies to detect anthropogenic climate change. Global near real-time observations from *in situ* and satellite instrumentation are being used to initialize and validate seasonal-to-inter-annual models of climate variability.
- Application of climate data, primarily from historical data banks and current weather observations has enabled nations to make informed decisions in climate-sensitive sectors of their economies. For example, early warning systems that continuously monitor environmental conditions and their impact on food crops have been developed and deployed in Africa, Asia and Latin America.
- Impacts of climate have been studied on a country basis, employing methodologies agreed upon at an international level. These studies have assisted governments in the identification and implementation of suitable policy options and response measures to mitigate adverse effects of climate variability and change. The FCCC requires that Parties carry out and report country studies and the pioneering work already carried out will be invaluable to countries in discharging their obligations under the FCCC and in determining how they might respond to future climate change.

It has been concluded that future international activities should be developed along four main thrusts:

- New frontiers in climate science and prediction
- Climate services for sustainable development
- Studies of climate impact assessments and response strategies to reduce vulnerabilities
- Dedicated observations of the climate system

with special attention given to issues that span all four thrusts, namely, capacity building through training and development; improved institutional arrangements and technology transfer; and public outreach to inform and educate.

(2) Thrust 1: New frontiers in climate science and prediction

Aim: The ability to simulate past, present and future climates, including the influence of human activity, at those spatial and temporal scales of interest to governments.

Lead agency: It is proposed that WMO, Unesco and its IOC and ICSU share joint responsibility for planning and coordinating this Thrust, under guidance affected through cross-membership of the Scientific Committees of the IGBP, WCRP and IHDP.

Requirements: The IPCC (Intergovernmental Panel on Climate Change) has identified requirements for future work arising from the need to:

- reduce the uncertainty in the predictions of the magnitude, timing and regional distribution of anthropogenic climate change;
- detect climate change.

Both these requirements necessitate improvements in the scientific understanding of the climate system, including both its natural variability and anthropogenic effects.

Overview: The achievement of the overall aim requires that two main lines of research are pursued. Firstly, understanding of the physical and dynamic processes of the atmosphere and oceans must continue to be developed but, in addition, there is the need to increase substantially knowledge of the chemical and biological processes that regulate the composition of the atmosphere and thereby regulate the climate system. Secondly, there must be continued development of the mathematical models which provide the basis for integrating knowledge of the climate system and which provide the tools for conducting the prediction of future conditions and determining the nature and extent of climate change and variability.

The WCRP will continue to determine the extent to which climate can be predicted and to develop the mathematical models to make such predictions. The IGBP addresses the interactions between the physical climate and the biosphere, including the Earth's surface, the ocean, and chemical changes in the atmosphere. It is these interactions which control the concentrations of greenhouse gases in the atmosphere and thereby the amount of global warming.

In addition to the above physical and biogeochemical research into the climate system, proper account must be taken of human activities when simulating the climate system. The IHDP of the International Social Science Council (ISSC) seeks to describe and understand the human role in causing global change and the consequences of these changes for society. Some research, such as that on

Table 2.1. International programmes contributing to Thrust 1

Objective	Focused programmes	Contributing programmes
1. Develop climate prediction and forecasting capabilities	WCRP, IGBP	IHDP
2. Understand the biological and chemical influences on atmospheric composition and climate	IGBP	WCRP
3. Understand the natural variability and the physical processes controlling the climate system	WCRP, IGBP	SCAR, IHP
4. Understand past climates	IGBP	WCRP, IHP
5. Understand human influences on the climate system	IHDP	IGBP, WCRP
6. Develop worldwide scientific capacity relating to the climate system	IGBP, WCRP, IHDP	IHP

perceptions and attitudes, technology transfer, or international negotiations are studied within the social sciences. Other areas, such as land and energy use have important climate effects and must be studied in close collaboration with natural scientists.

The specific objectives of this Thrust, together with the international programmes which are contributing to their realization, are shown in Table 2.1.

Thrust 1: Deliverables

Objective 1. Develop climate prediction and forecasting capabilities:

- realistic simulations of the present climate and reliable predictions of future global climate changes, from both natural and human causes, for up to a century ahead;
- ability to produce real-time climate forecasts for periods of several months, possibly several years;
- improved prediction of extreme events;
- regionally adapted information on transient climate anomalies incorporating information on regional physical, ecological, agricultural, economic and other relevant conditions for applications to economic activities and development.

Objective 2. Understand the biological and chemical influences on atmospheric composition and climate.

- improved methods for estimating the greenhouse warming potentials for radiatively active gases and aerosols, taking account of chemical couplings and other atmospheric effects;
- improved budgets for present-day sources and sinks of greenhouse gases, and for the biogenic production of aerosols;

- quantitative knowledge of the dynamics of biogeochemical processes affected by, and affecting, changes in atmospheric composition and climate change;
- improved climate predictions through better incorporation of biogeochemical processes in GCMs.

Objective 3. Understand the natural variability and the physical processes controlling the climate system

- better understanding of the magnitude of climate change through an improved understanding and accurate mathematical representation of the "fast processes" in the atmosphere and at the Earth's surface, which control the equilibrium climate of the Earth. This will permit correspondingly reduced uncertainty on the global climate response to external perturbations such as a change in the concentration of greenhouse gases;
- provision of more realistic predictions of anomalies in Earth surface radiation, evaporation and rainfall that result from global climate change and directly affect regional ecosystems and economic activities;
- increased understanding of low-frequency climate variability on decadal and multidecadal timescales;
- the ability to understand the rate of climate change;
- a global, high-precision description of the global oceanic circulation, and the instantaneous location and origin of the main water masses, suitable for establishing the baseline against which future changes can be detected;
- operational ocean models capable of accurately reproducing the present state of the global ocean and ocean–air fluxes and simulating possible future changes in these;
- determination of the current rate of change of the mass balance of polar ice-sheets and the corresponding effect on global sea-level and prediction of future changes in the polar ice-sheets and global sea-level.

Objective 4. Understand past climates

- better description of past climates and related ecosystem evolution;
- better discrimination of human-induced effects when considered in the context of contemporary natural variability;
- better understanding of the natural forces that can bring about major changes to the climate system, and hence their sensitivity to anthropogenic perturbation;
- improvements in GCM construction, and increased confidence in their ability to simulate changes in the Earth system behaviour, including scenarios for future conditions.

Objective 5. Understand human influences on the climate system

- realistic predictions of how land usage may change in the future and representation of these changes in GCMs;
- evaluation of the ways in which the production, allocation, distribution and consumption of human and material resources affect possible climate change;

- evaluation of the different forms of industrial transformation and energy use, on natural, regional and global scales and their consequences for the climate system;
- improved predictions of future emissions of greenhouse gases, based on increased understanding of national, regional and global trends in energy production and consumption.

Objective 6. Develop worldwide scientific capacity relating to the climate system

- effective capability in all nations to acquire and assess knowledge available on global and regional climate systems;
- effective capability of scientists in all regions of the world to participate in the development of Earth system science through regional scientific research activities relevant to the global climate system.

Thrust 1: Areas requiring specific/enhanced attention and lead programmes

- The basic understanding of climate processes (e.g., the hydrological cycle and clouds), necessary to provide scientifically well-founded quantitative predictions of natural climate variations as well as of anthropogenic climate change, requires the development and implementation of a series of dedicated research satellites over a period of several years [WCRP].
- There is a need to implement fully a range of other WCRP activities (e.g., to obtain an improved knowledge of ocean transport, the role of the land surface, development of climate modelling and provision of the computer resources required, and studies of the cryosphere) [WCRP].
- There is a need to implement fully IGBP's planned research programme studying biogeochemical processes in the climate system and to incorporate the results into climate models in order to improve their predictive capability [IGBP].
- There is a need for much closer collaboration between social and natural sciences particularly leading to additional research on the human influences on the climate system, for example, on the robust projection of GHG emissions. This work should be coordinated through IHDP with input from WCRP and IGBP [IHDP].

(3) Thrust 2: Climate services for sustainable development

Aim. The introduction, on an operational basis, of information and predictions on climate, climate change and climate variability into socio-economic decision-making on a worldwide basis.

Requirements. Agenda 21 is a comprehensive programme of specific actions and objectives towards global sustainable development. Central to many aspects of this theme is the need to produce and interpret, on a regular basis, climate information for regional and national purposes, in particular, in order to increase preparedness for drought and desertification and to provide knowledge relevant to sea-level rise through climate forecasts and application of these and other climate data through networks of climate service centres. Sustainable agriculture also requires applications of climate data and climate forecasts on an operational basis.

The need for climate forecasts and subsequent warnings of extreme climatic events, disseminated through regional centres, are all contained in the Yokohama strategy produced by the World Conference on Natural Disaster Reduction. This statement also emphasizes the importance of nations' self-reliance in order to increase preparedness for natural disasters.

Overview. The introduction of operational climate products and services in all nations of the world is needed, in order that nations may adapt their socio-economic development in light of climatic conditions. It is important to distinguish between climate variability and future climate change in the development of climate services—the relative importance of these two aspects of climate depends on the community under consideration.

The central, underlying theme for the development of operational climate services is *capacity-building* both in developing nations and collectively as a global community. International activities under this Thrust should contribute to the building of a technical capability to provide past and real-time climate information at the national and local level thus enhancing national efforts towards sustainable development.

The strategy for the introduction of operational climate services will be based upon the development of a network of national, regional and global centres functioning in near real-time to deliver operational climate services.

These operational climate services will be based upon the near real-time exchange of climate observations which meet international standards and those retained in computer data banks at local, regional and international locations. They will also be based on the experimental climate forecasts becoming available as a result of the effects of the research community.

The successful introduction of operational climate services requires the active engagement of the decision makers themselves; it cannot be done by the climate specialist alone. Therefore it is important to implement an effective marketing and training programme to inform decision makers of the climate services available to them and the benefits to be accrued through the use of such services. To be effective, such a programme must be supported by all the international agencies involved in the application of climate information to sustainable development.

The specific objectives of this Thrust, together with the international programmes which are contributing to their realization, are shown in Table 2.2.

(4) Thrust 3: Studies of climate impact assessments and response strategies to reduce vulnerability

Aim. To identify the means by which nations could reduce their vulnerability to climate, climate variability and both natural and anthropogenic climate change.

Requirements. The FCCC requires, *inter alia*, "stabilization of greenhouse gas concentrations in the atmosphere at a level that would prevent dangerous anthropogenic interference *with the climate system*". In practice, stabilization

Table 2.2. International programmes contributing to Thrust 2

Objective	Focused programmes	Contributing programmes
1. Deliver operational climate forecasts, and assess trends in climatic variables	WCASP	WCRP, IGBP, WCDMP, AgMP, HWRP, WWW, GCOS
2. Establish networks of national, regional and global climate services	WCASP	WCIRP, ICSU, WDCs, FAO programmes, WWW
3. Provide assistance to nations to enable them to respond to climate extremes and climate-related calamities, especially drought and desertification	WCASP	WCIRP, FAO programmes, WCDMP, AgMP, HWRP, IHP, MAB, IHDP, IGBP

requires an assessment of different emission paths, sinks, reservoirs and socio-economic factors. Determination of what constitutes dangerous anthropogenic interference with the climate system requires identification of measurable factors which are vulnerable to climate change in each relevant sector (e.g., the number at risk of sea-level rise); determination of relationships between these factors and climate change variables (e.g., rainfall and temperature) and thus an estimate of the maximum acceptable change in these variables; and determination of the stabilization level of greenhouse gas concentrations to limit climate change variables to acceptable levels. Clearly some of this work will also be carried under other Thrusts, particularly Thrust 1.

The FCCC also requires Parties to: publish emission inventories; formulate, implement, publish and regularly update programmes to address the anthropogenic emission by sources and removal by sinks of greenhouse gases; take measures to facilitate adaptation, cooperate and promote technology development; to conserve and enhance sinks and reservoirs of greenhouse gases. Vulnerability can be reduced by appropriate response strategies; there is therefore also a need for coordinated development of such strategies.

Agenda 21 calls for assessment of the environmental and socio-economic impacts of changes and fluctuations in the atmosphere; assessment of vulnerability and development of response strategies to sea-level rise; and assessment of vulnerability and development of response strategies with regard to water resources.

The Yokohama strategy on natural disaster reduction emphasizes the importance of disaster prevention through mitigation and preparedness, both of which require assessments of risk and development of response strategies.

Overview. Understanding the socio-economic and environmental implications of climate change and climate variability in the areas for which they are responsible is one of the prime interests of individual nations. This is coupled with their need to know what actions they can take individually and collectively to enhance the resilience of their socio-economic systems in areas of significant vulnerability.

Research in this area has today been sporadic and the subject of individual socio-economic professionals. It has not been the subject of an internationally coordinated programme of the type undertaken by the physical scientific community.

It is clear that much of the work will still have to be done by individual nations working with their own professionals and that this will form much of the basis for the assessments being conducted by the IPCC and required by the FCCC. At the same time, the tools by which such understanding is developed and applied within individual countries are only just beginning to emerge and could significantly benefit from the introduction of an international programme.

Global climate change has causes and effects which transcend national boundaries: energy production and consumption, irrespective of their origins, will influence global greenhouse gas emissions; a changing climate may trigger international migrations; and response and adaptation strategies will necessarily include international agreements.

An international framework will contribute to understanding these trans-boundary phenomena through the elaboration of a set of common research questions and agreed methodologies and data. It could also provide a focus for individual scientists working in this field, act as a "clearing house" for the exchange of information on techniques and methodologies to the large variety of different socio-economic systems which exist around the world.

It is proposed that such an international programme be built around the research activities of the IHDP and UNEP which support the application of climate impact study methods. The specific objectives of this Thrust, together with the international programmes which are contributing to their realization, are shown in Table 2.3.

Table 2.3. International programmes contributing to Thrust 3

Objective	Focused programmes	Contributing programmes
1. Identify environmental and socio-economic implications of both climate variability and change	WCIRP, IHDP	IGBP, WCASP, AgMP, HWRP, IHP, WHO programmes
2. Identify options for national response strategies to reduce vulnerability to climate variability and change	WCIRP, IHDP	FAO programmes, AgMP, IHP, MAB
3. Develop and publicize optimized methodologies, including appropriate datasets, for evaluation of climate impacts and response strategies.	WCIRP, IHDP	IPCC
4. Establish a coordinated network of national and regional climate impact and response programmes	WCIRP	IHDP, IGBP

(5) Thrust 4: Dedicated observations of the climate system

Aim. To provide on an operational basis systematic climate observations for climate system monitoring, including response and impacts; climate change detection; and research, modelling and prediction.

Requirements. Both Agenda 21 and the IPCC call for a coordinated effort to meet the observational requirements for climate system monitoring, climate change detection, research toward understanding and prediction and for data to apply toward national economic development.

The FCCC explicitly commits nations to promote and cooperate in systematic observation and the development of data archives related to the climate system to further understanding and to reduce or eliminate the remaining uncertainties regarding the causes, effects, magnitude and timing of climate change. The CCD also calls for national action programmes to, as a matter of priority, strengthen capabilities for systematic observation.

Overview. It is essential that the nations of the world be provided with adequate information on the current state of the climate and develop the capability to assess its future evolution so that governments and policy makers make well-informed decisions.

Continuous, global, well-calibrated data are required to:

- document climate changes;
- differentiate natural variability from anthropogenically forced change;
- reduce uncertainties about future climatic change;
- assess impacts on ecosystems.

Such data are also required as inputs to models being used to predict successfully seasonal-to-inter-annual climate variability (e.g., the El Niño phenomenon) and to validate longer-term climate models.

These data will also enable governments to enhance climate services for sustainable development (Thrust 2) and reduce vulnerability to climate variability and change (Thrust 3). To achieve this, the proposed strategy of climate observations will address the need for compatibility of climate data collection, where possible, with those socio-economic data being collected and made available in Thrusts 2 and 3.

Although many measurements are currently being made, a comprehensive system for observation does not now exist, but should be established with urgency to ensure the provision of necessary and appropriate climate observations.

The proposed strategy is to provide these observations via the Global Climate Observing System (GCOS). GCOS is intended to be an internationally coordinated, operational observing system that will build upon and augment where necessary, existing observing programmes in participating countries, including, for example, WWW, GAW, IGOSS and GLOSS, WCP—Water, GEMS and GRID.

It will integrate observations from both space-based (satellite) and surface-based data acquisition systems. It will include a comprehensive data and information management component (developed in concert with WCDMP) to ensure data availability and access to participating nations. It will provide a structure for continuing these observations into the next century where they will be of inestimable value to future societies.

GCOS is a relatively new initiative established in accordance with the recommendations of the Second World Climate Conference. Since its inception, an international Joint Scientific and Technical Committee (JSTC) and Joint Planning Office (JPO) have been developing the strategy and the central elements of an integrated plan for the establishment and operation of the global observing system.

The climate observing system should be an effective and efficient one based on sound scientific principles and be designed to accommodate past, current and future data.

It should provide accurate and representative observations from the atmosphere, ocean, land surface and ecosystem and monitor both the traditional climate parameters and critical ecosystem variables. It should have links to programmes with socio-economic information.

This Thrust outlines the strategy and objectives necessary to:

- specify the required components of the observing system based on scientific priorities;
- develop a comprehensive and feasible plan, including recommendations for the implementation of observations;
- propose an oversight and management structure to ensure the continuing efficacy of the observing system into the next century.
- The observing system consists of national activities and the necessary international planning and coordination to integrate them into a coherent programme.

The specific objectives of this Thrust, together with the international programmes which are contributing to their realization, are shown in Table 2.4.

Table 2.4. International programmes contributing to Thrust 4

Objective	Focused programmes	Contributing programmes
1. Design an effective operational climate observing system	GCOS	WCDMP, WWW, GAW, IGOSS, GLOSS, CEOS, GOOS, GTOS, WCP—Water, WCRP, IGBP, GEMS and GRID
2. Establish, coordinate and manage the Initial Operational System by integrating and enhancing existing components	GCOS	WCDMP, WCRP, WWW, GAW, IGOSS, GLOSS, CEOS, IGBP, WCP—Water, WDCs, GEMS, GOOS, GTOS
3. Develop new components to provide a comprehensive and responsible observing system to meet future needs	GCOS	WWW, GAW, GEMS, GOOS, GTOS, IGOSS, GLOSS, CEOS, WCDMP, WCRP, IGBP, WCP—Water

Specific capacity building initiatives

Regional networks play a key role in supporting climate research, capacity building, service provision and data and information exchange. Support to such networks, both by nations and by the international organizations developing the Climate Agenda, is required in order to meet the overall aims of the Agenda. Networking initiatives particularly relevant to this Agenda are:

- the global change System for Analysis, Research and Training (START) being developed by IGBP, IHDP and WCRP;
- the Inter-American Institute for Global Change Research (IAI), an intergovernmental initiative;
- the Asia-Pacific Network for Global Change Research (APN), an intergovernmental initiative;
- the European Network for Research in Global Change (ENRICH) being implemented by the European Commission;
- the Climate Impact and Response Strategies Network (CIRSNet) established and implemented under the WCIRP;
- the Climate Change Information Exchange Programme (CC:Info) and the Climate Change Training Programme (CC:Train), carried out by the Interim Secretariat for the FCCC in cooperation with UNEP and UNITAR.

Capacity-building activities on climate-related issues are also being implemented by Unesco and its IOC. In addition to relevant training activities in oceanography, hydrology, etc., regional information networks support climate-related research and strengthening of relevant national scientific capacity. Fig. 2.1 illustrates the interactions between various international organizations and their programmes related to the World Climate Programme.

Although the Climate Agenda is a very comprehensive document, it is clear at this preliminary stage of its development that further improvements are necessary. The problem of priorities and unsolved questions have not been precisely formulated as yet. Some thoughts on this subjects are mentioned below.

Perhaps, the key aspect of the Climate Agenda which requires further consideration and detailed analysis is the necessity to more clearly formulate the global climate change problem as an inseparable part of a much more general global change problem. As far as climate is concerned, an important achievement of the last years has been the realization of the need to investigate the whole coupled climate system "atmosphere – oceans – land surface – cryosphere – biosphere". Now the time has come to take into account interactions between the nature and the society (socio-economic development) because it is obvious that, eventually, these are specific features of the socio-economic development which determine the climate dynamics. We shall dwell upon this subject in more detail later on, but here we shall briefly mention as an illustration the problem of accomplishing the recommendations contained in the UN Framework Convention on Climate Change (FCCC). These recommendations are, no doubt, very important, but obviously cannot be fully realized on a global scale because they contradict the requirements of the socio-economic development of the Third

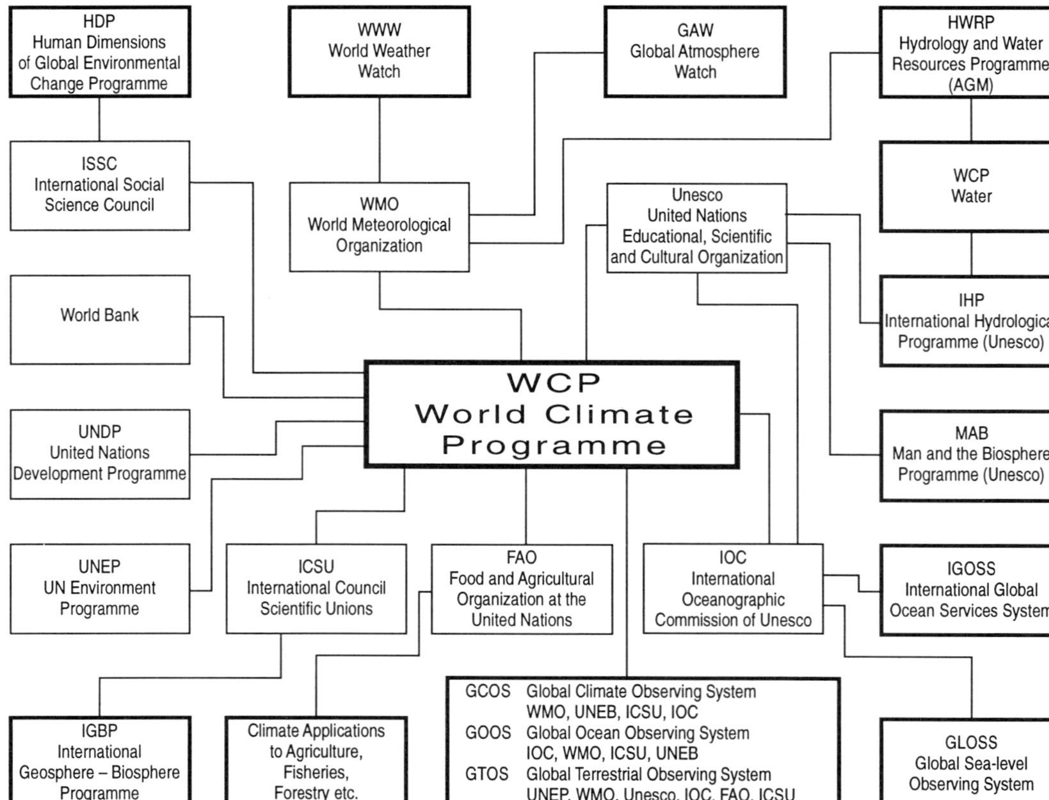

Fig. 2.1. International organizations and their programmes related to the World Climate Programme.

World. Such a situation has resulted from an insufficient account of human dimensions of the global climate change problems.

Climate change problems are numerous and therefore should be carefully prioritized, which is, of course, not easy to do. The "Summary for Policymakers" of the Climate Agenda (The Climate Agenda, 1995) contains the following conclusions which (in addition to that said in Thrust 1: Deliverables) may be used as an indication of scientific priorities: (1) greenhouse gas concentrations have continued to increase; (2) anthropogenic aerosols tend to produce the negative radiative forcing; (3) climate has changed over the past centuries; (4) the balance of evidence suggests a discernible human influence on climate; (5) climate is expected to go on changing in the future; 6) there are still many uncertainties.

These conclusions are, undoubtedly, very important, but still reflect certain overemphasis on the "greenhouse" climate warming. Future interdisciplinary research should be directed, however, towards solutions of the fundamental problems which embrace the whole complexity of global climate change processes. As for interactive physical, chemical and biological processes, a number of aspects of climate change deserve special attention, including:

(1) Atmosphere–ocean interaction, including gas exchange which regulates the atmospheric chemistry (CO_2, DMS, sulphur gases, etc.). A number of principal case studies: the Tropics (with special emphasis on El Niño/Southern Oscillation and the North Atlantic deep waters formation); higher latitudes (specific features of climatically significant processes in the Arctic and Antarctica).
(2) Atmosphere as a colloidal medium: aerosol – cloud – radiation interactions.
(3) Biosphere dynamics in its relevance to climate change. Coupled consideration of global biospherical cycles and climate dynamics.
(4) Solar impacts (solar activity and climate).

These are just a few illustrations to emphasize an urgent need for the problem-oriented planning of the global observing systems based on the combined use of both conventional (*in situ*, remote sensing) and satellite observations. Deficit of the problem-oriented planning results in obtaining incomplete information. The ongoing degradation of conventional observations (especially, networks of surface stations) requires special measures to continue the long-term data series. There is an obvious underestimation of the importance of ship, aircraft and balloon observations.

It is very important that the Framework Convention on Climate Change (FCCC) was in the focus of the Rio Conference in 1992. But to avert a danger of the anthropogenically induced climate change, it is not enough to accomplish the recommendations of the FCCC (which, unfortunately, cannot be fulfilled in case of developing countries). The problem is much more complicated and prohibitively difficult.Not less than future development of civilization has to be optimized. This is where human dimensions acquire their central position (Kondratyev, 1990a, 1996b; Kondratyev et al., 1997).

2.2 THE WORLD CLIMATE RESEARCH PROGRAMME (WCRP)

2.2.1 Climate observations data

Before discussing the World Climate Research Programme we shall briefly characterize the present state of the global climate diagnostics. The recent deep concern with respect to the problem of both observed and potential future climate changes has been expressed in numerous monographs and other publications (Alcamo, 1995; Allen et al., 1994; Baden, 1994; Crutzen and Ramanathan, 1996; Giambelluca and Henderson-Sellers, 1996; Hempel, 1994; Hobbs, 1995; Houghton, 1986, 1994; Houghton et al., 1996; Jones et al., 1996; Joussame, 1995; Kemp, 1994; Kondratyev, 1982, 1983a,b, 1985, 1988, 1990a,b, 1991a–c, 1992a–g, 1996; Kondratyev and Moskalenko, 1983, 1984; Kondratyev and Galindo, 1996; Kondratyev and Cracknell, 1996a,b; Le Treut, 1996; McCormick et al., 1995; Peixoto and Oort, 1992; von Storch and Navarra, 1995; Strzepek and Smith, 1995; The Global Climate, 1984; Viner and Hulme, 1994; Ye, 1993; Zhang and Fletcher, 1993, and many others). Very important are extensive reports (Climate Change, 1990, 1995, 1996a–c; Climate Change: Science..., 1991; Climatic Change News, 1995; Climate Impacts...,1995; CLIVAR, 1995; IPCC, 1992; IPCC Special..., 1994; Jackson, 1994) prepared by the Intergovernmental Panel on Climate Change (IPCC) with the participation of a great number of international experts. First of these reports had served the basis for estimates contained in the documents of the Second World Climate Conference held in Geneva in 1990 (Kondratyev, 1991b) and was of great practical importance in preparing the Framework Convention on Climate Change approved by the Second UN Conference on Environment and Development (Ferman, 1992; German Advisory..., 1996; Kon-dratyev, 1992i,j, 1993a; Program of Actions, 1993; Suzuki et al., 1996) and in working out the "Berlin Mandate" (The Berlin Mandate, 1995).

Taking as a basis the reports (Climate Change, 1990; Climate Change:Science..., 1991), Houghton (1994) answered in his monograph the following major questions: Does human activity affect the climate? Is the global climate warming a reality? How far may climate changes go in the future? Will they be followed by serious disasters and will the latter be more frequent? Can we adapt ourselves to climate changes or should we seek for means of reducing and, if possible, preventing the undesired changes?

According to the World Meteorological Organization's (WMO) definitions, climate variability is the variability of the stationary stochastic process which determines the climate on a scale of several decades, whereas climate change implies a difference between two stochastic processes modelling the climate for successive time intervals of several decades or, in other words, it is a difference of the climatic characteristics averaged over a long time period. Climate variability is characterized by extrema as well as by differences in mean monthly, seasonal and annual values compared to mean climatic estimates calculated based on the supposed stationary state and ergodicity.

Introduction of the two notions to describe the climate variability corresponds to representing the climate as stationary and non-stationary processes and, thus, the climate may be considered a non-stationary process which can be approximated by a stationary process over a shorter time period of several decades. This stationary approximation is connected with the notion of climatic norms considered as characteristics of averages with respect to which the climate fluctuates. Other definitions of climate variability and change have been proposed by Hasselmann et al. (1995).

An important consideration of global temperature deviations as a random walk has been accomplished by Kärner (1996) who has shown that a reasonable understanding of the contemporary mean climate (in terms of the surface air temperature) is attainable, assuming a random forcing to the climate system and treating the temperature deviation as a response to it. The forcings occur due to volcanic eruptions, redistribution of cloudiness, variations in snow- and ice-covered areas, changes in solar output, etc. A similar approach has been earlier suggested by Gordon (1991).

In this context, Mitosek (1995) exemplified some results from analysis of long-term series of mean monthly and mean annual values of the river run-off to describe the climate variability and climate change. The results obtained suggest the following conclusions: (1) in analysis of the series of hydrological characteristics one should take into account the difficulty in recognizing changes in the regime of a given phenomenon, since the possible trends are masked by noise; (2) the parametric tests based on the supposed normal distribution cannot be realized in the case of non-Gaussian processes (therefore the parametric tests are more appropriate); (3) the results of the Kruskal-Wallice tests (i.e., proceeding from the hypothesis of the stationary state and ergodicity of averages and of climate variability) should be rejected at the level of a 5% statistical significance; (4) according to the tests of the reliability of averages and variability after Mann-Kendall, an acception of the hypothesis of the stationary state and ergodicity makes it necessary to reject as doubtful (in the case of data on the mean annual river runoff) 42.0% (9.1%) of the cases in analysis of the river runoff averages (variability).

The unique and dynamic nature of the Earth is determined by an interaction of its components: the atmosphere, ocean, land, cryosphere, and the biosphere. The following five fundamental processes determine the major features of the climate dynamics: (1) convection, condensation nuclei, and cloud formation; (2) ocean circulation and its interaction with the atmosphere and cryosphere; (3) land hydrology and the interaction of the hydrological processes with vegetation cover; (4) biochemistry of the greenhouse gases (GHG); (5) circulation and chemical processes in the upper atmosphere. The major way to solve the related problems is to study the enumerated processes from the data of observations and numerical modelling. As for the first of the processes mentioned above, the main goal here is to develop more adequate models of formation, development and dissipation of clouds considering the effect of convection, cloud condensation nuclei (determining changes in the cloud particles size distribution and, respectively, optical properties of clouds), and the cloud–radiation interaction (Cloud–Radiation..., 1994). The atmosphere–ocean interaction is a key factor of climate formation and gas exchange (including the problem of the ocean as a component of the global carbon cycle).

The monograph "The Global Warming Experiment" (1995) gives a brief analysis of the role of most substantial climate-forming factors and presents an overview of the latest results of studies of climate variability based on numerical climate modelling and the diagnostics of ground and satellite observations. The analysis in question has shown that the well-known results of numerical modelling substantially overestimate the mean global "greenhouse" climate warming. New estimates suggest the conclusion that it can reach only 0.10°C over ten years but not 0.30°C over the same period, as has been considered earlier. In accordance with the data of Parker and Folland (1995), the mean global warming in 1994 (compared to the period 1951–80) constituted $+0.31 \pm 0.030$°C, and the warming over the last 135 years has reached 0.60°C, taking place, mainly, between 1910 and 1940 and during the last 15 years. The 1991 Pinatubo eruption had reduced the warming, but by the year 1994 its influence had practically ceased. According to IPCC-95 (Houghton et al., 1996), the global mean surface air temperature (SAT) has increased by between $\sim$0.3 and 0.6°C since the late nineteenth century.

A new important step in obtaining the satellite SAT information has been made by Kodera et al. (1996) who substantiated an empirical relationship between SAT, SST (sea surface temperature), wind speed, and humidity to retrieve SAT from satellite observations of the following parameters: SST, wind speed (microwave remote sensing data), and near surface water vapour mixing ratio (calculated from retrieved precipitable water with the use of an empirical relationship).

Kondera et al. (1996) has shown that an error of monthly mean satellite-derived air temperature is -0.3 ± 3.1°C and that of sensible heat flux is 10.0 ± 37.6 W/m^2. The estimated error of the inter-annual SAT variation is $0.0° \pm 1.5$°C, and that of sensible heat flux is 0.3 ± 19.0 W/m^2. In spite of the approximate nature of the method, its significance is undoubted because of the possibility of retrieving the components of the air–sea heat and moisture exchange.

Starting from the 1976–77 winter, a persistent anomaly has appeared in the tropical Pacific manifesting itself through the annual formation of the El Niño upwelling, with global-scale consequences (Kerr, 1995). The numerical modelling results show that regular positive SST anomalies contributed substantially to the record-level global climate warming during the 1980s and play an important role as a factor of interdecadal climate variations. The regular variability of the fields of wind and currents in the northern Pacific can also serve as one more factor in such variations.

The numerical climate modelling performed at the US National Meteorological Centre based on the observed SST variations revealed a warming over the land strikingly similar to that observed in the 1980s, especially in North America. Similar results have been obtained by the specialists of the Scripps Institution of Oceanography and the Geophysical Fluid Dynamics Laboratory, the results being practically similar, both with the prescribed global SST field evolution and with only the Pacific SST anomaly (these results will be discussed in more detail later on). It means that it was the Pacific Ocean that governed the process of global warming, whose mechanism consists in an enhancement of the latent heat release in the atmosphere through precipitation formation. The resulting warming of the tropical atmosphere affects the weather and climate on a global scale. It may be that the original cause of the Pacific warming lies in the increasing atmospheric greenhouse effect.

In the context of this problem, the necessity to take into account an impact of the processes in the Atlantic Ocean and in the extra-tropical zone of the Pacific Ocean on the general atmospheric circulation is not in doubt. Numerical modelling has shown, for instance, that regardless of the processes taking place in the tropical Pacific, a consideration of the interaction between the atmosphere and the whole Pacific Ocean in the northern hemisphere (even with the tropical belt excluded) reveals a 20-year circulation cycle determined, apparently, by a feedback between the winds blowing across the northern Pacific and a gigantic gyre around the Pacific Ocean bringing the heat from the south (the long duration of this cycle—20 years—reflects the great inertia of the cycle). All these results testify to the fact that it is difficult to distinguish between natural and anthropogenic climate changes.

Most serious sources of uncertainties in the numerical climate modelling originate from an inadequate account of the atmosphere–ocean interaction, cloud cover dynamics, and the cloud–radiation interaction. A parameterization of the processes connected with the snow- and ice-cover variability cannot be considered reliable.

Palaeoclimatic data demonstrating the presence of a correlation between the surface air temperature (SAT) and carbon dioxide concentration are an important source of information (Eddy and Oeschger, 1993). But, on the other hand, there have been periods when the CO_2 concentration has increased 4–10 times and the SAT has remained practically constant; on the contrary: substantial changes in temperature have taken place with almost constant CO_2 concentrations. So, for example, the warm climate in the middle of the Carboniferous period was, apparently, determined not by the greenhouse effect but by heat transport in the ocean.

An important source of information which has been broadly used to assess the greenhouse impact on climate is the ice core records of greenhouse gases. On the basis of such records, it has been concluded that the current atmospheric levels of CO_2 are higher by 25%, of CH_4 by 115%, and of N_2O by 19%, in comparison with the pre-industrial and ancient atmospheres. As pointed out by Jaworowski (1994), "during the last decade these records became the basic input parameters in the models of the global carbon cycle and a cornerstone of the man-made climatic warming hypothesis".

The results obtained by Jaworowski et al. (1992) and Jaworowski (1994) have clearly revealed that "no study has ever directly demonstrated that the contents of greenhouse gases in air recovered from bubbles in old polar ice, or even from recently deposited snow, represent air in its original atmospheric composition". In his survey paper Jaworowski (1994) has analysed a number of arbitrary assumptions which lead to a wrong interpretation of ice core data, especially due to an arbitrary selection of "reliable" data, ignoring fractionation of gases trapped in the ice sheets, the assumption that the air inclusions are much younger than the ice in which they are entrapped, etc. (i.e., the "age corrected" records of greenhouse gases have been widely used as a proof that the amounts of these gases were lower in the nineteenth century than today).

A note of caution introduced by Jaworowski (1994) concerning a poor understanding of the gas fractionation process is very important. It is true that "there is a need for experimental studies which would provide the essential physical and chemical parameters needed. Before such studies are carried out, conclusions on low pre-industrial atmospheric levels of greenhouse gases cannot be accepted when deduced from ice core data".

From the data of the ice core analysis, about 135 000 years ago, over the subsequent 20 000 years a sharp temperature drop occurred, with only small oscillations of CO_2 concentrations. The estimates showed that, for the "greenhouse" explanation of the Pliocene climate warming, the CO_2 concentration should be supposed to have increased up to 1300 ppm, that is, the concentration ought to have increased four times compared to the present level, whereas in fact, it was close to that observed nowadays. About 440 million years ago (in the Ordovician period), when the CO_2 concentration exceeded the present concentration by an order of magnitude, there was a glaciation (glaciers had covered vast regions of the supercontinent Gondwanaland). These examples illustrate the impossibility of explaining the palaeoclimate variations by changes in CO_2 concentrations (the greenhouse effect) only.

As Houghton (1994) notes, an analysis of various indirect climate indicators reveals their considerable variability over the last hundred years, in particular, periods of cooling and warming of unclear origin. The most probable cause is the internal variability of the climate system (e.g., the GHG effect could not be substantial, since their concentration had remained comparatively constant; the idea of a possible contribution of solar activity remains hypothetical, and the influence of the volcanic eruptions could have been episodic).

Extensive information on variations in atmospheric chemistry and climate during the last million years has been derived from the chemical and isotopic analyses of ice cores (note once more, however, that the need of further assessments of the reliability of these data has been convincingly demonstrated (Jaworowski, 1994; Jaworowski et al., 1992)). It was found out, in particular, that the last glacial period started about 120 000 years ago and completed approximately 20 000 years ago. A correlation has been established between temperature and concentrations of carbon dioxide and methane. Analysis of sea deposits suggested the conclusion that during the last million years there have been six or seven glaciation periods about 100 000 years apart. Apparently, the main cause of such climate variations (cyclicity) had been changes in the parameters of the Earth's orbit leading to a re-distribution of solar radiation, though the estimates show that a consideration of only this one factor is clearly insufficient. One should suppose the existence of feedbacks intensifying the effect of changes in the orbital parameters, due to CO_2 variations including (Berger et al., 1984).

It is important that, although the palaeoclimate variations were slow (e.g., during the last 8000 years the climate has been surprisingly stable), sometimes sudden and considerable changes in temperature had taken place (of 5–6°C) over a period of less than 100 years (e.g., during the glaciation period before 100 000 years ago). A mysterious drastic temperature drop (by 7°C over about 50 years) occurred in the period of the "Younger Dryas" (10 700–12 000 years ago), a possible cause of which could have been a change in the thermohaline circulation due to a glacier melting in North America and subsequent freshening of sea waters which had prevented the warm waters from flowing northward (the reverse process could have caused a climate warming).

Hughen et al. (1996) have shown that many abrupt sub-decade to century-scale oscillations in surface–ocean biological productivity took place that are synchronous with climate changes at high latitudes. The abrupt climatic changes are consistent with a North Atlantic circulation forcing mechanism due to thermohaline circulation dynamics.

An intensive discussion on abrupt climatic changes has arisen recently (Gwiazda et al., 1996; Revel et al., 1996) in connection with the discovery of the so-called Heinrich layers (this phenomenon will be considered later on). Possibilities of such drastic climate changes should be taken into account in future climate forecasts.

Noting the presence of a mean global warming trend within 0.3–0.6°C over the last century, Houghton (1994) emphasized the necessity of taking into account the inter-annual and interdecadal climate variabilities determined, apparently, by the internal variability of the climate system. In this connection Houghton notes that "in view of the observed natural climate variability, it is difficult to state with confidence that the trends of the mean global temperature increase over the last century and during the last decade have been determined by a global warming due to increasing GHG concentrations". So, for instance, the temperature increase between 1910 and 1940 was too fast to be explained by a comparatively slow increase of GHG concentrations during that period. The problem is that the temperature variations to be observed during future one–two decades should be carefully analysed to detect a "greenhouse signal" (in this context, the study by Ellsaesser (1992) is very important).

Solution of the problem of recognition of the anthropogenic climate changes "fingerprints" is one of the key aspects of present climate theory (see, e.g., Kondratyev, 1992d; Kondratyev and Cracknell, 1998a,b; Thaker, 1996). As Karl (1994) notes, a new step in this direction has been an analysis of data on variations in the mean seasonal temperature field in the troposphere and stratosphere from the results of microwave remote sounding of the atmosphere for the period 1963–88, which suggests the conclusion that the regularities observed confirm the results of the numerical modelling of the "greenhouse" climate changes: there was a tropospheric warming (particularly considerable in the upper tropical troposphere) and stratospheric cooling. According to statistical estimates, there are 9:1 chances that this congruence was not accidental (later publications (Christy and Spencer, 1995; Christy et al., 1995) drew attention to the difficulty of satellite data interpretation). On the other hand, this conclusion should be considered doubtful, since it is apparent that an increase of the greenhouse effect is only one of the factors governing climate change. So, for instance, the total ozone content (TOC) decline observed during this period could also cause a stratospheric cooling, and more frequent and powerful El Niño/Southern Oscillation events during the 1980s also contributed to the tropospheric warming.

The reliability of observational data should be further analysed. If, for instance, instead of the 1963–88 data we examine the results of observations during the period 1958–84, when a similar CO_2 increase was observed, it turns out that the stratospheric cooling in this period was weaker, and the tropospheric warming appeared close to zero.

In discussing the data of meteorological observations over the last 100 years it was emphasized (The Global..., 1995) that, for instance, the data of the satellite thermal sounding for 16 years (from 1979 to 1994, see Goldberg and Fleming, 1995) revealed a weak cooling trend of about -0.06°C/10 yr in the troposphere in this period, whereas, as the results of the numerical climate modelling show, a warming of about 0.3°C/10 yr should have taken place. With the Pinatubo contribution to the temperature change excluded (Kondratyev, 1993b, 1996a) the satellite data suggest the conclusion of a warming trend of 0.09°C/10 yr.

One of the most popular conclusions of the "greenhouse" climate warming theory consists in asserting a considerable enhancement of high-latitude warming. However, the satellite data for 16 years and their extrapolation over the period 40 years reveal the presence in the Arctic of a climate cooling trend reaching 0.37°C/10 yr (1950–90), that is, the Arctic temperature decrease reached $\sim$1.5°C over 40 years (the results of numerical modelling suggest a 40-year warming of $\sim$2°C). Thus the numerical modelling of climate changes does not guarantee reliable results in view of the inadequacy of the current climate models.

One more important conclusion can be drawn from the results of surface observations at the densest and reliably functioning US network: the SAT values for the last 100 years (1895–1995) averaged over the USA territory have only oscillated with an amplitude not exceeding 10°C. It follows from all these data that the mean global SAT increase over the last century reaching $\sim$0.5°C ought to have been determined mainly by the factors not related to the greenhouse effect.

The authors of the monograph (The Global..., 1995) believe that with the present inadequate understanding of the causes of climate change it would be unwise to invest huge finances to the solution of the problem of reducing the GHG emissions to the atmosphere without a clear understanding of the goals of these investments. To stabilize the CO_2 level in the atmosphere, the USA must halve its emissions, which may ruin the economy. And this is when the needs of the economic development needs of China alone determine the prospects of a CO_2 increase in this country by 600 million tons annually by the year 2025, which practically compensates the reduction mentioned above.

The results considered enable one to appreciate the expediency of postponing an accomplishment of recommendations of the Framework Convention on Climate Change for 10 years "to clarify the problem". The "price" of this postponement may be only a 0.2°C mean global SAT increase in the next century. Note should be taken, however, that this (quite rational, in our opinion) conclusion has been questioned by many specialists, which is illustrated by the rather contradictory discussions at the First Conference of the participants of the Framework Convention on Climate Change held in the spring of 1995 in Berlin (Lanchbery, 1995; Victor and Salt, 1995). The "Berlin Mandate" (1995) still marks the problem of reduction of GHG emissions as a first-priority problem, though does not contain any specific recommendations in this respect, which would limit the acceptable levels of emissions. The need to substantiate future levels of emissions resulted in the recommendation for industrial countries (as well as recommendations which would provide a GHG reduction due to the system of the GHG sinks) to fix the dates as 2005, 2010, and 2020.

The situation with regard to meeting the FCCC requirements concerning the reduction of GHG emissions still remains controversial, as is clearly demonstrated by the report published by the World Energy Council (WEC) in connection with the second annual conference of the parties of the climate convention (Masood, 1996).

Commitments by most industrialized countries to reduce emissions from fossil fuels to 1990 levels by the end of the decade are unlikely to be met. The fact is that global CO_2 emissions from fossil fuel burning rose by 12% in the first half of the current decade. Although the largest rise has come from developing countries (a three times stronger increase of emissions than for industrial countries), industrialized countries (except France, Germany and the United Kingdom) will not be able to meet the FCCC requirements.

During the period between 1990 and 1995, most member countries of the Organization for Economic Cooperation and Development (OECD) registered a 4% rise in CO_2 emissions. The countries in the Asia–Pacific region (except Australia, Japan and New Zealand) showed a 30% rise. Emissions in the countries in the Middle East increased by 35%, in Africa by 12.5%, and in Latin America by 8%.

In contrast, CO_2 emissions in eastern Europe and the former USSR have already dropped below their 1990 levels because of the economic decline. In 1995, emissions from the Commonwealth of Independent States (CIS) were 70% of their 1990 levels, and 75% for central and eastern Europe.

Therefore, M. Jefferson, deputy secretary-general of the WEC said that, on the basis of the latest survey, setting firm targets for further reductions would not be realistic at this stage (Masood, 1996).

Concerning an early detection of "CO_2 climate signal", the conclusion can be drawn that the use of any single "fingerprint" of anthropogenic climate changes cannot be productive. But, on the other hand, a complex analysis of the meteorological information is hindered by the "greenhouse" signal/noise ratio being different for various meteorological parameters, which complicates the unambiguous recognition of a "greenhouse fingerprint".

To more reliably solve this problem, it is necessary, in particular (Karl, 1994; Karl et al., 1994, 1995a,b,c): (1) to substantiate possibilities of identifying the anthropogenic climate changes determined by GHGs and aerosols as well as by other forcings (especially those resembling the greenhouse ones); (2) to reliably assess the contributions of the natural climate variability (in this context the long-term numerical climate modelling is important); (3) to select the longest and sufficiently reliable (homogeneous) observational data series (including information on indirect climate indicators); (4) to establish correlations between the changes in climate, biospheric parameters, and socio-economic development playing the most important role as factors of survival on the planet.

The IPCC-95 Report (Houghton et al., 1996) has summarized the most important results related to the issues of detection and attribution of an anthropogenic climatic impact in the following way:

- The limited available evidence from proxy climate indicators suggests that the twentieth century global mean temperature is at least as warm as any other century since at least AD 1400. Data prior to 1400 are too sparse to allow the reliable estimation of global mean temperature.
- Assessments of the statistical significance of the observed global mean surface air temperature trend over the last century have used a variety of new estimates of natural internal and externally forced variability. These are derived from instrumental data, palaeodata, simple and complex climate models, and statistical models fitted to observations. Most of these studies have detected a significant change and show that the observed warming trend is unlikely to be entirely natural in origin.
- More convincing recent evidence for the attribution of a human effect on climate is emerging from pattern-based studies, in which the modelled climate response to combined forcing by greenhouse gases and anthropogenic sulphate aerosols is compared with observed geographical, seasonal and vertical patterns of atmospheric temperature change. These studies show that such pattern correspondences increase with time, as one would expect as an anthropogenic signal increases in strength. Furthermore, the probability is very low that these correspondences could occur by chance and they are also inconsistent with those expected for solar and volcanic forcing.
- Our ability to quantify the human influence on global climate is currently limited because the expected signal is still emerging from the noise of natural variability, and because there are uncertainties in key factors. These include the magnitude and patterns of long-term natural variability and the time-evolving pattern of forcing by, and in response to, changes in concentrations of greenhouse gases and aerosols, and land surface changes. Nevertheless, the balance of evidence suggests that there is a discernible human influence on global climate.

An optimization of the global economic development remains a critically important and unsolved problem. As has been mentioned (Ziel der CO_2..., 1995), an intensive development of transport and the related increase of CO_2 emissions to the atmosphere creates a serious threat to the solution of the problem of preventing an undesirable global climate warming through reduced CO_2 emissions (Germany decided to reduce CO_2 emissions to the 1987 level by the year 2005).

Without resolute actions on reduced energy consumption, the global-scale annual CO_2 emissions will increase by 10 billion tons by the year 2010, which corresponds to about half the total emissions in 1991. The main contribution to this increase will be made by the countries of south-eastern Asia and by China. In addition, in 2010 the distribution of per capita CO_2 emissions in the countries will remain strongly inhomogeneous: from 1 ton per year in Africa and India to 22 tons per year in North America. The CO_2 concentration in the atmosphere will not stabilize by 2010, it will even almost double. Unfortunately, such a measure as global-scale afforestation cannot help to restore the acceptable CO_2 level (Nilsson and Schopfhauser, 1995).

At present the main burden of reducing GHG emissions to the atmosphere recommended by the Framework Convention on Climate Change lies on the industrial countries, but even in this case the only way to realize it is to bring back the 1990 level of emissions, which does not ensure a stabilization of the GHG concentration level (a 60% reduction of emissions is needed for this purpose). Naturally, such a situation is determined by economic factors.

In this connection a conception of "Joint Implementation" has been developed, the main idea of which is that the "donor" countries will finance "host" countries (credit them) to reduce the GHG emissions (Jackson, 1994; Jepma, 1995; Kuik et al., 1994). This GHG "trade" is supposed to ensure a minimization of expenditures on reducing the GHG emissions. The opponents to this conception believe, however, that its realization will not stimulate the developed countries to solve the problem of global warming and, moreover, may even intensify the GHG emissions. An accomplishment of the "Joint Implementation" concept will, no doubt, require a careful and differentiated approach.

2.2.2 Key aspects of the World Climate Research Programme

In 1979 the First World Climate Conference (WCC-1) was held which produced the World Climate Programme (WCP) that served the basis for subsequent multidisciplinary developments to get a deeper insight into the laws of climate formation and its changes, to assess the impact of climate on both human activity and ecosystems, as well as to analyse the socio-economic aspects of the problem (see the references).

The WCC-1 made a Statement on the problems of global climate (Kondratyev, 1991b, 1992c) where it was emphasized that, although people use now the advantages of the present favourable climate, human activity is strongly affected by climatic variability, especially by droughts and floods. First of all, it concerns the developing countries located either in arid and semi-arid zones or in excessively humid regions. The interdependence of the climates in different countries necessitates a coordinated global-scale strategy for a better understanding and rational use of climate.

At present, man causes inadvertent local- and (to some extent) regional-scale climate changes. There is a deep concern about the fact that further enhancement of human activity may lead to considerable regional and even global climate changes, which determines the urgency of international cooperation in solving this problem. Climate changes caused by natural processes will continue.

Though considerable progress has been achieved in understanding the causes of climate changes, the role of various factors is still unclear. However, an anthropogenic increase of atmospheric CO_2 can, probably, contribute to a gradual warming of the lower atmosphere, especially in high latitudes. Probably, some regional- and global-scale anthropogenic effects will have been detected by the end of this century, and by the middle of the next century they will become substantial. Anthropogenic effects such as various aspects of agricultural activity (especially the enhanced use of nitrogen fertilizers) and forestry, as well as emissions of halocarbon compounds (freons), and others claim serious attention.

Estimates of the present climatic trends show that, apparently, they will be determined mainly by natural factors, at least, till the end of this century. Since people in the future will, presumably, be able to cause advertent large-scale climate changes, it would be irresponsible to consider such actions before we understand the mechanisms that govern the climate, which would be sufficient to forecast the consequences of the forcings. It is important to make the appropiate international agreement.

The complexity of the climate changes problems is schematically illustrated in Figs 2.2 and 2.3 (The Climate Agenda, 1995).

The Eighth WMO Congress held in April–May 1979 decided to work out and accomplish a World Climate Programme (WCP) comprising the following four programmes: (1) the Climatic Data Programme (CDP), (2) the Applied Climatology Programme (ACP), (3) the Programme of Climatic Implications of Human Activity (CIAP), (4) the World Climate Research Programme (WCRP).

At the Second WMO Congress (1–23 May 1991) it was decided to specify the programmes enumerated above: (1) the World Climate Applications and Services Programme (WCASP); (2) the World Climate Data and Monitoring Programme (WCDMP); (3) the World Climate Impact Assessment and Response Strategies Programme (WCIRP); (4) the World Climate Research Programme (WCRP).

The following circumstances determine the necessity of the WCP: (1) an increasing role of climate as a natural resource and increasing scales of the climatic impacts on human activity in conditions of the growing global population and increased life standards; (2) possibility of anthropogenic impact on climate; (3) probability of a climate change in the near future; (4) improved means to study the climate and its implications on human activity.

The WCP is mainly aimed at: (1) studying the climatic mechanisms (determination of the relative role of natural and anthropogenic factors of climate); (2) enhancing and improving climate observational databases; (3) applying climatic knowledge to plan human activity; (4) investigating the impact of climate changes on various aspects of human activity.

The following directions of applied climatology are of high priority: (1) food production (agriculture and fisheries); (2) water resources; (3) energy production; (4) medicine and construction climatology; (5) transport and communication; (6) development of marine and coastal resources.

From the viewpoint of climate monitoring, the following data are of primary importance: (1) meteorological and aerological observations; (2) information about trace gases and aerosol components of the atmosphere; (3) oceanographic data (including data on sea-ice); (4) hydrological data (including data on snow and ice cover). The geophysical data base includes: (5) measurements of solar constant and radiation budget at the top and bottom of the atmosphere; (6) data on stratospheric dust loading; (7) geomorphological data.

Since studies of the atmosphere–ocean interaction are vital for the climate problem, regions of the World Ocean where this interaction is most active (the

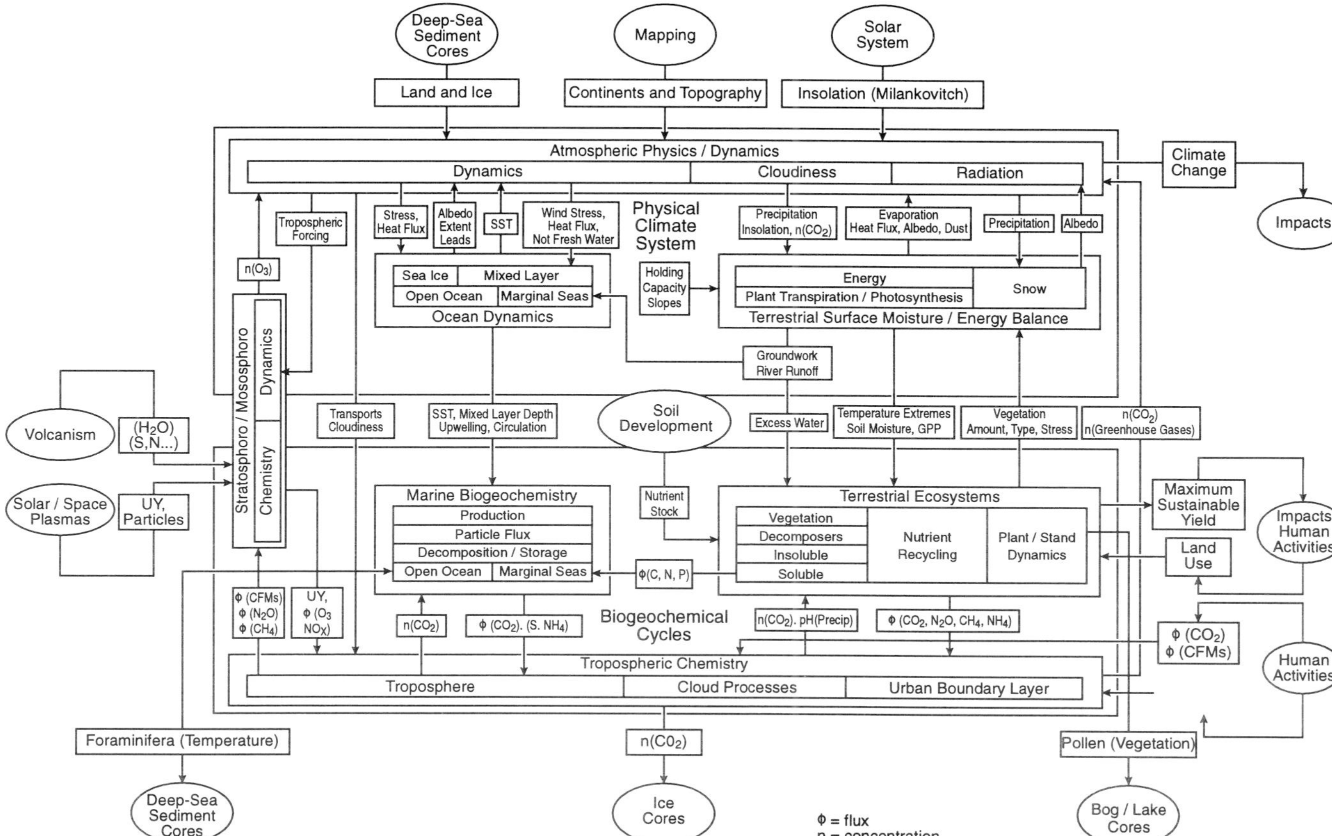

Fig. 2.2. Conceptual model of Earth system process operating on timescales of decades to centuries.

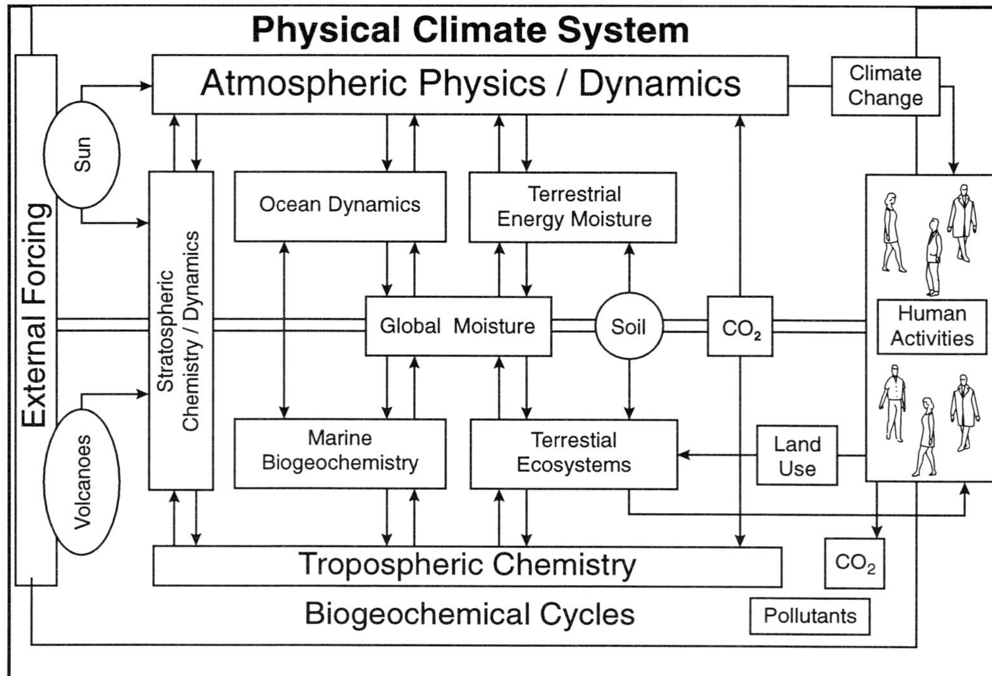

Fig. 2.3. The climate system.

regions of the Gulf Stream and Kuroshio, the zones of unstable stratification near the ice cover edge, the regions of upwelling, the zones of monsoon formation, etc.) must be studied in the first place as has been substantiated by the programme "Sections" (Marchuk, 1975; Marchuk et al., 1986; Energy-active zones..., 1989).

The problem of the meridional (towards higher latitudes) heat transport in the ocean and the transport of heat from one part of the ocean to another, varying under the influence of mesoscale eddies, must be particularly emphasized. Possible variations in the ocean albedo due to varying biological productivity must be analysed. Apart from meteorological, hydrological, geophysical, oceanographic and ecological information, the respective data are needed on agricultural activity, land use and industrial production. An adequate 30-year database on the climate system (atmosphere–ocean–continents–cryosphere–biosphere) must be a major goal.

Particular attention should be paid to monitoring of the anthropogenic climatic factors and their manifestations. This information should be obtained through enhancing the systems of conventional and satellite observations.

Major objectives of the WCRP aimed at climate studies in time scales from several weeks to several decades are: (1) the determination of the predictability of climate, and (2) the extent of man's impact on climate. To reach it, the following is needed: (1) a deeper insight into the laws of regional (a scale of about 1000 km and more) and global climates; (2) more reliable assessments of the trends of regional and global climates; (3) development and improvement of the physico-mathematical climate models able to simulate an actual climate, to assess its predictability; (4) studies on climatic sensitivity to natural and anthropogenic effects, that is, an assessment of climate changes caused by these effects.

Fig. 2.4 illustrates the interactions between global models and global observations (The Climate Agenda, 1995).

The WCRP developments are concentrated in six main directions (Hinsman 1995; Report..., 1992): (1) the Global Energy and Water Cycles Experiment (GEWEX) aimed at studying the cloud–radiation interaction as well as water cycles in the atmosphere and on land surface (GEWEX will be discussed in detail later on); (2) the programme of the tropical ocean and global atmosphere studies (TOGA) aimed at studies of the dynamics of the atmosphere–ocean interactive system in the tropics which seriously affects the inter-annual variability of the climatic system (El Niño/Southern Oscillation). The observational phase of this programme was completed in 1995; (3) the World Ocean Circulation Experiment (WOCE) focused on studies of "slow" climate changes determined by the ocean–atmosphere interaction; (4) studies of the Arctic climatic system (ACSYS) with account of specific high-latitude climate-forming processes; (5) processes in the stratosphere and their role in the formation of climate (SPARC) bearing in mind an analysis of the interactive chemical, dynamic and radiative processes in the context of global climate changes. The numerical climate modelling programme (including both "fast" and "slow" variability of the climate system).

These developments are mainly aimed at a deeper insight into the laws of climate variability which would create a reliable basis for climate forecasts in time scales from seasons to a century.

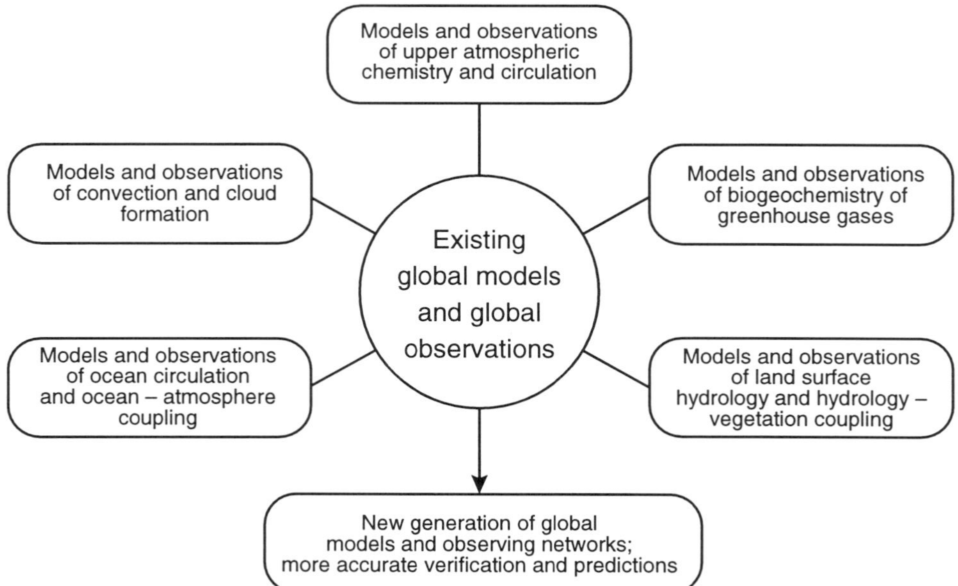

Fig. 2.4. Interactions between global models and global observations.

Studies should have the following order of priority: (1) the synthesis of observational data for regional and global climate systems; (2) the development of various theoretical climate models; (3) the role of the oceans in the climate system; (4) the biogeochemical cycles of radiatively active atmospheric gaseous components; (5) the formation of clouds, their spatial distribution and radiative properties; (6) types of aerosols, their optical properties and effects on clouds; (7) the hydrological cycle (evaporation, precipitation, etc.); (8) the properties of and processes on land surface and in the cryosphere, their impact on climate; (9) palaeoclimatic reconstructions, important for understanding the regional and global climate changes.

All the five interactive components of the climate system (atmosphere, ocean, land surface, cryosphere, biosphere) are objects of studies. A key and complicated problem is a study of relationships between the contributions to observed climatic variability in various time scales made by rapid internal fluctuations in the atmosphere and a slower variability determined by the boundary conditions at surface level and at the top of the atmosphere (variations in the snow and ice cover extent, SST, extra-atmospheric insolation, etc.), as well as other external factors (volcanic eruptions, CO_2 growth). The solution of this problem requires, first of all, complex diagnostic studies not only of the atmosphere but also of the ocean. Based on observations, an analysis must be made of the energy balances (in all its manifestations), momentum, water vapour and other trace gases on regional to global scales. The laws of the climate system can be understood only with the use of quantitative techniques in estimating the transport and transformation in the atmosphere of most important components of atmospheric and oceanic energetics. This determines a particular urgency of GEWEX (Climate Change ..., 1990, 1991, 1994, 1996a–c; Kondratyev, 1992f).

Still it is not clear which climate models will be most promising, and this complicates the planning of observations aimed at developing and testing the techniques for a parameterization of various processes. There are, however, processes whose importance raises no doubts. It is, for instance, processes of the radiation–cloud interaction (Marchuk et al., 1986). Studies of the climate system put forward rather high requirements for the accuracy of various quantities, which cannot be met yet. For example, extra-atmospheric insolation must be measured to an accuracy of $\sim 10^{-2}$ %, SST to $\sim 0.1°C$, and the heat transport in the ocean and precipitation to several per cent.

Of particular importance are complex studies of land surface processes with the consideration of albedo, evaporation, evapotranspiration, soil moisture, snow and ice cover changes related to the problem of anthropogenic impacts on climate (Henderson-Sellers and McGuffie, 1994; Henderson-Sellers et al., 1995a,b; Land-Use ..., 1995; Randall et al., 1995; Sellers and Hall, 1987; Sellers et al., 1996a–c; Viterbo and Beejaars, 1995).

Also important is the problem of the effect of aerosols on the atmospheric radiative regime and cloud formation (A Plan ..., 1996; Jennings, 1993; Kondratyev, 1991b, 1992a,b; Marchuk et al., 1986). In this case the development of some atmospheric aerosol models which could be used in numerical climate modelling plays a key role (Hobbs 1995; Kondratyev, 1982, 1983b, 1990b, 1991b , 1992d ; Kondratyev et al., 1983).

No doubt, the effect of clouds on climate formation is decisive, and this requires developments of the respective parameterization techniques with all most important feedbacks taken into account.

To concentrate on key aspects, the WCRP is sub-divided into three directions: (1) the first direction covers the problems of long-range (monthly-to-seasonal) weather forecasts. (2) The second direction is centred at the inter-annual variability. (3) The third direction is aimed at studies of the processes in time scales of decades and longer. Though, from the viewpoint of the WCRP, time scales from several months to several decades are of top interest, the palaeoclimatic reconstructions (especially to check the reliability of climate models) are also important. The spatial scale varies from regional ($\sim$1000 km) to global. The WCRP foresees three approaches to the solution of the formulated problems: (1) to study the climatically significant processes; (2) to develop the physico-mathematical climate models; (3) to perform a statistical analysis of the parameters of the climate observed.

A vitally important aspect of climate studies is to realize the chaotic nature of its changes. The difficulty in understanding the causes (forecasting, moreover) of climate change is determined by numerous feedbacks affecting climate formation and produced by chaotic nonlinearity of the climate system. In this context, Abarbanel (1985) analysed possible approaches to analysis of time series in conditions of chaotic nonlinear systems. The analysis was aimed at revealing a structure with available data seemingly lacking it. The data characterizing the behaviour of the chaotic deterministic structure are based on the geometric structure of fractal dimensionality in combination with a regular temporal variability. The structure is an attractor in the sense that to this structure asymptotically corresponds a large set of initial conditions for the system. Since such a structure has fractal dimensions, it was called a strange attractor.

The main goal of the overview (Abarbanel, 1985) was to discuss the problem of signals characterizing strange attractors. An examination of a chaotic orbit reveals the absence of any temporal regulation of evolution, and the Fourier analysis shows a wide set of frequencies. In this case the original task is to determine the space in which the time series should be analysed, or, in the assumed terms, to accomplish a phase-spatial reconstruction. In this connection it is important to search for motion invariants playing, in the case of nonlinear system, the same role as the characteristic frequencies in the examination of linear time series.

The next goal of the overview was to consider the parameterized models of the system's evolution within the reconstructed phase space and to find reliable information about the motion invariants. Concrete tasks were the following: (1) to establish the space of motion visualization and to analyse the chaotic time series; (2) to substantiate the classification of the system producing time series in terms of invariants of chaotic motion; (3) to formulate in the visualized space such models of motion which agree with the invariants and make the future behaviour forecast possible. Abarbanel (1985) emphasized that such a generalized consideration is directly related to the climate system, since it is, no doubt, chaotic.

A serious contribution to understanding the chaotic and "surprised" climate changes was made by palaeoclimatic evidence which helped to analyse the possibility to recognize the anthropogenic climate changes. As Gassmann (1993, 1994, 1995) emphasized, though the ideas of anthropogenic climate changes due to CO_2 growth have been widely recognized in the scientific community and caused concern, it does not mean that the formulated conclusions about the anthropogenic climate changes have been accepted.

This conclusion can be drawn with a consideration of new palaeoclimatic data (ice core analyses) testifying to a surprisingly substantial variability and instability of the past climate; for example, the preceding Emian interglacial period had been characterized by several drastic climate changes. The duration of the longest period of continuous climatic optimum (with a temperature 2 K higher than during the Holocene) had constituted only 2500 years. The climate had changed at least 20 times, the most drastic change (by the end of the Emian period about 115 000 years ago) having been followed by a 15-year transition to glaciation; the return to the original warm climate had taken only 70 years.

Thus there is a contradiction between the predicted gradual climate warming and the palaeoclimatic data on its sharp changes. This situation is explained by that climate changes used in climate modelling are treated as small variations with respect to the basic stable state. In addition, the possibility of the dynamics of the basic climatic state when one regime is replaced by another was ignored. Generally speaking, the spasmodic behaviour of the climate system with its nonlinearities and feedbacks is not unexpected. An adequate understanding should be based on realization not only of the chaotic nature of climate but also of its self-organizing capability. It is important that the transitions between various states of self-organization and between the chaotic state and self-organization are spasmodic: old structures are rapidly replaced by new ones. The old structures disappear at certain (critical) values of some parameters controlling the state of the structures, with small random fluctuations being able to become important factors of the subsequent evolution of the structures suffering a sharp phase transition. The problem of climatic multiequilibria will be discussed in this context in section 2.4.4.

In view of uncertain stochastic fluctuations and a weak cause-and-effect chain, even in the cases when the critical values of the parameters are well known, the time of transition to the future state cannot be predicted. It is only possible to describe probable metastable states. These circumstances are very typical of the global climate system with its numerous feedbacks and high nonlinearity (especially as it concerns the water cycle). A specific feature of the climate system consists in close interrelationships between various global regions due to the atmospheric and oceanic transport. All this can explain the climatic instabilities repeatedly happening during the last 200 thousand years, including the last two glaciation epochs and the Emian interglacial period. There is a paradoxical fact: still there is no answer to the question why during 5% of this whole period (the Holocene) the climate remained rather stable. While this situation remains unclear, it should be supposedly considered accidental and rather sensitive to external forcings.

The most natural situation is not a gradual warming but rapid climate changes.

Therefore along with the "classical scenario" of a gradual climate warming by ~0.3 K over 10 years (this scenario is an exception rather than the rule), it is worthwhile introducing the following scenarios (Gassmann, 1994): (1) a "sharp global change" characterized by a sudden temperature rise during a 1-to-10-year time period (the scenario of "strong uncertainty"); (2) a "sharp regional change" (the scenario of weak global but strong regional uncertainty).

Analysis of data on the oxygen isotopic ratio for several Greenland ice cores revealed considerable and sharp climate changes during the last ice period. Since this isotopic ratio depends on air temperature during the precipitation formation, such changes testify to climatic instability. The same conclusion about climatic oscillations has been drawn from analysis of marine depositions which revealed a stratified structure of several cores taken in the region of the North Atlantic. It was found out that the so-called "Heinrich layers" are observed in the 40–55°N latitudinal belt and refer to the period 7–10 thousand years BC.

These results enabled Paillard (1995) to propose the following scenario of palaeoclimate evolution. In the latitudinal belt 40–55°N an intensive release of icebergs by the Greenland glaciers to the Atlantic Ocean took place. From the data on isotopic ratio and foraminifera the temperature at that time was at a minimum. Immediately after the "Heinrich event" the temperature sharply rose by several degrees over several decades, followed by a gradual cooling, sometimes interrupted by episodic drastic temperature drops (the latter was clearly determined from analysis of ice cores). These secondary events form the major part of the "Dansghaard-Oeschger events" revealed from the data for ice cores and occur with periods of about 2000 years.

Though from these data the "Dansghaard-Oeschger events" are as substantial as the "Heinrich events", they are not clearly verified by analyses of marine depositions and land cores. This serves as the basis for distinguishing between "major events" connected with intensive releases of icebergs, and "small events" not confirmed by the data for depositions.

Such two-level climatic oscillations can be reproduced with the supposed existence of two different periodic factors affecting the ocean–atmosphere system: intensive releases of icebergs every 7–10 thousand years and weaker oscillations of the glaciers. However, with an account of the interaction with the environmental climate system, the hierarchical regime mentioned above can be described with the supposed existence of only one mechanism for the glaciers' impact on climate.

Since the Heinrich events are not likely to be a direct response to climate cooling, it is important that, according to the simplified climate model considered, its changes are determined by the glaciers' dynamics and not vice versa. The ocean intensifies the climatic variations: small changes in the freshwater supply cause strong climate changes. A powerful inflow of freshwater to the ocean suppresses for some time the thermohaline circulation, increasing thereby the temperature contrast between low and high latitudes. However, soon after this the thermohaline circulation is restored, and the climate at high latitudes sharply warms.

One may believe that an additional volume of freshwater goes to the ocean due to intensive releases of icebergs caused by ice melting at the foot of the glaciers at the expense of geothermal heat flow and friction. This supposition is confirmed by revealed instability of the Laurentian glaciers during the last glaciation and by the fact that the Heinrich detritic layers always contain a considerable amount of thaw water. The vertical temperature profile through the glacier thickness is determined by surface temperature as a boundary condition.

The enumerated regularities suggest the following scenario for the hierarchical climatic oscillations (Wuebbles et al., 1995): upon an intensive release of icebergs the high-latitude temperature sharply increases. At this time the height of the glacier sheets is at a minimum. Therefore the heat diffusion can provide a faster heating of the glacier foot and as the melting point approaches, a new release of icebergs takes place (even before the ice volume gets completely restored), but this secondary event is weaker and even negligibly small. The heating of the glaciers' foot weakens and, thus, the ice accumulates till the next release of icebergs. Of course, it should be borne in mind that this scheme is simplified and does not reflect the complexity of real processes.

Using isotope geochemistry data, Gwiazda et al. (1996) and Revel et al. (1996) have discussed the problem once more but reached opposite conclusions. Gwiazda et al. (1996) have found (from the data on the lead isotope composition for the second to last of the Heinrich events, called H-2, which occurred just before the height of the last glaciation, about 21 000 years ago) strong evidence that this Heinrich event was dominated by a massive iceberg discharge of the Laurentide ice sheet lobe located over Hudson Bay. These results support the model that invokes trapping of geothermal heat by the base of the ice cap and subsequent basal melting as the mechanism that triggered the Heinrich events. The data considered eliminate the possibility that during the Heinrich events, all ice caps were equally affected because in that case the Heinrich layer materials should consist of debris derived from more than one source, as in the case for the ambient glacial sediment, rather than being overwhelmingly composed of detritus from the Churchill province.

From the Sr and Nd isotopic compositions data on the lithic fraction of the last climatic cycle sediments from the North Atlantic (~40–60°N) Revel et al. (1996) have drawn the conclusion that during the last glacial period the Heinrich events (abrupt massive discharges of ice-rafted detritus, IRD) originated not only from the Laurentide ice sheet, but also from other sources. Three major North Atlantic ice sheets (Fennoscandian, British Isles, and Icelandic) probably surged simultaneously, discharging ice and IRD into the North Atlantic. Revel et al. (1996) disagree with the hypothesis mentioned above, which implies a unique, Laurentide origin of the H-2 driven by a internal mechanism. Thus, the Heinrich events, being a pan-North Atlantic phenomenon, require a common regional, external forcing.

The complicated correlation of the climates of the northern and southern hemispheres revealed by the palaeoclimatic data for Greenland and the Antarctic for the last interglacial period is well known. An idea was. suggested (Jouzel et al., 1995) that, in this case, of key importance was the formation of the North Atlantic deep waters (NADW) due to their effect on the SH sea-ice cover extent. The available data do not confirm that the correlation between the hemispherical climates mentioned above was determined by changes in the GHG content. The estimates showed, for example, that direct radiative forcing due to changes in the methane content cannot explain the temperature changes exceeding 0.1°C. It is also important that the warming in Antarctica had started (from the data of ice cores) much earlier than in Greenland (this reflects the fact that the initial stage of the interglacial period in the southern hemisphere was characterized by specific functioning of the interhemispherical correlation).

Analysis of the possible contribution of the Milankovich mechanism (variations in the orbital parameters) suggested the conclusion that the initial disturbing forcing must occur in this case in the northern hemisphere. To explain the early response of the southern hemisphere, an assumption was made about the rapid propagation to the southern hemisphere of the initial response to the insolation forcing which involves the Nordic Sea. The dynamics of the interglacial period from the data for Antarctica does not contradict the possible role of the Milankovich mechanism, and the smoother warming regime can be explained by the great thermal inertia of the oceanic southern hemisphere (however, the necessity of further testing of the reliability of indirect palaeoclimatic data should be borne in mind).

These circumstances (the role of the Milankovich mechanism and the processes in the North Atlantic) do not exclude a substantial contribution of the SH conditions to the formation of the SH climate, as well as the impact of the processes in the SH ocean on the NADW formation. Calculations have shown, for example, that the magnitude of the deep outflow from the Atlantic Ocean is proportional to the westerly wind stress at the latitude of the tip of South America. This conclusion is favoured by approximately synchronous strong changes in the atmospheric circulation reflected by intensified dust loading of Antarctica nearly 14 600 years BC and large-scale NADW changes revealed from observational data.

Though these considerations are mainly of a speculative character, there are no doubts about the empirical data testifying to the two-stage character of the last interglacial period clearly recorded in Western Europe, in Greenland, and in the North Atlantic. For example, in Greenland a gradual temperature decrease started in the Bölling (B) period about 14 500 years BC continued in the Alleröd (A) period, and then the Younger Dryas (YD) cooling event occurred, suddenly ceasing about 11 500 years BC.

It follows from the observational data that a succession of the BA/YD events had probably extended over the whole northern hemisphere, but information about the southern hemisphere is inadequate. New data on Antarctica (ice core drilled at Dome B in East Antarctica) were published, however (Jouzel et al., 1995), which point both to the global scale of the two-step regime of the last interglacial period and to the existence of substantial interhemispherical differences. In particular, it is possible that the coolest phase of an Antarctic cold reversal (ACR) in Antarctica, with a threefold weaker temperature decrease than that recorded in the YD period in Greenland, could have preceded the YD. Antarctica had not experienced abrupt climate changes, and the two periods of deglaciation had started there earlier than in Greenland.

An unusual antarctic phenomenon which has been recently discussed by Kapitsa et al. (1996) is an enormous lake, 200 km long, covering an area of 10 000 km^2 (comparable to Lake Ontario), and in some places at least 500 m deep located about 4 km under the ice of central East Antarctica. The mean age of this freshwater lake is about one million years. Undoubtedly, this lake should also be studied in the climatic context.

The warmest July of 1994 (compared to previous years) has again stimulated the question whether it was a manifestation of usual climate changes or of the long-term trend. In this context, of interest are the results of analysis of a Greenland ice core 3029 m long, considered by Fraedrich (1995). These results once more show that climate variations had not been continual but had been interrupted by sudden changes. Therefore the anthropogenic climate warming would not lead to the formation of the subtropical climate in Mid-Europe. Depending on the circumstances, one may expect a climate cooling in the future (as for the recurrent palaeoclimatic cooling, it will follow, no doubt). Analysis of the ice core mentioned above, revealed a temperature change reaching 10°C over several decades. In this connection an analysis has been made of the role of geological and astronomical factors of the global climate changes.

Normile (1995) has pointed out that it is hard to find a high-latitude warming in palaeoclimatic data, because seasonal and regional variations in the polar regions make it difficult to separate suspected greenhouse signals from the noise of natural fluctuations. Summarizing the discussions which took place during Wadati Conference on Global Change and the Polar Regions (Tsukuba, Japan, 7–10 November 1995), Normile (1995) quoted some of the participants.

R. Barry: "There is no evidence of a recent warming trend" (on the basis of observed snow cover and sea-ice dynamics).

M.J. Zwally: a search for long-term trends in ice cover "has given ambiguous results" (see also Johannessen et al., 1995, and Belchansky, 1995, who published certain evidence of the Arctic's shrinking sea-ice).

J. Walsh said he has found spotty matches from a comparison between observed Arctic temperature, precipitation, and extent of sea-ice predictions from GFDL climate model (observations do not show, for instance, retreating sea-ice).

The results obtained by Vaughan and Doake (1996) differ from what has been mentioned above. The analysis of time-series of observations of the areal extent of nine ice shelves on the Antarctic Peninsula has shown that five northerly shelves have retreated dramatically in the past 50 years, while those further south show no clear trend, which is believed to be an indication of a large-scale climate change.

Of course, as in other similar cases it is very important to identify contributions from internal variability of the climatic system and external impacts (including anthropogenic ones). In this context, of great interest are the results obtained by White and Peterson (1996) who discovered significant inter-annual variations in the atmospheric pressure at sea level, wind stress, sea surface temperature and sea-ice extent over the Southern Ocean. These anomalies propagate eastward with the circumpolar flow, with a period of 4–5 years and taking 8–10 years to encircle the pole. It has been suggested that such a system of coupled anomalies to be called the Antarctic Circumpolar Wave and that it probably plays an important role in climate regulation and dynamics both within and beyond the Southern Ocean.

As far as the Arctic is concerned, an important new feature mentioned by Macdonald (1996) is that a change has recently occurred in the Arctic Ocean's circulation: "A layer of water that derives from the Atlantic is warming up, and the exchange between the Arctic Ocean's two major basins—Canadian and Eurasian—has somehow altered... The core of the Atlantic layer (the region between about 200 and 800 m) is now 0.5 or 1°C warmer than in any historical observations.... No-one has suggested it to be a sign of climate change; rather it illustrates that the Arctic Ocean is dynamically coupled to the Atlantic and much more variable than previously suspected". There is no doubt that a displacement of surface water from the Arctic Ocean is of utmost importance for global ocean circulation and therefore for global climate. A new feature discovered in the Arctic is also much higher levels of chlorophyll concentration and biological productivities than previously reported (perhaps, by as much as a factor of ten).

Rahmstorf (1995) has accomplished a numerical modelling to assess the sensitivity of the North Atlantic conveyor (the thermohaline circulation) to the input of freshwater using a global ocean circulation model coupled to a simplified model atmosphere. The most important conclusion has been that the nonlinearity of the system results in the possibility that moderate changes in freshwater input (of less than 0.06 Sv; 1 sverdrup = 10^6 m^3/s) into the North Atlantic can induce transitions between different equilibrium states, leading to substantial changes in regional climate. Quite small perturbations to the present hydrological cycle may lead to sea-surface temperature changes of several degrees on timescales of only a few years.

In his comments on Rahmstorf (1995) results, Weaver (1995) has emphasized: "Combined with results from previous studies, it sends out a warning about the possible response of the North Atlantic conveyor to increasing atmospheric greenhouse gases. Nevertheless, a cornerpiece of the puzzle is missing. That is the corroboration of these results using fully coupled atmosphere–ocean models which resolve the annual cycle, do not drift away from the present-day climate and do not need to employ flux corrections to prevent this drift".

The major result of new studies consists in the conclusion that the complexity of the climate system and the difficulty of climate forecasts have been underestimated. New results substantially broaden the spectrum of possible responses of the global climate system to an anthropogenic CO_2 increase, which raises the level of uncertainty. In these conditions the perspective strategy of preventing undesirable consequences of climate changes must be robust with respect to the influence of sharp regional or global climate changes.

It would be premature, therefore, to believe that "the ability to predict the broad characteristics of regional climate—and the anthropogenic contribution to it—on seasonal to inter-annual timescales could be realized within the next few years" (Newton, 1995). It is obvious from what will be discussed later that the number and scale of uncertainties are so great that it is difficult to expect rapid solutions.

2.3 SOME RESULTS FROM THE SECOND WORLD CLIMATE CONFERENCE

The Second World Climate Conference (WCC-2) was held from 20 October to 7 November 1990 in Geneva with the participation of 747 experts from 120 countries; the main goal was to summarize the results from the work done during the previous decade and to discuss the prospects for both further scientific developments and the "human dimensions" of the problem, as well as the needed policy capable of preventing the harmful (indeed, catastrophic) impact of climate change on human activity and life (Kondratyev, 1990a,b, 1992d–j).

In this connection, WCC-2 consisted of the two stages: (1) a scientific discussion on the basis of the IPCC Report with the participation of about a thousand specialists from 70 countries, and summarizing the results of the WCP (29 October – 2 November, 1990); (2) a meeting at ministerial level to work out political recommendations (this second stage differs WCC-2 from WCC-1 which was mainly a scientific conference). The IPCC Report was unanimously approved, which was marked both in the final "Statement" of the participants of the scientific-technical stage of the conference and in the closing "Declaration" adopted by the meeting at ministerial level. These documents (like the IPCC Report) have played an important role at negotiations concerning the preparation of the "Framework Convention on Climate" approved by the Second UN Conference on Environment and Development.

Opening the WCC-2, Zhou Zhingmeng (China), President of WMO, emphasized that a keen interest in the problems of climate is determined by their first social priority and serious socio-economic consequences of possible climate changes in the future, and in this connection an increase of GHGs in the atmosphere and the anthropogenic impact on the ozone layer are of particular importance. Imperfection of the climate theory determined, first of all, by an inadequate account of the interactive processes in the "atmosphere–ocean" system as well as the dynamics of the cloud cover and its interaction with radiation needs further efforts to develop more reliable techniques of the numerical modelling and forecast of climate. An extremely important role will be played by the global climate observing system with the use of both conventional and satellite observation means to be created on an international basis.

Speeches of welcome were made by a representative of the Switzerland government (Ch. Emmenegger) and D. Fulmi, President of the State Council of Geneva, as well as by the chiefs of all the international organizations participating in preparations for WCC-2: G. O. P. Obasi, WMO, Secretary-General; M. K. Tolba, UNEP, Executive Director; F. Major, Unesco, Director-General; M. G. K. Menon, ICSU, President; H. de Heng, a representative of FAO. Everybody emphasized an urgency of the problem of climate and its changes affecting the human activity and health as well as the necessity of further intensive related studies and developments.

Note should be taken that with all the importance of speeches of the leaders of the international organizations mentioned, their considerations have not produced a clear conception of long-range (e.g., on a scale of a decade) prospects for studies and developments within the WCP and its scientific component, WCRP, as well as other related programmes (first of all, the IGBP). Such a situation is believed to result from the stereotypes established during the first decade of the accomplishment of WCP and connected with an unsubstantiated allegation of the decisive role of anthropogenic factors in the present climate change—an increase of GHGs concentrations and the resulting climate warming during the last century (some publications contain unsubstantiated conclusions about detecting the "CO_2 signal" from the data of observations).

It is this circumstance that has caused in a number of cases an unjustified enthusiasm of some experts in the numerical climate modelling about rather approximate calculations of climate changes due to an instantaneous CO_2 doubling (or, sometimes, quadrupling) to the detriment of some other numerical experiments aimed at analysing the sensitivity of the climate system to the effect of the factors such as cloud cover dynamics, cloud–radiation interaction, ocean–atmosphere interaction, and others.

At the second stage of the conference, Mrs Thatcher, then Prime Minister of Great Britain, was right to say: "So far, nobody can say with confidence that the mean global temperature increase has been caused by human activity", and further she emphasized an extremely inadequate understanding of the global carbon cycle ("It is still enigmatic, how the atmospheric carbon gets absorbed, including its small amount due to anthropogenic sources: whether most of it gets to the ocean as they usually believe?"). Mrs Thatcher stressed that the problem of climate cannot be considered in isolation from other present ecological problems, such as: an explosive growth of population, loss of soil fertility, increasing pollution of the ocean, intensive use of fossil fuel, deforestation (in the tropics, in particular).

To summarize the scientific and technical results of WCC-2, 12 working groups were formed to prepare the respective recommendations generalized then in the form of the "Statement". The assessments contained in this Statement are based on the conclusions of the IPCC Report. It has been emphasized, in particular, that the problem of climate goes beyond the atmospheric and oceanic sciences, affecting each aspect of life on the planet. An importance of this problem in the determination of the future state of the environment and the economic well-being is increasing. To avoid the risk of undesired climate changes, it is recommended that action be taken on reduction of GHG emissions. To stabilize the CO_2 concentration by the middle of the twenty-first century at a level exceeding the pre-industrial level by 50%, an immediate 1–2% annual decrease of the concentration would be needed. Otherwise, by the year 2025 an equivalent (taking account of other GHGs) concentration of CO_2, supposedly, will have doubled (like the energy consumption and the size of population). The conference stated that every country has both technical and economical capabilities to reduce CO_2 emissions. Of course, this statement raises doubts. There were no persuasive grounds for the conclusion about the necessity of the annual 1–2% reduction of emissions. The conclusions of the IPCC (International Panel on Climate Change) Report to be discussed later are based on a certain model of the global carbon cycle and do not even contain references showing the existence of other models which lead to quite different conclusions (see, for example,

Gorshkov, 1995; Gorshkov et al., 1992, 1994). No doubt, if there are contradictory assessments, the "monopoly" of any one model is absolutely unacceptable.

Unfortunately, unjustified opinions of some experts concerning the anthropogenic factors of the future climate (for example, some believed that by 2025 the NH polar ice will have completely melted) has created both in the public opinion and at the governmental level a stereotype of the "greenhouse effect", erroneously concentrating only on GHG emissions, to the detriment, for example, of the problem of biospheric protection, whose priority raises no doubts. It is because of this that the mass media discussed during WCC-2 only the problem of global warming and GHG, and during the two days of the conference at the ministerial level a picket of "greens" was organized in front of the entrance to the Geneva Congress Centre with posters against the leaders of those countries (USA, USSR, UK, Saudi Arabia, etc.) which, based on careful scientific assessments, were more cautious about the problem of reducing the emissions compared to the 12 countries—members of the European Economic Community and some other countries—ready to start reducing GHG emissions immediately. The inscription on the posters read: "Climatic criminals".

The "Statement" determines the priority of the following directions of climate studies: (1) cloudiness and the hydrological cycle; (2) greenhouse gases, global carbon cycle, and other biogeochemical cycles; (3) oceans: physical, chemical and biological processes; the atmosphere–ocean interaction; (4) palaeoclimatic investigations; (5) polar ice caps and sea-ice; (6) terrestrial ecosystems.

Of importance was the statement of the fact that "the existing observing systems to monitor the climate system are inadequate to the operative and research objectives". On this basis the conclusion has been drawn: "There is an urgent need for the global climate observing system (GCOS) on an operative basis similar to the WWW and including both space- and surface-based components". The GCOS should be designed to meet the following needs: (a) monitoring the climatic system to detect climate changes and monitoring the responses to climate changes, especially in the case of terrestrial ecosystems; (b) data to use in the economic development on a national scale; (c) studies to improve understanding, modelling and forecast of the climatic system.

The main components of the GCOS have been suggested: (1) the improved WWW; (2) the global ocean observing system (GOOS) to measure the physical, chemical and biological characteristics; (3) continued and broadened programmes to monitor other key characteristics of the climate system, such as the most substantial components of atmospheric chemistry (including those observed within the Global Atmospheric Watch), changes in the parameters of terrestrial ecosystems, cloud cover and hydrological cycle, ERB, ice cover and precipitation over the oceans (later the Global Terrestrial Observing System (GTOS) has been added). All the countries should provide a complete and open data exchange needed for studies of the climatic system and the processes taking place in it, as well as for the numerical modelling and assessment of the climate change forcings. Considerations with respect to the problems of the Global Climate Observing System testify to the complexity of this task and the need for concentrated efforts to execute it.

As for the socio-economic aspects of the climate problem, the "Statement" underlines that "Actions should be taken immediately which include: (a) analysis of the societal forcings of climate change on national and regional scales as well as study of possible actions on regulation and adaptation; (b) more close cooperation and contacts of the scientists in the sphere of natural and social sciences in order to take into account the information on climate in planning the socio-economic development; (c) considerably increasing financial support to studies of the climate forcings and adaptation to its changes. Of particular importance is an increase of the efficient use of energy and improvement of technologies of the use of non-fossil fuel, bearing in mind not only a reduction of GHG emissions but also steps towards more sustainable socio-economic development".

Unfortunately, all these correct considerations have not been supplemented with the needed analysis of the effect of political, social, economic and other factors on the environment and climate. In this connection a paradoxical situation has developed. On the one hand, the efforts of the IPCC WG-2 and 3 have been dedicated to these problems, and on the other hand, they have turned out to be concentrated practically only on a detailed analysis of the effects of climate changes on human activity (water resources, agriculture and food; oceans, fisheries and coastal zones, energetics, land use and planning urban building, forestry) and health, but not vice versa.

At the final stage of the first part of WCC-2 a great deal of attention (and heated discussion) was concentrated on GHG emissions and grounds (or their absence) for recommendations on their reduction. But the fundamental problem in which an interactivity of the system "human society–environment" , the effect of the socio-economic factors on the biosphere, environment and climate manifest themselves, had been practically neglected. Therefore the discussions (as well as the final "Statement" and the closing "Declaration" worked out on their basis) have been distorted towards the problem of reducing the emissions of CO_2 and other GHGs. Unfortunately, this tendency remains and has shown itself subsequently in the "Berlin Mandate" (The Berlin Mandate..., 1995).

Of course, the reasons are connected with an insufficiently careful approach to the preparation of the scientific–technical part of WCC. If the programme of the conference had included the reports on the biospheric dynamics and the socio-economic origin of the threat to the environment (the continuing growth of population, the growing impoverishment of many developing countries due to the enormous "pumping" of profits to the developed industrial countries, etc.), everything would have been in its place, including the problem of GHG emissions which, for all its importance, should not top the list.

The conference has approved support to the following three directions of the international activity: (a) continuation of the programmes of global observations and studies within the WCP, IGBP and other related international programmes; (b) continuation of the IPCC activity on the assessment of the state of the climate problems as the scientific basis for negotiations and preparation of the convention; (c) working out the convention on climate change as a guide for the respective policy of the governments.

The "Declaration of Ministers" (DM) adopted at the second stage of WCC-2 with the participation of the representatives of 137 countries confirmed the IPCC conclusions and highly estimated the effort on the WCP realization during the last decade. The global-scale strategy formulated in this document ran: "we believe that a decision on the global-scale response should be immediately made and realized, which would simultaneously provide the stable development of all countries based on the best knowledge available, following from the IPCC estimates". And further: "We emphasize that special effort should be made in the spheres of uncertainty defined by the IPCC. We state the need of an intensified scientific research in the field of the socio-economic consequences of climate change and development of the response strategy".

Then a most substantial statement followed: "We note that the consumption per capita in some parts of the globe as well as an expected population growth are factors reflecting on the expected increase of the emissions of the greenhouse gases" (unfortunately, the DM did not fix attention on the presently overpopulated planet (Gorshkov et al., 1992, 1994; Kondratyev, 1990a, 1992c; Kondratyev et al., 1996b)). Then the DM emphasizes: "lack of complete scientific clarity should not cause a delay in taking economically grounded actions on preventing the environment from degradation", but further the DM reads: "However, we agree that the developed countries with still a comparatively low energy consumption (measured either per capita or on another accepted basis) which will expectedly grow, as well as some countries whose economics is in the transitional phase, can create the objectives, programmes and (or) strategies in the interests of the socio-economic development, with the simultaneous improvement of energy usage in the economic activity. We are calling the developed countries to analyze the feasibility and options before the UN Conference on Environment and Development to be held in 1992 and in this context to work out the programmes, strategies and (or) objectives concerning a gradual reduction over the decades to come of emissions of the greenhouse gases not controlled by the Montreal Protocol, including carbon dioxide, methane and nitrous oxide". The actions to provide the socio-economic sustainability include: an increase of the energy use efficiency, maximum possible usage of ecologically safe technologies, conservation of forests (taking into account the role of forests in preserving the biodiversity, protecting the soil stability and the hydrological system).

The Declaration of Ministers emphasizes: "We call to immediately start negotiations on the Framework Convention on Climate Change after the decision will be made at the 45th session of the UN General Assembly recommending the ways, means and specialities of these negotiations". And then: "...it is extremely desirable that the Framework Convention on Climate Change containing the respective obligations and other principles be signed in Rio-de-Janeiro during the UN Conference on Environment and Development".

It would be fair to point out in conclusion that the WCC-2 was an outstanding event both from the viewpoint of an urgency of its problems and by the authoritative complement at its scientific–technical and intergovernmental stages. The final report of the IPCC deserves, on the whole, a positive opinion.

However, one is anxious about the fact (especially in connection with the preparation of the Framework Convention) that the socio-economic origin of the global ecological problems, including climate changes, has not been adequately analysed at the conference: there is no doubt that at present the problem of climate should be considered in the context of the ecological problems (Gorshkov et al., 1992, 1994; Kondratyev, 1990a, 1993c–e; Kondratyev et al., 1996b).

2.4 THE IPCC REPORTS AND RELEVANT STUDIES

An increasing urgency of the problem of global climate changes determines the necessity of the regular summarizing of the results of the relevant studies and developments. Such was the role, for example, of the International Conference held in Villach (Austria) in 1985. In 1988 the World Meteorological Organization (WMO) and the UN Organization on the Environmental Problems (UNEP) made a joint decision on the formation of the Intergovernmental Panel on Climate Change (IPCC). The IPCC was to prepare for the Second World Climate Conference the report characterizing the present understanding of the observed regularities of climate change and possibility to forecast the climate, its impact on the environment and human activity as well as the needed economic actions to prevent undesirable climate changes. To prepare the report, three working groups (WG) were formed: (1) WG-1. The scientific assessment of climate change (Chairman, Dr J. T. Houghton, at that time Director-General of the UK Meteorological Service); (2) WG-2. Climate change implications (Dr Yu. A. Israel, at that time Chairman of the USSR Goskomgidromet); (3) WG-3. Climate change policy (Dr F. Berntel, USA).

It should be emphasized that the WG-1 task was especially difficult for at least two reasons: (1) even an interpretation of the causes of the climate warming trend over the last 100 years is ambiguous, and the existing climate forecasts should be considered only conditional scenarios; (2) some experts and the mass media have been particularly successful in supporting the unjustified idea that the global climate warming observed in this century is determined by a CO_2 increase, and by the middle of the next century a catastrophic climate warming will take place followed by an increase of the World Ocean level.

Unfortunately, some judgements about the climate have been surprisingly incompetent. For example, the brochure "WMO and Global Warming" (WMO, Publ. No.741, Geneva, 1990, 24 pp.), which begins with an incorrect analogy between a normal greenhouse and the greenhouse effect of the atmosphere. As far back as the end of the last century a famous American physicist Wood, having replaced the usual glass with quartz in the laboratory model greenhouse and observing no changes in the greenhouse functioning, illustrated thereby that the point is not that being transparent to solar radiation the glass is opaque to the soil thermal emission. About 40 years ago in our publications (see, for example, Kondratyev, 1954, 1969, 1992d) dedicated to the greenhouse theory and experiments in conventional greenhouses, it was convincingly shown that the glass only "cuts off" the soil–atmosphere turbulent heat exchange. The same WMO brochure, for some reason, states that "the four most important greenhouse gases are carbon dioxide, methane, chlorofluorocarbons, and nitrous oxide" though it has been well known that water vapour contributes most. In addition, as has been mentioned already, the nature of this GHG which can be converted into liquid and solid phases, giving clouds, is most complicated, and it is an account of this fact that determines one of the main difficulties of the climate theory. But then, the authors of the brochure, contradicting their own statement, write: "Water vapor exhibits a strong greenhouse effect, but its concentration in the atmosphere is determined within the climatic system itself and *on a global scale the effect on this system is negligibly small*" (our italics). And this was written in 1990 when the anthropogenic impact on the global water cycle raised no doubts.

F. Cotti, the Federal Minister of Switzerland, writes (Geneva News and Environmental Report, November 1990, p. 19): "Now the IPCC firmly established that *the atmospheric warming determined by anthropogenic emissions of carbon dioxide and other so-called greenhouse gases* will seriously affect the ecosystems and the society", and further: "Apparently, carbon dioxide contributes much more to the enhancement of the greenhouse effect than other components". Even K. Tolba, the UNEP Executive Director, states in the same issue: "The IPCC WG-1 reported that an increase of GHG emissions has led to the global warming and the related climate changes". No comment needed. It should be emphasized, however, that such delusions are not typical since the IPCC Report contains clearly formulated correct conclusions in this respect.

Note that the WG-1 report states, first of all, the reality of the greenhouse effect. It manifests itself in that the atmosphere containing greenhouse gases determines an average temperature of the Earth surface which is 30°C higher than that with the GHG-free atmosphere. The most substantial GHGs are (in order of priority) water vapour, carbon dioxide, methane, halocarbons, tropospheric ozone, and nitrous oxide. Similar situation is observed on other planets (Kondratyev, 1983b; Kondratyev and Moskalenko, 1984; Kondratyev, 1992d). So, for example, on Venus the greenhouse effect reaches 523°C, whereas on Mars it is only 10°C.

Calculations with the current climate models on the assumption of a doubled CO_2 concentration have led to the conclusion (Climate Change..., 1990) that the mean global climate warming (SAT increase) will be within 1.5–4.5°C (apparently, the value 2.5°C is most probable) with the simultaneously intensified water cycle which will manifest itself through increasing precipitation (and evaporation) within 2-13%. With the GHG concentration increase equivalent to a CO_2 doubling assumed to take place by 2030, bearing in mind the slowing-down effect of the World Ocean, one may believe that by this time the mean global SAT may have increased by 1–2°C.

Global warming should be followed by diverse regional climate changes. Apparently, a faster heating of land compared to the ocean should be expected. Calculations have led to the conclusion that the warming in southern Europe and North America will be close to mean global and will be accompanied with decreased summertime precipitation and soil moisture. Apparently, the Asian monsoon will intensify. So far, it has not been possible to forecast the effect of the global climate warming on the frequency and intensity of catastrophic events (typhoons, hurricanes, floods, etc.). At the expense (mainly) of the thermal expansion and melting of the continental ice the World Ocean's level may increase within 10–30 cm (the catastrophic increase of the ocean's level by several metres due to the breaking and melting of the Western Antarctic ice sheet is improbable).

The report (Climate Change..., 1990; Kondratyev, 1991b,c, 1992c) rejects as incorrect the forecasts of the future climate based on the past climates as analogues (see also Shabalova and Konnen, 1995; Covey, 1995). However, the emphasis is justly laid on palaeoclimatic studies which may provide information on the climates of the past needed to test the climate model's reliability. The problem of the palaeo-analogues correctness brought forth a hot discussion at the meeting of the group of experts on analogues and global climate changes (a detailed consideration of this problem is contained in Eddy and Oeschger (1993). The members of the group were J. Mahlman (USA, Chairman), G. Boer (Canada), M. I. Budyko (Russia), K. Hasselmann (Germany), J. F. B. Mitchell (Great Britain), T. M. L. Wigley (Great Britain).

J. F. B. Mitchell noted in his report that the climate models have several advantages: (1) they ensure a comparatively complete and concrete account of the climate-forming factors; (2) they make possible a realistic prescription of external forcings; (3) they describe a wide range of climate characteristics; (4) their global scale; (5) they make it possible to test their reliability by comparing with the observed climate. The shortcomings of numerical modelling are: (1) sensitivity to the choice of the schemes of parameterization of the sub-grid processes; (2) low reliability of the calculated regional climate changes; (3) limited capability of validation from the data for other climates. Answering the rhetorical question about the acceptability of palaeo-analogues for climate forecasts, J. Mitchell compared the quantity of the radiative forcing equal to zero during the Holocene, and the value 4 W/m^2 with a CO_2 doubling. In these cases the annual temperature course turns out to be different. With a CO_2 increase, according to numerical modelling, a warming takes place in the tropics, whereas the palaeo-analogues reveal the cooling. Thus the palaeo-analogues cannot be used in climate forecasts, the more so that most of the climate change during the Holocene can be explained by variations in the orbital parameters, which differs radically from the greenhouse effect.

In a later publication, Skiles (1995) has justly emphasized complexities of model validation: "In the case of climate change, however, signals from the planet are noisy and difficult to decipher, and there has been a notorious lack of agreement on what they mean. Studies predicting what the hydrosphere and biosphere will be like in a changed world are speculations done in the absence of validation data" (see also Oreskes et al. (1994) and Hammit (1995)).

H. Oeschger, a well-known Swiss scientist, made a profound report on palaeoclimatology with a detailed analysis of the validity of various sources of information on palaeoclimate (ice cores, tree rings, sea and lake deposits, pollen, historic chronicles, etc.) and an overview of the results obtained. H. Oeschger emphasized that the dynamics of GHG (CO_2, first of all) in the nineteenth century was determined by the biospheric processes, and in the twentieth century by fossil fuel burning. The growth of methane concentration turned out to be in proportion to the population growth, which reflects its anthropogenic origin (rice paddies (Cao et al., 1995), domestic animals, etc.), though the role of natural processes in the northern regions is, of course, substantial (the problem of gas hydrates, etc.). To understand the GHG palaeotrends, an account of interaction of the marine and continental biospheres is important. Analysis of data on climate changes during the last glaciation has shown that they are similarly determined by the GHG dynamics and variations in the orbital parameters (it is interesting that the "signal" of the latter is different in phase in the northern and southern hemispheres). The atmospheric greenhouse effect enhancement during the last century was estimated at 2.2 W/m^2. Oeschger drew attention to the problem of the climate–biosphere interaction and emphasized the importance of analysis of the possible two states of the atmosphere–ocean system.

During recent years the climate changes have been simulated with the use of the latest coupled global models with the prescribed not suddenly doubling but continuously increasing (1% annually) CO_2 concentration (see Section 2.5). The results of such calculations testify to substantially different responses of the climatic system to instantaneous and transient increases of CO_2 concentration. New results of numerical modelling revealed the possibility of a very long-range internal climate variability with an amplitude of about 0.5°C (Mahlman, 1995) and a strong impact of mixing through the circumpolar oceanic thickness on the formation of climate (this is the reason why there has been no marked heating in the North Atlantic).

Analysis of the data of meteorological observations for the last 100 years revealed a mean global SAT increase within 0.3–0.6°C, with a most intensive warming observed in the 1980s, which, on the whole, agrees with the numerical modelling conclusions. There are no grounds, however, to consider it to be determined by the GHG concentration growth. This conclusion can also be drawn for other reasons and, specifically, because the spatial distribution of the warming in the two hemispheres is quite different and does not adequately correspond to the numerical modelling results. This can be explained by the effect of the inadequately simulated natural (internally induced) climate variability.

The fact is also important that the estimates of the greenhouse warming give values close to natural climatic variability. Of interest is the fact that the warming in the current century was concentrated in two periods: from the 1920s to the1940s and after 1975. From the 1940s to the early 1970s in the northern hemisphere there was a climate cooling (though it was in this period that industry developed, excluding the war years). At the same time, the SH mean SAT remained almost constant. The post-1975 spatial distribution of the warming was very inhomogeneous: in some regions (mainly in the northern hemisphere) the cooling (as before) took place, continuing up to the present day. Analysis of data on precipitation revealed a stronger inhomogeneity. While, for example, the rain rate over most of the former USSR territory increased, starting in the 1950s, south of the Sahara the rain rate dropped.

So far, there are no grounds to believe that the climate has become more variable during the last several decades. There is no doubt, however, that high-temperature episodes will be more frequent in the future simply due to the general temperature increase. Now we shall discuss certain basic issues of the climate change problem.

2.4.1 Trends of GHG concentration and the atmospheric greenhouse effect

From the viewpoint of assessing possible anthropogenic impacts on climate, the trend of GHG concentration growth is a starting factor which should be taken into account (and predicted). From the data on direct and indirect indicators one can judge with confidence about the presence in the atmosphere of not only current GHG growth but also its palaeovariations. Analysis of ice cores suggested, for example, the conclusion that the concentration of carbon dioxide, methane and nitrous oxide during the glaciations had been much lower than in the interglacial periods. One thousand years before the industrial revolution the concentration of these GHGs had been comparatively stable (over the past 1000 years there had been fluctuations 280 ± 10 ppmv). However, as the size of the population and industrialization grew, the content of GHGs in the atmosphere grew too. The analysis of new Greenland ice core records (Alley et al., 1996) confirms earlier findings about the increase of the atmospheric CO_2 concentration at the end of the last glaciation and a steady increase since the beginning of the industrial epoch. According to Etheridge et al. (1996), pre-industrial CO_2 mixing ratios were in the range 275–284 ppm with the lower levels during AD 1550–1800. The data in Table 2.5 show changes in the GHG concentration (estimates of the volume mixing ratio) compared to the pre-industrial period as well as the present estimates of the rate of concentration growth (Houghton et al., 1996).

Table 2.5. Observed trends of concentrations of the major GHGs in the atmosphere

	CO_2	CH_4	N_2	CFC-11	HCFC-22 (a CFC substitute)	CF_4 (a perfluoro-carbon
Pre-industrial concentration	~280 ppmv	~700 ppbv	~275 ppbv	zero	zero	zero
Concentration in 1994	358 ppmv	1720 ppbv	312[a] ppbv	268[a] pptv[b]	110 pptv	72[a] pptv
Rate of concentration change[c]	1.5 ppmv/yr 0.4% /yr	10 ppbv/yr 0.6% /yr	0.8 ppbv/yr 0.25% /yr	0 pptv/yr 0% /yr	5 pptv/yr 5% /yr	1.2 pptv/yr 2% /yr
Atmospheric lifetime (years)	50–200[d]	12[e]	120	50	12	50 000

[a] Estimated from 1992–93 data.
[b] 1 pptv = 1 part per trillion (million million) by volume.
[c] No single lifetime for CO_2 can be defined because of the different rates of uptake by different sink processes.
[d] This has been defined as an adjustment time which takes into account the indirect effect of methane on its own lifetime.
[e] The growth rates of CO_2, CH_4 and N_2O are averaged over the decade beginning 1984; halocarbon growth rates are based on recent years (1990s).

Now the causes of the anthropogenically induced growth of the concentrations of carbon dioxide, methane, chlorofluorocarbons (CFC) and tropospheric ozone are well known, but less so for nitrous oxide. A particular situation concerning GHG emissions exists in Africa, where the high proportion of emissions comes from agriculture, land-use change, and other sources which are difficult to assess (Fitzgerald et al., 1995).

Of interest is the long life-time in the atmosphere of the components such as carbon dioxide (in this case the uncertainty of estimates is determined by the difficult account of the processes of the global carbon cycle formation), CFC (especially CFC-12), and nitrous oxide (besides, in case of CO_2 the concept of a single characteristic timescale is of limited value because of the multiplicity of processes responsible for the carbon cycle formation). This testifies to the long-term consequences of these GHG emissions to the atmosphere: an equilibrium level of concentrations (even after the emissions cease) can be reached only over many centuries. So, for example, if in 1990 the CFC-12 emissions completely ceased, in 2100 the level of the concentration of this freon in the atmosphere would still reach approximately one-third of the present level. Another situation exists for some substitutes for CFCs and methane, for which the time to reach an equilibrium concentration constitutes only several decades. The estimates have shown that to stabilize the concentrations of various GHGs at the present level in the future, the following emissions should be relatively reduced (the respective estimates are given in brackets): carbon dioxide (60–80%), methane (12–20%), nitrous oxide (70–80%), CFC-11 (70–75%), CFC-12 (75–80%), CFC-113 (85–95%), HCFC-22 (40–50%).

The data in Table 2.5 testify to a more rapid (than in the case of CO_2) increase of the concentration of methane, nitrous oxide and chlorofluorocarbons, which reflects their increasing contribution to the formation of the greenhouse effect. Table 2.6 contains some details of the methane budget (Houghton et al., 1996).

Global average methane concentration has increased during the last 200 years. The increase was 6% over the decade starting in 1984. Methane concentration in 1994 was about 1720 ppbv, 145% greater than the pre-industrial concentration of 700 ppbv (Table 2.5). Over the last 20 years, there has been a decline in the methane growth rate: in the late 1970s the concentration was increasing by about 20 ppbv/yr, during the 1980s the growth rate dropped to 9–13 ppbv/yr. Around the middle of 1992, methane concentrations briefly stopped growing, but since 1993 the global growth rate has returned to about 8 ppbv/yr.

Individual methane sources are not well quantified. Carbon isotope measurements indicate that about 20% of the total annual methane emissions are related to the production and use of fossil fuel. In total, anthropogenic activities are responsible for about 60–80% of current methane emissions (Table 2.6). Methane emissions from natural wetlands appear to contribute about 20% to the global methane emissions to the atmosphere. Such emissions will probably increase with global warming as a result of greater microbial activity. In 1992 the direct radiative forcing due to the increase in methane concentration since pre-industrial times was about $+0.47\ W/m^2$.

Recently a number of attempts have been undertaken to explain methane cycle

Table 2.6. Estimated sources and sinks of methane for 1980 to 1990. All figures are in Tg[a] (CH_4)/yr. The current global atmospheric burden of CH_4 is about 5000 Tg (CH_4)

	Individual estimates	Total
(a) Observed atmospheric increase, estimated sinks and sources derived to balance the budget		
Atmospheric increase[b]		37 (35–40)
Sinks of atmospheric CH_4:		
Tropospheric OH	490 (405–575)	
Stratosphere	40 (32–48)	
Soils	30 (15–45)	
Total atmospheric sinks		560 (460–660)
Implied sources		
(sinks + atmospheric increase)		597 (495–700)
(b) Inventory of identified sources		
Natural sources		160 (110–210)
Anthropogenic sources:		
Fossil fuel related	100 (70–120)	
Total biospheric	275 (200-350)	
Total anthropogenic sources		375 (300–450)
Total identified sources		535 (410–660)

[a] 1 Tg = 1 teragram = 1 million million grams, which is equivalent to 1 million tonnes.
[b] Applies to 1980–90 average. Table 2.5 uses the average for 1984–94.

dynamics. Cao et al. (1996) have developed a methane emission model based on the hypothesis that plant primary production and soil organic matter decomposition act to control the supply of substrate needed for methanogens; the rate of substrate supply and environmental factors, in turn, control the rate of CH_4 production, and the balance between CH_4 production and methanotrophic oxidation determines the rate of CH_4 emission into the atmosphere. Coupled to data sets for climate, vegetation, soil, and wetland distribution, the model was used to calculate spatial and seasonal distributions of CH_4 emissions at a resolution of 1° latitude by 1° longitude.

Calculations made by Cao et al. (1996) indicate that the net primary production (NPP) of wetlands ranges from 45 gC/m^{2}yr for northern bogs to 820 gC/m^{2}yr for tropical swamps. Total CH_4 emission is 92 Gt/yr (previous assessments gave values within the range 47–111 Gt/yr). Sensitivity analysis showed that the response of CH_4 emission to climate change depends upon the combined effect of soil carbon storage, rate of decomposition, soil moisture and activity of methanogens. Cao et al. (1996) have emphasized that the focus of further field studies should move away from flux measurement per se, towards investigations of the mechanistic processes and quantitative relationships concerning the impact of environmental factors on CH_4 production and oxidation in soil. Methanotrophic oxidation especially warrants further research. Since oxidation consumes much of the CH_4 produced, it is probably the major determinant of the magnitude of CH_4 emissions into the atmosphere.

With the purpose of explaining recent CH_4 dynamics, Law and Nisbet (1996) have accomplished calculations of the sensitivity of the CH_4 growth rate to changes in CH_4 emissions from natural gas and coal. Various causes have been suggested to explain the decline of the CH_4 growth in 1991–92 (the Pinatubo eruption, biomass burning, etc.) but particular interest has been focused on the possible impact of a decline in the gas leakage rate from one of the world's major natural gas producers, namely, the former Soviet Union (FSU). During the 1980s and early 1990s the FSU contributed between 31% and 38% of the total global natural gas production. Much production was in Siberia and emissions of CH_4 from gas leaks at the wellhead and from pipelines could be considerable, although difficult to reliably assess.

Using a 2-D global model of the atmosphere, Law and Nisbet (1996) have assessed the impact of changes in the emissions of CH_4 from coal and natural gas, derived from production data and gas leakage rates, on the CH_4 growth rate during the period 1980 to 1994. Since the gas leakage is one of the least well known variables, particularly in the FSU, the sensitivity calculations were made for the FSU gas leakage range. The calculations indicate that during the period 1989 to 1993 the modelled CH_4 growth rate decreased by $\sim$1–2 ppbv/yr at northern midlatitudes (assuming constant gas leakage rates), showing that declining emissions, primarily in the form of natural gas from the FSU, may have contributed to the general reduction in the CH_4 growth rate from 1989 to 1991. However, this does not explain the large reduction which was observed between 1991 and 1992. A model with FSU gas leakage rates decreasing by 1% /yr from 5% in 1990 to 2% in 1993 shows a decrease in the CH_4 growth rate of $\sim$9–12 ppbv/yr (or $\sim$3–5 ppbv/yr^2) over this period which is similar in magnitude to the observed decline over the same period. Law and Nisbet (1996) have pointed out that the sensitivity calculations illustrate a large potential for significantly lowering the growth rate of atmospheric CH_4 by reducing fossil fuel emissions from coal and, in particular, by reducing the leakage rates from natural gas installations.

An important aspect of the estimates of the "greenhouse" climate warming are calculations of the contribution of different GHGs. This problem has been discussed in detail in monographs (Borisenkov and Kondratyev, 1988; Kondratyev and Moskalenko, 1983, 1984; Kondratyev, 1992d) as applied to the atmospheres of the Earth (Borisenkov and Kondratyev, 1988) and other planets (Kondratyev and Hunt, 1982; Kondratyev and Moskalenko, 1983). The main difficulty in assessing the contribution of various GHGs to the formation of the greenhouse effect and the related climate changes is explained by the presence of numerous feedbacks (both positive and negative), which determines an extremely complicated account of the respective interrelated processes (Borisenkov and Kondratyev, 1988; Houghton et al., 1996; Kondratyev, 1990b, 1991a,b, 1992c; Rosema et al., 1993). Naturally, with the feedbacks ignored one should expect a global climate warming directly determined by the enhanced greenhouse effect. In estimating the contributions of various GHGs to the formation of the greenhouse effect it is important that these contributions calculated for one GHG molecule differ greatly. For example, the effect of methane on the greenhouse effect calculated for one molecule is 25 times stronger than in the case of CO_2, and that of the CFC-11 molecule is stronger by a factor of 11 000. This circumstance is particularly important in connection with the fast growth of CH_4 and CFC-11. Therefore the increase of the equivalent CO_2 concentration calculated with an account of the contribution of other GHGs constituted by now about 50% whereas the growth of the real CO_2 concentration was half as much.

To assess the contribution of various GHGs to the formation of the greenhouse effect, it is worthwhile introducing the notion of the global warming potential (GWP) for GHGs, defined as a GHG contribution to the greenhouse effect calculated per 1 kg of an instantaneous emission of a given GHG with respect to the relevant contribution of CO_2 for the present atmospheric conditions.

If $c_i(t)$ is atmospheric concentration for the ith gas as a function of time and a_i - the instantaneous radiative forcing of a single molecule a_i, then

$$GWP_i = \frac{\int_0^\infty a_i c_i(t)\, dt}{\int_0^\infty a_r c_r(t)\, dt} \tag{2.1}$$

where subscript r refers to the reference molecule (CO_2). In case all a_is are calculated relative to CO_2 (and its radiative forcing is taken to be equal to unity), then the denominator becomes

$$T_{res} = \int_0^\infty c(t)\, dt \tag{2.2}$$

Table 2.7 (Houghton et al., 1996) contains results of calculations of relative cumulative contributions for three time intervals (see also Changnon, 1995; Daniel et al., 1995; IPCC Special..., 1994; IPCC, 1994; Pinnock et al., 1995; Wuebbles et al., 1995). The typical uncertainty is 35%, not including the uncertainty in the CO_2 reference. Table 2.7 demonstrates, for example, that in the decades to come methane will contribute much to the greenhouse effect, whereas the effect of the more long-lived nitrous oxide will manifest itself during a longer time period. It concerns the CFC compounds even more.

Though it follows from this table that CO_2 is a least efficient GHG, huge amounts of emitted CO_2 ($\sim$60%) determine its prevailing contribution to changes in the greenhouse effect over 100 years. Data such as those given in Table 2.7 are important in planning actions on GHG reduction to prevent undesirable climate changes. In this connection the priority of the problem of increasing concentrations of CO_2, CH_4, and CFC is quite natural.

Hydrofluorocarbons (HFCs) have been chosen as one of the major replacements for both CFCs and HCFs because they do not deplete stratospheric ozone. Since they are optically active in the IR region (i.e., contribute to the greenhouse effect),

Table 2.7. Global warming potential (mass basis) referenced to the updated decay response for the adopted carbon cycle model and future CO_2 atmospheric concentrations held constant at current levels. Typical uncertainties are about ±35%

Species	Chemical formula	Lifetime and reference	Global warming potential (time horizon) 20 years	100 years	500 years
CO_2	CO_2	Bern model, revised	1	1	1
HFC-23	CHF_3	>264[b]	9 100	11 700	9 800
HFC-32	CH_2F_2	5.6[b]	2 100	650	200
HFC-41	CH_3F	3.7[b]	490	150	45
HFC-43-10mee	$C_5H_2F_{10}$	17.1[b]	3 000	1 300	400
HFC-125	C_2HF_5	32.6[b]	4 600	2 800	920
HFC-134	$C_2H_2F_4$	10.6[b]	2 900	1 000	310
HFC-134a	CH_2FCF_3	14.6[b]	3 400	1 300	420
HFC-152a	$C_2H_4F_2$	1.5[b]	460	140	42
HFC-143	$C_2H_3F_3$	3.8[b]	1 000	300	94
HFC-143a	CF_3CH_3	48.3[b]	5 000	3 800	1 400
HFC-227ea	C_3HF_7	36.5[b]	4 300	2 900	950
HFC-236fa	$C_3H_2F_6$	209[b]	5 100	6 300	4 700
HFC-245ca	$C_3H_3F_5$	6.6[b]	1800	560	170
Chloroform	$CHCl_3$	0.51[b]	14	4	1
Methylene chloride	CH_2Cl_2	0.46[b]	31	9	3
Sulphur hexafluoride	SF_6	3 200[c]	16 300	>23 900	34 900
Perfluoromethane	CF_4	50 000[c]	4 400	6 500	10 000
Perfluoroethane	C_2F_6	10 000[c]	6 200	9 200	14 000
Perfluoropropane	C_3F_8	2 600[c]	4 800	7 000	10 100
Perfluorobutane	C_4F_{10}	2 600[c]	4 800	7 000	10 100
Perfluoropentane	C_5F_{12}	4 100[c]	5 100	7 500	11 000
Perfluorohexane	C_6F_{14}	3 200[c]	5 000	7 400	10 700
Perfluoro-cyclobutane	$c\text{-}C_4F_8$	3 200[c]	6 000	8 700	12 700
Methane**	CH_4	12.2 ± 3[b]	56	21	6.5
Nitrous oxide	N_2O	120[d]	280	310	170
Trifluoroiodo-methane	CF_3I	0.005[d]	3	1	1

[a] The GWP for methane includes indirect effects of tropospheric ozone production and stratospheric water vapour production, as in IPCC (1994).

[b] Based upon the revised CH_3Cl_3 lifetime from Prinn et al. (1995) and updated chemical kinetic data. Also includes updated information relating to radiative forcing per molecule based upon Pinnock et al. (1995), where appropriate.

[c] See Roehl et al. (1995) for discussion of radiative forcing per molecule for these gases. Lifetimes as in IPCC (1994).

[d] As in IPCC (1994).

Gierczak et al. (1996) have conducted a study to obtain more precise values of the GWP for $CF_3CH_2CF_3$ (1,1,1,3,3,3 - hexafluoropropane, HFC-236fa) and $CF_3CHFCHF_2$ (1,1,1,2,3,3 - hexafluoropropane, HFC-236ea). The global warming potentials for these gases are, respectively, 5610 and 2200 for a 20-year horizon and 5160 and 220 for a 500-year horizon (see relevant data in Table 2.7).

Table 2.8. contains a summary from IPCC-95 (Houghton et al., 1996) of lifetimes for radiatively active gases and halocarbons.

Stratospheric ozone depletion due to halocarbons is about -2% (globally) over the period 1979 to 1990 with half as much again occurring both immediately before and since; the total radiative forcing is thus now about -0.1 W/m^2. Tropospheric ozone appears to have increased since the nineteenth century over the northern mid-latitudes where few observational records are available; if over the entire northern hemisphere, tropospheric ozone increased from 25 ppb to 50 ppb at present, then the radiative forcing is about $+0.4$ W/m^2.

The concept of 'global warming potential' has been subject to various critical remarks (see IPCC-1994). Moore and Braswell (1994) have suggested instead of GWP a notion of the lifetime of excess atmospheric carbon dioxide. They have pointed out that there are two types of difficulties in the formula used (2.2): (1) most model-derived estimates of the relaxation of the CO_2 concentration reveal a function which is not always well approximated by weighted sums of exponentials although these are often used to calculate T_{res}; (2) the function $c(t)$ is quite sensitive to assumptions about the terrestrial biosphere and the relaxation.

As an interim step, Moore and Braswell (1994) suggested to introduce the single half-life time ($T_{1/2}$) as a simple indicator of the Earth's system ability to remove excess atmospheric CO_2. The single half-life time is defined as the time it takes for a model atmosphere to relax from its present-day value (or some future value, though it would alter $T_{1/2}$) half-way to its new equilibrium pCO_2 value. In this case a modified GWP is introduced in which the single half-lives of the radiatively important gases replace the integral in (2.1). If one assumes a terrestrial biosphere with a fertilization flux, then the best estimate is that the single half-life time for excess CO_2 lies within the range 19 to 49 years, with a reasonable average being 31 years. In the case of forest re-growth only, the average value for the single half-life time for excess CO_2 increases to 72 years, and if the terrestrial component is completely removed, then the single half-life increases further to 92 years.

Gaffin et al. (1995) have expressed certain comments and reservations concerning the concept of the single half-life time.

As has been mentioned above, the Framework Convention on Climate Change (FCCC) contains recommendations on reducing the CO_2 emissions to the atmosphere by the year 2000 to the 1990 level. In connection with preparing the IPCC-1995 report (IPCC Special..., 1994; IPCC, 1994), the levels of stabilization of the CO_2 concentration (first column of Table 2.9) were estimated for different cumulative emissions of CO_2 over the period 1990–2100 (in GtC/year). The same table shows the amounts of cumulative emissions corresponding to the known scenarios of emissions (IS 92a–f) (Climate Change, 1990). As is seen, the CO_2 concentration can be stabilized at an acceptable low level only with small emissions,

Table 2.8. Lifetimes for radiatively active gases and halocarbons

Species		Lifetime		Concentration (ppbv)		Current growth	Radiative forcing	
		Year	Uncertain	1992	pre-industrial	ppbv/yr	$W/m^2/ppbvq$	W/m^2
Natural and anthropogenically influenced gases								
carbon dioxide	CO_2	variable		356 000	278 000	1600	1.8×10^{-5}	1.56
methane	CH_4[a]	12.2	25%	1714	700	8	3.7×10^{-4}	0.47
nitrous oxide	N_2O	120		311	275	0.8	3.7×10^{-3}	0.14
methyl chloride	CH_3Cl	1.5	25%	~0.6	~0.6	~0	0	
methyl bromide	CH_3Br	1.2	32%	0.010	<0.010	~0		0
chloroform	$CHCl_3$	0.51	300%	~0.012		~0	0.017	
methylene chloride	CH_2Cl_2	0.46	200%	~0.030		~0	0.03	
carbon monoxide	CO	0.25		50–150		~0		[b]
Gases phased out before 2000 under the Montreal Protocol and its amendments								
CFC-11	CCl_3F	50	10%	0.268	0	+0.000[c]	0.22	0.06
CFC-12	CCl_2F_2	102		0.503	0	+0.007[c]	0.28	0.14
CFC-113	CCl_2FCClF_2	85		0.082	0	0.000[c]	0.28	0.02
CFC-114	$CClF_2CClF_2$	300		0.020	0		0.32	0.007
CFC-115	CF_3CClF_2	1700		<0.01	0		0.26	<0.003
carbon tetrachloride	CCl_4	42		0.132	0	−0.0005[c]	0.10	0.01
methyl chloroform	CH_3CCl_3	4.9	8%	0.135[#]	0	−0.010[c]	0.05	0.007
halon-1211	$CBrClF_2$	20		0.007	0	.00015		
halon-1301	$CBrF_3$	65		0.003	0	.0002	0.28	
halon-2402	$CBrF_2CBrF_2$	20		0.0007	0			
Chlorinated hydrocarbons controlled by the Montreal Protocol and its amendments								
HCFC-22	$CHClF_2$	12.1	20%	0.100	0	+0.005[c]	0.19	0.02
HCFC-123	CF_3CHCl_2	1.4	25%		0		0.18	
HCFC-124	CF_3CHClF	6.1	25%		0		0.19	
HCFC-141b	CH_3CFCl_2	9.4	25%	0.002	0	0.001[c]	0.14	
HCFC-142b	CH_3CF_2Cl	18.4	25%	0.002	0	0.001[c]	0.18	

Table 2.8 *(continued)*

Species		Lifetime		Concentration (ppbv)		Current growth	Radiative forcing	
		Year	Uncertain	1992	pre-industrial	ppbv/yr	$W/m^2/ppbvq$	W/m^2
HCFC-225ca	$C_3HF_5Cl_2$	2.1	35%		0		0.24	
HCFC-225cb	$C_3HF_5Cl_2$	6.2	35%		0		0.28	
Perfluorinated compounds								
sulphur hexafluoride	SF_6	3200		0.032	0	+0.0002	0.64	0.002
perfluoromethane	CF_4	50 000		0.070	0	+0.0012	0.10	0.007
perfluoroethane	C_2F_6	10 000		0.004	0		0.23	
perfluoropropane	C_3F_8	2600			0		0.24	
perfluorobutane	C_4F_{10}	2600			0		0.31	
perfluoropentane	C_5F_{12}	4100			0		0.39	
perfluorohexane	C_6F_{14}	3200			0		.46	
perfluorocyclobutane	c-C_4F_8	3200			0		0.32	
Anthropogenic greenhouse gases not regulated (proposed or in use)								
HFC-23	CHF_3	264	45%				0.18	
HFC-32	CH_2F_2	5.6	25%				0.11	
HFC-41	CH_3F	3.7					0.02	
HFC-43-10mee	$C_5H_2F_{10}$	17.1	35%				0.35	
HFC-125	C_2HF_5	32.6	35%				0.20	
HFC-134	CF_2HCF_2H	10.6	200%				0.18	
HFC-134a	CH_2FCF_3	14.6	20%				0.17	
HFC-143	CF_2HCH_2F	3.8	50%				0.11	
HFC-143a	CH_3CF_3	48.3	35%				0.14	
HFC-152a	CH_3CHF_2	1.5	25%				0.11	
HFC-227ea	C_3HF_7	36.5	20%				0.26	
HFC-236fa	$C_3H_2F_6$	209	50%				0.24	
HFC-245ca	$C_3H_3F_5$	6.6	35%				0.20	
HFOC-125e	CF_3OCHF_2	82	300%					
HFOC-134e	CHF_2OCHF_2	8	300%					
trifluoroiodomethane	CF_3I	<0.005					0.38	

Note. This table lists only the direct radiative forcing from emitted gases. The indirect effects due to subsequent changes in atmospheric chemistry, notably ozone (see below), are not included. The W/m^2 column refers to the radiative forcing since the pre-industrial, and the W/m^2/ppbv column is accurate only for small changes about the current atmospheric composition. In particular, CO_2 , CH_4 and N_2O concentration changes since pre-industrial times are too large to assume linearity; the formulae reported in IPCC (1994) are used to evaluate their total contribution.

A blank entry indicates that a value is not available. Uncertainties for many lifetimes have not been evaluated. The concentrations of some anthropogenic gases are small and difficult to measure. The pre-industrial concentrations of some gases with natural sources are difficult to determine. Radiative forcings are only given for those gases with values greater than 0.001 W/m^2.

[a] Methane increases are calculated to cause increases in tropospheric ozone and stratospheric H_2O; these indirect effects, about 25% of the direct effect are not included in the radiative forcings given here.
[b] The direct radiative forcing due to changes in the CO concentration is unlikely to reach a few hundredths of a W/m^2.
[c] Gases with rapidly changing growth rates over the past decade, recent trends since 1992 are reported.

Table 2.9. The level of stabilization of the CO_2 concentration

Stabilization CO_2 concentration (ppm)	CO_2 emissions (GtC)	Emission scenario CO_2 emissions (GtC)	IS 92
350	300–430	—	—
450	640–800	770	c
550	880–1060	980	d
650	1000–1240	—	—
750	1220–1420	1430	b
—	—	1500	a
—	—	1830	f
—	—	2190	e

which determines the difficulty in solving the problem of reducing the emissions. This problem was discussed in detail by Mintzer et al. (1994) as well as in Houghton et al. (1996).

Table 2.10 data (Houghton et al., 1996) characterize the profiles leading to stabilization at 450, 550, 650, 750, and 1000 ppmv levels.

The First International Conference of the FCCC participants (COP) was held in Berlin from 26 March to 7 April 1995 to analyse the convention adequacy and make arrangements with respect to further efforts to prevent undesirable climate changes. As before, the problem of reducing the GHGs emissions to the atmosphere was in the focus.

By the year 1995 in more than 24 countries the programmes had been developed aimed at reducing GHG emissions by 2000 to the 1990 level. According to the FCCC, the next phase of the problem solution should begin not later than in 1998. Therefore the Berlin Mandate was accepted calling to continue the work on the FCCC improvement to complete preparations of the Agreement by 1997. The emphasis is laid on the problem of GHG sources and sinks in the context of their

Table 2.10. Total anthropogenic CO_2 emissions accumulated from 1991 to 2100 inclusive (GtC)

Case				Accumulated CO_2 emissions 1991 to 2100 (GtC)
IS 92 scenarios		c		770
		d		980
		b		1430
		a		1500
		f		1830
		e		2190
		Concentration profiles A[a]	profiles B[b]	Concentration
Stabilization at	450 ppmv	630		650
	550 ppmv	870		990
	650 ppmv	1030		1190
	750 ppmv	1200		1300
	1000 ppmv	—		1410

[a] As in IPCC (1994).
[b] Profiles that allow emissions to follow IS 92a until at least the year 2000.

dynamics for the periods till 2005, 2010, and 2020. As for the results of the accomplished developments, they can be formulated as follows (Skea, 1995; Ten interesting..., 1995):

(1) The climate convention does not treat the problem of stopping the climate change but only limiting its scale.
(2) Whatever had happened with climate in the past (the Milankovich cycles, ice epochs, periods of rapid warming, etc.), the direction and the level of the human impact on the dynamics of natural processes is a relatively new phenomenon which can either stop or change the natural climate change.
(3) Climate changes are not reduced to an increase of mean global temperature. Direct and most serious impacts of climate changes on man are manifested mainly through extreme climatic events such as droughts, floods, abnormally high temperatures and wind speed, the increase of the ocean level, and others.
(4) Climate changes (including global warming) tell differently on people; they are characterized by strong spatial and temporal variabilities.
(5) The impact of climate change on man, his activity and the biosphere can be both negative and positive (the latter refers, for example, to fertilization due to CO_2 growth).
(6) Global warming does not automatically raise the World Ocean level as a result of glaciers' melting, since (at least, at the initial stage) the mass of polar ice can grow under the influence of snowfalls. The main factors determining the increase of the World Ocean level are: the thermal expansion of water masses

during the warming, the glaciers melting, and the geological processes of interaction between the continents and oceanic plates.

(7) Though the annual CO_2 emission of 6–7 GtC to the atmosphere (mainly, due to carbon fossil fuel burning and changes in land use) is still small (< 10%) compared to natural carbon reservoirs, it is sufficient to produce a substantial impact on the global carbon cycle in view of slow assimilation of anthropogenic CO_2.

(8) Even with the constant CO_2 concentration in the atmosphere maintained at the present level, the concentration will continue growing, at least, for two centuries, though slowing down in time.

(9) In the centre of the 15 μm CO_2 absorption band, the CO_2 absorption of IR radiation is so intensive that the effect of saturation occurs, but outside this interval (14–18 μm) the growth of CO_2 concentration intensifies the absorption and, hence, the greenhouse effect.

(10) The complicated chemistry of the atmosphere containing a great amount of optically active trace gases and aerosol determines the complex nature of the effect of the atmospheric chemistry variations on the climate, including the effects of both warming (greenhouse gases) and cooling (aerosol). Therefore, for example, actions on removing aerosol pollutants to improve the air quality may favour a climate warming.

Discussing the problems of the impact on the greenhouse effect and climate formation, one should not forget an optically active component, such as aerosols, the more so since increasing atmospheric pollution with gas components (e.g., sulphur dioxide) is connected with intensified gas-to-particle conversions in the formation of anthropogenic aerosols. Unfortunately, the fact has been ignored till recent time that the atmosphere is a dynamic colloidal medium. The development of the numerical modelling of the future climate should foresee an interactive account of the biogeochemical cycles including gas-to-particle reactions giving aerosols (see items 2.4.2 and 2.4.3).

The global numerical climate modelling carried out during the last several years taking account of not only the enhancing atmospheric greenhouse effect (due to the anthropogenic growth of GHG concentrations) but also the increasing content in the atmosphere of the anthropogenic sulphate aerosol (annual emissions of sulphur dioxide transformed then to aerosols are equal to 70–80 TgS, which is equivalent to emissions of 5 t sulphur dioxide per second) revealed a much more complicated pattern of climate formation than was initially supposed: the aerosol-induced climate cooling substantially compensates the greenhouse warming, and the spatial–temporal variability of the aerosol content determines the formation of the more variable geographical distribution of climate (Ackerman and Chung, 1992; Aerosol Forcing..., 1995; Alkezweeny, 1995; A Plan..., 1996; Bazhenov et al., 1994; Charlson et al., 1992; Chýlek, 1995; Chýlek et al., 1995a,b; Curry, 1995; Harries et al., 1995; Hsu et al., 1996; Jennings, 1993; Jones et al, 1994; Jones, 1995; Karl et al., 1995b; King et al., 1995b; Kondratyev, 1954, 1969, 1972, 1983b, 1991a–c, 1996a,b; Kondratyev et al., 1976, 1983; Kondratyev and Ivlev, 1995; Mitchell et al., 1995a,b;

Natural air ..., 1996; Penner et al., 1994a,b; Quinn et al., 1995; Rodhe and Langner, 1993; Rogers, 1994; and many others).

2.4.2 Tropospheric aerosols and climate

The impact of aerosols on climate manifests itself both directly, i.e. through increasing the Earth's albedo with the growing content of purely scattering aerosols, and indirectly, i.e., through influencing the cloud particles size distribution resulting in the albedo increase caused by the small droplets' concentration growth when the concentration of sulphate condensation nuclei increases. An account of the indirect aerosol impact on climate is a very complicated and still unsolved problem which determines the uncertainty of assessments of the contribution of aerosols to the formation of climate. Substantial difficulties appear in assessing the climatic implications due to the different sensitivity of the system to radiative forcings due to the greenhouse effect ($\sim 1.7°C/W/m^2$) and aerosol ($1.0°C/W/m^2$) (Karl et al., 1995b; Wigley, 1994). This determines, in particular, an inadequateness of simple energy-balance models to assess the global climate change. These circumstances point to the fact that to remove the existing uncertainties of the assessments of climate change, more effort and time will be needed than it has till now been supposed. There are a number of mechanisms through which aerosols affect climate; three of them are being of particular importance (King, 1993; King et al., 1993; Kondratyev, 1992d, 1996b; Penner, 1995; Penner et al., 1994a,b):

(1) The direct effect of aerosols on the radiation budget of the surface-atmosphere system through a redistribution of the shortwave solar radiation and infrared thermal emissions in the system due to scattering and absorption by aerosol particles, non-uniformly distributed in the atmosphere (Box and Trautmann, 1994; Charlson et al., 1987, 1991, 1992; Charlson and Lelieveld, 1994; Charlson and Wigley, 1994; Coakley and Cess, 1985; Engardt and Rodhe, 1993; Grassl, 1988; Hales et al., 1992; Hidy and Wolf, 1995; Hunter et al., 1993; Jennings, 1993; Karl et al., 1995b; Kaufmann and Chou, 1993; Kiehl and Briegleb, 1993; Kiehl and Rodhe, 1995; Kondratyev, 1991a; Langner et al., 1992; Liou, 1992; Marchuk et al., 1992; Meszaros, 1985; Novakov and Penner, 1993; Penner et al., 1994a,b; Prening, 1991; Schlesinger et al., 1992).

(2) The key role of aerosol particles in water phase transformations in the atmosphere and, in particular in cloud formation, which may be more important for atmospheric energetics than the direct effect of aerosols (Ackerman et al., 1993; Anderson et al., 1992; Andreae, 1994; Blanchet, 1994; Boers et al., 1994; Charlson and Wigley, 1994; Chýlek and Hallett, 1992; Falkowski et al., 1992; Fouquart and Isaka, 1992; Grassl, 1975, 1992; Hanson and Svenningson, 1994; Harshvardhan, 1993; Hobbs, 1993; Jones et al., 1993; Kim and Cess, 1993; King et al., 1995a; Kondratyev, 1972, 1991a, 1993g; Leatich and Isaac, 1994; Lelieveld and Heintzenberg, 1992; Marchuk et al., 1986; Novakov et al., 1994; Parungo et al., 1994; Platnick and Twomey, 1994; Raga and Jonas, 1993; Toon, 1995; Toon et al., 1994; Twomey, 1991). An issue of concern is the role of gas-to-particle reactions of aerosol formation (condensation nuclei formation) due to natural

and anthropogenic emissions to the atmosphere of the compounds, such as dimethylsulphide and sulphur dioxide (Charlson et al., 1987; Lawrence, 1993; Malin et al., 1993; Putaud et al., 1993).

(3) Heterogeneous chemical processes, in particular the disintegration of ozone molecules on the surface of aerosol particles, which affect the changes in the atmospheric gas composition (Kondratyev, 1989a,b). Complex field studies of the interaction between aerosols and radiation accomplished in different regions of the globe within the CAENEX and GAAREX programmes (Kondratyev, 1972, 1991a) show that the aerosols absorb a considerable amount of solar radiation and thus perturb the atmospheric radiation budget to a great extent.

The following problems are most urgent:

- Impact of dust outbreaks from arid regions on the radiative regime of the atmosphere over the continents and the oceans.
- Transport of aerosols to high latitudes leading to the formation of Arctic haze and extended cloudiness in the Arctic (Marchuk et al., 1986; Kondratyev et al., 1996a). Of special importance are the aerosol perturbations due to volcanic eruptions (Kondratyev and Galindo, 1997).

A very complicated problem is that of the role of radiation impact on the formation and evolution of the atmospheric aerosol field. Interactions should be studied between radiative, convective and turbulent heat and mass exchange in the processes of aerosol generation and formation, as well as the chemical composition, particle size distribution and optical properties of aerosols (Jennings, 1993; Kondratyev, 1991a).

Numerical modelling of aerosol formation and global distribution requires an essential advancement; further substantiation of the aerosol models is necessary.

2.4.2.1 Modelling of aerosol properties

Modelling of aerosol properties is extremely difficult because it involves a consideration of a large number of parameters affecting their variability, and also because there are no reliable criteria for choosing the respective parameters. For example, the spatial and temporal variability of aerosol properties is determined by a complex set of processes of the generation, transformation and evolution of an ensemble of aerosol particles, their propagation from the sources and removal from the atmosphere. Separate consideration of these processes, which has often been the case, is incorrect, since these processes are interrelated.

The problem of aerosol modelling may be subdivided into two tasks, namely:

(1) Global spatial distribution of aerosols (the aerosols mass concentration field), including consideration of average characteristics of the generation, propagation and removal of particles from the atmosphere (Kondratyev, 1992d; Kondratyev et al., 1983).

(2) Local aerosol properties resulting from the effects of transformation processes of size spectra and composition of atmospheric aerosols of different origin (Grassl, 1975; Ivlev, 1992; Meszaros, 1981).

The processes of particle generation determine, first of all, their physical and chemical properties, and their concentration in the atmosphere, and therefore, subsequent transformation of the size distribution spectra and particle removal rate from the atmosphere.

The following types of natural aerosol particles can be identified:

(1) products of disintegration and evaporation of sea spray;
(2) mineral dust, wind-driven to the atmosphere;
(3) products of volcanic eruptions: those directly emitted to the atmosphere and resulting from gas-to-particle conversion;
(4) particles of biogenic origin: those directly emitted to the atmosphere and resulting from the condensation of volatile organic compounds, e.g., terpenes, and chemical reactions taking place between them;
(5) smoke from biomass burning on land;
(6) products of natural gas-to-particle conversion, e.g., sulphate formed from sulphur compounds supplied by the ocean surface.

Basic types of anthropogenic aerosols are particles of industrial emissions (soot, smoke, cement, road dust, etc.) as well as products of gas-to-particle conversion. Besides, part of the aerosol matter is formed by heterogeneous chemical reactions, in particular, by photocatalytic reactions.

As a result of the comparatively long lifetime of aerosol particles, their transport, mixing and interaction with each other and with gaseous compounds of various substances, as well as differences in the physical and chemical properties of the particles coming from various sources are smoothed over. This results in the formation of background aerosols.

Characteristics of background aerosols vary in different atmospheric layers. This is relevant, in particular, to tropospheric and stratospheric aerosols which do not mix very well, and their particles differ considerably with regard to their composition and size distribution. There are regions in the troposphere where particles are very diverse, and this is due to relatively short lifetimes of tropospheric particles, especially in the surface layer, and also to the presence of some highly intensive sources of aerosols.

Processes of aerosol transport and removal from the atmosphere are governed by both the meteorological factors and the aerosol particle properties. Processes dependent on the meteorological conditions involve organized convective and advective particle transport, as well as mixing due to turbulent diffusion. Individual properties of particles govern some processes of particle removal from the atmosphere, e.g., dry and wet deposition.

In the course of transport of particles and their removal from the atmosphere, their size (mass) and composition are transformed due to coagulation and condensation growth, as well as to heterogeneous chemical reactions. These processes are responsible for the particle size spectrum and determine the complex refraction index value as well as the particle shape, i.e., optical properties of particles (Ivlev, 1992; Kondratyev et al., 1983).

Most essential gas-to-particle reactions affecting the formation of aerosol particles include (Jennings, 1993; Kondratyev, 1991a):

(1) Reactions between sulphur dioxide and hydroxyl radicals which eventually lead to the formation of easily condensed sulphuric acid molecules in the presence of water molecules.
(2) Reactions of hydrocarbons with ozone and hydroxyl radicals, with the subsequent formation of primary products with nitrogen oxides, such as peroxyacetylnitrate (PAN).

All processes of the formation and evolution of aerosol systems depend on periodical variations of solar radiation in the atmosphere and at the surface. The processes of atmospheric heating near the surface, water evaporation and chemical reactions of aerosol formation are characteristic of a pronounced diurnal variation resulting in variations of aerosol characteristics which in some cases exceed the annual variation. Considerable daily variations of anthropogenic aerosols have been observed.

Aerosol matter cycles are closely connected with hydrological processes in the atmosphere: on the one hand, clouds and precipitation play an important role in the formation, transformation of aerosol particles and their removal from the atmosphere; on the other hand, aerosols are condensation nuclei whose physico-chemical properties determine the microphysical processes in clouds. The fact that water vapour molecules and aerosol particles have approximately equal lifetimes has an explanation. Therefore, more reliable knowledge of interactions between aerosols and cloud elements is necessary to better understand the processes of aerosol formation and the variability of aerosol properties in time and space.

Although the available information on aerosol properties is inadequate, the volume of existing data is enormous and has been briefly summarized in a number of recent monographs (e.g., Jennings, 1993; Kondratyev, 1992a–g; Kondratyev et al., 1983, 1996a,b). The latest studies have been specifically devoted to sulphate aerosols.

For instance, a study has been carried out (Thornton et al., 1996) of sulphur dioxide as a source of cloud condensation nuclei in the upper troposphere of the Pacific Ocean within the NASA Pacific Exploratory Mission—West A (PEM-West A) experiment conducted during September–October, 1991. An important circumstance is that the experiment was accomplished in the conditions when the troposphere over the western Pacific was not impacted by continental outflows from Asia. The instrumentation on board the NASA Ames Research Center DC-8 aircraft allowed one to measure sulphur dioxide along with dimethylsulphide (DMS), carbon disulphide (CS_2) at altitudes up to 12 km.

The most significant result was an increase of SO_2 with altitude above the mixed layer both for the continental and central Pacific regions and for the equatorial and high-latitude regions (observations included a transect of the subtropical troposphere over the North Pacific Ocean). It was pointed out (Thornton et al., 1996) that the SO_2 distribution was apparently influenced by the long-range transport from convective activity over and near the continents, by injections of stratospheric air into the troposphere, and, possibly, by tropospheric emissions from the Pinatubo volcano. The SO_2 increase with altitude results in the formation of a reservoir of oxidized sulphur in the upper troposphere which can be a precursor of a significant number of condensation nuclei (CN) and cloud condensation nuclei (CCN). Undoubtedly, these CN and CCN could have a widely spread effect on cloud formation over the Pacific Ocean. Calculated H_2SO_4 production rates averaged 1600 molecules/cm^3 s above 8.5 km for the Central Pacific troposphere, that is, production of CN and CCN was quite significant to affect the cloud formation and, respectively, the radiation regime, producing the radiative cooling.

Specific attention has been paid (Arimoto et al., 1996) to studies of the sources of aerosol SO_4^{2-} and the long-range transport of continental materials to the North Pacific. For this purpose, the aerosol particle samples collected over Asia and the North Pacific were analysed to investigate relationships among the atmospheric sea salt, mineral aerosol, biogenic emissions (methanosulphonate, MSA) and several anthropogenic substances (sulphate, nitrate, and various trace elements). A comparison of samples for four coastal sites (Hong Kong, Taiwan, Okinawa, and Cheju) and three remote Pacific islands (Shemya, Midway, and Oahu) has shown that non-sea-salt (nss) SO_4^{2-} and MSA were uncorrelated at the East Asian sites because in these cases the pollution sources presumably overwhelm the biogenic emissions of nss SO_4^{2-} . At the coastal-continental sites, marine biogenic emissions accounted for only 10 to $<5\%$ of total nss SO_4^{2-} . In contrast, over the ocean the concentration of nss SO_4^{2-} and MSA were correlated, and higher percentage of biogenic nss SO_4^{2-} occurred: 155% and 70% at Oahu and Midway, respectively. Differences in the nss SO_4^{2-}/NO_3^- ratios for Cheju, Hong Kong, Taiwan, Okinawa, Midway, and Oahu suggest regional differences in the pollution component of the aerosol. The data considered revealed a strong seasonal pollution component in the sulphate aerosol. The meteorological conditions favouring the long-range transport of Asian dust to the North Pacific stimulate the transport of anthropogenic materials. It is important, however, that the composition of the air sampled at the coastal-continental stations may be quite different from the air transported to the remote ocean.

The study of Tegen et al. (1996) has reminded us of the significance of the flux of mineral aerosols from soils disturbed by erosion, over-cultivation and deforestation, mostly in arid and sub-arid regions. It is well known (Grigoryev and Kondratyev, 1993; Kondratyev, 1972; Kondratyev et al., 1976; Swap et al., 1996) that the most impressive cases of this kind are the long-range transport of desert aerosols from the Sahara desert and the Kara-Kum desert (Central Asia). As has been pointed out (Tegen et al., 1996), model calculations indicate that 5020% of the total atmospheric dust mass originates from disturbed soils. Using a radiative transfer model embedded in a general circulation model, the authors find that dust from disturbed soils causes a decrease of the net surface radiation forcing of about -0.96 W/m^2, accompanied by an atmospheric heating increase which may be a significant forcing for atmospheric dynamics. The global and annual mean forcing due to the increase in GHGs that occurred between 1850 and 1980 is about $+2.1$ W/m^2 with an uncertainty of about 10%. The contribution of sulphate aerosols is -0.28 W/m^2. If the contributions of anthropogenic GHGs and sulphate forcing are added, the resulting global mean flux at the surface is reduced to -0.6 W/m^2, compared to $+0.4$ W/m^2 for the case of only anthropogenic greenhouse gases and sulphate. Thus the dust from the disturbed sources actually changes the sign of the global mean anthropogenic forcing at the surface.

As for the top-of-atmosphere (TOA) radiative forcing due to mineral aerosols, it is close to zero because of mutual compensation of the shortwave and thermal components of the forcing in the longitudinal mean. Tegen et al. (1996) have correctly emphasized the preliminary character of their estimates and the necessity to develop more adequate models.

Li et al. (1996) have discussed measurements of the light-scattering properties of the North African dust transported across the Atlantic Ocean to Barbados by trade winds. The results of observations demonstrate that the African mineral dust may be the dominant light-scattering aerosol through the tropical and subtropical North Atlantic region (unfortunately, Li et al., (1996), have not referenced the CAENEX results (Kondratyev et al., 1976) which showed that dust aerosols are both scattering and absorbing components). Due to Li et al. (1996), the annual mean dust concentration in the Barbados trade wind air is 16 times that of non-sea-salt sulphate. Therefore the net scattering by mineral dust is about four times that by nss aerosol, although the latter has the mean scattering efficiency four times higher than in case of dust.

Important sources of aerosols are the aerosol emissions by the tropical forest and savannah biomass burning (Echalar et al., 1995; Chlek and Wong, 1995; Chlek et al., 1995), gas-to-particle conversion of marine atmospheric dimethylsulphide (Ayers et al., 1995), black carbon components (Chlek et al., 1995a). The estimates obtained by Chlek and Wong (1995) show that the global average value of the direct RF due to the biomass burning may vary between -0.2 and -1.1 W/m^2.

2.4.2.2 Aerosol climatic impact

The relationship between climate changes due to the enhanced greenhouse effect of the atmosphere and increased aerosol concentration is strongly dependent on differences in the behaviour of anthropogenic aerosols and greenhouse gases (GHGs), caused by the spatial–temporal inhomogeneity of the aerosol distribution (Kondratyev, 1996). These differences can be accounted for by the presence of powerful local sources, the short lifetimes of aerosols in the troposphere (of the order of two weeks), the photochemical particle growth processes in the presence of solar illumination, a strong interaction with natural aerosols, cloud elements and water vapour. It is the interaction between aerosols and water vapour that is first of all responsible for the limited lifetimes of both in the troposphere: about 10–15 and 6 days, respectively, as well as for their spatial–temporal inhomogeneity. The lifetimes of aerosols and water vapour are also dependent on the altitude and latitude of their location (Ivlev, 1992; Kondratyev, 1991a).

It is important that anthropogenic sulphate aerosols are mostly formed at the middle latitudes of the northern hemisphere (Europe, the USA, Japan). With the total emission level of sulphur-containing gases of about 10^{14} g/year, the contribution of anthropogenic sources constitutes about 70%, about 80% of this amount being relevant to the northern hemisphere, that is, 50–60% of the total emission (Kondratyev, 1991). However, the efficiency of these sources in the formation of sulphate aerosols is strongly variable: from 6% to 40%. Thus, the amount of anthropogenic sulphate aerosols in the global atmosphere varies from 0.4 mg/m^2 to 2.8 mg/m^2. With the backscattering coefficient for sulphate aerosols in the shortwave spectral region about 10 m^2g^{-1}, the relative contribution of backscattered solar radiation will be about $4 \times 10^{-3} - 2.8 \times 10^{-2}$.

According to the estimates of Lelieveld and Heintzenberg (1992), the sulphate concentration 0.5–2.0 mg m^{-3} is responsible for a solar radiation extinction of about 0.5–1.0 W/m^2. The attenuation of solar radiation by anthropogenic sulphate aerosols must be approximately 2–2.5 times greater. Owing to the spatial inhomogeneity of anthropogenic sulphate aerosols, this value may vary within two orders of magnitude.

Hegg et al. (1995) have pointed out on the basis of their airborne measurements of aerosol light-scattering efficiencies for a portion of the north-east Atlantic seaboard of the USA during July 1993 that sulphate light-scattering efficiency, b_{sp}, varied within 2.2–3.2 $m^2\ g^{-1}$, which is lower than the value used in recent model assessments of the climatic impact of aerosols (Charlson et al. (1992) assumed b_{sp} = 5 $m^2\ g^{-1}$ instead of 2.7 $m^2\ g^{-1}$ obtained in the observations mentioned above).

In case of pure sulphate aerosol a regression gives (Hegg et al., 1995):

$$b_{sp}\,(SO_4^{2-}) = (2.7 \pm 0.4)\,[SO_4^{2-}] - (6.2 \pm 3.8) \times 10^{-6}\,(m^{-1})$$

where $[SO_4^{2-}]$ is the non-sea-salt sulphate aerosol concentration (g/m^3).

On the other hand, direct observations give:

$$b_{sp} = (2.1 \pm 0.8)\,[SO_4^{2-}] + (7.37.3) \times 10^{-6}$$

that is, most of light-scattering was not associated with sulphate.

Although the role of sulphate submicrometre aerosols in the radiation regime of the lower atmospheric layers is doubtless very important and reveals itself as cooling, it should be noted that this effect is not quite general. Data of observations obtained in the course of implementation of complex CAENEX and GAAREX programmes (Kondratyev, 1972, 1991a; Kondratyev et al., 1976) have revealed the presence of a sufficiently large amount of aerosol particles absorbing shortwave solar radiation in the troposphere, which may lead sometimes not to cooling but to radiative heating of the lower tropospheric layers. In most cases such absorption is due to the presence of soot and iron oxides in the aerosol matter. In the polluted urban atmosphere the contribution of soot in the mass aerosol concentration may reach 10–30%, and that of sulphate components 20–40%.

It is important that a considerable portion of soot particles is represented by the small-size fraction (particle radius $r < 0.1$ μm). According to observational data, the modal radius of primary soot nuclei may reach 20–40 μm, whereas that of sulphate particles is in the size range of about 0.15 μm (Kondratyev, 1991a). The coagulation of primary soot nuclei leads to the formation of fractals having a lower sedimentation rate and anomalous optical properties. If there are mineral or other aerosol particles close to a soot source of high concentration (which is also relevant to sulphate particles), primary soot nuclei are condensed on these particles forming a cover layer. The specific absorption of shortwave radiation by soot may then increase by an order of magnitude or more (Kondratyev and Ivlev, 1995).

It should be also noted that in most cases the emissions of sulphur-containing gases, which are transformed into sulphuric acid and sulphates, are accompanied by simultaneous release of soot to the atmosphere. Soot is known to be an effective catalyst in the process of SO_2 oxidation giving SO_4. The optical properties of non-uniform particles thus formed differ substantially from the optical properties of homogeneous particles. To calculate the optical characteristics of spherical particles, their modelling with polydispersive two-layer spheres has been successfully performed (Ivlev et al., 1992; Kondratyev and Ivlev, 1995).

The formation of the two-layer cloud particles resulting from the condensation of water vapour on the soot-containing particles leads, as a rule, to an increase of albedo for shortwave radiation. In such cases clouds have a greyish-brown or brownish coloration. In particular, at sufficiently large optical thicknesses an enhancement of molecular absorption by trace gases may appear due to an increased photon path length in a cloud (Twomey, 1991), and as a result of the formation of a thin layer of adsorbed molecules which absorb radiation. The increase of cloud albedo may take place due to a change in the droplet size distribution with the growing concentration of CCN: the mean modal radius decreases, the number density of small droplets grows, and this changes the scattering function and increases the contribution of backscattering (Kondratyev and Ivlev, 1995).

Cloud formation processes are greatly affected by nitrogen oxides and their acids, whose source, to a large extent, is industrial activity. Their role is especially important in the processes of formation of polar stratospheric clouds. In particular, one of the types of polar stratospheric clouds is formed at the heterogeneous condensation of nitric acid on the surface of sulphuric acid droplets (Kondratyev, 1989a,b).

It is a well-known fact (Kondratyev et al., 1996a) that Arctic haze and clouds contain a considerable amount of anthropogenic impurities, including soot, which decreases substantially the albedo of polar regions and reacts against the global effect of atmospheric cooling due to anthropogenic sulphates.

It should be emphasized that heterogenic chemical reactions in the atmosphere related to anthropogenic activity are one of the most important factors of the impact of anthropogenic aerosols on climate.Firstly, these reactions stimulate a faster removal from the atmosphere of a number of gaseous impurities (sulphur dioxide, ozone, hydrocarbons, in particular, methane, nitrogen oxides, hydrochloric acid), thus reducing their lifetimes. Secondly, they influence the evolution of the aerosol particle size spectrum, and consequently, the albedo value. At the same time, photochemical heterogeneous processes can be a sensitive climatic regulator of solar–terrestrial relationships (Shaw, 1987).

Aerosols play an important role in the heterogeneous destruction of ozone molecules both in the stratosphere and the lower atmosphere (Kondratyev, 1989). Heterogeneous reactions of the ozone molecule destruction are mainly due to atmospheric dust particles containing metal oxides, and above all, aluminium oxides. In the presence of shortwave solar radiation ($\lambda > 320$ nm), the destruction of ozone molecules is possible in the process of photochemical oxidation of triplet-exited sulphur dioxide (Kondratyev and Ivlev, 1995).

Since the aerosol-induced radiative forcing (ARF) is comparable with the opposite influence due to the growing GHG concentration (GRF), an important problem appears of the more reliable estimation of the relationship between the ARF and GRF contributions to climate change.

Mitchell et al. (1995a) were first to apply a 3-D global climate atmosphere–ocean interactive model developed at the Hadley Centre for Climate Prediction and Research (UK) to simulate a combined impact of GHGs and sulphate aerosol on climate. The aerosol impact (cooling due to backscattering) was taken into account through a relevant increase of the surface albedo. Calculations made for the time period 1860–2050 show that the mean global SAT increase in the future will constitute 0.3°C/10 years when only GHGs are taken into account, and 0.2°C/10 years, if the combined influence of GHGs and sulphate aerosol is considered.The greenhouse radiative forcing was 0.4 W/m^2 in 1860; then it increased to 1.2 W/m^2 in 1960, and in 1990 it further rapidly increased to 2.5 W/m^2. After 1990 the rate of increase constitutes 0.6 W/m^2 per 10 years.

Taylor and Penner (1995) performed a numerical modelling of climate using the NCAR CCM-1 coupled model of the atmosphere–ocean system together with the Lawrence Livermore National Laboratory model of the chemical processes in the troposphere describing the gas-to-particle reactions (with the participation of hydroxyl OH) of the transformation of dimethylsulphide and sulphur dioxide to sulphate aerosol and its transport in the atmosphere (the Grand Tour model).

The CCM-1 model satisfactorily simulates the observed climate though it somewhat underestimates the SAT mainly due to substantial systematic errors in the upper troposphere (Table 2.11). The observed and calculated ERB (Earth's Radiation Budget) values agree well, but the annual course of the cloud-radiation forcing turned out to be underestimated. The prescribed emissions of sulphur dioxide take into account industrial sources, fossil fuel and biomass burning. Total emissions corresponding approximately to the 1980 level constitute 78 MtS/year.

The numerical modelling of climate for a period of 20 years has been carried out for the following four scenarios: (1) pre-industrial conditions (the CO_2 mixing ratio is 275 ppm); (2) the present concentration of CO_2 (345 ppm); (3) combined present anthropogenic emissions of sulphur compounds (at the level of 1980) and CO_2 concentration 275 ppm; (4) the same as in the third scenario but for the CO_2 concentration 345 ppm.

The data in Table 2.11 have been obtained by averaging the results of calculations over the last ten years. As can be seen, the mean global ARF value constitutes about -0.95 W/m^2, whereas the GRF contribution is equal to 1.26, which determines a weak total effect of climate warming. Of course, of importance is the fact that the indirect effect of aerosol on the optical properties of clouds making a certain contribution to the ERB variations is left out of account. This is why, in particular, Wigley (1995a) has emphasized the preliminary character of such numerical modelling. The assessment made by Jones et al. (1994) shows that an indirect aerosol effect at the top of the atmosphere is approximately -1.3 W/m^2 in the global annual mean.

Boucher (1995) obtained a GCM estimate of indirect aerosol forcing using the

Table 2.11. The radiative forcing, ΔF, and respective climate changes (T_s, SAT; P, precipitation; C, cloud amount; SI, relative ice cover extent)

Cases	ΔF (W/m^2)	T_s (°C)	ΔT_s (°C)	P (mm/d)	ΔP (mm/d)	C %	ΔC %	SI%	ΔSI%
Pre-industrial									
NH		12.5		3.40		56.6		4.86	
SH		12.5		3.54		62.4		6.64	
Globe		12.5		3.47		59.5		5.76	
Present CO_2 concentration									
NH	1.26	14.5	1.9	3.48	0.09	55.0	−1.7	4.13	−0.74
SH	1.25	14.8	2.3	3.61	0.08	61.1	−1.3	4.39	−2.26
Globe	1.26	14.6	2.1	3.55	0.08	58.0	−1.5	4.26	−1.50
Present concentration of sulphates									
NH	−1.60	11.3	−1.2	3.36	−0.04	56.9	0.3	5.5	0.67
SH	−0.30	11.7	−0.8	3.48	−0.06	63.1	0.7	7.24	0.59
Globe	−0.95	11.5	−1.0	3.42	−0.05	60.0	0.5	6.39	0.63
Joint impact of CO_2 and sulphates									
NH	−0.34	13.0	0.5	3.43	0.03	55.8	−0.9	4.85	−0.0
SH	0.95	13.6	1.1	3.56	0.02	62.1	−0.3	5.40	−1.24
Globe	0.31	13.3	0.8	3.49	0.02	58.9	−0.6	5.13	−0.63
Observed values									
NH		14.9		2.6		58.9		4.4	
SH		13.5		2.7		65.6		4.5	
Globe		14.2		2.7		62.2		4.5	

satellite-retrieved cloud droplet effective radii (from these data, r_{eff} in the NH is less than in the SH by about 0.7 μm: 0.8 μm over land and 0.4 μm over the ocean). The indirect radiative forcing by anthropogenic aerosols could be between ~ −0.6 and 1.0 W/m^2 averaged over the 0°–50°N latitudinal band. The uncertainty of these estimates is difficult to assess but it is, at least, 50%. Erickson et al. (1995) have found that the climate response due to indirect aerosol forcing is strongest in the northern hemisphere winter, with cooling over the North Atlantic and North Pacific Oceans of the order of 2–6°C.

Twohey et al. (1995) have used observational data to assess the effects of aerosol particles on the microphysics of the coastal stratiform clouds. Aerosol and cloud droplet size distributions were measured in 13 research flights in a stratiform cloud within 300 km west of the northern California coast. The chemical composition of droplets was also assessed. An analysis of observational data shows that most of the clouds were polluted. In these cases the particle and droplet number concentrations decreased, and the mean droplet size increased with the increasing distance from the coast. Compared to the unpolluted stratiform cloud, the polluted cloud had, on average, a sulphate concentration higher by an order of magnitude, a droplet number concentration higher by a factor of 6, the size of droplets smaller by a factor of 2, and the 3.7 μm reflectance higher by a factor of 2, but there was no significant difference in the visible reflection.

An important result obtained by Taylor and Penner (1995) is the conclusion that there is a greater climate sensitivity to GHGs than to aerosol forcing. Cox et al. (1995) have come to the conclusion, however, that these sensitivities are essentially equal and the previous result was an artifact of calculations. One can therefore believe that the global and annual mean temperature respond to the algebraic sum of longwave greenhouse forcing and shortwave aerosol forcing.

Analysis of data on the geographical distribution of ARF and GRF as well as the contributions of CO_2 and aerosol to climate changes has shown that a maximum of the aerosol-induced cooling is located over vast regions of the NH continents and not concentrated in the region of the ARF maximum over Europe, since the effect of aerosol is manifested through a complicated interaction with the atmospheric general circulation (of importance are, in particular, changes in the ice state of the Norwegian and Greenland Seas).

It is of interest that with the combined consideration of the impacts of CO_2 and aerosol the characteristic NH high-latitude greenhouse climate warming disappears. One more important feature is that the greenhouse and aerosol impacts on climate are not additive: the linear model foresees a global warming which is overestimated by 40%. Since the sensitivity of the applied climate models determined by the ratio of T_s and RF mean global values happened to be very high (6.4°C for a doubled CO_2 concentration), further analysis is needed on the basis of a more adequate model. The existing characteristic regional structures of the temperature field response to anthropogenic impacts necessitate a search for respective signatures to recognize the relevant disturbances.

The growing attention to assessments of the combined climatic impact of increasing CO_2 and aerosols has resulted in a number of recent relevant studies. Summarizing the results obtained, Mitchell et al. (1995b) have pointed out that the existing estimates vary widely. The estimates of the current global mean rate of cooling due to scattering by anthropogenic sulphate range from 0.3 to 0.9 W/m^2, with maximum values of 3 W/m^2 and more occurring over the industrial regions of the NH mid-latitudes. It means that locally the direct effect of sulphate aerosols may reverse the warming of ~ 2 W/m^2 due to greenhouse gases. Much more uncertain are the estimates of the indirect aerosol effect as CCN, which might lead to an increase of cloud brightness (i.e., albedo) and, hence, to a climate cooling. Recent estimates of the global mean cooling range within 0.4 to 1.3 W/m^2. It is important to take into account other radiative forcings due to volcanic eruptions (down to -4 W/m^2), stratospheric ozone depletion (down to -0.08 W/m^2, over the last decade), changes in tropospheric ozone and solar activity (estimates of the contributions of the last two factors are highly uncertain).

In continuation to the study of the combined effect of CO_2 and aerosols on climate, Mitchell et al. (1995b) have undertaken a simulation modelling using a low-resolution version of the 19-level coupled numerical weather prediction/climate model developed by the UK Meteorological Office (a 5° × 7.5°lat.–long. grid). The light-scattering by sulphate aerosols leads to an appropriate increase of the surface albedo. The atmospheric model is coupled to a 50-m mixed layer model of the ocean, with a prescribed seasonally varying heat convergence and a simple thermodynamic model of sea-ice. A comparison of the control simulation data (an integration for 75 years with the results averaged over the last 40 years) with observations gave quite satisfactory results. For instance, the model successfully simulates the patterns of SAT and their seasonal variations. The greatest errors occur over the NH high-latitude continents where the model is too cold, especially in December and February. The model has a warm bias in the tropics and in the SH in both solstitial seasons.

In case of $2CO_2$ the global mean SAT increases by 5.2 K. An integration with a 39% increase in CO_2 giving a change in radiative heating due to a GHG increase since 1990, produced an equilibrium warming of 2.3 K, which is substantially higher than the observed warming. Another important disadvantage of calculations is that they give a warming everywhere, whereas observations reveal coolings in some areas of the globe, including parts of the NH mid-latitude continents.

Mitchell et al. (1995b) pointed out that an addition of the effect of aerosol scattering (uncertain by a factor of, at least, 3) leads to a better agreement with the observed pattern of changes over the NH continents and reduces the global warming by about 30%. A doubled aerosol forcing produces patterns still compatible with observations, but a further increase (trebling the forcing) leads to an abnormally extensive cooling in mid-latitudes.

Recent observations testify to the fact that the diurnal SAT range narrows over most of the northern extra-tropics with a CO_2 increase. An addition of the present industrial aerosol has a small effect on the diurnal change in the model because the direct effect of reduced solar heating at the surface is approximately balanced by the indirect effects of cooling. Therefore the ratio of the narrowed range of the diurnal change to the mean warming increases getting in closer agreement with observations.

An important feature of the results is that the simulated SAT response to the aerosol forcing in the NH mid-latitudes is fairly uniform throughout the year, despite a several-fold variation in forcing, because the processes in high latitudes tend to amplify the temperature changes in winter and reduce them in summer.

Mitchell et al. (1995b) believe that the idealized numerical experiments they carried out show that the radiative forcing due to sulphate aerosols is likely to have had a substantial impact on temperature changes during this century. They emphasize that an adequate assessment of climate change over the recent decades will require an account of all changes in the radiative forcing, including the direct and indirect effects of industrial, volcanic, and other aerosols, as well as a reliable assessment of internal climatic variability.

Even in the case of sulphate aerosols the dependence of the climatic response on various highly variable aerosol properties (size distribution, complex refraction index, etc.) makes it difficult to reliably assess the aerosol impact on climate.

Boucher and Anderson (1995) have shown, however, that within a plausible set of assumptions (sulphate aerosol resides in the accumulation mode size range and only interacts with water vapour and ammonia vapour), the direct climate forcing sensitivity to the variable aerosol properties is fairly small (20%). This conclusion requires, however, further confirmation.

Estimates of the trend of the content of sulphates in the atmosphere obtained by Alkezweeny (1995) from the data of analysis of the South Greenland ice core revealed a distinct increase of the sulphate aerosol concentration. With an account of the effect of this trend of aerosol on the cloud particles size distribution consisting in the growth of the concentration of small droplets and in the subsequent increase of the cloud albedo, the mean global SAT may decreaase within 1.0–2.0°C (of course, this estimate should be considered approximate).

A serious oversight in the estimates of the effect of tropospheric aerosols on climate is the neglect of the real complexity of the optical properties of aerosol, an atmospheric component which not only scatters but also absorbs the shortwave radiation (Chlek and Wong, 1995; Kondratyev, 1972, 1991a, 1992b; Kondratyev and Moskalenko, 1983; Marchuk et al. 1986). This problem has been discussed (Kondratyev, 1995d) based on real (obtained from observations) optical properties of aerosol (Kondratyev 1996b; Kondratyev and Ivlev, 1995; Kondratyev et al., 1983).

2.4.2.3 Unsolved problems

Further studies of the atmospheric aerosol properties and their climatic impact should primarily include:

(1) Improving standard radiative models of the atmosphere to more adequately simulate the characteristics of various types of aerosols, including low humidity effects, particle size distribution, vertical concentration profiles, etc.
(2) Obtaining more adequate observational information about the spatial and temporal variability of aerosols on global scales, on the basis of both *in situ* observations and remote sensing. A special part of the GCOS should be developed to accomplish observations relevant to aerosols and climate problem (Kondratyev et al., 1996b).
(3) Developing techniques to identify the anthropogenic aerosol content and composition on both regional and global scales, with emphasis on the monitoring of the sulphate dynamics and soot components.
(4) Studying the dependence of real optical properties of aerosols and their chemical composition and size distribution, bearing in mind the effects of varying air humidity.
(5) Further development of measurement techniques which include both remote sensing and *in situ* observations.
(6) Undertaking both observational and theoretical studies of the internal aerosol particle structure, considering their multicomponent formation, as well as the role of aerosol particles as cloud condensation nuclei.

In view of the growing interest in the aerosol impact on climate, of vital importance is an accumulation of long-term data series characterizing the trends of aerosol properties. Andreae (1996) has justly emphasized that in contrast to the outstanding documentation of GHG growth in the atmosphere over the last two centuries, there is nothing of the kind for atmospheric aerosols.

2.4.3 Stratospheric aerosols and climate

2.4.3.1 Introduction

Powerful explosive volcanic eruptions are a unique type of natural calamity whose consequences are of great importance, not only on a regional but also on a global scale. Of special interest are impacts of volcanic eruptions on global climate and the ozone layer (Hansen et al., 1992, 1995; Kondratyev, 1985, 1988, 1991a, 1992d; Kondratyev and Galindo, 1996; McCormick et al., 1995). Every explosive volcanic eruption is a large-scale "experiment" in nature, which opens up an opportunity to study the sensitivity of nature (e.g., the climatic system, the ozone layer) to external disturbances. The El Chichón volcano eruption in 1982 was in fact the first "experiment" when it was possible to undertake a relatively complete programme of observations (including satellite remote sensing) to monitor relevant environmental consequences.

Various observational means have been further developed by 1991, when an eruption of the Pinatubo volcano took place on Luzon Island (the Philippines). By that time, a very broad complex of satellite, aircraft, balloon and ground-based instrumentation to monitor the environmental dynamics had been developed and applied for various purposes. It has become possible not only to monitor the propagation of volcanic products over the globe, but also to observe (with the use of both *in situ* and remote sensing techniques) various properties of eruptive aerosols, as well as their impact on the Earth's radiation budget and climate. The 15 June 1991 Pinatubo eruption caused the strongest aerosol disturbance in the stratosphere compared to other eruptions in this century (Table 2.12).

Only the eruptions of the volcanoes Tambora (1815, 100 Tg) and Krakatau (1883, ~50 Tg) in the last century were more powerful, provided ancient eruptions are not considered. For instance, during the Toba mega-eruption (~71 000 years ago) the total stratospheric loading over a 6-year period was assessed to be within 2200–4400 Tg of sulphur acid aerosols (Zielinski et al., 1996). The prolonged Toba eruption could have resulted in an enhanced climate cooling during the initial two centuries of this ~1000-year cooling event (specific interest in the Toba eruption is determined by its possible coincidence with the initial phases of glaciation around 72 000 years ago).

2.4.3.2 The Mount Pinatubo eruption

An explosive eruption of the Mt Pinatubo volcano (Luzon Island, the Philippines: 15.14°N; 120.35°E) of the andesite type took place in June 1991 and had the most powerful impact on the atmosphere compared with all the earlier eruptions monitored by making use of satellite information (the previous eruption of Mt Pinatubo occurred 635 years ago). Gigantic ejections of SO_2 and other volcanic products have led to the formation of an atmospheric aerosol layer up to an altitude

Table 2.12. Most powerful volcanic eruptions in the twentieth century

Volcano	Date of eruption	Total emission of aerosol Tg (10^{12}g)
Stratospheric background aerosol	The 1979 level	1
Katmai	June 1912	20
Agung	March 1963	16–30
Fuego	October 1974	3–6
El Chichón	April 1982	12
Pinatubo	June 1991	30
Sierro Hudson	August 1991	3

of 30 km and higher, which was registered by space-borne remote sensing techniques and documented (McCormick, 1992). Above 20-km altitude, the aerosol cloud moved quickly westward and its front edge made a complete revolution round the Earth in about three weeks.

From the data of aircraft, ship and surface lidar soundings, the volcanic aerosol was initially concentrated in the major 20–27-km layer, though it was also recorded near the level of the tropopause and at altitudes up to 30 km. The total mass of emitted sulphur dioxide (from the data of the space-borne TOMS instruments to map the total ozone content, TOC) constituted $\sim$20 Tg. This sulphur dioxide gradually (with a characteristic time of $\sim$30 days) transformed into H_2SO_4 aerosol. By the end of 1991 the total SO_2 mass in the stratosphere had reached 30 Tg (from the SAGE data).

Satellite observations

The current satellite remote sensing observations permit a comprehensive enough monitoring of volcanic eruptions and their environmental impact. These are not only digitized images in the visible and IR spectral regions regularly transmitted by the meteorological satellites but also various data on trace gases and aerosols in the stratosphere (TOMS and SAGE instrumentation, aerosol optical thickness (AOT), AVHRR data, etc.).

The first eruption cloud $\sim$100 km^2 in size was discovered on 12 June and contained 25 kt (25$10^3$ tonnes) of SO_2. By 13 June, the cloud shifted 1100 km westward (and was located over Vietnam), its mass reaching 110 kt. On June, 14 the cloud could be hardly distinguished over the Indian Ocean (15°N) and then disappeared altogether. On 13 June, a second similar aerosol cloud (9100 km^2, 15 kt) was discovered over the western edge of Luzon Island, but it could be observed only for one day. On 15 June, a third cloud was registered (7500 km^2, 450 kt) which propagated from the west of Mt Pinatubo to South Vietnam (a distance of about 1600 km).

The 15 June cataclysmic eruption started before noon and continued till next morning. This eruption was discovered from the TOMS data analysis on 16 June,

Table 2.13. Comparative characteristics of different volcanic eruptions (TOMS data in 1982 and later)

Eruption characteristics	Nevarupta/Katmai, 1912	ElChichón, 1982	Nevado del Ruiz, 1985	Pinatubo, 1991
SO_2 content (kg)	$(5.2–20) \times 10^3$	7.0×10^3	750	$\times 10^4$
Ash ejections (g)	3.4×10^{16}	3.0×10^{15}	4.8×10^{13}	1.0×10^{16}
Gas/ash ratio	$(2–6) \times 10^{14}$	2.3×10^{-3}	0.015	1.9×10^{-3}
Volcanic explosivity index (VEI)	6	4–5	3	5–6

(the day, when the satellite was over the zone of eruption). The total mass of eruptive SO_2 at that moment was estimated at 15.5 kt (this value might have been underestimated because of a certain instrument malfunction). By June, 23 the eruptive cloud covered an area of 15×10^6 km^2 and stretched at a distance of the order of 104 km, from Indonesia to Central Africa, being fairly homogeneous in space.

Two weeks after the eruption (30 June) the length of the cloud within the latitude band 10°S–20°N reached 16×10^3 km, and the SO_2 mass became 12 Mt. Thus, in a fortnight the cloud contained 60% of the initially ejected SO_2. This gaseous cloud could be observed during the whole period of its motion round the globe (22 days).

Table 2.13 contains comparative data which characterize different eruptions. Since the great power of the Pinatubo eruption is an obvious fact, one might naturally expect serious climatic consequences.

An important characteristic feature of the eruptive aerosol was its rapid propagation across the equator: two weeks after the eruption it reached 10°S, and during the following several weeks it propagated in the longitudinal direction covering the latitudinal band 20°S–30°N. The numerical modelling of the southward movement of the eruptive cloud has led to the conclusion that the movement was governed by the meridional circulation induced by local stratospheric heating and not by the global-scale transport. This heating, not only affecting the horizontal transport but also intensifying the vertical motions, was determined by the absorption of the upward longwave radiation by the newly formed aerosol and manifested itself as a lower tropical stratospheric warming in late 1991, ceasing during 1992 as the eruptive cloud dissipated.

Since a maximum of ozone concentration in the tropics is observed at an altitude of ~25 km, an enhancement of the low-latitude meridional circulation at lower altitudes, where ozone has a long lifetime (during 1–3 months it experiences a gradual photochemical transformation to the background level), caused the transport of the poor-in-ozone air masses to higher latitudes. It is presumably this mechanism that caused the TOC decrease by 6–8% observed over the equator during first several months after the eruption (Kondratyev et al., 1995).

A specific feature of the low-latitude explosive eruptions is a slow transport of the eruption products from the tropical zone to middle and high latitudes. This refers also to the Pinatubo eruption, with the most intensive meridional transport observed in the lower stratosphere. By the end of 1993 the global mass of the stratospheric eruptive cloud had decreased to 5 Tg (the time for e-fold decrease constituted about one year). After the eruption the zonally averaged value of the aerosol optical thickness (AOT) at the wavelength 1 μm gradually decreased from >0.2 by the end of 1991 in the tropics to 0.006–0.03 (in different global regions) in mid-1993 (the background AOT values from the data for 1989–90 varied within 0.001–0.003).

Minnis et al. (1993) have accomplished first direct satellite measurements of the volcanic radiative forcing on a large scale with simultaneous observations of aerosol optical depth.

The processing of the observational data on the ERB components (the ERBE satellite) has shown that in July–September 1992 there was an increase of the Earth albedo compared to 5-year averages. In August 1991 the mean global albedo was 0.250 and higher, which exceeded the background level 0.236 by five standard deviations. The relative albedo increase in cloudless regions, where it was at a maximum, reached 20%. An albedo increase was also observed in regions of thick convective cloudiness, where it could be determined by the impact of eruptive aerosol on the cloud particles size distribution.

The volcanically induced increase of albedo has led to a decrease of the ERB by 8 W/m^2 (the latitudinal band 5°S–5°N) and 4.3 W/m^2 (40°S–40°N), that is, to the cooling effect of the surface–atmosphere system. During the second half of 1991 and most of 1992 this effect exceeded the contribution of the "greenhouse warming" and remained considerable in 1993. With the mid-tropospheric warming due to the 1992 ENSO event excluded, the temperature decrease in 1992 (compared to the pre-eruption conditions) reached 0.7°C.

Aircraft observations

An extensive programme of aircraft observations (the "Electra" flying laboratory) of the plume of eruption products was started by the NASA scientists three weeks after the eruption (Valero and Pilewskie, 1992). Six flights conducted over the period 7–14 June 1991 covered a region: 34°N–4.5°S and 76°W–46°W. Since the flight ceiling of the aircraft was 5.5–7.0 km, this created serious complications in several cases for the remote sounding of the stratosphere throughout the upper troposphere.

The complex of the on-board instrumentation included, in particular, instruments to measure total fluxes of direct solar, diffuse and solar global radiation; a multichannel radiometer (with an oscillating shadow-ring to measure direct solar radiation); a multichannel flux radiometer (MFR). All types of spectral instruments had similar spectral channels (380, 412, 500, 675, 785, 865, and 1064 nm) and were zenith- and nadir-directed. Data of solar radiation measurements were used to calculate the AOT (the Rayleigh component was calculated using observed atmospheric pressure values; the extra-terrestrial spectral distribution of solar radiation was found through an extrapolation of measured values). The errors of irradiance measurements did not exceed 1–2%.

Analysis of observational data revealed the growth of AOT with the distance from the volcano decreasing (15.14°N), but maximum AOT values were obtained, however, in the vicinity of 10°N, which reflects a specific propagation of the eruptive cloud. A maximum of AOT was observed on 13 July at 412 nm and was equal to 0.49. The data of spectral measurements revealed a rapid AOT decrease with the growing wavelength in the near-IR and allowed one to retrieve the particles effective radius as a parameter of the aerosol size distribution. Analysis of relevant data has led to the conclusion about the existence of two types of aerosol: monomodal aerosol (data for 7, 8, 12, and 13 July) and bimodal aerosol (data for 10 and 14 July).

The effective radius r_{eff} of aerosol particles was 0.18 and 0.35 μm (data for 8 July and 13), and the columnar mass was 35 and 80 mg/m^2, respectively. The bimodal aerosol size distribution observed on 10 July was characterized by the r_{eff} values 0.1 and 0.8 m, while the columnar mass reached 40 mg/m^2. Apparently, the bimodal aerosol size distribution resulted from the gas-to-particle conversion processes of the submicrometre sulphuric acid aerosol fraction formation in the presence of large volcanic ash particles. Calculations of the surface–atmosphere system albedo (under various assumptions with respect to aerosol properties) gave a maximum of 0.12 for the surface albedo 0.05, solar height 10° and AOT equal to 0.5.

An aircraft lidar sounding performed in July 1991 (Nd:YAG lidar, 532 nm) revealed a multilayered structure and horizontal inhomogeneity of the aerosol plume in the stratosphere 3–4 weeks after the Pinatubo eruption (Winker and Osborn, 1992). As an indicator of the aerosol content, the scattering ratio was chosen and defined as a ratio of the sum of the aerosol backscattering and Rayleigh scattering coefficients to the coefficient of Rayleigh scattering (SR) (with the reference to the SR value at the tropopause level which was 16 km).

The thickness of the aerosol layers located within the 17–26-km altitude band varied between 1 km and several kilometres. The layers located north of 20°N were relatively weaker (SR values ~3 and smaller), but south of 15°N the SR values, as a rule, exceeded 10. Maximum SR values of the order of 80 and higher were observed on 12–13 July at an altitude of ~25 km, but north of 11°N the powerful aerosol layer suddenly disappeared.

Assessments of the columnar aerosol mass made under the assumption of pure sulphuric acid aerosols gave a value of 95 mg/m^2, which is equivalent to the global aerosol mass 8 Mt (this estimate is rather arbitrary, however). Presumably, by the moment of observations, only about half the ejected SO_2 had been transformed into the sulphuric acid aerosol, that is, the sulphate aerosol mass could have eventually reached ~16 Mt (after the El Chichón eruption the eruptive aerosol mass was ~1 Mt).

The polarization lidar carried by "Electra" produced a linearly polarized signal at the wavelength 532 nm, and the backscattering signal of the stratospheric volcanic aerosol was recorded at two orthogonal polarizations, with subsequent calculations of both the scattering ratios (SR) and the depolarization ratio (DR) defined as a ratio of signals for the perpendicular and parallel polarization components (these were first aircraft DR observations). Lidar sounding data with the 15-m resolution were smoothed to the 150-m resolution.

Analysis of the DR data revealed the presence of aerosol layers with considerably different polarization signatures, which was a result of the variable chemical composition, physical characteristics and particle shape (Winker and Osborn, 1992). The depolarization of backscattered light is known to indicate the presence of scattering by anisotropic particles. Possible causes of depolarization in the case considered could be the presence of crystallized sulphuric acid droplets, aggregates of frozen droplets and silicate particles of irregular shapes. The 12–13 July data demonstrate the presence of a layer with high depolarization near the 25-km altitudes. Apparently, in the latter case it was almost pure sulphuric acid aerosol with a small admixture of ash and other non-spherical particles, whereas in the former case, mixed aerosols were observed. For the reliable interpretation of these data (with due regard to the depolarization effect), *in situ* measurements of the aerosol size distribution, chemical composition and particle shape are needed.

Following Mt Pinatubo's eruption, the aerosol particle mixing ratios and size distributions were measured in the lower stratosphere at altitudes below 21 km from NASA's ER-2 high-altitude aircraft at various times and locations in the northern hemisphere (Jonsson et al., 1996). The ER-2 condensation nuclei counter (CNC) determines the concentration of particles larger than 0.008 m in diameter with an accuracy of a few per cent. A focused cavity aerosol spectrometer (FCAS) measures the size distribution of particles in the 0.06–$\sim$2 μm diameter range, producing a particle size spectrum every 10 s of sampling time. A forward scattering spectrometer probe (FSSP-300), which detects particles between 0.4 and 20 μm in diameter, was also carried by the ER-2 aircraft during the 1991–92 and 1993 flight series.

Regular ER-2 aerosol measurements allowed one to monitor the evolution and purge of the volcanic aerosol due to an interaction of aerosol mechanics with tropospheric–stratospheric exchange processes, transport, and mixing. Jónsson et al. (1996) have discovered that during first 5 months after the eruption the volcanic plume spread to higher latitudes in laminae and filaments, being inhomogeneous and producing steep spatial gradients in the properties of the stratospheric aerosol. At the same time, the concentration of the newly formed particles in the plume rapidly decreased towards background values as a result of coagulation, while the particle size and the aerosol surface area continued to increase. By December 1991 the particle number mixing ratio and the aerosol surface area mixing ratio had become spatially uniform over a wide range of latitudes above 18 km. The surface area mixing ratio peaked in this region of the stratosphere exceeding $\sim$35 times their background values in the winter of 1992. The corresponding condensed mass mixing ratio enhancement was by a factor of $\sim$200.

Jonsson et al. (1996) have pointed out that after the winter of 1992 a gradual removal of the volcanic mass began, and initially it was dominated by sedimentation above 18 km. The aerosol surface area mixing ratio thus decreased by an order of magnitude over 2.5 years, and the aerosol volume, or condensed mass, mixing ratio decayed by an order of magnitude over approximately 1.7 years. Below 18 km, the purging of the Pinatubo aerosol at mid-latitudes appeared sporadic and disorderly and was strongly influenced by episodic rapid quasi-isentropic transport and dilution by the tropical air of tropospheric origin having high condensation nuclei mixing ratios but low mixing ratios of aerosol surface area or condensed mass, compared to the volcanic aerosol.

Baumgardner et al. (1996) have flown a new instrument simultaneously measuring the aerosol diameter 0.4 to 10 μm and the refractive index between 1.30 and 1.60 on the NASA ER-2 aircraft. The observations during the time period from 18 March to 4 November 1994 (45 flights) were mainly made (in connection with studies of the evolution of the ozone depletion in the Antarctic and the subsequent transport of the ozone-depleted air masses from the polar vortex into mid-latitudinal and tropical regions) between 45° and 70°S during the flights south of Christchurch (New Zealand). Additional measurements were made in the tropics. The results show that average stratospheric aerosol refractive indices varied from 1.40 to 1.42 over the latitude range from 70°S to 50°N, and from 1.34 to 1.46 over the vertical range 4–20 km. Baumgardner et al. (1996) have emphasized that the measured stratospheric refractive indices do not agree well with theoretical predictions, and the vertical profiles suggest the presence of non-spherical or absorbing particles in the altitude range 7-9 km. As has been shown earlier by Chlek et al. (1995a), particles with even small soot inclusions can be highly absorbing. Blake and Kato (1995) have discovered heavy soot concentrations at commercial airliners altitudes (9–11 km) and mostly in the northern hemisphere, extending from 10° to 60°N.

A new aspect of the impact of stratospheric aerosols on climate through particle formation in the upper tropical troposphere has been discovered by Brock et al. (1995), who conducted atmospheric measurements and numerical calculations which indicate that binary homogeneous nucleation of H_2SO_4–H_2O particles occurs in the upper tropical troposphere. Such a mechanism demonstrates the presence of particles outside the stratosphere that provide the nuclei to maintain the Junge layer during the volcanically quiescent periods.

Brock et al. (1995) have accomplished aircraft observations (NASA U-2 and ER-2 aircraft reached altitudes of ~21 km) in mid-latitudes, polar regions, and tropics in both the hemispheres between 1987 and 1994. The principal instrumentation was the condensation nuclei counters (CNCs) measuring the number mixing ratio (NMR) of particles in the diameter range 0.008 to ~3 μm, which is broad enough to determine the total particle NMR. In 1993 and 1994, a second CNC measured the non-volatile particle NMR by heating the aerosol stream to 192°C before counting. Only those data were considered which were not affected by volcanic eruptions and by the particles formed in the polar vortices. Measurements made in the fall, winter, and spring dominate the extra-tropical data.

Two pieces of evidence suggest that the particles observed near the tropical tropopause are composed of H_2SO_4–H_2O droplets: (1) more than 95% of the particles were volatile at 192°C; (2) the electron microscope analysis data (size distribution was nearly monodisperse with diameters of ~0.015 μm; sulphur was the only component significantly above background). The numerical modelling results confirm the sulphuric acid nature of aerosols. The new H_2SO_4–H_2O particles are formed by a homogeneous binary gas-to-particle conversion in the upper tropical troposphere. This maximum in the particle NMR lies in a region of net upward mass transport of air through the tropical tropopause and into the middle stratosphere.

Analysis of the observation data made by Brock et al. (1995) indicates that particle concentrations decrease with the altitude increasing above the tropical tropopause as a result of coagulation during the upward air transport produced by stratospheric circulations. A maximum of NMR at the tropopause is about seven times larger inside the tropics than outside (on two occasions, high particle NMRs were found in the upper troposphere outside the tropics; thus the processes that lead to the high particle NMRs are not unique to the tropics but are more common and widespread there). During the extended quiescent time periods, particles formed in the upper tropical troposphere provide nuclei upon which oxidized sulphur gases condense in the stratosphere. Brock et al. (1995) have emphasized that this particle source, coupled with aerosol microphysical properties and atmospheric transport, governs the number concentration of particles in the lower tropical and mid-latitude stratosphere. The observations indicate that particles produced in the upper tropical troposphere are frequently transported to the lower mid-latitude stratosphere, where they may significantly affect the particle number and surface area mixing ratios.

Balloon observations

Balloon observations have been very important for studying the consequences of the Pinatubo volcano eruption. During the period 17 July–28 August seven dust-sonde flights were accomplished by the University of Wyoming experts from Laramie (41°N). These launchings were made both before and after the arrival of the volcanic aerosol cloud to the region of Laramie. Apart from standard instruments to measure the atmospheric pressure, temperature and ozone concentration, the dust-sonde (DS) instrumentation complex included the following basic sensors: condensation nuclei (CN) counter; photoelectric counter (the particle radius $r > 0.01$ μm); fast-response 8-channel aerosol particle photoelectric counter ($r > 0.15$, 0.25, 0.50, 1.0, 2.0, 3.0, and 10.0 μm), as well as special two-channel ($r > 0.15$ and 0.25 μm) particle counter to investigate the particle volatility. For a number of flights, a three-stage cascade impactor was used to sample aerosol for subsequent electron-microscopic analysis in laboratory conditions. The maximum flight altitude varied within 35.5–38.8 km.

Deshler et al. (1992, 1993) accomplished an analysis of all measured CN vertical profiles which revealed that the first arrival of the eruptive cloud took place on 16 July (apparently, the oil-field fires in Kuwait contributed to the increase of the CN concentration at the tropopause level). By 26 July the aerosol number density below the level of westerly transport in the upper stratosphere reached a record level of about 50 cm^{-3} at the altitude 17 km. This stable lower aerosol layer due to a rapid transport of particles to northern latitudes was also revealed during the subsequence DS flights.

First evidence of the layer above 20 km (within the westerlies) was observed as a thin layer at the 23-km altitude (this layer was also identified from the lidar sounding data in Boulder, Colorado, during the time period 29 July–2 August. By August, the layer had ceased to be visible but appeared again over Laramie on 28 August with a lower aerosol concentration after the eruptive cloud made a full revolution round the Earth. The end of August was characterized by the presence of an extended (vertically) aerosol layer at altitudes from the tropopause level to 22 km, and mid-September it had reached the altitude 30 km.

The CN concentration at 15–20-km altitudes was so high (due to gas-to-particle conversion processes of the formation of sulphate aerosol) that the CN counter saturation threshold was overcome. This meant that the CN concentration exceeded 500 cm^{-3}, that is, was approximately two orders of magnitude larger than the background values for latitudes between 18 and 20 km.

Experiments with heating the aerosol particles to 150°C revealed that 95–98% of aerosols were volatile, which makes it possible to identify the aerosol particles as high-concentration droplets of sulphuric acid water solution and to estimate the characteristic time of the SO_2 to H_2SO_4 transformation at 38 days (earlier estimates for smoke aerosols in Kuwait have given 50% of volatile aerosols).

The aerosol size distribution was characterized by the presence of large particles above the level of wind direction reversal in the stratosphere (at altitudes 22–23 km compared to 16–17 km). The presence of aerosol particles with a modal radius of $\sim$0.07 μm was typical of altitudes below 20 km, whereas above this altitude the radius increased to 0.35 μm (in case of single-modal lognormal size distribution). In the lower aerosol layer and at the edge of the upper layer, the aerosol size distribution was initially bimodal (data for 2 August). On 30 July at an altitude of 23 km the measured values of mass concentration, mass mixing ratio and total surface area of particles were, respectively, 28 $\mu g/m^3$, 0.48 ppm, and 84 $\mu m^2/cm^3$, that is, much higher than similar parameters measured after the El Chichón eruption. Since the same observations revealed a high concentration of small-size sulphuric acid droplets, one may assume that such droplets have been produced either in the process of homogeneous nucleation or from the condensation of small ions.

Surface observations

The results of surface observations have been a very important component of the whole measurement data set (Di Girolamo et al., 1994; Post et al., 1992; Uchino et al., 1993). Thus, for instance, lidar soundings at the Mauna Loa Observatory (the Hawaii: 19.53°N and 155.58°W, altitude 3400 m, a.s.l.) detected on 1 July 1991 the arrival of the post-Pinatubo volcanic plume. In this context, Goldman et al. (1992) used solar spectra measured with a Fourier spectrometer to retrieve the SO_2 columnar content in the atmosphere (Table 2.14). The retrieval has been accomplished via intercomparison of measured and calculated absorption spectra for the ν_3 absorption band of SO_2 centred at 1362.05 cm^{-1} (the stronger ν_1 band at 1157.71 cm^{-1} could not be used because of a heavy overlapping with the H_2O and CH_4 bands).

It can be seen from Table 2.14 that in August the SO_2 content decreased to the sensitivity threshold of the spectrometer ($\sim 1 \times 10^{13}$ molecules/cm^2). This may be interpreted as evidence that by that time the whole amount of volcanic SO_2 had practically transformed into sulphuric acid aerosol.

Naturally, the most widely applied observational technique to monitor the volcanic aerosol dynamics was the lidar remote sounding. twenty days after the Pinatubo eruption a volcanic aerosol cloud was detected by the lidar at $\sim$14 km from Frascati, Italy (Gobbi et al., 1992). Three-and-a-half months after the eruption the Pinatubo-induced aerosol perturbation reached and then exceeded the maximum loads recorded 11 months after the El Chichón eruption.

Table 2.14. Summary of data on the SO_2 columnar content

Date	Spectral resolution (cm^{-1})	SO_2 content (10^{16} molecules /cm^2)
11 May	0.004	<0.9
9 July	0.010	(5.1 ± 0.5)
10–24 August	0.004	<0.9

Jäger (1992) analysed the results of lidar soundings (Nd:YAG laser) for Garmisch-Partenkirchen, Germany (47.6°N and 11.1°E) and discovered the first arrival of the volcanic plume to this point on 1 July 1991, when backscattering (BS) exceeded the background level; however, such an increase was not observed on the following day. During several months, quite variable BS vertical profiles were recorded. The layered structure of the aerosol vertical distribution may be accounted for by specific features of the general atmospheric circulation. The aerosol concentrated in the 7–15-km layer in early July was moving along the shortest trajectory. The double-layered aerosol distribution observed in August agrees well with the presence of typical (of mid-latitudes) west-to-east transport at altitudes below 20 km and wind direction reversal above 20 km. After transition to a typical winter circulation regime (beginning with September), with the prevailing west-to-east transport through the volcanic aerosol layer thickness, a single powerful aerosol layer was formed between the troposphere level and the 28-km altitude, with an enhanced northward transport of volcanic ejecta (this is similar to what was observed after the El Chichón eruption (Galindo, 1992)).

Lidar aerosol soundings of the troposphere and stratosphere at three locations around Boulder, Colorado, were performed making use of three lidars at wavelengths: 10.591 m (a CO_2 laser); 0.694 m (a ruby laser); 0.574 m (Post et al., 1992). The first arrival of the volcanic plume was observed on 27 July (it is quite probable that the stable stratus cloudiness that prevailed during the previous 5 days prevented volcanic aerosols from being detected). In 1982, the volcanic plume in Boulder was first identified only 4 months after the El Chichón eruption, whereas the Pinatubo ejecta reached the same point in 1.4 months, despite a far greater distance between the volcano and the observation point. Doubtless, such differences may be explained by specific features of stratospheric circulation, as well as (or) volcanic emissions. One more difference consisted in that soon after the Pinatubo eruption an enhancement of backscattering by the atmosphere was observed, while in the El Chichón case, a similar phenomenon occurred only 8 months after the eruption.

An extensive programme of lidar aerosol soundings after the Mt Pinatubo eruption has been accomplished in Japan (Uchino et al., 1993; Nagai et al., 1993). Uchino et al. (1993) have discussed the results obtained by a lidar network of stations expanding from Naha (26.2°N) to Wakkani (45.4°N) during the time period from June to October 1991 (most of the lidars are Nd:YAG laser second harmonic generation system with the wavelength 532 nm, although 589-nm dye lasers (two stations) and 1064-nm Nd:YAG fundamental (two stations) were also used). The first arrival of the Pinatubo aerosol was observed on 28 June 1991, at approximately the 16-km level. The layer of volcanic aerosol was observed till the end of the period of observations. The upper layer located above 20 km was first observed on 15 July then it disappeared at all stations to re-appear again on 6 August. After the El Chichón eruption in 1982, two main layers were observed at heights of about 18 km and above 20 km for months after the major eruption.

Nagai et al. (1993) have discussed the results of lidar soundings (Nd:YAG laser; 532 nm) at Naha (26.2°N, 127.7°E) on Okinawa Island. Volcanic aerosols had already reached Okinawa by mid-September 1991, when the observations started. The aerosol layer was located between the tropopause and the 30-km level, but it rised higher (up to 33 km) in mid-November when the total amount of aerosol increased and the maximum aerosol backscattering ratio reached 23.1 at the 25.8-km altitude on 4 December 1991. The depolarization ratio observations demonstrated the presence of not only volcanic ash particles but also spherical particles, probably, sulphate droplets, from the very beginning of observations.The depolarization ratio decreased with altitude and reached a peak value of 0.22 at about 16.4 km on 18 November 1991.

Avdyushin et al. (1993) discussed lidar aerosol soundings performed between July 1991 and April 1992, with a lidar network which consisted of five Rayleigh–Mie lidars operating in the 532–589-nm wavelength range. Two Russian lidar stations operated by the Roskomgidromet were on-board two ships: "Professor Zubov" (a double-yttrium–aluminium laser, 539.5 nm) and "Professor Wiese" (a dye laser RGW, 589 nm). Three French lidar stations operated by the CNRS were located at the Observatory of Haute-Provence (OHP: 44°N, 6°E), at the Centre d'Essai des Landes et Biscarosse (CEL: 44°N, 1°W) and on board a military ship "Henri Poincaré" (LIMA). The observations covered the western sector of the northern hemisphere (8°S–60°N, 80°W–6°E).

The data obtained show a great deal of variability in the structure of the post-Pinatubo volcanic aerosol cloud. The aerosol extinction coefficient (AEC) data indicate the presence of aerosols primarily in the altitude range 17–25 km, although occasionally they were measured as high as at 32 km. Comparison of the generally low-latitude measurements on "Zubov" and "Wiese" with those performed at mid-latitudes (OHP, CEL, LIMA) shows the presence at low latitudes of a powerful high-altitude cloud at 17 km, whereas at mid-latitudes only a lower-altitude cloud is persistently present near and below 20 km.

Post et al. (1992) have accomplished analyses of a dense set of lidar observations of the volcanic clouds from the Mt Pinatubo and El Chichón eruptions, primarily over Boulder, Colorado (USA). Processing of the CO_2 lidar ($\lambda = 10.59$ µm) data (the ruby lidar, $\lambda = 0.694$ µm, was also used) has revealed noticeable similarities and differences. Post et al. (1992) confirmed the importance of the tropopause folding events for purging stratospheric debris into the troposphere, as well as the role of gravitational settling to bring debris into the grasp of the folding mechanism. The more important role (in comparison with the post-eruption time) both in the poleward transport and in purging stratospheric material downward belongs to seasonal influences. The lengthening of the decay process of the Mt Pinatubo

Table 2.15. Sulphur dioxide loading of the stratosphere by volcanic dust veils with aerosol masses of at least 4 Tg (values of the volcanic explosivity index VEI are given in brackets)

Period	Mass of SO_2, (Tg)	Volcanic eruptions (VEI)
1881–1894	28	Krakatau (6), unidentified (?)
1895–1908	19	Soufrière (4), Santa Maria (6), Ksudach (5)
1909–1922	5	Katmai (6)
1923–1936	2	Komagatake (4)
1937–1950	0	
1951–1964	10	Agung (4)
1965–1978	0	
1979–1992	22	El Chichón (5), Pinatubo (6)

eruptive clouds could take place due to either an equatorial reservoir of debris or the eruptions of Mount Spurr.

Stothers (1996) has recently accomplished an analysis of pyrheliometric and some other available observations for the time period 1881–1992 to assess major optical depth perturbations from volcanic eruptions as a function of year, month, and latitude band. Global stratospheric aerosol loadings have been computed from the peak turbidities. Table 2.15 contains some of the results of these computations.

As can be seen, the most recent interval is characterized by an unusually high level of volcanic activity similar to what had been observed at the end of the previous century. The average annual production rate was 0.8 Tg/yr, just half the average rate in the 1972–92 interval. During the whole 1881–1992 period the contribution of the two short intervals 1883–1902 and 1982–91 in the generation of all stratospheric aerosol reached $\sim 80\%$.

Sato et al. (1993) have analysed mean monthly aerosol optical thickness (AOT) perturbations due to volcanic eruption for the 1850–1990 period.

Herber et al. (1996) have processed the long-term measurements by sun-photometers of the spectral dependence of AOT for several sites in the Arctic and Antarctic for the time period January 1991 through December 1994. In the Antarctic a pronounced decrease of atmospheric turbidity was observed in late September 1991 due to the Cierro Hudson eruption in August 1991, and a second increase (an order of magnitude above background values) was registered due to the influx of the Mt Pinatubo aerosol. A similar AOT perturbation was observed in the Arctic in March 1992 (satellite observations had already revealed this Pinatubo aerosol at high northern latitudes at the end of 1991). By the end of 1994 in both polar regions a significant increase of the Pinatubo-induced perturbations was observed.

2.4.3.3 Climatic impact

The climatic implications of volcanic eruptions have recently become of great concern. To study this problem, a broad complex of observations of volcanic ejecta and their propagation in the atmosphere has been carried out making use of both the conventional means (ground-based, aircraft, balloon) and satellites. Considerable progress has been achieved in the development of techniques for the numerical modelling of climate and large-scale transport (Fiocco et al., 1996; Kolomeev et al., 1993; Kondratyev, 1984, 1988, 1996a).

Of principal interest are attempts to reveal the volcanic signal in observed climate changes. The solution of the problem is very difficult because the inter-annual variability of surface air temperature (SAT) is determined by various processes which may be connected with external forcings, internal regular oscillations and random processes, largely due to dynamic instability of the atmosphere. The most probable external factors are large-scale eruptions and increased carbon dioxide concentrations in the atmosphere (so far, there has not been convincing enough evidence in favour of the effect of solar activity on climate). Regular changes in the atmospheric general circulation (AGC) in the North American region have been related to ENSO. A considerable contribution to atmospheric variability is made by its internal instability. With the varying intensity and location of cyclones and jet streams, a random variability of heat fluxes occurs in the atmosphere, leading to a random variability in air temperature.

It is a well-known fact that the most definite climatic impact of volcanic eruptions is the stratospheric warming (Angell, 1993; Kondratyev, 1984, 1988, 1993b; Kondratyev and Galindo, 1997). In this connection, Labitzke and McCormick (1992) performed a comparison of aerological data on stratospheric temperatures in the northern hemisphere for the 30 and 50 hPa levels, starting from June, 1991, with monthly mean temperatures for 20 years (1965–84) and 26 years (1964–89). This comparison resulted in a conclusion about significant temperature rises in July, August, September, and October 1991, within the latitude band 0°–30°N.

Angell (1993) compared the low-stratospheric warmings (100–50 hPa or the 16–20-km layer) following the Agung (1993, 8°S), El Chichón (1982, 17°N) and Pinatubo (1991, 15°N) eruptions, on the basis of a 63-station radiosonde network and using the nine-season-averaged temperatures to minimize the influence of the quasi-biennial oscillation (QBO). Some earlier investigators used a post-eruption peak temperature as a measure of stratospheric warming. Angell (1993), however, has noted that the peak temperature is sensitive to the phase of the QBO in which the eruption occurs.

A recognition of the volcanic climatic signal on the basis of observations is not a simple problem, because the eruptions often happen during the El Niño/ Southern Oscillation (ENSO) events.

Portman and Gutzler (1996) have applied the superposed epoch method to search for volcanic and ENSO signals in the US surface climate records. For this purpose, anomalies of monthly mean SAT and total precipitation (TP) taken from the US Historical Climatology Network were composed (averaged) over years of major explosive volcanic eruptions, ENSO warm events, and ENSO cold events since the year 1900. It was assumed that volcanic eruptions and ENSO events occur independently. After assessing for significance of all composite anomalies with regard to several statistical and physical criteria, the composite ENSO-related anomalies were subtracted from the anomalies of temperature and precipitation associated with the volcanic eruption.

Table 2.16. Approaches and results of some recent investigations of volcanic eruptions and the United States temperature deviations

	Groisman (1985, 1992)	Lough and Fritts (1987)	Robock and Mao (1992, 1995)	Portman and Gutzler (1996)
Was any ENSO signal removed?	No	No	Yes	Yes
Shortest averaging interval	Season	Season	Month	Month
Earliest–latest eruption years	1815–1982	1631–1888	1875–1991	1902–1982
Number of eruption years	9–10	26	12	10
Summary of major results	Cool in summer, US East Coast	Warm (especially in winter), western US; cool (especially in summer), central and eastern US	Warm in first winter after tropical eruption (below 30° latitude) and in first or second winter after another eruption, continental US	Warm in first autumn and winter, eastern US; cool in first winter and spring, western US; cool in summer, southern US

After filtering out the ENSO-related anomalies it has been found out that volcanic signals are strongly pronounced east of the Continental Divide, for example, where positive monthly temperature anomalies exceeding 1°C occur during the first autumn and winter after the eruption. As Portman and Gutzler (1996) find, negative temperature anomalies occur west of the Continental Divide during the first winter and spring after the eruption and in the southern United States during the summer of the first post-eruption calendar year.

Positive monthly precipitation anomalies exceeding 15 mm in magnitude are found in the south-eastern United States during the first winter and spring after the eruption. Precipitation anomalies smaller in magnitude and yet significant, such as positive anomalies in the north-western United States, are found during the summer of the first post-eruption calendar year.

Table 2.16 data represent a partial summary of some recent studies of the impact of volcanic eruptions on SAT changes in the USA. An important difference between various sets of data is the removal of ENSO signals. The data under consideration demonstrate clearly the importance of filtering out of the ENSO signals.

Since in the study by Portman and Gutzler (1996) the composite variations in SAT anomalies associated with the largest volcanic eruptions closely resemble the composite variations associated with the ENSO cold events, the removal of signals associated with the coincident ENSO events (all are warm) was found out to strengthen rather than weaken the apparent volcanic signals over many grid regions.

Portman and Gutzler (1996) have emphasized difficulties in the characterizing of the volcanic aerosol forcing due to numerous variable parameters, such as eruption latitude, time of the year, plume height, and the amount of sulphur produced. Large variations of these parameters from one eruption to another, plus a small number of major volcanic eruptions during modern times seriously complicate an empirical analysis of observational data and illustrate the importance of the relevant climate modelling in the form of sensitivity calculations.

A most complete analysis of the volcanic climatic impact (with regard to SAT variations) has been conducted by Robock and Mao (1995) who examined the regional and seasonal patterns of SAT effects of large volcanic eruptions using the climate records of the past 140 years. This study was focused on the short-term (a few years) effect of the historic major volcanic eruptions, even though a long-term effect is possible, due to long-term scales of the oceanic response for a few post-eruption decades.

Robock and Mao (1995) have persuasively demonstrated that the major explosive volcanic eruptions produce clear temperature effects in different seasons and locations. An important aspect of data processing has been filtering out of the low-frequency variations and the ENSO signal. In general, the volcanic effects produce a cooling, but they give a winter warming signal over the high-latitude NH continents. For 2 years following powerful volcanic eruptions the surface cools significantly by 0.1–0.2°C (in the global mean) in each hemisphere, in the summer of the latitude band 0–30°S, and by 0.2°C in the summer in the latitude band 30–60°N. On the contrary, in the first winter after major tropical eruptions and in the second winter after major high-latitude eruptions, North America and Eurasia get warmed by several degrees, whereas northern Africa and south-western Asia get cooled by more than 0.5°C.

In the cases of simultaneous ENSO events, the ENSO-produced warming masked the volcanic cooling during the first post-eruption year. The time scale of the ENSO response is only one year while the volcanic response time scale is 2 years, so the cooling in the second year is evident whether the ENSO signal is removed or not.

The results mentioned, both the global cooling and the NH continental winter warming, agree with general circulation model calculations. Robock and Mao (1995) have emphasized that several issues involving volcanoes and climate remain to be investigated, including the possible cause-and-effect relationship with the ENSO. More efforts are needed to more reliably separate the ENSO and volcanic signals, in particular, over North America. It is still not quite clear, how much of the winter warming over North America is volcanic and how much is due to ENSO.

An important task is the numerical modelling of the volcanic climatic impact. Since the simulation modelling of the volcanic climatic impact has been mostly concentrated on assessments of a long-term impact of stratospheric sulphur acid aerosol resulting from gas-to-particle volcanic SO_2 conversion, Gerstell et al. (1995) have considered the radiative and dynamical perturbations associated with the short-lived but more strongly absorbing sulphur dioxide and ash clouds. Using radiative transfer model calculations as well as observational data from satellites, aircraft, and ground-based instrumentation, Gerstell et al. (1995) have estimated the amplitudes of the stratospheric radiative heating rate perturbations produced by each of these components during first few weeks after the El Chichón eruption, before the sulphur acid aerosol cloud fully developed.

A week after the 12 April 1982 eruption the net radiative heating rate perturbations exceeding 20 K per day were found at altitudes near 26 km, which were mainly due to solar radiation absorption by silicate ashes. The sulphur dioxide gas and sulphuric acid aerosols each produced net heating perturbations that never exceeded 3 K per day. Observations indicate, however, that in spite of the strong heating due to ash absorption, stratospheric temperatures never increased by more than a few degrees Kelvin. Gerstell et al. (1995) show that the radiative heating was largely balanced by upwelling and adiabatic cooling. The results considered demonstrate a necessity to take into account the radiative forcing by ashes and by the sulphur dioxide gas, in particular, in connection with its influence on the vertical and horizontal dispersal of the volcanic plume.

Of great importance for assessing the volcanic impact on climate is the simulation modelling of the global distribution of sulphate aerosol produced by a powerful volcanic eruption. Besides, volcanic eruptions play an important role in the global sulphate cycle of the Earth's atmosphere. A very complete modelling of this type has been accomplished by Pudykiewicz and Dastoor (1995). The methodology of this study (as applied to the case of the June 1991 Pinatubo eruptions) is based on the use of the dynamic global spectral model of the atmosphere coupled with a system of mass conservation equations for atmospheric trace species. The chemistry of SO_2 oxidation in dry and wet phases (the process of gas-to-particle conversion) and the emission field are represented by the right-hand sides of the tracer equations. Twenty-eight levels are used in the terrain-following coordinate system with the model top at 0.5 hPa.

An important feature of the dynamic model is an introduction of a relatively sophisticated parameterization of stable and convective clouds, which is crucial for the study because the interaction between clouds and aerosols is one of the most important mechanisms affecting the global distribution of atmospheric tracers. The verification of the model results has been accomplished through an intercomparison of calculated and observed aerosol optical thicknesses for the case of the Pinatubo eruption. The results of this intercomparison were positive.

The Pinatubo eruption case was simulated for a prescribed virtual source of SO_2 emission defined by a Gaussian function. Relevant assessments have shown that only $\sim 7\%$ of the total mass of sulphur dioxide were injected into the troposphere. This is why it was possible to neglect the liquid-phase oxidation of SO_2 to SO_4, which, in general, is thought to be far more important than the dry-phase oxidation. The latter has been simulated by the following reaction:

$$SO_2 + OH\ (H_2O,\ O) \xrightarrow{M} SO_4^{-2} + HO_2$$

The simplified chemistry scheme of gas-to-particle conversion includes only SO_2 and two forms of sulphates, one in the form of aerosol suspended in the atmosphere and the other dissolved in clouds. The tracer propagation and transformation equation in the operator form are as follows:

$$\frac{\partial \psi i}{\partial t} = \left(\sum_{j=1}^{3} L^i_j\right) \psi^i$$

where L^i_j are the operators representing advection, diffusion, and chemical reactions, and ψ^i is the vector of the mixing ratios of chemicals.

The last substep of modelling related to scavenging and chemical reactions is accomplished using either the analytical solution of the linear chemistry equation or numerical methods for stiff ordinary differential equations if the chemistry scheme is nonlinear. In this case the set of equations describing an oxidation of SO_2 could be solved analytically.

The model was integrated by Pudykiewicz and Dastoor (1995) using the initial conditions provided by an objective analysis valid at 0000UTC 12 June 1991, slightly before the first eruption, which was observed at around 1200UTC. The release of SO_2 from Mt Pinatubo took place as a series of massive eruptions, with the most powerful one occurring on 15 June 1991. The height reached by debris was $\sim$30 km, with some isolated towers of material observed even at an altitude of 40 km.

Model integration with the sulphur chemistry scheme was carried out for 50 days. The results show that following the atmospheric injection of about 20 million tons of SO_2, the sulphuric cloud was transported on a global scale. The predominant direction of transport of the material ejected above the tropopause was governed by stratospheric easterlies in the northern hemisphere and by stratospheric westerlies in the southern hemisphere. A fraction of material injected below the tropopause was driven mostly by tropospheric westerlies.

Pudykiewicz and Dastoor (1995) have emphasized that the methodology they developed could be used not only in further studies of the propagation of volcanic products and their climatic impact, but also for the numerical simulation of the interaction of stratospheric ozone with the sulphate aerosol layer.

Eluszkiewicz et al. (1996) have accomplished very detailed calculations of radiative transfer in the atmosphere. Using the multilevel δ-Eddington/adding method they considered a 61-level atmosphere between the ground and 0.01 hPa level, taking into account practically all importaant optically active components of the atmosphere: in the case of infrared radiation transfer, gases, such as H_2O, CO_2, O_3, N_2O, CH_4, O_2, NO_2, as well as clouds. An assessment of the radiation effects of the Pinatubo eruption indicates that the effect of volcanic aerosols rapidly diminished in time: already by July 1992 there was no aerosol heating in high northern latitudes and the peak tropical heating rates in the stratosphere were halved compared to those in January 1992. The radiative effects of the eruption are confined to pressures greater than 10 hPa with the altitude of maximum heating descending with time due to droplet sedimentation.

On the basis of a 3-D global climate model developed at the Goddard Institute for Space Studies (GISS), Hansen et al. (1992) made preliminary estimations of the climatic impacts of volcanic eruptions. For simulation modelling, three scenarios of volcanically induced stratospheric AOT have been prescribed:

(1) **El**: Aerosol properties correspond to the conditions of the El Chichón eruption (aerosol consists of sulphuric acid droplets); AOT decreases from 1 immediately after the eruption to zero 6 months later; volcanic aerosols are initially distributed homogeneously in the range 0–30°N (latitude belt), then they propagate over the globe with AOTs twice as high in the 30–90°N zone , compared to the 30°N–90°S zone.
(2) **2 × El**: a doubled AOT is prescribed , compared to **El**.
(3) **P**: The same temporal AOT variation as in the **El** and **2 × El** models (an exponential decrease during 10 months following the eruption, with the time constant 1 year), but AOT is 1.7 times greater than for **El**, and the AOT global distribution is more realistic than before (the hemispherically homogeneous aerosol distribution is reached by January 1992).

The results of numerical modelling show that in the case of scenario **P**, the global mean ERB change due to the Pinatubo eruption might have reached $\sim$4 W/m^2 by the beginning of 1992. This implies that the volcanically induced climate cooling might have exceeded the greenhouse warming due to an accumulation of all greenhouse gases since the beginning of the industrial revolution. The decrease of global mean SAT could have reached 0.5°C by the end of 1992, being three times greater than MSD for the global mean SAT.

Calculations of SAT changes for scenarios **El** and **2 × El** have demonstrated that in these cases the volcanically induced climate cooling was too low to change the 1991 climatological conditions (one of the warmest years in the current century) because of the gigantic thermal inertia of climatic system and the low level of initial forcing. However, in the case of scenario **2 × El** (just as in the case of scenario **P**), the SAT decrease reached 0.5°C; therefore, if this estimate is realistic, the predicted climate cooling must be much stronger than the 0.2°C warming typical of the ENSO impact.

Although the assessment of future regional-scale climate changes is not reliable enough, one can see that an account of the volcanic impact on climate brings about essential differences in estimates. For example, with the volcanic impact neglected, one obtains a decrease from 33% to 15–20% for the probability of unusually cold winters in Moscow (Russia) these years, compared to the 1950s; with the volcanic forcing taken into account, one obtains an increase up to 30–50%. With regard to the level of confidence in the numerical climate modelling results, it is limited by the uncertainties involved in the GISS model (one of which, for example, is a neglect of the interaction with deep layers of the ocean and the thermal inertia of the climatic system). A parameterization of the cloudiness–radiation interaction is equally inadequate.

There were suggestions that volcanic eruptions can trigger the development of El Niños, but Robock et al. (1995) have shown, for instance, that the strong 1982–83 El Niño and 1982 El Chichón was a coincidence.

2.4.3.4 Conclusion

After three powerful explosive volcanic eruptions which happened within approximately one decade (St Helens, 1980; El Chichón, 1982; Pinatubo, 1991), the development of new ground-based, aircraft, balloon and satellite means of observations has progressed substantially. The same refers to the development of new numerical models and to the level of understanding of what happens in the atmosphere under the influence of volcanic eruptions. This is especially true for the Pinatubo eruption case, when it became possible to use a unique observing system and new 3-D global models of the transport of ejecta and the climate change. The problem, however, is far from being solved.

The principal conclusion is perhaps as simple as that: the problem has proved to be far more complex than it was believed before. A persuasive proof of this simple truth is a discovery of an interaction between volcanic aerosols and atmospheric (both stratospheric and tropospheric) ozone. It has become clear that it is not only an impact of eruptions on ozone which matters, but also an effect of ozone changes on the climatic consequences of the eruptions (stratospheric aerosol–ozone interactions will be discussed in more detail later on). This implies that the accounting for the volcanic impact requires the development of new global climate models with coupled dynamics, radiation, and photochemistry, as well as an adequate simulation of the interaction between the stratosphere and troposphere. To verify the new modelling results, far more comprehensive observations are needed. Hence, a related international programme of both the observations and modelling is needed, since only a combination of these may lead to the solution of numerous existing uncertainties.

2.4.4 The atmosphere–ocean interaction

From the viewpoint of assessing the anthropogenic impacts on climate, of key importance is the numerical modelling of the atmosphere–ocean interactive system aimed at analysis of the processes determining the slowed-up response of the climate system to external forcings (Marchuk, 1975; Marchuk et al., 1986). Even with only an interaction of the atmosphere and the oceanic mixed layer 100 m thick taken into account the time of heating is about 100 years. A consideration of circulation taking place in deep waters of the ocean (in this connection of primary importance is the input of heat to the deep waters of the high-latitude North Atlantic and in the Antarctic) gave a characteristic time of several hundred years to reach an equilibrium heating.

Most powerful mechanisms for the transformation and re-distribution of energy are functioning in the ocean. Having a low albedo and, respectively, absorbing most of the solar radiation reaching the Earth's surface, the ocean re-distributes this energy via the currents. The energy then goes to the atmosphere where it is realized mainly as the latent heat of condensation. The oceanic currents are determined by the momentum, heat and water exchange between the ocean and the atmosphere. The currents have a complicated spatial structure determined by the spatial distribution of winds as well as by the distribution of the continents and the underwater relief.

The important role of the global water cycle in the formation of climate and its changes is not in doubt. This also explains the great interest in the problem of the atmosphere–ocean interaction and the effect of the oceanic processes on climate changes. In the past, outbreaks of freshwater to the North Atlantic could have supposedly stopped or, at least, attenuated the thermohaline circulation in the North Atlantic, which has led to changes in local and global climates (Tang and Weaver, 1985). A supposition was made that the glaciers of Greenland, Europe and Laurentides had served as the sources of freshwater, the inflow of which had broken the thermohaline circulation.

The recent analysis of the Greenland ice cores and other palaeoclimatic evidence showed, however, that the temperature fluctuations in the Eemian interglacial period could have also happened in the absence of massive ice sheets. Even in the last glaciation period the inflow of freshwater from the glaciers might not have been the major cause of sharp changes in the North Atlantic thermohaline circulation. The coupled modelling of climate suggested the conclusion that an alternative source of freshwater is its transport in the atmosphere from low latitudes.

So far, a reliable simulation of freshwater transport in the atmosphere using the coupled 3-D climate models has been a very complicated, and still unsolved, problem. In atmospheric models the SST field (at the lower boundary) or the mixing layer of the ocean were usually prescribed, which made it impossible to simulate the impact of freshwater on the thermohaline circulation in the ocean, whereas in the ocean models the climatic values of salinity near the surface or the freshwater inflow were prescribed. As for the coupled models of the atmosphere–ocean system, they are characterized by the prescribed freshwater inflow or a substantial correction of the freshwater volume.

In view of these circumstances, to simulate an enhancement of the freshwater transport in the atmosphere in conditions of climate warming, Tang and Weaver (1985) used a coupled box model in which the freshwater transport is determined by the poleward heat transport. The latter is determined by the sum of transports of: (1) the sensible heat in the ocean determined by the ocean dynamics and related to the thermohaline circulation; (2) the sensible heat in the atmosphere in proportion (as suggested) to the fourth power of the meridional temperature gradient; (3) the latent heat in the atmosphere. The total global heat transport $T(y)$ (y = latitude) is confined to the values of the radiation balance $R(y)$ of the surface–atmosphere system known from the satellite data:

$$dT(y)/dt = R(y)$$

Thus the latent heat transport can be found as a difference between the radiation balance and (respectively) the sensible heat transport in the atmosphere and in the ocean. This opens up a possibility of diagnostic estimation of the poleward freshwater transport from the data on the latent heat transport. According to this scheme, with the atmospheric warming and, hence, decreasing meridional temperature gradient the sensible heat transport decreases, and to compensate this decrease, the latent heat transport gets intensified, which agrees with the conclusions based on complete coupled climate models.

The main result of the discussed numerical modelling is the conclusion that the stability and variability of the climate system depend first of all on its mean climatic state (total heat content of the system) through the poleward freshwater transport. With the low heat content, when a weak freshwater transport is observed, and the thermohaline circulation provides a rapid advection from high latitudes of freshwater precipitated from the atmosphere, the present climatic stability occurs with the characteristic descending branch of circulation in high latitudes. On the contrary, with the high heat content and intensified freshwater transport in the atmosphere the thermohaline circulation weakens; an intensifying freshwater transport and a slower advection in the ocean lead to an accumulation of freshwater in the high-latitude ocean, which finally breaks the thermohaline circulation. An equilibrium state forms with the downward motions at the equator.

With the freshwater transport in the atmosphere not so intensive, there is a possibility of restoring the thermohaline circulation, and this generates oscillations of both the freshwater circulation and the thermohaline circulation. With sufficiently strong freshwater transport the restoration of the thermohaline circulation is impossible. If the transport gets more intensified, the effect of salinity becomes a dominating factor and the equatorial "overturning" of circulation with persistent downward motions at the equator takes place.

A numerical experiment for the climate warming conditions due to a CO_2 doubling ($2 \times CO_2$) for a period of 100 years has demonstrated a collapse of the thermohaline circulation with its subsequent recurrence (Tang and Weaver, 1985). The latter, however, does not apply to the $4 \times CO_2$ case. Despite a schematic character of the applied climate model not considering many feedbacks and even the sign of the freshwater transport at a global warming unknown, no doubt, the dynamics of the climate system changes substantially in conditions of a warmer climate. This is confirmed by: (1) the stable interglacial climate during the Holocene: there had been no substantial climate changes from the late YD period (10 000 years ago) in contrast to the climate dynamics during the Eemian interglacial period, when oscillations had taken place between the states of cold, moderate and warm climates (the problem of reliability of the Eemian climatic evidence still remains); (2) the warm climate of the Carboniferous period which can be considered the consequence of the "overturning" of the thermohaline circulation in the tropics.

Hense and Romer (1995) performed a statistical analysis of the results of the numerical climate modelling in the tropics using the ECHAM-3 model for perpetual January with two abnormal fields of the Atlantic Ocean surface temperature (SST) as a boundary condition: (1) low SST (1904–13); (2) high SST (1951–60). The choice of the tropics was determined by the need to validate the working hypothesis according to which the changes in the ocean deep water temperature due to SST variations determine the anomalies of the wind and pressure fields in the tropics.

The results obtained (Hense and Romer, 1995) testify to the presence of a clear "tropical signal" in the fields of wind and pressure manifested as a response of the applied climate model to the heating of the oceanic deep waters which takes place not near an SST anomaly but on the adjacent continents (especially in South America) with an opposite sign for remote regions between Indonesia and the International

Date Line. This signal can be generally characterized as an intensification of the east–west circulation in the tropics with the ascending branch over South America in the case of an abnormally high SST compared to an abnormally low SST.

In what follows various aspects of atmosphere–ocean interaction will be discussed in more detail in the context of climate modelling.

A very successfully progressing area of the atmosphere–ocean interaction research (in the context of climate change) is connected with studying the climatic impact of El Niño/Southern Oscillation (ENSO) as well as other action centres and composite extreme anomaly (CEA) distributions which have been (and will be) discussed in this survey in a number of cases. Composite extreme anomaly (CEA) is defined as the difference between extreme positive and negative composite anomalies.

The conceptual aspect of the problem is that new results more and more convincingly reveal the global scale of such an impact. A recent illustration of this conclusion is the study by Liang et al. (1995, 1996) who have conducted an intercomparison of observed and GCM simulated inter-annual teleconnection patterns in the northern hemisphere on a monthly basis (NCAR CCM Version 1 was used with an additional account of GHGs (Wang et al., 1996) for a 100-year simulation). The study was based on 1946–91 observations and two separate 100-year simulations corresponding to the present climate and a greenhouse warming climate.

Liang et al. (1996) have shown that, for the present simulation, the GCM reproduces observed spatial and temporal variations of the action centres of four principal teleconnection patterns: the North Atlantic Oscillation (NAO), the North Pacific Oscillation (NPO), the Pacific/North American pattern (PNA), and the Eurasian pattern (EUA). Substantial model biases exist, however, in the magnitude, in the regional structure and in the monthly transition of anomalies. In the greenhouse warming climate, the GCM continues to simulate the four inter-annual teleconnection patterns. Systematic changes, however, are found for the Pacific/North American and Eurasian patterns in winter, where the action centres shift to the east, and the CEAs weaken over land.

Liang et al. (1996) have pointed out that, when case 2050 (global warming due to GHGs increases) is compared with case 1990, no systematic differences are found in the duration of the four teleconnection patterns. However, the PNA in April and the EUA in October are not robust, while the NAO is clearly shown in September. In addition, the NAO centres are shifted to the west (by approximately 25°) and north during spring, while the PNA and EUA are both located 5–10° to the east of case 1990 during winter.

Liang et al. (1996) have emphasized that it is difficult to judge (at this stage) what mechanisms are responsible for the generations and maintenance of the observed and simulated inter-annual teleconnection patterns (for this purpose respective sensitivity experiments are necessary), but, tentatively, one can suggest that the likely dominant mechanism generating the inter-annual teleconnection patterns in the model are internally driven by instabilities and nonlinearities in atmospheric flow. In a dynamically coupled atmosphere–ocean GCM that incorporates comprehensive land surface processes it is necessary to improve the simulation of inter-annual teleconnection patterns.

Bitz et al. (1996) have applied a simplified single-column, energy balance coupled model of the atmosphere – sea-ice – upper ocean system to study the low-frequency natural variability of the arctic climate system. The variability in the system was induced by the prescribed forcing with realistic, random perturbations in the atmospheric energy transport and cloudiness. Since the purpose was to analyse the low-frequency variability, the model was integrated for approximately 1000 years. As has been mentioned earlier, an assessment of natural climate variability in the Arctic is of specific interest in the context of "CO_2 signal" detection under conditions of its high latitude enhancement. The model developed by Bitz et al. (1996) predicts that the volume of perennial sea-ice varies predominantly on decadal timescales, while other arctic variables vary mostly on intra-annual and inter-annual timescales. The variance of the simulated sea-ice volume is most sensitive to perturbations of the atmospheric forcing in late spring, at the onset of melt. Undoubtedly, the existence of significant low-frequency natural variability of sea-ice complicates the use of observed trends in sea-ice to detect a climate change associated with greenhouse warming.

An important result is that for the frequency range $f > (50\ \text{yr})^{-1}$ half of the variability of ice thickness is consistent with the features of a linear model, directly forced by perturbation of atmospheric energy flux convergence. But another half of the variability is due to high-frequency variability in the atmospheric energy flux convergence which induces a low-frequency variability through nonlinear processes. The calculations show, that a single anomaly in the melt onset date influences the sea-ice thickness for several decades. It appears that an anomalous melting associated with the onset of the melt season can induce a very broad, low-frequency response in the sea-ice thickness.

From the results obtained one can estimate how long one must wait to detect a change in the mean sea-ice thickness by using statistical decision theory. For example, sea-ice thickness must be monitored for 60 years for a 1.4-m change in the mean sea-ice thickness to be significant at the 95% confidence level.

An important consequence of the atmosphere–ocean coupling is a possibility of generation of multiple climatic equilibria. Using a climate model of an intermediate complexity (two-level atmosphere and two-dimensional zonally averaged ocean), Saravanan and McWilliams (1995) have demonstrated that large-scale fluid motions are primarily responsible for the nonlocal interactions between different parts of the globe, and contribute significantly to the nonlinearity of the system. Coupled integrations conducted for periods of several centuries (to identify the equilibrium states of the atmosphere–ocean system) have shown that there exist at least three types of equilibria, which are distinguished by whether they have upwelling or downwelling in the polar regions (the fluid dynamical processes considered include the horizontal transport of heat and moisture in the atmosphere, both by the mean meridional circulation and by the three-dimensional eddies, as well as the meridional and vertical transport of heat and salt by the oceanic thermohaline circulation).

Saravanan and McWilliams (1995) have shown that the dominant interdecadal mode of the internal variability of the climatic system is associated with the advection of oceanic temperature anomalies in the sinking regions. A rapid increase of the GHGs concentrations leads to a collapse of the meridional overturning in the ocean. Introduction of a large positive surface freshwater anomaly in the high latitudes leads to a temporary suppression of the sinking motion, followed by a rapid recovery, due primarily to the high latitude cooling associated with the reduction of oceanic heat transport. In this evolution, the secondary roles played by the atmospheric heat transport and moisture transport in destabilizing the thermohaline circulation are compared, and the former is found to be dominant. In this context, it should be mentioned that Manabe and Stouffer (1993) have discovered (on the basis of long integrations) that quadrupling the CO_2 concentration leads to a "halocline catastrophe", where the North Atlantic thermohaline circulation shuts off.

In their recent study Manabe and Stouffer (1995) have accomplished a simulation of abrupt climate change induced by freshwater input to the North Atlantic Ocean in view of new palaeotemperature records (Greenland ice cores and deep-sea cores) which suggest that large and abrupt changes of North Atlantic climate occurred frequently during both glacial and post-glacial periods (one example is the Younger Dryas cold event). It has been assumed that these changes result from rapid changes in the thermohaline circulation (THC) of the Atlantic Ocean, which were caused by the release of large amounts of melt water from continental ice sheets.

The GFDL coupled climate model has been used by Manabe and Stouffer (1995) to accomplish two types of numerical modelling: (1) the control experiment with a 1000-year time integration of the model (to prevent a rapid drift of the model state from the realistic initial condition, a flux correction was applied: the fluxes of heat and water obtained from the atmospheric component of the coupled model are modified by given amounts before they are imposed at the oceanic surface); (2) the freshwater (FW) experiment with the identical adjustments. In the FW experiment, a surface freshwater flux of 1 Sv (10^6 m^3/s) is released uniformly over the latitude belt between about 50° and 70° in the North Atlantic Ocean during the first 10-year period (the time integration is performed, however, over a period of 300 years). The initial condition for the FW experiment is the state of the coupled atmosphere–ocean model at year 501 of the 1000-year control integration. Manabe and Stouffer (1995) have pointed out that the melt water is applied to the simulated state of the current interglacial rather than the last glacial period. The response of the THC under conditions of a glacial environment should be studied in the future.

The numerical experiments accomplished by Manabe and Stouffer (1995) show that in response to a massive surface flux of freshwater to the northern North Atlantic of the model, the thermohaline circulation weakens abruptly, intensifies and weakens again, followed by a gradual recovery, generating episodes that resemble the abrupt changes of the atmosphere–ocean system recorded in ice and deep-sea cores. The associated change of surface air temperature is particularly large in the northern North Atlantic ocean and its neighbourhood, but is relatively small in the rest of the world.

The study by Manabe and Stouffer (1993) reveals that the simulated, multi-century response of the THC to a doubling of atmospheric CO_2 resembles the response to the freshwater input described above. Thus substantial changes in the intensity and distribution of the THC could also occur in response to future increases of greenhouse gases in the atmosphere.

In the context of palaeoclimatic data analysis, MacLeod and Huber (1996) have considered a reorganization of the deep ocean circulation accompanying a Late Cretaceous extinction event as an illustration of the fact that the deep ocean circulation may be a significant factor in determining the global climate. They have presented stable-isotope records of ocean temperature and salinity that indicate that bottom water in Late Cretaceous oceans of the southern hemisphere became cooler and less saline at the same time (about 70 Myr ago) as widespread biotic changes (mid-Maastrichtian bioevents include floral shifts in the American west, changes among benthic foraminiferal assemblages, and extinction among both tropical, reef-forming rudistid and widely distributed, deep-sea inoceramid bivalves).

Szilder and Lozowski (1995) have applied a zero-dimensional, time-dependent, two-component atmosphere–ocean model to investigate the influence of the primary climate feedbacks on the steady-state and time-dependent behaviour of the climatic system. Their model results show that water vapour and snow-ice albedo feedbacks may lead to multiple stable equilibria. A sensitivity study accomplished by Szilder and Lozowski (1995) have demonstrated that depending on the rate of change of cloud infrared emissivity and cloud albedo with temperature, several regimes with different stability characteristics may be identified. Under conditions of the model, the cloud water content and cloud altitude feedback, coupled with enhanced greenhouse longwave forcing, lead to the formation of a new warm stable equilibrium. A number of other important conclusions have also been obtained: (1) in the case of time-dependent simulation with a single-layer ocean, the global mean surface temperature rises gradually, then rapidly, finally reaching a value approximately 14 K warmer than at present, when an enhanced greenhouse flux of 4 W/m^2 is added; (2) even without additional shortwave or longwave forcing, the cloud amount feedback may lead to free oscillations with a period of several years (when a weak annual solar forcing is imposed, these oscillations become chaotic).

A very important area of coupled climate models applications are studies of the contribution of processes in tropical oceans to global climate change with the special rôle of ENSO phenomenon. A particular aspect of the problem is the interaction between the seasonal cycle and ENSO as an expression of inter-annual variability. It has been shown that the seasonal cycle is a critically significant interactive component of the inter-annual variability.

Chang et al. (1994) have suggested considering the irregular variability of ENSO as a low-order chaotic process, due to annual variations. To substantiate such an approach, an interactive model of the atmosphere–ocean system has been used which was able to reproduce the seasonal cycle as a response to an external perturbation. Basic perturbations are surface fluxes of latent and sensible heat which are controlled mainly by the annual course of solar radiation.

Calculations have demonstrated that the model adequately simulates SST variability within timescales from seasonal to inter-annual. In the process of steady enhancement of the disturbance (due to the annual variation) the dynamics of the coupled model is characterized by the presence of several transitions between periodic (with fixed frequency) and chaotic states before final dissipation of the internal oscillating mode takes place and the regular seasonal cycle forms.

An important result of the calculations is that the dynamics of the coupled atmosphere–ocean system may be very sensitive not only to the change of internal parameters, but also to external perturbations. Depending on the circumstances, the coupled system behaviour may vary from periodic to chaotic under the weak impact of external perturbations.

Kerr (1995) has pointed out that the interaction between the atmosphere and Pacific Ocean under conditions of positive SST anomalies during the 1980s could significantly contribute to global warming on land and interdecadal climate variability. The Pacific Ocean could regulate the whole process of global warming through the enhancement of evaporation from the ocean surface, increase of atmospheric water vapour content and latent heat (when precipitation is formed) and subsequent warming of the tropical atmosphere which could impact the global climate.

A very interesting analysis of an impact of the atmosphere–ocean interaction on global climate has been accomplished by Hurrell (1996), who has shown on the basis of the multivariate regression that nearly all of the cooling observed in the northwest Atlantic and the warming across Europe and downstream over Eurasia since the mid-1970s resulted from the changes in the North Atlantic Oscillation (NAO), and the NAO accounts for 31% of the hemispheric inter-annual variance over the past 60 winters. Over the Pacific basin and North America, the temperature anomalies result in part from tropical forcing associated with the ENSO phenomenon but with important feedbacks in the extratropics. The changes in circulation over the past two decades have resulted in a surface temperature anomaly pattern of warmth over the continents and coolness over the oceans.

Recent rapid development of coupled climate models and a necessity of a more adequate account of various climate-forming processes have resulted in a requirement to consider the climatic impact of the ocean optical properties on the solar irradiance within the ocean. Since the ocean optical properties are not well documented in space and time, Cohen-Solal and Le Treut (1996) have undertaken a sensitivity study to assess the corresponding SST uncertainties on a global scale using a model coupling the LDM (Laboratoire de Météorologie Dynamique, CNRS) AGCM with an integral mixed-layer model and a thermodynamic sea-ice representation. The principal objective of the study has been to assess the role of the solar radiation absorbed in the upper ocean more precisely, and in particular, the vertical distribution of the solar absorption. Although solar radiation is mostly absorbed within the uppermost metre of the ocean (up to 90% in the near-IR region) a substantial part penetrates deeper (more than 10% of solar radiation can reach a depth of 30 m in clear water). This penetration may be important when integrated over a long period of time.

To take into account the variable ocean water properties, Cohen-Solal and Le Treut (1996) have considered the five water types of the Jerlov classification (types I, IA, IB, II, and III). The numerical modelling results show that turbid waters exhibit, in general, a stronger seasonal cycle of the SST of about 2°C. The SST sensitivity to optical properties is at a maximum in the subtropics and the mid-latitudes of the summer hemisphere. It corresponds precisely to the area in which the observed optical properties present a large temporal variability which is therefore likely to have an action on the seasonal cycle of the ocean surface temperatures. Estimates have been obtained of the model sensitivity to the four terms: the direct impact of the water type change, feedback due to the mixed layer change, feedback due to the surface solar irradiance change, and feedback due to the non-solar heat fluxes change. The first two terms dominate the SST change. The direct effect tends to increase the warming of the mixed layer. In addition, the mixed layer depth diminishes because of a higher stabilizing effect of solar radiation on the total turbulent kinetic energy (TKE) budget. This tends to increase further the summer warming of the SST as well as their winter cooling.

A very important problem is to improve the accuracy of the optical properties of the global ocean which may be done in the near future with the use of satellite remote sensing data (SeaWiFS ocean colour scanner, ADEOS data), to retrieve such optically active ocean components as chlorophyll (phytoplankton), dissolved organic matter, and suspended matter (Bukata et al., 1995). An unsolved question, however, is how to obtain information on the bio-optical state in the deeper ocean, in particular, in the euphotic layer.

2.4.5 Cloud cover dynamics. Cloud–radiation interaction

2.4.5.1 Introduction

Clouds are a most dynamical component of the climatic system playing a key role in its energetics. This manifests itself, for example, through an important role of the latent heat released in cloud formation and the prevailing role of the cloud cover dynamics as a factor of the ERB spatial and temporal variability (Kondratyev et al., 1988; Wielicki et al., 1995). In this connection of crucial importance is an adequate parameterization of the processes of cloud formation in climate models (Arking and Ziskin, 1994; Arking et al., 1996; Cloud–Radiation . . . , 1994; Kondratyev, 1992d; Li et al., 1993, 1995; Li and Barker, 1994; Liao et al., 1995; Marchuk et al., 1986; Nakajima, 1993; Platt, 1994; Schweiger and Key, 1994; Senior and Mitchell, 1993; Simpson et al., 1988; Sinha and Shine, 1995; Tiedtke, 1993).

The undoubted inadequacy of the cloud parameterization schemes used now in climate models determines the necessity of checking the reliability of the available schemes and their further improvement, bearing in mind an accomplishment of an extensive programme of numerical experiments to analyse climatic sensitivity to cloud cover characteristics and various observational programmes.

A special feature of the processes of cloud formation is their numerous scales, which seriously complicates the solution of the problem of cloud parameterization. In this connection it is quite urgent to "telescope" the large-scale climate models to use (for individual regions) 3-D meso-scale models which realistically represent the processes of cloud formation in the interaction with the radiative and other processes (Giorgi et al., 1994; Kondratyev, 1992d; Li and Moreau, 1996; Liu et al., 1994; Marchuk et al., 1986; Walsh and McGregor, 1995). Of course, the mesoscale models should be also verified, and this necessitates the planning and accomplishing of the respective field experiments, especially in broken cloudiness conditions.

Summarizing the minimal requirements to be met to be confident in climate model predictions, Wielicki et al. (1995) have emphasized a necessity of the four basic elements:

- detailed dynamical and radiative cloud models verified against field and laboratory experiments for a wide range of cloud types and conditions;
- successful construction of simplified cloud model parameterizations using detailed cloud physics models;
- availability and verification of the accuracy of global satellite observations of radiative fluxes and cloud properties;
- agreement of climate models with satellite observations on a range of space and time scales (global, regional, annual, seasonal, monthly, diurnal).

These are, undoubtedly, very important elements, but, in accordance with the recommendation of the same authors (Wielicki et al., 1995): "the full observational strategy must include not only EOS satellite observations, but also a wide range of surface and field experiment data to support the global satellite measurements", it would be proper to point out here (as it has been later in the paper) that the basic aim is a combined system of satellite and conventional (surface, aircraft, balloon) observations. A non-trivial aspect of the problem is an optimization of the combined use of various observational means.

As far as cloud variables are concerned (in the context of the cloud–radiation impact on climate), Wielicki et al. (1995) recommend the following minimum set of cloud variables:

- cloud liquid water path LWP (or ice water path);
- cloud visible optical depth;
- cloud particle size distribution;
- cloud particle phase/shape;
- cloud fractional coverage;
- cloud temperature/height;
- cloud infrared emittance.

Since clouds never consist of pure water, it would also be important to add to this list a single scattering albedo for cloud droplets which would also emphasize the importance of cloud chemistry. One more complication is a spatial inhomogeneity of clouds which sometimes affects the radiative transfer very significantly.

Wielicki et al. (1995) have justly recommended four primary types of cloud measurements (Table 2.17) and made a reference to the successful effort of a synergistic approach to studying the cloud–climate processes within the First ISCCP (International Satellite Cloud Climatology Project) Regional Experiment (FIRE). Here, again, it must be remembered, however, that cloud chemistry should not be overlooked.

Studies of cloud-induced, climate-forming processes are aimed at an understanding of the dynamics (formation, development, dissipation) of cloud cover, taking with account of a parameterization of the sub-grid (macro- and meso-scale) morphology of clouds, as well as the key role of the cloud–radiation interaction

Table 2.17. Cloud and radiation measurement systems

Measurement system	Space scale	Time scale
Laboratory cloud microphysics	Micrometres to metres	Seconds to minutes
Aircraft, *in situ* and remote sensing	Metres to 100 km	Seconds to hours
Ground-based, *in situ* and remote sensing	Metres to 10 km	Minutes to years
Satellite global remote sensing	100 m to global	Hours to years

(Sundqvist and Starr, 1994). The qualitative description of the factors determining the type, phase state, evolution, lifetime, and radiative properties of clouds depending on the prescribed large-scale thermodynamical and dynamical conditions is decisive. So, for example, the type and the size of clouds depend on many factors, including atmospheric stratification and vertical motions, and their evolution is determined by the transformation of water into precipitation as well as by an interaction among clouds, radiation, diabatic heat flux divergences, and circulation.

Considerable progress has been recently achieved in understanding the climate-forming role of clouds (Marchuk et al., 1986; Sundqvist and Starr, 1994; Szilder and Lozowski, 1995; Wielicki et al., 1995; Zhang and McFarlane, 1995). This, in particular, refers to studies of the microphysical processes in maritime stratocumulus clouds, including:

- processes of their dissipation;
- assessment of the effect of atmospheric aerosols on microphysical processes.

Progress has been achieved in studies of the microphysical properties and dynamics of cirrus clouds (it turned out, for example, that the layers of cirrus clouds can be quite thin). Observations revealed, in particular:

- the presence in clouds of a considerable number of small (< 50 μm) ice crystals which exhibit a dominating effect on the cloud–radiation interaction;
- a substantial effect of volcanic aerosols on the microphysical processes in cirrus clouds.

The use of improved models of cirrus clouds has demonstrated the importance of the interaction of microphysical, radiative and dynamical processes in the formation of cloud properties. Serious progress has been achieved in studies of the precipitating cloud systems, but the multi-layered frontal cloudiness characterized by the mixed (water–ice) phase state has been explored inadequately.

Of primary importance is an accomplishment of the complex programme of field experiments and numerical modelling to study: cloud systems; cloud–radiation interaction; cloud–aerosol interaction. The field problem-oriented experiments should be, first of all, aimed at studies of the processes connected with cirrus, strato-cumulus and precipitating clouds, and with mixed-phase clouds. Of special importance is an analysis of the role of the spatial inhomogeneity of cloud cover.

In the context of the cloud–aerosol interaction the following problems claim special attention:

- the effect of condensation nuclei on the microphysics and radiative properties of clouds;
- solar radiation absorption by cloud aerosols;
- the formation of aerosols in clouds due to chemical reactions in clouds and in the atmosphere (it can turn out to be a major global source of aerosol formation).

Most urgent developments in the sphere of numerical modelling are (Sundqvist and Starr, 1994):

- the development of the high-resolution models on limited spatial scales useful for a parameterization of the cloud dynamics in climate models;
- assessments of contributions of the cloud-radiative forcing (including its diurnal course) and microphysical processes (with the role of aerosols taken into account) to the formation of the cloud cover properties and structure;
- improving the algorithms to retrieve the cloud characteristics from the data of remote sounding.

The urgency of the cloud–climate problems determines the keen interest in them demonstrated in various countries (see, for example, Cloud–Radiation..., 1994). For example, developments in Australia regarding the climate-forming role of cloud–radiation interaction are being carried out within several programmes (Platt, 1994). The major goal of the Southern Ocean Cloud Experiment (SOCEX) programme is to study the dependence of the albedo of strato-cumulus clouds over the SH ocean on cloud particle size distribution. The dominating factor determining the number, density and size distribution of cloud particles is, in the conditions considered, the concentration of the cloud condensation nuclei (CCN) which depends on natural emissions of dimethylsulphide (DMS) by the ocean to the atmosphere. The DMS emissions are characterized by a clear annual change with a minimum in mid-winter and maximum in mid-summer, which determines the CCN annual course. To study the processes mentioned above, surface and aircraft observations were carried out near the north-western coastline of Tasmania.

The main tasks of the Experimental Cloud Lidar Pilot Study (ECLIPS) programme are as follows:

- to accumulate a database on the height of cloud bases and their optical properties from the data of ground lidar sounding;
- to improve techniques of retrieving cloud cover characteristics from satellite data based on comparisons of satellite and surface lidar data;
- to develop techniques to retrieve surface heat balance components from satellite data;
- to accumulate a database on the optical properties of clouds in the interests of the ISCCP.

The Australian specialists also take part in studies within the American programme ARM (Proceedings of the Fourth..., 1995) consisting in studying upper- and middle-level clouds with the use of lidar and radiometric sounding.

In Japan, studies on cloud–radiation interaction were started in the mid-1980s within the WCRP programme (Asano, 1994). The related Western North Pacific Cloud-Radiation Experiment (WENPEX) project with the participation of the scientists of the Beijing University (P.R. China) was aimed at studies of climate, cloud radiative properties, and cloud–radiation interaction in the western sector of the North Pacific. The project had four major objects to study:

- cloud cover climatology in the western Pacific region;
- distribution of extended cloudiness and its bulk radiative characteristics;
- radiative and physical properties of stratiform clouds;
- formation of marine stratiform clouds in the atmospheric boundary layer (ABL).

The objective of one of the groups of specialists (the Meteorological Research Institute of Japan) was to carry out a field experiment in the region of Tsukuba-Hachijojima, and that of another (Tokyo University) to carry out a field experiment on the islands of the south-eastern Pacific (January, 1991). The task of the latter was to study the structure of the ABL covered with clouds when the cold air masses from the continent reach the warmer ocean surface, as well as relationships between the radiative properties and structure of the cloud cover.

Analysis of the results of aircraft measurements suggested the following conclusions:

- the strato-cumulus clouds were exclusively convective and were characterized by an intensive entrainment process;
- the main factor of the variability of the cloud radiative properties was their total water content which could be reliably estimated from the data of the airborne microwave radiometer at the frequency 37 GHz;
- the strato-cumulus clouds were very inhomogeneous both vertically and horizontally; the so-called "excessive" absorption of solar radiation was not observed.

The data of surface observations of the upper-level ice clouds suggested the following conclusions:

- the average size distribution of the cirrostratus cloud particles can be approximated by the power law function with a mean exponent -3.5;
- there is a clear correlation between shortwave radiation fluxes and the optical thickness of clouds in the visible;
- the optical thickness of clouds in the IR atmospheric transparency window is approximately half as much as in the visible (Brogniez et al., 1996; Gayet et al., 1995; Pincus et al., 1995).

Upon completion of the WENPEX programme, the Japanese Cloud–Climate Study (JACCS) programme planned for the period 1991–2000 was started. Its major objective is to provide a more adequate parameterization of the cloud–radiation interaction in climate models as well as to use satellite data in the pursuit of climate studies.

2.4.5.2 Cirrus clouds

Of special interest is the problem of an impact of cirrus clouds on climate. As has been mentioned above, the most substantial physical processes determining the formation, development and dissipation of cirrus clouds are the microphysical, dynamical and radiative processes (Starr, 1994). Cirrus clouds are characterized by a relatively wide cloud particles size distribution, with a large number of both small (10–100 μm) and large (100 μm–1 mm) ice crystals. As a rule, the form of crystals is rather diverse: columns, plates, and various aggregates.

Recent observations revealed a considerable effect of aerosols (volcanic, in particular) on the size distribution of cirrus cloud particles. Though the ice content, the particle size distribution and the form of cirrus particles depend on temperature, the strong variability of cirrus clouds seriously hinders a consideration (parameterization) of their effect on climate in numerical climate models. This variability is connected with the effect of the processes of various scales (from the size of an individual cloud to synoptical scales). A characteristic feature is the presence of cellular structures (with the size of cells from 100 m to 1 km), which are often organized as clusters 10–100 km in size within the large-scale systems of cirrus cloudiness.

In conformity with these circumstances, the radiative properties of cirrus clouds, affecting the dynamics of clouds and the microphysical processes, vary widely (in particular, owing to a nonlinear dependence of the radiative characteristics on the microphysical properties of clouds, especially with their small optical thickness).

It was found that cirrus clouds not only directly affect the radiative regime and climate but are also an important factor of the upper tropospheric water balance. For example, the downward transport of cloud water due to the presence of large crystals determines a re-distribution of water vapour between the upper and middle troposphere, which affects the atmospheric radiative regime and the formation of precipitation. While the vertical water transport is mainly governed by large crystals, small crystals mainly affect the radiative properties of clouds.

An important result of recent studies has been the fact that cirrus clouds form usually in rather shallow layers (100–500 m) and are often multi-layered. This fact illustrates the necessity of high vertical resolution in parameterizing the effect of cirrus clouds on climate. A key element of such a parameterization should be a consideration of mesoscale cloud systems (and the development of respective models).

As Liou (1994) notes, a parameterization of the radiative properties of crystal clouds requires an account of the following factors:

- the size distribution of crystals in terms of their mean (effective) size and ice-water content;
- scattering and absorbing properties of the irregular ice crystal shapes (such as hollow columns and bullet rosettes);

- vertical and horizontal inhomogeneity of cirrus cloud bands.

In recent years, considerable progress has been achieved in the first two directions. The results obtained point to a crucial importance of an account of single scattering albedo for non-spherical ice particles in calculations of the cirrus cloud albedo (Takano and Liou, 1995). In studies of feedback among the microphysical processes, radiation transfer and temperature field, establishing a correlation between cirrus cloud particle size distribution and the temperature field plays a fundamental role.

For an adequate understanding of the cloud–radiation climatic feedbacks, the dependence of the cloud particle size distribution on temperature and other factors must be taken into account. The effect of the small ice crystals less than ~20 μm (often the optical remote sounding does not recognize such particles) on the radiative properties of cirrus clouds and the effect of the spatial inhomogeneity of clouds on their albedo and emissivity claim special attention (Arnott et al., 1994).

Significant progress has been achieved in the development of remote sensing technique to study cirrus clouds (Atlas et al., 1995; Hutchinson et al., 1995; Schmidt et al., 1995). Using the data obtained from the Advanced Very High Resolution Radiometer (AVHRR) 3.7-μm and 10.9-μm channels, Ou et al. (1995) have developed a retrieval scheme to simultaneously infer the cirrus cloud optical depth and the mean effective crystal size based on the theory of radiative transfer and parameterization (Rao et al. (1995) have substantiated a technique to remove the solar component from the AVHRR 3.7-μm radiances when retrieving the cirrus cloud parameters). For the daytime satellite data processing a numerical scheme has been used to remove the solar component from the 3.7-μm radiance (the scheme is based on the correlation between the 3.7-μm (solar) and 0.63-μm reflectances). For validation purposes (to determine the microphysical and optical properties of cirrus clouds), the photomicrographic data collected by the balloon-borne replicator have been used. The principal result of the study is that the retrieved cirrus cloud temperature, mean effective ice crystal size, and optical depth closely match the *in situ* data.

The next stage of the study by Ou et al. (1995) was directed towards the utilization of the retrieved cirrus cloud properties to compute the surface radiative fluxes, which were then compared with the ground-based observations (the Coffeyville area 0.5° × 1.0°, Kansas, 26 November and 5 December 1991). Such a comparison indicates that the computed downward surface IR and shortwave fluxes are within 5 and 10 W/m^2 of the measured values, respectively, near the time of satellite overpass. In a later study, Ou et al. (1996) substantiated a technique of cloud classification which identifies cirrus, cirrus/low clouds, and low clouds. This technique has been successfully verified through a comparison with the ground-based observation.

Chýlek and Dobbie (1995) have accomplished the Monte Carlo simulations for the case of the radiative properties of finite inhomogeneous cirrus clouds to study the impact of various cloud morphologies (roughness, voids, waves and horizontal spreading) on the radiative properties of finite, thin model cirrus clouds. The results show that cloud roughness, voids, and waves generally decrease the cloud reflectance as well as cloud absorption. But, although the local horizontal variations in reflectance can be quite large (several hundred per cent), variations in the total reflectance, integrated over the top surface of a cloud, are in the range of a few per cent.

A very informative summary of the cirrus cloud optical properties measured from high-flying ER-2 aircraft (multispectral visible and infrared imaging radiometers and a lidar profiler) during the 1986 and 1991 FIRE (First—International Satellite Cloud Climatology Project—Regional Experiment) has been published by Spinhirne et al. (1996). The results have been compared to model calculations for three cloud phase function models: a spherical water droplet, an ice column crystal cloud, and a Henyey–Greenstein function. Spinhirne et al. (1996) have pointed out that the results for the observed shortwave to longwave relations for cirrus may be explained to a large degree by the results of direct bidirectional reflectance measurements at visible wavelengths. Real cirrus clouds are typically either of a significant vertical extent or are multilayered, and their microphysics is variable all through. The size and type of crystals are known to show extreme variations (this is why it is difficult to choose an adequate model, but it is clear that a spherical water droplet model is not acceptable).

A new direction of research in the remote sensing field is the development of the microwave remote sensing of cirrus clouds. In this context, Brown et al. (1995) have discussed the potential of a spaceborne 94-GHz (3.2 mm) radar for useful measurements of the vertical distribution and water content of ice clouds on a global scale.

To analyse the requirements for such remote sensing data, the sensitivity calculations of longwave (LW) radiation fluxes for a number of model ice clouds have been performed. This has made it possible to determine a minimum of the cloud optical thickness (COT) that will cause changes in the outgoing longwave radiation (OLR) or flux divergence within a cloud layer above 10 W/m^2, and in the surface downward LW flux greater than 5 W/m^2, compared to the clear-sky values. Such COT values have been used to define the notion of a "radiatively significant" cloud. Different "thresholds of radiative significance" have been calculated for each of the three radiation parameters and also for tropical and mid-latitude cirrus clouds. Then, extensive observational datasets of the ice crystal size spectra from mid-latitude and tropical cirrus have been used to assess the radar's capability of meeting these measurement requirements.

Brown et al. (1995) have also used the LW flux calculations to establish the required accuracy within which the COT should be known in order to estimate LW fluxes or flux divergence within specified limits of accuracy. Accuracy requirements are also expressed in terms of ice water content (IWC) because the cloud parameterization schemes in GCMs should be validated. Different schemes for the layered cloud parameterization in GCMs can lead to changes in the zonally averaged IWC, which exceed +100% and −50%, and a measurement accuracy in terms of IWC is required to match this.

On the basis of the numerical modelling results, Brown et al. (1995) have arrived at the following conclusions:

(1) Clouds, radiatively significant for LW flux divergence, have optical thickness exceeding 0.05 (ice water path, IWP = 0.63 g/m^2 for the effective radius of particles r_e = 10 μm) and require it to be measured to within +60% and −40% of the true value.
(2) Clouds, radiatively significant for the surface downwelling LW flux, have COT exceeding 0.20 (IWP = 2.5 g/m^2, r_e = 10 μm) and require it to be measured to within +100% and −50%.
(3) Clouds, radiatively significant for OLR, have COT exceeding 0.04 (IWP = 0.50, r_e = 10 μm) and require it to be measured to within +40% and −30% of the true value.

An analysis made by Brown et al. (1995) shows that a radar with a threshold of −30 dBZ should detect 99% (92%) of a "radiatively significant" cloud in mid-latitudes (tropics). This detection efficiency may be reduced significantly for tropical clouds at very low temperatures (−80°C).

Atlas et al. (1995) have also discussed an application of radar technique to retrieve the properties of ice clouds. Kent et al. (1995) have discussed the surface temperature related variations in the tropical cirrus cloud as measured by the SAGE II solar occultation satellite instrument over the altitude range 10.5–18.5 km. An analysis of a large dataset has made it possible to reveal the geographical distribution and the temporal variation of the cloud occurrence over all longitudes on timescales from less than one day to the El Niño–Southern Oscillation (ENSO) cycle. It has been found that a maximum of cloud occurs over the warm-pool region of the Pacific Ocean, with secondary maxima over the South American and Central African landmasses, where the percentage of cloud occurrence in the upper troposphere can exceed 75%. The cloud occurrence at all altitudes within the tropics, over both land and ocean, increases with the underlying surface temperature at a rate of approximately 13% /°C.

2.4.5.3 Lower tropospheric clouds. Cloud–radiation interaction

A method for the satellite remote sensing of the cloud optical thickness and effective particle radius from the AVHRR multispectral radiance data has been developed by Nakajima and Nakajima (1995). To derive the reflected solar radiation, undesirable radiation components (such as ground-reflected solar radiation and thermal emission) were removed with the use of the data for AVHRR channels 1, 3, and 4 (via subtraction from radiances in channels 1 and 3). The method developed can be applied to a broad range of water clouds, from semitransparent to thick clouds.

The AVHRR data processed by Nakajima and Nakajima (1995) cover the oceans during FIRE and the Atlantic Stratocumulus Transition Experiment (ASTEX). A comparison between the retrieved and *in situ* values indicates a good agreement. The expected retrieval error is less than 25%, and it is less than 10% for most cases of realistic clouds (cloud optical thickness $\tau_c > 5$, surface albedo $A_s < 0.3$). Such factors as non-plane-parallel cloud geometry, ice crystals, and aerosols require further quantitative investigation. The results show, in particular, that marine stratocumulus clouds were drastically enhanced by ship track effluents and dust-contaminated airflows from the continent. Observations indicate that maritime cloud particles (r_e ~20 μm) reduce their size (r_e = 5~15 μm) as a result of the cloud–aerosol interaction in the pair of ship tracks (see also Gayet et al., 1996).

Hayasaka et al. (1995a) have discussed the results of retrievals (from the AVHRR data) of geometrical thickness, liquid water content, and radiative properties of stratocumulus clouds over the western North Pacific. Asano et al. (1995) have developed a remote sensing technique to retrieve cloud properties from four channels of the airborne Multichannel Cloud Pyranometer (MCP): 500, 760, 938, and 1650 nm, within the ranges: $1 \leqslant \tau_{500} < 100$; $4 \leqslant$ μm $r_e \leqslant 25$; LWC $\leqslant$ 1 g/m^3; water vapour content $\leqslant$10 g/m^3 (in cloud water amount).

The still unsolved problem of cloud–radiation interaction (in the context of climate change) requires, first of all, a careful analysis of interrelationships among the cloudiness characteristics, outgoing shortwave (reflected) radiation (OSR), and outgoing longwave radiation (OLR). The problem is that a small difference (the Earth's radiation budget, ERB) of the two large values (OSR and OLR) has to be defined precisely enough. A difficult aspect of this problem, emphasized by Poetzsch-Heffter et al. (1995), is that both OSR and OLR depend not only on the amount of clouds but also on their geometrical (e.g., top height, thickness, and structure) and microphysical (e.g., optical thickness, droplet distribution, phase, and shape of ice particles) properties, which are very difficult to monitor.

The independently existing ISCCP data on clouds and the ERBE datasets do not permit a reliable analysis of cloud–radiation interrelationships. Therefore Poetzsch-Heffter et al. (1995) have accomplished a study, the principal purpose of which was to obtain a complex dataset on clouds, OSR and OLR, making use of both observations (ISCCP) and radiation calculations (a two-stream approximation for prescribed cloudiness conditions on the basis of the ISCCP data). Since the ISCCP cloud data have been obtained through a retrieval from the satellite narrowband radiances data, the method suggested can be considered as a conversion of narrowband radiances to broadband irradiances (fluxes). Such an approach results in obtaining both the radiation fluxes and the cloud type information.

To make the data reliable, Poetzsch-Heffter et al. (1995) undertook two calibration tests before the diagnostic studies: the modelled narrowband filtered radiances were verified against the ISCCP satellite observation data (internal tests) and the modelled broadband fluxes were compared with the ERBE data (external tests). After the successful completion of the tests, the OSR (Q) and OLR (F) values for each of the 35 ISCCP cloud classes were determined. A reduced set of nine cloud classes has been discussed (three bins for cloud top height, three bins for the optical thickness).

The global monthly and annual means of albedo and the greenhouse effect have been analysed for one month of each season (April, July, October, 1985; January, 1986). The analyses have shown that for the global annual mean, the cloud radiative cooling of the surface–atmosphere system in the shortwave spectral range dominates the cloud radiative warming in the longwave spectral range by a factor of 2. Seasonal variations of the radiative forcing are mainly due to changes in the shortwave cloud-forcing component. The variations range from −43.9 to −52.9 W/m^2 (C_{SW}, shortwave) and from 22.5 to 25.1 W/m^2 (C_{LW}, longwave). The main sensitivity β of the ERB to changes in cloud amount $\beta = \partial Q/\partial A_c - \partial F/\partial A_c = \partial C/\partial A_c \mp \partial C_{SW}/A_c + C_{LW}/A_c = C/A_c$ (A_c = cloud amount) varies from a minimum of -29.4 W/m^2 per a 100% change in the cloud cover in April to a maximum of -47.7 W/m^2 per a 100% change in the cloud cover in January (with the annual means: $\beta = -37.5$ W/m^2, $\partial Q/\partial A_c = 74.9$ W/m^2, and $\partial F/\partial A_c = 37.4$ W/m^2). This is also mainly due to the shortwave component.

A detailed analysis of the effects of the nine cloud types indicates the possibility of distinguishing among cooling, warming, and radiatively neutral cloud types. Low stratus and mid-level nimbostratus clouds contribute about 80% to the total cooling due to all clouds. Their effect on the OLP, however, is small (less than 20%), but they contribute almost 50% to the OSR. Deep convective clouds produce the largest single effect for both longwave (8.8 W/m^2) and shortwave (-13 W/m^2) forcings, but their net effect is small. Only cirrus clouds have a net warming effect. For the cirrus (type 7), the greenhouse effect dominates the albedo effect by a factor of 2.5 because they are nearly transparent in the solar spectral range but almost optically opaque for the longwave part.

The whole set of mean annual data on sensitivity coefficients is listed in Table 2.18. These coefficients depend on both the height and the optical thickness of clouds. Both properties are nearly equally important. As can be seen, it is possible to distinguish three categories of radiatively effective clouds:

- warming clouds: type 7 (optically thin cirrus);
- radiatively neutral clouds: types 4, 8 (altocumulus, optically thick cirrus);
- cooling clouds: types 1, 2, 3, 5, 6, 9 (all low, mid-level and high thick clouds).

The principal difficulty of the calculations considered is a reliable assessment of potential errors.

Greenwald et al. (1995) have investigated the large-scale spatial distribution and temporal variability of the cloud liquid water path (LWP) over the world's oceans (on a 2.5° × 2.5° grid over a 53-month period from July 1987 to December 1991) and the relationship of cloud LWP to temperature and the Earth's radiation budget using the satellite measurement data (SSM/I and ERBE) as well as the ISCCP data.

The highest LWP values (more than 0.13 kg/m^2) have been observed along the principal routes of the northern mid-latitude storms and in areas dominated by tropical convection. The distinctly trimodal zonally averaged profile is characterized by maxima located in mid-latitudes and near the equator. The global average marine cloud LWP is about 0.113 kg/m^2. The typical highest amplitude of annual variations is between 15% and 25%, but in certain locations it can exceed 30%. An intercomparison of cloud LWP with temperature for low clouds during JJA and DJF of 1990 shows significant positive correlations at colder temperatures and negative correlations at warmer temperatures. The correlations are characterized by strong seasonal and regional variabilities.

The observed albedo (for the surface–atmosphere system) to LWP relationships agree reasonably well with theory, where the average effective cloud droplets radius (r_e) is 11.1 μm and the standard deviation is 5.2 μm (the large r_e variability indicates

Table 2.18 Annual mean values (1985–86, global) of the sensitivity parameters of the nine cloud types (all values in W/m^2 per a 100% cloud cover change)

Cloud type	$\partial Q / \partial A_c$	$\partial F / \partial A_c$	β
Cumulus or stratocumulus	−28.2	−8.0	−20.2
Optically thin stratus	−70.7	−12.6	−58.1
Optically thick stratus	−115.9	−13.1	−102.8
Altocumulus or altostratus	−25.2	−21.4	−3.8
Optically thick altostratus	−77.3	−37.0	−40.3
Nimbostratus	−113.5	−36.1	−77.4
Optically thin cirrus	−20.5	−51.0	30.5
Optically thick cirrus or cirro-cumulus	−87.7	−90.4	2.7
Cirrostratus or cirrocumulus	−155.3	−105.5	−49.8

that other factors may be important in the albedo–LWP relationships). Greenwald et al. (1995) have found that a 0.05 kg/m^2 increase in LWP (for LWP < 0.2 kg/m^2) results in a -25 W/m^2 change in the net cloud forcing at a solar zenith angle of 75°.

Kyle et al. (1995) have discovered quasi-biannual global, mid-latitude, and tropics oscillations using the top-of-the-atmosphere outgoing longwave radiation (OLR), surface air temperature (SAT), and cloud amount for the period 1979 to 1989. In the detrended datasets (12-month running means), OLR and SAT strongly positively correlated with explained variances of 96.4 (global), 84.8 (tropics), and 75.9% (mid-latitude). The OLR and cloud amount correlated negatively; however, the results were less definitive with explained variances of 86.7% and 79.4%, respectively, for the globe and mid-latitudes, but only 17.4% in the tropics. The maximum–minimum ranges of the quasi-biannual global oscillations are relatively small: less than 1 W/m^2 for OLR, less than 0.3°C for SAT, and less than 1% for cloud amount.

2.4.5.4 Model–observations intercomparisons

Being one of the basic climate-forming factors, the re-distribution of the incoming solar radiation between the atmosphere and the surface is determined by a complicated set of feedbacks functioning in the climate system. In this connection, Li and Barker (1994) undertook a complex analysis of the data of satellite and surface observations and of the results of numerical climate modelling to compare them from the point of view of the specific features of the solar radiation redistribution. Table 2.19 contains a summary of mean global estimates of the shortwave radiation budget (SWRB) components from the following data:

- GEBA (Global Energy Balance Archive)—surface observations and approximate calculations using the empirical formulas;
- ERBE—satellite observations;
- ISCCP—data of the International Satellite Cloud Climatology Project;
- CCCII, CCM2, ECMWF - results for three models of global climate.

Table 2.19. Mean global mean annual values of the SWRB components (W/m^2)

Component	Source					
	GEBA	ERBE	ISCCP	CCC II	CCM 2	ECMWF
Solar constant	1365	1365	1365	1365	1370	1384
Outgoing SW radiation	101.2	101.2	101.2	111.3	94.3	96.0
SW radiation absorbed by the atmosphere	98.1	83.0	68.9	58.0	67.6	66.0
SW radiation absorbed by the surface	142.0	157.0	171.2	172.0	180.6	184.0

As can be seen, the data of numerical modelling are characterized by an underestimated value of SWR absorbed by the atmosphere and, respectively, overestimated SWR absorbed by the surface, with maximum differences exceeding more than ten times the greenhouse effect enhancement due to a CO_2 doubling. The difference between the GEBA and ISCCP data are explained, apparently, by insufficient representativeness of surface observations which were supplemented with results of calculations from empirical formulas. The ISCCP data on the surface SWRB are also overestimated.

A comparison of the ERBE and ISCCP data revealed serious regional differences connected, probably, with an inadequate technique in the ISCCP data processing. A comparison of the mean meridional profiles of the SWRB components from the data of ERBE and CCC II has shown that the values of the outgoing SW radiation agree best. The differences of estimates for the outgoing SWR, the SWR absorbed by the atmosphere and by the surface are approximately the same in conditions of clear skies and mid-level cloudiness. This points to the fact that the SWR absorption by clouds is not a major cause of differences. Thus one can believe that the major cause of differences is the absorption of SWR by the cloud-free atmosphere. In this connection, the choice of the calculation technique for the SWR absorption by water vapour is important as well as the content of water vapour and aerosol in the atmosphere.

Arking and Ziskin (1994) analysed the change in the relationship between cloud amount, water vapour content and the ERB as a function of SST from the 4-year observational data (February 1985 to January 1989) for the tropical band 25°S–25°N (in the longitude interval 150°E–90°W) and averaged over 2.5° lat. 2.5° long. squares. The observational data reveal a substantial difference between two locations of the strong cloud–radiation correlation in the band 10–25°S, where it is positive, but the signs of its LW and SW components are opposite. The latter reflects the fact that at the eastern location the low-level cloudiness is the prevailing factor of the feedback.

Wild et al. (1995) have compared calculated surface radiative fluxes of the ECHAM 3 spectral GCM with T21, T42, and T106 resolutions with the observation data from the GEBA. The ECHAM 3 models show a clear tendency to overestimate the global annual mean incoming shortwave radiation at the surface due to underestimated atmospheric absorption. In contrast to the conclusion just mentioned, Wild et al. (1995) believe that the underestimated SWR absorption by clouds is a major cause of differences.

Garrat and Prata (1996) have pointed out that global climate models are characterized by excess net radiation at land surfaces (up to 20–27% on an annual basis), apparently, due to overestimates in the downwelling shortwave radiation flux and underestimates in the upwelling longwave flux. Part of this excess, however, may be compensated for by an underestimate in the downwelling longwave radiation. They analysed the observational data on the downwelling longwave flux at the surface at several land stations in Europe, the United States, Australia, and Antarctica, and showed that in the cases of four climate models there was an underestimation of this component of the net radiation on an annual basis by up to 10 W/m^2, yet with low statistical significance.

Garrat and Prata (1996) believe in a number of causes for such an underestimation: the known underestimate in the boundary-layer air temperature, low model cloudiness and neglect of minor gases such as methane, nitrogen oxide, and the freons. Thus, the bias in the downwelling longwave flux, together with those found earlier for the downwelling shortwave and upwelling longwave fluxes, are consistent with the model bias found earlier for the surface net radiation. The following figures characterize the hypothetical observed values of mean annual global land fluxes inferred from model calculations (in brackets, in W/m^2): net radiation (65 ± 5); downwelling shortwave flux (176 ± 12); downwelling longwave flux (305 ± 8); upwelling longwave flux (374 ± 5); reflected shortwave flux (42); net shortwave flux (134); net longwave flux (-69).

Intercomparisons between numerical modelling results and observations of the Earth's radiation budget components (especially, outgoing longwave radiation) have been broadly used to validate climate models (a recent important study of this kind has been accomplished by Chen and Roeckner (1996)). Yu et al. (1996) have justly pointed out, however, that such an approach is based on an incomplete validation of some of the cloud properties which are the key parameters in determining the cloud feedback on climate, because the ERB's components represent the combined radiative effects of clouds, surface, and atmosphere. Therefore, tuning the model to obtain a good simulation of the ERB data neither guarantees a reliable surface RB (a key input to coupled atmosphere–ocean models) nor does it help significantly reduce the uncertainties associated with cloud feedback mechanisms. Therefore Yu et al. (1996) have undertaken an effort to verify the cloud schemes used in Laboratoire de Météorologie Dynamique GCM through an intercomparison with the ISCCP C1 dataset which contains global statistics of cloud properties with their spatial and temporal variability for several complete years. The results show that a direct comparison of simulated low cloudiness and that shown from satellites is not possible. They also reveal some model deficiencies concerning the cloud vertical distribution.

Analysis of the numerical climate modelling results revealed two important features characterized by the global distribution of the ERB difference in clear- and cloudy-sky conditions (cloud–radiation forcing, CRF):

- the summertime prevailing contribution of clouds in middle and high latitudes to the formation of the negative ERB difference (this also determines the negative mean global CRF);
- the formation of this ERB difference (CRF) in low latitudes as a small difference of large values of its shortwave and longwave components.

On the other hand, of interest are considerable (and opposite in sign) values of CRF for the atmosphere (naturally, in the presence of clouds the radiative heating of the atmosphere grows, that is, CRF > 0) and for the surface (clearly, the clouds produce the radiative cooling of the surface compared to the clear-sky conditions). The need for more complete information necessitates:

- use of the means of active (lidar, radar) satellite remote sounding to obtain data on the 4-D fields of characteristic of clouds and aerosol;
- substantiation of the techniques to validate the numerical climate modelling results with the use of complex data of satellite and conventional observations.

In the context of cloud parameterization, special attention should be paid to the hypothesis that an important role is played by dimethylsulphide (DMS) produced by marine phytoplankton (Restelli and Angeletti, 1992). The DMS, getting to the atmosphere, experiences the process of gas-to-particle conversion giving sub-micrometre aerosols (cloud condensation nuclei, CCN) affecting the microphysical and optical properties of clouds and, thus, the ERB and the climate. To check the validity of this hypothesis, complicated numerical and observational experiments are needed. It is of interest, in this connection, that an analysis of the data of satellite observations of the Earth's albedo in the region of the USA eastern coastline and the adjacent part of the Atlantic Ocean performed recently by R. Cess, revealed a marked increase of the cloud albedo over the ocean, which may be explained by the effect of industrial emissions of sulphur dioxide to the atmosphere.

As Stephens (1994b) noted, for a better understanding of the role of clouds in the formation of climate it is important to have the following information:

- the 4-D distribution of the optical properties of clouds and aerosol properties determining the cloud microphysics;
- global fields of the content in the atmosphere of liquid water, water vapour and ice;
- characteristics of the cloud–radiation interaction.

Because of a poor parameterization of the cloud dynamics and cloud–radiation interaction in climate models, further efforts are needed to improve the parameterization. In this context, Gultepe et al. (1996) have developed the cloud microphysical parameterization schemes for GCMs which are based on airborne observations conducted in maritime stratus clouds over the east coast of Canada during the North Atlantic Regional Experiment in the summer of 1995. Statistical techniques have been used to obtain regression equations for droplet number concentration (N_d), interstitial aerosol number concentration (N_i), temperature, vertical air velocity (w); and liquid water content (LWC), effective droplet radius

(r_{eff}), and total particle concentration ($N_d + N_i$). An important result is that multiple relationships are present among N_d, N_i and w; and r_{eff}, LWC, and N_D. For individual clouds, an inverse relationship between the interstitial aerosol concentration and droplet concentration is always observed.

Bintanja (1996) has suggested a parameterization of shortwave and longwave radiative fluxes for use in zonally averaged climate models which is based on detailed calculations of radiative fluxes at the top of the atmosphere and at the surface as a function of solar constant, solar zenith angle, cloud parameters (amount, optical depth, height, and emissivity), surface albedo, surface air temperature, surface temperature, and atmospheric CO_2 concentration to substantiate the respective regressions.

A more universal prognostic cloud water parameterization scheme for use in global climate models has been developed by Del Genio et al. (1996). A very important feature of the scheme is that it allows for life cycle effects in stratiform clouds and permits the cloud optical properties to be determined interactively. All important microphysical processes have been taken into account in the parameterization: autoconversion, accretion, Bergeron-Findeisen diffusional growth, and cloud/rain water evaporation. Small-scale dynamical processes, including a detrainment of convective condensate, cloud-top entrainment instability, and stability-dependent cloud physical thickness variations have been also considered. Cloud optical thickness is calculated from the predicted liquid/ice water path and a variable droplet effective radius estimated by assuming the droplet number concentration to be constant. Microphysical and radiative properties are assumed to be different for liquid and ice clouds, and for liquid clouds over land and ocean.

The parameterization developed by Del Genio et al. (1996) has been used in the baseline GISS GCM, Mode II ($4° \times 5°$ horizontal resolution with nine vertical levels). The principal aim of numerical modelling has been to validate the model. In this context, Del Genio et al. (1996) have justly emphasized a necessity for climate models which would predict a change, to concentrate the validation on the variability of the current climate, but not only the mean state. Successful results of the validation have been judged on the basis of the model's depiction of both the mean state; diurnal, seasonal and inter-annual variability; and the temperature-dependence of cloud properties.

As far as the results of the modelling are concerned, it has been shown that a realistic simulation of the tropical cumulus anvil radiative properties and their variation is crucial for a plausible climate sensitivity estimate. Two conclusions of general significance are that:

- climate sensitivity to greenhouse gas increases can be determined only in the context of actual climate change scenarios with coupled atmosphere–ocean models;
- the "thermostat" concept (to be discussed in detail in the next paragraph) is too simplistic to apply to global climate change, since the tropical anvil properties depend on the general circulation and thus the SST gradient rather than merely responding to local changes in SST.

A new simplified parameterization scheme for the penetrative cumulus convection suitable for use in GCMs has been developed by Zhang and McFarlane (1995). The scheme has been used in the Canadian Climate Centre GCM. There were a number of important results obtained including the conclusion that an increase of the albedo of cirrus clouds that occurs over warm pools of the ocean results in smaller global mean increases of surface temperature and evaporation. Also, it has been found that an introduction of penetrative convection alone enhances the global mean surface warming due to a CO_2 doubling but allowing an additional feedback associated with the enhancement of the cirrus cloud albedos over the warm ocean surface nearly offsets this. This is associated with a reduced radiative input at the surface, which is balanced mainly by a reduction in heat loss due to evaporation.

2.4.6 Excess absorption of solar radiation by clouds

2.4.6.1 Introduction

Li et al. (1995) have justly emphasized that "after more than 40 years of work, both theory and observation of the absorption of solar radiation by clouds are still fraught with uncertainties" (see also Crutzen and Ramanathan, 1996). A new illustration to this conclusion is a very intensive discussion of the problem of the so-called "excess or anomalous" absorption of solar radiation by clouds (Ramanathan et al. (1996) are correct in their statement that there is no reason to talk about "anomalous absorption" but "it is safe to consider this phenomenon as simply 'excess absorption' to point out that there is excess solar absorption in cloudy atmospheric columns, when compared with clear sky column absorption").

The basic problem under discussion is relationship between absorption of shortwave radiation (SWR) by the atmosphere and the Earth's surface which has been obtained on the basis of observations and numerical modelling. Cess et al. (1995) have come to the conclusion that "globally, GCMs that constrain their planetary albedos with satellite data, may overestimate the solar energy reaching the surface by as much as 8% or equivalently 25 W/m^2 (Ramanathan et al., 1996).

The principal purpose of this section is to answer the question why models may overestimate the solar energy reaching the surface. The answer to this question is not at all simple, especially because the problem has a number of aspects to be considered. It has been known for quite a long time (Arking, 1991; Kondratyev, 1969, 1972, 1983a,b, 1992b; Kondratyev and Binenko, 1984; Marchuk et al., 1986; Stephens et al., 1978; Stephens and Tsay, 1990) that models underestimate absorption of SWR by clouds, but it is just an aspect of the problem.

2.4.6.2 Observational data

Of course, the first aspect to be clarified is the reliability of observational data. The situation in this respect is fairly complex. What we need to know is SWR absorption (SWRA) by the atmosphere to then determine, how much solar energy is absorbed by the surface. SWRA (or what is often called atmospheric or cloud-radiative forcing (CRF) due to shortwave radiation) may be found from combined surface and satellite observations. While the latter are really of global scale, the surface solar radiation measurements are rather fragmentary and therefore existing global set of data cannot be considered as adequately representing the global distribution of SWR. Besides, different spatial scales of surface point observations and satellite data averaged over large territories complicate a combined analysis of such data (important comments in this respect have been made by Arking et al, 1996).

What we need to determine is CRF which is the difference of shortwave radiation balances—SWRB (net SWR)—at the top of the atmosphere (TOA net SWR values) and at the surface. SWRB itself is the difference between upward and downward SWR fluxes (irradiances). Thus, a necessity arises to twice differentiate measured values of SWR fluxes. Undoubtedly, it might lead to serious errors in the determination of CRF. Unfortunately, persuasive assessments of such errors do not exist yet. Certain efforts have been made to estimate errors in case of aircraft and balloon observations of SWR vertical profiles in the atmosphere (Kondratyev, 1969; Kondratyev et al., 1976; Kondratyev and Binenko, 1984; Stephens et al., 1978; Hayasaka et al., 1995a,b; and others).

In view of the absence of sufficiently reliable estimates of observational errors we would like to emphasize here the urgent necessity of careful consideration of this problem. An important aspect of the problem is the use of various assimilation techniques to process inhomogeneous (in space and time) data series, which also requires a critical analysis. An illustration of this problem is the determination of clear-sky RF values which is unavoidably based on various kinds of extrapolation (see Arking et al., 1996).

2.4.6.3 Effect of clouds on atmospheric absorption of solar radiation

It so happened that the discussion on the redistribution of solar energy between the atmosphere and the surface was focused on assessments of cloud impact on atmospheric absorption of solar radiation (Cess et al., 1995; Li et al., 1995; Ramanathan et al., 1995; and later publications), since it was obvious *a priori* that models underestimated cloud absorption. Ramanathan et al. (1996) have correctly noted that "some of the fundamental flaws of our radiation models, at least those used to assert the zero net cloud effect, is that they assume: (1) clouds are flat plates with horizontally homogeneous properties, (2) cloud drops and crystals are made of pure water, (3) the absorption is by Lorentzian lines with arbitrary specified wavelength cut-offs, and (4) poor treatment of aerosol effect". One more important feature to be mentioned is spatial inhomogeneity of cloud cover.

To assess SWR absorption by clouds, Cess et al. (1995) have undertaken an analysis of the data of simultaneous satellite and surface observations of the SWR fluxes at four locations: American Samoa (14.25°S, 170.56°W), Barrow (71.32°N, 157°W), Boulder (40.05°N, 105.01°W) and Cape Grim (40.67°S, 114.69°E), and at 11 stations located in the state of Wisconsin. The data of pyranometric observations of the upward and downward SWR fluxes make it possible to calculate the shortwave radiation budget (SWRB), and satellite data contain information on the outgoing shortwave radiation (OSWR). The SWRB and OSWR differences for real cloud conditions and clear skies characterize the shortwave radiative forcing $C_s(S)$ and $C_s(TOA)$ at the levels of surface and atmospheric top, respectively (the satellite data were averaged over $1° \times 1°$ or $2.8° \times 2.8°$ lat.-long. grid).

From the data for Boulder, the mean diurnal $C_s(S) = -92.6$ W/m^2 and $C_s(TOA) = -63.2$ W/m^2, that is, $C_s(S)/C_s(TOA) = 1.46$, whereas the calculated $C_s(S)/C_s(TOA) \approx 1$, from which it follows that the calculated SWR absorbed by the cloudy atmosphere is substantially underestimated (by $C_s(S) - C_s(TOA) = 30$ W/m^2).

In view of possible errors in the estimates, an alternative technique has been applied based on estimation (by linear regression) of the variable

$$\beta = -dA_{TOA} / d(Q_S / Q_{TOA})$$

where A_{TOA} is the albedo of the surface–atmosphere system and Q is insolation at the levels of surface (S) and at the top of the atmosphere (TOA). From the data of observations at Boulder, $\beta = 0.59$, whereas calculations with the ECMWF (the European Centre for Medium Range Weather Forecasts) and CCM2 (the National Center for Atmospheric Research Community Climate Model) climate models gave 0.70 and 0.81, respectively.

An agreement of the calculated β values with observations is only possible with the calculated absorption supposedly underestimated. It is easily seen that

$$C_s(S) / C_s(TOA) = (1 - A_s) / \beta$$

where A_s is the surface albedo. For the Boulder conditions ($A_s = 0.17$) the quantity $\beta = 1.41$ agrees well with the estimate given above. All the other locations are also characterized by underestimated calculated absorption (with an exception of the cases of high snow surface albedo).

Cess et al. (1995) believe that, though the water vapour content in the cloudy atmosphere is greater than with the clouds absent, this circumstance, as illustrated by respective estimates, cannot explain the growth of SWR absorption. The universal character of the considered "anomalous" ("excess") absorption by clouds (its independence on the location of the observation point) prompts one to reject also a possible role of the aerosol effect on the optical properties of clouds in view of the strong spatial and temporal variability of the aerosol concentration and properties.

Thus the problem consists, presumably, in inadequacy of the present ideas of the optical properties of clouds. With the mean global surface albedo assumed to be 0.1, the excess absorption of solar radiation by clouds not considered in present climate models, turns out to be about 25 W/m^2, that is, it exceeds the $2 \times CO_2$ -induced enhancement of the greenhouse effect by almost an order of magnitude (Borisenkov and Kondratyev, 1988). Therefore there is an urgent need to understand the nature of this absorption and its consideration in numerical climate modelling. In accordance with Cess et al. (1995) data, the mean global mean annual value of the shortwave cloud–radiation forcing $C_s(TOA)$ characterizing the effect of clouds on the shortwave radiation budget of the surface–atmosphere system (the difference in the SWRB in the system for mean conditions of clouds and clear skies) varies from -45 to -50 W/m^2 (of course, the clouds with their high albedo reduce the radiation budget). The value

$$C_s(\text{TOA}) = C_s(\text{S}) + C_s(\text{A})$$

is determined by the sum of the contributions from the surface (S) and atmosphere (A). If SWR absorption grows in the presence of clouds, then $C_s(\text{A}) > 0$.

To analyse the formation of $C_s(\text{TOA})$, the data of observations of the surface heat balance and heat transport in the ocean for the region of the "warm pool" (WP) in the western Pacific (140–170°E; 10°N–10°S) have been processed by Ramanathan et al. (1995). In this region the SST reaches a mean annual value of ~302.5 K, maximum for the World Ocean. The WP region is characterized by the humid and cloudy atmosphere with the frequent occurrence of cumulus clouds (60% of cases). An unexpected feature of WP turned out to be a small mean annual dynamical heat transport (D) out of the WP mixed layer due to horizontal advection and vertical diffusion: $0 \leqslant D \leqslant 20\ \text{W/m}^2$, which constrains the long-term annual mean net downward surface heat flux (H) to small values, since over an annual cycle no net heating of the mixed layer should take place. This condition

$$H - D = 0$$

is used to close the WP heat balance from the data of observations. It turned out that the closing is only possible in the case of strong reduction of global radiation by clouds (> 100 W/m^2) which, in its turn, requires high values of $C_s(\text{A})$ (~35 W/m^2) in the region of WP, that is, an anomalously intensive SWR absorption by clouds.

The resulting heat flux H at the ocean surface can be represented as

$$H = S_c + C_s(\text{S}) - F - E - \text{h} \qquad (1)$$

where S_c is the clear-sky SWRB, F is the net longwave radiation, E is the heat loss due to evaporation, h is the turbulent sensible heat flux. The mixed layer net heating $Q = H - D$ is determined by two components: horizontal advection in the mixed layer (D_a) and downward entrainment of heat into the thermocline below (D_e). The available data of ship and satellite observations make it possible to determine all the heat balance components of (1) except for $C_s(\text{S})$ which is calculated as a residual term.

Calculations made by Ramanathan et al. (1995) show that the mean annual values of the components of (1) turned out to be: $S_c = 275\ \text{W/m}^2$ (with possible variations within 270–280 W/m^2), $F = -45\ \text{W/m}^2$ (35–45 W/m^2), $E + h = -110\ \text{W/m}^2$ (90–125 W/m^2). In this case $C_s(\text{S}) = -100\ \text{W/m}^2$ (80–135 W/m^2) (for the region between 140°E to 170°E and 10°N to 10°S).

The following relationship corresponds to the optimal values of the components of (1)

$$f_s = C_s(\text{S}) \,/\, C_s(\text{TOA}) = 1.5$$

whereas calculations give $f_s < 1.2$. The results obtained by Ramanathan et al. (1995) reveal the fundamental gap in the present understanding of the effect of clouds on the SWR transport mentioned above. The value $f_s = 1.5$ means an increase of absorption by the atmosphere in the WP region by 35 W/m^2 (compared to 100 W/m^2 for the cloud-free atmosphere), and the oceanic mixed layer heating due to solar radiation absorption is, respectively, reduced by 35 W/m^2. All this means a change in the meridional energy transport in the ocean and in the atmosphere from the tropics to mid-latitudes within 25–50%, which, of course, is of fundamental importance from the viewpoint of climate formation.

To check these results from the data of direct observations, Pilewskie and Valero (1995) carried out simultaneous aircraft pyranometric measurements of total (the wavelength interval 0.3–4.0 μm) upward and downward SWR fluxes in the tropical Pacific (140–180°E; 0–15°S) using a DC-8 (Learjet) aircraft (top altitudes 8 to 12 km) as well as a stratospheric ER-2 aircraft flying at an altitude of ~20 km (near the tropopause). The goal of measurements was to obtain data on the contribution of clouds to the SWR absorption by the atmosphere between the two aircraft. Through an extrapolation, data for the whole atmosphere were obtained.

Analysis of the observational results revealed a strong SWR absorption by clouds: on the average, the SWR absorbed by clouds reached 165 W/m^2, whereas, according to calculations, it was negligible. These results agree well with the estimates from the data on the ocean heat balance mentioned above. Since the results of spectral measurements revealed only specific selectivity typical of water, a supposition of possible explanation of excess absorption by cloud pollution cannot be accepted. Therefore Pilewskie and Valero (1995) believe that the nature of absorption remains unclear (probably, the effects of the broken cloud morphology causing an increase of the length of the free path of photons play some role).

Evans et al. (1995) have accomplished surface measurements of the solar shortwave flux in Peterborough (Ontario, Canada) in January and February 1995, using equipment based on solar cell sensors and pyranometers. These observations have yielded some information concerning the excess cloud absorption through the comparison of SWR fluxes for clear and overcast skies. On the average, the column absorption under heavy clouds was about 50 W/m^2, up to a maximum of about 75 W/m^2.

In general agreement with the data by Cess et al. (1995), the CRF ratio was found to vary from 1.4 to 2.0.

The problem of "anomalous" absorption by clouds would not have been highlighted as has been done in some publications (Cess et al., 1995; Ramanathan et al., 1995) if their authors were acquainted with the results of complex studies of cloud–radiation interaction carried out under the programmes CAENEX (Complex Atmospheric Energetics Experiment), GAAREX (Global Atmospheric Aerosol Experiment), FGGE (First GARP—Global Atmospheric Research Programme—Global Experiment) (Kondratyev, 1972, 1983a,b, 1992d; Kondratyev et al., 1976, 1996; Marchuk et al., 1986), and with recent results of calculations of radiation transport for both overcast and partial cloudiness (Kondratyev et al., 1996). In this connection, here are some illustrations of the respective results.

The vertical soundings of the cloudy (5 October 1972) and cloud-free (6 October 1972) atmosphere over the Azov Sea revealed a strong transformation of both total and spectral radiative characteristics of the atmosphere on these days, determined by the emission of aerosol from the industrial zone of the cities Donetsk and Zaporozhye (the aerosol optical thickness of the atmosphere from the data of actinometric measurements $\tau_a = 0.35$). Analysis of the data of aircraft spectral measurements in the cloudy atmosphere carried out during the CAENEX (Kondratyev and Binenko, 1984) revealed: (1) the SWR absorption by clouds (10 points, S_c) in the visible is close to neutral; (2) an active role of the cloud top in the SWR absorption; (3) maximum wavelength dependence of absorption in the oxygen and water vapour absorption bands; (4) the cloud aerosol-induced strong absorption at $\lambda = 0.5$ μm (up to 0.15) compared to more "clean" clouds over the Black Sea (10 April 1971), when the relative absorption was 0.03 (the optical thickness of clouds was 25 and 19, respectively).

Simultaneous measurements of the attenuation coefficient ε and size distribution parameters of the stratified clouds 550 m (5 October 1972) and 450 m (10 April 1971) thick revealed: (1) the cloud top (within 50–100 m) has maximum ε, maximum values of water content, number concentration and modal radius of droplets; (2) the possibility of estimating the average droplet radius $\bar{r}$ with the use of the dependence $\bar{r} = 0.9\bar{w}/\rho\varepsilon$, where $\bar{w}$ is the mean water content, ρ is the water density, and $\varepsilon = \sigma + \kappa$ is the volume attenuation coefficient expressed as a sum of scattering and absorption coefficients; (3) the characteristic feature of marine clouds compared to clouds over land, consisting in a greater contribution of large particles to the size distribution spectrum and in water content which could result from the impact of industrial aerosol pollution of clouds over land.

Processing the data of measurements of the SWR total fluxes (0.3–3.0 μm) showed that the SWR absorbed by the layer 0.2–1.2 km in cloudless conditions (6 October 1972) for the sun elevation 38° was 35 W/m^2 but for the same layer in the presence of clouds and aerosol the absorbed SWR reached 91 W/m^2 (5 October 1972). Thus the contribution of absorption due to the cloud (as a difference of these values) constitutes 56 W/m^2. If we estimate the residual absorption by cloud, both total (0.3–3.0 μm) and in the visible (at $\lambda = 0.5$ m), in the climatological terms of other studies (Cess et al., 1995; Ramanathan et al., 1995) for the atmospheric layer 0.2–8.5 km, then $C_s(S)/Cs(TOA) = 140/118 = 1.16$, and for $\lambda = 0.05$ μm: $C_s(S)/C_s(TOA) = 45/18 = 2.5$, which points to an excess absorption by clouds in the visible. Since an object of the study was very horizontally homogeneous extended stratified cloudiness (it should be remembered that an element of the sub-grid climatological analysis is 250 × 250 km^2), the results of local aircraft soundings of the atmosphere could be used to assess the radiative forcing of clouds in the layer up to 8.4 km. The same estimates for more clean clouds over the Black Sea at the sun elevation 53° gave values $f_s = C_s(S)/C_s(TOA) = 1.11$ and 1.18, respectively. The radiation measurements performed on clear and cloudy days over the industrial town of Rustavi (Georgia) enabled one to estimate the ratio $C_s(S)/C_s(TOA)$ at 1.07 from the data of pyranometric measurements and 1.33 for $\lambda = 0.05$ μm.

Complex measurements carried out under the GAAREX programme over Zaporozhye (the sun elevation was 50°) and outside the city (on the windward side) have shown that the city intensifies the SWR absorption by clouds over the city, compared to conditions outside the city, within 21 W/m^2 to 77 W/m^2 (depending on the optical thickness of clouds at the wavelength 0.5 m varying from 16 to 38). Therefore the SWR absorption prevails over the cloud longwave cooling. The values of $C_s(S)/C_s(TOA)$ varied from 1.2 to 2.6, which reflects an intensifying effect of clouds and pollution aerosols on the SWR absorption. Analysis of the data of sounding of the cloudy atmosphere over the Ladoga lake (the sun elevation was 26°, $\sigma = 80$) gave 1.13 for the ratio $C_s(S)/C_s(TOA)$ within the wavelength interval 0.5–1.8 μm and 1.8 for $\lambda = 0.5$ μm.

In the case of the surface with a high albedo (ice or snow) the $C_s(TOA)$ differences for clear and cloudy weather are small and therefore the relative error in estimating the radiative flux divergence, cloud absorptance and the ratio $C_s(S)/C_s(TOA)$ can reach 60% and more. Thus, the estimates of the SWR absorption by clouds become unreliable.

Based on the use of measured values of spectral downward and upward radiation fluxes and asymptotic formulas of the theory of radiation transfer in clouds, the spectral dependences of the coefficients of scattering σ and absorption κ for a polydisperse turbid medium have been retrieved. With the use of analytical formulas the single scattering albedo $\omega_0 = \sigma/(\sigma + \kappa)$ and the imaginary part of the complex refraction index were estimated from the data of the aircraft soundings of the cloudy atmosphere mentioned above (Kondratyev et al., 1996b).

The calculated values of σ and κ agree with the measured values of the attenuation coefficients and cloud optical thickness. The results of calculations with account of multiple scattering both molecular and on droplets for $\sigma = 30\ km^{-1}$ as well as with account of the mean aerosol absorption coefficient in the cloud ($0.08\ km^{-1}$) make it possible to explain the excess absorption of solar radiation by clouds in the spectral interval 0.4–0.7 μm by the effect of multiple scattering which enhances aerosol and cloud droplets absorption (a certain contribution is, apparently, made by the changing optical properties of dirty clouds). An excess SWR absorption by clouds in the visible occurs always, and the presence of aerosol in clouds enhances this radiative effect depending on its scattering and absorbing properties (σ and κ) and on the place of its location: between droplets (interstitial aerosol), inside them, or on the surface.

Hayasaka et al. (1995b) conducted simultaneous observations of upward and downward total shortwave radiation fluxes below and above stratocumulus clouds in the western Pacific with the help of two aircraft equipped with pyranometers. The results of observations compared with Monte Carlo simulations show that as a result of horizontal inhomogeneity of cloud cover and respective divergence or convergence of radiation fluxes an additional absorption seemed existing which actually did not exist. Relevant corrections to remove the influence of horizontal inhomogeneity have been suggested resulting in complete coincidence of measured and calculated (for horizontally homogeneous cloud layer) absorption values. It may be concluded that:

(1) Aircraft measurements and Monte Carlo simulations suggest that horizontal convergence or divergence of radiation fluxes in the inhomogeneous cloud causes large apparent vertical flux absorptance.
(2) If apparent flux divergence is removed through a relevant correction of

observation data, which were obtained in synchronized aircraft observations, these data become consistent with the theoretical absorptance obtained by the plane-parallel cloud model approximation.

(3) Water vapour above and within the cloud seems to be important for the absorptance of clouds as well as cloud liquid water content.
(4) It is inferred that the anomalous absorption pointed out by aircraft observations in previous studies does not exist.
(5) It is expected that the cloud droplet size distribution is not so effective to the cloud absorptance in the usual size range.

Broken-cloud enhancement of solar radiation absorption has also been discussed by Byrne et al. (1996).

Summarizing some of the results discussed above and their own results of processing a 4-year global record of solar flux observed from both space and the Earth's surface, Li et al. (1995) have pointed out that f_s values are highly variable in the tropics with a median of about 1.1, and consistently less than 1.0 in polar regions. They have proposed that "large values and high variations of f_s may be related to the presence of absorbing aerosols and the uncertainties in both the observed and inferred solar flux data used here. Therefore, a substantial revision of our understanding of cloud absorption and its impact on the atmosphere's energy budget may not, after all, prove to be necessary if the effects of absorbing aerosols are properly incorporated" (unfortunately, like Cess et al. (1995); Ramanathan et al. (1995), and some others, Li et al. (1995) have failed to reference older publications on this very subject by Kondratyev, 1972, 1983, 1988, etc., as well as by Kondratyev et al., 1976, 1983; Kondratyev and Binenko, 1984, etc.).

The assumption made by Li et al. (1995) that $f_s > 1$ due to the impact of absorbing aerosol is based on the facts that in the tropics strongly absorbing aerosols produced by biomass burning could play an important role, whereas in mid-latitudes maximum f_s values were obtained near Hamburg and the Rhine Valley, that is, in heavily polluted areas. In all cases for a sizeable fraction of months $f_s < 1$ which may be explained by the influence of cirrus clouds (backscattering by cirrus clouds leads to a reduction of absorption by aerosols located below the clouds). Li et al. (1995) have pointed out that although the observed variation of f_s has been explained by changes in solar zenith angle and aerosol effects, the potential dependence of f_s on clouds structure (morphology) also deserves examination. They conclude that their study "does not rule out the existence of the cloud absorption anomaly, but rather indicates that its magnitude (if it exists) on a global scale may not be as large as suggested in some recent reports".

The problem of "excess absorption" has been further discussed by Stephens et al. (1996). Arguing against an approach to the interpretation of observational data, used by Cess et al. (1995) as well as by Pilewskie and Valero (1995), Stephens (1996) pointed out that "without convincing error analysis, without a reproduction of the results using different analyses and different data, and without an explanation of why other published results are inconsistent with their own, then it is difficult to evaluate the findings of Cess et al. (1995)".

In response to this statement, Cess and Zhang (in Stephens et al., 1996) have emphasized that: (1) Stephens reflects a traditional viewpoint in stating that cloud "absorption occurs in place rather than in addition to clear sky absorption", which cannot be accepted; (2) "Stephens interprets the data in the Report by Cess et al. (1995) with the use of an atmospheric radiation model that adopts conventional plane-parallel clouds".

As far as the impact of cloud cover horizontal inhomogeneity on assessments of "excess absorption" is concerned, Cess and Zhang pointed out that since they used satellite and surface observations, this argument (which may be true for aircraft measurements for isolated clouds when radiation escapes from the sides of the clouds as Hayasaka et al. (1995a,b) have shown) is not relevant. Both satellite–surface (Cess et al.) and aircraft (Pilewskie and Valero) measurements refer to cloud systems. In this context, Hayasaka et al. (1995a,b) concluded that the "simulation suggests the shortwave absorptance of inhomogeneous clouds can be evaluated reasonably by means of appropriate spatial averaging". Therefore Cess and Zhang conclude that "the model's clouds are underpredicting cloud SW absorption by overestimating cloud-sky surface insolation relative to the clear skies; we see no other plausible explanation".

Pilewskie and Valero (Stephens et al., 1996) find several errors in arguments suggested by Stephens.

First of all, in contradiction with Stephens' statement, the aircraft data are not consistent with simulation modelling results. Not all of the absorption by clouds but only some cloud absorption (in view of considerable overlap in the absorption bands of liquid water and water vapour) occurs in place of clear sky absorption. A two-aircraft technique and averaging procedure applied by Pilewskie and Valero made it possible to successfully solve the methodical problems of measurements mentioned by Stephens. Thus, a general conclusion is that the results obtained by Cess et al. (1995) and by Pilewskie and Valero (1995) are mutually consistent and correct.

As has been mentioned above, in connection with studying the role of aerosol–cloud interaction, Twohy et al. (1995) have conducted 13 research flights (September–October 1989) in stratiform clouds within 300 km of the northern Californian coast to measure aerosol and droplet size distribution as well as the chemical composition of cloud droplets. Microphysical and chemical properties indicated that most of the clouds were influenced by pollution from the North American continent, but pristine marine clouds were sampled on one flight during westerly flow conditions.

In the polluted case, particle and droplet number concentrations decreased, mean droplet size increased, and satellite-derived reflectance at 3.7 μm (AVHRR data) decreased with an increasing distance from the northern California urban region. Relative to the unpolluted stratiform cloud, the polluted cloud had, on the average, a sulphate concentration that was higher by an order of magnitude, droplet number concentration higher by a factor of 6, droplet sizes smaller by a factor of 2, and 3.7-μm reflectance that was higher by a factor of 2. These differences were due to large differences in CCN number concentration in clean (usually fewer than 100 cm^{-3}) and polluted (up to 1000 cm^{-3} and more) atmospheres. However, no significant difference in the visible reflectance was detected between the two cases, probably, a result of differences in liquid water path.

Twohy et al. (1995) have pointed out that the location of aerosol layers above clouds, particle distributions within clouds, and the chemical composition of the cloud water supported the hypothesis that anthropogenic pollution was the primary source of the enhanced aerosol absorption. The cloud microphysical properties sampled as far as 300 km downwind of San Francisco were apparently influenced by pollution aerosols.

Thus, the study accomplished by Twohy et al. (1995) verifies that anthropogenic aerosol particles can have substantial effects on droplet number and size in stratus clouds on at least a regional scale. The climatic impacts depend on cloud lifetime and liquid water path as well as droplet size. Since the influence of enhanced CCN on these quantities is less straightforward, further studies are necessary to reliably assess the climatic implication of the aerosol–climate interaction. Twohy et al. (1995) have emphasized that several key mechanisms (e.g., the physical and chemical controls of particle number concentration, the efficiency with which particles are incorporated as CCN, the factors affecting liquid water path, and the influence of CCN number on drizzle production and cloud lifetime) need to be better understood to establish a definitive and quantitative relationship between an enhanced pollution and atmospheric cooling effects.

In the context of a conclusion concerning underestimation of shortwave cloud absorption in the present-day climate models, Kiehl et al. (1995) have undertaken assessments of such an underestimation using the NCAR CCM2 and assuming that additional shortwave cloud absorption leads to an additional 20 W/m^2 global annual average absorption in the atmosphere, with an equivalent reduction of shortwave flux at the surface. Calculations made by Kiehl et al. (1995) (the enhanced absorption effects are calculated by tuning the cloud top single scattering albedo to reproduce the observations of Cess et al., 1994) show that additional shortwave cloud absorption leads to warming of the upper troposphere by as much as 4 K, reduction of the strength of the Hadley circulation by 12% and to lowering surface wind speeds in the tropics. On the other hand, these lower wind speeds result in a 25 W/m^2 reduction in surface latent heat flux. A consideration of mean zonal values shows that the dynamical and thermodynamical changes in the circulation due to enhanced cloud absorption lead to significant response in the surface energy fluxes and the atmospheric hydrological cycle. Later calculations with an improved CCM2 (with added diagnostic cloud water capability) indicate a much stronger response to enhanced absorption. For example, the warming of the upper tropical troposphere is 5–6 K, that is, models with interactive cloud water may show a stronger response to the enhanced shortwave cloud absorption than the CCM2.

As has been mentioned earlier, Boucher (1995) has accomplished a numerical climate modelling for prescribed effective cloud droplet radius r_{eff} changes due to satellite remote sensing data (r_{eff} in the northern hemisphere is less than in the southern hemisphere by about 0.7–0.8 μm over land and 0.4 μm over ocean). The assessment has indicated that the radiative forcing by anthropogenic aerosols could be between about -0.6 and -1.0 W/m^2, averaged in the 0–50°N latitude band.

The uncertainty of these estimates is difficult to estimate but it is at least 50%.

Due to the average global estimate by Boucher and Lohmann (1995), the global average indirect radiative forcing is about -1.0 W/m^2 and ranges from -0.5 to -1.5 W/m^2. The total uncertainty in the forcing is certainly larger than this range.

Arking (1996) and Arking et al. (1996) as well as Chen et al. (1995) have confirmed the conclusion that the present-day global climate models underestimate the amount of solar energy absorbed by the atmosphere within the range of up to 25–30 W/m^2 but suggested a quite different explanation for this fact. Arking (1996) has found on the basis of both observations and models that clouds have little (~ 5 W/^{2}) or no effect on atmospheric absorption. Therefore his conclusion is that "water vapour is the dominant influence on atmospheric absorption, and improvements in our models lie in improving the parameterization of water vapour absorption and, perhaps, discovering of additions absorption in the clear atmosphere".

The discussion on the nature of excess absorption of solar radiation by clouds still continues and has resulted (as we have already seen) in controversial judgements. It is obvious that further theoretical and, first of all, complex observational field studies are necessary with specific emphasis on 3-D inhomogeneity of cloud cover and cloud optical properties (it should be also reminded that even an old problem of IR continuum absorption by water vapour has not been solved as yet). New efforts in this direction have been started within the ARM (Atmospheric Radiation Measurement) Programme (Wiscombe, 1995).

The principal conclusion from the studies discussed above is that there is neither "anomalous" nor "excess" absorption of solar radiation by clouds. The reality is such that there is a number of processes which may be responsible for the disagreement between observed and modelled redistribution of absorbed SWR between the atmosphere and the surface, including: (1) multiple scattering in clouds with absorbing droplets (they never consist of pure water) which leads to the enhancement of cloud absorption; (2) specific features of radiation transfer in a 3-D inhomogeneous medium (broken clouds); (3) unknown contribution of absorption by water vapour (besides what is known); (4) peculiarities of radiation transfer in clouds consisting of a mixture of water droplets and aerosol particles. Last but not least, reliability of observations should be further assessed. It has been demonstrated by Marley et al. (1993) that water-soluble infrared absorbers which can contribute to the longwave radiative forcing of clouds, fog, and aerosols deserve serious attention.

2.4.6.4 Aerosol impact on atmospheric absorption of solar radiation

At the end of the 1940s, V. Kastrov from the Central Aerological Observatory (Moscow) started his long-term aircraft pyranometric observations of fluxes in the free atmosphere under various conditions over the European part of the former USSR territory (see Kondratyev, 1954, 1969). The most important result of his measurements has been a discovery that atmospheric aerosol absorption of solar radiation is close, by its magnitude, to water vapour absorption. Later on, balloon and aircraft observations within the CAENEX and GAAREX Programmes were conducted which confirmed Kastrov's results (Kondratyev, 1972; Kondratyev et al.,

1976; Kondratyev and Binenko, 1984). Of special significance were aircraft spectral measurements of SWR fluxes under clear sky and overcast cloudiness conditions with simultaneous measurements of aerosols and cloud properties (aerosol counters and filters were used which allowed one to obtain information not only on the number, concentration and size distribution of aerosol and cloud particles, but also on their chemical composition and, hence, optical properties).

Aircraft measurements made in various parts of the former USSR (European territory, Central Asia, the Arctic, Kamchatka and Chukotka) as well as over the tropical Atlantic confirmed the principal result mentioned above: aerosol absorption of solar radiation is always significant and on the average is approximately equal to absorption by water vapour. Observations in the presence of overcast (horizontally homogeneous) cloud cover in the polluted atmosphere of industrial regions revealed the existence of exceedingly strong SWR absorption by dirty clouds (see also Chlek and Wong, 1995; Stephens, 1994a).

It is quite clear, thus, that in the context of the problem discussed, the aerosol impact on atmospheric absorption of solar radiation requires serious attention (Kondratyev, 1996c).

Very specific aerosol dynamics has been observed in the Arctic. During the late winter–early spring period the Arctic Haze is formed (Shaw, 1987) while in summertime, when the high latitude atmosphere is not subject to the impact of pollution transport from the south, natural processes of marine dimethylsulphide formation and subsequent gas-to-particle reactions of sulphate aerosol formation in the atmosphere take place. To test the hypothesis that marine biogenically produced dimethylsulphide (DMS) gas can exert a significant global climatic control through the formation of sulphate aerosol and its subsequent impact on cloud properties, the Atmospheric Programme of the International Arctic Ocean Expedition (IAOE-91) was accomplished during the time period from 1 August to 6 October 1991 within the latitude range 75° to 90°N (Leck et al., 1996). A broad programme of aerosol measurements (number, concentration and size distribution in the diameter range 3 nm to 500 nm) as well as seawater DMS and distribution of phytoplankton biomass studies were conducted on board the Swedish icebreaker *Oden*.

Some of the results discussed by Leck et al. (1996) and Leck and Persson (1996) show that concentrations of seawater and atmospheric DMS decreased at about the same rate during the period of measurements spanning about three orders of magnitude. Methane sulphonate and non-sea-salt sulphate (nss) in the submicrometre particles, which may be derived from atmospheric DMS, also decreased similarly, suggesting that the first part of the hypothesis mentioned above was true. Three distinct modal sizes were usually present: the "ultrafine", Aitken and "accumulation" modes centred at 14.45 and 175 nm diameter, respectively. The presence of ultrafine particles, implying recent production, was more frequent than has been found in lower latitude remote marine areas. Evidence suggests that they were mixed to the surface from higher levels.

Leck et al. (1996) have emphasized that sudden and often drastic changes in aerosol concentration and size distribution were surprisingly frequent in view of the relatively slowly changing meteorology of the central Arctic during the expedition period and the absence of strong pollution sources. They were most common in particles likely to have taken part in cloud formation (> 80 nm diameter). Leck and Persson (1996) have pointed out that although the limited data obtained indicate the likelihood of a relationship between DMS and degraded phytoplankton material (phaeopigments), seawater DMS showed no simple correlation with phytoplankton standing stock over the large areas and different seasons covered. The area weighed summer and winter fluxes of DMS from the Arctic Polar Ocean to the atmosphere were estimated at 2.0 and 0.03 μmol /m^2 day, respectively. On an annual basis, winter biogenic sulphur emissions are negligible compared to the summer emissions in the region. The total emissions of marine biogenic sulphur from the northern hemisphere high latitudes was estimated at approximately 4 Gmol/yr. In view of the importance of the results discussed further, more complete (including cloud and radiation measurements) studies are necessary to assess climatic impact of natural sulphate aerosols in the central Arctic.

The most important general conclusion is simple and obvious: our knowledge of radiation transfer in real atmosphere is far from being adequate. Further studies are needed to reach a sufficiently adequate level of understanding. Only on such a basis necessary improvements of radiation parameterization in climate models will become possible.

2.4.7 Climate diagnostics

The key aspect of the climate change problem is connected with an assessment of the present-day basic observed features of global climate and its changes in the context of numerical modelling results. It is of interest to analyse the progress made recently in this field. This analysis is favoured by the recent publication of the " Supplement" to the IPCC report (IPCC 1992...,) as well as by some other studies summarizing the results available (Jastrow et al., 1990; Jones et al., 1996; Kodera, 1995, 1996; Kodera et al., 1996; Kondratyev and Galindo, 1997; Kondratyev and Cracknell, 1998a,b; Ratcliffe, 1995; Rind, 1996; Sreenath, 1993; The Global..., 1995; Wallace et al., 1995, 1996; Wang and Ye, 1993; Yatagai and Yasunary, 1994; Ye, 1993, and many others) and the work on the IPCC-1995 preparation (Climatic Change News, 1995; The Berlin Mandate..., 1995; The Climate Agenda..., 1995; The GCCIP Newsletter, 1995), including IPCC-95 Report itself (Houghton et al., 1996).

Of course, it would be unwise to expect that during several years important changes could take place in the fundamental ideas of climate, in general, and the role of the greenhouse effect, in particular. The main conclusions of the report (Climate Change, 1990) have remained basically unchanged (Fig. 2.5 presents a summary of observed climatic trends during the instrumental period of record from IPCC-95 (Houghton et al., 1996):

(1) There is a substantial increase of anthropogenic emissions of greenhouse gases: carbon dioxide, methane, CFCs, nitrous oxide (Etheridge et al., 1996; Raynaud, 1996; Rosema et al., 1993; Schlesinger, 1991).
(2) Since the late nineteenth century the mean global SAT has increased within 0.3–0.6°C, and by about 0.2 to 0.3°C over the last 40 years. Recent years have been

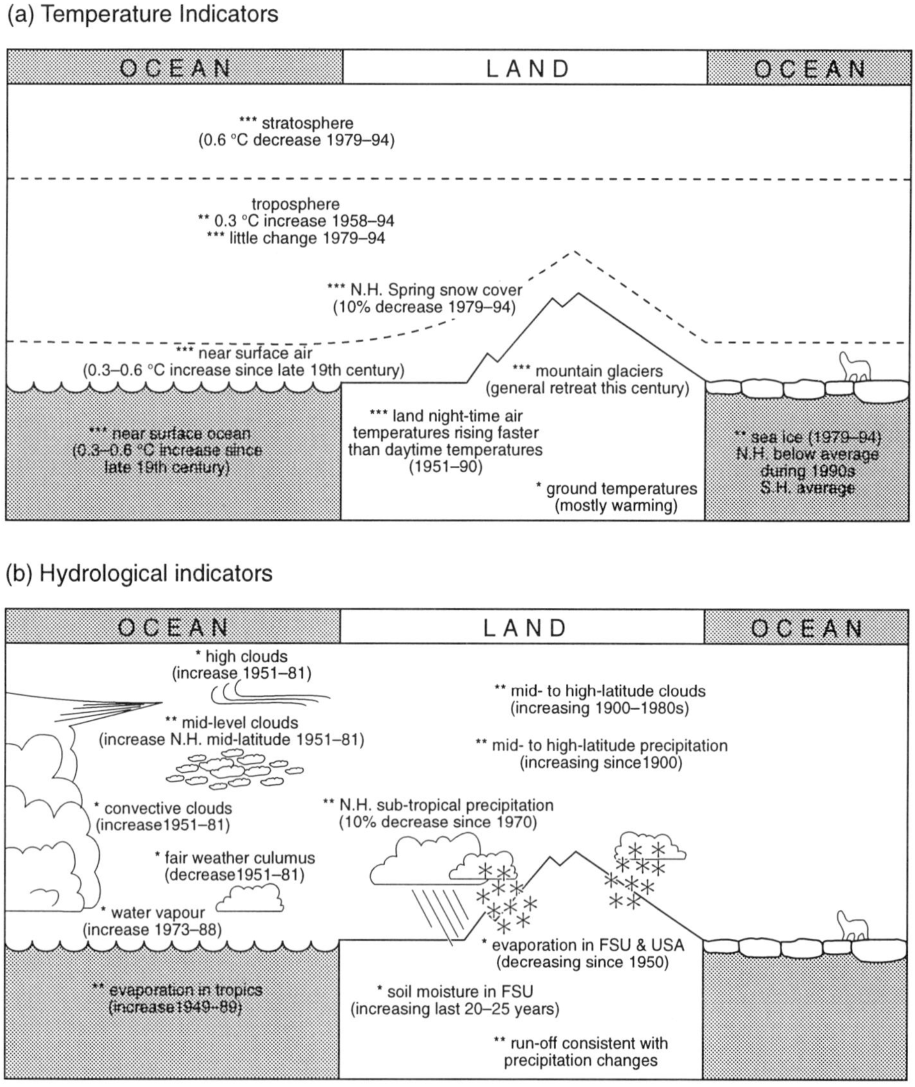

Fig. 2.5. Summary of observed climatic trends during the instrumental period of record.

among the warmest since 1860, that is, in the period of instrumental record. Fig. 2.5 data present a summary of observed climatic trends.

It is important, however, to remember about the inhomogeneity of SAT variations in space and in time (Kodera, 1995, 1996; Wallace et al., 1995, 1996). An illustration are the results discussed by Jones (1995) of recent variations in mean temperature and the diurnal temperature range in the Antarctic. All but three stations (from nearly 20 stations for the time period since IGY, 1957) show an increase of monthly mean SAT, amounting on the average to 0.57°C (1957–94). All of this warming occurred before the early 1970s. Since then there has been no change. An average continental diurnal temperature range shows no trend over 1957 to 1992. Jones et al. (1995) believe that the marked decrease in the mean SAT over Antarctica during 1993 and 1994 is unrelated to sea-ice variations which have shown little change since the early 1980s.

Later on, Jones (1996) has discussed new data on the climate dynamics for Oceania (Australia, New Zealand and the South Pacific). Analysis of the data for the past 150 years shows that a long-term warming is evident in New Zealand (1.1°C since the 1890s) and Australia (0.7°C since 1908) but the islands north of the South Pacific Convergence Zone only began warming in the late 1970s, which may be related to the increasing frequency of El Niño events.

Since many land-based climate records indicate a decrease in the diurnal temperature range through the present century, Balling and Christy (1996) have assessed the spatial and temporal trends in the diurnal temperature range over the period 1979 to 1994. Microwave sounding data (MSU) from NOAA satellites have been used to calculate the tropospheric temperature difference between the 19.30 LST and 07.30 LST temperature values as an indicator of the true diurnal temperature range. The results show that the diurnal temperature range is: (a) larger over land areas than over ocean areas; (b) high in summer, particularly in low latitudes, and low in winter, especially in the higher latitudes, and (c) decreasing substantially in the winter season for the land areas between 30°N and 60°N.

(3) The size of climate warming agrees, on the whole, with the forecasts based on climate models, but it is of the same magnitude as natural climate changes. Thus, the observed SAT increase could be determined by natural variability, but, on the other hand, such a variability and other anthropogenic impacts could compensate still more substantial (than was supposed) anthropogenic "greenhouse" warming. The conclusion of IPCC-95 (Houghton et al., 1996) is that "the balance of evidence suggests a discernible human influence on global climate".
(4) Estimates based on analysis of observational data and numerical climate modelling make it possible to consider as most probable the following limits to the mean annual mean global SAT increase with a CO_2 doubling: 1.5–4.5°C.
(5) Inadequate understanding of the laws of climate determines numerous uncertainties of the forecasts of its possible future changes from the viewpoint of magnitude, time and regional spatial distribution of climate change.

In this context, of critical importance are studies of the contribution of internal variability of the climatic system accomplished by Kodera (1995, 1996) and Wallace et al. (1995, 1996). Kodera (1996) has analysed linear trends in the northern hemisphere 500 hPa geopotential height, lower troposphere temperature, and zonal-mean zonal winds for the period 1965–93 after removing variabilities related to the equatorial SSTs as "known factors" by means of a linear regression. The results indicate that during the winter, decadal trends originate from two different sources of variability. Trends over the Pacific Ocean are related to changes in the equatorial SST. The decadal trends over Eurasian-Atlantic sector, however, are related to the stratospheric polar-night jet and meridional propagation of planetary waves, which can be considered internal modes of variability in the winter circulation. During the summer, trends in the subtropical lower-tropospheric temperature are explained as a direct response to changes in the equatorial SST. Trends over eastern Siberia, however, cannot be explained in a similar way. Instead, a relation to trends in the extent of the spring snow cover has been suggested by Kodera (1996). These results confirm a conclusion made by Wallace et al. (1995, 1996) that the major portion of recent decadal trends in the NH winter temperature is dynamically produced, in response to changes in the atmospheric circulation pattern. The origin of warming over the Eurasian sector stays unexplained, however (Yatagai and Yasunari, 1994).

(6) A reliable detection of the enhanced greenhouse effect from the data of observations is hardly possible sooner than in ten years and more.

Along with the consistency of the enumerated conclusions, the last few years have been marked (as has been mentioned above) by a deeper insight into both the observed features of the global climate variability and its determining factors. We shall start with the climate diagnostics.

A thorough analysis of the data of surface and ship observations of SAT as well as the temperature of the upper ocean layer confirmed the previous conclusion that from the late nineteenth century till nowadays the mean global SAT increased by $\sim 0.5°$. To substantiate this conclusion, a detailed analysis of various corrections has been made (Jones and Briffa, 1992; Jones, 1994; Jones et al., 1996; Parker et al., 1995) to provide the global data base homogeneity; the regularities of the spatial and temporal SAT variability have been considered over this century and especially the SAT changes between 1981 and 1990 compared to the period 1931–40. An approach to an estimation of the errors in the mean annual mean global SAT values has been substantiated (Shen et al., 1994).

The existing possibilities of analysis of the SAT observations are confined to the period after 1860. Between 1860 and 1990 the observational network had gradually broadened, and by 1920 only the inland regions of Africa, South America and Asia, the Arctic coastline and the whole Antarctic had not been covered with regular meteorological observations.This enabled one to obtain the mean monthly SAT anomalies (with respect to the period 1951–70) over the grid $5° \times 5°$ lat.-long.. A consideration of anomalies was determined by the need to minimize the observation uncertainties determined by the factors such as changes in the observation techniques and location of the stations, the effect of urbanization, the representativeness dynamics (as the number of stations grows), and others. According to the estimates obtained (Houghton et al., 1996), the contribution of urbanization and desertification to the overall global warming could not exceed a few hundredths of a degree. Practically, starting from 1890, the changing number of stations has not affected the mean hemispherical SAT over the land.

Most of the marine database constitute the results of ship SST observations from 1854 to nowadays (the COADS database; Oort et al., 1987; Parker et al., 1995). Though the SST–SAT difference over the ocean can reach 6–8°C, these data are highly correlated, and therefore the SST anomalies can serve a sufficiently reliable substitute for SAT anomalies. It is very difficult, however, to provide the SST data homogeneity in view of the past radical changes in the observational technique (this is connected, first of all, with the transition from measuring the temperature of the water sampled with a linen bucket to measuring the temperature of the water pumped to cool the ship engines).

With the total mean annual warming in both hemispheres reaching about 0.5°C starting from the end of the last century, the NH SAT dynamics is characterized by a strong spatial and temporal inhomogeneity. So, for example, between 1940 and 1970, a 0.2°C mean hemispherical cooling took place in some seasons (in the southern hemisphere the warming was more monotonic). The NH spatial SAT dynamics is characterized by a considerable variability in the form of warming and cooling periods (the seasonal differences are at a maximum in the northern hemisphere). Whereas the SAT level in the summers and falls of the 1980s was approximately the same as in the 1930s and 1940s, the winters and springs in the 1980s were the warmest. The summer seasons of the 1850s and 1970s turned out to be as warm as during the last two decades.

The warming trend determined by the mean annual data manifests itself in all the seasons except in summer. The cooling observed in the 1870s and 1880s was most clearly manifested in summer. The characteristic feature of the NH data is the coolings in 1870, 1910 and in late 1930, but a warming in 1920. A comparison of data for the periods 1901–45 and 1946–90 revealed a maximum warming in April (0.28°C), and a minimum one in August and October (0.06°C), with a clear annual change with maximum warming in spring and minimum warming in autumn. Analysis of the combined data on SST anomalies over the land and ocean (with respect to the period 1950–79) revealed the same but weaker trends as over the land, though with some small differences. So, for example, the data for the NH land revealed in some seasons the periods of cooling which practically do not show themselves in the combined data. The latter show approximately the same SAT variability in both hemispheres.

The correlation of data for NH and SH turned out to be unexpectedly high (0.79 between 1901 and 1990). The effects of El Niño (the "warm years") and La Niña (the "cold years"), which testify to the importance of filtering out the contribution of these events in analysis of the CO_2 signal, have been carefully studied. The data in Table 2.20 characterize the mean decadal anomalies (with respect to 1950–79), demonstrating that the decade of the 1980s turned out to be the warmest over the last 40 years, with the spatial–temporal dynamics of the SAT field in the northern hemisphere being fairly homogeneous.

Table 2.20. Trends of the mean hemispherical and mean global SAT (°C)

Period	NH	SH	Globe
1861–1870	−0.26	−0.36	−0.31
1871–1880	−0.27	−0.36	−0.32
1881–1890	−0.35	−0.31	−0.33
1891–1990	−0.27	−0.25	−0.26
1901–1910	−0.33	−0.37	−0.35
1911–1920	−0.34	−0.24	−0.29
1921–1930	−0.16	−0.26	−0.21
1931–1940	−0.04	−0.12	−0.08
1941–1950	0.04	−0.04	0.00
1951–1960	0.05	−0.05	0.00
1961–1970	0.02	0.00	0.01
1971–1980	−0.05	0.06	0.00
1981–1990	0.20	0.26	0.23

The warming formed mainly in winter (DJF) and in spring (MAM) with maxima over northern and central Asia and over the western sector of North America, whereas in some regions of Asia cooling trends were observed. The main NH regions with a relatively stable SAT below the norm covered the central Pacific and the northern Atlantic, including Iceland and Greenland. The SAT anomalies in the southern hemisphere were weaker, except in some regions of the Antarctic.

Schlesinger and Ramankutty (1994, 1995) have discovered a 65-to-70-year oscillation of observed SAT, which does not exist, however, everywhere over the Earth, but rather over the North Atlantic Ocean and its bordering continental regions of North America, Europe, and northern Africa. The proposed causes of the oscillations are: (1) random radiative forcing of the climate system, that is, climatic noise; (2) oscillatory forcing external to the climate system; (3) an oscillation internal to the climate system, perhaps not unlike that of the ENSO, but with a longer timescale. A critical examination of the first hypothesis has led Schlesinger and Ramankatty to the conclusion that this hypothesis can be rejected at a very high level of statistical confidence.

Thus the climate warming during the last decade was not homogeneous. This may also be demonstrated by sea surface temperature data.

Hansen and Bezdek (1996) have considered North Atlantic SST data from the Comprehensive Ocean–Atmosphere Data Set (COADS) to investigate the behaviour of temperature anomalies on multi-year timescales during the period 1948–92 in the latitude band 0°–70°N. From a 45-year sequence of January maps they have found out the already familiar phenomena of generally cold conditions prior to 1951, a long warm interval from 1951 through 1967, and again the cold period from 1968 through 1977. The years 1978 through 1982 were largely devoid of persistent strong anomalies, but moderate cold conditions returned during 1983–86. Warm conditions dominated the North Atlantic from 1987 onward. Hansen and Bezdek (1996) have emphasized, however, that within these thermal epochs, a total of five cold anomaly features and nine warm anomaly features have been identified. These features have individual lifetimes of 3 to 10 years. A typical size is 20° of latitude or longitude, but they range from barely detectable to spanning the width of the basin, the latter especially in low latitudes. Most of the anomalies move long distances along certain preferred paths. These paths generally follow the routes of the subarctic and subtropical gyres.

An important part of global climate diagnostics is connected with the analysis of sea level data. It has been pointed out in the IPCC-95 (Houghton et al., 1996) that over the last 100 years the global sea level has risen by about 10 to 25 cm (in accordance with tide gauge records). "It is likely that much of the rise in sea level has been related to the concurrent in the rise in the global temperature over the last 100 years.... The ice sheets remain a major source of uncertainty in accounting for past changes in sea level, because there are insufficient data about these ice sheets over the last 100 years".

An interesting scale analysis to sort the different causes of mean sea level changes has been accomplished by Marzocchi and Mulargia (1996) on the basis of the observations for the northern Adriatic Sea. They have discovered that the mean sea level in the region is inhomogeneous in both time and space because of a number of causes which determine the sea-level changes (ice sheets dynamics, tectonics, climatic effects, sediment compaction and subsidence, etc.). At some sites the rate of increase in mean sea level is equal to that of the Mediterranean basin, which in turn equals to the change in global mean sea level. But, at other sites, starting from about year 1950, the rate is an order of magnitude larger.

Hendricks et al. (1996) have accomplished an empirical orthogonal function (EOF) analysis of global TOPEX/POSEIDON radioaltimeter satellite data to assess global sea level dynamics using observational data for the first two years of the satellite functioning. The results obtained earlier from the tide gauge data indicated a rate of increase of 1.8 ± 0.1 mm/yr over the past century. Recent TOPEX/POSEIDON altimeter data have found out a rise in sea level 5.8 ± 0.7 mm/yr (Nerem, 1995) and 4.1 ± 1.0 mm/yr (due to various estimates), which are significantly higher than the tide gauge estimates. The satellite data may not be quite correct, however, because of the relatively short time series of the altimeter data.

Satellite altimeters measure sea surface height (SSH) relative to an Earth-centred coordinate system, providing an absolute reference frame for studies of the sea-level rise. The analysis accomplished by Hendricks et al. (1996) has demonstrated that a large portion of the rise in mean sea level estimated from 2 years of TOPEX/POSEIDON altimetry is caused by inter-annual SSH variations. Therefore a precise determination of the long-term rate of global sea level rise using satellite altimetry may require 10 years or more of accurate SSH measurements (for the time being the accuracy of SSH measurements is better than 5 cm RMS).

EOF decomposition of 69 cycles of the TOPEX/POSEIDON altimeter data, accomplished by Hendricks et al. (1996), indicates dominant annual and inter-annual timescales of variability. The inter-annual modes capture oscillations in the tropical Pacific characteristic of recent El Niño events. A 20-year history of the change in mean sea level reveals a rise of 5.2 mm/yr. The analysis of the contribution of each EOF mode to global mean sea level variations has shown that 82% of the rise in mean sea level is caused by a single inter-annual mode of variability. The length of the time series (2 years) is not enough, however, to resolve a complete ENSO cycle which dominates the inter-annual EOF modes. Therefore most of the rise in mean sea level derived from TOPEX altimetry is an artifact of incomplete temporal sampling of inter-annual variability.

Walsh et al. (1996a) have analysed decadal changes in the Arctic sea level pressures of the "buoy era", 1979–94. A significant decrease in the annual mean has been found out. The changes of the annual means are larger in the Central Arctic than anywhere else in the northern hemisphere. Walsh et al. (1996) have concluded that the pressure decrease, which has been compensated by pressure increases over the subpolar oceans, implies that the wind forcing of sea-ice contains an enhanced cyclonic component relative to earlier decades.

Palaeoclimatic proxy data based on sequence stratigraphy imply that the sea-level rose and fell 100 m or so in less than 1 million years several times during the Cretaceous. Since the Cretaceous climate has been thought to have been too warm for an extensive accumulation of continental ice, it was assumed that sequence stratigraphic data are recording apparent sea-level changes caused by regional tectonic activity. Stoll and Schrag (1996) have shown, however, that Lower Cretaceous bulk carbonate from deep-sea sediments records sudden inputs of strontium resulting from the exposure of continental shelves. These strontium data from an interval spanning 7 million years in the Berriasian–Valarginian imply that global sea-level fluctuated about 50 metres over the time scales of 200 000 to 500 000 years. Oxygen isotope measurements indicate that the growth of continental ice sheets caused these rapid sea-level changes. In view of such results it may be assumed that an Antarctic ice sheet may have existed despite overall climatic warmth, because high SSTs enhance water transport to the centre of Antarctica, where the temperature may remain below freezing.

An important component of climate diagnostic data are the results of the the Earth's radiation budget measurement, which have been discussed in detail earlier (Houghton et al., 1996; Kondratyev, 1983a,b, 1992d; Kondratyev et al., 1996b; Marchuk et al., 1986; Ward, 1995; Yamanouchi and Charlock, 1995). We shall discuss some of these results later on.

A most important problem is an analysis of the SAT time series and other climatic parameters to detect anthropogenic changes. In this connection a technique was developed and applied to filter out the climate signals (Hasselman et al., 1995; North et al., 1995; North and Kim, 1995; Thaker, 1996) to recognize the contributions of forcings determined by changes in the solar activity and in the atmospheric greenhouse effect; in both cases the principal possibility to obtain $S/N > 1$ was demonstrated. An analysis was made (Goody et al., 1995) of the possibility to detect the climatic "greenhouse" signals from the data of satellite observations of the spectral distribution of the outgoing longwave radiation. A perspective direction of a search is the ocean acoustic tomography (Munk et al., 1995).

A very important assessment of the dynamic contribution to hemispheric mean

SAT trends has been made by Wallace et al. (1995) who have shown (on the basis of land station data from the northern hemisphere) that roughly half of the temporal variance of monthly mean hemispheric mean anomalies in SAT during the period from 1900 through 1990 were linearly related to the amplitude of a distinctive spatial pattern (cold ocean–warm land: COWL) in which the oceans are anomalously cold and the continents are anomalously warm poleward of 40°N when the hemisphere is warm. Apart from an upward trend since 1975, to which El Niño has contributed, the amplitude time series associated with this pattern resembles seasonally dependent white noise. Wallace et al. (1995) have argued that the variability associated with this pattern is dynamically induced and is not necessarily an integral part of the fingerptint of global warming.

In a later study Wallace et al. (1996) have confirmed the results mentioned after consideration of monthly mean time series of: (1) SAT anomalies averaged over the NH based on data from land stations; (2) SST anomalies averaged over the NH oceans, and (3) 1000–500 hPa thickness anomalies averaged over the region poleward of 40°N. A principal conclusion is that the anomalous warmth of the winters of the 1980s is consistent with the strong positive bias of the COWL pattern index during the cold seasons of that decade.

The overall assessment of the detection and attribution issues has been formulated in section 1.1 in accordance with IPCC-95 (Houghton et al., 1996); (later on we shall discuss this problem in more detail).

The main problem of the climate signals detection consists in the absence of sufficiently long and reliable observation series, since, as has been mentioned earlier, only recently the development has been started on substantiation and practical application of the global climate observing system (GCOS) (Global..., 1995a–c; Goldberg and Fleming, 1995; Jones et al., 1993; Karl et al., 1994, 1995a–d, 1996; Kondratyev, 1992d,e,g,h, 1994a; Kondratyev et al., 1992a, 1996; Kondratyev and Cracknell, 1996a,b; MTPE/EOS..., 1995; Rango, 1994).

Because the central problem for the identification of the greenhouse climate signal has become a detection of trends in climatological data series, Visser and Molenaar (1995) have proposed in this context a structural time series model with which a stochastic trend, a deterministic trend and regression coefficients can be estimated simultaneously. The model meets the requirements which are to be satisfied to prove detection: (1) to make inferences on significant rises or declines in trends; (2) to take into account natural variability in climate series, and (3) to compare output from GCMs with the trends in observed climate data.

Visser and Molenaar (1995) have applied the model to trend estimations in cases of a single station in the Netherlands (1706–1990) and annual mean surface air temperature (SAT) at northern hemisphere land stations (1851–1990). To analyse potential cause-and-effect explanations, the latter time series has been regressed on four variables: Southern Oscillation Index (SOI), volcanic dust index (VDI), sunspot numbers (SSN), and a simulated temperature signal induced by increasing greenhouse gases (GHGs). The results of such a regression show that the influence of SSN on global temperatures is found to be negligible. The correlations between SAT and SOI as well as VDI appear to be negative. For SOI, this correlation is significant, but for DVI it is not, probably, because of a relatively small number of volcanic eruptions during the sample period. The relation between SAT and GHG is positive.

An attempt to apply the model for prediction purposes has not been successful. Visser and Molenaar (1995) have suggested possible explanations of such a result.

The data of satellite microwave remote sounding play an important role in the detection of climatic signals. Especially it refers to obtaining the data on the sea-ice cover dynamics reflecting the present climate changes. An accumulation of sufficiently long data series obtained with the Nimbus-5 ESMR radiometer, the Nimbus-7 scanning multichannel microwave radiometer (SMMR) and microwave radiometer SSM/I (the DSMP satellites of the US Department of Defense) has provided a reliable data set on the global distribution of the sea-ice cover concentration (SICC) and ice cover extent (ICE) (Johannessen et al., 1994; Kondratyev et al., 1996a).

Nomura (1995) undertook the SSMR data processing for the period January 1979–August 1987 and of the SSM/I data (July 1987–December 1993) to analyse the climatic trends and the short-periodic variability of ICE in the northern and southern hemispheres. Estimates of the relative mean square deviation (MSD) of these data have shown that it varies within 5–7% (in winter) and 10–15% (in summer). The main reason of the summertime MSD increase is an approximate account of ice melting (with an appearance of water over the ice surface it should be considered the water surface). The SICC retrieval becomes difficult in the presence of the optically thick clouds, precipitation and sea surface roughness. In these cases a false signal of the ice cover sometimes appears, especially near the coastlines (therefore the SICC was considered to be zero if the SST is above $+1°C$).

Since the ECMWF climate model needs data on the ice cover "mask", determining for each element of the spatial grid the presence or absence of the ice cover, the ICE was calculated from the SICC data. The location of the ice edge was determined by the threshold SICC value (50%). Validation of the choice of this criterion by comparing with the results of the manual processing of SMMR data for the Baltic Sea revealed an excellent agreement.

Analysis of the multi-year trends of the mean global ICE values for the period 1979–91 showed practical constancy in the southern hemisphere and a very weak decreasing trend in the northern hemisphere during the last several years of this period. The annual course in the Arctic and Antarctic determined by different geography turned out to be quite different (the ocean restricted by land from the south in the Antarctic and the more complicated land geography in the Arctic).

The main feature of the ICE annual change in the Antarctic consists in its large amplitude and, respectively, stronger "sharpness" than in the Arctic. In conditions of the Antarctic (Arctic) in the period April–July (September–January) the rate of ICE growth is about $(2.5–3.0) \times 10^6$ km/month and $(2.0–2.5) \times 10^6$ km/month, respectively, and the rate of ICE decrease due to ice melting in the period November–December (April–August) is 5.5×10^6 km/month and $(1.2–2.0) \times 10^6$ km/month.

An interesting feature of the Arctic is that here the rate of ice melting is less than the rate of its growth. A maximum of ICE is reached in February–March, but the

ICE dynamics is characterized by a strong geographical inhomogeneity. In the Antarctic a maximum of ICE is observed in August–October, an important feature being the formation of persistent polynyas, especially in the Ross Sea and the Weddell Sea. The sea-ice cover is characterized by the strong spatial and temporal variability in the form of inter-annual variations and geographical diversity.

Very strong inter-annual variations are observed in the Sea of Okhotsk: in 1979 it was almost completely covered with ice, and in 1984 only a small part of the sea near the western coastline was covered with ice.

Maslanik et al. (1996) have discovered the net trend in the Arctic Ocean summer ice cover over the 17-year period (1978–95) equal to -0.6% per year, with the extent of the perennial ice pack reduced by 9% during 1990-95 compared to 1979-89. The strongest reduction took place in the Siberian sector of the Arctic Ocean. The antarctic polynya in the Weddell Sea was very large in 1973–74, but in the period 1979–91 it could not be found. Thus against a fairly stable ice cover in the Arctic and Antarctic, on the whole, a strong local spatial and temporal ICE variability is observed, which, of course, must substantially affect the climate.

Simmonds and Jacka (1995) have studied the climatology of the Antarctic sea-ice extent based on 20 years of data (1973–92) with emphasis on inter-annual variability and extrema. In the first half of the year the greatest variability is found in the western hemisphere, particularly in the Weddell and Ross Seas regions, while in the second semester the variability displays a considerable degree of zonal symmetry. The possible links between the Antarctic sea-ice extent and the Southern Oscillation Index (SOI) have also been analysed. One of the strongest relationships is between the ice extent in the south-east Indian Ocean in the months April through October and the SOI in the previous 12 months.

Gloersen et al. (1996) have analysed the sea-ice cover concentration (SICC) dynamics using the Nimbus-7 SMMR 9-year record. The results show that the 9-year trends are most positive in the Beaufort and East Greenland Seas, whereas the most negative areas are in the Barents Sea and the sea-ice surrounding Svalbard.

Gloersen et al. (1996) have emphasized that after removing the average seasonal cycle and trend, the standard deviations of the second residual SICC run is as high as 15% in the central Arctic. Thus there is a significant activity in the ice concentrations of a region that is frequently considered to consist of nearly full consolidated ice cover. A substantial fraction of a second residual consists of oscillations in the sea-ice cover, with periods less than and also greater than the seasonal cycle. A tidal and the ENSO-like components have been identified in oscillations, but, in general, their nature is not yet clear.

Krabill et al. (1995) have accomplished precise airborne laser-altimetry surveys at locations on the Greenland ice sheet. The results indicate a thickening in the western Greenland ice sheet of up to two metres between 1980 and 1993.

Bitz et al. (1996) have studied the low-frequency natural variability of the Arctic climate system using a single-column, energy-balance model of the atmosphere, sea-ice, and upper-ocean system with the purpose to also analyse the implications of the low-frequency natural variability in sea-ice volume for detecting a climate change. The results indicate that a 1.3-m change of the mean sea-ice thickness would require measurements taken over 12 years in order to detect a climate change at a 95% confidence level.Rott et al. (1996) have pointed out that the dynamics of ice shelves which cover 11% of the total area of Antarctica may serve an indicator of climate change because of sensitivity of ice shelves to changes in the temperature and circulation in the atmosphere and/or the ocean. For example, probably, as a result of regional atmospheric warming, the Wordie Ice Shelf decreased from ~2000 to 700 km^2 during the time period from 1966 to 1989.

Rott et al. (1996) have accomplished an analysis of the dynamics of the northern Larsen Ice Shelf (LIS) during the phase of its collapse early in 1995—in comparison with earlier data, using sequences of synthetic aperture radar (SAR) images from the ERS-1 satellite. The satellite data analysis was complemented by field observations made during October and November 1994. The combined results of satellite and surface observations indicate that the two northernmost sections of the ice shelf fractured and disintegrated almost completely within a few days. This breakup followed a period of steady retreat that coincided with a regional trend of atmospheric warming. The part of the LIS north of Robertson Island has retreated slowly but constantly since the 1940s; the retreat accelerated after 1975, and finally the two parts of the ice shelf collapsed early in 1995. The observations imply that after an ice shelf retreats beyond a critical limit, it may collapse rapidly as a result of perturbed mass balance.

High-latitude climate dynamics have been discussed in detail elsewhere (Curry, 1995; Curry et al., 1995; Easterling, 1995; Kondratyev et al, 1996a; Lee and North, 1995; Nagurny, 1995; Vavrus, 1995; Verbitsky and Oglesby, 1995).

Since sea-ice cover instability is one of the most important climate-forming factors, this determines the necessity to develop adequate numerical modelling techniques to simulate it in climate models. From the viewpoint of large-scale geophysical processes, sea-ice is a broken thin layer of the polar ocean—whose concentration and thickness are governed by dynamic and thermodynamic processes. The sea-ice cover serves an interface between two much larger geophysical media—the atmosphere and the ocean, which determines the important role of ice cover in the processes of atmosphere–ocean interaction (Lemke, 1995).

The importance of sea-ice as a climate system component is determined by its impact on the surface radiation budget (due to high ice albedo) as well as on the momentum, heat and matter exchange between the atmosphere and the ocean. In the cooling period (ice cover formation) the process of brine emission takes place with subsequent convection favouring the mixed layer deepening and the formation of deep and bottom waters. In the warming period (sea-ice melting) the mixed layer becomes thinner, since a surface layer of relatively fresh water appears.

The sea-ice transport results in that the regions of its melting are removed, as a rule, from the place of ice formation. Therefore the annually averaged resulting rate of freezing in a certain region is seldom equal to zero, being usually either positive or negative depending on the presence of either divergence or convergence of the sea-ice flow. The quantity of the resulting rate of freezing characterizes the non-stationary ice-floes near the surface which strongly affect the ice concentration field. The sea-ice motion characteristics (especially a convergence and divergence and, hence, the resulting rate of freezing) depend strongly on the coastline geometry and atmospheric forcings (wind, in particular).

The dynamic sea-ice models usually consider the sea-ice as a 2-D continuum characterized by four quantities: fields of snow and ice cover thickness, concentration, and horizontal velocity. The prognostic equations result from the laws of preservation for snow and ice masses with account of the momentum balance (including the internal tangential tensions).

The dynamic-thermodynamic ice cover models include four major components:

(1) The surface energy balance equation considering the factors of atmospheric forcing (wind, temperature, humidity, global radiation) and heat conductivity through the ice cover, to calculate the surface temperature (the energy balance equation is based on its parameterized components, such as net radiation, longwave emission, latent and sensible heat fluxes).
(2) The model describing the ice cover heat conduction with prescribed surface temperature and heat flux from the ocean to the ice cover. The difference between the heat fluxes due to heat conduction in the ice cover and the ocean mixing determines the rate of freezing or melting. The ice starts melting only when the temperature of the ice surface reaches 0°. In these conditions the surface temperature is fixed as zero and the excess energy determines the snow and ice melting.
(3) The equation of momentum balance with account of Coriolis force, lateral stress in the atmosphere, in the ocean and inside the ice cover as well as the ocean surface state, which makes it possible to calculate the ice movement velocity. The internal tensions in the ice cover are determined by its rheology which relates the internal tensions with the deformation.
(4) The equation for the balance of snow and ice masses as well as for ice cover concentration, the use of which makes it possible to precalculate the concentration and thickness of ice cover for each cell of the spatial grid from the data on the velocity of ice motion and freezing/melting, found with the use of the model of heat conduction and energy-balance equations for the surfaces of ice and open water.

As a rule, in the models of the ice-covered ocean the atmospheric forcing is prescribed from climatic data, and the data of numerical weather forecast have been also successfully used. The numerical modelling with the use of the data of numerical weather forecast has demonstrated possibilities of an adequate simulation of the annual change of the ice-cover extent and of the ice-edge location. A comparison with the data of observations of the drifting sea buoys confirmed a sufficient reliability of reproduction of both large- and small-scale ice motions.

Further efforts are needed to provide a more reliable prediction of the ice-cover concentration (models overestimate it). As Lemke (1995) notes, experience of the ice dynamics numerical modelling has shown that the reliable results can be obtained only with the atmospheric forcing prescribed correctly. The respective estimates determined the following requirements for the errors in various parameters: 0.6°C (air temperature), 0.2 (cloud amount), 16% (humidity), 27 W/m^2 (global radiation), 18 W/m^2 (atmospheric longwave emission), 24% (wind speed).

To assess the importance of sea-ice thickness and extent changes for climate dynamics under conditions of control (1 × CO_2) and global warming (2 × CO_2) simulations, Rind et al. (1995) have undertaken numerical modelling experiments. First of all, through varying the sea-ice distributions and thicknesses in the control run the sensitivity to a doubled atmospheric CO_2 has been analysed. The calculations show that more realistic thinner sea-ice amplifies the atmospheric temperature sensitivity in these experiments by about 15% (to a warming of 4.8°C), because it is easier for thinner ice to be removed as the climate warms.

An experiment with sea-ice not allowed to change between the control and doubled CO_2 simulation illustrates that the total effect of sea-ice on surface air temperature change, including cloud cover and water feedback that arise in response to sea-ice variations, amounts to 37% of the temperature sensitivity to a CO_2 doubling, accounting for 1.56°C of the 4.17°C global warming. This is about four times larger than the sea-ice impact when no feedback is allowed. An important factor of sea-ice dynamics is the poleward ocean heat transport. Rind et al. (1991) have shown that an increase of this transport of just 15% globally (although greater locally at high latitudes) is sufficient in the GISS GCM to remove all sea-ice. The sea-ice changes in the different 2 × CO_2 experiments affect the ice mass balance over Antarctica, implying global sea level changes due to this effect alone, ranging from increases of 0.23 cm/yr to no change. The results discussed illustrate the importance of sea-ice changes for the Antarctic thermal and ice mass budgets (see also Rind and Overpeck, 1994).

The results considered highlight the importance of properly constraining the sea-ice response to climatic perturbations, necessitating the use of more realistic sea-ice and ocean models. Without better sea-ice parameterizations (including the effects of finer vertical resolution), the sea-ice contribution to climate sensitivity cannot be fully understood. For instance, it may well be very different on the longer time sales when North Atlantic Deep Water (NADW) production changes can propagate to the Southern Ocean than for transient or even 2 × CO_2 climate changes.

Rind (1995) has pointed out that in accordance with prevailing scientific opinions, the high-latitude environment will be subject to a substantial impact of global warmig, while the tropics are considered as a fairly stable part of the globe. This is not true, however, because potential global temperature rise during the next century will also result in serious consequences for the broad latitude band from southern Europe to South America.

Observation data show that during the warmest decade of the 1980s the temperature increase took place mainly in the tropics: the tropical ocean was 0.25–0.75°C warmer than in the period 1951–80. Although the causes of this warming are still not clear, the numerical modelling results show that there must be a temperature rise in the tropics within 1–4°C due to a CO_2 doubling. In this context some palaeodata have been reconsidered. More careful analysis shows that, during the last glaciation, mountain glaciers reached lower levels (by about 900 m). Traces of glaciers have been discovered on Mauna Kea (Hawaii) where at the present time the zero isotherm is located a few hundred metres above the top of the mountain. Surface air temperature in the tropics decreased by 5°C, not being stable at all. A similar conclusion concerning the instability is presumably true for periods of warmings as well.

It is doubtful, however, whether CO_2 concentration in the tropics 3 million years ago was as high as has been suggested. It is more probable that the concentration was close to the present one. The principal cause of the climate warming could have been an enhancement of the meridional heat flux directed to the pole. Thus, in accordance with numerical modelling results, one can believe that during periods of global cooling and warming the tropics were not stable.

An important discovery from palaeoclimatic data has been that "the climate system is capable of switching between significantly different modes, suggesting that climatic surprises may lie ahead" (Overpeck, 1996). Changes of several degrees in temperature within years to decades were registered over the high northern latitudes which have been attributed to abrupt shifts in oceanic thermohaline circulation and are probably linked to abrupt shifts at low latitudes.

In this context, the recent analysis of the two Greenland ice cores (just 30 km apart), which are more than 3 km deep and extend back 110 000 years, deserves serious attention (Alley et al., 1996). These results show that the Earth has experienced large, rapid, regional-to-global climate oscillations through most of the last 110 000 years, on a scale that agricultural and industrial humans have not faced. These millennial-scale events represent large climatic deviations that, probably, include changes in temperature of many degrees Celsius, twofold changes in snow accumulation, order-of-magnitude changes in wind-blown dust and sea-salt loading, and large changes in methane concentration, etc., with cold, dry, dusty, and low-methane conditions correlated. Changes during these events commonly occur over decades or less (see also Bodri and Cermák, 1995). Similar conclusions on rapid changes in global palaeoclimate during the period 10 000 to 16 000 years ago have been obtained by Mayewski et al. (1996), who concluded, in particular, that rapid climate change events occurred more frequently in Antarctica than previously demonstrated.

Another important part of the globe are the world's drylands which comprise about 40% of the Earth's land suface (over 1 billion people live here). In this context, Hulme (1996) has studied precipitation and temperature trends in nine dryland regions for the period 1900 to 1994. Although all drylands have warmed during this period (and this warming was greater than the average warming over global land areas), no widespread desiccation in drylands climate was found except in the African Sahel, where the mean linear trend, which is equivalent to a 21% decline in precipitation over the course of 100 years, has been observed (much of the drying has occurred since the 1960s). It is to be noted that "desiccation" is defined as an aridification resulting from a dry period lasting a decade or more. As far as "desertification" is concerned, it is (in accordance with Agenda 21) "land degradation in dryland areas resulting from various factors, including climatic variations and human activities". "Drought" has been defined as a period of two years or more with below-average rainfall (Hulme, 1996).

The processes of drylands degradation are multifaceted and, for instance, the precise contribution of climate change to desertification is still a subject of controversial judgements, although it is quite clear that climate change alters the frequency and severity of drought and can cause desiccation in various regions of the world. Hulme (1996) has emphasized that "it does not necessarily follow, however, that drought and dessication will by themselves induce, or even contribute, to desertification . . . a variety of social, economic and political factors may operate—independently of climate—to cause desertification and, consequently, either out-migration or social disintegration in dryland communities. More common is the co-existence of social and natural pressures on the "dry land" which leads to the desertification phenomenon".

Hulme (1996) has chosen nine dryland areas to assess the recent climatic change in the world's drylands and has shown that observed climate warming may have contributed to a worsening of the ratio of the mean annual precipitation, P, to the potential evapotranspiration, PE, as an indicator of desertification (dry lands are defined as having a P/PE ratio of between 0.05 and 0.65) in many of the dryland regions. There is an anomaly in the three African dryland regions (the Sahel, southern Africa and North Africa) where warmer years/decades are significantly correlated with drier years/decades. Causes of the correlation require further careful analysis. Other dryland areas (six) are characterized by the absence of significant wetting or drying. As has been demonstrated by climate modelling, the whole land surface of the Earth is wetter under warming conditions, which mainly results from an increased oceanic evaporation.

To conclude, it is important to point out that a new IPCC-1995 Report (Houghton et al., 1996) contains, in particular, a new assessment of the existing climate diagnostics data. In this context, Wigley has said (see Monastersky, 1995): "In 1990, we said that on the basis of the observable record alone, we cannot claim to have detected the greenhouse effect. We can now see the fingerprint of man in the past temperature record. That's a pretty radical change".

Karl et al. (1996) have investigated climate dynamics within the contiguous United States through the development and analysis of two complex indices of climate change, a Climate Extreme Index (CEI) and a US Greenhouse Climate Response Index (GCRI). They have found out that the CEI, defined by an aggregate of conventional climate extreme indicators, supports the notion that the climate of the United States has become more extreme in recent decades. Yet the magnitude and persistence of the changes are not now large enough to conclude that the climate has systematically changed to a more extreme state as opposed to fluctuating about a near-stable state.

The US GCRI is calculated from the annual arithmetic average of the following five indicators of the percentage of the conterminous US area:

(1) The percentage of the United States with much above normal mean temperature (minimum temperature times 0.525 plus maximum temperature times 0.475).
(2) The percentage of the United States with much above normal precipitation during the months of October through April (the cold season).

(3) The percentage of the Unites States in extreme or severe drought during the months of May through September (the warm season).
(4) The percentage of the United States with a much greater than normal proportion of precipitation derived from extreme 1-day precipitation events (exceeding 2 in or 50.8 mm).
(5) The percentage of the United States with much below normal day-to-day temperature differences.

Each of these five indicators defines an anticipated response of the US climate related to increases of atmospheric greenhouse gases derived from the Intergovernmental Panel on Climate Change (1990, 1992, 1995). In addition to its role in monitoring for anticipated climate responses, another reason for developing a US GCRI is to obtain additional information by analysing multiple, mostly independent parameters, each of which is expected to respond to increases of greenhouse gases and/or temperatures.

Karl et al. (1996) have discovered a positive trend of the US GCRI during the twentieth century which is consistent with an enhanced greenhouse effect (there is only a 5% chance that the increase could arise due to a chance alone). The increase of the GCRI is not large enough, however, to unequivocally reject the possibility that this increase could result from other factors, including natural climatic variability. Both indices increased rather abruptly during the 1970s, a time of major circulation changes over the Pacific Ocean and North America (these changes are not, however, a dominating factor, as has been shown by the statistical analysis).

Further monitoring of the indices over the next decade is necessary as well as studies of an impact of the enhancing greenhouse effect on the circulation changes (with special emphasis on the ENSO phenomenon) to attribute the dynamics of the indices to the "greenhouse" impact.

2.5 NUMERICAL CLIMATE MODELLING

The present-day assessments of the critically important climatic factors such as the atmosphere–ocean interaction, cloud cover dynamics and its interaction with radiation, are still rather approximate (Kondratyev, 1992d; Marchuk et al., 1986; Kondratyev and Cracknell, 1998a,b). Despite certain progress in understanding that the atmosphere is a colloidal medium containing the suspended optically active aerosol, the climate theory treats this fact inadequately (A Plan ..., 1996; Houghton et al., 1996; Kondratyev, 1991a–c, 1992a–d, 1996a–c; Kondratyev et al., 1996a,b). Further development of the climate theory will require huge efforts to better understand and take into account these and other basic climate-forming processes. Certain progress made during several years after the IPCC Reports (IPCC ..., 1992, 1994, 1995) refers only to a part of substantial problems. First of all, it concerns studies of the role of the atmospheric greenhouse effect and assessments of the contribution of aerosol to the climate formation. Without repeating ourselves, consider the basic subjects.

2.5.1 Greenhouse gases

Naturally, the reliable simulation of the climatic implications of the greenhouse effect is only possible with adequate climate models for interactive forecasts of the climate and greenhouse gases (GHGs) dynamics (Schlesinger, 1991). Without such models one should base oneself on the *a priori* GHGs dynamics. In this case of great importance is an analysis of GHGs trends from the observational data and respective forecasts (preferably, on the basis of analysis of the global biogeochemical cycles of various elements, and first of all, carbon, nitrogen, sulphur). In this connection, certain progress has been made, in particular, in specifying the data on GHGs trends.

As has been mentioned earlier, for the long-lived GHGs such as CO_2, methane, CFCs, nitrous oxide and tetrachlormethane (TCM), the anthropogenic increase of their concentration continues during the last decade at a rate either remaining constant or increasing. There are, however, exceptions, such as methane and some halocarbons, whose rate of concentration growth has slowed (this also happened to CO_2 and N_2O in the beginning of the 1990s). In the case of methane it decreased from $\sim$20 ppb/year in late 1970 to $\sim$10 ppb/year in 1989 (the reasons for such dynamics are still unclear).

The general growth of CO_2, CH_4 and N_2O reached (by 1992) about 30%, 145% and 15%, respectively. These trends may be attributed largely to human activities, mostly fossil fuel use, land-use change and agriculture (Beran, 1995; IPCC WG1 ..., 1996; Rosema et al., 1993). According to Gunarathan and Young (1995) assessment, at the present time about 77% of CO_2 emissions and 54% of all GHGs emissions are due to fossil fuel burning.

Palaeodata from the analysis of the air trapped in polar ice (Raynaud et al., 1996) indicate fluctuations of about 10 ppmv for CO_2 and 70 ppbv for CH_4 over the last millennium. The CO_2 and CH_4 records extended to the Holocene period show significant variations (up to about 40 ppmv for CO_2 and 150 ppbv for CH_4).

Other GHGs are characterized by the following trends: (1) a substantial TOC (total ozone content) decrease during the last one-two decades taking place in all latitudes (except the tropics) in spring, summer and winter, the trend being stronger in the 1980s compared to the 1970s; the TOC decrease has been mainly determined by the decreasing content of ozone in the lower stratosphere (at altitudes below 25 km), where the latitude-depending rate of decrease reached 10% for 10 years (the estimates have shown that even if all governments meet all the recommendations of the London Supplement to the Montreal Protocol, the concentration of chlorine and bromine in the stratosphere will grow for several years to come, and this means, in particular, an inevitable regular formation of the TOC springtime minimum in the Antarctic); (2) in the NH mid-latitude troposphere (at altitudes below 10 km) the ozone content during the last 20 years increased at a rate of $\sim$10% for 10 years (the data of ozone-sondes for several surface stations); (3) apparently, the content of carbon monoxide in the northern hemisphere increased at a rate of 1% per year; (4) so far, there has been rather fragmentary information concerning the ozone precursors such as nitrogen oxides NO_x and non-methane hydrocarbons (NMHC).

Information on the relationships between the components of the global budget and carbon cycle such as emissions to the atmosphere due to fossil fuel burning, changes in land-use and deforestation (i.e., destruction of the continental biosphere), bioproductivity of marine biosphere (the "oceanic pump"), cement industry, and others remains contradictory. While the estimates obtained by Gorshkov (Gorshkov, 1995; Gorshkov et al., 1992, 1994) from indirect data on carbon radio-isotopes have made it possible to close the global carbon cycle, direct estimates of the contributions of various components have failed to close the cycle.

According to Gorshkov (1995), the terrestrial biosphere is now functioning as a source of CO_2 producing 6 GtC/year—at the same level as CO_2 emissions due to fossil fuel burning (6 GtC/year). In the atmosphere reside 4 GtC/year, whereas the CO_2 assimilation by the ocean reaches 6 GtC/year (due to biological processes—the "biological pump") and 2 GtC/year (CO_2 diffusion from the atmosphere to the ocean). These estimates provide the closing of the global carbon cycle but stay controversial. From the data of the Supplement (IPCC 1992..., 1992), the most reliable level of CO_2 emissions due to fossil fuel burning in 1989–90 was 6.0 ± 0.5 GtC/year. Model calculations by Winguth et al. (1994) predict a temporary uptake of 0.6 GtC during the ENSO year 1983 and assess the concurrent release of CO_2 from the land biosphere of about 2 GtC. The global ocean annual primary production ranges between 36.5 and 45.6 GtC/yr (depending on the assumptions made). Heimann and Maier-Reimer (1996) have obtained the oceanic CO_2 uptake rate 2.1 ± 0.9 GtC/yr. Due to Friedlingstein (1995) the present total amount of CO_2 emissions is about 7 GtC/year. Using a gridded biospheric carbon model, Friedlingstein et al. (1995) have come to the conclusion that about two-thirds to three-fourths of the 1850–1990 integrated missing sinks is due to the CO_2 greening of the biosphere. The remainder may be due to the increased level of nitrogen deposition starting around 1950. Total emissions of CO_2 due to fires at oilfields in Kuwait in 1991 reached only 0.065 GtC, that is, 1% with respect to global anthropogenic emissions.

Foley (1995) has obtained the following estimates of the carbon storage potential of the terrestrial biosphere: 800 GtC in vegetation, 150 GtC in litter and 1370 GtC in soil organic matter.

In accordance with the estimates discussed by Woodwell et al. (1995), the amount of carbon released through burning fossil fuels in 1992 is thought to have been about 6 GtC. In addition to that the net release from land-use changes (including deforestation) is 1.6 ± 1 GtC. An amount of the carbon which is accumulated in the atmosphere varies from year to year. Between 1988 and 1991 the accumulation ranged between about 5 and 2 GtC annually and averaged about 3 GtC/yr. The difference ($7.8 - 3 = 4.6$ GtC/yr) would be expected to be the amount absorbed by the oceans. Woodwell et al. (1995) have emphasized, however, that the amount accumulating in the oceans, as estimated by several techniques, seems to be limited to about 2 ± 1 GtC/yr. The rest is what is called "missing carbon".

This unexplained sink of carbon may be attributed to an accumulation of carbon by forests. There are many processes, however, which make it difficult to explain it this way, including: the toxification of extensive areas by industrial wastes, ozone and acid rain; the continued rapid deforestation and impoverishment of land in the tropics; the increasing rates of deforestation on harvesting of primary forests and their replacement by young, secondary forests with low carbon stocks in the northern hemisphere; the increasing frequency of forest fires in northern forests, etc. Warming over the past century may also have increased the decay of soil organic matter, releasing CO_2 into the atmosphere in a positive feedback. Therefore Woodwell et al. (1995) conclude that there is "no evidence that the forests are accumulating carbon at a rate sufficient to resolve the issue of imbalance in current accounts".

Still there was a number of arguments expressed in favour of a potential rôle of the terrestrial biosphere processes as contributors to the explanation of "missing carbon". A very important step forward in this context has been made by Keeling et al. (1996) who analysed an extensive O_2 /N_2 concentration ratio in both northern and southern hemispheres to show that, between 1991 and 1994, much of the carbon was trapped in the growing forests of the NH. The decrease of O_2 concentration during this time period was slower than it should be due to fossil fuel burning (which takes oxygen to form CO_2 and water). The difference may be explained only as the result of oxygen production through photosynthesis under conditions of the net growth of the land biosphere.

In accordance with Keeling et al. (1996) data of the total global anthropogenic CO_2 emission of about 6 Gtc/yr the land biosphere sequestered 2.0 ± 0.9 GtC/yr over a 4-year period. Since the atmospheric CO_2 content increased by 2.3 GtC/yr, the ocean uptake must have been 1.7 ± 0.9 GtC/yr. The O_2/N_2 ratio is lower in the northern hemisphere which may be interpreted as the dominating role of the NH in CO_2 sequestering. The results considered have been confirmed by Bender et al. (1996) who found a global land biotic sink of 3.3 ± 1.6 GtC/yr for the 1991–93 period on the basis of O_2/N_2 data, as well as by Ciais et al. (1995) who discovered a land biotic sink from 30° to 60°N of about 3.5 GtC/yr in the 1992–93 period based on the north–south gradient in CO_2 and $^{13}C/^{12}C$ isotopes.

Goulden et al. (1996) have emphasized the urgency of studying the impact of inter-annual climate variability on the annual CO_2 exchange in the northern hemisphere which may change by 1 GtC/yr, if their results on exchange of CO_2 by a deciduous forest in New England may be considered as typical of northern biomes. Data in Table 2.21 from IPCC-95 Report (Houghton et al., 1996) characterize the annual average anthropogenic carbon budget for 1980 to 1989.

The scheme of the global carbon cycle due to these data is presented in Fig. 2.6. Data in Table 2.22 illustrate different assessments obtained by Bender et al. (1996).

To further clarify the "missing sink" question, Hudson et al. (1994) used a global carbon cycle model (GLOCO) for relevant assessments. They pointed out that because temperate and boreal forests are N limited, CO_2 fertilization of these forests is less than predicted by short-term CO_2 response factors. Modelling of temperate/boreal forest fertilization by anthropogenic N deposition suggests that CO_2 is initially sequestered at a C:N ratio of ~100, rather than the steady-state value for the ecosystem of ~30. If N deposition is to account for 40–70% of the fertilization of the terrestrial biosphere not explainable by CO_2 fertilization and temperature

Table 2.21. Annual average anthropogenic carbon budget for 1980 to 1989. CO_2 sources, sinks and storage in the atmosphere are expressed in GtC/yr

CO_2 sources	
(1) Emissions from fossil fuel combustion and cement production	5.5 ± 0.5[a]
(2) Net emissions from changes in tropical land-use	1.6 ± 1.0[b]
(3) Total anthropogenic emissions = (1) + (2)	7.1 ± 1.1
Partitioning amongst reservoirs	
(4) Storage in the atmosphere	3.3 ± 0.2
(5) Ocean uptake	2.0 ± 0.8
(6) Uptake by northern hemisphere forest regrowth	0.5 ± 0.5[c]
(7) Inferred sink: 3 − (4 + 5 + 6)	1.3 ± 1.5[d]

[a] For comparison, emissions in 1994 were 6.1 GtC/yr.
[b] Consistent with Chapter 24 of IPCC WGII (1995).
[c] This number is consistent with the independent estimate, given in IIPCC WGII (1995), of 0.7 ± 0.2 GtC/yr for the mid- and high-latitude forest sink.
[d] This inferred sink is consistent with independent estimates of carbon uptake due to nitrogen fertilization (0.5 ± 1.0 GtC/yr), plus the range of other uptakes (0–2 GtC/yr) due to CO_2 fertilization and climatic effects.

Table 2.22. Anthropogenic carbon fluxes during 1991–93 inferred from Bender et al. (1996)

Rate term	Flux or change in concentration
Observed $d[CO_2]/dt$	1.5 Gt/yr (0.7 ppmv/yr)
Fossil CO_2 production	6.1 GtC/yr (2.9 ppmv/yr)
Fossil O_2 consumption	4.1 ppmv O_2 /yr[a] (20 per meg/yr)
Observed $d[O_2]/dt$	12 ± 4 per meg/yr
Land biosphere O_2 production	8 ± 4 per meg/yr (1.7 ± 0.8 ppmv/yr)
Land biosphere CO_2 uptake	3.3 ± 1.6 Gt/yr[b] (1.6 ± 0.8 ppmv/yr)
Oceanic CO_2 uptake	1.1 ± 1.6 Gt/yr[c]

[a] Fossil O_2 consumption is 2.9 ppmv CO_2/yr 1.4.
[b] Land biosphere CO_2 uptake is ((1.7 ± 0.8) ppmv/yr) / 1.05.
[c] Oceanic CO_2 uptake is 6.1 Gt/yr − 3.3 (±1.6) Gt/yr − 0.7 Gt/yr.

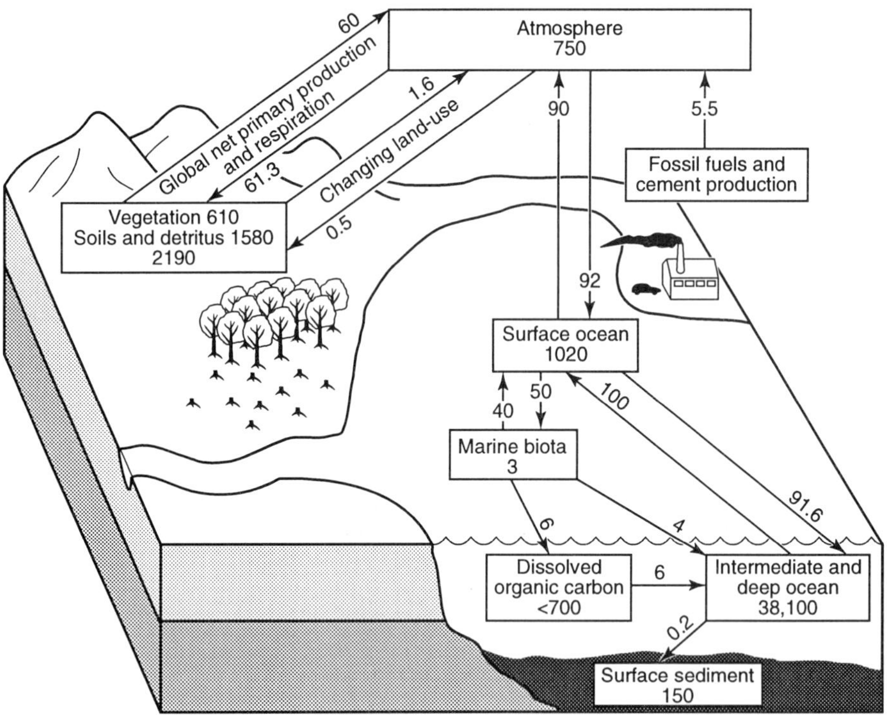

Fig. 2.6. The global carbon cycle, showing the reservoirs (in GtC) and fluxes (GtC/yr) relevant to the anthropogenic perturbation as annual averages over the period 1980 to 1989. The component cycles are simplified and subject to considerable uncertainty. In addition, this figure presents average values. The riverine flux, particularly the anthropogenic portion, is currently very poorly quantified and so is not shown here. Evidence is accumulating that many of the key fluxes can fluctuate significantly from year to year. In contrast to the static view conveyed by figures such as this one, the carbon system is clearly dynamic and coupled to the climate system on seasonal, inter-annual and decadal time-scales (e.g., Schimel and Sulzman, 1995).

increases, then Hudson et al. (1996) assessed that 26-30 GtN/yr of anthropogenic deposition in the temperate and boreal zones would be required. Recent anthropogenic NO_x and NH_3 deposition fluxes at northern temperate latitudes have been estimated at 20–28 GtN/yr. Thus fertilization by anthropogenic N emissions probably constitutes a significant portion of the missing CO_2 sink. In this context of great interest is a study by Matthews (1994) to map the global distribution of consumption and associated emissions of nitrous oxide and ammonia on the basis of information on the use of nitrogenous fertilizers.

As for the estimates of the contribution of deforestation to the global carbon budget, they remain very approximate. Calculations of the average rate of deforestation in the Amazon basin (Brazil) between 1978 and 1989 gave 2.1 million ha annually (Mha/year). The rate of deforestation increased till mid-1980 and then it started decreasing; in 1990 it constituted 1.4 Mha / year. From the data of FAO the rate of cutting the global tropical forests between 1981 and 1990 was equal to 17 Mha/year and exceeded the rate of deforestation in 1976–80 by 50%. The absence of new reliable global data on deforestation does not give grounds for changing the estimate 1.6–1.0 GtC (1980) given in the IPCC Report.

During the last several years serious attention has been paid to attempts to specify the contributions of the continental and marine biospheres to the global carbon budget (some of the attempts have been mentioned above). These contributions are characterized by inputs of small amounts of carbon to the atmosphere in the equatorial belt determined by CO_2 outgassing by warm tropical waters and by the continental biospheric component—a small difference of two large quantities: sources (including, deforestation) and sinks. Probably, a powerful sink exists in the northern hemisphere (due to both the oceanic and the continental components of the biosphere) and a weak sink in the southern hemisphere. For the oceanic sink the estimate 2.0 ± 0.8 GtC/year given in the IPCC-95 Report can still be considered most reliable. In the continental biosphere which supposedly serves a sink for carbon, the compensation processes take place determined by afforestation and fertilization due to increasing concentrations of CO_2 and nitrogen compounds; these processes still cannot be reliably estimated quantitatively. Still unsolved (and controversial) is the problem of the "missing sink" which is of the order of 1–2 GtC/year (due to Friedlingstein et al. (1995) the CO_2 imbalance is 1.8 ± 1.2 GtC/yr which, being integrated over the industrial period, amounts to about 100 GtC). This "imbalance" is an important factor which hinders the CO_2 dynamics forecasts and reliable assessments of the global warming potential (GWP).

According to Joos and Siegenthaler (1995) and Sundqvist (1993), between 1980 and 1989 the CO_2 concentration in the atmosphere increased at an average rate of about 1.5 ppm/year (which is equivalent to 0.4% or 3.2 billion tons of carbon) and constituted approximately half all the anthropogenic emissions for this period.

The carbon cycle is formed with contributions from the atmosphere, the oceans, the terrestrial biosphere and (in geological time scales) marine deposits and rocks, the following processes playing the key role (Joos and Siegenthaler, 1995): (1) the CO_2 exchange between the atmosphere and the ocean; (2) the CO_2 exchange between the surface and deep layers of the ocean (the latter being a gigantic reservoir for

Table 2.23. Mean annual components of the global carbon cycle over the period 1980–89 (GtC/year)

CO_2 sources	
(1) Emissions due to fossil fuel burning and cement industry	5.5–0.5
(2) Changes due to land use in the tropics	1.6–1.0
(3) Total anthropogenic emissions	7.1–1.1
Distribution of carbon by reservoirs	
(4) Accumulation in the atmosphere	3.2–0.2
(5) Assimilation by the ocean	2.0–0.8
(6) Assimilation due to afforestation in the northern hemisphere	0.5–0.5
(7) Additional land sinks (fertilization due to increasing concentrations of carbon dioxide and nitrogen, effect of climate changes: [(1) + (2)] − [(4) + (5) + (6)]	1.4–1.5

carbon); (3) assimilation or emissions of CO_2 due to land-use (e.g., deforestation); (4) photosynthetic assimilation of CO_2 by the land vegetation cover; the long-range accumulation of this carbon in wood and soils affecting the level of CO_2 concentration and the climate; CO_2 emissions to the atmosphere in the process of vegetation and soil respiration.

The data in Table 2.23 contain estimates of anthropogenic impacts on the global carbon cycle, including a "missing sink" reaching 1.4–1.5 GtC/ year (Joos and Siegenthaler, 1995).

The process of "fertilization" (stimulation of primary production by the growing concentration of CO_2 and nitrogen) is particularly interesting and open to discussion (Boehmer-Christiansen, 1994a,b; Joos and Siegenthaler, 1995). Probably, during the 1980s the CO_2 sink due to fertilization, constituting 0.5–2.0 GtC/year, was functioning. An additional fertilization due to nitrogen could reach 0.2–1.0 GtC/year.

As has been noted (Archer et al., 1995), during the last two centuries an increase of the biomass density and shrubland expansion has been observed in many regions of the south-western USA as well as in other arid and semi-arid global regions, the reasons being unknown. Traditionally it is explained by the effects of climate changes, forest fires, pastures. As an alternative, a supposition has been made that the tree and grass dynamics in the ecosystems of deserts, prairies and savannahs is determined by "fertilization" due to a CO_2 increase in the post-industrial epoch.

This hypothesis is favoured by the following circumstances: (1) trees refer to the photosynthetic category C3 whereas the grass cover (in the south-east USA and in the tropical regions) refers mainly to the category C4; (2) a CO_2 increase stimulates the photosynthesis mainly (but not always) in the vegetation of the type C3; (3) an intrusion of trees to the grass-covered regions (C4) during the last 200 years was followed by a 30% increase of CO_2 (from 270 ppm to 350 ppm).

Also, the hypothesis is attractive in the sense that CO_2 grew on global scales, which had created conditions for comparatively synchronous intrusion of the tree vegetation to the regions new to it. In this connection of interest are palaeodata testifying, for example, to the fact that the shifting of the regions of grass (C4) and tree (C3) vegetation recorded in the Chihuahuan Desert 7–9 thousand years ago had, apparently, resulted from changes in CO_2 concentration (there are, however, data which do not show any relation to CO_2).

In this context an analysis has been made (Archer et al., 1995) of the hypothesis of fertilization due to a CO_2 increase which suggested the conclusion that the propagation of the tree vegetation observed in the past had not been connected with the CO_2 increase. This propagation (shift) should be considered to have resulted from a joint impact of a set of factors (climate change, CO_2 increase, fires, pastures), and it is impossible to select the contributions of individual factors whose ratio had changed in time. There is no doubt, however, that in the post-industrial epoch the CO_2 increase (and, probably, climate changes) had been dominating factors of changing the grass cover for trees, though they not always can explain the local dynamics which can be connected, for example, with eating up of vegetation by cattle.

Summarizing their assessments of CO_2 fertilization, Okamoto et al. (1995) arrived at the conclusion that CO_2 fertilization exists on the global scale, but there are still large uncertainties and therefore further investigations are necessary. King et al. (1995a) have pointed out in this context that the hypothesis that the CO_2 enhancement of terrestrial NPP explains the historical missing carbon sink of the global carbon cycle budget is not supported by the model consideration.

As has been mentioned already earlier, the recent years have been marked by a number of new attempts to assess various components of global carbon cycle with special emphasis on biosphere–atmosphere interaction. The World Ocean has been considered till recently as the principal sink for CO_2. But later studies have demonstrated that the contribution from the terrestrial biosphere is also very significant. To undertake relevant quantitative estimates, Friedlingstein (1995) has developed a geographically referenced global SLAVE model (Scheme of Large-Scale Exchange between Vegetation and Atmosphere) with 5° lat. 5° long. resolution which allows one to simulate the global vegetation dynamics (nine types of vegetation) as well as dynamics of the principal reservoir of carbon and a number of nutrients (first of all, nitrogen) on the basis of available climatic information. The basic purpose of this simulation modelling has been a better understanding of anthropogenic impacts on the biosphere through climate change, CO_2 growth and nitrogen deposition on land.

The assessment shows (Friedlingstein, 1995) that effects of CO_2 fertilization, nitrogen deposition and climate warming lead to the following changes of the annual carbon budget components: 1.64, 0.62, and -0.9 GtC, respectively.

A number of studies have been devoted to an assessment of the boreal forests contribution to global carbon budget. For instance, it has been shown that the sink strength due to the temperate and boreal forest (data for Canada, the USA, Europe, former USSR) amounts to about 300 MtC excluding soil changes. It will either remain or become weaker within the following 100 years. Kurz et al. (1995) have shown that the carbon budget of Canada's forest (1920–89 time period) indicates that these forest ecosystems have been a carbon sink of approximately 0.2 GtC/yr.

Turner et al. (1995) have obtained estimates which show that the US forests sequestered carbon in the 1990s at a rate of 80 GtC/yr but would come close to carbon equilibrium by the 2020s. The carbon sink on the forest land base currently offsets 6% of the US fossil fuel emissions but that proportion is likely to decrease over the coming decades.

Heimann (1995) has pointed out that an important contribution to the global carbon cycle dynamics for time scales from a few years to decades belongs to El Niño/Southern Oscillation (ENSO) phenomenon. The analysis of the carbon isotopes ($^{13}C/^{14}C$) data shows that during strong ENSOs (like in 1983 or 1987) the ocean was an important CO_2 sink because of upwelling weakening and relevant cessation of CO_2 emissions by equatorial waters of the Pacific. At the same time, the presence of an anomalous CO_2 source on land has been revealed. Such a situation was not observed, however, during the 1992 ENSO because of cooling in mid-latitudes of the northern hemisphere due to the Pinatubo eruption (presumably, it was also the cause of certain slowdown of the CO_2 concentration rate of growth in 1992–93). Similar assessments have been considered by Meyers and O'Brien (1995).

Heimann and Maier-Reimer (1996) have reviewed recent assessments of the oceanic uptake of CO_2 using the $^{13}C/^{12}C$ isotope ratio, or the air–sea $^{13}C/^{12}C$ isotopic disequilibrium, and ^{13}C measurements. These three methods yield inconsistent oceanic CO_2 uptake rates for the time period 1970–90, ranging from 0.6 to 3.1 GtC/yr. Heimann and Maier-Reimer (1996) have shown, however, that using a nonlinear estimation procedure, a consistent scenario with an oceanic CO_2 uptake rate of 2.1 ± 0.9 GtC/yr can be established.

Keeling et al. (1995) have also used the carbon isotopes data (since 1978) to assess the relationship between contributions of terrestrial and marine biospheres to global carbon cycle. It has been shown that in the periods of CO_2 concentration growth the contribution of terrestrial biosphere as a source of carbon was enhanced, whereas during the periods of CO_2 concentration decline the opposite effect occurred. If one can remove short-time inter-annual variations, then during the 12.5-year period (1978–91) of isotope observations the World Ocean was a carbon source of about 3.0 GtC/yr, which agrees with the simultaneous warming of surface waters during the same period. On the other hand, the land biosphere sink of about 1.2 GtC/yr took place which could only partly compensate the oceanic source. During the next two years the latter was close to zero, while the land biosphere sink reached 3.0 GtC/yr.

As has been mentioned above, one can believe that during ENSOs the ocean becomes a more powerful carbon sink than usually (the opposite is true for the land biosphere). The volcanically induced cooling, if it was also accompanied by a precipitation enhancement in the northern hemisphere, could significantly stimulate the growth of vegetation and substantially reduce CO_2 emissions by the terrestrial biosphere.

In the context of global carbon cycle dynamics, an important rôle belongs to assessments of potential changes of carbon reservoirs through an accomplishment of a global afforestation programme. Nilsson and Schopfhauser (1995) have conducted a detailed analysis of the results of potential realization of such a programme with various constraints due to political, economic and technical circumstances taken into account. Data in Table 2.24 characterize the geographical distribution of forests over the globe which occupy about 3,417 billion ha, that is, approximately 26% of the whole land territory.

The forest systems mentioned in Table 2.24 in their above-ground parts contain about 500 to 800 GtC, which is approximately equal to the mass of carbon located in the atmosphere in the form of CO_2. There are various opportunities to intensify an assimilation of CO_2 by forests (and, thus, to reduce the CO_2 concentration growth): reduction of forest cutting, increase of forest productivity, afforestation. The Nordwijk Declaration (1989) calls for an increase in the annual growth of forest cover to 12 million ha by the year 2000. If we take into account that every year about 17 million ha of tropical forests are lost, this means that the annual forest increase must reach 30 million ha.

Earlier assessments of the scale of afforestation which is necessary to stabilize CO_2 concentration at the present level, gave values within the range 3.4 to 700 million ha to compensate carbon emissions within the limits 0.05–5 GtC. Nilsson and Schopfhauser (1995) have obtained new estimates which show that the global land territory available for afforestation is only 345 million ha, which is significantly below the estimates obtained earlier (1.0–1.5 billion ha). Table 2.25 contains the results of numerical modelling to assess possibilities of the carbon sequestration through afforestation during the time period 1995–2050.

A principal result of the calculations is that the impact of new forests on the carbon budget may become significant in only 40–50 years. The maximum rate of carbon sequestration of $\sim$1.5 GtC/yr will be reached in about 60 years and is less than 50% of the present growth of the carbon reservoir in the atmosphere (3.8 GtC/yr). Total mass of sequestered carbon for the whole 100-year time period will be about 104 GtC (Gunarathan and Young (1995) mentioned the range 50–170 GtC). It is clear thus that the realistic afforestation programme will not lead to the CO_2 stabilization: only about 30% of the anthropogenic carbon emissions may be compensated. Other measures are also necessary (development of new sources of energy, increase of efficiency of energy use, enhancement of marine phytoplankton production, etc.). Tans and Bakwin (1995) believe that new reliable enough assessments of the CO_2 assimilation by the World Ocean and terrestrial biosphere are necessary (see Gorshkov (1995) in this context).

Since CO_2 exchange between the atmosphere and terrestrial biosphere depends on the productivity of ecosystems, an important problem is an assessment of net primary productivity (NPP). The use of two different approaches for such an assessment has led Bouger (1995) to the following two NPP values: 51 and 68 GtC/yr. In the first case the value of gross primary productivity (GPP) which is determined by carbon fixation is 125 GtC/yr.

Myneni et al. (1995) have estimated potential (i.e., unstressed) GPP of terrestrial vegetation from 1982 to 1990 using monthly mean global green leaf area and photosynthetically active radiation obtained from observations made from several geostationary and polar orbiting satellites. Mean annual GPP value was estimated at about 179.9 $\pm$ 1.6 GtC. It should be mentioned that NPP is the difference between GPP and autotrophic respiration R_a. Due to Bouger (1995), $R_a = 74$ GtC/yr and therefore NPP $= 125 - 74 = 51$ GtC/yr.

Table 2.24. The world's forested areas in 1990, by continent and ecological zone

Ecological zone		Africa	America	Asia[a]	Pacific	Europe[b]	Total
Temperate (including boreal, Mediterranean, sub-temperate)	Temperate and boreal forests (million ha)	8.1	511.4	188.4	47.3	904.3	1659.5
	% of temperate land area	1.1	22.8	10.6	5.9	32.3	19.9
Tropical	Tropical forests (million ha)	600.1	839.9	274.9	42.6	1757.5	
	% of tropical land area	26.7	50.1	30.7	78.6	36.3	
Total	Total forested area (million ha)	608.2	1351.3	463.3	89.9	904.3	3417.0
	% of total land area	20.5	34.7	17.3	10.8	34.4	26.1

[a] Includes the Middle and Near East but excludes the Asian part of the former Soviet Union.
[b] Includes all of the former Soviet Union.

Table 2.25. Estimates of annual carbon-fixation rates by above- and below-ground biomass to be achieved with the proposed global plantation programme

Year	Carbon in above-ground biomass (million t/yr)	Carbon in below-ground biomass (million t/yr)	Total carbon in biomass (million t/yr)
2015	407	122	529
2025	635	191	826
2035	864	259	1123
2045	991	297	1288
2055	1140	342	1482
2070	1092	328	1420
2085	1058	317	1375
2095	1084	325	1409

Table 2.26. Mean and standard deviation of biome GPP (GtC/year) during 1982–90

Biome	Mean	Standard deviation
Ice / polar desert	0.0607	0.0035
Semidesert	2.3720	0.1156
Tundra	2.2149	0.0150
Taiga	7.7962	0.0204
Cold deciduous forest	2.0017	0.0104
Cool grass / shrub	5.6423	0.1480
Cool conifer forest	2.6424	0.1480
Cool mixed forest	5.6987	0.0454
Temperate deciduous forest	7.7283	0.1245
Evergreen / warm mixed forest	14.9989	0.2285
Warm grass / shrub	8.9113	0.3037
Hot desert	8.9016	0.4989
Xerophytic woods / shrub	17.1024	0.2321
Tropical rain forest	23.1533	0.1015
Tropical seasonal forest	24.0256	0.1323
Tropical dry forest / savannah	45.8432	0.3112

Table 2.26 data obtained by Myneri et al. (1995) characterize average annual GPP of the various biomes. As one can see, the three biomes account for about 52% of the total annual carbon assimilation: the tropical dry forest/savannah, tropical seasonal and rain forest.

It is important that their productivity is stable from year to year, with an inter-annual variability of 0.55 Gtc/yr. On the other hand, the biomes with high inter-annual variability are the hot deserts and semideserts, cool and warm grasses / shrubs, and xerophytic woods / shrubs. Their annual GPP is 43 $\pm$ 1.3 GtC, or about 24% of the global total GPP. The inter-annual variability of GPP (1.6 GtC) is quite small compared to the mean annual GPP $\sim$180 GtC.

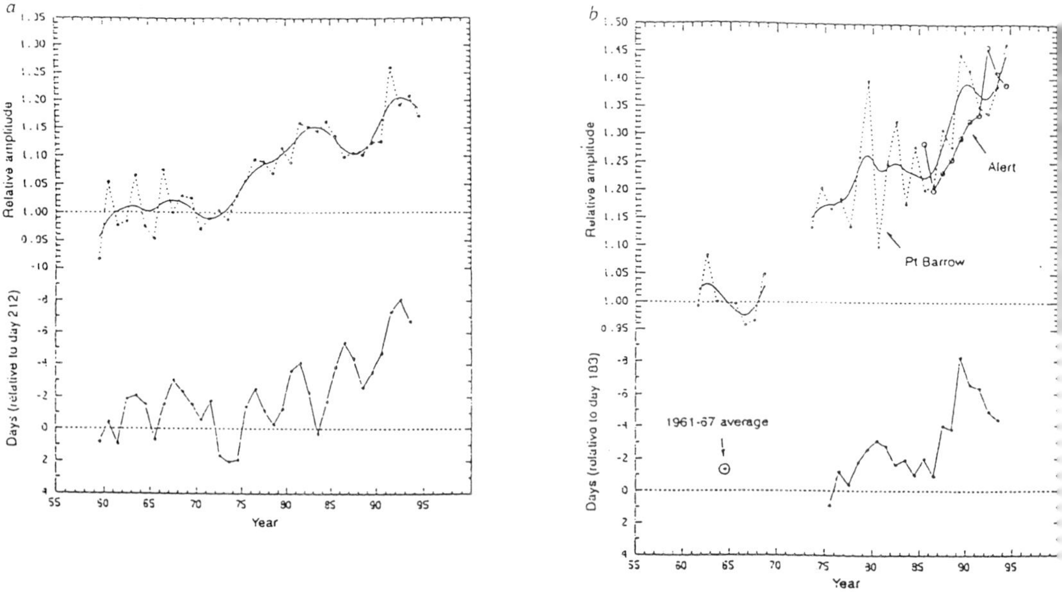

Fig 2.7. Trends in relative amplitude and timing of the seasonal cycle of atmospheric CO_2. *a*, At Mauna Loa Observatory, Hawaii. Annual values of the amplitude (upper plot, dots connected by dashed lines) were determined from weekly averaged continuous concentration data fitted annually to a phase-locked 4-harmonic seasonal function. This function, together with a linearly increasing gain factor, was first established by a fit to the full record, after which the gain was redetermined separately for each year. The linearly increasing gain factor, referenced to 1964, was computed to be $0.675 \pm 0.045\%$ per year. A smoothing spline (solid curve, with standard error, σ_A, of 0.028%) shows quasi-decadal changes in the relative amplitude. The timing of the downward zero crossing of the seasonal cycle (lower plot, dots connected by solid line segments), was obtained from 3-year fits to the CO_2 record and plotted relative to the 212th day of the year. The zero crossing is with respect to the seasonally adjusted CO_2 concentration. *b*, As *a*, but for Point Barrow, Alaska. The data from 1973 to 1994 are from flask samples collected 2–4 times per month, augmented from 1961 to 1967 by continuous CO_2 measurements, also calibrated by our laboratory. The fitting procedure (which spans a record gap from 1968 to 1973) yields a linear gain factor of $1.318 \pm 0.100\%$ per year, again referenced to 1964. A smoothing spline (s_A of 0.050% as in *a*), shows quasi-decadal changes. Also shown is the trend in amplitude at Alert. Canada (connected open circles in upper plot), plotted so that the sum of the squared annual differences form Point Barrow data, is minimized. The zero crossing time at Point Barrow (lower plot) is plotted relative to the 183rd day of the year. The circled dot in 1964 was determined from 3-year fits for 1961–63 and 1965–67, averaged.

Keeling et al. (1996) have analysed CO_2 concentration observations at Mauna Loa (since 1958) and Point Barrow (since 1961) and found out the annual amplitude of the seasonal cycle has increased by 20%, as measured in Hawaii, and by 40% in the Arctic. These increases are accompanied by phase advances of about 7 years during the declining phase of the cycle, suggesting a lengthening of the growing season. Data in Fig. 2.7 illustrate the results obtained by Keeling et al. (1996). The annual amplitude observed at Mauna Loa since 1958 shows almost no trend until the mid-1970s; then it decreased almost steadily until 1984; decreased for several years, and increased again to a record maximum in 1991. Thus, a quasi-decadal oscillatory pattern is seen with peaks in 1983 and 1992. Since the timing of the minimum in the seasonal CO_2 cycle advanced irregularly after 1980 by about 3 days while the maximum and the upward zero crossing showed no significant long-term changes, it may be interpreted as resulting from a lengthening of the growing season during the same time period in which the annual amplitude has increased.

Similar results have been obtained in the Arctic (Point Barrow, Alaska) where the total gain in the annual amplitude from the mean for 1961–67 was 39.5 ± 3.0, approximately, twice that at Mauna Loa. For another Arctic station, at Alert, Canada, with a record beginning in 1985, only small year-to-year fluctuations were observed, but an upward trend was similar to that of Point Barrow.

With the aim to find out the causes of the variability observed, Keeling et al. (1996) have considered the correlation between changes in the annual amplitude and land temperature (Fig. 2.8). On the decadal timescale (when short-term inter-annual fluctuations have been filtered out) a quasi-decadal pattern of temperature with extremes in warmth near 1981 and 1990, about two years before decadal maxima in CO_2 amplitude, was observed. This correlation indicates large-scale responses of the carbon cycle to climate change, but the nature of such responses is not obvious. Keeling et al. (1996) have examined various potential causes and concluded that the increase of the observed annual amplitude can be explained neither by direct short-term responses of plants to temperature nor by other factors unrelated to temperature. They pointed out that decadal maxima in amplitudes occurred, approximately, when the annual NPP (net primary production) has maximally exceeded the annual HR (heterotrophic respiration), and vice versa. One can therefore believe that NPP, on a decadal timescale, has been stimulated when it warms and diminished when it cools, with the response lagging temperature.

Keeling et al. (1996) have concluded: "These striking increases over 30 years could represent unprecedented changes in the terrestrial biosphere, partially in response to some of the highest global average temperatures since the beginning of the global records, and partially in response to plant growth being stimulated by the highest concentrations of atmospheric CO_2 in the past 150 000 years".

Analysis of existing versions of the global carbon cycle with the changing dynamics of CO_2 has shown that with the CO_2 emissions maintained at the present level, by the end of the twenty-first century CO_2 concentration will have reached 500 ppm. The maintenance of stable CO_2 concentration at the present level is only possible with emissions reduced to a level much lower than in 1990. Reaching the prescribed stabilization level depends mainly on the cumulative volume of emissions and not on the chosen scenario of the temporal variations in emissions. The following ranges of variability of the cumulative emissions correspond to various levels of CO_2 stabilization (given in brackets): 300–430 GtC (350 ppm), 880–1060 GtC (550 ppm), 1220–1420 GtC (750 ppm). For comparison purposes, one can

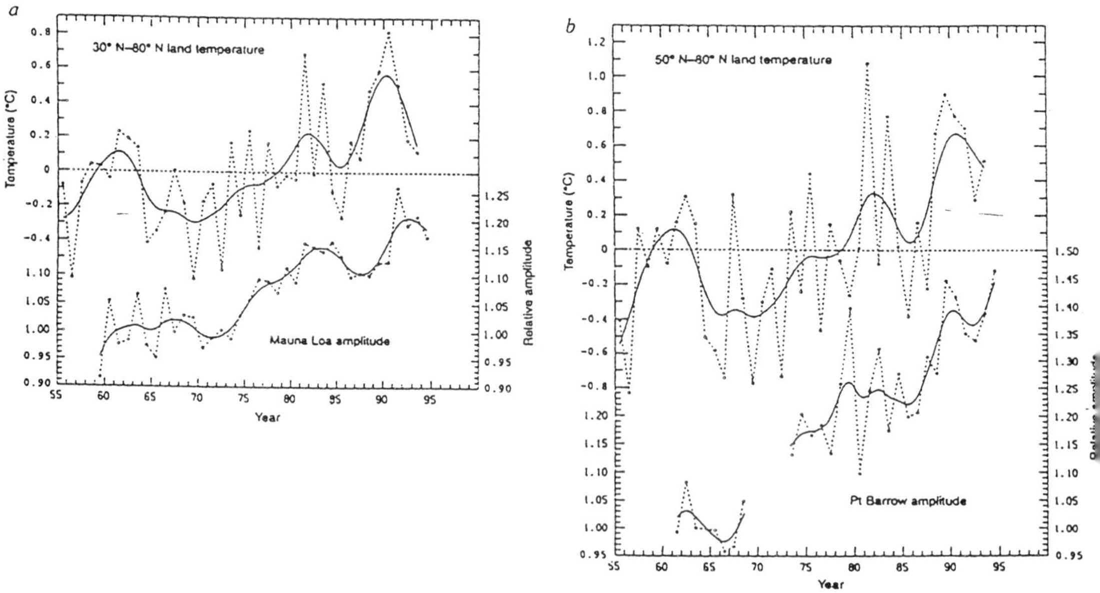

Fig. 2.8. Comparison of trends in the relative amplitude of the seasonal cycle of atmospheric CO_2 with trends in annual average land temperature. For clarity, the two curves are plotted with visually equivalent scales of linear increase, *a*, Trend in temperature in the zone from 30°N to 80°N (upper plot, annual dots connected by dashed line segments), also shown by a smoothing spline (standard error, σ_γ, of 0.215°C) to reveal quasi-decadal variability. Lower plot, the trend in relative amplitude of CO_2 at Mauna Loa Observatory, as in Fig. 2.7. The average increase of amplitude with respect to temperature was found by linear regression of annual data, allowing for errors in both variables, represented by σ_A and σ_γ, respectively. Best-fit estimates based on the lowest χ^2 for the period 1973–94 (for comparison with Point Barrow) showed nearly identical results for lags of one and two years in amplitude relative to temperature, and gave an increase of 19.0 $\pm$ 4.0% per °C (reduced χ^2 = 1.2) for the two-year lag. (If run from 1966 to 1994 for a two-year lag, the result was 22.7 $\pm$ 3.7% per °C.) *b*, Similar comparison of the amplitude at Point Barrow, Alaska, as in *a*, but using the trend in temperature between 50°N and 80°N (σ_γ of smoothing spline 0.320°C). The best-fit regression in this case was for a lag in amplitude of one year and gave an increase of amplitude with temperature for 1973–94 of 25.5 $\pm$ 5.6% per °C (reduced χ^2 = 0.9). Linear correlation coefficients confirmed the lags for both stations.

mention that to the IPCC IS92 scenario (Table 2.27) corresponds to the range of cumulative emissions 770-2190 GtC.

The rate of increase of the methane concentration decreased during the last decade, most rapidly in 1991–92, but in 1993, again, it started growing. During 1980–90 the average trend constituted ~13 ppb (0.8% or 37 million tons of nitrogen annually).

Christensen and Cox (1995) have pointed out that tundra ecosystems between 50°N and 70°N are estimated to contain 14% of the global soil carbon and account

Table 2.27. Some ecological parameters and GHDs dynamics corresponding to scenarios 1992a–f; TREP stands for total requirements for energy consumption; the carbon cycle intensity is defined as a ratio of carbon mass per unit TREP; CFCs include CFC-11, 12, 113,115, 115; GNP stands for gross national product

					Tropical deforestation		Annual emissions					
Scenario	Years	TREP/GNP decrease (mean annual change, %)	Decrease of the carbon cycle intensity (mean annual change, %)	Cumulative emissions due to fossil C, C, GtC	Tropical forest cutting, million hectares	Cumulative emissions of C, GtC	Year	CO_2	CH_4, Tg	N_2O, TgN	CFC, Tg	SO_2, TgS
IS92a	1990–2025	0.8	0.4	285	678	42	1990	7.4	506	12.9	827	98
							2025	12.3	659	15.8	217	141
	1990–2100	1.0	0.2	1386	1447	77	2100	20.3	917	17.0	3	169
IS92b	1990–2025	0.9	0.4	275	678	42	2025	1168	659	15.7	36	140
	1990–2100	1.0	0.2	1316	1447	47	2100	19.0	917	16.9	0	164
IS92c	1990–2025	0.6	0.7	228	675	42	2025	8.8	589	15.0	217	115
	1990–2100	0.7	0.6	672	1343	70	2100	4.6	546	13.7	3	77
IS92d	1990–2025	0.8	0.9	249	420	25	2025	9.3	584	15.1	24	101
	1990–2100	0.8	0.7	908	651	30	2100	10.3	567	14.5	0	87
IS92e	1990–2025	1.0	0.2	330	678	42	2025	15.1	692	16.3	24	163
	1990–2100	1.1	0.2	2050	1447	77	2100	35.8	1072	19.1	0	254
IS92f	1990–2050	0.8	0.1	311	725	46	2025	14.4	697	16.2	217	151
	1990–2100	1.0	0.1	1690	1686	93	2100	26.6	1168	19.0	3	204

for between 20% and 25% of the natural methane emissions (total contribution of CH_4 to global warming is about 15% at the present time). Numerical modelling has been performed for arctic tundra to assess the enhancement of methane emissions due to global warming (2 × CO_2 versus 1 × CO_2 cases). Depending on soil moisture conditions the increase in methane emissions could amount to anywhere between 0 and 5% of the current global emissions.

With the present emissions of methane preserved, its concentration will stabilize in 50 years at the level 1900 ppb, that is, it will exceed the present level by 11%. With the annual methane emissions reduced to a level below the present one by ~35 million tons (about 10% with respect to total anthropogenic emissions), the concentration of methane may remain at the present level (these estimates proceed from the fact that natural sources and sinks of methane are not affected by climate changes and changing chemistry of the atmosphere).

The rate of increase of some ozone-destroying gases (halocarbons) has been markedly reduced, which reflects an efficiency of the Montreal Protocol and its Supplements. In 1992 the total concentration of chlororganic compounds in the troposphere increased only by 1.6%, that is. almost half as much compared to 1989 (2.9%).

Preserving the present emissions of nitrous oxide should stabilize its concentration at a level ~400 ppb in several centuries (this exceeds the present level by 30% and the pre-industrial concentration by 50%). With the present emissions reduced by more than 50%, the concentration of N_2O may stabilize at the present level (310 ppb).

A comparison of data on the sinks of methane and its accumulation in the atmosphere has given the total (natural and anthropogenic) annual emissions of CH_4 ~500 Tg, the quantities of the total emission components being still uncertain. Recent specifications of the data on the rate of reactions of CH_4 removal by atmospheric hydroxyl OH (the rate of reaction decreased) has made it possible to more reliably assess the contribution to the methane budget of the sources such as rice paddies, and to take into account some additional sources (e.g., wastes and cattle). Analysis of the data on CH_4 isotopes shows that about 100 Tg CH_4 (20% of the total source of CH_4) are produced by fuel burning (coal, oil, natural gas). Emissions of methane by the rice paddies vary widely depending on the type of soil and vegetation conditions. New estimates gave a lower level of these emissions compared to the estimates of the IPCC Report (Climate Change, 1990) as well as a more reliable lifetime of methane in the atmosphere: 11 years.

So far, both sources and sinks of nitrous oxide remain unclear, which hinders an explanation of the observed increasing trend of nitrous oxide and its forecast. The global-scale consumption of CFCs (freons) CFC-11, 12 and 113 dropped by 40% in the beginning of 1990 compared to the level of 1986, which is much below the level foreseen by the Montreal Protocol. The Supplements to the Montreal Protocol adopted in 1990 in London will promote further reduction of the level of freons consumption (respectively, emissions to the atmosphere). With their complete exclusion, the scales of the substitutes such as HCFC and HFC will increase.

Aerosol is a most variable and difficult to consider (and forecast) optically active atmospheric component. Its spatial and temporal variability is determined by a complicated set of natural (biospheric dynamics, dust storms, volcanic eruptions, etc.) and anthropogenic processes (industrial emissions of sulphur compounds, biomass burning, subsonic aviation, etc.) (Kondratyev, 1992d). In this connection, most interesting are gas-to-particle processes of the sulphate aerosol formation, an analysis of which has shown that the role of natural factors is not so important as has been supposed earlier, which makes one concentrate on analysis of anthropogenic processes.

In view of impossibility of reliable forecasts of the GHGs and aerosol dynamics in future, one may speak only about developing scenarios of probable changes with account of rather contradictory assumptions on the trends of socio-economic development (naturally, the reliability of the scenarios decreases with their forecast range). Based on specified scenario A (SA90) proposed in IPCC Report (Climate Change, 1990), six new alternative scenarios have been substantiated (IS92a–f) which foresee various versions of the socio-economic development. The data in Table 2.27 illustrate the respective versions of the dynamics of the greenhouse gases which differ greatly. If we compare these six scenarios with SA90, we shall see that smaller emissions are typical of IS92, compared to SA90. This is explained by the mutually compensating variability. For example, the forecast of the more intensive population growth and, hence, intensified emissions of GHGs are compensated by excluded emissions of halocarbons and more optimistic possibilities to use the renewable energy sources.

Maximum GHGs emissions correspond to scenario IS92e, which is based on the combined account of the factors such as moderate population growth, intensive economic development, wide accessibility of fossil fuel, and hypothetical (in the future) exclusion of the use of nuclear energy. Minimum GHGs emissions are foreseen by scenario IS92c (first an increase and then a decrease of the population size by the middle of the next century, weakly growing economics, limited supplies of fossil fuels). On the whole, the data in Table 2.27 point to an inevitable increase of GHGs concentrations in the next century, provided the respective preventive actions are not taken.

Sarmiento et al. (1995) have estimated anthropogenic carbon emissions required to stabilize future atmospheric CO_2 at various levels ranging from 350 ppm to 750 ppm. Table 2.28 data present results of 5450 and 5750 simulations. The combination of the atmospheric increase and uptake by the ocean and terrestrial biosphere gives an estimate of the allowable anthropogenic emissions in the scenarios considered.

The most important result is that the estimated uptake by the terrestrial biosphere and ocean are of comparable magnitude and that, in combination, these two sinks greatly exceed the amount of CO_2 that remains in the atmosphere. In both the terrestrial biosphere and ocean the fractional terrestrial and oceanic uptake capacity is significantly reduced as the CO_2 level increases and the rate of CO_2 growth increases.

Sarmiento et al. (1995) have emphasized that a great deal of research is needed in order to reduce the uncertainty of the carbon sinks and concluded: "The large future terrestrial uptake predicted by our models should be viewed primarily as an

Table 2.28. Results of S450 and S750 Simulations

	Annual increase (Gt yr^{-1})	Total 9 (Gt)				
	Mean, years 1980–89	Year 2000	Year 2100	Year 2200	Year 2300	Years 1990–2300
S450 scenario						
Atmosphere	3.38	3.34	0.00	0.00	0.00	205
Ocean	1.83	2.32	1.80	1.16	0.90	497
Terrestrial biosphere	2.23–2.65	2.54–3.38	1.14–1.95	0.40–0.98	0.26–0.73	322–554
Total	7.4–7.9	8.2–9.0	2.9–3.8	1.6–2.1	1.2–1.6	1024–1256
S750 scenario						
Atmosphere	3.38	4.05	4.70	0.95	0.00	842
Ocean	1.83	2.42	3.98	2.83	1.92	956
Terrestrial biosphere	2.23–2.65	2.61–3.51	2.40–4.41	0.86–2.61	0.23–1.53	507–1029
Total	7.4–7.9	9.1–10.0	11.1–13.1	4.6–6.4	2.2–3.5	2305–2827

The range in terrestrial biosphere increases is for the models with the three following different temperature forcings: constant at the initial value, equilibrium warming due to the CO_2 increase, and equilibrium warming due to twice the CO_2 increase. The mean total increase of 7.4-7.9 Gt/yr in 1980–89 agrees within the uncertainty with the estimated carbon release of 7.0 ± 1.2 Gt/yr from fossil fuel burning and land use changes. The agreement is even better when the actual temperature record prior to 1989 is used to force the terrestrial biosphere model.

indication of the need for additional global change research on the terrestrial biosphere, rather than as a firm forecast".

Craig and Holmén (1995) have obtained assessments of the CO_2 fertilization and temperate forest regrowth which indicate that the observed CO_2 record can accommodate a carbon budget balanced by CO_2 fertilization but that the balance with forest regrowth is more difficult. Due to the sensitivity tests, CO_2 concentration by the year 2100 may vary within the range 604–695 ppm while net uptake (by the year 1990) changed from 2.57 to 2.67 GtC/yr. Future CO_2 projections are sensitive to how the carbon budget is balanced.

A model has been developed to estimate the column-integrated oceanic primary production using chlorophyll concentration retrieved from satellite observations. This information will open a new possibility to assess the ocean contribution to the global carbon cycle. An important assessment of the glacial–interglacial change in the terrestrial carbon storage within the range 750–1050 GtC has been obtained by Crowley (1995) through the analysis of a new pollen database.

The main factors governing the dynamics of future CO_2 emissions are: the growth of population and economics, structural changes in economics, price of energy, progress in technologies, accessibility of various energy sources (fossil fuel and renewable energy sources), prospects of nuclear energetics. Of great importance are the development tendencies in the former USSR and in the countries of East Europe, which have been considered in these scenarios. The prescribed trends of GHG enable one to assess the possible dynamics of the atmospheric greenhouse effect.

2.5.2 Radiative forcings

In this context, most substantial progress is connected with assessments of the contributions of the ozone dynamics in the atmosphere and the troposphere, the climate-forming role of sulphate aerosol and analysis of an adequacy of the global warming potential (GWP) concept. Calculations have shown that the TOC decrease observed during the last two decades led to a decrease of radiative forcing (RF) for the mid- and high-latitude "surface–atmosphere" system. Being globally averaged over 10 years, this decrease turns out to be equal (but opposite in sign) to a RF increase due to CFC growth during the same time period. Since in this case the regional inhomogeneity of the RF distribution is quite essential (especially in high latitudes), the estimates must be specified with the use of the global 3-D models. With the prescribed homogeneous global 10% increase of the ozone content in the tropics (with account of the data of fragmentary observations mentioned above), the RF should increase by 0.1 W/m^2. As has been mentioned above, assessments of the possible effect of anthropogenic sulphate aerosol in the northern hemisphere (in the southern hemisphere this effect is negligibly small) for clear-sky conditions have led to the values of cooling for the last century of the order of 1 W/m^2. No doubt, the estimates of the climate-forming contribution of aerosol are conditional (see Kondratyev, 1992d) and should be further verified, the more so that the role of aerosol as a factor affecting the radiative properties of clouds can be very important.

Raynaud et al. (1996) have compiled Table 2.29 data which characterize the present-time greenhouse radiative forcing and its dynamics in the past. Rind (1996) has accomplished an intercomparison of various global-scale forcings (Table 2.30).

Table 2.29. Estimates of greenhouse trace gas radiative forcings (absolute values) for different key-periods of the last 10 000 years (Holocene)

Time period	Trace gas	Atmospheric mixing ratio change	ΔF (W/m^2)
Between pre-industrial times and 1992	CO_2	278 → 356 ppmv	1.56
	CH_4	700 → 1714 ppbv	0.47
	N_2O	275 → 311 ppbv	0.14
	O_3		0.2–0.6
Last millennium (pre-industrial period)	CO_2	$\leqslant$ 10 ppmv	$\leqslant$ 0.22
	CH_4	$\leqslant$ 70 ppbv	$\leqslant$ 0.04
	N_2O	< 10 ppbv	< 0.04
Holocene (without including industrial era)	CO_2	$\leqslant$ 40 ppmv	$\leqslant$ 0.97
	CH_4	$\leqslant$ 150 ppbv	$\leqslant$ 0.09

Table 2.30. Uncertainties in global forcing and response

Forcing	Comment	Uncertainty in radiative forcing (W/m^2)	Δsurf. temp. (°C) due to radiative uncertainty (with equilibrium sensitivity of ~1°C/W/m2)[a]	Time period (years)
Volcanic aerosols	$\Delta\tau = 0.22$	4	1	few years
Solar variability	Total irradiance without background 0.05% with background 0.4%	0.8	0.6	centuries
	UV effects on strat. and trop. dynamics/ radiation	?	response may not be proportional to forcing	
Tropospheric aerosols	Without cloud effects	0.6	0.5	centuries
	With cloud effects	1.2	?	
Trace gases	10 ppmv for CO_2	0.15	0.15	decades–centuries
	? for ozone	?	?	
Desertification/ deforestation	48% land affected max of 5% change	0.8	0.6	several hundred
Ocean variability	El Niño		0.1–0.2	several years
	NADW		0.2	decades

[a] Note that regional changes can be three times the global effects, and even higher (5–20 times) for more local forcing (oceans, land surface).

Two conclusions are especially important: (1) many of the forcing uncertainties (potential RF changes) are of the same order of magnitude, that is, 0.5–1 W/m^2; (2) the forcing for climate models has many uncertainties, although the possibility exists that the more uncertain variations are quite small.

In connection with the results mentioned, Rind (1996) have emphasized that coupled atmosphere–ocean models produce intra-decadal variability of 0.3–0.4°C in the global mean SAT without any changes in forcing. Differences in thermohaline circulation intensity in the coupled model produce decadal average SST variations of 0.5–1°C in the North Atlantic (Delworth and Manabe, 1995), of the same order of magnitude as some observations (Kushnir, 1994). Then Rind (1996) concludes: "While there is no guarantee that this model 'noise' represents 'natural variability' of the actual system, it is reasonable to assume that climate variations of this magnitude may in fact be 'unforced'. This is an important concept to be kept in mind, since the proxy data often indicate fluctuations of a similar order of magnitude to this model variability".

Table 2.31. The global warming potential

Gas		Indirect effect
CO_2	1	None
CH_4	11	Positive
CFC-11	270	Uncertain
CFC-12	3400	Negative
HCFC-22	7100	Negative
HFC-134a	1600	Negative

Cai (1996) has specifically considered the response of the present-day thermohaline circulation to a greenhouse gas induced global warming under different surface thermohaline conditions. The results indicate that the global thermohaline circulations under the weak relaxation show an initial weakening and shallowing, continued weakening upon the cessation of the atmospheric warming, and eventual re-establishment of their strengths.

As has been mentioned, to compare the contributions of various GHGs to the formation of atmospheric greenhouse effect with respect to CO_2, the GWP conception has been proposed consisting in assessing the contributions of GHGs to the greenhouse effect per 1 kg. Table 2.31 gives the GWP values obtained with account of new data on the lifetime of various GHGs with the prescribed period of forcing 100 years. Since, apart from direct impact on the formation of the greenhouse effect, a possibility of indirect effect through chemical reactions producing other greenhouse gases should be taken into account, the third column of the table contains the related remarks (the respective quantitative estimates are unreliable).

New RF estimates showed (MacCracken, 1995) that during the period from the pre-industrial era to nowadays the contributions of various GHG constituted: CO_2 (the concentration increased from 280 to 356 ppm) 1.56 W/m^2; CH_4 (from 0.7 to 1.7 ppm) 0.5 W/m^2; N_2O (from 275 to 310 ppb) 0.18 W/m^2. The total contribution of CFC-11, 12, 113, 114, 115, methylchloroform and tetrachlormethane was $\sim$0.3 Wm^2, and that of freon substitutes >0.05 W/m^2.

With an account of the ozone increasing trend in the troposphere, its mean global RF is estimated at 0.2–0.6 W/m^2. The CFC-induced depletion of the ozone layer should result in the negative RF -0.1 W/m^2, whereas the contribution of CFCs as greenhouse gases constitutes 0.1-0.2 W/m^2. The mean global RF due to sulphate aerosol varies from -0.2 to -0.9 W/m^2, and due to biomass burning from -0.05 to -0.6 W/m^2. Note also that measurements of the solar constant over the period from 1850 determined the RF value within 0.1–0.5 W/m^2, and the Pinatubo eruption determined RF = -4 W/m^2 over several years.

In accordance with the recent IPCC summary (Houghton et al., 1996), the direct radiative forcing of the long-lived greenhouse gases (2.45 W/m^2) is due primarily to increases in the concentrations of CO_2 (1.56 W/m$_2$), CH_4 (0.47 W/m^2) and N_2O (0.14 W/m^2). The direct RF due to the CFCs and HCFCs combined is 0.25 W/m^2. However, their net radiative forcing is reduced by $\sim$0.1

W/m^2 because they have caused stratospheric ozone depletion which gives rise to a negative RF.

At present some long-lived greenhouse gases (particularly HFC (a CFC substitute), PFCs and SF_6) contribute little to radiative forcing but their projected growth could contribute several per cent to RF during the twenty-first century (see Daniel et al., 1995; Pinnock et al., 1995).

The tropospheric ozone RF is not yet well characterized, but it is estimated at about 0.4 W/m^2 (15% of that from the long-lived GHGs).

Total tropospheric aerosol negative RF is about -0.5 W/m^2, as a global average, and an indirect aerosol–cloud forcing is possibly of a similar magnitude. As is seen, the new assessments mentioned are close to estimates considered above.

Sinha and Toumi (1996) have accomplished a comparison of climate forcings due to chlorofluorocarbons and carbon monoxide. Estimates of RF due to CO, CFC-11 and CFC-12 show that global-mean anthropogenic climate forcing by CO is determined to be 32% of that by CFC-11, 12% of that by CFC-12, and 9% of that by CFCs combined. Even if the contemporary global-mean CO concentration is tripled, the climate forcing by CO is still only 75% of that due to CFC-11. Thus, the contribution from CO is of only minor significance.

In view of the principal contribution of water vapour to the greenhouse effect, Sinha and Harries (1995) have assessed the impact of far-IR absorption by water vapour on the greenhouse effect for the sub-arctic winter (SAW) and tropical (TRP) atmospheres (the far-IR is defined as wavelengths longer than about 20 μm). The calculations show that the sensitivity of the normalized greenhouse effect $g = (E-F)/E$ (E is the emission of the surface; F is the outgoing longwave radiation) to water vapour concentration is largest in the far-IR for the SAW atmosphere, and in the window region for the TRP. The total greenhouse effect $G = E - F$ is equal to 170 and 50 W/m^2, respectively. The proportion of G in the far-IR is 27% for the TRP and 35% in the SAW case; g is, respectively, 0.372 and 0.202. It is of interest that water vapour change giving the same forcings as a CO_2 doubling is only +17% and +43% (fractions of forcing from water vapour change in the far-IR are 19% and 53%). One can conclude from the assessments obtained by Sinha and Harries (1995) that the contribution of the far-IR absorption by water vapour to the formation of the greenhouse effect is very significant and therefore deserves further careful studies on the basis of observations and numerical modelling. The problem of the greenhouse Earth as observed from space has been discussed by Harries (1996).

Frey et al. (1996) have analysed the variability of the spectral greenhouse parameter g_λ using collocated Advanced Very High Resolution Radiometer (AVHRR) and High-Resolution Infrared Sounder-2 (HIRS/2) observations during the months of July 1993 and January and July 1994. The principal result of this study is that the spectral greenhouse parameters at wavelengths sensitive to middle and upper atmospheric water vapour content is dependent on SST via its connection to the large-scale atmospheric circulation pattern.

The phenomenon of the atmospheric supergreenhouse effect (Meehl and Washington, 1995) originating in the tropical Pacific has recently become of particular interest. The western tropical Pacific is characterized by most strongly variable SST and atmospheric moisture content. This points to the fact that in this region the level of the water vapour-induced atmospheric greenhouse effect (GE) should be at a maximum.

The SST in the region considered is known to never exceed 305 K. To explain this situation, the hypothesis of "thermostat" has been proposed (Ramanathan and Collins, 1991, 1993): a SST increase is limited by the self-regulating mechanism of the formation of cirrus clouds. In this context, of importance is the formation of the atmospheric supergreenhouse effect (SGE): the upward longwave radiation flux, G_a, absorbed by the atmosphere increases more rapidly with the growing SST than the thermal emission of the ocean surface. This happens because of the growing moisture content of the troposphere. In response to SGE, highly reflective cirrus clouds are produced which act like a thermostat, shielding the ocean from solar radiation. Such an effect of cirrus clouds may limit SST by values less than 305 K. The concept of the "thermostat" has produced, however, a fairly controversial discussion.

If the hypothesis of "thermostat" holds (the solar radiation determines an increase of T_s, which leads to an increase of the upper-level cloudiness and, thus, favours a decrease of the solar radiation input, that is, prevents further T_s increase), the following should be observed: (1) positive correlation between cloud amount and T_s in conditions when the T_s increase leads to an upper clouds enhancement; (2) negative correlation, if clouds reduce the input of solar radiation. An expected time shift between the cloud development and T_s decrease (due to the ocean's heat capacity) makes it possible to distinguish between these correlations.

The observational data show that such a correlation is always positive (for any time shift) reaching a maximum with a 1-month time shift. Therefore the effect of the upper clouds on the incoming solar radiation should be considered weak, and, if clouds do really regulate the SAT, it is not due to their radiative forcing of the ocean surface. In accordance with Arking and Ziskin (1994) results, the Pacific warm pool is the product of large-scale atmospheric and ocean circulation patterns and air–sea interaction.

Analysis of the data on the dependence of the outgoing longwave radiation (OLR) on T_s in clear-sky conditions revealed an unexpected situation: for $T_s >$ 300 K the OLR decreases with T_s increasing in conditions of the normal clouds regime (with the constant relative humidity and increasing T_s one might expect an OLR increase). This happens, apparently, because the moisture content of the upper troposphere increases strongly at $T_s >$ 300 K. These conditions reflect the presence of an unstable situation when a T_s increase intensifies the heat input to the atmosphere, which promotes further heating of the ocean surface and continuing T_s increase. Since in conditions of the thick upper clouds regime this instability does not occur, one may believe that these clouds explain the phenomenon of a "warm pool" observed in the western tropical Pacific: a convergence in the lower troposphere together with an increasing moisture content in the upper troposphere and an enhancement of the greenhouse effect, causes a T_s increase, which determines the thickening of the upper cloudiness ensuring the stability and restraining the T_s increase. This phenomenon is similar to the "thermostat" mechanism suggested earlier but it does not imply the clouds radiative forcing of SST, which is not confirmed by the data of observations.

Arking and Ziskin (1994) have concluded that "although cloud radiative effects do not regulate sea surface temperatures on the monthly time scales considered here, they may be important when considering long-term climate effects, where cloud feedback can involve changes in height, optical thickness, and the microphysical properties of clouds, altering the balance between shortwave and longwave effects. Also, our results do not preclude the possibility that clouds and their radiative effects play a role in the dynamics of the Pacific 'warm pool' region on time scales shorter than 1 month or by way of indirect processes connected with atmospheric and oceanic circulation. Further analysis is needed, especially with cloud data that have higher spatial and temporal resolution and with concomitant data providing the temperature, humidity and velocity fields, so that one can incorporate non-radiative processes into the analysis".

Meehl and Washingtom (1995) have emphasized the necessity of studying the balance between the two competing feedbacks: the cloud albedo feedback and the supergreenhouse effect.

Lubin (1994) analysed the results of the Central-Equatorial Pacific Experiment (CEPEX) carried out from 7 March to 7 April 1993 to obtain quantitative estimates of the SGE forcing of the heating of the atmosphere and ocean surface. The processing of the earlier satellite data has shown that the formation of SGE starts when SST > 300 K and is determined by two factors: (1) non-linear increase of G_a at a high SST; (2) more rapidly increasing G_a with the growing SST compared to the thermal (absolutely black) emission of the ocean surface. The SGE existence had been confirmed by satellite measurements of the upward longwave radiation at the level of tropopause.

In the CEPEX period, the ship spectral measurements of the atmospheric thermal emission were undertaken in zenith in the wavelength interval 5–20 μm (500–2000 cm^{-1}) with the resolution 1 cm^{-1}. The data of these measurements reveal the water vapour-induced SGE over the equatorial ocean. With the SST increasing from 297 K to 303 K, the radiative cooling (negative longwave radiation budget) of the ocean surface decreased by 30–50 W/m^2. Hence, the SGE originally detected from satellite data directly contributes to the ocean surface heating; the water vapour-induced continuum absorption (800–1250 cm^{-1}) and a powerful tropical convection play an important role in the formation of SGE.

Fu et al. (1996) have examined connections between the large-scale inter-annual variations of clouds, deep convection, atmospheric winds, vertical thermodynamic structure, and SSTs over global tropical oceans over the time period July 1983–December 1990. The principal conclusion of this study has been that the coherence between clouds and SST tendency shows that SST tendency leads cloud variations in the equatorial Pacific. Thus the change of clouds does not dominate the sign of SST tendency even though the cloud change was at a maximum during the 1987 El Niño. In some areas of the Indian, subtropical Pacific, and North Atlantic Oceans, cloud change leads the SST tendency. Therefore, cloud change might affect the SST tendency in these regions.

The "thermostat" hypothesis suggested by Ramanathan and Collins (1991, 1993) has produced an active discussion, including certain reservations and objections. For instance, Hartmann and Michelson (1993) have come to the conclusion that there must be an increase of latent heat flux with SST of about 7.2 W/m^2 K, which exceeded the atmospheric longwave supergreenhouse effect. They conclude therefore that evaporative cooling is a dominant limiting factor in determining tropical SST. Zhang and McPhaden (1995) have shown, however, that the relationship between latent heat flux and SST is more complex than the simple monotonic increase as suggested by Hartmann and Michelson (1993). On timescales of several days, convective activity enhances surface evaporation, which together with the decrease of surface solar radiation, acts to cool the sea surface. However, on scales of climatic interest, convection acts to reduce surface evaporation. High SST gives rise to more convective activity, which through interaction with the large-scale circulation, increases the low-level large-scale convergence and decreases the surface wind, leading to evaporation in spite of the increased surface–air humidity difference. Therefore, although individual convective events can significantly enhance surface evaporation on short timescales, the long-term average effect of convection is to suppress surface evaporation at high SST due to its interaction with the large-scale circulation. Thus, evaporative cooling may not provide the negative feedback required to restore the SST. An important circumstance is that the results considered are based on the data from the equatorial Pacific. More data are needed (i.e., from the Indian and the Atlantic Oceans) to verify these results.

In a continuation of the discussion of the "thermostat mechanism" Pierrehumbert (1995) has once more reconsidered the question of the regulation of tropical SST. The reconsideration has been based on the analysis of the tropical heat budget through consideration of an idealized radiative-dynamic model of the tropical general circulation. Pierrehumbert (1995) has argued that evaporation on its own cannot provide an effective regulating mechanism although it makes up a large part of the surface energy budget of tropical oceans. Clouds also cannot serve as regulators unless there are substantial departures from the observed near cancellation between cloud greenhouse and cloud albedo effects. It has been shown, in particular, that the prediction by Ramanathan and Collins (1991) of highly stable tropical climates is based on an inconsistent set of assumptions about the behaviour of the atmospheric heat transport. When the heat transport has been treated by Pierrehumbert (1995) in a consistent manner, clouds (in equilibrium) are found to have little impact on the net column atmospheric heating, the atmospheric circulation, and the warm pool SST. The tropical climate can be quite sensitive, however, to radiative perturbations.

Pierrehumbert (1995) has found out that the main determinant of tropical climate is the clear-sky water vapour greenhouse effect averaged over the entire tropics. In the absence of dry "radiator fins" maintained by subsidence, the tropical temperature would tend to fall into a runaway greenhouse state that could be stabilized only by heat export to the extratropics. The warm pool atmosphere cannot get rid of its heat, because of the strong water vapour greenhouse effect; this heat must be exported via zonal and meridional heat fluxes, to drier regions where it can be radiated to space. These dry, non-convective regions act like "radiation fins" stuck into side of the warm pool atmosphere. The lateral heat flux required to redistribute energy within the tropics is carried largely by the Hadley and Walker circulations. The warm pool SST is very sensitive to the optical thickness of the dry pools and to the relative area of the dry and subsiding versus convective regions, A_2/A_1. Presumably, there is an undiscovered stabilization principle at work, which involves an adjustment of A_2/A_1. Key unsolved problems are determination of A_2/A_1 in the tropics and of the degree of dryness of the subsiding regions. In this context Pierrehumbert (1995) has criticized the "thermostat" hypothesis.

To better understand the mechanisms that provide the upper limit on SST, Waliser (1996a,b) has studied four-dimensional atmosphere and ocean conditions which are associated with the formation of hot spots in the oceans. The hot spots are defined as regions where SST exceeds 29.75°C and that have an area greater than 1×10^6 km^2. The study indicates that formation of hot spots is largely confined to the region within the 28°C isotherm of the long-term mean SST, with greatest concentration occurring in the western Pacific warm pool. The analysis of the warm-pool hot spots (0°–10°S, 156°E–176°W) shows strong influences from inter-annual, annual, and 30–60-day timescales (the typical lifetime of an ocean hot spot is about 1 month, with several lasting 2–3 months, and only a couple lasting more than 3 months) with La Niña conditions appearing to inhibit the formation, southern summer favouring the formation and the descending (ascending) phase of the Madden–Julian oscillation (MJO) favouring formation (the decay). The interaction with the MJO indicates how internal, or remotely forced, atmospheric variability, in addition to local feedbacks, may be playing a role to help limit the SST. The out-of-phase relationship between SST and deep convection associated with the formation and decay of the ocean hot spots suggests the possibility of a positive feedback mechanism for MJO. The ocean hot spots tend to form in the western Pacific when convection is suppressed over this region in favour of enhanced convection to the west, over the Indian Ocean. With respect to the surface heat budget, Waliser (1996a,b) has shown that the data indicate that during the hot spot evolution, the convective perturbations to the surface shortwave radiation exceed those for evaporation by at least a factor of 2.

The analysis of the composite ocean conditions revealed that the rapidly varying atmospheric conditions associated with the hot spot evolution induce significant changes below the surface layer of the ocean. Removing the El Niño time period from the composite data considered demonstrates that the composite ocean is more dependent on the inter-annual state than the composite atmosphere. Concerning the "thermostat" hypothesis the results considered indicate that the cooling effects due to cloudiness appear more important than those due to evaporation on limiting the very warm SSTs associated with hot spots in the western Pacific warm pool. However, there are indications that atmospheric variability uncoupled to the local SST (e.g., MJO) provide significant perturbations to the local convective environment that are important to both the formation and limiting mechanisms of very warm SST. Waliser (1996) has emphasized that such results indicate that the ocean should not be rendered too simple with respect to understanding the limiting mechanisms of high SST.

The problem of climatological interactions between deep convection, upper tropospheric humidity, and the greenhouse effect has been further studied by Soden and Fu (1995) through the combined analysis of satellite and conventional observations data. The study combined the TOVS 6.7-μm water vapour channel with coincident cloud and radiation budget data from ISCCP and ERBE. The results obtained by Soden and Fu (1995) for the tropics reveal a positive interrelationship between deep convection, upper-tropospheric humidity, and the greenhouse effect. Regional variations in the frequency of deep convection (FDC) are strongly correlated ($r = 0.82$) with variations in an upper tropospheric humidity index (UTH), indicating that locally (i.e., on the scale of 2.5° grid box), an increased convection is associated with an increase in the upper-tropospheric relative humidity. Such a conclusion is also true for regional scales: on both local and regional scales, areas of increased convective activity are associated with increased upper-tropospheric humidity that is the net effect of deep convection is to moisten the upper troposphere. Even for the entire 30°N–30°S tropical belt there is a positive relationship between the FDC and UTH.

The analysis of ERBE measurements of the outgoing longwave radiation (OLR) has shown that the increased upper-tropospheric humidity is associated with an enhanced greenhouse trapping, demonstrating the importance of moisture transport by deep convection upon the relative energy balance.

In extratropical regions, temporal variations in upper-tropospheric humidity exhibit little relationship to variations in deep convection, suggesting the importance of other dynamical processes (such as large-scale quasi-isentropic transport) in determining changes in the upper-tropospheric moisture for this region.

An intercomparison between observations and simulations using GFDL GCM, conducted by Soden and Fu (1995) has demonstrated qualitatively successful results.

Kent et al., (1995) have used SAGE II satellite data to study the properties of tropical cloud over the altitude range 10.5–18.5 km. At all altitudes, most observed cloud is subvisual, this fraction increasing from just over 50% of the total cloud at lower altitudes to almost 100% above 16 km. The geographical distribution and temporal variation of the cloud occurrence have been examined over all longitudes on timescales from less than 1 day to that of the ENSO cycle. Significant variations in cloud occurrence have been found on each of these scales and have been compared with the underlying surface temperature changes. Maximum cloud occurs over the warm pool region of the Pacific Ocean, with secondary maxima over the South American and Central African landmasses, where the percentage of cloud occurrence in the upper troposphere can exceed 75%. More tropical cloud is found in the latitude band "equator–20°N" than in the band "equator–20°S". Cloud occurrence at all latitudes within the tropics over both land and ocean, increases with the surface temperature at a rate of approximately 13%/1°C. The ENSO cycle changes in the cloud occurrence have been observed not only in the tropics but also over the subtropics including a difference in the meridional Hadley circulation between ENSO warm and cold years.

An important step forward in studying the warmest waters (the warm pool, WP) in the equatorial Indo-Pacific Ocean has been made by Schneider et al. (1996) on the basis of global climate simulation modelling using the coupled model developed at the Max-Planck-Institut für Meteorologie (Hamburg) at T42 (2.8° × 2.8°) and T21 (5.6° × 5.6°) resolutions. The coupled model ECHO consists of the atmospheric GCM (ECHAM3) and ocean GCM(HOPE). An important feature of the model is that it does not employ a flux correction. The analysis of the results shows that the model simulates a warm pool, whose temperatures are consistent with observations for a model atmosphere at T42 resolution. At T21 resolution, the model underestimates surface temperatures by 1 K. For most regions of the warm pool the surface heat gain is by vertical advection. The advection cooling is activated by inertial oscillations forced by synoptic atmospheric variability. At T42 resolution the seasonal migration of the warm pool is simulated reasonably well, as are the surface heat fluxes, winds, and clouds. However, simulated precipitation is too small compared to observations, implying that the surface density flux is dominated by fluxes of heat.

As Schneider et al. (1996) have discovered in the Pacific portion of the warm pool, the average net heat gain of the ocean amounts to 30–40 W/m^2. In the northern branch, this heat gain is balanced by vertical advection, while in the southern branch, zonal, meridional and vertical advection cool the ocean at approximately equal rates. At the equator, the surface heat flux is balanced by zonal and vertical advection and vertical mixing. The Indonesian and Indian Ocean portions of the warm pool receive from the atmosphere 30 and 50 W/m^2, respectively, and this flux is balanced by vertical advection. Schneider et al. (1996) have emphasized that the model produce a surprisingly realistic simulation of the tropical region, although, of course, further intercomparisons with observations are necessary. It will become possible after an accomplishment of the Tropical Ocean and Global Atmosphere Coupled Ocean–Atmosphere Response Experiment (TOGA COARE) field experiment.

A quite different (alternative) "thermostat" mechanism has been suggested by Sun and Liu (1996). Using a three-box model for the tropical Pacific Ocean with a simple atmosphere, they have demonstrated that the dynamic coupling between the atmosphere and the ocean may serve a thermostat mechanism. In this context, Sun and Liu (1996) have pointed out that the ocean currents that connect the western Pacific Ocean with the eastern Pacific Ocean are driven by surface winds. Surface winds are in turn driven by SST differences between these two regions. This kind of dynamic coupling between the atmosphere and ocean (which plays a central role in the ENSO phenomenon) may limit the SST in the tropical Pacific Ocean to below 350 K even in the absence of cloud feedbacks.

To summarize the discussion concerning the "thermostat" hypothesis, one can say that the results obtained are fairly controversial which simply means that further studies are necessary. The well-balanced conclusions have been formalized by Waliser (1966): "Given the drastic difference in complexity between the real climate system and the thermostat hypothesis, which attempts to describe one aspect of it, it is difficult to ascribe validity, particularly when (1) factoring in remote effects on the local high SST environment; (2) assessing the 'thermostat's' role in regulating, versus limiting, the present climate's SST; or (3) extending its applicability to future climate changes. Of less difficulty, however, is assessing the qualitative contributions made by the hypothesis. Chiefly, the hypothesis has underscored the importance of cloud-forcing feedbacks in high SST regimes relative to the more conventional mechanism of evaporation. By doing so, it has challenged much of the tropical climate community to reevaluate their understanding of the surface energy budget in these regions and reconsider the relative strengths of the feedbacks that keep this system at the observed equilibrium. On the practical side, the controversial nature of the hypothesis has provoked a host of alternative propositions and motivated the acquisition of data that can be used to test these propositions as well as future models and theories. Finally, from a pedagogical standpoint, the thermostat hypothesis has and can continue to serve as a useful paradigm in our efforts to further unravel the complex nature of our climate".

The western tropical Pacific is of special interest for climate change studies because it is characterized by the warmest open-ocean surface temperatures found on Earth. It may be assumed therefore that a significant CO_2-induced warming would result in warm SST conditions unlike any found in the Earth's present climate. This is why Knutson and Manabe (1995) have conducted an investigation of the time-mean response over the tropical Pacific region to a quadrupling CO_2 concentration using a global coupled atmosphere–ocean general circulation model developed at GFDL The atmospheric component is a global nine-level spectral GCM (7.5° long × 4.5° lat. computational grid, rhomboidal truncation at zonal wave number 15). The ocean component is a 12-level gridpoint GCM with horizontal resolution 3.75° long. × 4.5° lat.

The analysis of the numerical modelling data has shown that tropical Pacific sea surface temperatures rise by about 4–5°C. The zonal SST gradient along the equator decreases by about 20%, although it takes about one century (with CO_2 increasing at 1% per year compounded) for this change to become clearly evident in the model. Over the central equatorial Pacific, the decreased SST gradient is accompanied by similar decreases in the easterly wind stress and westward surface currents and by a local maximum in precipitation increase.

2.5.3 Numerical climate modelling and its reliability

2.5.3.1 Introduction

The gradual improving of the numerical climate models through an adequate parameterization of the climate-forming processes (of particular importance is a development of the coupled models of the atmosphere–ocean system with a consideration of the biospheric processes) and an increase of the spatial and temporal resolution of the models raises the reliability of the numerical modelling results. Use of the coupled models has made it possible to simulate the climatic variability on the scales of decades, which in some respects agrees with the observed one (the oceanic component of the models demonstrates the presence of long-term fluctuations connected with changes in the thermohaline circulation). The mechanism of the feedback functioning due to water vapour has been better understood, though serious uncertainties of the precalculated changes in the moisture content of the upper troposphere in the tropics still remain; the problem of an adequate simulation of the cloud–radiation interaction is still unsolved; only first steps have been made towards an account of biospheric processes. As has been mentioned above, of special interest is the problem of consideration of the climate-forming role of aerosol (A Plan..., 1996). All these circumstances explain rather a low reliability of the simulation and prediction of regional climate changes, a possibility of unexpected changes is quite possible.

A new important step forward in the climate theory is developing a theory of the time-dependent climate dynamics (instead of, for example, changes in the equilibrium climate with an instantaneous CO_2 doubling). By the present time, 23 coupled global circulation models (CGCM) have been developed in eight countries (Proceedings of Workshop..., 1995) which have been used to simulate global climate dynamics for time periods up to 1000 years (Manabe and Stouffer, 1996) with a continuously improving parameterization of various climate-forming processes (i.e., sea-ice cover dynamics, land surface and ocean processes, biosphere dynamics, chemical processes, etc.). Unfortunately, a realization of such models is still accompanied, as a rule, by the need to remove the climate drift through the so-called "flux correction" consisting in matching the calculated fluxes of heat and moisture with the observed ones (Colman et al., 1995; Meehl, 1995). Such an " adaptation" of the model ensuring a realistic simulation of the present climate may substantially distort the response of the model to small forcings such as, for example, an increase of GHG concentrations (as a rule, in modelling a time-dependent climate, the CO_2 transient increase by 1%/year is prescribed), although Colman et al. (1995) concluded from the results of a non-flux corrected transient CO_2 experiment with the coupled GCM that, despite the secular climate drift, the major patterns of atmospheric and oceanic temperature change are similar to changes from flux corrected models.

The internal variability of the climate system dominates as a factor of geographic variability during first several decades of numerical experiments with prescribed increasing CO_2 concentrations. Upon establishing the geographic distribution it varies very little and turns out to be similar to that corresponding to the equilibrium climate models, including special features, such as: (1) stronger SAT increase over the land compared to the ocean; (2) an increase (on the average) in high latitudes, in the monsoon region of Asia, and in the wintertime mid-latitudes; (3) a summer-time decrease of soil moisture (on the average) in some regions of the mid-latitude continents. An account of the gradual CO_2 increase suggested, however, the conclusion that in the northern band of the North Atlantic and in the southern regions near the Antarctic the warming is 60% or more weaker than the equilibrium one, by the moment of a CO_2 doubling.

Reliable climate forecasts can be obtained only under two conditions: (1) interactively simulated (not prescribed *a priori*) GHGs concentrations; (2) the climate models should take into account all substantial natural and anthropogenic climate-forming processes based on consideration of the coupled climate model "atmosphere–hydrosphere–lithosphere–cryosphere–biosphere" (the contribution of solar activity to the formation of climate is still unclear). Since until recently the use of climate models was concentrated, mainly, on an account of radiative forcings due to a GHGs increase, apparently, the results thus obtained characterize mainly the greenhouse component of climate changes.

In connection with preparations for the first conference of the participants of the Framework convention on climate change (COP-1) held in Berlin from 27 March to 7 April 1995, the experts of the UK Meteorological Office undertook an overview of the results of the use of the numerical climate model developed at the Hadley Centre for Climate Prediction and Research (Modelling Climate..., 1995). The main objective of the numerical modelling was to simulate the observed climate changes and to forecast the climate for the period 1860–2050, with account of basic natural and anthropogenic climate-forming factors (the latter include emissions of greenhouse gases and sulphur dioxide transformed into sub-micrometre sulphate aerosol through gas-to-particle reactions).

The most important result of the numerical modelling consists in simulating the mean global SAT increase by 0.5°C observed over the last 100 years: while a consideration of the "greenhouse" warming determines a SAT increase within the range 0.6–1.3°C, an additional account of the "aerosol" cooling reduces the warming to the observed level 0.5°C (0.45 ± 0.15°C). The forecast of the long-term variations of the global SAT (with the prescribed transient annual CO_2 increase by 1%) gave the warming trend 0.2°C/10 years (apparently, this rate of SAT variations doubles the limit accepted for most sensitive ecosystems). With the "aerosol" cooling excluded, the pure "greenhouse" warming will be 0.3°C/10 years, which agrees with the results of other calculations (the IPCC estimates, in particular).

Calculations of the global distribution of SAT variations over the period from 1990 to nowadays compared with observational data revealed a considerable inhomogeneity due to local sources of aerosol which determine the formation of local regions of climate cooling (naturally, the calculated inhomogeneity of the field of SAT variations is considerably smoothed compared to the observed one, and the differences between the locations and amplitudes of SAT anomalies are rather substantial).

The forecast of averaged SAT variations for the period from 1970–90 to 2030–50 (the 20-year averages were calculated) revealed practically in all cases a climate warming with a much greater warming in the NH high-latitudes. Calculations with the aerosol contribution left out of account showed that with increasing GHG concentrations the following should take place: (1) enhancement of precipitation in middle and high latitudes (especially in winter) and its cessation in many sub-tropical regions (the distribution of soil moisture follows, on the whole, the distribution of precipitation, but the effect of specific features of the spatial structure of the river runoff and evaporation is also manifested; the forecast for the nearest 50 years determines a possible increase of soil moisture over most of North America and northern Asia, which is mainly determined by the wintertime enhancement of precipitation; the distribution of soil moisture in summer is very complicated and should be further analysed); (2) the rise of the World Ocean's level by about 4 cm per 10 years.

The numerical modelling with account of aerosol confirmed qualitatively these conclusions. Still limited computer potential does not permit an accomplishment of the global numerical modelling with a high spatial resolution sufficient for the regional analysis. This prompts one to use the "nested" models (Cubasch et al.,

1995c; Giorgi et al., 1994; Giorgi, 1995; Walsh and McGregor, 1995). At the Hadley Center (Modelling Climate..., 1995) a "nested" model was developed (within the global model with a resolution of ~270 km) with the resolution 50 km. This model was used for a period of 10 years for the territory of Europe (the boundary conditions for the "nested" model were prescribed from the data of the global numerical modelling).

A comparison of the wintertime and summertime precipitation fields calculated with the use of the global and "nested" models has shown that in the second case one could reach a much more realistic spatial distribution of the precipitation field and more reliable reproduction of rain rate (this is confirmed by comparison with the observational data). The regional model provides, for example, an orographically induced enhancement of precipitation in Scotland, Norway, and the Alps, adequate to that observed. The earlier estimates of the effect of the $2 \times CO_2$-induced climate changes (with aerosol neglected) on the global distribution of the yield of most important crops showed that this effect should be mainly negative in low latitudes and positive in high latitudes (different climate models give, however, substantially different results). No doubt, an account of the aerosol effect should markedly change the results obtained earlier, especially from the viewpoint of the regional-scale manifestations of the climatic impact on agricultural crops.

Of paramount importance are the data of observations of the frequency of occurrence of weather anomalies during the last century (in conditions of global climate warming). An analysis of these data suggested the conclusion that the frequency of occurrence of weather anomalies did not increase. It is essential, however, that a 1.6% increase of the mean global rain rate took place. Numerical modelling of global climate warming gave similar results. From the data of Houghton (1994), in the global warming conditions, the frequency of occurrence of floods and droughts as well as the intensity of tropical cyclones should increase.

On the whole, the global warming should intensify the existing negative changes in the environment. For example, a possible rise of the World Ocean level will create difficulties for Bangladesh and for many island countries, and floods and droughts will cause the loss of productive land, deforestation, unfavourable changes in the water regime, and others. Even favourable (in some cases) consequences of the warming will require an adaptation and, hence, serious expenditures. Though in the warming conditions the global level of agricultural production may be preserved (due to adaptation), the contrast between the developed and developing countries will deepen.

Serious negative consequences will occur in connection with the impact on the ecosystems, especially in middle and high latitudes (forests, primarily). An increased duration of the periods of heat stresses in low latitudes will be harmful to human health and will drive some tropical diseases (such as malaria) to higher latitudes.

Approximate estimates of economic losses gave the mean global value of the order of 1% GNP, which may be twice as large in the case of developing countries. These estimates, however, do not take into account a number of factors (e.g., possible social and political conflicts). About 3 million new "ecological" refugees may appear annually, that is, more than 150 million people by the middle of the next century. All these estimates have been obtained on assumption of a CO_2 doubling. With further CO_2 increase, the consequences of climate change will be more painful (see Climate Change..., 1996c)

2.5.3.2 Early detection of an anthropogenic climate signal

Santer (1995) undertook a numerical modelling of the global climate to assess the contributions of increasing concentrations of carbon dioxide and aerosol to changes in the SAT field, to further analyse possibilities of detecting the anthropogenic climate signal by assessing correlations between the initial and subsequent SAT fields. The results of calculations show that with a separate consideration of the effect of CO_2 and that of aerosol, the SAT change cannot be reliably attributed to any one cause.

Solution of the problem of anthropogenic climate change detection is also difficult with a combined account of both these factors; here the centred pattern correlation statistic technique should be used. Calculations have shown that in summer and in autumn the change in the structure of the SAT field under the combined influence of CO_2 and aerosol is similar to that observed, and this similarity grows with time (over the last 50 years). With only the aerosol effect taken into account the location of the region of maximum response does not coincide with the region of maximum effect.

The results obtained by Santer (1995) do not permit one to exclude the possibility that the signal of the combined effect of CO_2 and aerosol may be actually a manifestation of the long-range (in time scales of many decades) natural climate variability. These results agree qualitatively with the hypothesis of partial compensation of the greenhouse climate warming by the aerosol cooling. The large contribution of sulphate aerosol to the formation of climate revealed by the numerical modelling testifies to the importance of considering chemical processes in the numerical modelling of the past and future climates.

The necessity of solving the problem of the early detection of anthropogenic climate signal has stimulated numerical modelling efforts with the aim of assessing the greenhouse-induced climate changes.

One of the consequences of the potential impact of atmospheric CO_2 increase on climate assessed from numerical climate modelling has been a distinct tendency for soil moisture to be substantially reduced during the summer over extensive continental regions in middle latitudes. To improve understanding of the mechanisms responsible for the dryness mentioned, Wetherald and Manabe (1995) performed several numerical experiments (for a CO_2 quadrupling) using a GCM with idealized geography developed at the Geophysical Fluid Dynamics Laboratory (GFDL).

The results show that the simulated reduction of soil moisture in middle latitudes begins in late spring and is caused by an excess of evaporation over precipitation. The increase of carbon dioxide and the associated increase of atmospheric water vapour enhances the atmospheric thermal emission at the continental surface at all latitudes. However, owing mainly to the CO_2-induced change in mid-tropospheric relative humidity, the increase in thermal emission is greater in the equatorward side of the rain belt, making more energy available there for both sensible and latent heat.

Since the saturation vapour pressure at surface increases nonlinearly with surface temperature, a greater fraction of additional radiative energy is realized as latent flux at the expense of sensible heat. Therefore, evaporation increases more than precipitation over the land surface in the equatorial side of the rain belt during spring and early summer and initiates the drying of the soil there. As the rain belt moves poleward from spring to summer, the soil moisture decreases in middle latitudes, reducing the rate of evaporation. This reduction of evaporation, in turn, causes a corresponding decrease of precipitation in middle latitudes, keeping the soil dry throughout the summer.

A tendency to increased summer dryness also occurs in high latitudes, but mainly results from the earlier removal of highly reflective snow cover in spring, which enhances the evaporation in the late spring, lengthening the period of drying during the summer season (the contribution of such a mechanism in middle latitudes is small). The drying of soil is also enhanced by the land surface–cloud interaction in both middle and high latitudes. Owing to the reduction of cloud cover that results from the decrease of relative humidity in the lower troposphere, solar radiation absorbed by the continental surface increases, thereby enhancing evaporation and further reducing the soil moisture in summer. Wetherald and Manabe (1995) have pointed out simplifications of their model and a necessity (for simulations of real climate) to apply a GCM with a more realistic parameterization of land surface processes.

To consider a fundamental problem of identification of internal and externally forced variability of the climatic system (the most popular aspect of this problem is the detection of CO_2 signal), Harzallah and Sadourny (1995) have applied an updated version (Cycle 5) of the LMD AGCM to analyse the variability of atmospheric flow by separating it into an internal part due to atmospheric dynamics only and an external (or forced) part due to the variability of SST forcing. The principal purpose of the study has been to demonstrate that middle and extratropical atmospheric responses to SST anomalies can be separated from internal variability using a set of long runs. The simulations to solve this problem are a set of seven long-term integrations (with the LMD AGCM) all forced by 19 years of observed SSTs (1970–88). The forced atmospheric variability is obtained as the average over the seven simulations (ensemble mean). The internal part, on the other hand, is obtained from the deviation between individual simulations and the ensemble average used to calculate the forced one.

In the case study considered (northern hemisphere winter: December–January–February) Harzallah and Sadourny (1995) put emphasis on internal variability of sea level pressure (SLP) and 500-hPa geopotential height. The results of the numerical modelling show that although statistical significance is not reached in some extratropical regions, large significant extratropical responses are found at the North Pacific–Alaska sector for SLP and over western Canada and the Aleutians for 500-hPa geopotential height.

The investigation of the influence of SST variations on internal variability has been accomplished by using a 7-year simulation with the prescribed climatological SST seasonal cycle. It has been found out that inter-annual SST changes strongly influence the geographical distribution of internal variability; particularly, it tends to increase it over the oceans. The use of EOF decompositions has demonstrated that the model realistically simulates the leading observed variability modes.

Harzallah and Sadourny (1995) have found that the geographical structure of internal variability patterns is similar to that of total variability, although similar modes tend to evolve rather differently in time. The zonally symmetric seesaw dominates the internal variability for both observed and climatologically prescribed SST. The Pacific–North America and western Pacific patterns, on the other hand, are the dominant modes associated with patterns of SST variability; the latter is related to Atlantic anomalies, while the former responds to both El Niño events and extratropical forcing. Adding the inter-annual variability of SST is found to shift the internal atmospheric variability from continents to oceans, with northern Europe and northern Central Asia showing the most important decrease of variations.

Kawamura et al. (1995) have made a significant contribution to studies of the internal variability of the global climatic system. They have shown that the tropical SSTs of interdecadal variability contribute a great deal to the decrease and increase in the northern hemisphere land surface temperature observed in recent decades (these results of simulation modelling are based on the use of a T42 GCM version of the Japan Meteorological Agency (JMA-GSM89) global climate model). The interdecadal variability of the northern hemisphere land surface air temperature, at least in summer, is controlled by the tropical SSTs of interdecadal variability. The decadal changes of the hemispheric land surface temperature seen in recent decades, in particular, are primarily caused by tropical forcing.

To assess whether the global positive SAT trend should be forecast for the future, Woodward and Gray (1995) have discussed the generic problem of determining whether an observed trend in a time series realization is a random (i.e., short-term) trend or a deterministic (i.e., permanent) trend. In this context, an autoregressive integrated moving average (ARIMA) model and a "deterministic forcing function + autoregressive (AR) noise" model have been considered as possible random trend and deterministic trend models, respectively, for realization displaying the trending behaviour. For classifying an observed time series realization as ARIMA or "function + AR", a bootstrap-based classification procedure has been applied using linear or quadratic forcing functions. The principal conclusion of the simulation study is that the temperature anomaly series have to be classified as ARIMA (i.e., having random trends). Woodward and Gray (1995) have emphasized, however, that their results should not be interpreted to indicate that there is no deterministic component in the data, but that these data have a preference for the ARIMA model when compared to models with linear or quadratic trends and stationary autoregressive residuals.

Since a wavelet transfer (WT) has been suggested as a tool to study multiscale, nonstationary processes occurring over finite spatial and temporal domains, this methodology has been broadly applied for various purposes, including detection, chaos and fractal research, and many others. Lau and Weng (1995) have applied WT methodology, which, in some ways, is a generalized form of Fourier transform (FT) and windowed FT (WFT), to solve the problem of the detection of a climate signal as the culmination of interactions among physical processes operating on a wide range of spatial and temporal scales. Since two completely different time series with different local information may result in very similar mean states, it is important to recognize that it is the totality of both the local and the global information that constitutes a true climate signal. Lau and Weng (1995) have emphasized that the WT is a tool that is ideally suited to solve such a problem. Unlike the FT that maps a 1-D time series to a 1-D spectrum, the WT maps a 1-D time series to a 2-D image that portrays the evolution of scales and frequencies with time.

Lau and Weng (1995) have applied the WT to solve the climate signal detection, which most important aspect is the determination of trends in the presence of a large climate variability. The four illustrations of the WT use have been considered: (1) amplitude modulation (this is specifically relevant to a climate system involving nonlinear interactions between various scales of interference of a frequency components from its sidebands; (2) frequency modulation (which is relevant to the fundamental physical properties of the climate system undergoing secular changes such as the increase in atmospheric moisture due to global warming); (3) abrupt change in frequency (a catastrophic event that has a long-term impact); (4) abrupt change in time (a case of an impact of volcanic eruption on the global temperature variations).

The WT methodology has been applied to two widely used real climate time series: (1) 2.5 million-year (Myr) deep-sea sediment record of $\delta^{18}O$ at site 607 in the North Atlantic, which provides information on the variation of global ice volume; (2) the monthly northern hemisphere surface temperature (NHST) averaged from 5×5 grid dataset over the period January 1854–July 1993. The principal conclusions substantiated by Lau and Weng (1995) show that in case of palaeoclimatic data the presence of a 40-kyr (for the period from 0.7 to 2.5 Myr) and a 100-kyr (during the last 0.7 Myr) quasi-periodic oscillations and an abrupt transition between the two frequency regimes at about 0.7 Myr before the present, consistent with previous study, took place. The two dominant cycles (40 and 100 kyr) coincide approximately with frequencies for two of the major components of orbital forcings associated with obliquity and eccentricity variations.

The second case (the contemporary NHST variations) is characterized by a myriad of oscillatory modes from inter-annual (2–5 yr), interdecadal (1–12 yr, 20–25 yr, and 40–60 yr), and century (180 yr) scales for different periods of data record. An interesting feature of the present century is the scale interaction across the inter-annual and interdecadal boundaries, which indicates a frequency modulation from a 5.6-yr cycle in the early 1990s to a quasi-10-yr cycle in the 1970s. Also emerging out of the early 1990s is a 40–60-yr oscillation that is maintained up to the present. The common features in the time-frequency characteristics of both time series (in spite of large differences in time scales) suggest that the variations of the Earth's climate are consistent with those exhibited by a nonlinear dynamical system under external forcing.

The recent search for an anthropogenic climate signal has been undertaken by Hansen et al. (1995b, 1996), Hasselman et al. (1995), Karoly et al. (1994), Mitchell et al. (1995a,b), Santer (1995), Santer et al. (1995a,b, 1996) and Taylor and Penner (1995). In a most recent study (Santer et al., 1996) the four new steps have been made to achieve a higher statistical significance of the results: (1) the relative detectability of the vertical temperature-change signals due to individual and combined changes in atmospheric CO_2 and sulphate aerosols (zonal mean annually averaged meridional cross-sections of the temperature field in the troposphere and lower stratosphere were considered); (2) signal data from two different models were used in order to explore the sensitivity of the results to model-dependent uncertainties in the definition of an anthropogenic signal; (3) modification of a combined CO_2 + aerosol vertical temperature-change signal due to observed changes in stratospheric ozone was considered; (4) information from three long control runs performed with coupled atmosphere–ocean GCMs (CGCMs) was used to assess the significance of trends in the modelled-versus-observed pattern similarity. Such integrations allow one to analyse the internally generated climate variability on a range of space and time scales. In all the cases the model data were used with fixed levels of anthropogenic pollutants ("equilibrium experiments") and from integrations in which the atmospheric concentrations of GHGs and aerosols increase over time ("transient experiments").

A most clear feature of modelled and observed patterns of annual-mean zonal-mean temperature change as a function of latitude and height is (with only CO_2 taken into account) is tropospheric warming and stratospheric cooling, which is the well-known direct radiative signature of the growth of CO_2. Maximum warming occurs in the tropical upper troposphere, and temperature changes are hemispherically symmetric. Consideration of either aerosols only or CO_2+aerosols reveals a hemispherically asymmetric response, with an increased cooling and reduced warming in the northern hemisphere (due to aerosol impact). Stratospheric O_3 reduction is characterized by stratospheric cooling, with maximum cooling at high latitudes in both hemispheres (but the response is not hemispherically symmetric: statistically significant cooling occurs over a wider latitude range in the NH than in the SH, primarily because of a hemispheric asymmetry in the observed ozone changes).

An intercomparison between calculations and observations has been accomplished with the use of the so-called "centred correlation statistics", $R(t)$:

$$R(t) = [\sum_{x=1}^{n} (\Delta D(x,t) - \Delta\hat{D}(t)(\Delta M(x) - \Delta\hat{M}]/[ns_D(t)s_M]$$

where ΔD and ΔM are the temperature-change fields for observed data and model output, respectively, and x and t are discrete indices over space ($x = 1,2,\ldots, n$, the combined latitude–height dimension of the radiosonde data) and time ($t = 1963,\ldots,1987$). Observed changes are anomalies relative to the average over 1963–73, and model changes are differences between time averages of perturbation and control experiments. The ^ indicates a spatial average. The observed spatial variance is $S^2{}_D(t)$

$$S^2_D(t) = [\sum_{x=1}^{n} (\Delta D(x,t) - \Delta\hat{D}(t)]^2 \,/\, (n-1)$$

with the model spatial S^2_M variance defined similarly. The $R(t)$ positive trend defines that the observed time-varying patterns of temperature change are becoming increasingly similar to the model-predicted responses.

In the case of the 50–850 hPa layer the behaviour of $R(t)$ is similar for all signals except S-TP (consideration of aerosol impact only): trends are positive over the full 25-year period, indicating an increasing expression of the model-predicted patterns in the observed annually averaged data. Santer et al. (1996) have emphasized that these similarities are due partly to the large-amplitude pattern of stratospheric cooling / tropospheric warming common to all signals that incorporate CO_2 effects. The modelled–observed comparisons over 50–850 hPa have demonstrated that the positive 25-year $R(t)$ trends for all signals considered (CO_2 only, CO_2+aerosols) differ considerably from unforced trends which characterize the internal climate variability (this conclusion is based, however, on the assumption that the GCMs used provide credible estimates of the spectrum and patterns of internal natural variability on timescales ranging from 10 to 25 years). On shorter timescales (10 and 15 years) the signal/noise ratios decrease.

The general conclusion made by Santer et al. (1996) is that the results considered suggest that the similarities between observed and model-predicted changes in the zonal-mean vertical patterns of temperature change over 1963–87 are unlikely to have resulted from the natural internally generated variability of the climate system: "The degree of pattern similarity between models and observations increases through this period. It is likely that this trend is partially due to human activities, although many uncertainties remain, particularly relating to estimates of natural variability". Santer et al. (1996) have pointed out that the main uncertainties relate to:

(1) Estimates of the relative magnitudes and spatial and temporal evolution of the different forcings including both anthropogenic factors which have not been considered (indirect aerosol effects, non-CO_2 GHGc, etc.) as well as purely natural forcings (solar activity effects, volcanic eruptions, etc.).
(2) Still non-adequate parameterization of some climate-forming processes in the atmosphere and ocean.
(3) Reliability of model-derived natural internal climate variability on decadal and longer timescales.
(4) Certain time-varying instrumental biases in the radiosonde data, and their incomplete spatial coverage.

In view of the uncertainties mentioned, Santer et al. (1996) have concluded: "Although we have identified a component of the observational record that shows a statistically significant similarity with model predictions, we have not quantified the relative magnitude of natural and human induced climate effects".

In addition to what has been concluded by Santer et al. (1996), it would be proper to mention that the assumption of pure scattering aerosols represented by changes in the surface albedo is far from being realistic, because aerosol number density, chemical composition and, respectively, optical properties are strongly variable in space and time and in most cases aerosols contain absorbing components (this is especially important from the viewpoint of aerosol–cloud interaction), as has been discussed in Section 2.4.2. Since aerosol effects have happened to be very important in reaching a closer agreement with observations, these circumstances may become of critical significance. Undoubtedly, numerical climate simulation experiments with aerosol as a varying optically active (scattering and absorbing) component of the atmosphere are necessary, although for the time being, prohibitively difficult (Kondratyev, 1972, 1982, 1983b, 1988, 1991a, 1992a,b, 1995a,d, 1996b). It is equally clear that further assessments of a contribution of natural internal climate variability are most important, especially in the light of a number of recent results (Bengtsson et al., 1996a; Graham, 1995; Meehl and Washington, 1996; Stendel and Bengtsson, 1996) which will be discussed later on.

An important re-analysis of combined land SAR and marine SST data for the recent half-century has been accomplished by Hansen et al. (1996) who produced what they call a land–ocean temperature index (LOTI) for 1950–95 (LOTI anomalies have been considered defined as deviations from the 1951–80 mean). These data have confirmed once more the reality of global warming (Fig. 2.9). The 1995 LOTI (0.39°C) exceeds the previous high LOTI (in 1990) by 0.01°C. The most prominent

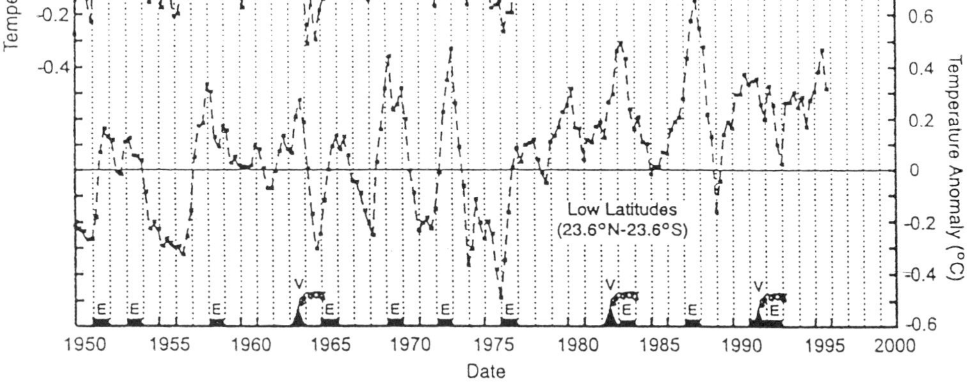

Fig. 2.9. Change of land–ocean temperature index at seasonal resolution. The index combines SSTs between 59°N and 45°S and surface air temperatures of meteorological stations. E and V mark the years of major El Niños and volcanoes.

signals in low latitudes are El Niños, but they also leave a signature on global LOTI. It is quite obvious that by 1995 the global LOTI has returned to the pre-Pinatubo level.

Hansen et al. (1996) have pointed out that the temperature change of recent decades can be described as a jump occurring in the late 1970s or as a fairly steady increase since the mid-1960s with superposed fluctuations. They believe that the latter interpretation seems more plausible, but either conclusion is valid until a longer record is available.

The geographical and seasonal distribution of LOTI change for 1965–95 is characterized by the greatest warming in the NH winter over Eurasia and North America. Intense cooling occurs in the mid-latitude central Pacific and in a region between Hudson Bay and Greenland stretching into the North Atlantic. The magnitude of the cooling decays in the summer and fall. At lower latitudes there is an extensive warming over the oceans, and, with more seasonal variation, warming over continents. The tropical warming of recent decades is both sustained and geographically pervasive, that is, it represents more than an increase of the frequency or intensity of El Niño.

The fact that the recent global warmth occurs near a minimum of the 11-year cycle of solar irradiance variations, and the time of record ozone depletion supports the contention of an underlying global warming trend. Hansen et al. (1996) believe that observed temperature change patterns do not suggest a driving of climate by ocean circulation changes during the past three decades (later on we shall consider the results which are in the contradiction with such a conclusion). Since the recent temperature anomalies increase the meridional temperature gradient and should tend to increase the strength of the jet stream across North America and the Atlantic, one can assume, for example, that this will fuel powerful storms that can move rapidly into Europe.

In the context of the latter conclusion, the study by Landsea et al. (1996) of downward trends in the frequency of intense Atlantic hurricanes during the past five decades should be mentioned. Landsea et al. (1996) have pointed out that the most recent years of 1991 through 1994 have experienced the quietest tropical cyclone activity on record in terms of frequency of tropical storms, hurricanes, and intense hurricanes. This was followed by the 1995 hurricane season, the busiest in the past 50 years. But, despite 1995's activity a long-term (five decades) downward trend continues to be evident primarily in the frequency of intense hurricanes. Besides, the mean maximum intensity attained by the strongest hurricane each year has not shown a significant change.

Landsea et al. (1996) have emphasized that they do not suggest that the Atlantic basin is necessarily representative of global activity, but it is useful for long-term analysis because of the unique reliable record of length.

Carnell et al. (1996) have applied the first United Kingdom Meteorological Office (UKMO) coupled ocean–atmosphere model to assess the storminess with the present-day and gradually increasing CO_2 concentrations, with special emphasis on the North Atlantic. Their principal conclusions are as follows: (1) the detection of significant changes in extreme events in the observed record is unlikely to occur in the near future; (2) the size of predicted changes is small and would be unlikely to be statistically detectable given the limited length of the observational record.

Hasselmann et al. (1995) have emphasized in conclusion of their study that the probability that the observed SAT increase during recent decades is of natural origin is estimated at less than 5%. However, this numbcr is dependent on the estimated natural variability level, which is still subject to some uncertainty.

2.5.3.3 Further development of global climate modelling

Though the present coupled (atmosphere–ocean) climate models (CGCM) can correctly simulate some features of atmospheric variability, they are still far from reality in view of unsatisfactory parameterization of the cloud cover dynamics, cloud–radiation interaction, biospheric processes (the latter is particularly important for the coupled forecast of GHG dynamics). Therefore the reliability of the simulation (even more the forecast) of regional climates remains low; there is, for instance, no convincing evidence of possible changes in the frequency of the storm conditions. Inadequacy of CGCM determines the possibility of unexpected climate changes ("surprises").

Uncertain forecasts of the time and magnitude of climate change in the future (particularly concerning the regional climate changes) are explained by inadequate understanding of the role of a number of the climate-forming factors: (1) sources and sinks of GHG and aerosol, the processes of their transformation in the atmosphere; (2) cloud cover dynamics, cloud-radiation and cloud–aerosol interactions, global water cycle; (3) the atmosphere–ocean interaction (the delaying impact of the deep-water circulation, in particular); (4) polar ice cover dynamics; (5) land surface processes and feedbacks determined by hydrological and biospheric processes governing the interactivity of regional and global climate changes.

The gradual removal of the enumerated and other causes of inadequate reliability of numerical climate modelling as well as solution of the problem of global climate changes, on the whole, are possible with a deeper insight into the laws of climate, including: (1) an improvement of the global climate observing system (with an emphasis on observations of cloud properties, GHGs, aerosols and extra-atmospheric insolation); (2) further studies of the climatically significant processes, especially those connected with the dynamics of the cloud cover, the World Ocean, and global carbon cycle as well as palaeoclimatic variations; a consideration of the socio-economic development producing unfavourable environmental changes, climatic implications including. Intensified international cooperation should make an exclusively important contribution to the solution of the enumerated and other problems.

Concerning the key problem of an account of cloud cover dynamics in climate models, it should be mentioned that, first of all, it consists in the necessity for cloud dynamics parameterization determined by the sub-grid scales of both horizontal and vertical dimensions of clouds. An experience of such a parameterization has convincingly demonstrated the strong dependence of the sensitivity of climate to cloud cover variability on specific features of the chosen parameterization technique (Bintanja, 1996; Del Genio et al., 1996; Gultepe et al., 1996). Progress in satellite observations of the cloud cover characteristics have determined the possibility of accomplishing the International Satellite Cloud Climatology Project (ISCCP), which created the basis for comparisons of the results of the numerical climate modelling with the observational data. An important result of realization of the Earth Radiation Budget Experiment (ERBE) has been an application (to test the model adequacy) of the concept of the cloud-radiation forcing (CRF).

Nevertheless, progress in improving the parameterization technique for cloud cover dynamics and cloud–radiation interaction made during the last decade has been slow, especially from the viewpoint of considering the partial cloudiness and mesoscale structures of the cloud cover. Progress has been made in developing the prognostic schemes with a separate account of liquid-water and ice clouds, including (Geleyn, 1994): (1) use of the prognostic equation for cloud amount based on statistical relationships between cloud amount and the rate of changing the saturating specific humidity as a function of the liquid-water and ice content of clouds; (2) substantiation of equations describing the microphysical processes of the formation and dissipation of liquid-water and ice clouds; (3) giving an account of the interaction of the processes in convective and large-scale clouds.

An important step forward has been made by Hahmann et al. (1995) who have explored the land-surface temperature and radiative fluxes response of the NCAR CCM2/Biosphere–Atmosphere Transfer Scheme to modifications in the optical properties of clouds. For this purpose, climate simulations have been compared with observation data on the Earth's radiation budget, surface radiation budget and cloud amount data with an emphasis on global and regional spatial scales and the seasonal timescale. The analysis of the reliability of the control simulation has revealed reasonable overall patterns, but also serious discrepancies. In particular, in the mid-latitude summer, the model reflects too little solar radiation (by as much as 50–100 W/m^2) in comparison with satellite observations, has surface shortwave fluxes that are too large (by as much as 40–80 W/m^2) in comparison with surface observations, and is low in cloud amount (by as much as 10%) compared to ISCCP climatology. Such discrepancies result in an overestimation of the land-surface temperature (by up to 15°C).

To improve numerical modelling results, Hahmann et al. (1995) have revised the parameterization of both shortwave and longwave radiative transfer through more reliable account of the cloud optical properties which has led to a much better agreement with observation data. It has been shown, thus, that there is a large sensitivity of summer surface temperatures to cloud optical properties.

The first version of the general circulation model which was developed at the Center for Climate System Research/National Institute for Environmental Studies (CCSR/NIES) (a detailed consideration is contained in the paper by Nakajima et al. (1995)). The horizontal resolution of this spectral model is T21 (about 600 km). The 60-layer model has approximately 500 m vertical resolution in the upper troposphere and lower stratosphere. The top level of the model is located at about 0.5 hPa (approximately 50 km). The moist convective adjustment scheme has been used as the cumulus parameterization.

The CCSR/NIES model has been applied to simulate the stratospheric quasi-biannual oscillation (QBO). The principal result is that the simulated oscillation occurs in the middle stratosphere, being slightly higher than that observed in the equatorial lower stratosphere. The amplitude of the zonal-wind oscillation is about 20 m/s, which is comparable to that observed. However, the calculated easterly phase is weaker than the westerly phase, which is opposite to what has been observed.

From the point of view of an account of various climate-forming factors, considerable progress has been made in the parameterization of the land-surface processes with the biospheric dynamics taken into account, in improving the parameterization of the shortwave and longwave radiation transfer which made it possible to use more complete information on the spectral dependence of the optical properties of clouds and aerosol (an adequate parameterization of radiative processes is still hindered by limited resources of the computer time and insufficient reliability of the input parameters).

Some climate models still overestimate the net radiation at the surface level and underestimate the absorbed atmospheric shortwave radiation (SWR) (even in the absence of clouds). Serious errors are connected with calculations of the SWR absorption by clouds (see Section 2.4.6). The problems of account of the spatial structure ("geometry") of the cloud cover and nonlinear correlation between cloud properties and radiation fluxes remain practically unsolved. The latter means that the averaged (for the spatial grid elements) values of radiation fluxes obtained from the averaged cloud cover characteristics may differ substantially from the observed ones. In this context the conception of "effective" characteristics of clouds requires a thorough analysis. Solution of the problem of validation of the numerical climate modelling results from the data of satellite and conventional observations needs further effort.

No doubt, the main ways to study climate are connected with the development of coupled models. Though the coupled ocean–atmosphere models were first used in numerical modelling in the late 1960s, only by the end of the 1980s did progress in computational technology make possible calculations of the evolution of climate over time intervals of the order of one century and more. Naturally, such models are most useful for the numerical modelling of the present climate variability and for assessments of possible future anthropogenic climate changes.

To summarize the initial results from such developments, a workshop was held at the Scripps Institution of Oceanography (USA) on 10–12 October, 1994 with the participation of 45 leading specialists from eight countries representing almost every scientific body dealing with the development and application of the coupled climate models. As has been mentioned above, during recent years, 23 coupled global climate models have been developed in 16 institutions of eight countries (Meehl, 1995; Proceedings of Workshop ..., 1994). Meehl (1995) has summarized the results of the workshop in the following way:

1. Most first-generation global coupled models are termed coarse-grid models due to the approximate 5° × 5° grid resolution in the atmosphere and ocean. This coarse resolution is necessary for computational efficiency for the decadal and longer timescale integrations. Several groups are now formulating somewhat higher-

resolution coupled models with approximate horizontal resolution of about 2.5° × 2.5°. So far no coupled model has been run with a global eddy-resolving ocean general circulation model (GCM), because the computational requirements are prohibitive. However, improvements in components and formulations have been made since the late 1980s.

For example, many groups are using some type of sea ice dynamics with fractional ice cover-age in addition to the more traditional thermodynamic and/or free-drift ice schemes. Several groups have added more complex land-surface process schemes (over and above the simple bucket soil moisture formulation). In addition to the Bryan-Cox-type ocean GCM, there are now a number of other ocean model formulations, some with isopycnal coordinates, some with sigma coordinates, and others with a variety of improved mixing parameterizations. Several groups are experimenting with stretched or rotated ocean grid schemes to add greater resolution (about 0.5° latitude) in the equatorial oceans and to provide a more suitable grid over the North Pole. There are also several efforts to address the computational problems associated with global coupled models using massively parallel computer architectures. Such activities are still largely experimental, with most groups continuing to use more conventional supercomputer hardware. Other future coupled model plans include the introduction of more components and processes in the climate system such as interactive land vegetation schemes, biogeochemistry, and atmospheric chemistry, among others.

2. Many present-generation global coupled models showed a consistent error of warmer than observed sea surface temperatures (SSTs) off the west coasts of the subtropical continents. This error has generally been assumed to be associated with the poor simulation of the boundary-layer stratocumulus clouds in those locations. There are also indications that the surface wind must have a sufficiently large alongshore component to produce upwelling strong enough for cool SSTs. Generally in coarse-grid coupled models, possibly as a consequence of the smoothed topography, surface winds in those regions are somewhat onshore with upwelling that is too weak. Additionally, there was the suggestion that thermoclines that are too deep and diffuse in the coarse-grid models could inhibit upwelling of water of the proper temperature (i.e., upwelled water would be too warm). Coarse-grid ocean models continue to have problems in properly simulating western boundary currents. In spectral atmospheric models, Gibbs ripples from the spectral topography result in errors in surface wind and precipitation fields.

3. A number of groups are verifying not only mean climate quantities but also climate variability on various time and space scales. Even though in most cases the variability is of lower amplitude than observed (e.g., ENSO-type variability with amplitudes typically about half that observed), numerous examples were presented of global coupled models' ability, both with and without flux correction, to simulate inter-annual, interdecadal, and longer timescale variability that qualitatively resembles such fluctuations seen in the observed climate system. This suggests that some modes of coupled variability in the climate system are robust, since the present generation of global coupled models are not only coarse grid but have clear systematic errors and often use correction terms at the air–sea interface (see item 5).

4. A typical systematic error in the ocean components of the coupled models was a too diffuse tropical thermocline with consequent warming of the water below the thermocline and in the deep ocean. This seemed to be due to a combination of at least the coarse vertical resolution and vertical diffusion. Presumably, improved diffusion schemes, higher resolution, and alternative vertical coordinates (e.g., isopycnal, sigma) will help to remove this deficiency.

5. Coupled model components (atmosphere, ocean, and sea ice) contain many systematic errors at present. When coupled together, they produce sizeable errors in the coupled simulation. Present approaches to deal with this situation have led some groups to either leave the errors in the coupled simulations, to correct freshwater and heat fluxes, or to correct freshwater flux, heat flux, wind stress, and SST. Subsequent model versions have reduced simulation errors in the component models, thus lessening the problem of systematic errors in the coupled simulations, but it is likely that many groups will continue to employ some form of flux correction (or flux adjustment) in the future until the model components are significantly improved. In spite of the various types of errors and flux adjustments or lack thereof, climate sensitivity is remarkably similar in the different global coupled models (e.g., the patterns of climate change from an increase of CO_2 in the models are qualitatively similar). This suggests again that some aspects of the climate system are fairly robust in terms of their response to increased CO_2 forcing.

6. Variability in atmospheric GCMs coupled to nondynamic mixed-layer oceans was compared to variability in global coupled models. The results suggest that on short (inter-annual or less) timescales such variability in the mid-latitudes is similar between the two classes of models, but that on longer timescales (e.g., decadal and longer) ocean dynamics and the coupling between ocean and atmosphere probably play a more active role. This seems to support the original hypothesis of Bjerknes.

7. Even though initial attempts at global coupled model simulations were and still are on the decadal and century timescales, some groups have run multicentury integrations and have examined very long timescale variability. Analyses have shown that there are regions of fluctuations of the amplitude of climate variability on century and longer timescales. Some groups have also completed integrations of the time history of climate from about 100 years before present to 100 years in the future with appropriate CO_2 concentrations and probable sulphate aerosol concentrations. The simulation of the observed climate record in global mean surface air temperature is improved if the past record of sulphate aerosol forcing is represented. The latter experiment also suggests that the CO_2 climate change signal may have been masked until now by sulphate aerosol effects, with possible alterations to the geographical pattern of climate change. Some experiments are being planned with the inclusion of additional greenhouse gases.

8. Ensemble-type experiments (with a CO_2 transient increase starting from various times in a control integration) point to the importance of the initial state for the response of the climate system and raise the fundamental issue of initialization and spinup. These types of ensemble experiments could prove to be useful in the CO_2 climate change detection problem.

Despite the diversity of the models, an important feature is their similar sensitivity to external forcing (e.g., to a CO_2 increase), that is, in this respect the models are sufficiently robust. Much attention has been paid to analysis of CGCM capabilities from the point of view of simulating the internal (natural) climatic variability. Though in most cases the variability amplitude is underestimated compared to that observed (e.g., in the case of the variability of the EL Niño type it is almost halved), the existing CGCM (both with the flux correction and without it) provide a qualitatively reliable reproduction of inter-annual, interdecadal and longer climatic variations. This reflects the robustness of a number of modes of climatic variability.

A comparison with simplified models in which only a fixed oceanic mixed layer is prescribed, has shown that the results are similar in case of short-term (inter-annual and shorter) variability, but differ greatly in case of long-term (interdecadal and longer) variations. Some attempts of numerical modelling for a period of several centuries revealed fluctuations of the amplitude of climate change, even on a scale of hundred years and longer.

New efforts to simulate climate for a hundred years ahead with the prescribed transient CO_2 increase (1%/year) gave results agreeing with those discussed in the IPCC-1992 Report. Use of CGCM to simulate the course of the mean global SAT during the last century, taking account of increasing CO_2 and sulphate aerosol, gave estimates agreeing well with the known global warming value $0.45 \pm 0.15°C$.

The "ensemble" numerical experiments taking account of continually growing CO_2 concentration, starting with different time moments (in the case of control integration), revealed the important role of the prescribed initial state in the formation of the climate system response to a forcing. These results revealed a fundamental problem concerning the initialization and "speeding up" of the models. (Such experiments can be also very important in solving the problem of detecting the anthropogenic climate signal.)

The initial state of the ocean plays the main role in CGCM when determining the climate evolution over the period of, at least, one hundred years, and impacts substantially on the climate drift. At present, three approaches are used to initialize the ocean through prescription of: (1) climatic 3-D fields of temperature and salinity; (2) climatic fields "speeded up" for a period of ~ 100 years with prescribed climate forcings at the ocean surface level; (3) calculated equilibrium state of the ocean.

An accomplishment of the Global Ocean Circulation Experiment (WOCE) has made it possible to obtain a most reliable observational database to prescribe the initial state of the ocean in climate simulations. A search for adequate approaches to consider the interaction among the climate system components is an important and difficult problem.

Within the Atmospheric Model Intercomparison Project (AMIP) supported by the US Department of Energy and accomplished by the Working Group on Numerical Experimentation, 30 different atmospheric models were compared for a 10-year period (1979–88), with prescribed boundary conditions determining the SST, ice cover distribution, CO_2 concentration, and solar constant (Gates, 1996; Proceedings..., 1995; Slingo et al., 1995). The goal of this Project is to compare the mean monthly geographical distributions of the parameters such as atmospheric pressure at the sea level, the surface fluxes of heat, moisture and momentum, the surface air temperature, the total cloudiness, and the temperature, geopotential, humidity and wind at 850, 500 and 200 hPa , upper tropospheric velocity potential and zonal wind, precipitation and the Earth's radiation budget components to assess the models' capability to simulate the mean seasonal state of the atmosphere and the tropical intraseasonal oscillations.

The AMIP diagnostic subprojects are important: they are aimed, in particular, at studying low-frequency variability, the ocean surface currents, monsoon circulations, cloud cover dynamics, stratospheric circulation, extreme climatic variability, and processes in polar regions.

The first overview of the results of developments within the AMIP diagnostic subproject "Variability in the tropics in time scales from synoptic to intraseasonal" (Slingo et al., 1995) demonstrated the important role of the discussed numerical modelling in analysis of systematic uncertainties and revealing their determining factors. The goal of this subproject was to compare 15 models of atmospheric general circulation (AGC) based on statistical analysis of the time series of daily values of the speed and zonal wind potential in the upper stratosphere averaged over the equatorial belt. Calculated values were compared with the data of an objective analysis for the period 1982–91.

The results obtained show that the models under discussion can adequately simulate the AGC intraseasonal oscillations. Most of the models reveal the eastward propagation of the upper tropospheric velocity potential anomaly, though, according to other models, there is a greater tendency to form the standing oscillations, and in one of two cases the field of velocity potential turned out to be quite chaotic (without any prevailing direction of propagation).

With the eastward propagation of the signal, the values of periodicities were quite acceptable, though, as a rule, the calculated periods were shorter than the observed ones (about 50 days). The spatial–temporal spectral analysis showed that none of the models could reveal the dominance of the intraseasonal oscillations. Though some models showed maxima on intraseasonal time scales, a relatively greater amplitude of oscillation corresponded to maxima at higher frequencies (<30 days) compared to those observed.

Almost every model underestimates the strength of the intraseasonal variability. The observed intraseasonal oscillations are characterized by a clearly cut annual course of the frequency of their occurrence with a maximum in the NH winter–spring period. Most of the models did not reproduce this regularity. Though the considered decade is too short to analyse the inter-annual variability of intraseasonal oscillations, an analysis of this type was performed and showed that some models do simulate a "suppression" of oscillations in the period of El Niño revealed from the observational data.

A study of relationships among the modelled intraseasonal variability, its annual change and characteristics of the basic state of the climate has shown that a weak annual change is typical of the models with a weak intraseasonal variability. The results obtained demonstrate that a reliable description of the basic climate is an important condition for an adequate simulation of the intraseasonal oscillations. It turned out, in particular, that the models most realistically simulating the intraseasonal oscillations are characterized by the distribution of precipitation which correlates with an increased SST. In such models, as a rule, the convection is parameterized, closed with respect to buoyancy and not the moisture convergence.

Existing cloud observation and numerical modelling data are fairly controversial even in case of the global area-weighted averages and standard deviations of annual mean total cloud amount for each of the models and observational datasets (Weare et al., 1995). For instance, the means of the ISCCP C2 and Meteor observations (for the common period July 1983 through June 1988) are both near 62%, which is nearly 10% greater than that of Nimbus-7 (for the 6 years of its record). The model means tend to be near 55% but vary from a low of less than 28% to a maximum of over 78% (it is important, however, that various definitions of "total cloud amount" were used). Most models have standard deviations from the global annual mean that are 5% to 10% smaller than the C2 and Meteor observations, suggesting much smaller spatial variability than that observed. There are substantial differences between observed and calculated global distributions of the annual mean cloudiness. The root-mean-square differences (rmsd) between the annual means of the model calculations and the C2 observations after global means are removed are all greater than 14%. The largest rmsd are more than 28%, whereas the difference between the C2 and Meteor observations is less than 8%.

In connection with the Atmospheric Model Intercomparison Project (AMIP), an intercomparison has been undertaken between calculated (29 models) and observed (ISCCP C2 as well as Nimbus-7 and Meteor observational estimates) data on the mean-annual global distribution of total cloudiness and on the mean seasonal cycle (Weare et al., 1995). The results reveal that the root-mean-square differences between the annual means of the model calculations and the C2 observations after global means are removed vary from about twice to nearly four times the difference between C2 and Meteor observations. The large differences are in some cases due to the fact that although a model qualitatively has patterns of spatial variations similar to those of the observations, the magnitude of those variations is much too small. In other cases the models have produced the approximate magnitude of the spatial variability of the observations but display sizeable errors in the pattern of that variability.

Weare et al. (1995) have also discovered pronounced deficiencies with respect to the model simulations of the mean seasonal cycle. For instance, the differences between the zonal averages of total cloudiness for contrasting seasons suggest that near 60° most models predict minima in cloudiness in summer, whereas observations strongly suggest the opposite. The C2 and Meteor data show a modest signal in total cloudiness for the only important inter-annual variation during the July 1983 through June 1988 observation period—the 1986/87 ENSO event. A few models reproduce this event about as well as do the Meteor observations, whereas many models fail to show any evidence of it. In general, those models that have more sophisticated physical processes tend to better simulate the cloud observations. Chen et al. (1995) have accomplished a specific analysis of the numerical modelling results for the northern and southern polar climates for 1979–88 for 14 global climate models (in the context of the AMIP intercomparisons). The results suggest that the accuracy of model-simulated climate features in high latitudes primarily depends on the horizontal resolution and the parameterization of physical processes in the GCMs. A more detailed discussion is contained in Proceedings... (1995).

A new assessment of global climate model simulations of Arctic air temperatures has been accomplished by Tao et al. (1996) as a subproject of the Atmospheric Model Intercomparison Project (AMIP) discussed above. Nineteen GCMs have been compared with prescribed forcing as observed SST and sea ice from a 10-year period (1979–88) with the purpose of analysing the model biases as well as the models' simulations of the inter-annual variations contained in the observational data. The principal results have been that:

(1) The models capture the latitudinal and seasonal variability of Arctic temperatures, although a cold bias of 1–3°C is apparent over the northern continents. The continental bias is strongest over Eurasia during spring and is consistent with the absence of vegetative masking of the high-albedo snow.
(2) Over the Arctic Ocean, a warm bias of 3°C during spring is comparable to the radiative effect (in the absence of albedo-temperature feedback) of a CO_2 doubling in climate model experiments.
(3) The largest model biases, which are consequences to the models' smoothing of the orography, are found over Greenland. Improvements in model resolution will probably reduce these biases.
(4) Although the magnitudes of the inter-annual variations are similar both in the models and in the observational data, the simulated and observed variations show little correlation.

Tao et al. (1996) have emphasized that the model used suffers various limitations arising from both the models and the observations. For instance, the differences of up to several degrees Celsius among historical climatologies suggest that observational uncertainties may not be much smaller than the differences among models. What is quite clear from the results obtained by Tao et al. (1996) is that the highest priorities for an improved simulation of Arctic temperatures are parameterizations of cloud–radiation interactions and of local surface–atmosphere interactions, especially with regard to surface albedo and vegetative masking.

Washington and Meehl (1996) have integrated a global AGCM coupled to a global 1-degree, 20-level ocean GCM with dynamic and thermodynamic sea ice with CO_2 increasing at 1% per year compounded for 75 years (CO_2 doubles at about year 70). Flux correction was not used in the experiment. The increase of global average mean-annual SAT at the time of CO_2 doubling is 3.8°C.

Washington and Meehl (1996) have found that the warm subsurface Atlantic layer at intermediate depths in the Arctic is maintained mainly by the sinking and intrusion of water from the West Spitsbergen Current in the model and observations. With increased CO_2 in the model, the warmer surface waters are intruded in the upper portion of the Atlantic layer producing an anomalous warming in the model at depths between 200 and 400 m. This resembles an anomalous warm layer near those depths recently observed in the Arctic. As the climate warms and sea ice retreats, low clouds increase over the newly exposed water. Yet the consequent increase of cloud albedo over these regions is more than compensated for by the decrease of surface albedo due to the melting of sea ice. This produces a net decrease of planetary albedo in the Arctic that contributes to a strong ice-albedo feedback and comparatively high sensitivity of the model.

An important aspect of the results obtained by Washington and Meehl (1996) relates to the question whether the warm layer recently observed at intermediate depths in the Arctic (and revealed by the modelling considered) is due to climate changes associated with increased CO_2 or whether it is associated with some kind of inferential decadal timescale fluctuations. The answer to this question will require enhanced observations in the Arctic over a longer time period.

Washington and Meehl (1996) have pointed out that they are at present conducting a set of sulphate aerosol experiments with the present global coupled model. The preliminary results show that the high-latitude effects in the sulphate experiments are qualitatively similar to the experiment discussed (with the consideration of CO_2 only) but with less warming. Simulation experiments will also be conducted with ozone changes as well as alterations in other forcings.

As has been mentioned already, a very important specific feature of the Arctic are leads and polynyas which have a great impact on the energy budget of the polar ocean and atmosphere. In this context Grötzner et al. (1996) have studied the impact of sub-grid scale sea-ice inhomogeneities on the performance of the ECHAMN3 atmospheric model. The results indicate that a sub-grid scale distribution of sea ice is a dominant factor controlling the exchange processes between ocean and atmosphere in the Arctic. In the presence of leads and polynyas the heat and water transports are strongly enhanced leading to a significant warming and moistening of the polar troposphere. This affects the atmospheric circulation in high and mid-latitudes; e.g., the stationary lows are modified and the transient cyclonic activity over the subpolar oceans is reduced. Grötzner et al. (1996) have emphasized that a pronounced impact of sub-grid scale sea-ice distribution on the model climate can only be obtained when the nonlinear behaviour of the surface exchange processes is considered by a proper, physically based, averaging of the surface fluxes. A simple linear averaging of surface parameters is not sufficient.

The overall accuracy of atmospheric GCMs as of the early 1990s may be characterized by the data of Table 2.32 data (Proceedings..., 1995). A number of preliminary findings can be summarized as follows:

- The models provide a generally credible simulation of the mean seasonal structure of the large-scale circulation, although some models are outliers for particular variables.
- Outliers aside, no single model performs best in all respects.
- The models display a characteristic spread in their simulations for each variable that is nearly independent of latitude and season.

Table 2.32. Hemispheric mean seasonal root-mean-square differences between observations and the mean of the AMIP models

Variable	DJF		JJA	
	NH	SH	NH	SH
Mean sea-level pressure (hPa)	1.4	1.4	1.3	2.4
Surface air temperature (over land) (°C)	2.4	1.6	1.3	2.0
Zonal winds at 200 hPa (m/s)	2.4	1.8	1.8	2.4
Precipitation (mm/day)	0.80	0.71	0.62	0.77
Cloudiness (%)				
Outgoing longwave radiation (W/m^2)	2.8	3.2	2.9	5.5
Cloud-radiation forcing (W/m^2) 9.1	20.5	16.2	6.5	
Surface heat flux (over ocean) (W/m^2)	22.5	27.3	30.5	17.2

- The models generally overestimate the variance due to the seasonal cycle and underestimate that due to inter-annual variability.
- There are evident systematic errors in some variables, although the observational data are of uncertain reliability in some cases.
- The model errors are complex, and a suite of statistical measures is required for their proper characterization.
- The reasons for a particular model's behaviour are difficult to determine from the AMIP data alone, although there does not appear to be a strong dependence on resolution.
- While models on the whole appear to be improving, the lack of systematic performance data prior to AMIP makes estimation of the rate of improvement difficult.

The primary recommendation to emerge from AMIP is that the project should be expanded and maintained on a continuing basis.

Thompson and Pollard (1995) have investigated the sensitivity of the equilibrium climate to a doubled atmospheric CO_2 using the GENESIS (Global Environmental and Ecological Simulation of Interactive Systems) coupled global climate model version 1.02. The atmospheric component of this model is a heavily modified version of the NCAR CCM1 and is coupled to a multicanopy land-surface model (LSX); multilayer models of soil, snow, and sea ice; and a 50-m slab ocean mixed layer. New features of the model include explicit subgrid convective plumes, PBL mixing, a diurnal cycle, a complex land-surface model, sea ice dynamics and semi-Lagrangian transport of water vapour.

The principal result of the simulation modelling conducted by Thompson and Pollard (1995) is that the global mean-annual SAT increase due to a CO_2 doubling is 2.1°C, with global precipitation increasing by 3.3%. Over most land areas, most of the changes in precipitation are insignificant at the 5% level compared to inter-annual variability. Decreases of soil moisture in summer are not as large as in most previous models and only occur poleward of ~55°N in Siberia, northern Canada, and Alaska. Sea ice extent in September recedes by

62% in the Arctic and by 43% in the Antarctic. The area of the NH permafrost decreases by 48%, while the total snow cover extent in the NH in January decreases by only 13%. Thompson and Pollard (1995) suggest that the greater global warming (3.2°C) found in an earlier version of the model was due to more detailed interactions that no longer occur in the new version of the climate model.

The analysis of the effects of various modifications to the model physics have shown, in particular, that, despite radical changes to the convective and cloud schemes in the current model, the global warming due to a doubled CO_2 varied between only 1.73° and 2.38°C. Thompson and Pollard (1995) have emphasized that their result that penetrative convection in itself does not increase CO_2 warming support an idea suggested by Lindzen (1994): the CO_2-induced warming will intensify penetrative convection in the tropics and increase thereby the Hadley circulation. This is meant to produce drier and stronger subsidence in low latitudes and drier and warmer air above $\sim$5 km, which greatly diminishes the water vapour feedback and has a negative rather than positive feedback on global warming at the surface.

In the context of the climate predictability problem, an analysis has been made (Bengtsson et al., 1996a) of atmospheric response to the evolution of the SST global field between 1979 and 1992 with the use of the global spectral (T42) 19-level model of atmospheric general circulation (ECHAM-3) developed in the Max Planck Institute for Meteorology (Hamburg). Results were considered from five numerical experiments for a period of 14 years both for each of realizations and for the ensemble, on the whole, compared with the data for the control experiment for 30 years (the climatic mean monthly SST fields were prescribed).

These results have shown that the model can reliably reproduce the observed response of the atmosphere to the El Niño and La Niña events. During the El Niño the subtropical jet streams get intensified and move towards the equator in both hemispheres, and also a trend of formation of a weak eastward transport appears in the upper troposphere over the equator. The Southern Oscillation event was reliably reproduced in all cases. The inter-annual variability in mid- and high latitudes is determined by the prevailing forcing of chaotic dynamics, and the SST effect only weakly modulates the atmospheric radiation.

The capabilities of models to forecast the climate have been considered for quite different regions. The signal-to-noise ratio was high, as a rule, in the tropics, moderate in the western hemisphere and usually low over Europe. The forcing (signal) of the El Niño / Southern Oscillation (ENSO) event is clearly expressed in the NH spring.

A particularly strong signal in the precipitation field is observed at this time in the southern USA and south-western Asia. As a rule, during the ENSO warm phase a warming occurs in western Canada, whereas in western Europe the warming forms in the cold phase. This is determined by an advection of warm air masses by the Pacific cyclones, whose intensity grows during the ENSO warm phase, and, on the other hand, by the effect of the Iceland minimum intensifying during the ENSO cold phase. A supposition has been made (Bengtsson et al., 1996a) that the SST inter-annual variability in extra-tropical latitudes is governed, at least, partially by winds and the ENSO events.

Based on the use of the coupled climate model developed at the Japan Meteorological Research Institute, numerical experiments have been carried out to analyse the anthropogenic climate changes with the prescribed 1% / year increase of CO_2 concentration (Noda, 1995). The following two special features of the model are important: (1) a relatively high spatial resolution of the ocean unit in the model in low latitudes to provide a sufficiently reliable reproduction of the El Niño events; (2) a detailed model of the ice cover dynamics making possible an adequate description of the annual change of the ice concentration and thickness.

Calculations were made for a period of 70 years. By this time, the CO_2 concentration had doubled, and, as calculations showed, the mean annual, mean global SAT increased by 1.6°C. The atmospheric response to a CO_2 doubling was slow in the southern hemisphere and over the ocean, as had been obtained earlier. However, (in contrast to previous results) the SAT increase in the Arctic and in the adjacent continental regions was not clearly manifested during first 50 years of integration. Only in the period 50–70 years did a sharp increase of the Arctic SAT begin. Changes in the sea ice cover extent on a scale of interdecadal variations were found out from the data of both the experiment with prescribed CO_2 growth and the control experiment.

Around year 50 of integration there was a phase of transition from positive to negative anomalies in the course of interdecadal variations. This transition coincided in time with a fast warming due to a CO_2 increase due to enhanced radiative (greenhouse) forcing.

The results considered show that the spatial distribution of the atmospheric response to a CO_2 increase depends not only on spatial coordinates but also on time; in addition, it is strongly affected by natural climate changes. An interactive mode most clearly expressed in the "atmosphere–ocean" system is very close to that revealed from SST observations. This mode is characterized by inter-annual variations in the Pacific with the prevailing period $\sim$6 years, which is close to the characteristic time of the El Niño variations and interdecadal variations. The latter reflects a possibility to forecast for a period of several decades.

Murphy (1995a,b) and Murphy and Mitchell (1995a,b) used the coupled climate model developed at the Hadley Centre for Climate Forecasts (UK Meteorological Service) for the numerical modelling of global climate for a period of 75 years with the transient 1°C/year increase of CO_2 in the atmosphere. The results obtained show that in the case considered there was no SAT increase in the southern hemisphere at all during first 45 years of integration. By the moment of CO_2 doubling (year 70) the NH average SAT had increased only by 1.1°C, which intensified the upward fluxes north of the equator and the downward fluxes south of the equator, and, respectively, intensified (reduced) precipitation in the tropics of the northern (southern) hemisphere.

The SH ocean cooling with respect to the control integration reaching a maximum ($\sim$0.6°C) by year 45 of integration is a source of temperature contrasts. This cooling explains a weak warming from the data of integration for periods $>$60 years. The cooling is determined by the increasing difference "precipitation–evaporation" resulting in freshening the upper ocean layer, which, in its turn, reduces the convection. (Note should be taken, however, that the value of the difference "interactive calculation–control" may be affected by climatic drift, since a gradual warming in the SH ocean takes place both in the "interactive" and in the "control" cases.)

An increase of the Antarctic sea-ice cover extent determined the formation of a negative feedback, favouring a weakening of the "efficient" climatic sensitivity during the first half of the numerical experiment, which increased a delay of the initial mean global SAT growth. Estimates with the use of the energy-balance climate model "tuned" against the coupled model have shown that this delay is reduced if the feedback mentioned above is excluded.

Analysis of the response of the climate system to a CO_2 increase revealed considerable "land–ocean" contrasts. By year 70 of integration the mean temperature increase on land reaches 2.8°C, whereas over the ocean it constitutes only 1.3°C. It turns out that the role of direct forcing of the ocean's thermal inertia on the formation of the contrast is relatively small: a stronger warming on land is mainly determined by the presence of a stronger positive feedback due to shortwave radiation. This is determined by the contrast conditions of the lower cloud cover formation: over the ocean, the lower cloudiness thickens as a result of replacing the ice clouds by longer-lived liquid-water clouds in high latitudes, and the atmospheric stability in the low and middle latitudes of the southern hemisphere, whereas over land the lower cloudiness decreases due to decreasing soil moisture and evaporation. Calculations with the use of the energy-balance model considering the CO_2 increase over the period from 1965 to nowadays have shown that the calculated values of the "land–sea" contrast exceed those observed. This difference is substantially reduced with the cooling due to anthropogenic sulphate aerosol taken into account.

Murphy (1995a,b) has studied the transient response of the Hadley Centre coupled atmosphere–ocean model (AOGCM) to increasing carbon dioxide with the main purpose of investigating and quantifying the factors responsible for the magnitude, land–sea contrasts, and time-dependence of the global-mean response, focusing on the radiative feedbacks and delaying effect of oceanic thermal inertia. The principal tool of the study is a globally averaged four-box (land, ocean, troposphere, stratosphere) energy balance model (EBM) and the oceanic box-diffusion model, both tuned to the simulated AOGCM response. The divergence between EBM and AOGCM results may be used to assess the role of nonlinear feedbacks (not considered in the EBM) in determining the response of climate. The 75-year AOGCM control (contrast CO_2) and anomaly (increasing CO_2, 1% /year) integration were initialized from the end point of a preliminary 150-year "calibration" experiment.

Murphy (1995a,b) has shown that the land–sea contrast in the surface warming is explained almost entirely by the shortwave radiative feedbacks (SWF) associated with changes in cloud and surface albedo. The SWF has been decomposed into three components: CLOUD, ALB, and ABS due to, respectively, changes in the reflection to space by clouds, changes in surface albedo, and changes in the atmospheric absorption by water vapour and cloud. Since the ABS component is similar in magnitude over land and sea alike, the land–sea contrast in the globally averaged response is driven by differences in the cloud and albedo feedbacks (the latter is chiefly due to the retreat of the NH snow cover). The oceanic thermal inertia delays the response. However, the initial delay is enhanced by increases in the Antarctic sea-ice cover, which substantially reduce the effective climate sensitivity of the model in the first half of the 75-year experiment. The EBM overestimates the warming over land in the case when the observed anthropogenic greenhouse forcing from the pre-industrial period to the present day is prescribed. However, an inclusion of the direct forcing due to anthropogenic tropospheric sulphate aerosol eliminates the land/sea contrast in the response at 1990, leaving the simulated warming over land slightly below the observed value, although the rapid warming observed during the 1980s is well reproduced. An approximate estimate of global average 1990 aerosol forcing is as follows: -0.29 W/m^2 (TOA), -0.61 W/m^2 (land surface), and -0.16 W/m^2 (ocean surface), compared to a TOA greenhouse forcing of about 2.2 W/m^2. The vertical penetration of the oceanic response is small below 1000 m.

An intercomparison of the monthly mean shortwave surface radiation budget (SRB) under near-global scale retrieved from satellite data (the Pinker algorithm was applied) with calculations using the NCAR CCM2 climate model (a 10-year control simulation), accomplished by Ward (1995) revealed large differences. The largest discrepancies are found in the summer mid-latitude regions where CCM2 overestimates the absorbed surface solar fluxes relative to retrieved values by 50 to over 100 W/m^2. Large positive differences and some negative differences also occur elsewhere. Such large differences will seriously affect the ability of the model to perform accurate climate simulations. These discrepancies are nowhere near the 10 W/m^2 accuracy for SRB recommended for climate studies by the WCRP. Most of the differences are associated with deficiencies in CCM2's prediction of cloud optical properties and cloud amount. However, surprisingly significant differences also occur in clear-sky fluxes and surface albedo. The CCN2 was found to have larger clear-sky surface insolation than Pinker (retrieved) over nearly all land areas by more than 60 W/m^2 in some locations. Omission of aerosols in CCM2 is certainly a contributing factor, but aerosols alone are unlikely to account for the large differences in clear-sky surface insolation. The magnitudes of the differences in SRB for overcast conditions are much larger than those for clear-sky conditions. Thus the intercomparison considered has identified substantial deficiencies in the NCAR CCM2 model associated with the representation of cloud optical properties and cloud cover fractions. The most difficult problem to be solved in the future will be a proper accounting for aerosols, including their influence on cloud properties as condensation nuclei.

Liang et al. (1995) have applied a climate ensemble technique to simulate climate change using the CCM1 model with special emphasis on identifying model biases and climate change signals due to the enhanced greenhouse effect. Several aspects of observed inter-annual variability have been successfully reproduced by the model, including: global SAT, Arctic sea-ice extent; and regional SAT variability, sea-level pressure and 500 hPa height over about one-quarter of the observed data domains.

Substantial biases have been found, however, over broad regions, where strong seasonality and systematic links between variables were identified. When GHGs concentrations increase, atmospheric moisture variability is substantially larger over areas that experience the greatest surface warming. This corresponds to intensified hydrological cycle. The results obtained by Liang et al. (1995) indicate that regional changes in inter-annual variability due to the enhanced greenhouse effect are associated with mechanisms that depend on the variable and season.

Sloan and Rea (1995) have accomplished an assessment of an impact of enhanced greenhouse effect on climate for the early Eocene, which was the warmest time interval for the past 65 million years. The results obtained with the use of the NCAR GENESIS model with prescribed CO_2 increase (300, 560 or 1680 ppm) indicate the following response of the climate system to increased atmospheric CO_2: (1) increased mean annual global SAT by 2–6°C, and decreased latitudinal surface temperature gradients; (2) maximum warming at high latitudes, especially in the winter season; (3) increased mean global precipitation; (4) increased high-latitude clouds. A comparison with palaeoclimate data has shown that the climate response to increased CO_2 does not appear to explain completely the early Eocene latitudinal surface temperature gradient. For example, in the 1680 ppm increase case, the high-latitude oceans and continents are minimally warm enough to match palaeoclimate data, while low latitude regions are overly warm when compared to SST reconstructions. A consideration of the increased poleward oceanic transport and/or other factors may be required to produce model results that more closely agree with marine and terrestrial palaeoclimate data, especially at low latitudes.

A similar conclusion has earlier been made by Chandler et al. (1994) who used the GISS GCM to examine the Pliocene climate (about 2 million years ago when climate was nearly as warm as predicted for the coming century). The simulation based on specifying SST and vegetation distribution resulted in the annual average 1.4°C warming of the northern hemisphere. The warming was greatest at high latitudes and therefore the equator-to-pole temperature gradient decreased by 11.5°C. The Pliocene SST pattern implicates an increased ocean heat flux as a component behind the middle Pliocene warmth, since levels of CO_2, large enough to cause extreme high-latitude temperatures, would generate more tropical warming than is indicated by palaeorecords (due to simulation results, at low latitudes, temperatures were mostly unchanged).

Campbell et al. (1995) used empirical orthogonal function (EOF) analysis to examine the monthly variance structure of several general circulation model simulations to look for possible systematic changes of variability, not only due to increased CO_2 concentration in the atmosphere but also due to model configuration. A principal conclusion is that different model configuration has a stronger effect on simulated inter-annual variability globally than does an altered CO_2 forcing. The second important conclusion is that strongly different seasonal cycles appear in the high latitudes of the northern hemisphere in the coupled model under different CO_2 conditions. This phenomenon is mostly due to snow and ice reductions with increased CO_2 that contribute to a smaller-amplitude annual harmonic and larger-amplitude half-year harmonic of surface air temperature.

A new stage of the further development of global climate models has been marked by the Workshop on Cloud Microphysics Parameterizations in Global Atmospheric Circulation Models (Workshop..., 1995), of which the principal purpose has been to discuss perspectives of cloud-resolving models (CRMs). The most complete study of this kind belongs to Fowler et al. (1996) and Fowler and Randall (1996a,b).

Fowler et al. (1996) have developed a new version of the Colorado State University General Circulation Model (CSU GCM), including a parameterization of microphysical processes responsible for the formation and dissipation of water and ice clouds in order to: (1) yield a more physically based representation of the components of atmospheric moisture budget; (2) link the distribution and optical properties of the model-generated clouds to the predicted cloud water and ice amounts, and (3) produce more realistic simulations of cloudiness and the Earth's radiation budget through the consideration of coupled cloud dynamics, cloud microphysics and radiative processes.

The five prognostic variables for the mass of water vapour, cloud water, cloud ice, rain, and snow describe the bulk cloud microphysical properties (graupel and hail are neglected). An important feature of the model is that interactive cloud optical properties provide a link between the cloud microphysics and radiation parameterizations; the optical depth and inferred emissivities of large-scale stratiform clouds are parameterized in terms of the cloud water and cloud ice paths. At the present stage the model does not include a prognostic equation for the number concentration of cloud water droplets and cloud ice crystals.

The results have been presented by Fowler et al. (1996) in terms of January and July monthly averages for two annual cycle numerical simulations to assess the impact of cloud microphysics on the hydrological cycle: (1) in the "EAULIQ" case, large-scale moist processes and cloud optical properties are driven by the cloud microphysics parameterization; (2) in the "CONTROL" run, condensed water is immediately removed from the atmosphere in the form of rain, which may evaporate as it falls through subsaturated layers. Stratiform ice clouds are not considered in CONTROL. Comparisons between the simulated vertically integrated cloud water and the columnar cloud water retrievals from satellite microwave measurements over the global oceans indicate a reasonable agreement. Fowler et al. (1996) have pointed out a number of realistic features of the results. For instance, the sources of simulated cloud water and cloud ice are located where they should be expected (in the tropical convective activity centres over land and oceans and in the mid-latitude storm tracks). In the model, the chief source of cloud water is low-level large-scale supersaturation, whereas a detrainment at the tops of cumulus towers is most important for cloud ice, especially in the tropics.

A comparison between the climates, simulated by the CSU GCM, with and without liquid and ice cloud microphysics, and interactive cloud optical properties, which has been accomplished by Fowler and Randall (1996a), has revealed large differences between EAULIQ and CONTROL in terms of the longwave and shortwave CRF. Comparisons with observation data from the ISCCP and the ERBE show that EAULIQ yields improved simulations of the geographical distribution of the simulated cloudiness, ERB as well as longwave and shortwave CRF. Differences between EAULIQ and CONTROL are largest in the tropics and are mostly due to a decrease, in the EAULIQ run, in the amount and optical thickness of upper-tropospheric clouds. In comparison with CONTROL, EAULIQ yields a warmer troposphere except below 8 km between 30°N and 30°S. The meridional circulation is significantly weakened in the EAULIQ run. Fowler and Randall (1996a) have emphasized that one key result is that interactions among the cloud microphysics, cumulus, and radiation parameterizations lead to an increased stability of the atmosphere in EAULIQ, especially in the tropics. Multiple deficiencies remain, however, in the simulation of cloudiness and longwave and shortwave CRF. For convectively active regions, these problems partially result from an overestimation of upper-level clouds.

Fowler and Randall (1996a) have pointed out two major weaknesses of the present version of EAULIQ: (1) fractional cloudiness has not been considered; (2) there is no adequate parameterization of planetary boundary-layer clouds. Both these problems are being studied.

The problem of the CSU GCM model sensitivity has been the subject of the further study by Fowler and Randall (1996b). The purposes of 120-day sensitivity experiments corresponding to perpetual January conditions have been to: (1) determine the importance of advecting cloud water, cloud ice, rain, and snow at the temporal and spatial scale resolutions at present used in the model; (2) study the importance of the formation of extended stratiform envils at the tops of cumulus towers; (3) analyse the role of mixed-phase clouds in determining the partitioning among cloud water, cloud ice, rain, and snow and, hence, their impacts on the simulated cloud water properties; (4) evaluate the sensitivity of the atmospheric moisture budget and precipitation rates to a change in the fall velocities of rain and snow; (5) determine the model's sensitivity to the prescribed thresholds of autoconversion of cloud water to rain and cloud ice to snow, and (6) study the impact of the collection of supercooled cloud water by snow as well as accounting for the cloud optical properties of snow. The results of these sensitivity experiments allow one to identify the most important factors and processes as well as to substantiate the possibilities of the use of simplified approaches. Fowler and Randall (1996b) have underlined in conclusion the need for in-depth analysis of the spatial and temporal scale-dependence of the different microphysical parameters of the cloud parameterizations used in climate models.

A number of other studies have been devoted to various aspects of cloud-radiation parameterization in climate models. For instance, Tiedtke (1996) has developed an extension of cloud-radiation parameterization in the ECMWF model to take into account subgrid-scale variations of optical depth instead of the use of plane-uniform clouds in radiative transfer calculations. A consideration of cloud inhomogeneities and heterogeneities associated with convective and stratiform cloudiness has resulted in substantially increased and more realistic net downward shortwave radiative fluxes at the top of the atmosphere: over the tropical and subtropical oceans (30°S–30°N) the increase due to inhomogeneities and heterogeneities is on average 9 and 7 W/m^2, respectively.

Ghan and Leung (1995) have discussed the possibility of determining cloud droplet number concentration using the droplet number conservation equation to simulate cloud-aerosol interactions:

$$\frac{\partial N_c}{\partial t} = -\nabla(N_c V) + N_{vc} - E_{cv} - A_{cr} - C_{cr} - C_{ci} - C_{cs} - F_{ci}$$

where N_c is the droplet number concentration, N_{vc} represents the droplet nucleation rate, E_{cv} is droplet evaporation, A_{cr} is the droplet sink due to droplet coalescence, C_{cr}, C_{ci}, and C_{cs} are sinks due to collection by rain, ice, and snow, and F_{ci} is the sink due to droplet freezing to form ice crystals.

The two simulations for cases of marine and continental ammonium sulphate aerosol particles concentrations equal to 200 cm^{-3} and 1000 cm^{-3}, respectively, yield substantially different cloud droplet number concentrations, which results in changes of the planetary albedo. A five-fold increase in aerosol number leads to three-fold increase in droplet number, and, respectively, to a 1.1% increase in the mean planetary albedo, corresponding to the 3.6 W/m^2 increase in the shortwave CRF.

Most recent attempts to verify the climate modelling results have been focused on intercomparisons between simulations and observations of various parameters. Yu et al. (1996) have accomplished in this context a methodology study of the validation of clouds in GCMs using ISCCP satellite observations emphasizing that tuning models to obtain a good simulation of the Earth's radiation budget do not guarantee a good surface radiation budget, a key input for coupled atmosphere–ocean models. A consideration of cloud water content 3-D distributions and their evolution opens up much more adequate perspectives of the verification.

Yu et al. (1996) have compared cloud properties, such as cloud-top height and cloud optical thickness, analysed by ISCCP and simulated by Laboratoire de Météorologie Dynamique (LDM) GCM. The results indicate that a direct comparison of simulated low cloudiness and that shown from satellites is not possible. They also reveal some model deficiencies concerning the cloud vertical distribution. High convective clouds appear to be too thick. A large cirrus overestimation and large altocumulus and cumulus underestimations have been found. For middle and low clouds, optically thick clouds are overestimated, while the optically thin clouds are almost missed by the simulation. The principal conclusion is thus that the model is far from being realistic.

Airey et al. (1996) have evaluated simulations of terrestrial precipitation in four successive UK Meteorological Office/Hadley Centre climate model versions. A general conclusion is that the global precipitation fields (mean monthly precipitation patterns) from the control integrations of the four experiments improved with successive simulations. The largest single improvement comes with the increase in resolution and with changes in model physics in more recent versions.

Senior and Mitchell (1993) have studied the impact of changing the horizontal resolution (5°lat. × 7.5° long. and 2.5°lat. × 3.75°long. resolutions were compared) of a version of the Hadley Centre GCM in present-day and 2 × CO_2 climate simulations. This study has demonstrated generally favourable changes of the use of higher resolution. The main improvements are in the southern hemisphere in June, July, and August (JJA). Aspects of the model that show no improvements include the general coldness of the troposphere in all seasons and the pressure patterns in the northern hemisphere winter. The global mean annual mean warming due to a CO_2 increase is similar (3.4 K and 3.3 K), but there are differences in the latitudinal distribution of the zonally averaged temperature response: (1) the winter high-latitude amplification of the warming is damped at higher resolution in both hemispheres; (2) there are large differences in JJA between precipitation and tropical circulation; (3) changes in climate variability are quite different, especially in winter.

Srinivasan et al. (1995) have analysed monthly precipitation fields produced by a subset of 19 currently available AMIP model experiments for the tropical region using (for comparisons) a land-only observed dataset for the period 1980–88. The models show large variations in their ability to reproduce observed tropical precipitation, although spatial correlations indicate that some of the models simulate the pattern of observed precipitation fields fairly well. The correlations are strongest during boreal winter (DJF) and weakest during the boreal summer (JJA). A comparison between model and observed precipitation time series for two Central Pacific locations show that most models are unable to reliably simulate the inter-annual precipitation variability in this region.

Intercomparisons of GCMs and remotely sensed datasets (Nimbus-7 SMMR data) on snow cover and snow mass have been accomplished by Foster et al. (1996). The results show that both the models and SMMR represent seasonal and year-to-year snow distributions fairly well. The passive microwave data and several of the models, however, consistently underestimate snow mass, but other models over-estimate the mass of snow on the ground.

An extensive experience in developing the climate models (the coupled models, during the last years) has been gained at the National Centre for Atmospheric Research (NCAR), USA (Boville, 1995). The goal of a new project is to develop a climate system model (CSM) on the basis of more efficient cooperation of meteorologists, oceanologists and other specialists, bearing in mind the numerical modelling of the global climate in time scales from the inter-annual variability to centuries.

The long-range objectives of CSM aimed, first of all, at studies of climatic predictability, are as follows: (1) development of the continually improved model which would provide a thorough account of the climate-forming factors; (2) maintenance of a set of less complicated models; (3) numerical modelling to analyse the inter-annual and interdecadal climatic variability; (4) application of the obtained results to substantiate the respective policy at national and international levels; (5) a wide accessibility to the whole system of CSM is provided by a simple conceptual structure, it consists of individual models of the atmosphere, ocean, land surface and sea-ice cover, the coupling of which is ensured by the "flux coupler". This makes it possible to perform calculations for individual component models with the help of independent computers and to use a set of models of different complexity.

Bonan (1995) has made interesting assessments of the role of inland water basins

Table 2.33. Differences in 2-m height air temperature (*T*), latent heat flux (*LH*), sensible heat (*SH*), and total precipitation (*P*) between with lakes and without lakes simulations for January and July. Data were averaged for 60 grid cells with more than 5% inland water in a region of northwest Canada between 50° and 70°N and 130° and 95°W, including Great Bear Lake, Great Slave Lake, Lake Athabasca, and Lake Winnipeg. July data were broken into 41 grid cells with inland water between 5% and 20% and 19 grid cells with inland water greater than 20%. Significant differences from the without lakes control at the 90% and 95% levels are indicated by * and **, respectively, based on 5-year standard deviations of the area averages

	January	July	
	5–100%	55–20%	20–100%
ΔT (°C)	−5.3	−2.6**	−3.4**
ΔLH (Wm^{-2})	−0.3	−1.8	19.6*
ΔSH (Wm^{-2})	4.6	−1.4	−15.1**
ΔP (mm day^{-1})	−0.1	0.1	0.1

(lakes, swamps, marshes) in climate formation. Simulation calculations have been made with the use of the NCAR CCM2 which is a spectral general circulation model configured at T42 resolution (approximately 2.8° × 2.8° transform grid). The model has 18 vertical levels with the model top at 2.9 hPa, a 20-min time step (radiation calculations are only performed every model hour). The standard model uses prescribed climatologically varying SST, prescribed surface albedos, and prescribed surface wetness. The coupled climate model was run for 5 years with and without inland water subgrid points to determine the importance of inland water for global climate simulation. Data in Tables 2.33 and 2.34 illustrate some of the results.

As can be seen from Tables 2.33 and 2.34, the inclusion of lake, swamp and marsh subgrid points had a significant effect on the surface climatology of regions with high inland water area: in July this resulted in a spatially consistent signal in which high inland water regions were 2–3°C cooler, had increased latent heat flux (10–45 W/m^2), and decreased sensible heat flux (5–30 W/m^2) compared to the simulation without these water bodies. These changes were statistically significant in the lake region of north-west Canada, the Great Lakes region of North America, the swamp and marsh region of the Siberian lowlands, and the lake region of East Africa, but were not significantly different in the swamp and marsh region of Finland and north-west Russia.

In tropical lake regions (East Africa), the response to lakes was less in the rainy season (January) than in the dry season (July). Precipitation was unchanged in both months except for the Great Lakes region where precipitation increased in January. These changes in temperature, precipitation, and surface fluxes were consistent with mesoscale modelling studies of the effects of lakes on climate (cooler summers, warmer winters, increased latent heat flux and decreased sensible heat flux due to the lakes) and tended to bring the model closer to observations. The lakes had little effect on atmospheric moisture, radiation, or zonal circulation.

Table 2.34. As in Table 2.33 but for 44 grid points, in the Siberian lowlands between 55° and 70°N and 60° and 90°E. July data were broken into 15 grid cells with inland water between 5% and 20% and 29 grid cells with inland water greater than 20%. See Table 2.33 for the explanation of asterisks

	January	July	
	5–100%	5–20%	20–100%
ΔT (°C)	1.5	−1.8*	−2.0*
ΔLH (Wm^{-2})	−0.1	12.3*	19.5**
ΔSH (Wm^{-2})	3.3	−8.0	−6.5
ΔP (mm day^{-1})	−0.3	0.0	−0.3

At the Centre for the Ocean–Land–Atmosphere studies (COLA) various versions have been developed of the coupled global spectral (R15) climate model with 9- and 18-level atmospheric units and ocean units with moderate and middle resolutions, bearing in mind problems of the numerical climate modelling and analysis of climatic predictability (Schneider, 1995; Schneider et al., 1996). The model was also used to study the ENSO forcing of climate on scales of annual and inter-annual variabilities.

Calculations have shown that the modelled climate drift is sensitive to a parameterization of the cloud–radiation interaction and can be, apparently, removed almost completely by matching the precalculated parameters of clouds with those observed. Serious difficulty in the coupled modelling is explained by the low resolution of the coastal orography. As the spatial resolution of the model increases, the role of tuning the calculated climate to the observed one (an account of drifting) is reduced, as well as the problems of "speeding up" the model.

A comparison with the observational data has demonstrated the model's capability of reliably simulating the spatial–temporal variability of most substantial observed modes of climatic variations. An experience of ENSO forecasts suggests a preliminary conclusion that a specification of the average state and annual course by considering the interactivity of anomalies may raise the reliability of the forecast, though does not solve the problem (in particular, from the viewpoint of excluding the need of flux corrections). An account in the initial conditions of the characteristics of the tropical Pacific makes a certain contribution to the reliability of the forecasts of the inter-annual variability.

Very important progress in developing CGCMs has been achieved at the Geophysical Fluid Dynamics Laboratory, GFDL (USA). A preliminary analysis of a model's potential to simulate climate variability with the range from inter-annual changes to several decades has confirmed a good coincidence between calculations and observations (Delworth and Manabe, 1995).

To analyse the observed increase of the mean global SAT, Stouffer (1995) and Manabe and Stouffer (1996) undertook calculations of climate change for a period of 1000 years with the use of a GFDL global coupled climate model. None of the earlier calculations of natural climate change has revealed any SAT increase over 100 years reaching the level of the observed warming ~0.5°C.

The supposedly realistic results of the numerical climate modelling suggest the conclusion that the observed trend of SAT did not result from the atmosphere–ocean interaction but could be caused by external forcings, such as an increase of CO_2 and aerosol in the atmosphere. It is very difficult, however, to determine whether the numerical modelling results are realistic, since the length of the SAT observation series constitutes only ~ 110 years. The model is only known to reliably simulate the inter-annual and interdecadal SAT variations, excluding the characteristic time scale (3–5 years) of the ENSO event.

Calculations with the use of the low-resolution coupled global model have shown that the calculated fluctuations constituted only one-third compared to those observed in the tropical Pacific, but over the rest of the globe both the value and the spatial distribution of the standard deviation (SD) of the mean annual surface temperature is close to that observed, the SD being greater over the land than over the ocean, and in high latitudes than in low latitudes (again, with the exception of the tropical Pacific).

Analysis of the results of the numerical SAT modelling with the use of CGCM has shown also that they differ little from those obtained with the simplified coupled model with a fixed mixed layer of the ocean. If the results of calculations of the natural climate variability with the use of CGM are realistic, then this agreement may be important from the point of view of analysis of the ocean forcing on the natural climatic variability.

Delworth and Manabe (1995) analysed the capability of the GFDL coupled climate model to simulate the climate variability on scales from inter-annual to decadal variations, with the numerical experiments for the North Atlantic region as an example. The obtained results show that on a scale of inter-annual variability the spatial distribution of the first empirical orthogonal function (EOF) characterizing the SST field in winter agrees well with the data of observations.

Both cases considered reveal the alternating zonal bands of SST anomalies in the North Atlantic (the calculated data reveal, however, a more clearly expressed latitudinal structure). This structure of the bands in the SST field is connected with the similar structure of the anomalies of the height of the 500 hPa surface level. A comparison with the calculated results for the simplified model with the fixed oceanic mixed layer revealed a close agreement pointing to the fact that the dynamics of the ocean is not a critically substantial factor of the inter-annual climatic variability.

The observed variability of the SST field on a scale of decades is characterized by the presence of a dipole mode with the regions of strongly manifested positive (negative) anomalies near the Newfoundland coast and more negative (positive) anomalies near the USA south-eastern coastline. The structure of variability similar to that observed is typical of the second EOF of the precalculated SST field.

Manabe and Stouffer (1996) have emphasized that neither complete CGCM nor the approximate mixed layer model generates a sustained, long-term warming trend of significant magnitude such as that observed since the end of the last century. During the entire time series, a temperature trend as large as 0.5°C /century was never sustained over the period longer than 60 years. Therefore Manabe and Stouffer (1996) concluded that the results mentioned suggest that the observed warming trend is not caused by the interaction among the atmosphere, ocean, and continental surface. Instead, it may have been induced by a sustained trend in thermal forcing such as solar irradiance, GHGs, and aerosol loading.

A number of full CGCMs have been developed at the Max Planck Institute for Meteorology in Hamburg (Bengtsson et al., 1996a). One of them (ECHAM-3 which has been mentioned earlier) was used to simulate an atmospheric response to the evolution of the global SST field during the time period from 1979 to 1992. It has been found out that the model successfully simulates such phenomena as El Niño and La Niña as well as such a stable feature of the global general circulation as Southern Oscillation. The numerical modelling has demonstrated that the inter-annual variability of the atmosphere in mid- and high latitudes is determined by chaotic dynamics, whereas the SST impact is limited by a weak modulation of atmospheric radiation. An analysis has been made by Bengtsson et al. (1996a) of potential predictability for various regions of the globe. It has been shown in this context that signal-to-noise ratio is, as a rule, large in the tropics, has a moderately high level in the western hemisphere, and is usually small over Europe.

In a later study Stendel and Bengtsson (1996) further developed these results with the purpose of analysing an opportunity of monitoring the temperature of the troposphere by means of a general circulation model. Newell and Weare (1976a,b) have demonstrated earlier that SST anomalies are the driving force for global temperature anomalies. Recently, Graham (1995) has confirmed this general conclusion and shown from observational data (1970 to 1982) that the global average tropospheric temperature is determined by observed ocean surface temperature to a large extent and with an accuracy that is rather close to that achieved by available observing systems like radiosondes. According to Graham (1995), "In agreement with a diverse suite of controversial observational evidence from the past 40 years, the upward trend in simulated tropospheric temperatures is caused by an enhancement of the tropical hydrological cycle driven by increasing tropical ocean temperatures. Although it is possible that the observed behaviour is due to natural climate variability, there is a disquieting similarity between these model results, observed climate trends in recent decades, and the early expressions of the climatic response to increased carbon dioxide in numerical simulations".

The estimates of the simulated global atmospheric heat budget components obtained by Graham (1995) show that the heat budget is dominated by changes in the latent heat flux (that is, precipitation), which shows swings of ± 2 W/m^2 in the global average during the simulation. The second important component is outgoing longwave radiation (OLR), acting to counter the effects of the increased latent heating. Between 1970 and 1988, OLR increased by about 0.7 W/m^2 (about 0.3% of the global average). One more significant component is upwelling (from the surface to the atmosphere) thermal radiation (i.e., surface net longwave radiation). Because of the increase of the tropospheric water vapour content, net longwave radiation decreased by about 0.2 W/m^2 during the time period mentioned. An interesting conclusion is that changes in sensible heating and solar absorption has little effect on the low-frequency temperature changes.

In continuation of earlier studies, Stendel and Bengtsson (1996) have investigated the sensitivity of the ECHAM-4 model to forcing by observed SST in different regions and validated tropospheric and stratospheric temperature with the MSU (satellite Microwave Sounding Unit) brightness temperatures for the period 1979 to 1993 when three fully developed warm events (El Niño) and three fully developed cold events (La Niña) were observed (there was a special effort undertaken to verify the reliability of the MSU datasets).

Stendel and Bengtsson (1996) have pointed out that there is a remarkable overall agreement of observed and simulated temperature anomalies in the troposphere, whereas the correlation in the stratosphere is zero. Tropospheric temperature anomalies agree well in the regions that lack conventional observing systems, that is, over the oceans and in the tropics. This means a possibility to assess the effects of climate change with a greater degree of confidence by means of general circulation models. Differences between observed and simulated tropospheric temperature anomalies can be attributed to stratospheric aerosol due to two major volcanic eruptions (El Chichón and Mt. Pinatubo).

Fig. 2.10 data illustrate a successful simulation of monthly global anomalies of lower tropospheric (T_{2R}) MSU brightness temperature by the ECHAM-4 model (a synthetic channel, 2R, has a maximum of the weighting function near 700 hPa). Table 2.35 shows a regional statistical analysis for selected regions of the Earth. As is seen, the correlation exceeds 0.8 in large parts of the tropics with largest values in the eastern Pacific. However, correlations are considerably smaller in regions with a large temperature variance, that is, the extra-tropical mainlands, especially North America and Europe. Stendel and Bengtsson (1996) have emphasized that this is due to the fact that the variability at middle and high latitudes is dominated by nonlinear dynamics, whereas in the tropics the response to the boundary conditions is essentially linear. It is also natural to expect (Bengtsson et al., 1996a) lower correlations over the eastern hemisphere, since the ENSO signal is generally weak there. In spite of that, there are extra-tropical regions with a large correlation of T_{2R} and ECHAM-4 temperatures, suggesting that SST anomalies play an important role also in parts of the middle and high latitudes.

Meehl and Washington (1996) have applied a second-generation, global coupled, general circulation climate model to simulate climate change in the Pacific region with transient (1%/year) CO_2 increase. The model was integrated for 75 years and the last 20 years of the experiment have been analysed (years 56–75; the CO_2 level doubles at approximately year 70). The results (Table 2.36) obtained without flux corrections were compared to a control integration averaged over the same time period with the present-day amounts of CO_2.

Table 2.36 shows that the changes in SST near the time of CO_2 doubling reveal the eastern tropical Pacific area warming more (3.49°C) than the western Pacific (2.21°C). Absorbed solar radiation at the surface decreases in the western Pacific area (-3.92 W/m^2) and increases in the eastern Pacific area ($+2.23$ W/m^2) due to cloud albedo feedback effects and changes in cloud amount (shown in the table).

There are small changes in the net IR (surface net longwave radiation) and sensible heat flux, but there is a large increase of latent heat flux (positive sign denotes heat removed from the ocean surface) in the eastern Pacific area compared

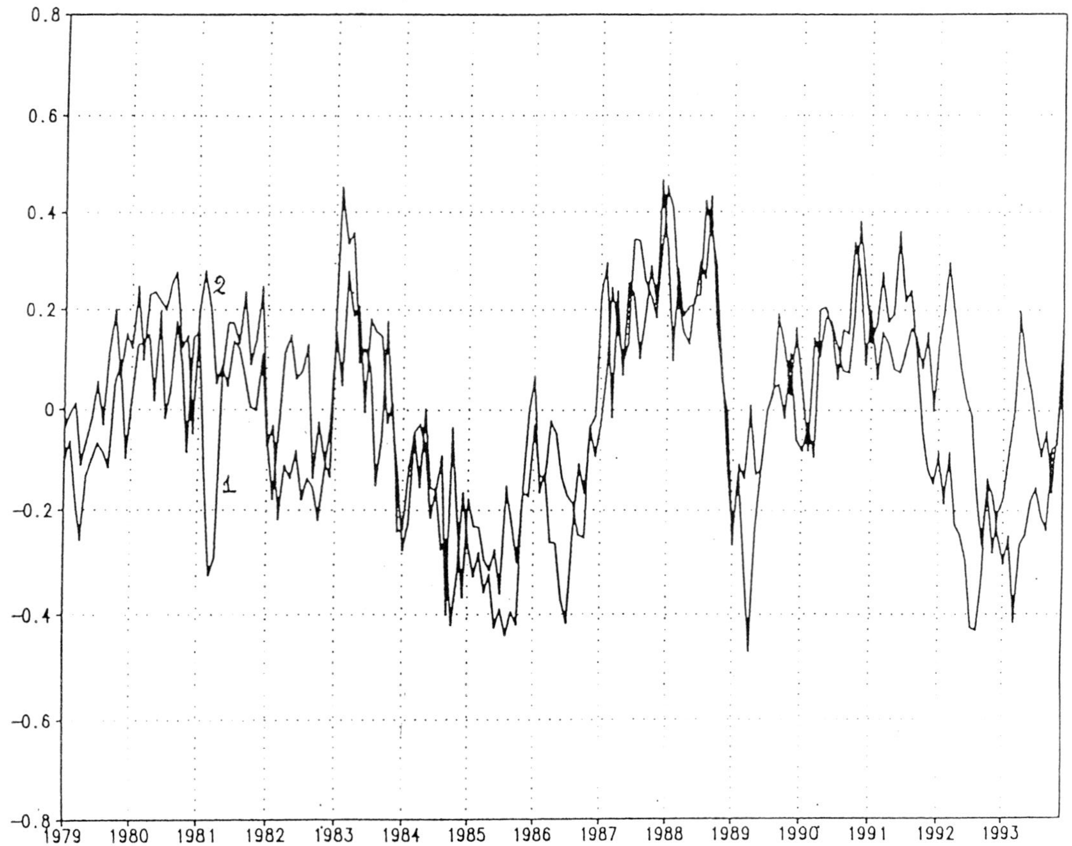

Fig. 2.10. Monthly global anomalies of lower tropospheric (T2R) MSU brightness temperatures (curve 1) and ECHAM-4 simulation with global observed SST 1979–93 (curve 2). Deviations [K] from mean 1979–93.

Table 2.35. Statistical regional comparisons of temperature anomalies between MSU_{2R} and ECHAM-4 model simulations. Left columns: correlations (1979–93) with AMP observed SST run. Right columns: variance (%) explained by experiments TAN and PAN, respectively. Centre column: annual rms differences (*K*) between ECHAM-4 and MSU_{2R}. The experiment TANO is based on observed SST from 1983 to 1991 everywhere between 20°S and 20°N, climatological SST poleward of 30° and a linear transition between both from 20° to 30° on both hemispheres. The relative importance of the tropical Pacific was examined in the experiment PANO with observed SST only in the tropical Pacific

Area	Boundaries				Corr. ECHAM4/MSU_{2R}					Explained variance [%]			
	Latitude		Longitude		Annual average		Monthly average			TANO		PANO	
	S	N	W	E	unmodified	adjusted for volcanic effects	unmodified	adjusted for volcanic effects	rms difference [*K*]	unmodified	adjusted for volcanic effects	unmodified	adjusted for volcanic effects
Gobe					0.82	0.93	0.63	0.78	0.18	78.5	76.1	80.2	92.0
Land	−90	90	−180	180	0.56	0.78	0.29	0.47	0.36	85.0	82.5	68.1	70.9
Ocean					0.84	0.91	0.69	0.87	0.16	64.0	59.6	60.6	64.6
N. hem.	0	90	−180	180	0.72	0.89	0.51	0.68	0.23	80.4	78.3	83.4	87.6
S. hem.	−90	0	−180	180	0.86	0.93	0.58	0.78	0.20	70.4	67.1	59.6	62.7
Tropics					0.83	0.87	0.73	0.78	0.21	93.1	92.2	89.4	89.1
Trop. cont.	−30	30	−180	180	0.72	0.82	0.55	0.65	0.31	87.7	86.0	79.5	77.4
Trop. oceans					0.86	0.89	0.77	0.81	0.19	94.6	94.0	90.9	91.2
Niño-3	−4	4	−150	−90	0.92	0.88	0.92	0.87	0.43	97.7	97.3	98.1	97.9
N. Pacific	30	50	150	230	0.78	0.86	0.83	0.89	0.55	42.4	57.2	65.6	73.1
N. Atlantic	20	50	−60	−20	0.72	0.80	0.64	0.79	0.52	76.1	74.0	62.2	60.3
N. America	30	70	−125	−75	0.27	0.35	0.21	0.43	1.13	22.8	14.7	24.2	19.1
Eurasia	40	70	−10	140	−0.23	−0.01	−0.08	0.13	0.97	0.4	20.5	0.1	0.0

Table 2.36. Area averages from the coupled model

	SST (°C)	Abs. solar (Wm^{-2})	Net IR (Wm^{-2})	Sensible (Wm^{-2})	Latent (Wm^{-2})	Advection (Wm^{-2})	Q-con. (%)	Q (%)	TCLD (%)	Albedo (%)
E. Pacific	+3.49	+2.23	+0.92	+2.11	+8.67	+7.65	+62	+22	−13	−4
W. Pacific	+2.21	−3.92	+0.28	+0.88	+0.81	+5.28	+3	+16	−5	+1

Notes: The western tropical Pacific area is defined for 5°S–5°N, 140°E–170°E, and the eastern tropical Pacific area is defined for 5°S–5°N, 120°W–90°W. Computed minus observed area averaged SST differences for the control case are 1.89°C for the eastern Pacific area, and 0.91°C for the western Pacific area, with the coupled model somewhat systematically warmer than observed as noted in the text and elsewhere. To compare the effects of the cloud albedo feedback on the longitudinal SST response across the tropical Pacific to an increase of CO_2, we note that the eastern Pacific area-averaged SSTs in the first column warm 58% more than the western Pacific area-averaged SSTs. In contrast, the eastern Pacific area warms only 30% more than that the western Pacific area in a different version of the coupled model without cloud albedo feedback. In a version of the atmospheric model coupled to a non-dynamic slab ocean the east Pacific area warmed 20% more than the western Pacific area when cloud albedo feedback was included, but only 5% more without cloud albedo feedback. Using a sample from April from the present coupled model both with and without the cloud albedo scheme, the change in total absorbed solar radiation at the top of the atmosphere for the western Pacific area with cloud albedo feedback is -5.1 W/m^2. The contribution due to changes in cloud albedo alone from the cloud albedo feedback scheme (the difference in C_s at the top of the atmosphere with and without the cloud albedo feedback scheme) is -11.9 W/m^2. Thus the contribution due to cloud decreases is $+6.8$ W/m^2. In the eastern Pacific area, decreases in cloud fraction (increased convective cloud with smaller grid fraction coverage) contribute to increased absorbed solar radiation ($+13.8$ W/m^2), but there is also a contribution from increased cloud albedo of -7.2 W/m^2 (the difference in C_s with and without the cloud albedo feedback scheme). Thus the contribution due to cloud changes is $+21.0$ W/m^2. There is 65% more solar radiation reflected by the clouds in the west compared to the east solely due to the cloud albedo feedback in the model, but there are also contributions from changes in cloud amount and type in response to the dynamical coupling between ocean and atmosphere. Without cloud albedo feedback and ocean dynamics, the major contributor to decreasing east–west SST gradient is temperature-evaporation feedback that can produce similar El Niño-like climate-change signals associated with tropical Pacific warming. Shown in this table are area-averaged results for increased CO_2 experiment minus control for the present version of the coupled model, December–January–February, 20-year averages. Total cloud and albedo for the eastern Pacific area control case are 0.61 and 0.23, respectively, while those quantities for the western Pacific area control case are 0.79 and 0.28, respectively. Abbreviations used: absorbed solar at the surface (abs. solar), net infrared at the surface (net IR), latent and ocean heat advection (adv.); surface moisture convergence (Q-con.), surface moisture (Q), total cloud amount (TCLD), and albedo.

to the western Pacific area ($+8.67\ W/m^2$ versus $+0.81\ W/m^2$). This is due to greater avaporation in the eastern Pacific associated with large SST increases, as well as relatively large increases of surface moisture (+16%) in the west Pacific that contribute to inhibiting larger evaporation increases there. There is also large increase in low-level moisture convergence in the eastern Pacific area (+62%) compared to the eastern area (at has been observed).

Meehl and Washington (1996) have concluded: "We cannot definitely attribute the recent warming in the Pacific (and associated global warming of the 1980s and early 1990s) to increased levels of CO_2 in the atmosphere. Yet the model results presented here and elsewhere, as well as some observational results, point to the possibility that CO_2-induced climate change in the Pacific region could have this signature. However, there may also be decadal-timescale variability in the tropical Pacific region that could complicate the interpretation of this signal".

In the context of separating the global warming trend and natural variability in observed long-term series, the association of inter-annual and interdecadal variations of global-mean temperature with tropical Pacific SST appearing in a model and observations has been discussed by Kang (1996) on the basis of long-term temperature variations using data from GFDL coupled atmosphere–ocean model 139-year simulation and the observed datasets of surface air temperature (SAT) over the globe for 139 years (1854–1992) and tropical Pacific SST for 42 years.

Kang (1996) has shown that global-mean SAT of both the model and observation are characterized by interdecadal oscillations with timescales of about 20 years and 10 years and inter-annual fluctuations with timescales of 3–6 years. It has been demonstrated that both inter-annual and interdecadal variations of the global-mean temperature anomaly (GMTA) are associated with the corresponding variations of tropical Pacific SST. An intercomparison between calculated and observed SAT has revealed that the model is capable of adequately simulating many important features of SAT variability over the globe and SST fluctuations over the tropical Pacific, except that the inter-annual variations of the model with timescales of 3–6 years are relatively weak.

The analysis of a potential impact of sampling errors of SAT and SST observations has shown that the long-term GMTA variations with timescales longer than few years are not much affected by the sampling problem

To assess potential changes in the temperature seasonal cycle due to an enhanced greenhouse effect, Mann and Park (1996a) have compared trends in the amplitude and phase of the seasonal cycle in observational temperature data in the NH with the response of two general circulation models (GFDL and CCM2) to CO_2 increase. Sizeable amplitude decreases are observed in both models and observations. Significant phase delays (i.e., later seasonal transitions) are found in the simulations, opposite to the phase advances isolated in the observations. The retreat of winter sea ice in high-latitude regions appears to explain the models' response to CO_2 increase. The most important conclusion made by Mann and Park (1996a) is that much of the variability in the observational data is not predicted by the models. They conclude: "It is possible, probably likely, that the observed trends in the seasonal cycle represent a combination of internal variability, enhanced greenhouse effects and external forcings".

2.5.3.4. Climate–biosphere interaction

No doubt, a consideration of the effect of the GHG dynamics on climate should be interactive. This determines an urgency of development of the climate models which would provide an interactive simulation of the biospheric dynamics (Alexandrov et al., 1994; Archer et al., 1995; Betts et al., 1996; Chase et al., 1996; Claussen, 1996; Folly et al., 1994; Kondratyev, 1994a–d, 1995a–e; Kunina, 1993; Nemani et al., 1996; Sellers et al., 1995, 1996; Sud et al., 1996a–c; Tiedtke, 1993; Walker et al., 1995). In this context, Henderson-Sellers and McGuffie (1994) undertook a numerical climate modelling over the grid $4.5° \times 7.5°$ lat.-long. with the use of the NASA CCM1-Oz model, based on the schemes of parameterization of the atmosphere–biosphere interaction (BATS 1E 16 model) and the oceanic mixed layer 50 m thick (with an account of the three-layer component of the ice cover).

The biosphere was parameterized based on a simplified schematization of types of vegetation cover after Holdridge. The vegetation cover distribution was parameterized as a function of precipitation, temperature and possible evapotranspiration (the effect of fertilization due to a CO_2 increase and other factors was not taken into account), and the climatic effect of vegetation cover was manifested through its albedo and roughness (a minimum time scale of 1 year was taken). The unit of the biospheric parameterization was "tuned" against the condition of the present climate simulation.

Such a coupled "climate–vegetation cover" model was used to calculate the $2 \times CO_2$-induced climate changes for an instant doubling (calculated for a period of 15 years) and for gradually increasing concentration (for 45 years). A comparison of the results obtained revealed some differences, but they are similar in that in conditions of the global warming the vegetation cover extent, denoted as agricultural, increases through ousting the broad-leaved forests and grass cover (the tropical forests remain unchanged). On the other hand, the tundra shrinks, while the areas of deserts and broad-leaved trees broaden. On the whole, the pre-calculated regions of the vegetation cover are insensitive to initial conditions, though over 1–2 years the effect of differences in initialization may be appreciable.

Analysis of the calculated climate sensitivity to an account of various special features of the interactive biosphere has shown that the use of the annually renewable characteristics of land surface does not change the numerical modelling results, but produces variations in the mean zonal temperature and precipitation as well as evaporation intensification on the continents. The latter determines an intensification of the Hadley circulation, especially in July, which causes an increase of mean global evaporation.

The results obtained (Henderson-Sellers and McGuffie, 1994) (provided they are verified by further calculations) point to the fact that the climate model considered can be used for the vegetation dynamics diagnosis even with an incorrectly prescribed initial state of the climate. Naturally, this coupled climate model should be considered very approximate. Its special shortcomings are: different characteristic times of variability for the climate and the biosphere (of course, the vegetation cover response to climate changes is much slower than that modelled) and a conditional definition of agricultural vegetation. It is important, however, that the coupled climate model turned out to be much more realistic and gave, for example, a much stronger evaporation on the continents and, on the whole, a substantially different climate in conditions of a CO_2 doubling than the respective models with the fixed characteristics of vegetation cover.

A new version of the Météo-France climate model has been developed by Mahfouf et al. (1995) using an improved land surface parameterization: Interactions between Soil, Biosphere, Atmosphere (ISBA). The model includes such improvements as an incorporation of spatial variability of convective rainfall in canopy drainage estimation and subsurface gravitational percolation. A 3-year integration for the present-day climate with a T42L30 version of the climate model has been performed. Results obtained compare favourably well with available observed climatologies related to various components of the continental surface energy and water budgets. Certain differences with observations are due mostly to a poor simulation of the precipitation field and suggest specific improvements in the surface scheme concerning a representation of the bare soil albedo, the surface runoff, and the soil moisture initialization.

Sellers et al. (1995) have discussed early results from the 1994 field year (August 1993 through September 1994) of the Boreal Ecosystem Atmosphere Study (BOREAS) which is a large-scale international field experiment (covered most of Saskatchewan and Manitoba, Canada) with the goal to improve our understanding of the exchanges of radiative energy, heat, water, CO_2, and trace gases between the boreal forest and the lower atmosphere. The principal aims of the study have been: (1) sensitivity of the forest biome to changes in the physical climate system; (2) the carbon cycle and biogeochemistry in the boreal forest; (3) biogeophysical feedbacks on the physical climate system.

Sellers et al. (1995) have pointed out that the early results can be summarized as follows:

(1) The photosynthetic machinery of the boreal forest has considerably less capacity than the temperate forests to the south. In the mature forests, the growing season uptake of CO_2 by photosynthesis seems to be approximately balanced by a sustained CO_2 loss by respiration over the rest of the year.
(2) The coniferous vegetation in particular follows a very conservative water-use strategy. The vegetation transpiration stream is drastically reduced by stomatal closure when the foliage is exposed to dry air, even if soil moisture is freely available. This feedback mechanism acts to keep the surface evapotranspiration rate at a steady and surprisingly low level (less than 2 mm/day over the season).
(3) The low evapotranspiration rates coupled with high available energy during the growing season (the albedos are among the lowest observed over vegetated regions) can lead to high sensible heat fluxes and the development of deep planetary boundary layers, particularly during the spring and early summer, with intense mechanical and sensible heat-driven turbulence. Aerosols from forest fires can often greatly reduce the incident surface shortwave radiation flux during the growing season.
(4) From what is known from scientific literature on the current climate and

numerical weather prediction models, they all grossly overestimate evapotranspiration from the region.

An intensive development of global climate models with a coupled biosphere has stimulated an accumulation of factual information on biosphere dynamics. For instance, Nemani et al. (1996) have undertaken an effort of quantifying the anthropogenically induced global land cover changes from the coarse resolution satellite data on vegetation in combination with long-term climate and soil data. Differences between potential leaf area index (LAI), derived from climate–soil–leaf area equilibrium, and actual LAI obtained from satellite data have been used to assess changes in land cover. Changes in LAI between potential and actual conditions have been linked to climate by expressing them as possible changes in radiometric surface temperatures (T_r) resulting from changes in surface energy partitioning.

As expected, areas with high population densities, such as India, China and Western Europe, showed large reductions in LAI. Changes in global land cover expressed as summer, mid-afternoon T_r ranged from $-8°$ to $+16°C$ (deforestation resulted in an increase in T_r, while irrigated agriculture reduced the T_r). Nemani et al. (1996) have pointed out that many of the current climate models use potential vegetation maps to represent global vegetation. The results mentioned indicate, however, that there are widespread changes in global land cover due to deforestation and agriculture below the resolution of many models, and these changes could have a significant effect on climate.

Chase et al. (1996) have analysed the sensitivity of climate to global changes in LAI. For this purpose the NCAR CCM2 climate model has been used coupled with the biosphere–atmosphere transfer scheme (BATS). Global distributions of maximum LAI were prescribed as surface boundary conditions. Results from 10-year ensemble averages for January and July indicate that the largest effects of the decreased LAI in the actual LAI simulation occur in the northern hemisphere winter at high latitudes, despite the fact that direct LAI forcing is negligible in these regions at this time of the year (it could perhaps happen because of LAI forcing in the tropics which has long-ranging effects in the winter of both hemispheres).

An assessment made by Chase et al. (1996) for the Asian monsoon region for the month of July shows a decreased latent heat flux from the surface, increased surface temperature, and decreased precipitation with the actual LAI distribution. Although the statistical significance of the results has not been unambiguously established in the simulations considered, one can assume that an effect on modelled climate dynamics has occurred because of changes in maximum LAI, suggesting that further attention needs to be paid to accurate designation of vegetation parameters. Chase et al. (1996) believe that the incorporation of concomitant changes in albedo, vegetation fractional coverage, and roughness length is a first priority task for further research.

Betts et al. (1996) have discussed the land surface–atmosphere interaction in a broader context of the GEWEX Continental Scale International Project (GCIP) using observations from two North American field experiments (First International

Satellite Land Surface Climatology Project Field Experiment (FIFE) and Boreal Ecosystem Atmosphere Study (BOREAS) to improve parameterizations of land surface and boundary layer processes in the European Centre for Medium-Range Weather Forecast (ECMWF) model. Results obtained from a new ECMWF model with four predicted soil layers illustrate the interaction between the soil moisture reservoir, evaporation and precipitation on different time scales and space scales. These results suggest that some predictability exists in the extended range as a result of the memory of the soil moisture reservoir.

An important further step has been made by Sud et al. (1996a,b) to study the influence of the current level of tropical deforestation on the regional circulation, rainfall, and global climate in the case of the Amazonia region. As is well established, the influences of forests on land surface fluxes are: (1) low surface albedo that leads to an increase in the solar radiation absorbed by the biosphere; (2) significantly larger momentum fluxes through enhanced boundary-layer turbulence (as a consequence of larger surface roughness-length of forests) and, most importantly, (3) large evapotranspiration.

Using the Goddard Laboratories for Atmospheres (GLA) GCM with a simple biosphere model (SiB), Sud et al. (1996a) have accomplished two sets of 5-year long integrations for: (1) the control case (normally forested); (2) the partially deforested Amazonia–savannah (until the year 1988). Results obtained have revealed the following: (1) There was a large decrease in evapotranspiration (1.15 mm/day, roughly 33%) in the deforested region, while the basinwide decrease was only 0.24 mm/day (about 8%). (2) The associated decrease of precipitation in the deforested (basinwide) regions was about 0.30 (0.168) mm/day. Sud et al. (1996a,b) have pointed out that a precipitation decrease that is smaller than evapotranspiration decrease reflects a mitigating influence of increased moisture convergence, particularly within the boundary layer which gets warmed by the increased surface fluxes of sensible heat and longwave radiation. This implies that precipitation decrease in response to the 1988-level of Amazonian deforestation could be reversed (in contradiction with earlier assessments), that is, reforestation is likely to emerge if the anthropogenic pressures were eliminated and deforestation were forestalled. (3) The changes in the surface wind and rainfall patterns together with increased runoff and surface temperature indicate potential hydrological as well as ecological consequences of deforestation. In the case of simulations considered, only small regions of the Amazon Basin produced precipitation and circulation changes at a 90% or better statistical significance level. This is a consequence of a more realistic prescription of deforestation in Amazonia as compared to the total deforestation scenario used in previous studies and resulted in an overestimated response. Sud et al. (1996a,b) believe that the inference of an irreversible destruction of the biosphere remains an open issue. The answer depends upon the spatial extent and intensity of deforestation. Obviously, further studies are necessary with a strong observational support for verification purposes.

Using the NCAR CCM1-Oz model which includes a mixed-layer ocean model, a three-layer sea-ice model and BATS (the Biosphere–Atmosphere Transfer Scheme), McGuffie et al. (1995) have simulated regional to global-scale climate sensitivity to tropical deforestation. A 14-year control integration was followed by a 6-year deforestation experiment in which the tropical moist forest throughout the Amazon Basin, south-east Asia and tropical Africa was replaced by scrub grassland. An important conclusion made by McGuffie et al. (1995) is that the three deforested regions sustain different impacts on their regional climates. The largest disturbances occur in the Amazon Basin where total precipitation decreases by −437 mm/yr, evaporation decreases by −231 mm/yr and a marked decrease in moisture convergence is evident, although the surface temperature increases by 0.3 K. The African region is least affected by deforestation, although surface net radiation decreases year-round and there is a detectable decrease in moisture convergence in the dry season. The reason why regional responses differ is that regional circulation patterns are affected differently. For example, while ground surface temperatures increase in the southern Amazon and over this basin as a whole, the northern Amazon, south-east Asia and Africa all exhibit decreases in surface temperatures. A modification of atmospheric circulation patterns over deforested tropical regions results in climate responses distant from the disturbance, including the Asian monsoon dynamics and small but statistically significant changes in climate in the middle and high latitudes.

Illustrating a necessity to develop coupled biosphere–atmosphere models (in this case SiB2-GCM), Sellers et al. (1996a–c) have shown that the physiological response of terrestrial vegetation when directly exposed to an increase in atmospheric CO_2 concentration could result in a warming over the continents in addition to that due to the conventional CO_2 greenhouse effect. Since this problem of CO_2 fertilization has been discussed above, it is enough to mention that due to SiB2-GCM calculations, for $2 \times CO_2$ conditions, evapotranspiration will drop and air temperature will increase over the tropical continents, amplifying the changes resulting from atmospheric radiative effects. The responses of SAT and terrestrial carbon uptake are inversely related in the tropics year-round, but only during the growing season elsewhere.

An important conclusion made by Sellers et al. (1996a–c) is that non-local effects generated by the strong year-round physiological responses in the tropics, can affect the high-latitude climate and, thus, the radiative and physiological responses to $2 \times CO_2$ combine in a non-additive way. The calculations show that without a consideration of the physiological effect, the total increase in mean global SAT due to a CO_2 doubling (ΔT) is 1.0°C. The combined consideration of physiological and radiative effects resulted, however, in significant temperature changes: mean ΔT value for the tropics increased from 1.7°C to 2.1–2.6°C (depending on what kind of physiological response was considered). The higher limit (2.6°C) is close to calculated SAT increase due to Amazon deforestation (2–4°C).

A special aspect of climate–biosphere interaction is the problem of biogenic marine emission of dimethylsulphide which has been already discussed earlier. In this context, Andreae et al. (1995) conducted measurements of dimethylsulphide in the air (DMS_a) as well as the number concentration, size distribution, and chemical composition of atmospheric aerosols, including the concentration of cloud condensation nuclei (CCN), during February–March 1991 over the tropical South

Atlantic along 19°S. The results indicate that DMS_a is closely correlated with the sea-to-air DMS flux calculated from DMS concentrations in seawater and from meteorological data. Aerosol (CN and CCN) number and non-sea-salt sulphate concentrations were significantly correlated with DMS concentration and flux. It may be assumed therefore that DMS oxidation followed by aerosol nucleation and growth in the marine boundary layer is an important, if not dominating, source of CN and, possibly, CCN. Thus, the results obtained by Andreae et al. (1995) provide strong support for several aspects of the CLAW hypothesis (named after the initials of its authors, Charlson, Lovelock, Andreae, and Warren) which proposes the existence of a feedback loop linking DMS emission from marine phytoplankton to sulphate aerosol and climate (see Lovelock, 1995).

Claussen (1996) has considered the problem of the sensitivity of a biome model to simulated climate variability to assess a possibility of distinguishing realistic vegetation shift from that which is due to the uncertainty in the estimate of equilibrium vegetation patterns from finite time interval climatologies. For this purpose, global biome patterns from various, equally likely numerical realizations of the present-day climate have been compared. Claussen (1996) has found that the changes in global biome patterns to be expected from interdecadal variability of the atmosphere affect 9–12% of the continental surface (Antarctica excluded). There is no unique difference pattern, although changes are mainly induced by the variability of annual moisture availability and, to a lesser extent, by winter temperatures. This variability of biome patterns reflects the uncertainty in the estimate of equilibrium vegetation patterns from finite time interval climatologies. Changes in biome distributions due to SST and CO_2 increase are larger than and different in kind from the changes due to interdecadal variability (boreal biomes are the most sensitive). Roughly, 30% of the land surface is affected by these changes.

As Woodwell (1995) has shown, a productive perspective to assess the role of the climate–biosphere feedbacks is connected with analysis of palaeoclimatic evidence. An analysis of an ice core from the station "Vostok" in the Antarctic revealed synchronous variations in temperature during the last 160 000 years (periods of warming and cooling) as well as in the concentrations of methane and carbon dioxide in the atmosphere. Though these data do not permit one to establish the cause-and-effect chain in these variations, they point to the fact that the temperature variability has been determined by the dominating feedback between an increase (decrease) of H_2O and CH_4 concentrations and temperature increase (decrease). During the cooling phases, the temperature variations had caused variations in the concentration of the trace gases considered: a 1°C change of temperature had caused variations in the CO_2 mixing ratio by $\sim$7.5 ppm, which is equivalent to the change of total CO_2 content in the atmosphere by 1 Gt.

Woodwell (1995) believes that though the temporal resolution of the palaeodata does not exceed 1000 years, in the context of the present global warming the problem appears of possible analogy in the functioning of mechanisms for climate warming both in the past and now. So far, there has been no convincing evidence that a GHG increase is leading now to the global warming (for this, much more complete observational databases are needed), but the following aspects of the problem are now clear: (1) an accumulation of GHG in the atmosphere accelerates the phase of slow climate changes; (2) such accelerating changes cannot stop themselves (without man's interference); (3) for the last 160 000 years, positive correlations between temperature and CO_2 and CH_4 concentrations had been a dominating factor of such palaeochanges in time scales of hundreds to thousands of years, as illustrated by the data for the Little Ice Age (such correlations are also manifested now); (4) according to the data of the "Vostok" ice core analysis, a 1°C increase of temperature is followed by the CO_2 increase by 7–10 ppm; (5) at present, this correlation is characterized by a time lag of days to weeks between temperature changes and subsequent changes in CO_2 concentration; (6) though the temperature is determined by a complicated set of positive and negative feedbacks, the positive feedback plays, of course, the dominating role: the warming entails a CO_2 increase favouring, in its turn, further warming; (7) knowledge of the formation of global carbon cycle is still uncertain; it is connected with difficulties in assessing the anthropogenic impacts on forests and confirming the fact that these impacts cause carbon emissions to the atmosphere which exceed an assimilation of carbon by the ocean and its accumulation in the atmosphere; (8) a most likely explanation of the "missing" sink for carbon consists in that carbon emissions by land are less than it follows from the available estimates due to unmeasured (and, probably, unmeasurable) increase of the resulting primary production, which determines a reduction of carbon losses by forests; (9) however, there are grounds to doubt that this situation will remain in conditions of the global climate warming which may suppress the development of boreal forests and tundra due to various factors (the effect of temperature increase and humidity decrease, damage by insects, deseases and fires, etc.); (10) all this can cause an increase of the annual CO_2 and CH_4 emissions of the order of 10^{15} gC and more; (11) though the contribution of the enumerated factors is partially compensated by carbon supplies in the remaining forests, no doubt, the total effect will remain an accumulation of carbon in the atmosphere; (12) it is highly probable that the global warming will be followed by a positive feedback consisting in intensified emissions of CO_2 and CH_4 by the NH forests and tundra (the assumption of the conserved present rate of CO_2 and CH_4 accumulation during the warming is, at least, naïve); (13) the atmospheric state can be stabilized under conditions of a 60% reduction of CO_2 emissions, reduced deforestation, and afforestation; (14) postponing the actions on emissions reduction may accelerate the emissions due to biospheric sources to a level which will be impossible to reduce by cutting the anthropogenic emissions of carbon.

Graham et al. (FAUNMAP . . . , 1996) have discussed the faunal dynamics due to climate change, using for this purpose data on the faunal history of the United States over the past 20 000 years (mammal faunas from 2945 localities in the contiguous United States have been considered). The results demonstrate that the geographic ranges of individual species shifted at different times, in different directions, and at different rates in response to late Quaternary environmental fluctuation. Thus, the faunal response to climate changes was very complex. The geographic pattern of faunal provinces was similar to the late Pleistocene and late Holocene, but different environmental gradients resulted in dissimilar species composition for these biogeographic regions. Modern community patterns emerged only in the last few thousand years, and many late Pleistocene communities do not have modern analogues. Faunal heterogeneity was greater in the late Pleistocene.

Thus, the principal conclusion of FAUNMAP analyses is that the simple assumption of faunal associations as individual members bound together in equilibrium by links of adaptation and competition is wrong. The reality is that species are not team players (as Gee (1996) pointed out)—they tend to look after themselves, not their associations, and respond to climate change in idiosyncratic way. The implications of such a situation are twofold: (1) the uncoordinated movement of species is response to climate change produces novel combinations; (2) ecological stability we see today is illusory because modern communities emerged only in the past few thousand years.

The global-scale "experiment" on the anthropogenic impact on the environment stimulates further development of techniques of the numerical modelling of global climate, the more so that, so far, reliable forecasts of future climate changes (on regional scales, in particular) have been impossible. The main difficulty of numerical modelling consists in the need to take account of numerous interactive factors in the global-scale climatic system, affecting the climate. The complexity of solution for such a problem and the huge volume of calculations needed prompt one in many cases to use simplified climate models which leave out of account the climate system complexity and do not provide the spatial and temporal resolution needed.

In this connection, within an extensive programme of development of adequate climate models, the climate system model framework (CSMF) has been substantiated (Mirin et al., 1994) which can provide a sufficiently complete and flexible account of the processes in the global climate system with the use of the most powerful computers a configuration of file parallelism. The CSMF realization has been discussed with the numerical modelling of two components of the climate system taken as an example: the atmosphere and the ocean. The parallelism of calculations (applied a task-decomposition approach) is realized with respect to both individual climate-forming processes and spatial domains with the subsequent matching via message passing. The results of initial tests have been considered, obtained with the use of almost all the parallel processors available as well as strongly efficient working stations and their clusters (Intel Paragon, TMC CM-5, Meiko CS-2) as well as Cray C-90. In the case of the ocean model, calculations were also made for the system "ocean–sea ice". The results of calculations of the Antarctic ice cover extent agree well with observations.

2.6 SOLAR ACTIVITY AND CLIMATE

The subject of this paragraph is especially difficult in spite of numerous efforts and abundant scientific literature (see some references in the surveys by Kondratyev (1989b, 1992d), Nesme-Ribes (1994), and Roederer (1996)). The solar activity–climate problem remains mysterious, since there are still no proven mechanisms explaining how solar activity might impact climate. Recent studies by Foukal (1994),

Kukla et al. (1994), Schönwiese et al. (1994), Sklyarov (1994), Sofia and Fox (1994), Tinsley and Heelis (1993) illustrate this conclusion. An extensive recent survey has been published by one of the leading experts in the field—Roederer (1996). In connection with the earlier effort, studies by Roberts (1975) and Eddy (1976) have to be mentioned.

Roederer (1996) has emphasized the necessity of considering three classes of solar variations: (1) an 11-year sunspot cycle (and, eventually, the related 22-year Hale cycle); (2) long-term secular changes of the sunspot cycle amplitude (the Gleissberg "cycle"); (3) short-duration, sporadic events, such as solar flares and the sun-controlled solar wind shocks are reversals of the interplanetary magnetic field.

There are two principal aspects of the problem of the solar activity impact on climate to be considered: (1) climate change due to variations of the extra-terrestrial solar irradiance (solar constant, SC); (2) solar activity variations (determined by various indices) and potential climatic consequences (we shall not consider here "Milankovitch mechanism" explaining palaeoclimatic variations as a result of orbital parameter changes and relevant re-distribution of solar radiation over the globe—see Berger (1992)).

2.6.1 Solar extra-atmospheric irradiance variations

Simple energy-balance calculations which have been later confirmed by 3-D climate modelling show that from the viewpoint of the order-of-magnitude assessments, a 1% solar constant change is approximately equivalent of a 1°C variation of mean global mean annual surface air temperature (SAT). One can also remind in this context an empirical regression between SAT (T_s), SC (S_o) and CO_2 concentration [CO_2] which has been substantiated by Ardanuy (1983) through the analysis of the Earth's radiation budget satellite measurements:

$$T_s = 0.26 + 0.76\, S_o + 0.11\, [CO_2]$$

Crowley and Kim (1996) have compared the estimates of solar variability since 1600 with the two estimates of the NH temperature change. Sun–climate correlations vary between 0.57 and 0.74 and are significant at a 95% level. One can therefore believe that solar variability significantly contributes to climate change on decadal to centennial timescales.

Thus, the problem is: to what extent does SC vary. Fortunately, during the last 10–15 years (i.e., for about one 11-year solar cycle) reliable enough satellite solar constant measurements have been made (before that only indirect assessments on the basis of the surface, aircraft and balloon SC measurements existed, which were rather controversial: (Kondratyev, 1983)).

According to Hartmann et al. (1993), current active cavity radiometer (ACR) designs are able to measure the total solar irradiance with an absolute accuracy of about 0.35% and the relative precision of the order of 0.01%.

Willson (1984, 1993; Willson and Hudson, 1988, 1991) was one of the pioneers of SC satellite observations. For this purpose, three high-precision cavity pyrheliometers were installed on board the Solar Maximum Mission (SMM) satellite launched in February 1980. The results for the first 153 days of observations gave a value of 1368.31 W/m^2 (with an accuracy of about $\pm 0.5\%$). The longest continuous record is the 14-year record of the Nimbus-7 ERB, which began in late 1978 and ended in early 1993.

Overlapping this record are the Solar Maximum Mission (1980–89) and the Earth Radiation Budget Experiment (1983 to the present) SC measurements. An ACR was also operating on the UARS satellite. In the beginning of 1994 a new ACR developed by Sklyarov (1994) was installed on board METEOR–3 satellite and is now successfully functioning. The observed mean SC values differ by 7 W/m^2 (about 0.5%), but all indicate the SC variations over solar cycle of about 0.1%. Future SC observations are connected first of all with the EOS programme.

Fröhlich (1993) has summarized the results of SC observations since the early 1980s. The following conclusions are most important:

- There are temporal SC variations within timescales from a few minutes to decades. A very important component of these variations for the range of timescales from a few days to several months is connected with the activity of the photosphere. Roederer (1996) has pointed out that the 11-year SC modulation comes from two sources of mutually counteracting effects: enhanced emissions from bright flares during solar maximum and enhanced blocking by sunspots (the former wins the latter). The relative amplitude of SC variations for timescale of the order of few months is about 0.1%. The amplitude of the 11-year periodicity is about the same.
- Now the existence of SC 11-year periodicity can be considered as a proven fact. Solar constant reaches maximum values during periods of maximum solar activity.
- If an average SC value equal to 1367 ± 2 W/m^2 is assumed, it means that the level of mean global extra-atmospheric solar irradiance is 240 W/m^2. Thus, a 0.1% change in SC leads to an irradiance change of 0.24 W/m^2. A SC increase by 2% is approximately equivalent (from the viewpoint of climatic impact) to a CO_2 doubling. Since the greenhouse climate forcing calculated over the period 1750 to 2000 is about 2 W/m^2, this is about eight times the magnitude of the climate forcing associated with an 11-year solar cycle (Hartmann et al., 1993). Larger solar variations on longer timescales have been suggested by empirical studies (Friis-Christensen and Lassen, 1991; Hoyt and Schatten, 1993).

Using such indicators of solar activity as solar diameter, solar cycle length and umbra/penumbra variations, Hoyt and Schatten (1993) have derived a possible increase in forcing of about 0.2 to 0.4 W/m^2 between 1850 and the present day which could non-negligibly contribute to climate change, but still is much smaller than the calculated greenhouse forcing for the same time period of more than 2 W/m^2. Lockwood et al. (1992) have mentioned, however, a possibility of future solar-induced climate fluctuations exceeding those of the historical past.

Hartmann et al. (1993) have pointed out that indices of solar activity (such as photometric sunspot index, the 1083 line with Mg e/w radio, CaII emission) are well correlated with total solar irradiance over the 11-year solar cycle, but the precision of these indices and their statistical relationships with SC are probably inadequate for

Table 2.37. Solar constant data obtained during the ATLAS 2 flight

Experiments Wm^{-2}	9 April 1993 08:00 to 17:30	11 April 1993 09:00 to 18:00	13 April 1993 11:00 to 19:00	15 April 1993 07:30 to 20:00	Mean
SOVA1-L (1)	1364.92	1365.23	1365.21	1365.24	1365.150
SOLCON-R (2)	1366.25	1366.65	1366.61	1366.53	1366.510
SOVA1-R (1)	1366.37	1366.64	1366.58	1366.68	1366.568
SOLCON-L (2)	1367.15	1367.44	1367.39	1367.31	1367.323
ERBS (3)	1365.71	1365.76	1365.66	1365.476	1365.650
SOVA2a (1)	1366.09	1366.42	1366.41	1366.39	1366.503
SOVA2b (1)	1366.22	1366.58	1366.61	1366.60	1366.503
ACRIM II (4)	Data not made available yet, mean value estimated through ratio of ACRIM II to SOVA1-L				1365.757

long-term trend detection. The problem is that the physical relationships between solar activity indices and SC are not understood (Foukal, 1994).

Outstanding studies of solar constant variability have been accomplished by Crommelynck (1993) and Crommelynck et al. (1994, 1996) during the ATLAS 2 shuttle mission period and later. A very important result of this study is also that it provides the basis for the definition of the Space Absolute Radiation Reference which has been materialized as a set of instruments for which a set of adjustment factors have been found, using such observational conditions when (April 1993) the sun did not have a single sunspot for several days and eight of the radiometric channels were all within 0.1%. This makes it possible to reduce all earlier SC measurement results to one single level. The confidence level of the accuracy is determined to be between 0.05% and 0.02%.

Table 2.37 (from Crommelynck et al., 1994b) illustrates the results of SC measurements based on the utilization of eight different instruments: two channel SOVA 1 (solar variability experiment) and SOVA 2 (two monochannel active and backup radiometers on board the European Retrievable Carrier (EURECA), ERBS (Earth's Radiation Budget Satellite), SOLCON (similar to SOVA (1) and Active Cavity Radiometer ACRIM II (ATLAS (2), ACRIM II (UARS). Here are taken: (1) Experiment on board EURECA; (2) ATLAS 2; (3) ERBS; (4) UARS.

An intercomparison of the results allowed one to substantiate the values of the adjustment factor for various radiometers (Table 2.38, from Crommelynck et al., 1994b) to reduce all SC values to the Space Absolute Radiation Reference Scale.

The mean absolute value of 1366.22 W/m^2 of the solar irradiance reduced to 1 AU (astronomic unit) deduced from eight radiometric channels during the ATLAS 2 mission and the ratios of the measurements to this value provided the definition of the Space Absolute Radiation Reference (SARR) with a confidence interval of 2.5×10^{-4} at the 0.900 probability level and 5.5×10^{-4} at 0.999.

The instrument complex of the SOVA experiment included five Sun photometers to monitor spectral SC variations at the wavelengths 335, 480, 500, 546, and 865 nm (Crommelynck et al., 1994a,b). These data characterize the temporal variability of

Table 2.38. Ratios of solar constant measurements obtained during ATLAS 2 and before, relative to SOVA1-L and to the MEAN during ATLAS 2

Experiments	Ratio relevant to SOVA1-L	Ratio relative to MEAN	Adjustment factor to MEAN
SOVA1-L	1.000000	0.999215	1.000786
ERBS	1.000366	0.999580	1.000420
ACRIM II	1.000444	0.999658	1.000342
MEAN	1.000786	1.000000	1.000000
SOVA 2a	1.000863	1.000077	0.999923
SOVA 2b	1.000991	1.000205	0.999795
SOLCON R	1.000996	1.000210	0.000790
SOVA1-R	1.001038	1.000252	0.999748
SOLCON L	1.001592	1.000805	0.999195
Operational during ATLAS 2, identified as out layers			
ACR 502	1.005956	1.005166	0.994861
ACR 504	1.006432	1.005642	0.994390
Not operational during ATLAS 2, indirect ratio to SOVA1-L			
ACRIM I	1.002248	1.001461	0.998541
ERB	1.005392	1.004602	0.995419

standard deviations for spectral solar constant at the SOVA photometers' wavelengths. Some cyclic variations are obvious with apparent differences in the phases between each wavelength. The power spectral density of the standard deviation (relative amplitude, normalized to maximum amplitude) characterizes prevailing frequencies. For all the photometers (wavelengths) there is a frequency component with a period of approximately 50 days with its first harmonic at a 25-day period.

As could be expected, the frequency spectra for spectral and total SC are rather similar (the prevailing periods are about 46 and 23 days). An important conclusion from all these data is that basic solar rotation period is not observed in the power spectral density. The observed periods are about 55, 46, 39, 25, and 23 days. These periods are not stationary and the power spectral content depends on the wavelength. Similar results were obtained by Willson (1984).

Crommelynck et al. (1994, 1996) believe that the results obtained may be interpreted in the sense that the Sun has: (1) non-spherical shell, dipole-like, rotating inside the Sun near the photospheric surface; (2) rotating sources in nonlinear interaction.

Chandra et al. (1995) have discussed some results of continuous measurements of the solar UV flux in the spectral range 115–420 nm since October 1991 which have been made with the use of two instruments on board the UARS satellite: Solar Ultraviolet Spectral Irradiance Monitor (SULIM) and Solar Stellar Irradiance Comparison Experiment (SOLSTICE). The analysis of observational data on relative changes in the solar Lyman-α (121.6 nm) and F (200–205) fluxes which cover the time period of about 3 years from solar maximum to near solar minimum conditions of solar cycle 22 indicate the presence of substantial changes. The Lymanflux decreased by about 45% from a mean solar maximum value of about 9 mW/m^2, and the integrated solar flux between 200–205 nm decreased by about 5% from a mean value of about 47 mW/m^2.

An important problem concerning long-term solar constant variations is a possibility to extrapolate SC variations in the past and in the future (Lean, 1991; Schatten et al., 1996). As has been mentioned above, various solar indices may be used to solve this problem. There were many attempts to find such a solution. One can illustrate the results by a few examples.

Hoyt and Schatten (1993) examined several different solar indices measured over the past century that are potentially proxy measures for the solar constant. They believe that the three solar indices allow the solar constant to be modelled back for the 1770–1992 time period: (1) the length of the solar cycle; (2) the normalized decay rate of the solar cycle; and (3) the mean level of solar activity (for all the proxy models considered the solar constant has varied by less than about $\pm 0.2\%$ in the present century; this is why it is difficult to detect solar constant variations).

A comparison with the NH SAT departures for the same time period show that approximately 71% of the decadal variance in the present century can be modelled with those solar indices, although this analysis does not include anthropogenic or other variations which would affect the results. Over the entire three centuries, $\sim$50% of the variance is modelled. In accordance with what has been mentioned before, Hoyt and Schatten (1993) note that "To understand our present climate variations, we must place the anthropogenic variations in the context of natural variability from solar, volcanic, oceanic, and other sources".

Lean et al. (1991) have assessed the change in the output of solar ultraviolet radiation (130–300 nm) since the time of the Maunder Minimum (1645–1715) till the present time. Their estimates show that between the Maunder Minimum and 1986 (the present-day solar activity minimum between cycles 21 and 22) reductions took place of 64% in the irradiance of the Lyman-α line, 8% at 200 nm, and 3.5% in the wavelength range from 210 to 250 nm. The reduction in the solar output for the entire spectral band between 120 and 300 nm has been estimated at 0.17 W/m^2, which is approximately 6% of the earlier estimated change in the total solar irradiance of 2.7 W/m^2 over the same time span.

Nesme-Ribes et al. (1993) have substantiated an approach to reconstruct the solar constant over the past four centuries (with the use of magnetic solar cycle variations as proxy data) and then accomplished climate change simulation experiments on the basis of an atmospheric general circulation model developed at the Laboratoire de Météorologie Dynamique (LMD AGCM: 7.5° long. × 3.2° lat. horizontal resolution, 11 layers). The response of the model to magneto-convective solar forcing during the Maunder Minimum, when the mean value of the top in solar constant reached 0.5%, has been studied to simulate the Little Ice Age phenomenon. The simulated cooling for the solar constant decrease by 1 W/m^2 is compatible with observational data.

An important conclusion is also that variations of globally homogeneous external forcing parameters, like solar constant or greenhouse gas loading (of equivalent forcing), yield climate responses with very similar geographical patterns. These responses may be distinguishable only in the stratosphere. Such responses are determined mainly by the amplitude of the external perturbation.

Nesme-Ribes et al. (1993) pointed out that there are a few conspicuous features of the results: the stratospheric warming from mid-latitudes to poles, contrasting with tropospheric cooling, which may be attributed to a secondary greenhouse effect by water vapour; relatively strong warming of the upper tropical troposphere due to a strongly positive cloud feedback in this version of the LMD AGCM, and the unexpectedly weak seasonal contrast in high-latitude cooling. The global average cooling is near 1.5°C with little seasonal dependence.

The most important conclusion made by Nesme-Ribes et al. (1993) is that the strong sensitivity of the results to parameterization approaches (especially in the case of cloud microphysics which makes the model highly sensitive), the possibility of other mechanisms acting simultaneously (such as volcanic perturbations) and the presence of internal climate variability due to the transient component of ocean dynamics (omitted in the numerical modelling considered) should be a reminder of the complexity of climate system dynamics and the difficulty to reach definite conclusions.

Kondratyev and Nikolsky (1995a,b) have discussed high mountain observations of spectral solar radiation to estimate energy contributions of ultraviolet (UV),visible (VIS) and near infrared (NIR) regions of the spectrum to changes in climatic parameters in the presence of solar activity variations of different periodicity (from a few minutes to an 11-year cycle). The observations of spectral fluxes and spectral transparency of the atmosphere were carried out over the period beginning from the maximum of the 21st cycle (1981) of solar activity through the subsequent minimum and maximum phases of activity till the middle of the decrease phase in the 22nd cycle (1994). In more than ten cases, occurrences of extremely anomalous high values of transparency of the atmosphere in UV and violet-blue ranges of the spectrum were revealed.

When analysing relevant circumstances, it has been discovered that, as a rule, anomalies in transparency arise when an active region is passing through the central meridian and are observed at local noon. One should keep in mind that when measuring through the atmosphere, the increase in extra-atmospheric solar spectral irradiance (ETSSI) is revealed by the spectral transparency excess over the calculated (model) value characteristic of the clear and dry atmosphere with air pressure and temperature profiles in the troposphere observed over the station at the moment of observations.

Further study showed that an apparent increase in transparency occurs when the Earth is illuminated by radiation coming from an active region (AR) and is characterized by the intensity maximum near 410 nm and by specific spectral variations. This kind of solar radiation bursts cannot be explained by an actual increase in transparency because in this case the transparency of the atmosphere for wavelengths in the spectral ranges mentioned above would have coincided with the values characteristic of the ideal atmosphere at the station level (2100 m) and in some cases would have reached values corresponding to the ideal atmosphere at the 3400-m level.

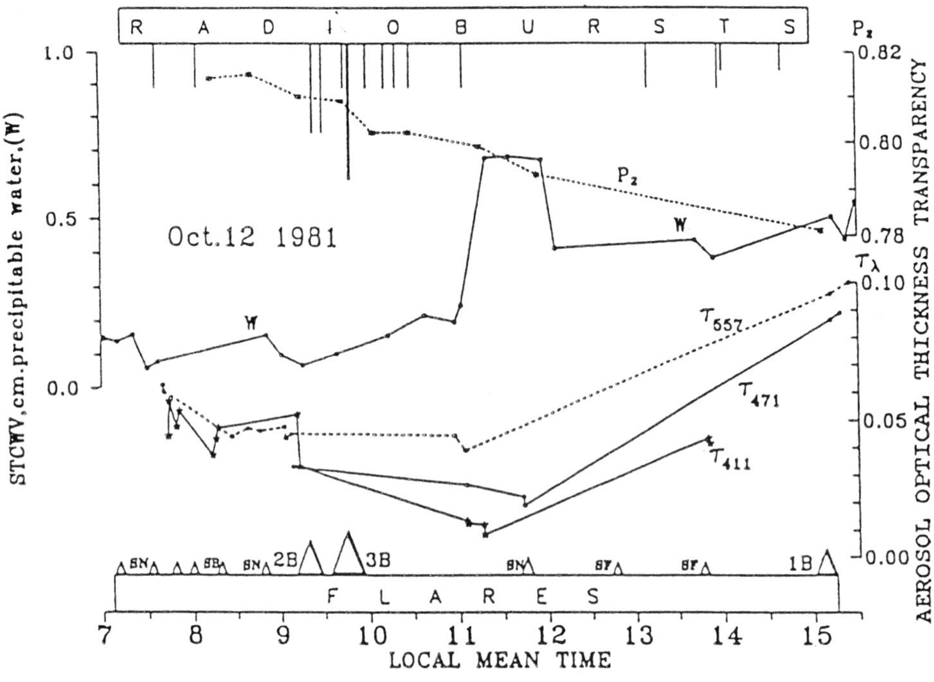

Fig. 2.11.

Comparison of the energy contribution of the "explosive" burst variations of solar spectral radiation (SSR) mentioned above with total solar constant variations due to Nimbus-7 data (diurnal mean values) showed that the excess radiation is generated by a very small part (1×10^{-7}) of the Sun's disk surface.

Before the analysis of main results of the observations let us consider an example of an anomalous situation displayed by variability of the optical parameters of the atmosphere in the period of high solar activity in October 1981. Figure 2.11 presents the daily variation (12 October 1981) of: (a) the total content of water vapour W (according to IR-hygrometer data); (b) spectral optical thickness τ of the atmosphere for wavelengths 411, 471, and 557 nm; and (c) total solar radiation transmission through the atmosphere (P_2). On the abscissa the symbols of duration and power of flares (below) and those of radioemission bursts from ARs (above) are plotted.

Data presented in Fig. 2.11 indicate the presence of the strong transformation of the optical characteristics of the atmosphere in the first half of the day which was the result of increasing power of the Sun's flares and radioburst intensity. The first feature to be mentioned is an extremely low value of W (0.1-0.25 cm precipitable water) for the interval from 7 to 11 LMT, then its unusually fast increase to the normal mean level (0.6 cm) and consequent sharp decrease 40 min later.

Of great importance is the situation never encountered before: an inverse (to normal) sequence of values of residual (aerosol) optical thickness (τ_{411}, τ_{471}, τ_{557}) as

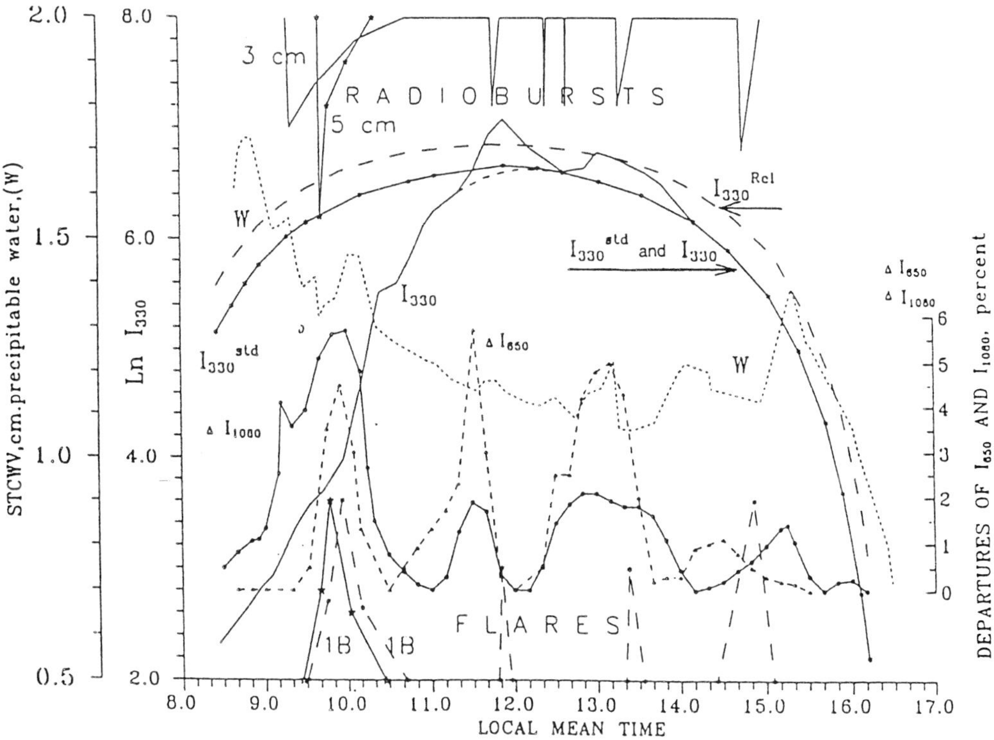

Fig. 2.12.

well as very low τ_{411} values ($\sim$0.01) and τ_{471} at noon. Such anomalously high transparency in the blue range of the spectrum should be evidently connected with a 25% burst of ETSSI, the maximum of which is at 411 nm; at shorter wavelengths there is a long wing of increasing transparency.

We now need to focus our attention on the most impressive and well-traced event of 2 November 1991. Fig. 2.12 shows the development of specific changes in solar radiation following two flares (1B and 1B) with an 8-min interval between their peaks, and radiobursts at some frequencies (a peak amplitude of 3125 sfu at $\lambda = 5$ cm at 9h 46m LMT (local mean time)). The beginning of the response of the optical parameters to the flares was noted as early as 9h 46m LMT.

The increase in solar irradiance in near-IR region (650 nm) reached 4.5% and, at the wavelength 1080 nm, about 5.5%. It is interesting to note that an intensification of solar radiation at $\lambda = 1080$ nm began 16 min before the first flare and 10 min before the beginning of the radioburst. The response of the total (according to spectroscopic data) content of water vapour (*W*), as it is plotted in Fig. 2.12, took place at 9h 58m LMT, that is, 11 min after the maximum of the first flare. *W*-change continued over about 30 min and reached 8.5%. For the flare 3B on 12 October 1981, W-change reached up to 45% and lasted about 60 min.

As can be seen from Fig. 2.12, bursts of radiation were continuing till 12h LMT. At 650 nm the photometer displayed two more 5% bursts and a final 1% burst. Three more bursts with 1.5–2% amplitude were recorded by the photometer at 1080 nm.

On the basis of the observations considered, one can make some suggestions as to the nature of sources of spontaneous bursts of solar radiation on 2 November 1991. A morning burst at wavelengths 650 and 1080 nm should be undoubtedly related to a pair flare at 9h 55m LMT while subsequent bursts, as follows from the work by Dyakonova and Chistyakov (1993), represent a manifestation of oscillating changes in the area of the faculae rings and in the Sun's diameter. Midday anomalies are due to the occurrence of a powerful nonequilibrium source of radiation (see further).

Now let us dwell on the main event of the day: the bursts of solar radiation in UV region (Fig. 2.12). From the data on I_{330} variations near noon it follows that 35 min before the upper culmination there appeared at the Sun an additional source of radiation characterized by essentially selective spectral variation. The main feature of this source was a strong increase of radiation in the blue spectral region. The data of observations show that a peak in I_{330} increase took place exactly at local noon. The burst lasted from 11h 30m till 12h 30m LMT. From Fig. 2.12 one can see that instead of following the preceding tendency of the increase along part of the dotted curve and coinciding with the $I_{330}{}^{std}$ curve which corresponds to a normal daily variation (the latter was constructed on the basis of data for the undisturbed atmosphere over the period 14h 20m to 16h 15m LMT), the values of I_{330} lay outside the standard (normal) levels and reached levels corresponding to an ideal atmosphere ($I_{330}{}^{Rel}$) and even higher ones. The increase in I_{330} was about 50%. The second increase in I_{330} lasting from 13h 00m till 14h 30m LMT did not locate outside the "Rayleigh" level but nonetheless was anomalous (+18%). After the second increase the I_{330} values followed closely the $I_{330}{}^{std}$ course (in accordance with Bougher's law).

Fig. 2.13 presents spectral variations in ETSSI, which took place at about noon on 2 November 1991, 10 September 1986, and 12 October 1981, as well as maps of location of active regions on the Sun for the two first dates. We should notice that for the events on 2 November 1991, $\Delta I_o\lambda$, variations in UV- and near-IR spectral regions do not coincide in time; the maxima of ΔI_{650} and ΔI_{1080} variations are 25–30 min ahead of the I_{330} maximum. That's why ΔI_{650} and ΔI_{1080} were not taken into account when we made the calculation of a noon integral using the ETSSI-curve for 2 November 1991. One may suggest that a very close location of active regions to the central meridian of the Sun (CMS) was a most significant feature on that day which provoked repeated occurrences of radiation bursts in different parts of the spectrum.

The duration of the entire process—the burst of solar radiation at 330–430 nm and in the ranges 650 and 1080 nm—was about 4 hours. At noon, when the variation

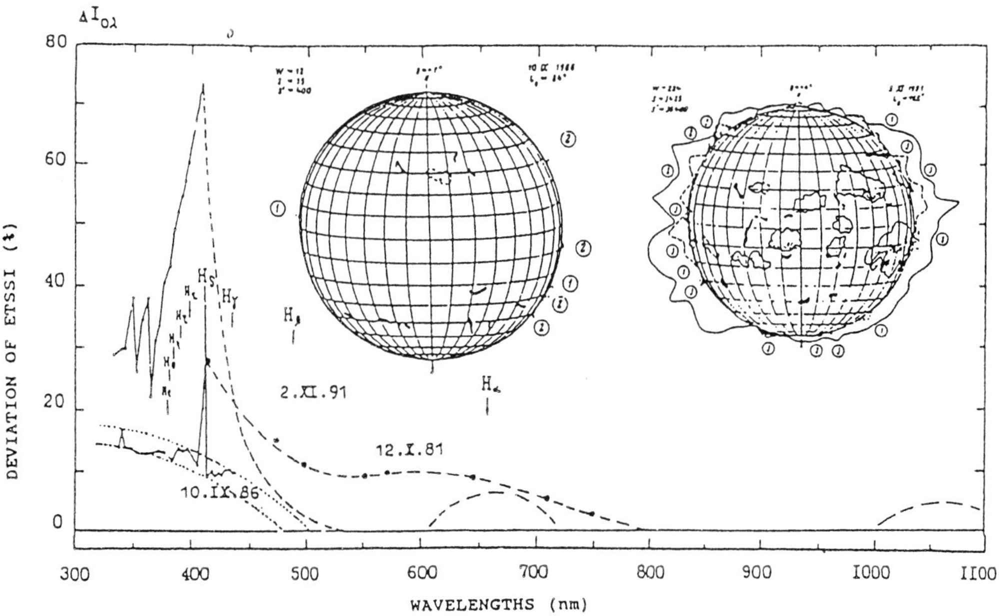

Fig. 2.13.

was at a maximum, the integral in the range 330–430 nm was about +100 W/m^2 (~8% of total solar constant I_o).

From the variation in the spectral distribution for 10 September 1986 towards 500 nm one may see that on that day there was also probability of occurrence (a band limited by a small-dot line) of low values of $\Delta I_{o\lambda}$ for $\lambda > 500$ nm. The solar disk for that day is shown on the left-hand map. The only AR at 30°N is located near the CMS; it is undoubtedly that very object which contains a source of ETSSI-burst (its duration was about 7 min and it was observed just at local noon).

The data for 12 October 1981 point to a longer period of ETSSI-increase (up to 6 hours) which is, however, not related to the effect of the noon-time illumination of the Earth mentioned above. On that day we observed some power flares (2B, 3B) which were responsible for excess radiation occurrence (from 28% at 411 nm to 10% in the 560–650-nm range, and dropping to zero at 800 nm). The integral for the range 410–800 nm was about 80 W/m^2. It should be taken into consideration that this integral does not include radiation with $\lambda < 410$ nm. This portion was approximately estimated at ~10 W/m^2.

To make statistical significance of parameters characteristic of ETSSI-burst occurrences more reliable, the spectrometer was supplemented with a two-channel filter photometer for 405- and 630-nm wavelengths. This instrument proved to be necessary as well as for estimating the degree of spectral compensation of bursts in the total flux of solar radiation. On 10 November 1992 at about 9h 30m UT a sharp

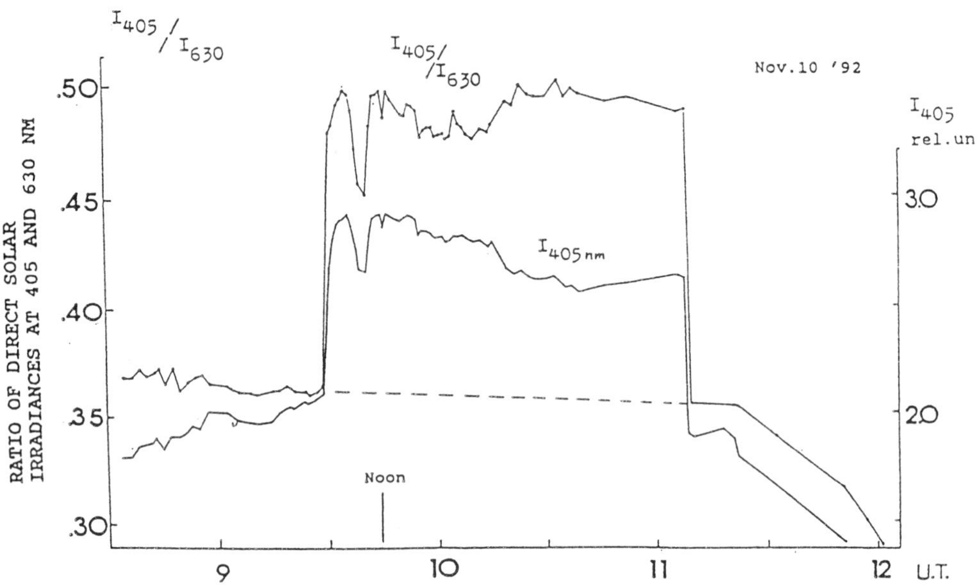

Fig. 2.14.

rise in the intensity of solar radiation for the 405-nm channel was recorded, but the data for the 630-nm channel did not confirm the occurrence of this variation, as can be clearly seen from Fig. 2.14. Since measurements in each channel were made every 2 minutes and the irradiance rise was very short, the precise determination of its beginning proved to be impossible—it was defined as 9h 33m $\pm$ 1 min UT. The intensity drop at 11h 07m $\pm$ 1 min UT was also very sharp.

To summarize the results discussed, one has to emphasize that the most significant events of solar spectral radiation bursts over more than 12-year observation period at a high-altitude station have been discovered. The number and amplitudes of solar spectral bursts were increasing when the solar activity (SA) maximum was approaching. A plausible mechanism of short-term variations of extra-atmospheric solar spectral irradiance (ETSSI) has been suggested. It appears that a burst of ETSSI arises when the Earth is sporadically irradiated by a strong flux of induced violet-blue light coming out of magnetic flux tubes in the active region of the Sun. Kondratyev and Nikolsky (1995a) confirm earlier conclusions that on a timescale of decades there is a close relationship between variations in faculae areas, the solar constant, and surface air temperature. On the basis of these results it may be suggested that at the end of the 1930s, when the Sun was very active, its effective output was about 0.4% and therefore the surface air temperature in the northern hemisphere about 0.4°C higher than in the first decade of the twentieth century.

2.6.2 Potential solar activity impact on climate

There were many mechanisms suggested to explain, how solar activity characterized by various indices might impact climate. A number of such mechanisms have been discussed by Kondratyev and Nikolsky (1995a,b). Here we shall confine a discussion to a consideration of the most recent attempt with emphasis on climate modelling.

Fig. 2.15 schematically presents a complexity of interrelationships between the most important energy input channels from the Sun, their continuous or sporadic variability and the regions in the terrestrial atmosphere that can be affected by this variability (Roederer, 1996). Solar magnetohydrodynamics regulates all forms of variability, including that of the intensity of cosmic ray flux, the configuration of the high-latitude ionospheric electric field and precipitation of energetic trapped electrons, all of which are controlled by the interplanetary magnetic field (IMF). The variability of the two photon channels on the left is controlled by surface structures, such as factual and pages (radiation emitters) and sunspots (blockers), whereas the particle channels are mainly controlled by processes in the coronal plasma. The Earth's internal magnetic field, which is variable on a secular scale, also influences the fluxes of the charged particles penetrating the atmosphere.

When discussing the impact of solar activity on climate it is important to briefly mention relevant empirical findings. It is a very difficult subject, however, because there have been numerous studies which discovered empirical correlations between various climate parameters and solar activity. We shall briefly mention here only most recent results which may be considered reliable enough. Investigations by Labitzke (1987), Labitzke and Van Loon (1992, 1993), Van Loon and Labitzke (1993) have been mentioned already. They were concentrated on the analysis of the correlation between the temperature of 30 hPa level and solar activity expressed by the 16.7-cm radio emission flux. Mendoza et al. (1991) have come to the conclusion that 63% of ENSO events occur during the descending phase of the solar cycle, and that there are twice as many events occurring one year after the maximum rate of sunspot decline (maximum negative gradient) than would be expected by random occurrence.

Friis-Christensen and Lassen (1993) have discovered a remarkable correlation between mean global mean annual surface air temperature and 11-year solar cycle length, which leads by 10–15 years. Reid (1991) has derived an empirical relationship between solar constant and sunspot number and, using a one-dimensional columnar model of the coupled atmosphere–ocean system, computed SST variations during the time period from the late 1600s. Calculated SSTs fit the observed temperature variations during the last century (of course, such a good coincidence produces a suspicion concerning the reliability of the results because the observed SST variability was undoubtedly not only due to solar activity changes). Andersen (1992) has mentioned a possibility of connection between surface winds, solar activity and the Earth's magnetic field.

As far as an application of climate models to assess solar activity effects is concerned, a majority of efforts have been concentrated on simulations of the solar constant variations impact (see Kondratyev, 1992d). For instance, the NCAR CCM1 model has been applied to simulate climatic consequences of

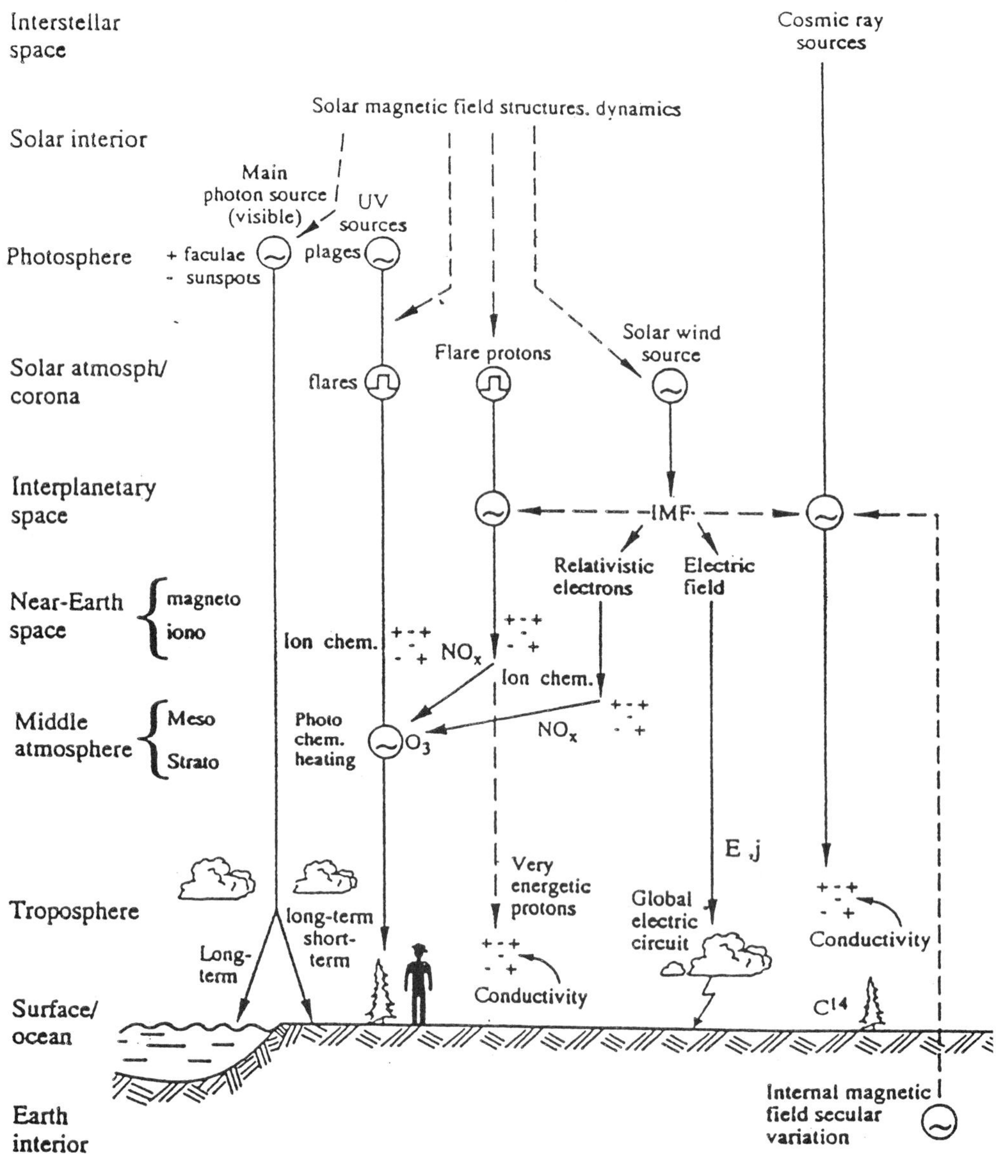

Fig. 2.15. A scheme of complex interrelationships between the most important energy input channels from the Sun.

CO_2 increase from 100 to 1000 ppm with various prescribed values of solar constant changes: −5, −2, 0, +2, +5%. The results show that: (1) the model climate reveals a large response in many aspects to both CO_2 and solar constant changes, with the response greatest at high latitudes (and related to sea-ice changes); (2) the response of most climate variables to changing CO_2 is strongly logarithmic over both continent and ocean. The response of SC changes is strongly linear over continents, but equally well described by either a linear or a log fit over the ocean; (3) perhaps, most importantly, the qualitative patterns of geographical response show a high degree of correspondence between CO_2 and SC changes, despite the very different nature of the forcings. This suggests that the nature of the geographic response may be an inherent characteristic of the climate system rather than dependent on the specific forcing.

It is obvious that only systems approach (interactive consideration of solar activity impact) may lead to a solution. It has been recently demonstrated by the results obtained by Jenkins (1993) who considered the sensitivity of a general circulation model (the NCAR CCMOA model has been chosen) to a faster rotation, enhanced CO_2 concentration, and reduced solar constant. Four sensitivity studies have been performed under the assumptions of zero land fraction and reduced solar constant conditions: (1) increased rotation rate of the Earth (shorter day length); (2) eight times the present CO_2; (3) three with a 15% reduction in the solar constant.

These numerical modelling results directed towards palaeoclimatic assessments indicate the following changes in the global mean SSTs compared to the control simulation (it has a solar constant reduced by 10% the present amount): +6.6 K; +12 K; +18.5 K, and +6.6 K relative to the control global mean temperature of 264.7 K. The warm SSTs are associated with a 20% reduction in clouds for the rotation rate experiments and higher CO_2 concentrations in other experiments. A very important conclusion is that in contrast with the previous studies that have used energy balance and radiative-convective models, new results require a much smaller atmospheric CO_2 increase to prevent the Earth from being covered with ice. This demonstrates a potentially important role of cloud feedback for climate dynamics of early Earth as well as lead to the conclusion that very large atmospheric CO_2 concentrations may not be necessary to counterbalance the lower solar constant that existed early in the Earth's history.

Observed variations of the total solar constant are too small to substantially contribute to climate change, but energy input from various potential solar activity impacts on climate are even much smaller. Therefore, the only possibility to explain the solar activity impact is connected with a search for some trigger mechanisms operating in the atmosphere. The most probable part of the atmosphere where such trigger mechanisms can function is the middle atmosphere, especially the stratosphere. This is why a number of recent studies have been devoted to a simulation of chemical–dynamical–radiative processes in the stratosphere.

Brasseur (1993) and Huang and Brasseur (1993) have developed a two-dimensional chemical–dynamical–radiative model of the middle atmosphere to investigate the potential changes of temperature, ozone, and other chemical constituents in response to variations in the solar ultraviolet flux, associated with the solar rotation (27 days) and the solar cycle (11 years) (see also Ko et al., 1993). The observations showed that the variation between the early 1980s (solar maximum condition) and the mid–1980s (solar minimum condition) is of the order of 50–60% at Lyman-, 5% at 180–200 nm and 6–7% at the wavelength 205 nm which is often used as an index to represent the intensity of the solar forcing on the stratosphere (these are SMW satellite data).

Brasseur (1993) has pointed out that the model simulates satisfactorily the response (amplitude and phase) to the 27-day forcing of ozone and temperature in the stratosphere but does not properly explain the ozone and temperature response of opposite sign observed near 70 km altitude. The change in the ozone column abundance associated with the 27-day solar forcing is estimated at less than 0.5%. Variations in middle and upper atmospheric ozone concentrations and temperatures induced by solar variations on the 11-year timescale are not negligible compared to changes produced as a result of human activities over the same period of time. The calculated change in the total ozone from solar minimum to solar maximum conditions is of the order of 1.1–1.3% in the tropics and increases with altitude, especially in winter, to amount up to 1.5–1.7% in the polar regions.

Brasseur (1993) has emphasized the presence of large unexplained differences between the calculated and observed stratospheric responses. These discrepancies could involve chemical, radiative, as well as dynamical processes which are not yet fully understood. In this context three-dimensional numerical modelling is specifically required. It should be mentioned in this connection that according to Hood et al. (1993), a positive solar cycle variation of independently measured ozone and temperature occurs with maximum amplitude near the low-latitude stratosphere. This conclusion has been made on the basis of the analysis of the Nimbus–7 SBUV data.

A new effort to investigate the role to stratospheric ozone in modulating the solar radiative forcing of climate has been undertaken by Haigh (1994). The main aim of this study has been to substantiate a physical mechanism for the explanation of observed solar activity—climate correlations (Friis-Christensen and Lassen, 1991; Labitzke, 1987; Labitzke and Van Loon, 1992, 1993; Lean, 1991; Reid, 1991; Schlesinger and Ramankutty, 1992; Van Loon and Labitzke, 1993, and others). For this purpose a global (from pole to pole) two-dimensional radiative–chemical–transport model has been developed which shows that the spectral composition of the solar variations and the photochemical production of stratospheric ozone together lead to a highly nonlinear relationship between the extraterrestrial and cross-tropopause solar radiative flux.

Two model runs have been carried out to assess the solar activity impact: (1) a control run for conditions of minimum activity in the 11-year solar cycle; (2) a perturbation run for observed solar irradiance during the maximum activity. This perturbation corresponds to a 0.08% increase of the total SC between solar minimum and solar maximum.

The change of spectral solar irradiance is most pronounced in equatorial latitudes. At the wavelengths $\lambda < 242$ nm the increase of irradiance results in additional photodissociation of molecular oxygen, and then the resulting atomic oxygen combines with O_2 to produce higher concentrations of ozone. This in turn absorbs more radiation especially between 200 and 350 nm but also, importantly, in the visible portion of the spectrum. The result is a substantial change of the solar radiation spectral distribution in the lower stratosphere, depending on latitude (on the varying solar zenith angle, i.e., atmospheric mass).

The influence of the larger incident solar radiation produced higher concentrations of ozone at all altitudes and an increase of total ozone content between 1.2% and 1.8%. Because of this increase the enhancement of the incident solar radiation flux is only 0.10 W/m^2, or 0.025%, compared with the increase in actually less solar radiation at 100 hPa at solar maximum than at solar minimum, because of extra absorption under conditions of larger atmospheric mass (e.g., at 65°N there is a 0.05% reduction of solar flux).

With higher ozone concentrations the stratospheric temperatures have also increased, and in response to these changes in temperature and ozone concentration, the downward fluxes of thermal emission are also amended. The higher temperatures lead to greater emissions mainly by CO_2, H_2O and O_3, whereas increases in ozone concentration will also tend to increase the emission in the 9.6-μm band; these result in an increased downward flux of the order of 0.05 W/m^2 at 100 hPa.

While changes of solar radiation flux are inhomogeneous, in the case of thermal emission the assessments mentioned may be considered as typical on a global scale. The increase of the downward thermal emission flux at all latitudes will result in that at 100 hPa level the net radiative forcing decreases from ~0.2 W/m^2 at the summer pole fairly uniformly to ~0.05 W/m^2 at the winter pole. Such changes may be followed by a modulation of planetary-wave activity and stratospheric wind structure which has been observed (Labitzke and Van Loon, 1993). It is also important to point out that the magnitude of changes in radiative fluxes is of the same order as it has been discussed in the context of the greenhouse-induced changes (Section 2.5). Haigh (1994) has emphasized that this will tend to make an extraction of the different components from data records rather difficult. Although the latter conclusion is undoubtedly true, a comparison of contributions to climate changes due to the greenhouse effect and solar variability requires further analysis. An important aspect of the problem is the thermal inertia of the atmosphere–ocean system which might lead to time lags between radiative forcing and climate response that are much longer than a decade. Therefore, continuous accumulation of GHGs may be more important than the cyclic solar changes which are, on the average, close to zero over the whole 11-year cycle.

According to Lean (1991), there is an increase of about 0.04% of solar irradiance in the wavelength region 750–1100 nm (which contains about 25% of the total solar constant) from solar minimum to solar maximum but a compensating reduction of down to 0.03% in most of the remaining quarter of the SC. This means that changes in the half of the solar spectrum ($\lambda > 730$ nm) which were not taken into account in the estimates made by Haigh (1994) does not contribute significantly to changes in the radiative forcing of climate on the solar-cycle timescale (this conclusion requires, however, further confirmation). Another factor which has been ignored by the model discussed are changes in tropospheric composition and temperature, in response to variations in solar fluxes, which can affect the upward thermal emission flux at the tropopause level.

The principal conclusion from simulation modelling accomplished by Haigh (1994) is that the solar radiative forcing of climate depends crucially on the spectral composition of the solar variability. The large observed changes at UV wavelengths cause variations in ozone concentration which can mask the lower atmosphere from variations in the overall flux and even, at winter middle and high latitudes, reduce the radiative forcing at solar maximum relative to its solar minimum value. The degree of this masking depends sensitively on solar variability in the wavelength regions 340 nm $< \lambda <$ 500nm and $\lambda >$ 750 nm which are little affected by ozone. The ozone and concomitant temperature changes in the lower stratosphere also affect the longwave radiation field.

In connection with observed correlation between solar-cycle length and atmospheric parameters (Friis-Christensen and Lassen, 1991) it would be important to explore potential interrelationship between solar spectral variability and cycle length. Haigh (1994) has emphasized that in the light of the results considered, two more conclusions follow: (1) simple energy-balance calculations could well be misleading; (2) there is a clear requirements for monitoring the entire solar spectrum over periods of decades.

To summarize the present-day knowledge concerning the impact of solar activity on climate, it is enough to quote Roederer (1996): "The topic 'Sun–weather relationships' has followed a long and fortuitous history of scientific speculation and controversy since the continuous observation of sunspots began in the 17th century. It was not until recently that more systematic studies with long-term data bases of meteorological, climatological and solar parameters led to an increasing, statistically robust, body of evidence for a causal connection between some manifestations of solar variability and changes in the troposphere and the climate system ... Most likely, several trigger mechanisms are at work simultaneously. A scientific understanding of possible external influences, however minor at first sight, is of crucial importance. We are still very far from achieving this goal".

2.7 ON A PERSPECTIVE CONCEPT OF CLIMATE STUDIES

A most important result from preparations of the IPCC Report and WCC–2 is a deviation from the stereotypes mentioned above, a clear understanding of serious uncertainties of the numerical climate modelling (whose initial results have served the basis for these stereotypes), and the need for radical improvements of the existing climate models. Since there is no substantiated conception of the development of climate studies for the period of the subsequent one to two decades, here are some ideas proceeding from the following key directions.

2.7.1 Introduction

An accomplishment of a new stage of the numerical modelling of the dynamics of the coupled climate model "atmosphere—ocean—land—cryosphere—biosphere" considering the fact that this system is open, that is, it is subject to external cosmic forcings (first of all, the solar forcings due to variabilities of the solar constant and solar activity is most difficult. In this connection, it is important here to consider the biosphere as an interactive component of the climate system. First steps in this direction have been made already (Kondratyev, 1992d,e; Kondratyev and Cracknell, 1998a,b): approaches have been proposed and realized to parameterize the biosphere from the viewpoint of taking into account the heat, moisture, and gas exchange between the atmosphere and the vegetation-covered surface. It was important to accomplish the programme of the complex field experiment FIFE together with a "simple" model of the biosphere (Sellers et al., 1996a–c).

The next "biospheric" stage is to generalize the climate theory through a coupled consideration of the global carbon cycle. Such a development of the numerical climate modelling might serve a sound basis for the CO_2 variability forecasts for the future, which at present are reduced to a set of rather conditional "scenarios" of the concentration growth. Of course, this generalization of the climate theory is very complicated. Solution of this problem is favoured, however, by a number of key projects within the WCRP and IGBP programmes discussed in Chapter 1 (first of all, the Joint Global Ocean Flux Studies, JGOFS) aimed at investigating the global carbon cycle may serve a sound and organic basis for the interaction of two large-scale programmes: WCRP and IGBP.

Further development of climate theory should foresee a coupled account of the cycles of sulphur and nitrogen. As for the sulphur cycle, of special interest is a simulation of gas-to-particle reactions of the conversion of gaseous sulphur compounds into sulphate aerosol (Kondratyev, 1991, 1992d). With reference to the stratospheric aerosol, it is, first of all, the problem of conversion of the volcanic sulphur dioxide to H_2SO_4 aerosol. The varying emission of dimethylsulphide by the ocean (due to phytoplankton dynamics) puts forth quite a new problem of taking into account not only the gas-to-particle mechanism for the formation of the sulphate condensation nuclei but also their further impact on the dynamics and optical properties of the cloud cover which may be an important factor of the ERB variability (hence, the climate). Here, again, there is a basis for concrete WCRP/IGBP interaction within the problems of biogeochemical cycles—the thematic gist of the IGBP. Thus, the first component of the perspective conception of the WCP (and, mainly, WCRP) is a step-by-step accomplishment of the "biospheric" stage of the numerical climate modelling, which, in its turn, will provide the development of realistic models of the biosphere.

2.7.2 Problems of observing systems

A substantiation and accomplishment of the Global Climate Observing System (GCOS) (Kondratyev, 1990a, 1992c,d; Kondratyev et al., 1992a, 1996). The WCC–2 participants supported unanimously the need to create the GCOS as a system to monitor climate, to detect climate changes, to monitor the responses (first of all, of the surface ecosystems) to climate changes. Also, the goals of the GCOS are to serve the needs of economic development: to provide it with climate data; to get a deeper insight into and to improve the techniques for numerical modelling and forecasts of the climate system's dynamics. The main components of the GCOS should be: (a) an improved programme of the World Weather Watch (WWW); (b) Global Ocean Observing System (GOOS) which would provide observations of the characteristics of the physical, chemical and biological processes in the ocean: (c) Global Terrestrial Observing System (GTOS); (d) carried out and improved programmes to monitor various key components of the climate system, which provide data on changes in the chemistry of the atmosphere and terrestrial ecosystems, cloud cover and water cycle, the ERB, ice cover and precipitation over the World Ocean, etc.

In this connection, of special interest is the Global Atmospheric Watch (GAW) functioning from 1989 under the auspices of the WMO on the basis of combined global ozone observing system and a network of the background monitoring of atmospheric pollution. The main objective of GAW is to provide regular observations of: (a) the content in the atmosphere of most important GHGs (carbon dioxide, CFCs, methane, nitrous oxide, tropospheric ozone); (b) total ozone content, its concentration near the Earth's surface and its vertical profile, as well as the content of gases—the ozone precursors—with the use of various approaches and instruments (spectrometers, sondes, lidars, microwave radiometers, including satellite observations); (c) surface radiation budget and its components, optical thickness (or transparency) of the atmosphere, meteorological visibility, UV radiation, total content of aerosol and water vapour in the atmosphere; (d) chemistry of liquid and solid precipitation; (e) the content of chemically active components, such as sulphur dioxide and various forms of nitrogen compounds, carbon monoxide; (f) number density and composition of aerosol (including mineral aerosol); (g) the content of radionuclides, including krypton–85, radon and tritium, as well as the isotopic composition of individual atmospheric components. A realization of GAW should make an important contribution to studies of the dynamics of the atmospheric greenhouse effect, global ozone layer, and acid rains.

A substantiation and accomplishment of GCOS is an extremely complicated problem. In this connection, one of the first-priority objectives is to detect and attribute anthropogenic climate changes (Borisenkov and Kondratyev, 1988; Kondratyev, 1992d). Here the difficulty is explained by lack of the information needed: comparatively long observation series are available only for SAT, but in this case an interpretation of data is sometimes contradictory. The anthropogenic SAT variations can be recognized only when either of the two conditions is met:

- if the SAT variation reaches an amplitude comparable with (or exceeding) the range of natural SAR variations;
- if progress in understanding the mechanisms for natural variability provides their reliable recognition and forecast, that is, a reliable assessment of the climatic "noise".

An accumulation of reliable global information on SAT and SST remains an important problem. The combined conventional and satellite observations of the cloud cover characteristics, parameters of the ERB and the atmosphere, bearing in mind the special role of the cloud–radiation interaction as a cloud-forming factor are one of the key GCOS components. It is this task that is in the focus of the EOS system and CLIMSAT satellites (Kondratyev, 1992h) together with the objectives of the GEWEX programme mentioned above. The key tasks include also observations of solid and liquid precipitation as well as obtaining diverse hydrological and oceanigraphic information. Serious attention has been paid to the idea suggested by J. Woods (Great Britain), J. Baker (USA) and R. Stewart (Canada) to develop and create a Global Ocean Observing System (GOOS). The main idea of this system consists in the combined use of satellite and robotized conventional means of observations covering the World Ocean (Allison and Moritz, 1995; Scientific Design . . . , 1995). The allocation of this system is planned to be completed by 2010, after which it will be able to function in the operational regime providing climate forecasts (of course, it should be borne in mind that the problem of climate predictability remains unsolved) in the interests of navigation, fisheries, and other economic activities (the zone of shelf claims particular attention).

Though the GCOS foresees a sufficiently complete monitoring of the global climate system "atmosphere—ocean—land—cryosphere—biosphere", a large group of specialists has initiated a substantiation of the global observing system to detect and monitor the response of ecosystems to global environmental changes (Heal et al., 1993; Kondratyev, 1992d–h; Kondratyev et al., 1988, 1992a, 1996a,b). The main goal of discussions relevant to GTOS was to reach a consensus with respect to: main principles which would serve the basis for the design of the Global Terrestrial Observing System; approaches to choose 50–100 first-priority (key) regions to be monitored, which would characterize a representative set of natural and cultivated ecosystems; strategies of an optimal combination of the periods of routine and intensive observations using *in situ* measurements and remote sounding techniques; observational databases needed for an adequate description of the dynamics of ecosystems and respective forcings; assessment of data quality and data management; organization and structure of the management systems at the national, regional and global levels.

Heal et al. (1993) believe that the GTOS should be realized under the auspices of UNEP, with an active participation of FAO and Unesco. The GTOS should be a main component (with respect to land surface observations) of GCOS, in close contact with developments within the IGBP core projects, such as GCTE and START (at present, the problems of the global environmental monitoring are of keen interest for more that 80 international organizations and programmes). The GTOS should be accomplished on the basis of a system of 311 biospheric reserves existing within the Unesco programme "Man and Biosphere" (their areas vary from 70 000 km^2 in north-eastern Greenland to local scales $< 5\ km^2$), the main tasks of which are: (1) to preserve the biodiversity; (2) to substantiate the stable use of land and water resources; (3) to substantiate the network for studies and monitoring on an international basis.

The main factors of the impact on terrestrial ecosystems are changes in: (1) atmospheric chemistry (CO_2 concentration, in particular); (2) climate; (3) land use (under the influence of demographic, economic, technological and social changes); 4) environmental pollution. An important basis for an optimal planning of the monitoring systems should be results from the numerical modelling of the ecosystems' dynamics over a wide range of spatial and temporal scales: leaf—plant—vegetation cover—landscape—biome—biosphere. For all these scales, studies of the biogeochemical cycles are of particular importance. A serious problem is the validating of global biosphere models, since empirical data are used for model calibration and, thus, the models are "tuned" to the present biosphere (this is typical of the climate models). Special observations at "independent" key sites (KS) can provide a partial solution of the problem. To obtain the data needed, a three-level monitoring network is planned:

(1) The KS for the intensive monitoring of many parameters of the ecosystems with a high temporal resolution for detailed studies of the processes and substantiation of their adequate models. The KS of this type should be based on the available observing points for the respective studies.
(2) The KS (50–100 in number) for less intensive monitoring, chosen at locations of strong contrasts (gradients) of the environment (at distances 100 km and farther), which will make it possible to obtain data to validate the models of the terrestrial ecosystems. The contrasts (ecotones) may be exemplified by the transitions "tunra–boreal forests", "savannah–desert", and others. Observations should be planned with account of the factors such as phenology, local climate, soil state, heat and moisture exchange, and others.
(3) The KS for the extensive monitoring at as many points as possible to extrapolate (interpolate) the results of detailed observations at the KS of the first and second levels.

In all cases, studies of biomes, ecotones and different types of gradients should be of first priority. In this connection, it is important to substantiate the requirements for the remote sounding data. Each KS should meet the following requirements:

(1) Possibility of carrying out observations for 50 years to come.
(2) The organization responsible for the KS should provide the needed complex of observations needed and the possibility of working on the following one or several basic priorities:
    (2.1) Biodiversity and changes in the composition of the ecosystems.
    (2.2) Changes in agricultural practice.
    (2.3) Changes in ecological fuctions.
    (2.4) Changes in landscapes and land use.

The requirements determining the priority of KS are as follows:

(1) The novelty of environmental conditions (climate, soil, land use, biodiversity) compared to the existing KS.
(2) Minimum expenditures.
(3) Importance for solving the problems of the socio-economic development (Earth resources management, including) and support at a local level.
(4) Location in a region of first priority for the participating organizations.

The GTOS programme (Kondratyev, 1992d, 1994a,c,d) contains a detailed substantiation of the requirements for three categories of the observational data (bearing in mind, mainly, the KS of the second level) aimed at: (1) general characteristics of KS (this requires only sampled and episodic observations); (2) solving the major problems of the programme; (3) obtaining various additional information. The main database of the second category includes the following sections (with a detailed enumeration of the parameters to be measured and acceptable errors of measurements): general characteristic of the KS and environmental conditions typical of it; climate; soil; "underground waters, river runoff"; vegetation cover; nutrients dynamics; the animal kingdom; the CO_2 exchange between the soil-vegetation cover and the atmosphere; the landscape composition. The emphasis is laid on the agroecosystems and biodiversity. In this context, close cooperation is needed within the programme DIVERSITAS and GCTE.

The main objectives of the international programme DIVERSITAS worked out in 1991 and supported by SCOPE and Unesco are connected with studies of the ecosystem functions of biodiversity: origin, maintenance and loss of biodiversity as well as with its inventory. Whereas the main interests of the programme DIVERSITAS are in studies of the groups of organisms (plants, animals, microorganisms), the main purpose of the key project of IGBP—GCTE (Global Change and Continental Ecosystems)—is an analysis of the effects of global changes on the biodiversity dynamics. Thus, these programmes can be considered complementary. +

Naturally, of practical importance for the successful accomplishment of the Programme is collecting, processing, storing and distributing the observational data. This will require local, regional and main data centres.

By the end of 1992, 53 institutions from different countries had participated in the accomplishment of the programme GIM (Global Integral Monitoring) under the auspices of UNEP. The conception of a set of global observing systems supported by FAO, Unesco, ICSU, WMO, UNEP, and IOC (International Oceanographic Commission) will be based on an interaction of three global systems: GCOS (basic system): atmosphere, ocean, land, cryosphere; GTOS: land (biodiversity, cultivated ecosystems, structure and functions of continental ecosystems, land hydrology); GOOS: ocean (processes in the coastal zone, pollution, marine bioresources, navigation and fisheries, mining).

An assessment of these programmes suggests the following conclusions:

1. As has been mentioned above, diverse developments to substantiate the global-scale programmes of observations of the biospheric and environmental dynamics are still not concrete enough from the point of view of their problem orientation. This refers also to GTOS which does not foresee, for example, a solution for the cardinal problems such as closed nature of global biogeochemical cycle and the role of biodiversity as the basis of global ecological sustainability. The programme GTOS justly lays emphasis on developing remote sounding techniques, but it must be remembered that only an optimal combination (and relevant planning) of the complexes of conventional and satellite observational means will lead to the accomplishment of the programme. In connection with the justified importance of

the problem of biodiversity in this programme, the prospects for using the remote sensing techniques to monitor the biodiversity deserve special attention (Kondratyev, 1993e; Kondratyev et al., 1990, 1993).

2. An aspect of conceptual importance is a multitude of various programmes of the global ecological monitoring proposed recently and supported by various international organizations. This suggests the conclusion on the needed optimization of both the available programmes and (which is more important) a system of respective international organizations, including the governmental (at the level of specialized UN organizations) and non-governmental (the latter refers, first of all, to the International Council of Scientific Unions, ICSU). Estrangement and poorly coordinated efforts, and excess scientific conferences at different levels to (sometimes) the detriment of the work, lead to unjustified expenditures of both the forces and the means. Of course, an optimization is an extremely complicated problem and requires cooperation to search for means of optimization.

Probably, a good model of optimization (with the concentration of the intellectual resources) could be the World Climate Programme (bearing in mind, first of all, the WCRP) in which the problem-oriented developments are most profound, which also determines a more concrete substantiation of the GCOS. The next stage could be the solution of the problems of the biospheric dynamics which (to some extent) are part of the WCRP.

2.7.3 Climate modelling problems

In conclusion, once more about climate change problems. Since the climate prediction is considered as a solution of the problem with initial data, it should give the time series of the statistical characteristics of climate for each region of interest (for instance, with a spatial averaging over $5° \times 5°$ lat.-long., with a 1-month interval for a period of 100 years using the regularly reviewed initial data). A simplified approach consists in prescribing a database for one year (with a characteristic annual change). In this case one could issue an annual climate forecast 100 years ahead, similar to the regularly published annual economic forecasts. This analogy is not accidental, since the data of the socio-economic forecasts serve the basis for assessing the contribution of anthropogenic forcings. The "anthropogenic" boundary conditions should contain information on the dynamics of land properties and water use, the rate of GHG emissions and the atmospheric pollutants reacting with them.

As for the limits of climatic predictability, they have been poorly studied and, apparently, depend on special features of the predicted parameters. Probably, the useful predictability of the climate system 100 years ahead is possible. Apparently, the accuracy of the forecasts will depend on an adequacy of the models and reliability of the initial data. The initial data should include: (1) an improved land surface monitoring system; (2) a renewed WWW (together with GAW); (3) a completely new system to monitor the World Ocean. If the errors in the prescribed initial state are sufficiently small and an icreaase of the errors in the dynamics of the ocean is not too fast, this will make it possible to obtain useful forecasts. The respective estimates have shown, for example, that to provide $S/N = 5:1$, the atmosphere—ocean heat exchange averaged over a year for the North Atlantic should be assessed with an error not more than 1 W/m^2, which can be reached only with the use of new measurement approaches (similar estimates made earlier had an accepted error of about 10 W/m^2).

The need to resolve the mesoscale structure of the oceanic dynamics (synoptic eddies, etc.) greatly complicates direct measurements over the whole ocean with the use of various *in situ* techniques, necessitating the choice of key regions and quantities, including: the seasonal boundary layer, gyres and currents; areas with concentrated kinetic energy, fluxes of heat, water and chemical components, inter-basin fluxes. Unfortunately, the methodology of recognition of the World Ocean's energy-active zones is not yet known enough (Marchuk, 1975; Energy-Active Zones..., 1989). Therefore, for the time being, the following regions were recommended to concentrate the observation means to study the seasonal boundary layer: (1) the Arctic Ocean (monitoring the processes which control the ice cover dynamics); (2) the Norwegian and Greenland Seas (formation of deep waters); (3) the high-latitude North Atlantic (formation of deep waters); (4) the Antarctic sea belt (formation of bottom waters); (5) the equatorial Pacific (El Niño); (6) the tropical Indian Ocean (the Somalia current and monsoon processes). Studies of the following key mesoscale processes claim special attention: (1) an interaction among the non-stationary eddies, persistent currents and bottom topography for horizontal scales of the order of several tens of kilometres; (2) an interaction between the elongated eddies in the non-stationary jet streams and the biological "pump" on the same scales. So far, there have been no adequate techniques to parameterize such processes. However, the tests of first high-resolution (with a grid step of ~25 km and 70 levels vertically) simulation models of the ocean have demonstrated that, on the whole, they successfully simulate an interaction among the eddies, persistent currents and bottom topography, though somewhat higher resolution would be worthwhile. An account of the nonlinear dynamics of the eddies has provided a much more adequate description of circulation in the ocean.

A principally new step towards further development of the OGCM consists in accounting for the chemical and biological processes with emphasis on the processes with participation of plankton and hydrosol which regulate the atmosphere—ocean carbon exchange. An important task is to provide the coordinated relevant developments within the programmes GOOS, TOGA, WOCE (World Ocean Circulation Experiment) as well as the IGBP core projects JGOFS (study of the oceanic component of the global carbon cycle) and GOEZS (study of the processes in the euphotic zone). The total programme of observations can be classified into five components: temperature and salinity (the hydrographic component); nutrients, concentrations of dissolved oxygen and geochemical tracers (the chemical component); the phase transformations of water: ice, precipitation, evaporation (the water component); an external forcing of wind stress, the atmosphere–ocean heat exchange, insolation, heat flux through the sea bottom (the "flux" component); the phytoplankton concentration in the surface layer (the biological component).

In connection with the problems discussed, of interest is an experiment (headed by W. Munk) on the acoustic sounding of the World Ocean proposed by the Scripps Institution of Oceanography (USA) with the participation of some other institutions, with the final aim of detecting global warming and the CO_2 climatic signal from the data on changes in the velocity of sound. During a 9-day phase of the experiment it was planned to use a small ship lying at anchor near Herd Island (an Australian island in the southern part of the Indian Ocean), to immerse a source of sound to the depth of $\sim$250 m at frequencies of 50–70 Hz with an intensity of 209 dB. The geographical location of this source is such that its waveguide-propagated sound can be recorded in different regions (the sub-water microphones supposed to be mounted on the station Mawson in the Antarctic, near the Bermudas and San Francisco, along the coastlines of South Africa, India, Australia, new Zealand and Canada). Based on analysis of the results of the tests, further decisions will be made.

An urgency of the accomplishment of GCOS contrasts with the lack of a quantitative and complete enough substantiation of the requirements for the observational data from the point of view of the priority of the parameters to be measured, acceptable errors, needed spatial and temporal resolutions, optimal combination of conventional and satellite observational means. When developing the WCP and WCRP programmes, serious efforts have been made to solve these problems (Kondratyev, 1982, 1992d; Kondratyev et al., 1992a, 1996b).

An important work on substantiation of the requirements for satellite data has been made by the COSPAR WG–6 (see Kondratyev, 1983a, 1992d; Kondratyev et al., 1996b), certain progress has been achieved in solving the problem of optimization of a set of spectral channels of the space-borne instruments (Kondratyev et al., 1992a, 1996b). Of topical value is the conception of the energy-active zones proposed by G. Marchuk (Marchuk, 1975; Energy-Active Zones . . . , 1989) as a methodology to reveal the global regions most substantial for the formation of climatic variability, as well as to optimize the combination of conventional and satellite means of observations. However, on the whole, the problem remains unsolved.

The difficulty of the situation is vividly illustrated by practically total lack of numerical experiments on climatic sensitivity to various factors and processes, whose results could have been used, in particular, to substantiate the priorities and to estimate acceptable errors. The related ideas are fragmentary and, as a rule, unconvincing. Of course, the discussed numerical experiments are expensive and computer-time consuming. But an incomparably enormous material outlay is connected with the creation of space-borne observation means (by initial assessment, the cost of the American EOS programme—the Earth Observing System—could reach about 30 billion dollars), and their non-optimal character will lead to more serious losses compared to expenditures on developments of the methodology of planning the observing systems. Therefore the scientifically grounded approach to the GCOS optimization is an extremely important problem.

2.8 CLIMATE DYNAMICS IN THE CONTEXT OF GLOBAL CHANGE: THE SOCIO-ECONOMIC ASPECTS

2.8.1 Introduction

As has been mentioned above, the Second UN Conference on Environment and Development (UNCED) was completed in June 1992 in Rio de Janeiro. Preparations for this conference tookn a lot of effort and expense. Particular working meetings were held on developments of framework conventions on climate (United Nations..., 1992) and biodiversity, as well as four sessions of the UNCED Preliminary Committee. This effort produced voluminous materials and documents. Of great importance is an analysis of the results of UNCED which has been an unprecedented event both in the range of the problems discussed and in the top-level representation of the countries—at the level of Presidents or Heads of Government.

The remarkable role of UNCED as a World Forum raises no doubts: it drew attention to urgent problems of global ecology characterized by various closely interrelated aspects connected with conditions of the socio-economic development and environmental dynamics. Closing UNCED, Dr Maurice Strong, Secretary-General of the Conference, noted (The Spirit of Rio..., 1992) that "first, the greatest international conference in the world has been a success. Needed people participated in this conference, and peoples of the world kept up with the conference with a keen interest. After this conference the world will not be the same. Diplomacy will not be the same. United nations will not be the same. And prospects for out Earth cannot ... should not ... be the same ... Finally, this decade should become a transition period, which will lead us on a new road, to a new economics—to the creation of an eco-nomic system".

It has been pointed out that the principal achievement of UNCED was the acknowledgement of the fact that the problems of the environment and development cannot be considered separately. The conference has convincingly demonstrated an interrelationship between the problems of poverty and a low level of development, on the one hand, and ecology, on the other hand, considering different standards of living in the North and in the South. The energy consumption and climate, the trade in tropical wood, the problem of poverty, and the increasing effect of population growth and desertification—all these aspects of global and regional ecology have claimed unprecedented attention. Extremely important has been the Global Forum held at the same time, supported by almost 9000 non-governmental organizations (29 000 participants, 450 000 visitors, and about 1000 meetings).

The Earth Charter ought to have become the fundamental document of the conference. Unfortunately, it has been replaced with another, rather vague and rhetorical document—the Rio Declaration (Kondratyev, 1993a). The reason is that the Preparatory Committee has failed to reach a consensus on the content of the Charter, whose principal objective should have been a well-reasoned statement on priorities in the field of ecology.

The second, most important document "Agenda 21", rather voluminous and expensive from the viewpoint of its realization, has been aimed at formulating practical measures needed to ensure a stable socio-economic development. Of course, such a document could not have been prepared properly without a constructive and clear Charter. This is the reason why such most important problems as the dynamics of population, the structure of consumption, the debt of Third-World countries, export of harmful wastes and deforestation have not been actually touched. Note, that the annual expenses of the realization of the "Agenda 21" recommendations have been estimated at 21 billion dollars. Such has also been the fate of the conventions on climate and biodiversity (Kondratyev et al., 1994).

The concern for the situation had been expressed before UNCED. This had been reflected, for example, in the documents prepared by non-governmental organizations of the conference "The Roots of the Future" held on 17–20 December 1991 in Paris called "Justice among people—justice among generations" and "Ya. Vananchi Agenda", with emphasis on interrelationships among the problems of ecology, economy, law, and policy. Spivy-Weber (1992) explained this interrelationship as applied to the problem of deforestation and convention on biodiversity. Marton-Lefevre (1992), summarizing the results of the International Conference on the UNCED Scientific Aspects (ASCEND) held in late November 1991 in Vienna, emphasized the necessity of a system's approach, enumerating the following key aspects of the problem: the dynamics of population; per capita consumption; degradation of soil and agricultural resources; social inequality and poverty; climate changes; a decrease of biodiversity; industrial pollution of the environment; freshwater resources shortage; energy production and consumption. Similar considerations have been expressed by Hadj-Sadok (1992).

It is clear that an extreme urgency of the global-scale importance of some ecological problems stimulate discussion on ways of raising an efficiency of international institutions which provide a coordination and cooperation of efforts. Radical ideas on a reorganization of available international organizations within the UN system have been expressed by Vavrousek (1992), who suggests, in particular, that a World Environment Organization be set up. Note that similar considerations have been expressed earlier (Kondratyev, 1991b).

As for the Framework Convention on Climate Change, it will not be difficult to reach an adequate consensus, since its development had been preceded by successful efforts of the IPCC, producing an extensive report which later was discussed and approved by WCC–2 (Climate Change..., 1990; Kondratyev, 1991a–c, 1992d). Unfortunately, the final outcome has been substantially biased by certain overemphasis on the problem of the greenhouse climate warming. Now we shall discuss this problem in more detail, although it should be mentioned that recently fundamental Climate Change 1995 reports have been published (Climate Change ..., 1995; 1996a–c).

2.8.2 Global climate change and the problem of GHG reduction

Speaking about the global climate we shall first mention the stable stereotype of "global warming" which has been persistently propagated both in scientific literature and in the mass media over the last 10–15 years. This is particularly strange because the IPCC Reports (Climate Change..., 1990; Climate Change: Science..., 1991; IPCC Reports, 1992, 1994) have convincingly dispelled this stereotype, noting that:

(1) the causes of the global warming taking place in this century are still not clear enough (this explains the priority of studies of the internal variability of the climate system instead of the one-sided exaggeration of the role of anthropogenic factors); (2) forecasts of the future climate based on palaeo-analogues are incorrect. The latter means that so far, there has been no reliable methods of climate forecasts. In this connection of particular interest is the problem of unexpected climatic dynamics (and the environment in general) and analysis of climatic "surprises" in the past (Kondratyev, 1992d).

Wielicki et al. (1995) have justly pointed out: "In the next century, the earth faces the potential of rapid environmental change. The magnitude, regional distribution, and timing of global environmental change could have a profound societal impact, but it is not yet possible to provide policy makers with unequivocal answers to vexing questions about these effects, due in large part to a lack of knowledge of the interdependent processes that affect regional and global climate".

The basic conclusion made recently by Vinnikov et al. (1996) is that "the climate change of the past three decades cannot be attributed solely to natural internal climatic variability, and is likely to have been caused at least partly by human activities".

Wallace (see Kerr, 1996) has justly pointed out: "I am a bit skeptical of the idea that every time we see something in the recent record that we have not seen before, it must be due to greenhouse warming".

It is obvious, thus, that "the balance of arguments" (terminology mentioned earlier in connection with the controversial statement in the IPCC–95 Report) demonstrates that what we need is to study climate change in all its complexity with an adequate account of both natural internal variability and various external impacts (including an anthropogenic contribution).

Now, we shall discuss, however, various approaches to solving the problem of the reduction of the GHGs emissions which has attracted a great deal of attention.

To substantiate the countermeasures and related climate policy to prevent undesirable GHGs induced climate changes, Jain and Bach (1992) developed a 1-D energy-balance climate model which combines an account of energy-economic and biochemical-climatic aspects of the problem. Calculations for 240 years (1860–2100) have been made with a time step of one year.

For the period 1860–1985, actual data on GHGs trends were taken, and calculations for the period 1985–2100 were based on prescribed different scenarios of changes in the GHG concentrations. Comparisons were based on a case of extrapolation of the present GHGs trends (BAU: business-as- usual scenario). The GHGs under consideration include: CO_2, CH_4, N_2O, CFC-11, -12, -113, -114, -115, HCFC-22, halon-1301, CCl_4, CH_3CCl_3, tropospheric ozone and stratospheric water vapor. Maximum acceptable values of the global climate warming 0.1°C/10 years or 2°C over the period 1860–2100 were prescribed as criteria for countermeasures to reduce the emissions of GHGs..

The climate model has been supplemented with models of the global carbon cycle (three reservoirs: atmosphere, mixed layer and deep layers of the ocean with account taken of interaction with the biosphere); a simple balance model to pre-calculate concentrations of CO_2, halons and N_2O; a model of chemical reactions determining the concentration of methane. The models enumerated provide the input data on GHGs concentrations for the climate model. One more model is designed to estimate the World Ocean's level with account of: (1) thermal expansion; (2) melting of mountain glaciers; (3) ablation of the Greenland ice sheet; (4) ablation or accumulation of the Antarctic ice sheet.

Testing the reliability of this climate model by comparing calculated and observed secular trends of SAT as well as the sea surface level has given, on the whole, positive results. An acceptable agreement has been obtained when comparing with data for a 3-D global model developed in the Max Planck Institute for Meteorology (Hamburg). Thus, the conclusion can be drawn that the model can be used to substantiate countermeasures to reduce the GHGs emissions and to carry out the respective climate policy (of course, in case the simulation modelling results are adequate).

According to the results obtained, unreduced levels of GHGs emissions will lead to a rapid increase of mean global SAT, which can be prevented only through countermeasures to reduce the GHG emissions. A reduction of emissions of CFCs and halons is most expensive. Of importance, though not decisive, is the preservation of tropical forests and afforestation, which will enable one to reduce total CO_2 emissions to the atmosphere.

A most efficient countermeasure would be to reduce industrial CO_2 emissions expected, first of all, from industrial countries. An interesting plan to reduce the GHG emissions has been suggested by the Enquête-Commission (Germany) which foresaw an individual approach to various countries depending on the level of their development (Table 2.39). As is seen, the developing countries are not supposed to reduce the emissions, but eventually, by the year 2100, a 50% cut of global GHGs emissions is planned. It means that with respect to the BAU scenario of preserved volumes of emissions, the strategy of emission reduction by this scenario ensures a reduction of the level of mean global SAT within 0.9–1.7°C (depending on the assumed sensitivity of the climate system to a CO_2 doubling). A realization of the countermeasures under discussion will also prevent a dangerous increase of the level of the World Ocean. Of course, it must be remembered that the results considered have been obtained on the basis of a highly simplified model.

A new approach to the analysis of the conflicting requirements of the FCCC has been suggested by Enting (1995) who has pointed out that:

- The requirement of limits to the acceptable amount of climatic change may imply a restriction on the maximum allowable CO_2 concentrations (C_{max}).
- The requirement that stabilization should "allow ecosystems to adapt naturally to climate change" (as the FCCC states) represents a potential constraint on the rate of growth of CO_2 (G_{max}).
- Alternatively, the constraint on CO_2 growth may be most appropriate as a constraint on the rate of change of radiative forcing;
- Economic development may require avoiding reductions of anthropogenic emissions below some prescribed minimum (E_{min}).

Table 2.39. Recommendations on a differentiated reduction of GHGs emissions (%) by various countries

Year	Industrial countries—level of economy				Other global	Developing	Mean
	High	Middle	Low countries	Average countries			estimates
1990	+5	+5	+5	+5	+5	+11	+6
1995	+5	+7	+8	+6	+10	+24	+10
2000	−10	−4	+5	−4	+15	+37	+4
2005	−30	−15	−5	−20	+25	+50	−5
2010	−37	−21	−11	−27	+15	+53	−10
2020	−50	−35	−25	−40	−5	+60	−20
2030	−60	−50	−43	−54	−22	+63	−30
2050	−80	−80	−80	−80	−55	+70	−50
2100	−80	−80	−80	−80	−55	+70	−50

- In addition, economic requirements are likely to constrain the achievable rates in reduction in emissions expressed in either absolute terms or as a constraint on the proportional reduction rates.

In this context, Enting (1995) has presented some illustrations indicating which combinations of upper bounds on concentrations and lower bounds on emissions are mutually inconsistent and which are consistent.

IPCC documents (IPCC, 1992, 1994; Houghton et al., 1996) indicate that the stabilization of CO_2 concentration on an acceptable level, defined by the Framework Convention, requires an eventual and sustained reduction of emissions substantially below current levels. Assessing relevant results, Wigley et al. (1996) have pointed out that the first conclusion of the IPCC (WG1) analysis, that meeting any of the prescribed targets will require emissions to decline eventually to levels well below today's, is robust. One cannot conclude from this analysis, however (as has been stated), that an immediate reduction in emissions is required if we are to stabilize CO_2 concentrations at 750 ppmv level or below.

In this context, Wigley et al. (1996) have examined alternative pathways for meeting the prescribed concentration targets and then considered both the economic (costs) and environmental (benefits) implications of choosing one concentration trajectory over another. New stabilization profiles (pathways of CO_2 dynamics) have been substantiated that explicitly (but only qualitatively) incorporate considerations of the global economic system, estimate the corresponding anthropogenic emission requirements, and assess the significance of the profiles in terms of global-mean temperature and sea level change.

A case study for the 550 ppmv stabilization conditions indicates striking differences between the WG1 analysis and new results in terms of the implied carbon intensity of the global energy system in the early decades of the next century. The revised pathways following BAU (business-as-usual) for 10, 20 and 30 years show that different cumulative emissions pathways diverge initially and then become nearly parallel as one approaches and moves beyond the stabilization point (AD 2150 in this case). Cumulative emissions are noticeably higher in the cases that follow IS92 scenario initially. Therefore Wigley et al. (1996) pointed out that "later emissions reductions allow greater total CO_2 production, particularly for higher stabilization levels. These cumulative emissions differences, not considered by IPCC, may have important economic implications".

As far as the economic aspect of the problem is concerned, Wigley et al. (1996) have emphasized that there are a number of reasons why concentration pathways with higher near-term emissions have lower overall mitigation costs. Such results should not be interpreted, however, as suggesting a "do nothing" or "wait and see" policy, because: (1) all stabilization pathways still require future capital stock to be less carbon intensive than under a BAU scenario; (2) new supply options typically take many years to enter the market place; (3) any available "no regrets" measures for reducing emissions should be adopted immediately.

Wigley et al. (1996) have calculated, how such climate indicators as global-mean temperature and sea-level rise are affected by differences in the pathways to stabilization at an atmospheric CO_2 concentration of 550 ppmv. The results indicate a noticeable impact of the choice of pathway. This conclusion depends critically, however, on how SO_2 emissions are assumed to change in the future. A potential interaction between CO_2 -induced warming and SO_2 -induced cooling still remains a highly uncertain aspect of the problem.

Peck and Teisberg (1995) have substantiated a specification of warming cost which was incorporated into an integrated assessment model of global climate change. In contrast to the conventional representation of warming costs as a determinist function of temperature rise, they represent the cost of global warming as large losses which occur with small probabilities that depend on the amount of temperature rise.

A new approach not to provide quantitative monetary estimates of costs and benefits of optical CO_2 emission strategies to assist decision makers, but to clarify the basic input assumptions and cause-and-effect relations has been substantiated by Hasselmann et al. (1996). An important conclusion of the study is that, since the global warming response for CO_2 emissions extends over several hundred years, the costs associated with the climatic impact of present and future CO_2 emissions must be optimized over the horizons far beyond normal economic discounting time scales. Hasselmann et al. (1996) have emphasized that the implementation of a long-term monitoring and continually returned regulatory policy requires more realistic modelling tools than are at present available. In this context the conclusion made by Zwick (1995) deserves serious attention: "The occurrence of natural oscillations on an interdecadal/inter-annual time scale shows that the climate system is more complex than thought of. The natural variability had been underestimated".

A most serious gap in the IPCC Report (1990), the second and third parts of which contain a detailed consideration of probable impacts of the potential climate warming on different aspects of human activity and natural environment, is the lack of an adequate analysis of the socio-economic reasons for changes in the environment and in the biosphere, which have caused an increase of GHG concentrations in the atmosphere and an enhancement of the greenhouse effect able to cause serious changes in the global climate and subsequent ecosystems changes (Hänninen (1995) has emphasized certain problems in assessing the latter). Another important oversight in the IPCC Report is connected with an isolated consideration of the problem of climate change, apart from the global changes of the environment, on the whole. At present, such a one-sided approach is unacceptable (Kondratyev, 1982, 1990a, 1991a, 1992d). Largely because of this, as well as because of neglecting the socio-economic factors of the present global changes, the role of the increasing GHG concentrations has been exaggerated, and, in this connection, a reduction of GHG concentrations has been put forward as a most important countermeasure to prevent undesirable global changes of the environment, whose realization would badly affect the economics of the developing countries. Consider the closely related natural and socio-economic aspects of this problem.

As has been shown earlier (Gorshkov, 1995; Gorshkov et al., 1992, 1994; Kondratyev, 1993c,d) an account of the level of the consumption per capita of the biospheric resources and the fact that the acceptable level of global consumption had been exceeded soon after the beginning of the industrial revolution are decisive for the solution of the problem of ecological security. A global-scale contradiction has appeared and grown with time, consisting in non-equivalent per capita consumption in the more developed countries (MDC) and less developed countries (LDC), and without its solution it is impossible to provide the global ecological (hence, economic, political, military) security. One can agree with Agarwal and Naraian (1991), that there are grounds to believe in the existence of ecological (environmental) colonialism consisting, on the one hand, in exploitation of the LDC natural resources and, on the other hand, in desire to equally distribute the responsibility for the impact on the global biosphere and the environment among MDC and LDC. Naturally, this evokes a great concern of many developing countries, where coal is a major energy-carrier, and rice-growing and stock-breeding are sources of methane.

As has been mentioned above, the main ecological paradigm nowadays is the interdependence of countries, and in this connection, of crucial importance is the concept of a unified world which must but now cannot resist the present increase of the global level of consumption, with enormous contrasts of consumption in the developed and developing countries (interesting ideas concerning the possibility to forecast the socio-economic development of various components of the unified world in terms of Kondratyev's 54-year cycles have been expressed by Marchetti (1991)).

Ehrlich and Ehrlich (1990) noted that the birth of one child in the USA determines a 100-fold increase of the load on natural resources compared, for example, to a child born in Bangladesh. The socio-economic contradictions are in that the reason for cutting the tropical forests is the need for economic survival of the rapidly growing poor population of the developing countries, but, on the other hand, the wood obtained provides a high standard of life in the developed countries. The latter means that a rich man living at a distance of many thousand kilometres from the zone of tropical forests can (indirectly) produce greater destruction than the local population.

To understand the present global ecological situation, it is important, in particular, to answer two questions: (1) how reliable is the available information; (2) what are contributions of the trends of anthropogenic effects on the environment in different countries (Clark and Munn, 1986; Gore, 1993). Let us consider the answers to these questions with the data on tropical deforestation and GHG emissions as an example, confining ourselves to the concept of "greenhouse" warming.

As Agarwal and Naraian (1991) note, the available data on tropical deforestation and emissions of carbon dioxide and methane are contradictory. So, for example, the scales of cutting of tropical forests which reached a maximum in 1987, substantially decreased in the subsequent years. A similar process took place in India, where in the 1970s the forests were annually cut over the area 1.5 million ha, and in the 1980s the size of the annual cutting decreased to about 50 000 ha. In the industrial countries the forested areas decreased not because of cutting but owing to the damage by acid rains, the size of damage being still unclear, though for some regions such data are available. By the year 1986, more than one-fifth of the forests in western Europe had been damaged.

As for GHG emissions, from the data of the Indian Centre for Science and the Environment (CSE), there are no grounds for accusing India or China of being responsible for global emissions of carbon dioxide and methane to the atmosphere, since their contribution remains relatively small. On the other hand, proceeding from the fact that stock-breeding is a considerable source of methane, it is important to understand what measures should be taken to reduce the emissions of methane. Of course, one of the ways is to reduce the cattle population; but how is it possible?

The average per capita consumption of meat in the industrial countries is six times higher (78 kg/year) than in the developing countries (14 kg/year); in the former it has increased by 20% during the last 15 years, and in the latter it has remained unchanged. Thus, apparently, a decrease of the cattle population is only possible in the developed countries, which will lead to a 40% decrease of CH_4 emissions (due to the cattle) and a 50% decrease of the beef consumption (the norm corresponding to the balanced non-vegetarian diet is 30 g/day, whereas an actual level of meat consumption is several times higher), with the preserved level of the dairy produce consumption. An important resource for reducing the emissions of methane is pork production (pigs generate very little methane).

In this connection, it is important to note that, for example, Western Europe imports more than 40% of fodder corn from the Third-World countries, and deliveries of beef from the countries of Central America to make hamburgers in the USA are connected with an intensive tropical deforestation (and, respectively, an increase of CO_2 emissions). It means that limitations on beef consumption in the industrial countries can produce not only important positive ecological but also positive socio-economical consequences.

As for the developing countries, it is difficult for them to reduce the cattle population, not only because the cattle in these countries perform other functions apart from milk and meat production (first of all, the cattle produce organic fertilizers and are used as draught animals).

Estimates of CH_4 emission by the rice paddies are complicated and not reliable, being based on several data of experiments on some fields in the developed countries. Naturally, the data for Spain and Italy cannot be extrapolated to the developing countries. Observations in India show, in particular, that emissions of methane depend strongly on soil properties. If, according to the data of the World Resources Institute (WRI) based on extrapolation, the annual emissions of methane in India should reach 18 million tons (the IPCC Report-1990 suggests the range 7–49 million tons), the estimates obtained by Indian specialists turned out to be in the range 3–9 million tons. Of course, to obtain more reliable estimates, further studies are needed.

Explaining his dissatisfaction with the Framework Convention on Climate Change (first of all, from the viewpoint of its declarative character), Pachauri (1992) drew attention to the enormous potential of GHG reduction through improved technologies. Table 2.40 characterizes capabilities of various countries on the assumption that these countries will reach the level of effective use of energy existing in Japan.

Analysing the global situation with GHG emissions, it would be correct to obtain the differentiated estimates of the sources and sinks of GHG for each country, which would create the basis for the fair solution of the problem of emission reduction. Of course, such estimates should consider the national quotas in the GHGs balance with respect to the planetary-scale resources, such as the troposphere and the World Ocean. So far, there have been no persuasive attempts to solve this difficult problem. There are only some estimates, which are, however, of great interest.

Table 2.40. Potential of CO_2 reduction by reaching the level of effective use of energy existing in Japan (million tons of carbon)

Country	Level of emissions	Industry production	Energy	Cars	Total
USA	Now	202.17	448.14	324.84	975.16
	Future	73.39	410.60	280.74	764.73
Canada	Now	22.22	24.07	27.62	73.91
	Future	6.95	21.96	23.88	52.79
France	Now	26.05	3.52	28.44	58.06
	Future	17.25	3.07	25.22	45.34
Germany	Now	40.78	46.29	34.93	122.00
	Future	28.31	41.35	30.72	100.38
Italy	Now	22.09	24.74	23.95	70.78
	Future	14.13	23.25	21.07	58.46
Great Britain	Now	25.63	50.73	28.53	104.89
	Future	14.27	47.17	25.09	86.55
Others	Now	71.40	64.89	116.10	252.39
	Future	32.80	58.01	102.10	192.91
Total	Now	410.34	662.44	585.41	1657.19
	Future	187.10	605.41	508.62	1301.13

Table 2.41. The effect of the uniform reduction of CO_2 emissions on the economics of the developing countries. (MTC = million tons of carbon)

Country	Emissions in 1990 (MTC)	Expected emission in 2000 (MTC)	Reduction of annual emissions by 2000 (MTC)	Annual expenditures on emission reductions ($\$10^6$)	Annual national investments ($\$10^9$)
Bangladesh	4.24	8.87	1.18	17.7	2.42
India	160	248	20	300	56.24
China	588	964	89	756	150.12
Thailand	37.8	74.4	9.2	65	21.6
Brazil	66	94	95	95	70.21

So, for example, from the WRI data, India takes fifth place by total GHG emissions, but with the size of its population (16.2% with respect to global population) taken into account, it turns out that total emissions of CO_2 and methane constitute in this case, respectively, only 6% and 14.4% with respect to the amounts of GHG absorbed by the global ecosystem. The respective estimates for the USA (4.73% of global population) are 36% (CO_2) and 20% (methane). It follows from these data that the USA and Japan, the industrial countries, bear responsibility for emissions of GHG and, besides, they are powerful sources of CFC.

If all the countries equally reduce the emissions, then, as Pachauri (1992) notes, the poor countries will have to pay most for it. The data in Table 2.41 illustrate the estimates obtained on the assumption that during the 1990s the annual reduction of CO_2 emission should reach 20%.

Proceeding from the validity of the per capita estimates of GHG emissions, Agarwal and Naraian (1991) formulated the following principal solutions of the problem of responsibility for emissions, harmful as sources of substantial global climate changes;

(1) The per capita account (for each country) of natural sinks for CO_2, methane and other GHG, which will determine an acceptable size of emissions of various GHGs for each country (since no natural sinks exist for CFCs, their production must be forbidden).
(2) Comparison of total emissions of carbon dioxide and methane obtained at WRI with the permissible limits of GHG emissions (from the CSE data) to estimate a surplus (or deficit) of emissions with respect to the acceptable level.
(3) Commercialization of relationships among the countries with excess and deficit emissions (the latter refers, first of all, to China and India) on the basis of per capita estimates.

Calculations of the CSE scientists have shown that the industrial countries exceed their total quota of acceptable emissions of CO_2 by 2839 million tons of carbon (MTC) (carbon equivalent), which constitutes 58% with respect to excess emissions of carbon dioxide. As for the developing countries, failing to use their quota, they create a "reserve" reaching 1459 MTC due to such an "economy" (1015 MTC "belongs" to China and India).

Contributions of the more developed countries to the GHG concentration increase differ greatly. While the ecologically efficient technologies used in France and Italy provide their low per capita contribution, for example, the share of Australia, with its 3% of global population, in the resulting emissions of CO_2 and CH_4 reaches, respectively, 1% and 7%. During the 200-year existence of this country-continent, the forested areas have been almost halved, which, naturally, has produced considerable "biospheric" emissions of CO_2. Only two countries—the USA and the former USSR, with their 10% of global population—produced CO_2 emissions constituting $\sim$40% with respect to the global emissions, and the contributions of the USA and Canada to methane emissions reaches about two-thirds of the global emissions of methane.

As for the LDC, even the total contribution of China and India to global emissions of CO_2 does not reach 0.5%, the per capita contribution of India being the least in the world: each American is equivalent (by per capita emission) to 8150 Indians. Fifteen countries, including nine industrial and six developing countries, determine 83% of the global resulting (with account of sources and sinks) emissions of CO_2 in the atmosphere (Kondratyev, 1992d).

Due to the recent report published by the World Energy Council (WEC), which was summarized by Masood (1996), during the first half of the current decade CO_2 emissions from burning fossil fuels rose by 12%. The largest rise comes from developing countries (the rate of increase was 2–3 times more than in industrialized countries).

A very important conclusion is that even developed countries (except France, Germany and the United Kingdom) are unlikely to meet the goal of returning CO_2 emissions to 1990 levels before the end of the century, as it is required by the FCCC. A specific situation has developed in Eastern Europe and the former Soviet Union because of the economic decline. In this part of the globe CO_2 emissions have already dropped to below their 1990 levels. In 1995, emissions from the Commonwealth of Independent States (CIS) were 70% of their 1990 levels, and 75% for Central and Eastern Europe. Between 1990 and 1995, most member countries of the Organization for Economic cooperation and Development (OECD) registered a 4% rise in CO_2 emissions. But the countries in the Asian-Pacific region (except Australia, Japan and New Zealand) showed a 30% rise. Emissions in the countries in the Middle East increased by 35%, in Africa by 12.5% and in Latin America by 8%. CO_2 emissions from burning oil in Malaysia increased by 66% between 1990 and 1995.

In the context of the situation mentioned, Abdul Bar al Gain, director of Saudi Arabia's environmental agency said: "Reducing CO_2 emissions because of climate change has placed developing countries in a really serious bind ... it's like telling a poor trader who saved for 20 years to buy a van to carry on making deliveries using his donkey cart. It's not fair. We need to find some other solution for those developing countries that will become increasingly dependent on fossil fuel burning".

M. Jefferson, deputy secretary general of the WEC said, on the basis of recent survey, that "setting firm targets for further reductions would not be realistic at this stage".

Naturally, the fair distribution of responsibility for the reduction of GHGs emissions requires a differentiated approach to various countries, which would exclude an "equalization" and foresee both penalty and reward. A system should be developed which would stimulate, for example, the countries such as India to maintain the emissions at a low level, and the USA to rapidly reduce them. The CSE specialists think that a system of emission quotas should be worked out taking account of per capita emissions for each country in order that the sum of emissions be equal to the power of natural sinks of GHG. Besides, the "resources" of emissions may be sold (at a fixed price) by the countries which partially use their emission quota. The countries with excess emissions should be fined, and the respective profit may form the Global Protection Fund. The finances of this Fund can be used to render help to the countries subject to the destabilizing forcing of climate changes and to develop and distribute technologies providing a reduction of GHG emissions.

Approximate estimates obtained within the International Project on Secure Energetics Perspectives (IPSEP) developed in the USA showed (Agarwal and Naraian, 1991) that an acceptable fine for the emission of 1 thousand tons of carbon to the atmosphere (it is equivalent to 3.7 thousand tons of CO_2 and 0.54 thousand tons of methane) might be \$15. Such a level of fine is equivalent to an increase of the raw oil price by $\sim 10\%$ (from the data for 1986). On the other hand, it follows from these estimates that, for example, India would have the right to get annually approximately 8.3 billion dollars from the polluters-countries (this sum is approximately equal to half the investments of India into the development of energy production according to a 7-year plan).

For the USA this situation means a fine for excess emissions of 6.3 billion dollars annually. A group of 20 LDC would obtain annually about 30 billion dollars from the polluters-countries: for example, China (11.31), India (8.3), Pakistan (2.08), Nigeria (1.45), Bangladesh (1.06). A higher fine for emissions, \$25 for 1 thousand tons of carbon (carbon equivalent) would mean an increase of the annual contribution to the Global Protection Fund to about 90 billion dollars (only the contribution from the USA would in this case reach 38.3 billion dollars).

Proceeding from the criterion that a maximum acceptable global climate warming is 0.1°C for 10 years, that is, an acceptable upper limit of cumulative warming by 2100 is 2°C, we obtain (from the IPSEP data) that an acceptable volume of total emission over the period 1985–2100 must not exceed 300 billion tons of carbon (BTC) (hence, the annual "norm" of emissions is ~ 2.6 BTC). It follows from the IPSEP estimates that measures on reforestation and control of forest cutting are able (at the best) to reduce to zero the input of CO_2 to the atmosphere due to deforestation. Therefore, of decisive importance (from the point of view of removing the danger of global warming and its consequences) is the control of the development of the hydrocarbon energetics and, in this connection, a fair per capita distribution of the resource of emissions 300 BTC mentioned above.

The respective estimates obtained with account of the present levels of population and demographic forecasts have shown that in this case the MDC will have spent their total emission quota (48 BTC) by the year 1999 if emissions remained at the level of 1986, whereas in LDC the emissions can remain at the level of 1986 till 2169. These estimates show that the industrial countries should stop emitting GHG right away. This situation is determined by that the North (developed countries) with its 20% of global population consumes 85% of the global production of energy with the use of non-renewable energy sources. Therefore the North has spent most of the "ecological capital" of the planet.

To prevent a global ecological catastrophe, the structure of production and consumption must be radically changed and, in particular, the consumption of some key natural resources (e.g., fossil fuel) must be reduced, bearing in mind the necessity to provide the socio-economic development of the South. An important political problem is that the North has never been ready to pay the true ecological price for consumed products and goods manifesting itself both directly (e.g., deforestation) and indirectly (e.g., CH_4 emissions in Algeria are connected with an export of natural gas to Europe). Therefore the working out of a new (in the interests of the Third-World countries and all the global community) concept of sustainable development will be a difficult problem and will require, first of all, a reliable scientific substantiation (of course, the estimates mentioned above are rather conditional: further efforts are needed in this direction). Unfortunately, the Second UN Conference on Environment and Development was only a modest step towards this objective.

The threat of the global GHG-induced climate warming and of the associated impact on human activity and life necessitates a development of a complex strategy of response and, in particular, answers to the following questions (Parikh et al., 1994). (1) What level must GHG emissions be reduced to? (2) What levels of GHG concentrations may be accepted? (3) Who must be responsible for actions on reducing the emissions at different stages of the realization of the reduction programme? (4) Should this programme be started immediately or it should be postponed until more reliable information will make it possible to substantially reduce the level of uncertainties (which uncertainties are most substantial)? (5) What are possible provisional strategies (before the more complete information is available)?

Considerations have been expressed (Parikh et al., 1994), with respect to the answers to the questions enumerated, to substantiate the decisions on provision of GHG reductions worthwhile both now and in the distant future, bearing in mind the following aspects: (1) the nature and origin of the problem (the conceptual approach to its solution, bearing in mind, in particular, that the North is responsible for 70% of CO_2 emissions and for 77% of total emissions); (2) the correlation of the problems of equality, poverty and global warming (the developing countries and poor people are affected most by climate changes); (3) the ethical problems (anthropocentrism and the ecocentric approach taking into account the effects on nature); (4) uncertainty in climate forecasts and subsequent forcings of the ecosystems.

In making decisions, it is important to take into account the risk and uncertainties at all stages and to substantiate possible alternative approaches. Of importance are economical assessments of potential consequences of climate change—expenditures on countermeasures on GHG reduction, as well as analysis of the feasibility of these measures in different countries.

Pillet (1994) undertook an analysis of the problem of decision making at governmental and parliamentary levels in connection with potential anthropogenic climate changes, under "strong" uncertainty and collective risks, which requires in this case a substantiated specific methodology of decision making.

In usual conditions, extended knowledge enables one to reduce (or remove) the uncertainty, and the successively realized learning is introduced to the scheme of decision making. However, in the case of climate change an accumulation of knowledge does not remove uncertainties, but, on the contrary, favours their increase, as has been illustrated by the recent results, which testify to an underestimated complexity of the global climate system in the past. In these conditions an account of the ecologo-economic dynamics, with the economic factors being of first priority, plays a decisive role.

Climate changes are connected with collective risk which is defined as a risk to which a great number of people are equally subject. To solve the problem of such a risk it is important to use the market mechanisms (in this connection, Pillet (1994) proposed a scheme of ecologo-economic correlations). Pillet expressed his ideas about the quantitative estimates of the risk connected with climate changes based on account of correlations between different aspects of the socio-economic development and environmental dynamics, as well as a "cascade" of uncertainties of the estimates, beginning from the forecast of the dynamics of the size of population.

Haites (1994) notes the possibility of analogy between climate change decision making and decisions on the use of a portfolio of investments. In the case of the climate problem, various measures are needed, most proper for a given country. Such measures include: (1) studies aimed at a deeper understanding of the causes of climate changes; (2) development of energy-saving and low-waste technologies as well as non-traditional energetics; (3) handing over the advanced technologies and joint (on an international basis) efforts on reducing the GHG emissions; (4) development of effective economical systems (cancellation of subsidies for fossil fuel extraction, etc.); (5) search for methods of adaptation to climate changes; (6) combination of countermeasures on reducing the GHG emissions and intensifying the sinks for GHG. Of special importance are estimates of the economic efficiency of actions on preventing the undesirable climate changes.

Awareness of the danger (from the point of view of potential global climate changes and their consequences) of intensified emissions to the atmosphere of CO_2 and other GHG has determined an urgency of the problem of emission reduction. As Harvey (1995) noted, one of the most radical ways to solve this problem is connected with the development and adoption of hydrogen energetics: electricity production using hydrogen and oxygen of electrolysed water: their subsequent recombination produces electricity (direct use of hydrogen is also possible).

In this connection, Harvey (1995) undertook a comparison of expenses and profits from the point of view of CO_2 emission reduction by the period 2000 to 2060: (1) of improved centralized production of electricity with the use of fossil fuel; (2) of the applied hybrid method with the use of fossil fuel and solar energy (photoelectric generators); (3) of the use of only solar energy and hydrogen energetics as an energy accumulator.

In conditions of the low-level insolation (the average annual net radiation is 160 W/m^2) the price of the "solar" electricity will be 5–13 cents/kWh by 2000–10, and in the case of the high-level insolation (260 W/m^2) it will vary within 4–9 cents/kWh. Improved power stations using fossil fuel will be able to function with an efficiency of 45–55% (coal) and 50–60% (natural gas) and will make it possible to reduce CO_2 emissions 2–3 times compared to the present power stations (in industrial countries). In the case of hybrid systems using solar energy and fossil fuel (as an additional source of energy when needed), at a low level of insolation the price of electricity will increase in the range 0–2 cents/kWh, but, as a rule, it will decrease at a high-level insolation, the share of the "solar" electricity constituting 20% and 30%, respectively (the price of electricity from power stations using gas and coal is, respectively, 3.5–7 and 4.5–7 cents/kWh). Solar-hydrogen electricity generation will raise the price by 0.2–5.4 (1.0–4.5) cents/kWh, compared to power stations using natural gas (coal).

To provide the compatibility of the solar-hydrogen technology, a tax should be imposed on the use of fossil fuel (the "carbon" tax) from $70 to $660 /ton (depending on the price of technology and fuel in the future). A negative ecological consequence of the hydrogen technology is an escape of hydrogen which, getting to the stratosphere, produces water vapour, which may affect the ozone layer and climate. Rough estimates have shown that with the annual escape 0.5% (of total production of hydrogen) the effect on climate may be comparable with the effect of the hybrid solar-fuel system (as a result of CO_2 emissions during one year), and with account of CO_2 emissions over 100 years the escape of hydrogen will constitute about 1% (with respect to the hybrid system). The effect on the ozone layer ought to be negligible, too.

As has been mentioned elsewhere (Gorshkov, 1995; Gorshkov et al., 1992; Kondratyev, 1990a, 1993c,d; Kondratyev et al., 1997), the development of the problem of interaction between society and the environment has led to the concept of the development of a single system "nature–society" based on the following general principles:

— the biosphere, a sphere of man's living and activity, experiences now the transition from natural evolution to the evolution controlled by the society, which produced unacceptable perturbations of the biosphere;
— biotic and abiotic components of the natural media (mineral resources, air, water, living organisms, and soils) used in the sphere of social production and consumption as resource categories and development conditions are the components of a single dynamic object that develops in accordance with its specific laws; the principal law is that life existed under conditions of acceptable biospheric perturbations as a self-supporting system;
— material production is associated with complicated mechanisms, artificially inserted into the environment, coupled with the environment and violating the environmental equilibrium;

— the sustainable socio-economic development is impossible without taking account of the environmental changes caused by this development, without the control of the environmental dynamics, provision of an optimal conformity of the environmental state and the long-term objectives of development, and, finally, without continuous improvement of the system of the development management itself.

The system should be oriented to a sustainable relationship between society and the environment, between the public's needs, the development of productive forces and production relations, and the state of the environment; it should be aimed at improving the nature–society system, on the whole; it should be an ecologically sustainable development based on the principle of the biotic regulation of the environment (Gorshkov, 1995).

In the present conditions of intensive development exhausting reserves, it is particularly important to coordinate production with social needs. Of great importance are: orientation of management towards increasing efficiency and quality; speeding-up scientific-technical progress; developing interest in the outcome, the initiative and efficiency in each link of the economic chain. An urgent need for transition to the rational use of natural resources will require a specification of the categories mentioned above, which will serve as the basis for methodology and practical management of the socio-economic development.

Pure air and water, continually broadening resource bases, fertile soils, highly productive natural complexes, natural recreation zones—all these can be considered the principal social needs.

Aggravation of conflicts in nature—society relationships is explained by lack of proper progress in the development and application of conceptual–methodological support for the improvement of the development management (Glantz, 1994). The present practice of development does not adequately reflect the existing "environment–society" interrelationships. So far, there has been no progress in improving socio-development as an interactive constituent of natural processes.

Considering the present ideas about the processes in the geosphere–biosphere, the basic conceptions of ecological studies should proceed from the necessity to take into account:

(1) an approach to the problem from the point of view of coupled environmental dynamics and development based on the concept of biotic regulation of the environment;
(2) substantiation of such strategies of the control of geospheric–biospheric processes which provide the harmony of human activity and nature, preserving (restoring, if necessary) the closed nature of coupled global biogeochemical cycles.

It is important to emphasize that global and regional-scale environmental problems are interconnected. In the first case, we speak about an assessment of the present state and predictions of the geospheric–biospheric evolution within the next decades and centuries. Regional problems require an analysis of the consequences of the impact of man's activity on the environment (reclamation measures, hydrotechnical structures, environmental pollution, etc.) considering their priority, the development and application of simulation modelling techniques to assess impacts on the environment, and the substantiation of requirements for ecologically safe technologies. In all the cases the multidimensional nature of the problems should be remembered.

2.9 GLOBAL DYNAMICS OF TOTAL OZONE CONTENT, ITS IMPACT ON THE VARIABILITY OF SURFACE UV RADIATION AND ECOSYSTEMS

Since the problem of the climatic impact of ozone has been discussed above, we shall concentrate here, after a brief summary of recent stratospheric and tropospheric ozone studies, on the implications for ecosystems of increasing depletion of the total ozone content.

2.9.1 Stratospheric and tropospheric ozone

2.9.1.1 Stratospheric ozone

Observation data. An accumulation of long data series of satellite observations of the total ozone content (TOC) has provided a database characterizing the global dynamics of TOC over the period of the last several decades. The most important result from these observations consists in detecting the mean global TOC decrease and the event of "ozone hole" in the Antarctic attracting a great deal of attention, as well as short-term but sharp TOC decreases in the NH high latitudes (Chandra et al., 1996; Hofmann et al., 1996; Jiang and Yung, 1996; Kondratyev, 1989, 1990a, 1995e; Manney et al., 1996; Miller et al., 1996; Prather et al., 1996; Varotsos and Kondratyev, 1995a,b; Wang et al., 1996; Zhao and Kondo, 1996). Naturally, the TOC decreasing trend intensifies the biologically active UV radiation (UV-B), which determines various forcings of the ecosystems and man. On the other hand, an increase of the surface ozone concentration is known to affect negatively both the ecosystems and human health: irritation of eyes and nasopharynx, impeded breathing, cough, etc. (Biggs and Joyner, 1994; Calvert, 1994; Lefohn, 1992).

Since the stratospheric ozone is an intensive absorber of solar radiation, and the tropospheric ozone is a greenhouse gas (Wang and Isaksen, 1995), there is no doubt that changes in the content of ozone in the atmosphere must affect the climate.

A WMO-supported new report on ozone (1994 Ozone..., 1995) notes the following most important aspects of the problem:

— Beginning from 1979, the rate of TOC decline in the mid-latitudes of both hemispheres constituted 4–5% over 10 years.
— The lowest values of TOC were observed in 1992–93 in the presence of the most clearly manifested "ozone hole" in the Antarctic and the lowest TOC values in

the whole observational period over the densely populated regions of the northern hemisphere.

— The earlier conclusion has been confirmed about the substantial role of methylbromide as an ozone-destroying component, the main sources of which are the fumigating of the crops, biomass burning, and cars.
— The planned countermeasures on the control of emissions of the ozone-destroying substances will not greatly change the maximum level of TOC decline to be reached during next 10 years.
— Conclusions have been confirmed that the effect of anthropogenic emissions of chlorine and bromine compounds together with heterogenic chemical processes on polar stratospheric cloud particles is the main reason for TOC decline in high latitudes, the effect being even stronger.
— The decrease of the stratospheric ozone content between 1980 and 1990 has led to a compensation (by $\sim$20%) of the greenhouse forcing which favoured the global climate warming. However, the increase of the content of tropospheric ozone in the same period has intensified the atmospheric greenhouse effect in the northern hemisphere, also by $\sim$20%. The problem of the ozone trends with opposite signs in the stratosphere and in the troposphere claims special attention also in the context of the effect of TOC variations on UV radiation. From the published data (Michaels et al., 1994), the TOC decline in 1992–93 was not followed by any increase of UV-B radiation. Important ideas on the necessity of a thorough analysis of data on the trends of TOC, UV-B and related biological consequences have been expressed by Ellsaesser (1994). Feister and Grewe (1995) have observed, however, higher UV radiation inferred from low ozone values at northern midlatitudes in 1992 and 1993 and pointed out that this could lead to adverse biological effects.

In spite of the cautions against an overemphasis of ozone impacts on the environment and ecosystems, there are no doubts that this problem deserves very serious attention.

Prather et al. (1996) have posed a number of important questions in this context. (1) What if chlorofluorocarbon (CFC) use had followed a free-market growth into all sectors and countries until the observation of stratospheric ozone depletion itself were the first warning? (2) What if the ozone hole had been discovered in 1985 (as indeed it was) but no prior suspicion had focused on chlorine from CFCs? (3) When we now expect to see recovery of the ozone layer as provisions of the Montreal Protocol show their effects in the global atmosphere? (4) How much stratospheric chlorine and associated ozone depletion would the global atmosphere have been committed to in a likely alternative scenario?

Answering the questions mentioned, Prather et al. (1996) have emphasized that: (1) if we had ignored the scientific evidence, or not had it at all, ozone depletion would be dramatically worse than that we are controlling today; (2) if CFCs had followed free-market growth until 2002, the Antarctic ozone hole would be a permanent fixture through the twenty-first century, instead of disappearing by 2050 as predicted in the Copenhagen '92 scenario; (3) ozone depletion is expected to reverse and recover measurably in the first decades of the next century; unambiguous detection of the recovery will take a decade or more.

In this context it is proper to quote the Policy Statement on Atmospheric Ozone accepted by the American Meteorological Society Council, 28 January 1996:

(1) The AMS recognizes ozone as an atmospheric constituent that has a number of important beneficial as well as detrimental effects on the atmosphere, on surface ecosystems, and on humankind. These include effects on both ultraviolet and longwave radiation, resulting effects on atmospheric wind systems, and direct impacts on plants and animals. Moreover, ozone is recognized as the dominant progenitor of much of the trace-gas chemistry that occurs in both the troposphere and the stratosphere. The AMS also notes that extensive couplings exist among these chemical, radiative, and dynamical components of ozone's behaviour, and that these couplings act to complicate ozone's composite behaviour and add substantial uncertainty to many of the currently available assessments of ozone's impact.

(2) Despite many uncertainties, ample evidence exists to substantiate that atmospheric ozone has been affected in important and even critical ways by human activity. In the case of the stratosphere, the existence of the Antarctic ozone hole is unquestionable, and our evidence that it results from human-produced halocarbons is overwhelming. Ozone depletion in the midlatitude stratosphere is less dramatic; however, the general agreement of the large numbers of available column and profile measurements make it reasonable to speculate that such depletion is indeed occurring and that it is largely human-induced.

   Anthropogenic activities also significantly influence tropospheric ozone. Humankind is directly responsible for the excessive ozone levels that often occur near the surface in and down-wind from populated areas. These effects are also felt throughout significant regions of the free troposphere.

   The effects in both the stratosphere and troposphere are sufficiently profound to mandate substantial concern, on both a local and a global basis.

(3) While the above manifestations of human impact on ozone are clear, there remain important gaps in our understanding of ozone's complex behaviour. For example, currently we cannot quantitatively explain the midlatitude observations of insignificant surface UV increases. Impacts of tropospheric ozone buildup on the free troposphere's chemistry are also in a highly speculative state. These issues and others must be resolved in order to satisfactorily forecast potential future manifestations and provide a firm basis for policy analysis and associated policy actions.

(4) Many of the noted uncertainties arise because of the strong couplings between chemistry, radiation, and atmospheric dynamics. Because of this, their resolution will require substantial combined effort by scientists having chemical and meteorological backgrounds. The AMS intends to actively provide one component of a continuing forum for this scientific interaction, and act in conjunction with other like-minded organizations for this purpose.

(5) Current ozone-control legislation (such as the US Clean Air Act, with 1990

amendments) and international agreements (such as the Montreal Protocol) tend to reflect the uncertainties noted above. The State Implementation Plan approach of the current US Clean Air Act is burdened by older and largely invalid concepts of chemistry and atmospheric transport, which make ozone-standard attainment difficult if not impossible at many locations. Future amended versions of this act should incorporate more manageable and scientifically valid approaches to ozone control, and substantial current meteorological and chemical research should be directed to determining effective strategies for this purpose.

With regard to the Montreal Protocol, it is encouraging to note that halocarbon limitations under this agreement appear to have resulted, recently, in decreases of some of the shorter-lived halogen-containing species. Owing to the noted uncertainties and complexities associated with stratospheric ozone depletion, however, the effects of this and other international agreements must be monitored continuously and carefully to ensure their effectiveness and to establish the basic understanding required for more effective maintenance during future years.

The recent years have been marked by very intensive studies of global total ozone dynamics on the basis of both observations and simulation modelling. As far as observations are concerned, an especially important contribution has been made by the analysis of the total ozone mapping spectrometer (TOMS) data.

Chandra et al. (1996) have used 13 to 14 years of the Nimbus-7 TOMS total ozone measurements to demonstrate that the TOC trends derived from satellite observations at northern and mid latitudes during winter and spring seasons appear to be significantly larger than predicted by recent 2-D photochemical models. There is a substantial longitudinal spread in TOC trends at mid-latitudes due to dynamical perturbations in the atmosphere. For example, decadal TOC trends during February over the 10° latitude band centred at 40°N vary from about -9% to about -4% with errors bars of 4% to 6% at the 2σ level. During the same month the zonally averaged trend is about $-6 \pm 3\%$.

Jiang et al. (1996) have analysed the decadal evolution of the Antarctic ozone hole (within the polar vortex area) during the austral winter and spring for 1980–91 TOMS data. The vortex-average TOC during September and October has declined at a rate of 11.3 Dobson Unit (DU) /yr (3.8%) from 1980 to 1987 (90 DU over 8 years) and at a smaller rate of 2 DU /yr (0.9%) from 1987 to 1991 (10 DU over 5 years, excluding the anomalous year 1988). Jiang et al. (1996) have interpreted the year-to-year trend in the ozone depletion rate during the earlier part of the decade as being due to the rise of anthropogenic chlorine in the atmosphere. The slower trend at the end of the decade indicates saturation of ozone depletion in the vortex interior. In subsequent years, increases in stratospheric chlorine may have enhanced the wintertime chemical loss of ozone in the south polar vortex even before major losses during the Antarctic spring.

Upper Atmosphere Research Satellite (UARS) Microwave Limb Sounder (MLS) data were used by Manney et al. (1996) to study evidence of ozone-hole analogues in the Arctic through the analysis of ozone concentration changes at various levels in the stratosphere. During the unusually cold 1994–95 Arctic winter the MLS data indicated enhanced chlorine monoxide (ClO) in late December and throughout February and early March. Between late December 1994 and early February 1995 the vortex-averaged ozone concentration at the 50 hPa level decreased by about 15% with local decreases up to 30%; additional local decreases of about 5% were observed between early February and early March.

Wirth and Renger (1996) have conducted airborne lidar ozone soundings during the Arctic winter 1994/95 which indicate that ozone mixing ratio declined by about 50% throughout the polar vortex during the period from the beginning of February to mid-March in a sharp defined altitude band from 420 K to 520 K potential temperature. For a number of reasons a major contribution to this ozone deficit inside the Arctic vortex by dynamical processes can be excluded. The Arctic winter 1994/95 with extraordinary low stratospheric temperatures, and a stable, long-lasting Arctic vortex with strong polar stratospheric clouds (PSC) activity was very similar to typical Antarctic winters, accompanied by ozone depletion due to chemical processes with subsequent ozone hole formation in spring.

Zhao and Kondo (1996) have investigated winter stratospheric ozone changes over Japan since 1991 from total ozone and ozonesonde data obtained in Sapporo (43°N), Tsukuba (36°N), Kagoshima (32°N) and Naha (26°N). They discovered sizeable TOC anomalies during the winters of 1992/93 and 1994/95 at the four stations. Substantial impacts of the Pinatubo eruption and Quasi Biannual Oscillations (QBO) on TOC were observed.

The record low TOC level at Mauna Loa Observatory during winter 1994/95 has been registered by Hofmann et al. (1996). During two days in late December 1994 the TOC dropped below 200 DU for the first time since ozone measurements began at this site over 30 years ago. UV solar radiation measured at Mauna Loa on clear days during this period increased inversely with ozone, as expected, more than tripling in relative intensity at 295 nm. Hofmann et al. (1996) believe that the unusually low TOC values were related to ozone transport from the tropics, and the time of phase transitions of the QBO. The overriding cause of a TOC reduction of about 3% since 1976 appears to be, however, the downward trend in ozone.

The necessity of increasing the reliability of observational data has led to a number of studies with the purpose of intercomparisons of various measurement techniques.

An intercomparison of TOC data from a Dobson spectrophotometer, TOMS, visible wavelength spectrometer, and ozonesondes for the period 1991 to 1994 at Arrival Heights/McMurdo, Antarctica, showed that while the Dobson, TOMS and ozonesonde measurements show good agreement, there is a seasonal dependence between data from the visible wavelength spectrometer and each of the other three instruments.

Miller et al. (1996a) have examined the non-seasonal and the seasonal trend behaviour of the ozone profile data from both ozonesonde and umkehr measurements in a consistent manner, covered the time period 1968–91. The results confirm the observation of significant negative ozone trends in both the lower stratosphere (15–20 km), about −6% per decade, and upper stratosphere (35–50 km), about

−6% per decade, separated by a nodal point in the region of 25–30 km. Miller et al. (1996) have pointed out that the upper stratosphere decrease is due to the classic gas-phase chemical effect of CFCs, whereas the cause of the lower stratospheric decline is still under investigation, but may well be associated with the chlorine and bromine chemistry in this region.

Miller et al. (1996) have studied ozone trends and solar effects on ozone (responses to the 11-year solar cycle represented by the 10.7-cm radio emission) from ground-based umkehr data as well as satellite solar backscattered ultraviolet (SBUV) measurements (Nimbus–7 and NOAA–11) for the period June 1977 to June 1991 at northern midlatitudes (30°–50°N) at altitudes between 25 and 45 km. The data indicate that the trend is approximately zero near 25 km but becomes more negative in the upper stratosphere, reaching nearly −7% per decade in the 40–45 km region which is evidently associated with the 10.7 cm solar flux demonstrate near-zero solar-induced ozone variations in the 25–30 km layer and statistically significant in-phase variations at higher altitudes which reach about 4.5% at 40–45 km from solar cycle maximum to minimum. Due to Wang et al. (1996) who analysed corrected ozone SAGE data, the solar cycle ozone dependence in the upper stratosphere is characterized by a local maximum response in the ozone concentration of approximately 3% at midlatitudes at approximately 41 km altitude. In the lower stratosphere the SAGE results indicate a sensitivity of about a 1.5% change over the solar cycle but this change is mostly occurring just below the ozone concentration peak.

Gregory et al. (1996) have examined Pacific Exploratory Mission (PEM-West A) aircraft ozone data concurrent with TOMS observations to assess, for relatively remote Pacific regions (110°–180°E longitude, 0°–40°N latitude), what TOC fraction can be attributed to lower-tropospheric ozone (surface to 9 km altitude) and how this fraction varies with time. The observations show that within the 30°–40°N latitude band the surface to 9 km ozone column contributes about 10% of TOC. Of this 10%, about half is attributed to the surface to 3 km layer. The contribution of tropospheric ozone is less near the equator (0°–15°N): about 5% (0–9 km) and 1% (0–3 km) of TOC. At 40°–50°N, ozone variations within the columns can result in changes in each column's contribution to the TOC that approach more than 50% of the column's contribution to the TOC.

In the context of stratospheric ozone variability, Russel et al. (1996) have discussed the results of a four-year (1991–95) global time-series of satellite observations of hydrogen chloride (HCl) and hydrogen fluoride (HF) in the stratosphere made within the Halogen Occultation Experiment (HALOE) programme of the UARS. These observations conclusively indicate that fluorocarbon releases—rather than other anthropogenic emissions—are responsible for the recent global increases of stratospheric chlorine concentrations by the level which is more than five times exceeds the "background" level. Therefore Russel et al. (1996) have concluded that chlorofluorocarbons beyond reasonable doubt are dominating ozone depletion causation in the lower stratosphere (all but a few per cent of observed stratospheric chlorine amounts can be accounted for by known natural and anthropogenic tropospheric emissions).

Since methyl bromide (CH_3Br) is considered to be a major source of bromine to the stratosphere and important contributor to the destruction of stratospheric ozone, Pilinis et al. (1996) have assessed the role of the oceans as a sink or a source of methyl bromide. They arrived at the conclusion that the oceans are a strong net source of methyl bromide to the atmosphere. Because this conclusion is sensitive to the assumptions about the production and distribution rates for CH_3Br, Pilinis et al. (1996) have emphasized a necessity of further research.

Ozone holes and polar stratospheric clouds (PSCs). It has been reliably established that a decisive contribution to the formation of Antarctic ozone hole belongs to heterogeneous chemical reactions on surfaces of PSC's particles (Kondratyev, 1989; Tolbert, 1996). Similar processes develop in the Arctic stratosphere. There are still many uncertainties, however, concerning such processes. This is why many studies of PSC's properties have been recently accomplished through field observations and simulation modelling.

Within the Airborne Southern Hemisphere Ozone Experiment / Measurements for Assessing the Effects of Stratospheric Aircraft (ASHOE/MAESA) project, Dye et al. (1996) measured particle volumes in a type I polar stratospheric cloud near the Antarctic polar vortex. The observations from board the ER-2 aircraft were made with the help of the Multiangle Aerosol Spectrometer Probe (MASP), which counts and sizes individual particles in the size range of 0.4 to 10 μm diameter by measuring light scattered by individual particles. The particle refraction index can also be estimated over a size range of 0.4 to 0.7 μm by measuring the ratio of light scattered in the forward and backward directions (during the same mission very important gas chromatographic measurements of minor gas components were made).

The results of observations in a PSC in the outer region of the Antarctic vortex obtained by Dye et al. (1996) show similarity with previous observations in the Arctic. The existence of PSCs composed of ternary droplets has been discovered in the Antarctic, thus unifying the understanding of PSC formation processes in the two polar regions. The onset of the cloud occurred near 193 K, 3 degrees colder than nitric acid trihydrate (NAT) saturation (the same conclusion has been made by Larsen et al. (1996)). Dye et al. (1996) have pointed out that the onset temperature, the smooth increase of volume with decreasing temperature, the inverse correlation of particle volume and enhanced NO_y (HNO_3 in the particles) with temperature, and comparisons of observations with an equilibrium model of ternary droplet growth (Koop and Carslow, 1996)—all support the notion that much of this type I PSC was ternary solution (H_2SO_4–HNO_3–H_2O) droplets (this provides confirmation of previous findings in the northern hemisphere). The ternary model does not fit, however, the observations in all regions. This may be due to the presence of some solid phase growth in agreement with impactor observations. Hamill et al. (1996) have shown that the mechanism of ternary solution droplets is heteromolecular condensation in which the particle is in equilibrium with environmental water vapour.

De Rudder et al. (1996) and Tie et al. (1995) have accomplished a model study of polar stratospheric clouds and their effect on stratospheric ozone. Describing the model, De Rudder et al. (1996) have emphasized that detailed microphysical processes accounting for the PSCs formation (in terms of heterogeneous nucleation, condensation and sedimentation) have been included in a global chemical/dynamical 2-D model to study the effect of heterogeneous reactions occurring on the surfaces of PSCs on stratospheric ozone. Three types of particles are considered in the model: (1) diluted sulphuric acid particles which serve as cloud condensation nuclei (CCN) for the PSC particles; (2) type I PSC particles (PSCI) composed of a NAT shell surrounding a small sulphate core, which act (as temperature decreases further) as condensation nuclei for, (3) type II PSC (PSCII) composed of an ice shell coating a PSCI core. The calculations show that PSC's particles distribution in the Arctic and Antarctic is quite different. Over Antarctica, nitric acid trihydrate particles (type I PSCs) are formed from early June to late September with a maximum surface area of 15 $\mu m^2/cm^3$, while in the Arctic, type I PSCs are formed only in January with a maximum surface area of 5 $\mu m^2/cm^3$. Ice crystal clouds (type II PSCs) are present over Antarctica in August, but are not seen in the Arctic. The results of simulation modelling considered by Tie et al. (1996) indicate that the heterogeneous reactions (conversion of $ClONO_2$ into ClO under sunlight conditions) occurring on the surface of PSCs are the likely cause of the ozone decrease observed from 1980 to 1990 in both Antarctica and the Arctic. The calculations successfully simulate the development of the springtime ozone hole over Antarctica. The largest ozone depletion starts in September and reaches a maximum in October (polar sunrise). The maximum ozone depletion is located at the top of the PSC layer (20 km), where the Cl_x concentration is highest. In the Arctic, the ozone depletion due to PSCs is much smaller because of the shorter occurrence period and the smaller surface area available for chemical conversion.

The formation rate of type I PSCs in both Antarctica and the Arctic has enhanced after the eruption of Mount Pinatubo in 1991 because of the dense eruptive sulphate aerosol cloud. However, the effect on the ozone depletion in the Antarctic was limited due to the fact that the conversion from $ClNO_2$ to ClO on PSC particles was almost "saturated" under non-volcanic conditions. Tie et al. (1996) have concluded therefore that, for the potential ozone depletion, the enlargement in the area covered by PSCs may be more important than the increase in PSC density.

Chipperfield et al. (1996) have developed a 3-D chemical transport model to investigate the potential for chemical ozone depletion in the Arctic polar vortex for the past four winters. Isolating the effect of meteorological variability on polar ozone loss (prescribing the same chemical initial conditions for each winter) they have shown that during 1994/95 the conditions in the Arctic polar vortex were most conducive to ozone depletion in the recent years. The calculations indicate that at 475 K (around 18 km) in the lower stratosphere the chemical ozone destruction between late December and mid-March was greater than 30% over a large area of the polar vortex. Larger percentage losses, reaching 50%, occur at slightly lower altitudes. The average model depletion in the entire region poleward of 60°N over the same period was around 22% at 475 K. The large depletion for 1994/95 is a consequence of the extreme meteorological conditions during this winter.

To study the effects of heterogeneous chemistry upon anthropogenic ozone depletion at northern midlatitudes, Solomon et al. (1996) have used a 2-D model with input data on stratospheric aerosol surface area distributions inferred from satellite 1-m extinction measurements. The principal conclusions made by Solomon et al. (1996) are as follows: (1) the short-term (inter-annual) and longer-term (decadal) changes in aerosols very probably played a substantial role along with trends in anthropogenic chlorine and bromine in both triggering the ozone losses observed at northern midlatitudes in the early 1980s and increasing the averaged long-term ozone depletions of the past decade or so; (2) the use of observed aerosol distributions enhances the calculated ozone depletion due to halogen chemistry below about 25 km over much of the past decade, including many periods not generally thought of to be affected by volcanic activity; (3) direct observations confirm the key aspects of the model chemistry that is responsible for this behaviour and demonstrate that aerosol changes alone are not a mechanism for ozone losses in the absence of anthropogenic halogen inputs to the stratosphere; (4) aerosol-induced ozone changes could be confused with 11-year solar cycle effects in some statistical analyses, resulting in an overestimate of the trends ascribed to solar activity; (5) the timing and magnitude of future ozone losses at midlatitudes in the northern hemisphere are likely to be strongly dependent upon volcanic aerosol variations as well as on future chlorine and bromine loading; (6) there is still a substantial quantitative disagreement between model predictions and observations: observed ozone changes are about 50% larger than calculated (this may be due to shortcomings in modelled photochemistry or dynamics or both).

Zhao et al. (1996) have applied a general circulation model which has been coupled interactively to a 3-D stratospheric photochemistry model to study the dynamical response of the Arctic polar vortex to ozone depletion catalysed by volcanic aerosols. It has been found out that temperatures may be depressed by as much as 3–7 K in late March in the lower stratosphere at northern mid- and high latitudes owing to ozone-radiation-dynamics feedback effects. In the model, the reduced solar heating in late winter and early spring associated with aerosol ozone depletion acts as a trigger for the net cooling that is observed. However, a major part of the cooling is related to anomalies induced in the circulation of the lower stratosphere. Zhao et al. (1996) have emphasized the complexity of the coupling between the dynamics and chemistry of the lower stratosphere as well as of relevant assessments of cause and effect related to perturbations in this region.

Randeniya et al. (1996a,b) have included the hydrolysis of $BrONO_2$ on sulphate aerosols ($BrONO_2 + H_2O_{aerosol} \rightarrow HNO_3 + HOBr$) in a 2-D model calculations of ozone changes over the period 1980–95. It has been found that the inclusion of this reaction has only a marginal impact on the calculated TOC trends between 1980 and 1995. The addition of this reaction had, however, a major impact on the calculated transient ozone loss under enhanced aerosol conditions (after the Pinatubo eruption) and approximately doubled the calculated ozone loss between 60°S and 60°N in 1992 (when maximum loss of ozone occurred). In contrast to the effects of the hydrolysis of N_2O_5, the effects of the hydrolysis of $BrONO_2$ are most marked at high latitudes in summer.

Model calculations made by Tie and Brasseur (1995) suggested that the ozone decrease observed a few years after the eruptions of Mt. Pinatubo and El Chichón may have been unique in the Earth's history, and is directly linked to the emission in the atmosphere of industrially manufactured CFCs. These calculations show that the response of ozone to a volcanic eruption similar to that of Mt. Pinatubo is very sensitive to the atmospheric chlorine loading. For chlorine loadings typical of the pre-industrial era, the TOC increased after the eruption, with the largest response occurring at high latitudes during the winter season. For chlorine loading representative of the 1980s, the response of total ozone becomes negative in winter at mid- to high latitudes. Tie and Brasseur (1996) have pointed out that this transition in the ozone response suggests that, as a result of CFC production and emission, the "NO_x effect" associated with volcanic eruptions (which tends to enhance stratospheric ozone) is counterbalanced by a "ClO_x effect" (which tends to deplete ozone). In the future (year 2015), the response of ozone is expected to become positive again, if the CFC production is sufficiently reduced (according to Montzka et al. (1996), the amount of reactive chlorine and bromine will reach a maximum in the stratosphere between 1997 and 1999 and will decline thereafter).

On the basis of laboratory simulations of the growth of nitric acid trihydrate, $HNO_3 \cdot 3H_2O$ on sulphuric acid tetrahydrate $H_2SO_4 \cdot 4H_2O$, Zhang et al. (1996) have come to the conclusion that, in the stratosphere, PSC particles may experience repeated cycles of evaporation and condensation of HNO_3 on pre-existing background frozen sulphate aerosols. Hence, growth of $HNO_3 \cdot 3H_2O$ on preactivated aerosols provides one important mechanism for polar stratospheric cloud formation and denitrification.

Dessler et al. (1996) have tested, for the first time, the photochemical theory of partitioning between the chlorine species $ClONO_2$ and ClO in the lower and middle stratosphere using simultaneous measurements of ClO, NO_2 and $ClONO_2$ from UARS satellite. The results provide no evidence for any problems in the present-day understanding of the partitioning between the chlorine species mentioned over the range of pressures (46 to 10 hPa) and latitudes (60°S to 60°N) considered.

Through the analysis of satellite and ozonesonde TOC data, Ziemke et al. (1996) have assessed the contribution of biomass burning into the variability of total ozone. The results indicate that zonal asymmetries observed in total ozone in the tropics originate primarily from tropospheric effects (dynamics coupled with biomass burning), while outside the tropics, zonal patterns in TOC result mainly from stratospheric dynamics. Due to Jiang and Yung (1996), there was an increase of tropospheric ozone concentration of $1.48 \pm 0.40\%/\text{yr}$ (or 0.21 ± 0.06 DU/yr) over South America and the surrounding oceans because of biomass burning during the time period from 1979 to 1992.

Impact of high-flying aircraft and space shuttles. Initial predictions of a potential catastrophic destruction of the ozone layer resulting from supersonic aviation (SSA) flights in the stratosphere have not materialized because it has so happened in reality that the number of stratospheric aircraft is still very limited (Kondratyev, 1989). Nevertheless, in view of future perspectives, the last few years have been marked by renewed interest in the problem of whether aircraft emissions significantly perturb the chemical and radiative balance of the atmosphere and if these perturbations can alter global climate (Brasseur et al., 1996; Flatøy and Høv, 1996).

In this context, Kärcher et al. (1996) have considered the aircraft plume chemistry in the jet regime, that is, during the first few kilometres behind the aircraft, thereby providing necessary input data for follow-on modellers. The principal aim was to study the relaxation of the strongly perturbed plume chemistry to background conditions. The evolution of hydrogen, oxygen, nitrogen, and sulphur species has been considered in detail and compared with observations with special emphasis on the coupling between mixing and chemistry. The case study considered is related to the impact of a B747 airliner cruising at around 10-km altitude. An important conclusion is that gas-to-particle conversion depletes the produced high levels of gaseous H_2SO_4 and results in the generation of large numbers of new binary aerosols.

Kärcher (1996) has accomplished model calculations for the formation of aerosols and the interaction between themselves and with exhaust gases in plumes of cruising airliners to assess the role of the current aircraft fleet as a source of new particles within the upper troposphere and lower stratosphere. Such particles may contribute to the observed increase of the stratospheric background aerosol surface area in the past decade, and thus be partly responsible for the measured ozone loss at northern midlatitudes between 10 and 25 km altitudes.

Weisenstein et al. (1996) have pointed out the importance of including aerosol formation within aircraft wakes due to small-scale chemical processing of emitted SO_2 in the estimates of the possible ozone impact of sulphur emissions from high-flying aircraft. In this context, Kärcher (1996) has calculated the dynamics of liquid and solid particles in the plumes of B747 and a stratospheric aircraft (HSCT), cruising at pressures 216 hPa and 73 hPa, respectively, and at ambient temperatures less than 223 K. It has been shown that a subset of *in-situ*-nucleated H_2SO_4/H_2O aerosols that stray liquid can transform into an almost pure HNO_3/H_2O mode because of enhanced uptake of HNO_3. Soot embedded in a liquid coating containing H_2SO_4 and HNO_3 triggers heterogeneous freezing of water ice and leads to visible contrails. The ice particles may acquire an $HNO_3 \cdot 3H_2O$ (NAT) coating upon evaporation of H_2O, which can potentially lead to a lengthening of their lifetime. If NAT coatings develop at the surface of contrail ice, the resulting persistence of such submicrometre particles could result in efficient heterogeneous chemical processing that may affect ozone. In view of current model uncertainties, it is very important to undertake an experimental verification of the results.

Hanning-Lee et al. (1996) have accomplished laboratory experiments to assess the rates of stratospheric ozone decomposition on aluminium oxide (alumina) as components of launch vehicles' solid rocket motor exhaust. The results indicate that the calculated ozone holes formed because of the ozone decomposition are rather narrow. In the worst case, ozone levels are within 5×10^{-5} of ambient in the centre of the plume. An approximate estimate of the global impact of alumina particles on ozone decomposition indicates a potential steady-state daytime depletion of less than 2.6×10^{-8} at present launch rates.

Assessing the polar stratospheric clouds effect on stratospheric ozone, Tie et al. (1996) have shown that, in the future, the density of PSCs in the Arctic could be enhanced owing to the potential emissions of water vapour and nitrogen species by projected high-speed civil transport (HSCT). The increase in PSC density could lead to a maximum of ozone depletion of 10% at the northern high latitudes in winter.

Randeniya et al. (1996a,b) have estimated (using a 2-D model) the impact of heterogeneous hydrolysis of $BrONO_2$ via sulphate aerosols in case of projected fleets of supersonic aircraft. The results show that the addition of this reaction has a major effect when the heterogeneous hydrolysis of N_2O_5 has reached saturation in the regions where the night length is short.

A simulation has been made to assess the space shuttle's impact on the stratosphere of a launch scenario of nine shuttles and three Titan rockets per year, using a 2-D photochemistry and transport model which includes heterogeneous reactions on a stratospheric sulphate aerosol (SSA) layer and polar stratospheric clouds (PSCs). As is well known, the launch of NASA's space shuttle and similar rockets results in injections of hydrogen chloride, carbon monoxide, water vapour, molecular nitrogen and aluminium oxide directly into the stratosphere. The calculations indicate that maximum total inorganic chlorine enhancements are about 12 pptv (which is about 0.4% of a 3 ppbv background value) in the middle to upper stratosphere at northern middle to high latitudes. Maximum ozone decreases associated with these chlorine increases are about 0.14% in the middle to upper stratosphere at northern middle to high latitudes in the early spring. The simulations made under various assumptions gave the following mean annual global TOC decreases: (1) 0.0056% (the consideration of gas phase reactions only); (2) 0.010% (taking account of gas phase reactions and heterogeneous reactions on the SSA layer); (3) 0.014% (the same as (2) plus heterogeneous reactions on PSCs). Jackman et al. (1996) have pointed out that stratospheric effects from heterogeneous reactions promoted by alumina emitted from the rockets could be larger than those predicted from chlorine emissions but need to be investigated further.

Volk et al. (1996) have emphasized that ozone at midlatitudes appears to be more sensitive to elevated levels of industrial chlorine than is currently predicted. Nevertheless, about 45% of air in the tropical ascent region at 21 km is of midlatitude origin, implying that emissions from supersonic aircraft could reach the middle stratosphere.

Calculations made by Flatøy and Høv (1996) who applied a 3-D model to study the effect of NO_x emissions from subsonic aircraft on ozone in the upper troposphere over Europe and the North Atlantic, show that during the period 1–10 July 1991 over North Europe, chemical ozone formation up to 1.8 ppbv/hr could occur, which is about factor of 50 higher than the average for 10.5 km over continental Europe during the 10-day period. The high production rates are not sustained, however, for more than a few hours. Calculations show that 10–25 molecules of O_3 (0.03-0.05 ppt/h) are generated on the average per NO_x molecule emitted from aircraft (2–3 ppt/h).

Using a 3-D model, Brasseur et al. (1996) predicted that in the year 2050 (with further growth in aviation) the NO_x concentration at 10 km could increase by 30–60%

at midlatitudes, and the concentration of ozone could be enhanced by 7% and 2% in summer and winter, respectively.

Thus, one can make a general conclusion that due to existing estimates neither subsonic or supersonic aviation nor rocket launches are a serious danger for the stratospheric ozone layer.

2.9.1.2 Tropospheric ozone

As it is well known (Hales, 1996; McKee, 1994; Prinn, 1994a,b), tropospheric ozone originates from two sources: intermittent engulfment of stratospheric ozone into the troposphere and *in situ* formation from chemical processes among tropospheric trace gases. An increasing number of ozone observations on the ground and in the free troposphere has resulted in a substantial improvement of the tropospheric ozone database. Numerous recent studies in the field of simulation modelling has led to a much better understanding of tropospheric ozone dynamics. Since relevant results have been discussed in a number of recent fundamental publications (Lefohn, 1992; McKee, 1994; Prinn, 1994a; Wang and Isaksen, 1995), only some of the most recent studies will be mentioned here.

Theodorsen et al. (1996) have considered surface ozone changes at high latitudes due to atmospheric pollution on the basis of observational data for Tromsø (70°N) and Longyerbyen Svalbard (78°N) from November 1992 and throughout 1993 using ozonometers with high temporal resolution. Significant ozone concentration variability has been discovered within a few seconds. It has been shown that at high latitudes in Norway, car exhaust and air pollution essentially destroy surface ozone.

Naja and Lal (1996) have conducted surface ozone concentration observations at Ahmedabad (23°N) during 1954–55 and 1991–93 periods. The results indicate that higher ozone concentrations including larger amplitudes of diurnal and seasonal variations have been observed during 1991–93 as compared to the measurements made during 1954–55. The observed seasonal maximum during 1954–55 is in spring months whereas during 1991–93 it is in winter months. The average ozone concentration increased from 14.7 ppbv during 1954–55 to 25.3 ppbv during 1991–93, resulting in a linear increase of 1.45% /yr.

Reichardt et al. (1996) have discussed Raman lidar measurements of ozone, water vapour, and cirrus clouds optical properties over northern Germany (Geesthacht: 53.4°N, 10.4°E) between October 1995 and January 1996. Pronounced minima in the ozone vertical profiles with values close to zero have been found several times in the presence of ice cloud layers in contrast to smooth ozone profiles with mixing ratio values of 50 to 100 ppbv in a cloud-free and dry upper troposphere. A cause of such minima may be heterogeneous chemical reactions on ice particles and changed conditions for homogeneous chemical processes in the presence of cirrus clouds. Further studies of cause and effect interrelationships are necessary, however.

Hudson et al. (1995) have developed a new algorithm to retrieve tropospheric column ozone from satellite TOMS data with the purpose to apply this algorithm to assessments of a contribution of biomass burning into the tropospheric ozone budget. They have demonstrated that over the region bounded by 20°W and 60°E longitude and 20°S and 0°S latitude during the 1989 biomass burning season tropospheric ozone content approached 20% of total ozone.

To assess impacts of increased emissions in Asia on tropospheric ozone and climate, Berntsen et al. (1996) have used a global 3-D chemical transport model (CTM) developed at NASA Goddard Institute for Space Studies (GISS). An extended set of chemical compounds (including such principal ozone precursors as hydrocarbons, nitrogen oxides, carbon monoxide) has been taken into account in the model, of which 25 are transported. The choice of the Asian region has been motivated by the fact that anthropogenic emissions are rapidly increasing in this region and because there is a large potential for future increases.

Three model experiments have been conducted to study the impact of tropospheric ozone from surface emission changes: (1) a reference run (R) with global emissions of NO_x and hydrocarbons representative for 1980 (Table 2.42); (2) two additional experiments with increased emissions (E_{A2}, E_{A4}): the anthropogenic sources have been increased by factors of 2 and 4, respectively. New emissions applied in the Asian region are then defined by equations:

$$E_{A2} = 2 \times F_a \times E_R + (1—F_a) \times E_R$$

$$E_{A4} = 4 \times F_a \times E_R + (1—F_a) \times E_R$$

where F_a (Table 2.42) defines the fraction of the emissions assumed to be of anthropogenic origin for each component. The Asian region is defined as the region between the equator and 80°N, and between 57.5°E and 172.5°W. The factor F_a gives the fraction of the emissions which are perturbed in the runs with increased emissions relative to the reference run (A2 and A4).

Table 2.42. Global and Asian surface emission rates applied in the reference run in the model

Species	Emissions (E_R)		
	Global (Tg/yr)	Asian (Tg/yr)	F_a
NO_x	34.6 (N)	7.5 (N)	1.0
CO	1575	294	0.50
C_2H_4	36.7	6.3	0.61
C_2H_6	37.0	7.4	0.61
C_3H_6	27.0	4.8	0.73
C_4H_{10}	29.9	5.7	0.52
C_6H_{14}	24.3	5.0	0.70
m-xylene	22.7	5.8	0.68
HCHO	1.11	0.2	0.61
isoprene	205	27	0.0
CH4	394	169	0.0

Table 2.43. Calculated surface concentrations of NO_x and total reactive carbon for the three model runs, and corresponding changes in surface ozone concentrations

Region	Case	NO_x (pptv)	TOT.C (ppbv)	ΔO_3 (ppbv%)	PE_{O3} (ppbv/Tg(N)yr^{-1})	$\Delta O_3/\Delta NO_x$ (ppbv/ppbv)
Japan	R	970	16.3			
	A2	1710	24.2	13.2/28	23	18
	A4	3260	40.4	22.5/47	8.4	6.0
China	R	1160	13.8			
	A2	2110	17.3	7.9/29	17	8.4
	A4	4060	25.1	14.0/52	6.5	3.1
India	R	290	16.6			
	A2	490	20.7	12.8/43	119	64
	A4	790	29.3	3.2/114	100	71

The change (A2 versus R and A4 versus A2) in the ozone production efficiency $PE_{O3} = \Delta E_{NOx}$ (ppbv/Tg (N) yr^{-1})), and in the ratio between increases in ozone and NO_x concentrations ΔO_3 (ppbv)/ΔNO_x (ppbv) are given in the last two columns.

The basic model case (R) for approximately 1980 emissions is run for 15 months. The perturbations are then run for four months each starting 1 May with initial conditions equal to the concentrations in the reference case at that time. The final results are averages for the month of August (after three months spin-up period).

Berntsen et al. (1996) have pointed out that the reference results lead to surface NO_x concentrations in the Asian region with maxima over Japan and north-east China of about 1 ppbv as a monthly averaged concentration in August. This level is about 45% and 30% of the corresponding levels predicted over the eastern United States and Central Europe, respectively. Over south-east Asia and India, the NO_x concentrations in the model are significantly lower (300 pptv or less).

The results obtained by Berntsen et al. (1996) indicate that in the 3-D model calculations, the adopted perturbations lead to significant increases in surface ozone in the Asian region (up to 70–90% in the case of A2). Table 2.43 data illustrate some of the results.

It is important that enhanced short-term ozone values during episodes are becoming more pronounced in the cases with high emissions. Significant ozone increases have also been obtained in the upper troposphere due to efficient convective transport. A doubling of NO_x emissions results in ozone increases up to 30% in the upper troposphere. Estimates of the regionally positive radiative forcing due to the increases of tropospheric ozone give values up to 0.5 W/m^2, which is 30–50% of the negative RF due to the direct effect of sulphate aerosols in the region. The averaged RF values for the northern hemisphere, the southern hemisphere and total are calculated to be 0.13, 0.016, and 0.073 W/m^2, respectively. The longwave RF dominates the total RF which is highly inhomogeneous. These conclusions are a very important demonstration of the role of tropospheric ozone as a greenhouse gas (Wang and Isaksen, 1995).

The calculations presented in Table 2.43 have been made to demonstrate the importance of the background concentrations impact on the increase of emissions.

Table 2.44. Contribution to ozone changes in the troposphere from various processes in the model for the whole northern hemisphere

	R (Tg/month)		A2 (Tg/month)		Diff.(A2 − R)(Tg/month)	
Process	May	August	May	August	May	August
strat. flux	96.7	33.4	95.8	31.9	−0.9	−1.5
deposition	80.5	71.0	85.2	77.8	4.7	6.8
net chem. prod. (PBL)	50.7	68.5	60.5	79.8	9.8	11.3
net chem. prod. (MFT)	−66.4	−74.2	−66.5	−81.8	-0.1	-7.6
net chem. prod. (UFT)	18.3	30.1	21.9	35.2	3.6	5.1
upw. flux top of PBL	−24.6	5.5	−20.4	10.0	4.1	4.4

PBL: planetary boundary layer (σ = 1.0–0.87, surface to appr. 1.5 km); MFT: middle free troposphere (σ = 0.87–0.39, appr. 1.5–7.5 km); UFT: upper free troposphere (σ = 0.39–0.14, appr. 7.5–13.7 km).

The value of the ozone production efficiency defined by the term $PE_{O3} = \Delta O_3/\Delta E_{NO3}$ is maximum over north-east India because of the significantly lower NO_x concentrations in this region. Table 2.44 shows the contribution of various processes to the change in tropospheric ozone in the northern Hemisphere.

As can be seen, in the middle free troposphere (approximately 1.5–7.5 km altitude) there is an increase in the flux of ozone from the boundary layer of about 4.4 Gt/month for August when anthropogenic emissions are doubled over Asia. There is also an increase in the downward transport from the upper free troposphere.

Berntsen et al. (1996) have pointed out in conclusion that with a sustained economic growth in the Asian region with increased use of coal as the primary source of energy, local pollution effects from sulphur emissions will probably lead to emission controls on sulphur emissions. The emissions of NO_x, CO and hydrocarbons leading to ozone production, are more difficult to control and will probably continue to increase. The net effect of anthropogenic emissions of sulphur dioxide and ozone precursors on the radiation budget over Asia is therefore likely to become positive in the future.

Flatøy et al. (1996) have applied a 3-D Mesoscale Chemistry Transport (MCT) model driven by meteorological data from a Numerical Weather Prediction (NWP) model to calculate ozone and other chemical species over Europe over a 10-day period (1–10 July, 1991) which is characterized by warm weather and frequent occurrence of cumulus convection. Since the photolysis rates are highly dependent on cloud amount and cloud optical depth, the meteorological model uses cloud water as one of the prognostic variables and has an advanced treatment of stratiform and convective cloud and precipitation processes which allows a reliable enough simulation of cloud properties. The results obtained by Flatøy et al. (1996) show that: (1) physical processes, especially convection, may dominate in the vertical distribution of ozone in the free troposphere; (2) sinking air which compensates for convective updrafts is important for the tropospheric ozone budget; (3) transport of ozone precursors from the boundary layer into the free troposphere by convection enhances the rate of O_3 photochemical production significantly.

A case study has been considered of the photochemical development and transport of ozone and other chemical components in the troposphere over Europe for the time period from 0000 UT on 1 July 1991 to 0600 UT on 11 July 1991. Model values have been compared to ground measurements and vertical soundings. The comparison with ground level measurements shows that the model performance is reasonable, but some phase and magnitude errors can be found (there are various causes, why this has happened). Photochemical production rates as high as 1.5 ppb/h was calculated when the boundary layer contained about 3 ppb NO_x, 130 ppb CO, and 20 ppb NMHC (non-methane hydrocarbons) which agrees with some earlier results.

Flatøy et al. (1996) have emphasized the necessity of further improvement of the model as well as more thorough comparisons with observation data. Future modelling will include a study of how far North American tracer emissions of NO_x, NH_3, SO_2, and VOCs (volatile organic compounds), and the secondary products formed can be identified, and in what chemical form. In addition, the chemical impact of subsonic aircraft emissions of NO_x over the North Atlantic on the upper tropospheric NO_x and ozone budgets will be calculated. New assessments will be obtained of how the flux of stratospheric ozone into the troposphere in tropopause folding events can influence the oxidation of NO_x and VOCs.

2.9.2 Effects of UV-B radiation on terrestrial ecosystems

A limited number of studies have been accomplished, so far, on the effect of UV-B radiation (wavelengths 280–320 nm) on various crops (mainly in conditions of special chambers and greenhouses and only in some cases in the field conditions); mainly individual plants have been studied and not populations, communities or ecosystems, though there is no doubt that long-term investigations of the response of the integrated ecosystems to an UV-B increase are needed (Effects..., 1992, 1993; Kondratyev et al., 1995). Interactions and feedback between individual species of ecosystems cannot be assessed reliably with the data available only for individual species. Therefore, study of individual species should be considered only as an initial stage of the subsequent, more complete investigations. One of the unexpected and surprising results of the considered consequences consists in detecting a very strong variability of the sensitivity of various species (different samples of plants within a given species) to changes in UV-B radiation. The characteristics such as vegetation growth and biomass distribution were used as the quantitative criteria of sensitivity. As a rule, the sensitivity of plants to UV-B changes is weaker in the field conditions than in chambers and greenhouses.

A decrease in biomass due to the ceasing processes of photosynthesis when the UV-B radiation intensifies is one of the concrete manifestations of the plants' sensitivity. In some cases the biomass of plants shrinks as a result of the decrease of the leaf cover extent, but in other cases some morphological changes appear, manifesting themselves as an increasing number of branches, changing thickness of leaves and growing foliage, or shortened trunks. The response to a UV-B increase may weaken owing to various interacting chromophores (e.g., nucleic acids, phytohormones and pigments). In some cases the functioning has been discovered of specific UV-B receptors, which absorb light mainly at wavelengths ~290 nm.

A few studies of tree species have shown, for example, that the increasing UV-B radiation suppresses the development of pines (*Pinus taeda*), with an accumulation of negative consequences for the subsequent three years.

Since one may expect in the future not only a UV-B increase but also an increase of CO_2 concentration and a climate warming, a combined account of these factors becomes urgent. From the data available, an increase of CO_2 concentration tells mainly on species C3, whereas species C4 are much less sensitive from the point of view of their photosynthetic reaction. Experiments with the combined account of these factors revealed in the case of C3 species such as rice and wheat that when CO_2 concentration increases, the biomass does not grow provided the plants are additionally irradiated with the UV-B lamps, whereas soy-beans do not react to CO_2 when the UV-B increases. With the solar irradiation, a temperature increase in the case of maize (C4) or CO_2 increase in the case of sunflower (C3) could compensate the negative forcing due to increasing UV-B radiation. If, however, the UV-B increase is combined with that of CO_2 over the saline marshes, these two factors practically do not interact. Thus, the response of plants to the combined forcing of CO_2 and UV-B depends substantially on plant species.

The effect of UV-B increase on the plant chemistry shows itself mainly through the increasing concentration of the UV-B—absorbing pigments (first of all, flavonoids) whose function is to protect the plants from the harmful effect of the UV-B. So, for example, an accumulation of such pigments in the epidermis of wheat protects the leaves from photosynthetic suppression. An increase of UV-B may also lead to changes in the concentrations of minor components such as alkaloids and coumarines, which can tell on the quality of agricultural produce, the rate of organic disintegration, and the intensity of emissions of trace gases to the atmosphere.

The effect of UV-B on the suppressed plants from the ecosystem is a separate aspect of the problem. It has been established, for example, that plants developing in conditions of the moisture deficit react weaker to a UV-B increase than those well moistured. The plants suffering from deficient phosphorus are also less sensitive to a UV-B increase (the effect of other deficient biogens has not been studied). Of importance is an account of interactions between the effects of soil contamination and UV-B increase. The following unsolved problems connected with the effect of increased UV-B radiation should be considered of first priority:

- Assessment of the impact on primary production of various agricultural and non-agricultural ecosystems.
- Assessment of the impact on the morphological changes in the architecture of the vegetation cover and biomass distribution as well as the respective consequences.
- Assessment of direct and indirect impacts on the users of the ecosystems' produce and of the possible effect of feedbacks on the ecosystems.
- Analysis of the genetic basis of resistance to the effects of UV-B changes.

2.9.3 Effects of UV-B radiation on aquatic ecosystems

The aquatic ecosystems cover a wide spectrum of habitat from the oceans to large and small ponds, rivers, estuaries and wetlands whose state varies from the level of undisturbed ecosystems to practically complete anthropogenic transformation. With all diversity, the aquatic systems have a feature in common: they are involved in the cycles of energy and substance, with the decisive role of the varying incoming solar radiation (as a source of energy) and its spectral distribution. The primary production whose level is bound up with the dynamics of nutritional chains and biochemical cycles of various compounds determining a wide range of phenomena, from chemistry of aquatic domains to regional and global weather and climate, is the focus of the processes of photosynthesis taking place due to solar energy. The UV-B—induced changes in the spectral regime signify an appearance of substantial environmental disturbances (first of all, connected with photoprocesses in aquatic ecosystems), which have existed over the long period of the evolution of aquatic ecosystems.

Numerous facts have been found out which testify to the destructive impact of UV-B radiation on aquatic organisms. In particular, the effect of UV-B increase on primary production in the Antarctic has been revealed. This not only affects phytoplankton resistivity but also breaks the polychromatic balance of the light field, which affects the interaction of the processes at molecular, biochemical, physiological and trophic levels and, apparently, causes changes in the evolution of natural communities. However, the consequences of the UV effects on regional and global aquatic primary production and related processes in the ecosystems remain unclear. In this connection, of crucial importance are data of satellite and surface observations which can be obtained in the near future: they will provide information about the global UV-B distribution over the ocean surface and enable one to estimate the depth of penetration. In this context the problem of obtaining the data on the photosynthetically active radiation (PAR) is also urgent.

The scarcely solved problem is an analysis of the effect of UV-B increase on the communities and ecosystems (a community is defined as part of the ecosystem, an ensemble of organisms which exist within a certain ecological region), though it is clear that the response of aquatic ecosystems may vary in a wide range, from zero to catastrophic. This determines the need to:

- obtain more complete and reliable databases on a UV-B increase on local, regional and global scales;
- reveal the aquatic ecosystems potentially more sensitive to UV-B forcings;
- investigate the relationships between the response of ecosystems to the forcings and their subsequent recovery;
- make quantitative estimates of the effects manifested on various spatial–temporal scales;
- reveal the processes in ecosystems affected by increasing UV-B radiation;
- investigate possible global positive feedback taking account of changes in cloud cover, temperature and processes of biological adaptation which can be determined by global changes in the level of the UV-B radiation reaching the Earth's surface.

2.9.4 Effects of UV-B radiation on global biogeochemical cycles

Cyclic variations of solar radiation manifested as daily and annual courses as well as the solar activity-induced long-range variability (an 11-year cyclicity is its most clear manifestation) substantially affect the biogeochemical cycles determined by a complicated interaction of biological, chemical and physical processes of the substance and energy transformation. A well-studied example of the cyclicity of the biogeochemical cycle is the annual course of the atmospheric CO_2 concentration determined by the global biomass cyclicity under the influence of the annual change in the incoming solar radiation and, respectively, vegetation processes. In this connection an important contribution is made by the anthropogenic CO_2 emissions which, however, despite their considerable volume ($\sim$6GtC/year), are by an order of magnitude less than the CO_2 exchange between the atmosphere and terrestrial biosphere ($\sim$60 GtC/year), and the close-in-value CO_2 exchange between the atmosphere and the ocean. The latter means that even small variations in the biospheric fluxes can cause substantial changes in the atmospheric CO_2 concentrations.

An analysis of the mean global CO_2 trend reveals an interesting fact that despite the slowing-down rate of the anthropogenic CO_2 emissions, the ratio of the observed rate of CO_2 increase to that due to fossil carbon fuel burning has substantially increased. This can be interpreted as a result of increasing UV radiation at the Earth surface level during the last 15 years, causing a suppression of photosynthetical activity and, hence, a decrease of primary production and weakening an assimilation of atmospheric CO_2 (the effect of other factors, e.g., changes in land use in the tropics, is not excluded).

A UV-B increase can not only suppress the photosynthetic activity but also intensify the disintegration of non-living organic substance. Over long time periods, variations in the productivity of ecosystems can affect the dynamics of the biogenic cycles. In particular, the rate of the cycle of the compounds resisting the disintegration processes (lignin, humus) can increase. The observational data show, for example, that an increase of UV-B radiation favours a disintegration of the aquatic humus substances.

The close interaction of the biogeochemical cycles determines inevitable accompanying changes in the cycles of nitrogen, oxygen, sulphur, halogens and metals playing an important role in the cellular metabolism. Three major manifestations are possible of the effect of a UV-B increase on the cycles of the enumerated elements:

— Direct effect on the biological processes can happen due to an intensified destruction of RNA and other biological molecules (first of all chlorophyll), and as a result, changes take place in the photosynthetic activity and respiration of terrestrial vegetation as well as in the processes either directly or indirectly connected with marine phytoplankton, such as microbe respiration, nitrification, sulphur cycle and metallic redox reactions with the participation of metals. These effects may be connected with such phenomena as global climate warming and CO_2 increase, and PAR variations.

— A UV-B increase affects strongly the geochemical processes in the sea and on land. The processes taking place both in the sea and on land include: producing the "greenhouse" and chemically active gases (CO_2, CO, NO, COS (carbonylsulphide), NMHC (non-methane hydrocarbons)); mineralization of nitrogen and phosphorus; transformation of organic matter to biologically simple organic compounds with a smaller molecular weight; a deeper penetration of UV radiation to the sea water thickness; changes in the redox state of the upper oceanic layer, and others.

— The UV-B—induced chemical processes in the atmosphere manifest themselves through changes in: the oxidizing property of the troposphere; the formation (more intensive) of greenhouse gases, such as methane and ozone; concentrations of aerosol and cloud condensation nuclei; concentrations of the ozone-destructing gases (including various halogenized compounds). Such phenomena should strongly interact with chemically active gases emitted by the burnt biomass (organic substances) and other anthropogenic processes.

All the enumerated changes may be followed by an appearance of feedback which either intensifies or reduces the formation of greenhouse gases and aerosol in the atmosphere, which will tell on the global climate dynamics.

2.9.5 Dynamics of atmospheric chemistry: the role of remote sounding

To obtain reliable estimates of the regional and global consequences of the effect of an UV-B increase on the biogeochemical cycles and respective feedbacks, it is necessary to develop numerical simulation models and to obtain adequate data of observations of the dynamics of atmospheric chemistry. A deeper insight into the correlation of climate changes and the biogeochemical cycles of the atmospheric trace gases (TG) such as CO_2, CO, N_2O, DMS, and OCS determines the increasing urgency of developments in the sphere of numerical climate modelling taking account of the remotely sensed data on the cycles of respective TG obtained from satellites. As an example, Erickson and Eaton (1993) discussed the possibility of using the CZCS data to estimate the global distribution of the phytoplankton-induced input of carbonylsulphide (OCS) from the surface waters to the atmosphere.

Results of retrieving the content of phytoplankton (chlorophyll) in the surface waters are used to determine the "photochemical lability" (PL) of dissolved organic substance in the surface layer of the ocean. Large values of PL are assumed to correspond to the regions with a large content of chlorophyll. The OCS flux from the ocean to the atmosphere can be calculated as

$$F_{\mathrm{OCS}} = K_{\mathrm{OCS}}\,([\mathrm{OCS}]_{\mathrm{SO}} - [\mathrm{OCS}]_{\mathrm{eq}}),$$

where K_{OCS} is the coefficient of diffusion (the rate of transmission), $[\mathrm{OCS}]_{\mathrm{SO}}$ the OCS concentration in the surface layer, and $[\mathrm{OCS}]_{\mathrm{eq}}$ the equilibrium OCS concentration in the ocean, to which the atmospheric mixing ratio 500 ppt (by volume) corresponds.

It is assumed that $[\mathrm{OCS}]_{\mathrm{eq}} = 10 \times 10^{-6}$ mol/l, and $[\mathrm{OCS}]_{\mathrm{SO}} = B\,[\mathrm{OCS}]_{\mathrm{max}}$, where $[\mathrm{OCS}]_{\mathrm{max}}$ is a maximum possible concentration of OCS in the oceanic surface layer, determined by the relationship: $[OCS]_{max} = 115\,[CHL] + 25$. The quantity [CHL] is the concentration of chlorophyll (mg/m^3) from the CZCS data. The quantity $[OCS]_{max}$ characterizes a maximum input of sulphur-organic compounds to the oceanic surface layer which can be photooxidized. The quantity $B = R_{SfC}/R_{max}$, where R_{SfC} is the mean monthly net radiation for a square 2.8° × 2.8° lat.-long.; $R_{max} = 375\ W/m^2$. Calculations with the NCAR CCM1 global spectral (T42, which corresponds to 2.8° × 2.8°) climate model provide data on the global fields of net radiation, surface wind speed, and SST every 24 h over 5 annual cycles (the data for year 4 of integration were used to estimate the OCS flux).

An analysis of the results of calculations for January and July has shown that the concentration of OCS in the oceanic surface layer varies from 10×10^{-6} mol/l in oligotrophic waters with a low level of insolation to 100×10^{-6} mol/l in the highly productive zones with a high level of insolation, never exceeding 200×10^{-6} mol/l (probably, this is connected with the "saturation" of the CZCS sensitivity range). The mean annual, mean global value of the concentration constitutes 27×10^{-6} mol/l and agrees well with the earlier estimate $32 \times 10^{\times 6}$ mol/l. Mean monthly values of K_{OCS} vary from 0 to 1500×10^{-9} mol OCS (in some NH regions the ocean sometimes turns out to be a sink for OCS). The NH Atlantic and Pacific Oceans are characterized by large OCS fluxes (> 1000) in July. The mean global, mean annual OCS flux is 0.25 TgS (as OCS) per year. This quantity should be considered the lower limit.

The interactivity of the chemical processes in the atmosphere affected by increased UV radiation determines the need to accumulate a global database characterizing the dynamics of atmospheric chemistry determined by the anthropogenic data. At present, it has been stated that the anthropogenic emissions of nitrogen oxides (NO_x) and sulphur gas due to industrial emissions, burning of fossil fuel and biomass affect substantially the chemical processes in the global troposphere.

In this connection, a project has been initiated (Penner et al., 1993) on drawing up an inventory of global emissions (GEIA) within the key IGBP project on the global atmospheric chemistry (IGAC) bearing in mind the data archiving over the grid 1° × 1° lat.-long. Such an archive, together with the models of photochemical processes in the atmosphere, can be used to forecast the dynamics of most important atmospheric trace gases (TG).

A comparison of the results thus obtained with the data of observations will make it possible to assess the adequacy of the present models of photochemical processes in the atmosphere as well as numerical modelling of the budget, transport and other processes. All this can provide the basis for reliable estimates of possible consequences of anthropogenic TG emissions to the atmosphere.

As has been mentioned above, two aspects of the discussed problems are of particular interest:

(1) The oxidizing property of the atmosphere and its possible changes under the influence of emitted methane, nitrogen oxides, carbon oxide and non-methane hydrocarbons (NMHC). The oxidizing property of the atmosphere is known to be determined mainly by the concentration of hydroxyl (Table 2.45), but ozone and hydrogen peroxide also play an important role.

Table 2.45. Atmospheric trace gases either removed or oxidized through chemical reactions in the presence of hydroxyl

Component	Value of the component considered
CH_4	Greenhouse gas; affects the content of tropospheric and stratospheric ozone and OH
CO	Urban atmospheric pollutant; affects the tropospheric O_3 cycle and OH
CH_3CCl_3 CH_3Br CH_3Cl	Greenhouse gases; their disintegration in the stratosphere gives free chlorine which destroys ozone
HCFCs	Hydrochlorofluorocarbons are greenhouse gases; the CFCs substitutes giving free chlorine and destroying ozone in the atmosphere
NMHCs	Urban atmospheric pollutants; determine an increase of tropospheric ozone in the regions with increased concentrations of NO_x, responsible for the sink reactions of tropospheric OH
DMS, H_2S SO_2	Lead to the formation of sulphate aerosol which contributes to the formation of acid rains and affecting the climate (due to increased backscattering of solar radiation); the formation of sulphate condensation nuclei affects the optical properties of clouds

If the hydroxyl concentration decreases on global scales, the concentration of the TG removed by hydroxyl may increase even in the absence of changes in the intensity of their sources. As for hydroxyl, it forms from O_3 photolysis (under the influence of solar radiation at wavelengths <310 nm) giving $O(^1D)$. The subsequent reaction of $O(^1D)$ with a water vapour molecule gives two hydroxyl radicals. Since the OH concentration in the background atmosphere is mainly determined by concentrations of CO, CH_4, NO_x and H_2O, to assess the effect of these TG on O_3 and OH, data are needed on the power not only of their natural but also of their anthropogenic emissions.

(2) The effect on climate of aerosol formed from gas-to-particle reactions with the participation of anthropogenic emissions of sulphate compounds. Various types of thus-formed anthropogenic aerosols are possible depending on their chemical composition (dissolved in water non-organic components, elemental or soot carbon, mineral dust). An important fact is that the processes of the formation of sulphate aerosol depend on the oxidizing property of the atmosphere, and this determines, in particular, the importance of data on the anthropogenic emissions of NO_x and SO_2, as well as CH_4 and CO.

The global database characterizing the spatial–temporal variability of NO_x and SO_2 emissions (over the grid $1^\circ \times 1^\circ$), the estimates of the global propagation of these TG, as well as the formation of sulphate aerosol and the effect of nitrogen oxides on tropospheric ozone have been discussed (Penner et al., 1993).

Calculations of the global field of the concentration of sulphate aerosol were made with due regard to five natural and anthropogenic sources of sulphate

Table 2.46. The power of global sources of chemically active sulphur compounds and their possible changes (MtS/year)

Source	Total emissions	Probable range
Fossil fuel and industry	66	60–100
Biomass burning	2.2	1–4
Soils, forests, crops	0.9	1–10
Volcanoes: degassing	3.4	
eruptions	6.4	
Oceans	23.7	12–40

compounds: the formation (mainly) of dimethylsulphide (DMS) by the oceanic phytoplankton, volcanic emissions of sulphur dioxide (SO_2); the terrestrial emissions of DMS and sulphurous hydrogen (H_2S) by soils, forests and crops; SO_2 emissions due to fossil fuel and biomass burning as well as industrial emissions of SO_2. The data in Table 2.46 characterize the global emissions of sulphur compounds.

The prescribed emissions in the model of the global transport and chemical transformation have led, on the whole, to a realistic reproduction of the observed concentration of sulphates in the atmosphere and their wet and dry deposition onto the underlying surface. In the numerical modelling of ozone concentration in the troposphere determined by oxidation of CO, CH_4 and NMHC in the presence of NO_x (and insolation), the sources of chemically active nitrogen compounds have been prescribed as shown in Table 2.47.

The tropospheric ozone number density turned out to be overestimated, though the near-surface concentration is close to that observed. This determines the necessity for further specified calculations (e.g., NMHCs have not been specified and the formation of PAN has not been considered). A most important problem is to

Table 2.47. The power of global sources of the chemically active nitrogen compounds and possible ranges of its changes (MtN/year)

Sources	Total emissions	Possible range
Anthropogenic		
Fossil fuel burning	22.4	15–5
Biomass burning	10	2–20
Aircraft wastes	0.23	0.1–0.7
Forest fires	0.2	0.1–0.8
Natural		
Emissions due to microbiological processes in soils	5	1–20
Lightning discharges	10	2–100
Formation in the stratosphere	1.0	0.5–1.5
Total	48.8	20–168

increase the reliability of the initial database on emissions, both natural and anthropogenic. Means of space-borne remote sounding, such as those used on the satellite UARS and planned for EOS, open up new prospects for obtaining a representative global database characterizing the dynamics of atmospheric chemistry.

A comparison has been made (Mote et al., 1993) of the data on stratospheric methane and water vapour observed with the UARS occultation sounding instruments HALOE and calculated with the use of the NCAR CCM–2 global climate model (for the system "troposphere–middle atmosphere"). This spectral model, developed in September 1992, includes the improved units to parameterize the radiation transfer and the dynamics of clouds and convection, with the use of the semi-Lagrangian scheme to calculate the transport of atmospheric trace gases. The version of the model corresponding to the "middle atmosphere" covers the atmospheric thickness 0–76 km and has a horizontal resolution of about $3.75° \times 3.75°$ lat.-long.

To increase the reliability of calculations of water vapour, the transport of methane as a tracer and its oxidation to H_2O were taken into account (the oxidized methane gives β of water molecules per one molecule of methane, the value of β being 1.5–2.0; since β grows with height and, taking account of the conversion of H_2 to H_2O, it is assumed that $\beta = 2$; the mixing ratio for methane in the troposphere is assumed to be 1.5 ppm). In the initial integration over 4 years (initial conditions are prescribed from the data as of 1 September) a low vertical resolution was used which then was raised to 75 levels (resolution ~ 1 km in the middle atmosphere).

Despite a somewhat simplified parameterization of the photochemistry of methane and prescribed climatological information, the numerical modelling has reliably reproduced a number of important observed regularities of the fields of concentrations of methane and water vapour: (1) sedimentation within an extensive stratospheric thickness in the circumpolar vortex of the southern hemisphere; (2) intensive dehydration within the vortex; (3) the existence of a region of low values of the water vapour mixing ratio stretching from the Antarctic to the NH tropics (this testifies to the fact that dehydration taking place in the Antarctic substantially contributes during part of the year to a decrease of the moisture content in the moderate and tropical latitudes). These results illustrate the prospects for satellite remote sounding from the point of view of monitoring the dynamics of the chemical composition of the global atmosphere and checking the reliability of the results of its numerical modelling.

2.9.6 Conclusion

Following the structure of the review (Kondratyev et al., 1995a), we shall briefly formulate the most important conclusions. Undoubtedly, the main unsolved problem is poorly studied effects of UV-B radiation on ecosystems. Therefore it is very important to investigate the effect of UV-B radiation on:

1. Terrestrial ecosystems, including:
   1.1. Primary productivity of individual agricultural and non-agricultural ecosystems.

- 1.2. Morphology (architecture) and the spatial structure of distribution of the vegetation biomass.
- 1.3. The structure of populations and its stability.
- 1.4. The effect on the chemical composition and related studies at the level of ecosystem.
- 1.5. The effect of the long-term evolution of vegetation taking account of the genetic dynamics.

2. Aquatic ecosystems, including:
   - 2.1. Variability of UV-B radiation and of the relationship UV-B : UV-A : PAR at different depths and the respective effects on the ecosystems.
   - 2.2. The spatial–temporal sensitivity of the ecosystems (primary productivity, balances of substance and energy, adaptivity) and UV-B changes.
   - 2.3. The effect of physical, chemical and biological factors on the manifestation of the impact of an UV-B increase on the ecosystems.
   - 2.4. Revealing the role of potential feedbacks (determined, for example, by changes in clouds and temperature) which can show itself in the process of the effect of increased UV-B radiation on the ecosystems.
3. Biogeochemical cycles and global changes (Table 2.48), including:
   - 3.1. The cycles of carbon and nitrogen in terrestrial ecosystems.
   - 3.2. The cycles of carbon and nitrogen in aquatic ecosystems.
   - 3.3. The cycle of sulphur in aquatic ecosystems.
   - 3.4. The cycles of biologically important metals and phosphorus in the aquatic domain.
   - 3.5. Formation and degradation of organogenes.
   - 3.6. Disturbances introduced to atmospheric biogeochemical cycles, which is of interest from the viewpoint of the effect on climate (CO_2, CH_4, N_2O, SO_2, DMS, O_3), the quality of tropospheric air (NO_x, SO_2, CH_4, CO, O_3, NMHC) and stratospheric chemistry (COS, CH_4, halogenometals).
   - 3.7. Development and application of remote sensing techniques to assess the effect of increased UV-B radiation on ecosystems on global scales with account of other related processes.
   - 3.8. Development and application of techniques of the simulation numerical modelling of the environmental dynamics (first of all, global biogeochemical cycles) in conditions of anthropogenic effects with a complex account of the interactively affecting factors such as a decrease of the total content of stratospheric and an increase of tropospheric ozone, an increase of UV-B radiation, an increase of the concentration of CO_2 and other greenhouse gases, an intensification of the greenhouse effect of the atmosphere, climate warming, and others.

Table 2.48. Direct and indirect forcings of UV-B radiation on the biogeochemical cycles

Carbon	Terrestrial ecosystems	(1) Photosynthesis, shifting of plant species. (2) Emission of non-methane hydrocarbons. (3) Composition of the substratum, disintegration of organic substance.
Carbon	Aquatic ecosystems	(1) Degradation and cycle of organic substance. (2) Spectral properties of aquatic domain. (3) Synthesis of organic carbon and CO_2 assimilation by phytoplankton and macrophytes.
Carbon	Atmosphere	(1) Exchange of carbon-containing gases. (2) The rate of chemical transformation, especially in the formation of radicals.
Nitrogen	Terrestrial ecosystems	(1) Fixation of nitrogen. (2) Nitrification, emissions of N_2O and NO. (3) Composition and disintegration of the substratum. (4) Emissions of inorganic N as a result of degradation of organic matter and subsequent effect on soil fertility, CH_4/CO sinks.
Nitrogen	Aquatic ecosystems	(1) Nitrification, emissions of N_2O and NO. (2) Formation of inorganic N as a result of degradation of organic matter and subsequent effect on accessibility of biogens. (3) Formation of radicals from NO_3^- and NO_2^-.
Nitrogen	Atmosphere	(1) Exchange of nitrogen-containing gases (N_2O, NO, NH_3) between the atmosphere and other natural domains. (2) The rate of chemical transformation, especially in the formation of radicals.
Sulphur	Terrestrial ecosystems	(1) Assimilation and emission of COS by vegetation.
Sulphur	Aquatic ecosystems	(1) Assimilation and emission of the precursors DMS, COS and S by aquatic organisms. (2) Formation and consumption of DMS. (3) Photooxidation of DMS.
Sulphur	Atmosphere	(1) Exchange of sulphur-containing gases (COS, DMS, SO_2) between the atmosphere and other natural domains. (2) The rate of chemical transformation, especially in the formation of radicals.
Metal–phosphorus compounds	Terrestrial ecosystems	(1) Emissions of inorganic P as a result of degradation of organic matter and subsequent effect on soil fertility.
Metal–phosphorus compounds	Aquatic ecosystems	(1) Existence of small admixtures of metals (e.g., Fe, Cu, Mn). (2) Formation of OH through reaction in the presence of Fe.

Table 2.48. *continued*

Oxygen	Terrestrial ecosystems	(1) Photosynthesis. (2) Disintegration of the substratum.
Oxygen	Aquatic ecosystems	(1) Photosynthesis and oxidation of DOM. (2) Formation and disintegration of H_2O_2. (3) Formation of HO_2 and OH.
Oxygen	Atmosphere	(1) Transformation $O_2 \rightarrow O_3$.
Halogens	Terrestrial ecosystems	(1) Formation of organohalogens.
Halogens	Aquatic ecosystems	(1) Formation of biogens and consumption of organohalogens.
Halogens	Atmosphere	(1) Disintegration of methylhalides, chlorofluorocarbons and their substitutes with the subsequent destruction of ozone.

2.10 PROSPECTS FOR SATELLITE MONITORING OF ATMOSPHERIC TRACE GASES AND AEROSOLS

2.10.1 Introduction

Studies on changes in the chemistry of the atmosphere from the point of view of the content of the optically active trace gases (TG) and aerosol are now in the focus of the environmental problems (Kondratyev, 1990a, 1992d; Kondratyev and Galindo, 1994; Kondratyev et al., 1996b). This is connected both with the fundamental problem of the closed nature of global biochemical cycles (Gorshkov, 1995; Gorshkov et al., 1992, 1994), the importance of which is still underestimated, and with the discussions on changes in global climate (Houghton et al., 1996; Kondratyev, 1992d) and in total ozone content in the atmosphere (Kondratyev, 1989) attracting a great deal of attention. Therefore developments have been carried out for a long time of various techniques for the satellite TG monitoring (Kondratyev et al. 1992a) which have been recently activated within the programme "Mission to Planet Earth" (MTPE) on the basis of the programme EOS. Jacob and Leovy (1995) were right to draw attention to the following questions:

(1) What are the anthropogenic impacts on stratospheric ozone (total ozone content (TOC) in the atmosphere) and how do they tell on the UV radiation flux at the surface level which affects both the biosphere and man?
(2) To what extent does the climate near the Earth's surface depend on the content of greenhouse gases (GHG) and aerosol in the atmosphere and which factors determine the variability of the optically active TG in the atmosphere?
(3) How does the oxidizing property of the atmosphere, governing its chemistry, vary (among other reasons, under the influence of anthropogenic factors)?
(4) How does air quality vary (especially in densely populated industrial regions) under the influence of anthropogenic emissions of pollutants to the atmosphere?

To answer these questions, we need a more adequate observational database on changes taking place in the atmospheric chemistry as well as deeper knowledge of the physical, chemical and biological processes which determine the atmospheric chemistry. The problems claiming special attention are:

(1) Factors determining the large-scale trends of TOC.
(2) Processes which control the variability of main GHG, including water vapor, carbon dioxide, methane, nitrous oxide, chlorofluorocarbons (CFC) and tropospheric ozone.
(3) Mechanisms regulating the concentration of ozone and other oxidants in the troposphere.
(4) Sources and sinks (global distribution), chemical and optical properties of aerosol.

In this context, the systems of conventional (surface, aircraft, balloon) and satellite observations planned for the year 2000 and beyond will be very briefly discussed.

2.10.2 Systems of observations of trace gases and aerosol

2.10.2.1 Surface observations

Programmes of surface observations foresee, first of all, broadening the existing networks of routine direct measurements of the concentration of stable gases, radicals and aerosol; dry and wet depositions; vertical profiles, composition and structural parameters of the atmosphere using lidar and microwave sounding, an accomplishment of complex observational programmes both on a regular basis and through respective expeditions (e.g., the expeditions undertaken in the Arctic to study TOC dynamics) being most urgent. Naturally, coordinated surface observations and satellite remote sounding (RS) gain importance.

2.10.2.2 Aircraft observations

Over the last several years, aircraft have been intensively used as platforms to study the atmosphere at various altitudes. The former USSR gained rich experience in using for this purpose both jet and turbo-jet (IL-18, AN-30) aircraft. Data in Table 2.49 illustrate the type of aircraft used in the USA to study the atmospheric chemistry (UAF, Uncontrolled Air Force). These aircraft carried various instrumentation for direct measurements of the chemical composition of the atmosphere and for remote sounding (especially with lidars). Clearly, simultaneous surface, aircraft and satellite observations are most urgent, bearing in mind the control of reliability and calibration of the RS data.

2.10.2.3 Satellite observations

The history of using satellites for the remote sounding of the atmosphere started soon after the launch of the first satellite, an application of manned space vehicles playing a special role at the initial stage (Kondratyev, 1995e; Kondratyev et al., 1992a, 1996b). At present, wide prospects for the remote sounding of the atmosphere are connected with the EOS programme within which six types of instruments are planned to be used (Table 2.50).

Table 2.49. Types of aircraft used in the USA to observe the atmospheric chemistry

Type of aircraft	Ceiling (km)	Range (km)	Payload (kg)
ER–2	15–20	5500	2000
DC–8	1–12	9000	5000
P–3	0.3–7	7000	7000
WB–57F	1–19	4000	2000
UAF	0.3–28	200–7000 +	100–400

Table 2.50. The EOS complex for the remote sounding of the atmospheric chemistry

Component	Instruments	Altitude range (km)
O_3, vertical profile	MLS, HIRDLS, SAGE-III	15–70
O_3, total content	ODUS	
O_3, profile in the troposphere	TES	2–15
OH	MLS	20–25
H_2O	MLS, HIRDLS, SAGE-III	5–60
CH_4, vertical profile	HIRDLS	15–70
CH_4, total content	MOPITT, TES	
CO, total content	MOPITT, TES	
ClO	MLS	15–50
$ClONO_2$	HIRDLS	20–35
HCl	MLS	15–50
CFCs	HIRDLS	15–35
NO_2	HIRDLS, SAGE-III	15–50
HNO_3	MLS, HIRDLS	15–40
N_2O, profile	HIRDLS	15–70
N_2O, total content	TES	
Temperature	MLS, HIRDLS	15–80
Tropospheric TG (NO, HNO_3, H_2O, etc.)	TES	
Aerosol	SAGE-III, EOSP, MODIS, MISR	

The MOPITT instruments will be installed on the first of the polar platforms EOS AM-1 to be launched in July 1998 (see Chapter 1 for details).

The platform CHEM-1 carrying the instruments HIRDLS, MLS, TES and ODUS planned to be launched in December 2002 will provide information on the global fields of a large set of TG in the stratosphere and the troposphere. In addition, the satellite Meteor-3M-1 planned for August, 1998 will carry the instruments SAGE III for the 3-D mapping of the content of ozone, water vapour, aerosol, nitrogen dioxide, and some other TG. The polarization observations with EOSP to be carried by the polar platform EOS AM-2 will be aimed to determine the optical thickness and polarization characteristics of atmospheric aerosol in the stratosphere and the troposphere. For this purpose, information on the aerosol optical thickness (AOT) and average size of particles will be obtained from the data of multichannel radiometer MODIS (AM-1,2; PM-1) with the same retrieving technique which was used to process the data of the standard instrumentation AVHRR carried by the NOAA satellites. To retrieve the AOT and average size of particles, multispectral and multi-angle MISR data (AM-1,2) will be used. Very interesting RS programmes to retrieve TG and aerosols are being developed in Japan on the basis of ADEOS satellites (the first ADEOS was launched on 17 August 1996) (Kondratyev and Tanaka, 1997).

2.11 KEY UNSOLVED PROBLEMS IN THE FIELD OF ATMOSPHERIC CHEMISTRY

Naturally, observational studies in the field of atmospheric chemistry should be based on the combined use of conventional direct measurements and satellite remote sounding (the problem of an optimized global observing system remains unsolved). From the point of view of both the specific character of the problems of atmospheric chemistry and the observation techniques, it is worthwhile to separate the discussions of the perspective developments connected with studies of the processes in the stratosphere and the troposphere.

2.11.1 Stratosphere

In this connection, two problems claim first priority.

Dynamics of the ozone content in the lower stratosphere and its determining factors. This complicated problem connected with the necessity of explaining numerous dynamic and chemical processes determining the changes in the total ozone content (TOC) in conditions of anthropogenic forcings due to emissions to the atmosphere of the ozone-destroying gases (ODG) such as CFC, requires, apart from chemical processes (first of all, heterogenic chemical reactions on the atmospheric aerosol particles), an account of the stratosphere—troposphere interaction, as well as the air exchange between high and middle, middle and tropical latitudes.

From the point of view of obtaining the observation data, this signifies the need for: (1) monitoring of ozone with high spatial and temporal resolutions and of most important ODG (ClO, NO_x), aerosol, water vapour, temperature and other meteorological parameters; (2) obtaining adequate information to meet the needs of the numerical modelling of the stratospheric variability. The problem of global dynamics of the ozone layer should be solved in the context of varying concentrations of CFCs and their substitutes (HCFCs), possible intensification of stratospheric aviation, as well as probable changes in temperature and circulation in the stratosphere (stratospheric climate) under the influence of increasing GHG concentrations (of course, the interactivity of the climatic and ozone-layer dynamics should be borne in mind).

Solution of these and other problems should be supported by the programme EOS CHEM, an accomplishment of which will provide global databases on the concentrations of O_3, ClO, NO_3, $ClONO_2$, HCl, HNO3, N_2O_5, CFCs, H_2O, CH_4, as well as on aerosol and temperature with the vertical resolution 2–3 km (MLS) and 1 km (HIRDLS). The stratosphere–troposphere exchange will be studied with the use of the combined data of aircraft measurements with the vertical resolution $\sim$100 m along the horizontal routes from 100 km to 500 km and of the satellite remote sounding.

In connection with an accomplishment of the EOS programme CHEM, two additional aspects of developments appear: (1) observations of the UV radiation flux and substantiation of requirements for these observations (Kondratyev et al., 1995a, 1996b); (2) expedience of including to the programme the measurements of concentrations of OH or other HO_x radicals through some modification of the MLS instruments. The urgency of the problem of hydroxyl is determined by its important role in the atmospheric chemistry for several reasons: (1) OH controls the conversion of CH_4 to H_2O; (2) reactions in the presence of HO_x are most important as mechanisms for ozone losses in the upper and lower stratosphere; (3) reactions in the presence of OH determine the rate of oxidation of gaseous sulphur compounds (SO_2 , OCS) to give sulphate aerosol; (4) OH competes with heterogenic chemical reactions with respect to control of the transformation of radicals and "reserve" components in the systems of families NO_y and Cl_x. So, for example, the reaction $OH + NO_2 \rightarrow HNO_3$ competes both with the inverse reaction ($OH + HNO_3$) and with reactions

$$ClO + NO_2 \rightarrow ClNO_2 \text{ and } OH + HCl \rightarrow Cl + H_2O$$

Since in the models of chemical processes the content of OH is expressed, as a rule, through concentrations of other gaseous components, the data of measurements are needed which characterize the dependence of OH on the concentration of such components in a wide range of real conditions.

Factors determining the ozone concentration in the middle and upper stratosphere. Though the ozone residing in the middle and upper stratosphere contributes relatively little to TOC and therefore weakly affects the near-surface UV radiation flux, it is very interesting as one of the factors responsible for the formation of the temperature field and atmospheric circulation in the stratosphere which may affect the structure and composition of the lower stratosphere. In this connection, the following two unsolved problems are most urgent: (1) the remaining differences between the data of observations and the results of numerical modelling for the zone of dominating photochemical processes at altitudes above 40 km; (2) poorly studied heterogenic chemical processes in the lower stratosphere (25–35 km). In this context, the results of observations with the instruments MLS, HIRDLS and SAGE-III will play an important role (of special interest are the MLS data on hydroxyl).

2.11.2 Troposphere

The combined use of the data of direct measurements and remote sounding will open up prospects for solving most urgent problems.

The fields of concentration of most substantial GHG (H_2O, CO_2, CH_4, N_2O, HCFCs, O_3) and their determining factors. While for the long-lived gaseous components such as CO_2, CH_4, N_2O and HCFCs, rather uniformly mixed up in the atmosphere, the representative information can be obtained from the data of direct surface observations, in the case of water vapour and tropospheric ozone characterized by strong spatial and temporal variabilities the data of remote sounding are important.

The fields of concentration of oxidants in the troposphere, ozone and hydroxyl, in particular. The oxidizers such as ozone, hydroxyl and others (H_2O_2, in particular) form in the troposphere as a result of a complicated set of photochemical reactions in the presence of nitrogen oxides, hydrocarbons and water vapour, the relative role of which remains poorly studied, though the primary importance of heterogenic chemical reactions on aerosol and cloud particles raises no doubts. To solve the complicated problems of the heterogenic chemistry of the atmosphere, not only the observational data on numerous TG and aerosol are needed but also the problem-oriented field experiments, bearing in mind, in particular, studies of the gas exchange between the atmosphere and the biosphere, the atmosphere and the ocean (in this context the correlation approaches of direct measurements of TG fluxes in the atmosphere can play a special role). The satellite RS should become an important source of information on the global fields of the precursors (and indicators) of oxidants in the troposphere such as NO_x, CO, hydrocarbons, water vapour, which are short-lived and strongly variable (Table 2.50).

A substantial source of ozone and NO_x for the troposphere is their transport from the stratosphere, the quantitative data on which are disputable. Since the gas exchange between the stratosphere and the troposphere is realized, apparently, via mesoscale processes, direct aircraft measurements with a high spatial resolution are very important. The satellite RS data will supplement them with respect to the spatial representativity and account of the role of large-scale processes.

Sources and properties of tropospheric aerosol. The recently achieved understanding of then urgency of the problem "greenhouse warming/ aerosol cooling" of climate (Kondratyev, 1991, 1992d, 1996) determines the special role of complex studies of atmospheric aerosol, including a wide application of surface (lidar) and satellite remote sounding techniques.

REFERENCES

Abakumova G.M., Feigelson E.M., Russak V., Stadnik V.V. (1996). Evaluation of long-term changes in radiation, cloudiness and surface temperature in the territory of the former Soviet Union. J. Climate, Vol. 9, N6, p.1319–1327.

Abarbanel H.D.I.(1985). Prediction in chaotic nonlinear systems: time series analysis for aperiodic evolution. Glob. Clim. and Ecosyst. Change: Proc. NATO Adv. Res. Workshop Model Ecosyst. and Their Changes. Maratea, Sept. 4–8, 1985. New York, London. p.185–237.

Acker J. G., Buselacchi A. (1995). Workshop on satellite and in situ observations for

climate prediction. Bull. Amer. Meteorol. Soc., Vol. 76, N2, p.241–250.
Ackerman S. A., Chung H. (1992). Radiative effects of airborne dust on regional energy budgets at the top of the atmosphere. J. Appl. Meteorol., Vol. 312, N2, p.223–233.
Ackerman A.C., Toon O.B. and Hobbs P.V. (1993). Dissipation of marine stratiform clouds and collapse of the marine boundary layer due to the depletion of cloud condensation nuclei by clouds. Science, Vol. 262, p.226–229.
Adamenko V.N. (1990). Ecological Programme of the USSR. Objectives and Means of Realization. Leningrad, Znanie Publ. 32 pp. (in Russian).
Adger W.N., Brown K. (1994). Land Use and the Causes of Global Warming. John Wiley & Sons, Chichester UK, 271 pp.
Aerosol Forcing of Climate. R. J. Charlson and J. Heintzenberg (Eds.), John Wiley and Sons, Chichester, UK, 1995.
Agarwal A., Naraian S. (1991). Global Warming in an Unequal World. A case of environmental colonialism. Centre for Science and Environment. New Delhi. 36 pp.
Airborne Particulate Matter in the United Kingdom. Third Report of the Quality of Urban Air Review Group (1996). The University of Birmingham, Birmingham, U.K., 176 pp.
Airey M.J., Hulme M., Johns T.C. (1996). Evaluation of simulations of terrestrial precipitation in UK Meteorological Office/Hadley Centre climate change experiments. Geophys. Res. Lett., Vol. 23, N13, p.1557–1560.
Alcamo J.(1995). Integrated Modelling of Global Climate Change. Kluwer Acad. Publ., Dordrecht. 328 pp.
Alexandrov G.A., Golitsyn G.S., Mokhov I.I., Petukhov V.K. (1994). Global climate changes and the regulating role of marshes. Izv. RAN, Ser. Geogr., N2, p.5–15 (in Russian).
Alkezweeny A. (1995). Trend analysis of sulphur dioxide emissions and sulphate concentrations and their application to global cooling. Atmosfera, Vol. 8, N2. p.91–99.
Allen J.R.L., Hoskins B.J., Valdes P.J., Sellwood B.W., Spicer R.A. (1994). Paleoclimates and Their Modelling. With Special Reference to Mesozoic Era. Chapman & Hall, Andover, Hants. 160 pp.
Alley R., Mayewski P., Peel D., Stauffer B. (1996). Twin ice cores from Greenland reveal history of climate change more. EOS, Vol. 77, N22, p.209–210.
Allison I., Moritz R.E. (1995). Sea Ice in the Global Climate System: Requirements for an Ocean Observing System. OOSDP Background Report. Univ. of Washington, Seattle. N7, 28 pp.
Andersen R.Y. (1992). Possible connection between surface winds, solar activity and the Earth's magnetic field. Nature, Vol. 358, p.51–53.
Anderson T.L., Wolfe G.V., Warren S.G. (1992). Biological sulphur, clouds, and climate. Ency. Earth Syst. Sci., Vol. 1, p.363–376.
Anderson T.L., Wolfe G.V., Warren S.G. (1995). Biological sulphur, clouds, and climate. In: Ice Core Studies of Global Biogeochemical Cycles. Ed. R.S. Delmas (NATO ASI Series. Vol. 30). Springer-Verlag, Berlin. p.139–165.
Andreae M.O. (1994). Climate effects of changing atmospheric aerosol levels. In:

World Survey of Climatology, Vol. XX. Future Climate of the World (Ed.: A. Henderson-Sellers).
Andreae M.O. (1996). Raising dust in the greenhouse. Nature, Vol. 380, p.389–390.
Andreae M.O., Elbert W., de Mora S.J. (1995). Biogenic sulphur emissions and aerosols over the tropical South Atlantic. 3. Atmospheric dimethylsulfide, aerosols and cloud condensation nuclei. J. Geophys. Res., Vol. 100, ND6, p.11335–11356.
Angell J.K. (1993). Comparison of stratospheric warming following Agung, El Chichón and Pinatubo volcanic eruptions. Geophys. Res. Lett., Vol. 20, p.715–718.
A Plan for a Research Program on Aerosol Radiative Forcing and Climate Change (1996). National Research Council, National Academy Press, Washington, D.C. 161 pp.
Archer S., Schimel D.S., Holland E.A. (1995). Mechanisms of shrubland expansion: land use, climate or CO_2? Climatic Change, Vol. 29, N1, p.91–99.
Ardanuy P.E. (1983). Determinability of interannual global and regional climatic changes of the earth radiation budget. First Conf. on Atmospheric Radiation, Oct. 31-Nov. 4, 1982. Boston, AMS. p.410–413.
Arefyev V.N., Kamenogradsky N.E., Kashin F.V. (1995). Carbon dioxide in the continental atmosphere. Meteorology and Hydrology, N4, p.87–96 (in Russian).
Arimoto R., Duce R.A., Savoie D.L., Prospero J.M., Talbot R., Cullen J.D., Tomza U., Lewis N.F., Ray B.S. (1996). Relationship among aerosol constituents from Asia and the North Pacific during PEM-West A. J. Geophys. Res., Vol. 101, ND1, p.2011–2023.
Arking A. (1991). The radiative effects of clouds and their impact on climate. Bull. Amer. Meteorol. Soc., Vol. 71, N6, p.795–818.
Arking A. (1996). Absorption of solar energy in the atmosphre: discrepancy between model and observations. Science, Vol. 273, N5276, p.779–782.
Arking A., Ziskin D. (1994). Relationship between clouds and sea surface temperatures in the western tropical Pacific. J. Climate, Vol. 7, p.988–1000.
Arking A., Chou M.-D., Ridway W.L. (1996). On estimating the effect of clouds on atmospheric absorption based on flux observations above and below cloud layer. Geophys. Res. Lett., Vol. 23, N8, p.829–832.
Arnott W.P., Dong Y.Y., Halett J. (1994). Role of small ice crystals in radiative properties of cirrus: A case study, FIRE-II, 22 November, 1991. J. Geophys. Res., Vol. 99, ND1, p.1371–1381.
Arola A., Lettenmaier D.P. (1996). Effects of subgrid spatial heterogeneity on GCM-scale land surface energy and moisture fluxes. J. Climate, Vol. 9, p.524–536.
Asano S. (1994). Cloud and radiation studies in Japan. Cloud Radiation Interactions and Their Parameterization in Climate Models. WCRP–86 (WMO/TD No. 648).Geneva,WMO, p.72–73.
Asano S., Shiobara M., Uchiyama A. (1995). Estimation of cloud physical parameters from airborne solar reflectance measurements for stratocumulus clouds. J. Atmos. Sci., Vol. 52, N20, p.3556–3576.
Atlas D., Matrosov S.Y., Heymsfield A.J., Chou M.-D., Wolff D.B. (1995). Radar and radiation properties of ice clouds. J. Appl. Meteorol., Vol. 34, N11, p.2329–2345.
Atmospheric Ozone. Policy Statement (Accepted by the AMS Council 28 January 1996). Bull. Amer. Meteorol. Soc., Vol. 77, N8, p.1247–1248.

Augstein E. (1991). The role of the Arctic regions in global climatic changes. Energy Environ. 1991: Int. Symp., Espoo, Aug. 1991. Atlanta, p.25–28.

Ausbel J.H., Victor D.G., Wernick I.K. (1995). The environment since 1970. Consequences, Vol. 1, N3, p.3–15.

Avdyushin S.I., Tulinov G.F., Ivanov M.S., Kuzmenko B.N., Mezhnev I.R., Manchecorne A., Chanin M.L. (1993). Spatial and temporal evolution of the optical thickness of the Pinatubo aerosol cloud in the Northern Hemisphere from a network of ship-borne and stationary lidars. Geophys. Res. Lett., Vol. 20, p.1963–1966.

Ayers G.P., Ivey J.P., Schäfer B., Gabric A. (1995). Short-term variability in marine atmospheric dimethylsulfide concentration. Geophys. Res. Lett., Vol. 22, N18, p.2513–2516.

Bacastow R.B., Keeling C.D., Lucker T.J., Wahlen M. (1996). The ^{13}C Suess effect in the world surface oceans and its implications for oceanic uptake of CO_2: Analysis of observations at Bermuda. Global Biogeochemical Cycles, Vol. 10, N2, p.335–346.

Baden J.A., Ed. (1994). Environmental Gore. A Constructive Response to "Earth in the Balance". Pacific Res. Inst. for Public Policy, San Francisco, CA, 270 pp.

Balling R.C., Jr. (1995). Analysis of German climatic variations during the period of instrumental records. Geophys. Res. Lett, Vol. 22, N3, p.223–226.

Balling R.C., Jr., Christy J.R. (1996). Analysis of satellite based estimates of tropospheric diurnal temperature range. J. Geophys. Res., Vol. 101, ND8, p.12827–12832.

Barker H.W., Li Z. (1995). Improved simulation of clear-sky shortwave radiative transfer in the CCC-GCM. J. Climate, Vol. 8, p.2213–2223.

Barker J.R. (Ed.) (1995). Progress and Problems of Atmospheric Chemistry. World Scientific, Singapore. 941 pp.

Barron E.J. (1994). Chill over cretaceous. Nature (Gr. Brit.), Vol. 370, N6489, p.415.

Barry R.G., Serreze M.C., Maslanik J.A. (1995). The Arctic sea-ice climate system: observations and modelling. Revs. of Geophys., Vol. 31, N4, p.397–422.

Baumgardner D., Dye J.E., Gandrud B., Barr K., Kelly K., Chan K.R. (1996). Refractive indices of aerosols in the upper troposphere and lower stratosphere. Geophys. Res. Lett., Vol. 23, N7, p. 749–752.

Bazhenov O.E., Kasyanov E.I., Titov G.A. (1994). Effect of aerosol on the absorption and average albedo of broken clouds. Izv. RAN. Ser. FAO, Vol. 30, N6, p.319–327 (in Russian).

Belchansky G.I. (1995). To the question of global change indications from trends of sea-ice cover extent with the use of satellite data. Doklady Russian Acad. Sci., Vol. 343, N1, p.130–135 (in Russian).

Bender M., Ellis T., Tans P., Francey R., Lowe D. (1996). Variability in the O_2/N_2 ratio of Southern Hemisphere air, 1991–1994: Implications for the carbon cycle. Global Biogeochemical Cycles, Vol. 10, N1, p.9–21.

Bengtsson L., Arpe K., Roeckner E., Schulzveida U. (1996a). Climate predictability experiments with a general circulation model. Climate Dynamics, Vol. 12, p.261–278.

Bengtsson L., Botzet M., Esch M. (1996b). Will greenhouse gas-induced warming over the next 50 years lead to higher frequency and greater intensity of hurricanes? Tellus, Vol. 48A, N1, p. 57–73.

Berger A.K., Imbrie J., Hays J., Kukla G., Saltzman B. (Eds.) (1984). Milankovitch and Climate. NATO ASI Series C: Mathematical and Physical Sciences. Vol. 126. D. Reidel Publ. Co., Dordrecht. 510 pp.

Berger A.L. (1992). Astronomical theory of palaeoclimates and the last glacial–interglacial cycle. Quarternary Sci. Rev., Vol. 11, p.571–581.

Berner E.K., Berner R.A. (1996). Global Environment: Water, Air, and Geochemical Cycles. Prentice Hall, Upper Saddle River, N.J. 376 pp.

Berntsen T., Isaksen I.S.A., Wang W.-C., Liang X.-Z. (1996). Impacts of increased atmiospheric emissions in Asia on tropospheric ozone and climate. A global 3-D model study. Tellus, Vol. 48B, N1, p.13–32.

Betts A.K., Ball J.H., Beljaars A.C.M., Miller M.J., Viterbo P.A. (1996). The land surface—atmosphere interaction: A review based on observational and global modelling perspectives. J. Geophys. Res., Vol. 101, ND3, p.7209–7225.

Biggs R.H., Joyner M.E.B., (Eds.) (1994). Stratospheric Ozone Depletion/UV-B Radiation in the Biosphere. NATO ASI Series 1: Global Environmental Change. Vol. 18. Springer-Verlag. Berlin. 358 pp.

Bintanja R. (1996). The parameterization of shortwave and longwave radiative fluxes for use in zonally averaged climate models. J. Climate Res., Vol. 100, ND4, p.7195–7202.

Bitz C.M., Battisti D.S., Moritz R.E., Beesley J.A. (1996). Low-frequency variability in the Arctic atmosphere, sea-ice, and upper-ocean climate system. J. Climate, Vol. 9, N2, p.394–408.

Blake D.F., Kato K. (1995). Latitudinal distribution of black carbon soot in the upper troposphere and lower stratosphere. J. Geophys. Res. D, Vol. 100, N4, p.7195–7202.

Blanchet J.-P. (1994). Mechanisms of direct and indirect climate forcing by aerosols In: Proc. Dahlem Workshop on Aerosol Forcing of Climate, Berlin, April 1994.

Bodri L., Cermák V. (1995). Climate changes of the last millennium inferred from borehole temperatures: Results from the Czech Republic—Part 1. Global and Planetary Change, Vol. 11, p.111–125.

Boehmer-Christiansen S. (1994a). Global climate protection policy: the limits of scientific advice. Part 1.Glob. Environm. Change, Vol. 4, p.140–159.

Boehmer-Christiansen S. (1994b). Global climate protection policy: the limits of scientific advice. Part 2. Glob. Environm. Change, Vol. 4, p.185–200.

Boers R., Ayers G.P., Gras J.L. (1994). Coherence between seasonal variation in satellite-derived cloud optical depth and boundary layer CCN concentrations at a mid-latitude Southern Hemisphere station. Tellus, Vol. 46B, p.123–131.

Bonan G.B. (1995). Sensitivity of a GCM simulation to inclusion of inland water surfaces. J. Climate, Vol. 8, N11, p.2691–2704.

Bond G.C. (1995). Climate and conveyor. Nature, Vol. ,377, N6548, p.383–384.

Bony S., Duvel J.-P., Le Treut H. (1995). Observed dependence of the water vapour and clear-sky greenhouse effect on sea surface temperature: comparison with climate warming experiments Climate Dynamics, Vol. 11, N7, p.307–320.

Borisenkov E.P., Kondratyev K.Ya. (1988). Carbon Dioxide and Climate. Leningrad: Gidro-meteoizdat. 320 pp. (in Russian).

Borrman S., Solomon S., Dye J.E., Luo B. (1996). The potential of cirrus clouds for heterogeneous chlorine activation. Geophys. Res. Lett., Vol. 23, N16, p.2133–2136.

Bosilovich M.G., Sun W.-Y. (1995). Formulation and verification of a land surface parameterization for atmospheric models. Boundary-Layer Meteorol, Vol. 73, N4, p.321–341.

Boucher O. (1995). GCM estimate of the indirect aerosol forcing using satellite-retrieved cloud droplet effective radii. J. Climate, Vol. 8, N5, p.1403–1409.

Boucher O., Anderson T.L. (1995). General circulation model assessment of the sensitivity of direct climate forcing by anthropogenic sulphate aerosols to aerosol size and chemistry. J. Geophys. Res., Vol. 100, ND12, p.26117–26134.

Boucher O., Le Treut H., Baker M.B. (1995). Precipitation and radiation modelling in a general circulation model: Introduction of cloud microphysical processes. J. Geophys. Res., Vol. 100, ND8, p.16395–16414.

Boucher O., Lohmann U. (1995). The sulphate–CCN–cloud albedo effect. A sensitivity study with two general circulation models. Tellus, Vol. 47b, N3, p.281–300.

Bouger B. (1995). Role dela vegetation terrestre dans le cycle global du CO_2 . C.R. Acad. Agr. Fr., Vol. 81, N5, p.109–123.

Boville B.A. (1995). NCAR climate system model. Proc. Workshop on Global Coupled General Circulation Models. Scripps Inst. Oceanogr., La Jolla, USA, Oct. 1994, WCRP–87 (WMO/TD-No.655). Geneva, p.46–47.

Box M.A., Trautmann T. (1994). Computation of anthropogenic sulphate aerosol forcing using additive perturbation theory. Tellus, Vol. 46B, p.33–39.

Brasseur G. (1993). The response of the middle atmosphere to long-term and short-term solar variability: a two-dimensional model. J. Geophys. Res., Vol. 98, ND12, p.23079–23090.

Brasseur G.P., Müller J.-F., Granier C. (1996). Atmospheric impact of NO_x emissions by subsonic aircraft: A three-dimensional model study. J. Geophys. Res., Vol. 101, ND1, p.1423–1428.

British Government Panel on Sustainable Development. First Report. Department of the Environm. London. 1995. 20 pp.

Brock C.A., Hamil P., Wilson J.C., Jonsson H.H., Chan K.R. (1995). Particle formation in the upper tropical troposphere: a source of nuclei for the stratospheric aerosol. Science, Vol. 270, N5242, p.1650–1653.

Broecker W. (1996). Glacial climate in the Tropics. Science, Vol. 272, p.1902–1904.

Brogniez G., Buriez J.C., Giraud V., Parol F., Vanbauce C. (1995). Determination of effective emittance and radiatively equivalent microphysical model of cirrus from ground-based and satellite observations during the International Cirrus Experiments: The 18 October 1989 case study. Mon. Weather Rev., Vol. 123, p.1025–1036.

Brown L.F., Ed. (1996). Vital Signs. World Res. Inst., Washington, D.C. 169 pp.

Brown P.R.A., Illingworth A.J., Heymsfield A.J., Mc Farquhar G.M., Browning K.A., Gosset M. (1995). The role of spaceborne millimeter-wave radar in the global monitoring of ice clouds. J. Appl. Meteorol., Vol. 34, N11, p.2346–2366.

Brown R.D., Goodison B.E., Nakamura M. (1996). Interannual variability in reconstructed Canadian snow cover, 1915–1992. J. Climate, Vol. 9, p.612–624.

Bruce J.P., Lee H., Haites E.F. (Eds.) (1996). Climate Change 1995. Economic and Social Dimensions of Climate Change. Cambridge Univ. Press. 448 pp.

Buffle J., van Leeuwen H.P., (Eds.) (1992). Environmental Particles. Lewis Publ., Boca Raton. 554pp.

Bukata R.P., Jerome J.H., Kondratyev K.Ya., Pozdnyakov D.V. (1995). Optical Properties and Remote Sensing of Inland and Coastal Waters. CRC Press, Boca Raton. 345 pp.

Bye J.A.T., Byron-Scott R.A.D., Gordon A.H. (1996). Climate: the elements. J. Climate, Vol. 9, N7, p.1546–1560.

Byrne R.N., Somerville R.C.J., Subasilar B. (1996). Broken-cloud enhancement of solar radiation absorption. J. Atmos. Sci., Vol. 53 (in press)

Cai W. (1996). Surface thermohaline forcing conditions and the response of the present-day global ocean climate to global warming. J. Geophys. Res., Vol. 101, NC1, p.1079–1093.

Callis L.B.., Baker D.N., Natarajan M., Blake J.B., Mewaldt R.A., Selesnick R.S., Cummings J.R. (1996). A 2-D model simulation of downward transport of NO_y into the stratosphere: Effects on the 1994 austral spring O_3 and NO_y . Geophys. Res. Lett., Vol. 23, N15, p.1905–1908.

Calvert J.G. (1994). The Chemistry of the Atmosphere: Its Impact on Global Change. Blackwell Sci. Publ., London. 394 pp.

Campbell G.G., Kittel T.F.G., Meehl G.A., Washington W.M. (1995). Low frequency variability and CO_2 transient climate change. Part 2: EOF analysis of CO_2 and model configuration sensitivity. Global and Planet Change, Vol. 10, p.201–216.

Cao M., Dent J.B., Heal O.W. (1995). Modeling methane emissions from rice paddies. Global Biogeochemical Cycles, Vol. 9, N2, p.183–195.

Cao M., Marshall S., Gregson K. (1996). Global carbon exchange and methane emissions from natural wetlands: Application of a process-based model. J. Geophys. Res., Vol. 101, ND9, p.14399–14414.

Carnell R.E., Senior G.A., Mitchell J.F.B. (1996). An assessment of measures of storminess: simulated changes in Northern Hemisphere winter due to increasing CO_2 . Climate Dynamics, Vol. 12, N7, p.467–476.

Cautenet G., Legrand M., Cautenet S., Bonnel B., Brogniez G. (1992). Thermal impact of Saharan dust over land. Part 1: Simulation. J. Appl. Meteorol., Vol. 31, p.166–180.

Cess R.D., Zhang M.H., Minnis P., Corsetti L., Dutton E.G., Forgan B.W., Garber D.P., Gates W.L., Hack J.J., Harrison E.F., Jing X., Kiehl J.T., Long C.N., Morcrette J.-J., Potter G.L., Ramanathan V., Subasilar B., Whitlock C.H., Young D.F., Zhou Y. (1995). Absorption of solar radiation by clouds: observations versus models. Science, Vol. 267, p.469–499.

Cess R.D., Zhang M.H., Ingram W.J., and co-authors. (1996). Cloud feedback in atmospheric general circulation models. J. Geophys. Res.., Vol. 101, ND8, p.12791–12794.

Chan Y.-H., Chan D., Higuchi K. (1996). On a two-dimensional energy balance climate model coupled interactively to a land carbon cycle model. Tellus, Vol. 48B, N3, p.329–346.

Chandler M., Rind D., Thompson R. (1994). Joint investigation of the middle Pliocene climate-II: GISS GCM Northern Hemisphere results. Global and Planetary Change, Vol. .9, p.197–219.

Chandra S., Lean J.L., White O.R., Prinz D.K., Rottman G.J., Brueckner G.E. (1995). Solar UV irradiance variability during the declining phase of the solar cycle 22. Geophys. Res. Lett., Vol. 22, N18, p.2481–2484.

Chandra S., Varotsos C., Flynn L.E. (1996). The mid-latitude total ozone trends in the Northern Hemisphere. Geophys. Res. Lett., Vol. 23, N5, p.555–558.

Chang P., Wang B., Li T., Ji L. (1994). Interaction between the seasonal cycle and the southern oscillation—frequency entrainment and chaos in a coupled ocean-atmosphere model. Geophys. Res. Lett., Vol. 21, N25, p.2817–2820.

Changnon S.A. (1995). State roles in the global climate change issue. Bull. Amer. Meteorol. Soc., Vol. 76, N2, p.227–254.

Charlson R.J., Lovelock J.E., Meinrat O., Warren S.G. (1987). Oceanic phytoplankton, atmospheric sulphur, cloud albedo and climate. Nature, Vol. 326, p.655–661.

Charlson R.J., Langner J., Rodhe H., Leovy C.B., Warren S.G. (1991). Perturbation of the Northern Hemisphere radiative balance by backscattering from anthropogenic surface aerosols. Tellus, Vol. 43A, p.152–163.

Charlson R.J., Schwartz S.Z., Hales J.M., Cess R.D., Coakley J.A., Hansen J.E., Hofmann D.F. (1992). Climate forcing by anthropogenic aerosols. Science, Vol. 255, p.423–430.

Charlson R.J., Lelieveld J. (1994). Multiphase atmospheric chemistry: implications for climate. In: Global Atmospheric-Biospheric Chemistry. Plenum Press, New York. p.57–69.

Charlson R.J., Wigley T.M.L. (1994). Sulfate aerosol and climatic change. Scientific American, Vol. 1270, p.28–35.

Chase T.N., Pielko R.A., Kittel T.G.F., Nemani R., Running S.W. (1996). Sensitivity of a general circulation model to global changes in leaf area index. J. Geophys. Res., Vol. 101, ND3, p.7393–7408.

Chen B., Bromwich D.H., Hines K.M., Pan X. (1995). Simulations of the 1979–88 polar climates by global climate models. Ann. Glaciol., Vol. 21, p.83–90.

Chen C.-T., Ramaswamy V. (1996). Sensitivity of simulated global climate to perturbations in low-cloud microphysical properties. Part I: Globally uniform perturbations. J. Climate, Vol. 9, N6, p.1385–1402.

Chen C.-T., Roeckner E. (1996). Validation of Earth radiation budget as simulated by the Max Planck Institute for Meteorology general circulation model ECHAM4 using satellite observations of the Earth Radiation Budget Experiment. J. Geophys. Res., Vol. 101, ND2, p.4269–4287.

Chen C.-T., Roeckner E., Soden B.J. (1996). A comparison of satellite observations and model simulations of column-integrated moisture and upper-tropospheric humidity. J. Climate, Vol. 9, N7, p.1561–1585.

Chipperfield M.P., Lee A.M., Pyle J.A. (1996). Model calculations of ozone depletion in the Arctic polar vortex for 1991/92 to 1994/95. Geophys. Res. Lett., Vol. 23, N5, p.559–562.

Chou M.-D., Arking A., Otterman J., Ridgway W.L. (1995). The effect of clouds on atmospheric absorption of solar radiation. Geophys. Res. Lett., Vol. 22, N14, p.1885–1888.

Christensen T.R., Cox P. (1995). Response of methane emission from arctic tundra to climatic change: results from a model simulation. Tellus, Vol. 47B, N3, p.301–309.

Christy J.R., Spencer R.W. (1995). Assessments of precision in temperatures from the microwave sounding units. Climatic Change, Vol. 30, p.97–102.

Christy J.R., Spencer R.W., McNider R.T. (1995). Reducing noise in the daily MSU lower tropospheric temperature data set. J. Climate, Vol. 8, p.888–896.

Chlek G.M., Johnson B., Damiano P.A., Taylor K.C., Clement P. (1995b). Biomass burning record and black carbon in the GISP–2 ice core. Geophys. Res. Lett., Vol. 22, N2, p.89–92.

Chlek P., Dobbie J.S. (1995). Radiative properties of finite inhomogeneous cirrus clouds: Monte Carlo simulations. J. Atmos. Sci., Vol. 52, N20, p.3512–3522.

Chlek P., Videen G., Ngo D., Pinnick R.G., Klett J.D. (1995a). Effect of black carbon on the optical properties and climate forcing of sulphate aerosol. J. Geophys. Res., Vol. 100, ND8, p.16325–16332.

Chlek P., Wong J. (1995). Effect of absorbing aerosols on global radiation budget. Geophys. Res. Lett., Vol. 22, N8, p.929–931.

Ciais P., Tans P.P., White J.W.C., Trolier M., Francey R.J., Berry J.A., Randall D.R., Sellers P.J., Collatz J.G., Schimel D.S. (1995). Partitioning of ocean and land uptake of CO_2 as inferred by measurements from the NOAA Climate Monitoring and Diagnostics Laboratory Global Air Sampling Network. J. Geophys. Res., Vol. 100, ND3, p.5051–5070.

Clark W.C., Munn R.E. (Eds.) (1986). Sustainable Development of the Biosphere. Cambridge Univ. Press. Cambridge. 491 pp.

Claussen M. (1996). Variability of the global biome patterns as a function of initial and boundary conditions in a climate model. Climate Dynamics, Vol. 12, p.371–379.

Clement A.C., Seager R., Cane M.A., Zebiak S.E. (1996). An ocean dynamical thermostat. J. Climate, Vol. 9, N7, p.748–759.

Climate Change. The IPCC Scientific Assessment. Eds.: J.T. Houghton, G.J. Jenkins, J.S. Ephraums. Cambridge Univ. Press, Cambridge. 1990. 365 pp.

Climate Change: Science, Impacts and Policy. Proc. Second World Climate Conf., Eds.: J.Jager, H.L. Ferguson. Cambridge Univ. Press, Cambridge, 1991. 567 pp.

Climate Change 1994: Radiative Forcing of Climate Change and an Evaluation of the IPCC Emission Scenarios. Ed. by J.T. Houghton et al. Cambridge Univ. Press, Cambridge. 1995, Vol. VII, 339 pp.

Climate Change 1995. (1996a). The Science of Climate Change. Ed. by J.T. Houghton, L.G. Meira Filho, B.A. Callender, N. Harris, A. Kattenberg, and K. Maskell. Cambridge University Press, Cambridge. 572 pp.

Climate Change 1995. (1996b). Impacts, Adaptation and Mitigation of Climate Change: Scientifical-Technical Analysis. Ed. by R.T. Watson, M.C. Zynyowera, R.H. Moss, and D.J. Dokker. Cambridge University Press, Cambridge. 878 pp.

Climate Change 1995. (1996c). Economic and Social Dimensions of Climate Change. Ed. by J.P. Bruce, H. Lee, and E.F. Haites. Cambridge University Press, Cambridge. 488 pp.

Climatic Change News. GCC Newsletter. 1995, Vol. 2, issue 3, p.1–5.
Climate Impacts. LINK Newsletter. 1995, N10, p.1–4.
CLIVAR. A Study of Climate Variability and Predictability. Science Plan. WCRP–89 (WMO/TD-No.690). Geneva.1995. 157 pp.
Cloud–Radiation Interactions and Their Parameterization in Climate Models. WCRP–86 (WMO/TD-No.648) Geneva. 1994. 147 pp.
Coakley J.A., Jr., Cess R.D. (1985). Response of the NCAR community model to the radiative forcing by the naturally occurring tropospheric aerosol. J. Atmos. Sci., Vol. 42, p.1677–1692.
Cohen-Solal E., Le Treut H. (1996). Impact of ocean optical properties on seasonal SST: results with a surface ocean model coupled to the CDM AGCM. Climate Dynamics, Vol. 12, p.417–433.
Collins W.D., Valero F.P.J., Flatau P.J., Lubin D., Grassl H., Pilewskie P. (1996). Radiative effects of convection in the tropical Pacific. J. Geophys. Res., Vol. 101, ND10, p.14999–15012.
Colman R.A., Power S.B., McAvaney B.J., Dahni R.R. (1995). A non-flux corrected transient CO_2 experiment using the BMRC coupled atmosphere–ocean GCM. Geophys. Res. Lett., Vol. 22, N22, p.3047–3050.
Cook E.R. (1995). Temperature histories from tree rings and corals. Climate Dynamics, Vol. 11, N4, p.211–222.
Coughlan M., Avissar R. (1996). The Global Water Energy and Water Cycle Experiment (GEWEX) Continental-Scale International Project (GCIP): An overview. J. Geophys. Res., Vol. 101, ND3, p.7139–7147.
Covey C. (1995). Using palaeoclimates to predict future climate: how far can analogy go? An editorial comment. Climatic Change, Vol. 29, N4, p.405–407.
Cox S.J., Wang W.-C., Schwartz S.E. (1995). Climate response to radiative forcings by sulphate aerosols and greenhouse gases. Geophys. Res. Let., Vol. 22, N18, p.2509–2512.
Craig S.G., Holmen K.J. (1995). Uncertainties in future CO_2 projections. Global Biogeochemical Cycles, Vol. 9, N1, p.139–152.
Crommelynck D. (1993). L'exérience SOLCON. Ciel et Terre, Vol. 109, N4, p.99–105.
Crommelynck D., Domingo V., Fichot A., Lee R.B. III (1994). Irradiance observations from the EURECA and ATLAS experiments. In: The Sun as a Variable Star: Solar and Stellar Irradiance Variations. Eds.: J.M. Pap, C. Fröhlich, H.S. Hudson, S.K. Solanki, Cambridge University Press.
Crommelynck D., Fichot A., Lee R.B. III, Romero J. (1996). First realisation of the space absolute radiation reference. (in press)
Crowley T.J. (1995). Ice age terrestrial carbon changes revisited. Global Biogeochemical Cycles, Vol. 9, N3, p.377–389.
Crowley T.J., Baum S.K. (1994). General circulation model study of late Carboniferous interglacial climates. Paleoclimates Data and Modeling, Vol. 1, N1, p.3–22.
Crowley T.J., Kim K.Y. (1996). Comparison of proxy records of climate change and solar forcing. Geophys. Res. Lett., Vol. 23, N4, p.359–362.
Crutzen P.J., Ramanathan V. (Eds.) (1996). Cloud, Chemistry, and Climate. Springer-Verlag, Vol. 1, N1, p.3–22.

Cubasch U., Hegerl G.C., Hellbach A., Hock H., Mikolajewicz U., Santer B.D., Voss R. (1995a). A climate change simulation starting from 1935. Climate Dynamics, Vol. 11, N2, p.71–84.

Cubasch U., Santer B.D., Hegerl G.C. (1995b). Klimatmodelle—wo stehen wir? Phys. Bl., Vol. 51, N4, p.269–276.

Cubasch U., Waszkewitz J., Hegerl G., Perlwitz J. (1995c). Regional climate changes as simulated in time-slice experiments. Clim. Change, Vol. 31, N2–4, p.143–174.

Cullen M.J.P. (1993). The united forecast climate model// Meteorological Magazine, Vol. 122, N1449, p.81–94.

Cundary V., Colombo T. (1992). Atmospheric temperature time series at Mt.Cimone observatory, Italy, 1946–1991. Il Nuovo Cimento, Vol. 15C, N4, p.381–390

Cundary V., Colombo T., Ciattaglie L. (1995). Thirteen years of atmospheric carbon dioxide measurements at Mt. Cimone station, Italy. Il Nuovo Cimento, Vol. 18C, Ser.1, N1, p.33–48.

Curry J.A. (1995). Interactions among aerosols, clouds, and climate of the Arctic Ocean. The Science of the Total Environ, Vol. 160/161, p.777–791.

Curry J.A., Schramm J.L., Ebert E.E. (1995). Sea ice-albedo climate feedback mechanism. J.Climate, Vol. 8, N2, p.240–247.

Curry J.A., Rossow W.B., Randall D., Schramm J.L. (1996). Overview of arctic cloud and radiation characteristics. J. Climate, Vol. 9, N7, p.762–774.

Daniel J., Solomon S., Albriton D. (1995). On the evaluation of halocarbon forcing and global warming potentials. J. Geophys. Res., Vol. 100, ND1, p.1271–1285.

De Gaetano A.T. (1996). Recent trends in maximum and minimum temperature threshold exceedences in the northern United States. J. Climate, Vol. 9, N7, p.780–790.

Del Genio A.D., Yao M.-S., Kovari W., Lo K.K.-W. (1996). A prognostic cloud water parameterization for global climate models. J. Climate, Vol. 9, N2, p.270–304.

Delmas R.J., Ed. (1995). Ice Core Studies of Global Biogeochemical Cycles. Springer-Verlag. Berlin. 475 pp.

De Luisi J.J., Mateer C.L. (1995). Comparison of observed ozone trends in the stratosphere through examination of Umkehr and balloon observation data. J. Geophys. Res., Vol. 100, ND6, p. 11209–11217.

Delworth T., Manabe S. (1995). North Atlantic interannual to decadal variability in the GFDL coupled model. Proc. Workshop on Global Coupled General Circulation Models. Scripps Inst. Oceanogr., La Jolla, USA, Oct. 1994, WCRP–87 (WMO/TD-No.655). Geneva, p.48.

Dendeney D. (1990). The case against linking environmental degradation and national security. Millennium, Vol. 19, N3, p.461–476.

Denne T.(1995). CO_2 limitation in the European Union policies and projections. Climate Change and Energy Use. Status and Options for European Policy. Contractor Reports. Part A. Luxembourg, the European Parliament. 73 pp.

De Rudder A., Larsen N., Tie X., Brasseur G.P., Granier C. (1996). Model study of polar stratospheric clouds and their effect on stratospheric ozone. 1. Model description. J. Geophys. Res., Vol. 101, ND7, p.12567–12574.

Deser C., Alexander M.A., Timlin M.S. (1996). Upper-ocean thermal variations in the North Pacific during 1970–1991. J. Climate, Vol. 9, N7, p.812–824.

Deshler T., Hofmann D.J., Johnson B.J., Rozier W.R. (1992). Balloon-borne measurements of the Pinatubo aerosol size distribution and volatility at Laramie, Wyoming during the summer of 1991. Geophys. Res. Lett., Vol. 19, p.199–202.

Deshler T., Johnson B.J., Rozier W.R. (1993). Balloon-borne measurements of Pinatubo aerosol during 1991 and 1992 at 41°N: vertical profiles, size distribution, and volatility. Geophys. Res. Lett., Vol. 20, p.1435–1438.

Dessler A.E., Kawa S.R., Considine D.B., Waters J.W., Froidevaux L., Kumer J.B. (1996). UARS measurements of ClO and NO_2 at 40 and 46 km and implications for the model "ozone deficit". Geophys. Res. Lett., Vol. 23, N4, p.339–342.

Dessler A.E., Kawa S.R., Douglass A.R., Considine D.B., Kumer J.B., Roche A.E., Mergenthaler J.L. Waters J.M., Russel J.M.,III, Gille J.C. (1996). A test of the partitioning between ClO and $ClONO_2$ using simultaneous UARS measurements of ClO, NO_2, and $ClONO_2$. J. Geophys. Res., Vol. 101, ND7, p.12515–12521.

Dettinger M.D., Ghil M., Keppenne C.L. (1995). Interannual and interdecadal variability in United States surface-air temperatures, 1910–87. Climatic Change, Vol. 31, N1, p.35–66.

Di Girolamo P., Cacciani M., di Sarra A., Fiocco G., Fua D. (1994). Lidar observations of the Pinatubo aerosol layer at Thule, Greenland. Geophys. Res. Lett., Vol. 21, p.1295–1298.

Dingmin Li, Shine K.P., Gray L.S. (1995). The role of ozone-induced diabatic heating anomalies in the quasi-biennial oscillation. Quart. J. Roy. Meteorol. Soc., Part B, Vol. 121, N524, p.937–944.

Disparity and hope. Tiempo, 1995, Issue 16, p.11–14.

Dixon K.W., Bullister J.L., Gammon R.H., Stouffer R.J. (1996). Examining a coupled climate model using CFC–11 as an ocean traces. Geophys. Res. Lett., Vol. 23, N15, p.1957–1960.

Doubleday N.C., Douglas M.S.V., Smol J.P. (1995). Paleoenvironmental studies of black carbon deposition in the High Arctic: a case study from Northern Ellsmere Island. The Science of the Total Environ., Vol. 160/161, p.661–668.

Dudek M.P., Liang X.-Z., Wang W.-C. (1996). A regional climate model study of the scale dependence of cloud–radiation interactions. J. Climate, Vol. 9, N6, p.1221–1234.

Dyakonova V.D., Chistyakov V.F. (1993). On oscillations of the solar radius for the duration of one day. In: V.F. Chistyakov (Ed.), Global Variations of the Sun and Physics of Active Regions. Dal' Nauka, Vladivostok, p.55–77 (in Russian).

Dye J.E., Baumgardner D., Gandrud B.W., Drdla K., Barr K., Fahey D.W., Delnegro L.A., Tabazadeh A., Jonsson H.H., Wilson J.C., Loewenstein M., Podolske J.R., Chan K.R. (1996). In-situ observations of an Antarctic polar stratospheric cloud: similarities with Arctic observations. Geophys. Res. Lett., Vol. 23, N15, p.1913–1916.

Easterling D.R., Peterson T.C., Karl T.R. (1996). On the development and use of homogenized climate datasets. J. Climate, Vol. 9, p.918–932.

Easterling W.E. (1995). Long-term climate monitoring by the Global Climate Observing System (GCOS). Climate Impact. Climatic Change (manuscript).

Echalar F., Gaudichet A., Cachier H., Artaxo P. (1995). Aerosol emissions by tropical forest and savanna biomass burning: characteristic trace elements and fluxes. Geophys. Res. Lett., Vol. 22, N22, p.3039–3042.

Eddy J. (1976). The Maunder Minimum. Science, Vol. 192, N4245, p.1189–1202.

Eddy J.A., Oeschger H. (1993). Global Change in the Perspective of the Past. John Wiley & Sons, Chichester. 383 pp.

Effects of Increased Ultraviolet Radiation on Biological Systems. (1992). SCOPE/UNEP. Paris. 47 pp.

Effects of Increased Ultraviolet Radiation on Global Ecosystems. (1993). SCOPE/UNEP. Paris. 47 pp.

Ehrlich P.R., Ehrlich A.H. (1990). The Population Explosion: Why Overpopulation is our No.1 Environmental Problem. Simon & Schuster, New York. 320 pp.

Ellsaesser H.W. (Ed.) (1992). Global 2000 Revisited. Mankind's Impact on Spaceship Earth. Paragon House, New York. 436 pp.

Ellsaesser H.W. (1994). The unheard arguments: a rational view of stratospheric ozone. 21st Century Science and Technology, Vol. 7, N3, p.37–45.

Eluszkiewicz J., Crisp D., Zurek R., Elson L., Fishbein E., Froidevaux L., Waters J., Grainger R.G., Lambert A., Harwood R., Peckham G. (1996). Residual circulation in the stratosphere and lower mesosphere as diagnosed from Microwave Limb Sounder Data. J. Atmos. Sci., Vol. 53, N2, p.217–240.

Energy-Active Zones: Conceptual Aspects. Progress in Sci. & Technol., Atmosphere, Ocean, Space—Programme "Sections".G.I. Marchuk, K.Ya. Kondratyev, V.V. Kozoderov, et al. Moscow, VINITI. 1989. Vol. 10, 250 pp.; Vol. 11, 368 pp. (in Russian).

Engardt M., Rodhe H. (1993). A comparison between patterns of temperature trends and sulphate aerosol pollution. Geophys. Res. Lett., Vol. 20, p.117–120.

Engelman R. (1995). Population and consumption. Tiempo, Issue 16, p.17–21.

Enting I.G. (1995). Analyzing the conflicting requirements of the Framework Convention on Climate Change. Clim. Change, Vol. 31, N1, p.5–18.

Erickson D.J., Eaton R.E. (1993). Global biochemical cycling climates with CZCS satellite data and general circulation models. Geophys. Res. Lett., Vol. 20, N8, p.683–686.

Erickson D.J., III, Oglesby R.J., Marshall S. (1995). Climate response to indirect anthropogenic sulphate forcing. Geophys. Res. Lett., Vol. 22, N15, p.2017–2020.

Erickson D.J., III, Rasch P.J., Tans P.P., Friedlingstein P., Ciais P., Maier-Reimer E., Six K., Fischer C.A., Walters S. (1996). The seasonal cycle of atmospheric CO_2: A study based on the NCAR Community Climate Model (CCM2). J. Geophys. Res., Vol. 101, ND10, p.15079–15097.

Estupiñán J.G., Raman S., Crescenti G.H.., Streicher J.J., Barnard W.F. (1996). Effect of clouds and haze on UV-B radiation. J. Geophys. Res., Vol. 101, ND11, p.16807–16816.

Etheridge D.M., Steele L.P., Langenfelds R.L., Francey R.J., Barnola J.-M., Morgan V.I. (1996). Natural and anthropogenic changes in atmospheric CO_2 over the last 1000 years from air in Antarctic ice and firn. J. Geophys. Res., Vol. 101, N6 D2, p.4115–4128.

Evans W.F.J., Puckrin E. (1996). Near-infrared spectral measurements of liquid water absorption by clouds. Geophys. Res. Lett., Vol. 23, N15, p.1941–1944.

Evans W.F.J., Reinhart C., Puckrin E. (1995). A ground based measurement of the anomalous cloud absorption effect. Geophys. Res. Lett., Vol. 22, N16, p.2135–2138.

Falkowski P.G., Kim Y., Kodler Z., Wilson C., Wirick C., Cess R.D. (1992). Natural versus anthropogenic factors affecting low-level cloud albedo over the North Atlantic. Science, Vol. 256, p.1311–1313.

Farla J.C.M., Hendriks C.A., Blok K. (1995). Carbon dioxide recovery from industrial processes. Clim. Change., Vol. 29, N4, p.439–461.

FAUNMAP Working Group (1996). Spatial response of mammals to late Quaternary environmental fluctuations. Science, Vol. 272, p.1601–1606.

Feister U., Grewe R. (1995). Higher UV radiation inferred from low ozone levels at northern mid-latitudes in 1992 and 1993. Global and Planetary Change, Vol. 11, N1, p.25–34.

Ferman C. (1992). Japan in the Greenhouse: Responsibilities, Policies, and Prospects for Combating Global Warming. Report 1992/13, Fridtjov Nansen Institute.

Fiocco G., Fuà D., and Visconti G. (Eds.) (1996). The Mount Pinatubo Eruption Effects on the Atmosphere and Climate. John Wiley & Sons, New York., 310 pp.

Fitzgerald J.F., Braatz B.V., Brown S., Isichei A.O., Odada E.O., Schoels R.J. (Eds.) (1995). African Greenhouse Gas Emission Inventories and Mitigation Options: Forestry, Land-Use Change, and Agriculture. Kluwer Acad. Publ., Dordrecht. 219 pp.

Flatøy F., Høv Ø. (1995). Three-dimensional model studies of exchange processes of ozone in the troposphere over Europe. J. Geophys. Res., Vol. 100, ND6, p.11465–11481.

Flatøy F., Høv Ø. (1996).Three-dimensional model studies of the effects of Nox emissions from aircraft on ozone in the upper stratosphere over Europe and the North Atlantic. J. Geophys. Res., Vol. 101, ND1, p.1401–1422.

Flatøy F., Høv Ø., Smit H. (1996). Three-dimensional model studies of exchange processes of ozone in the troposphere over Europe. J. Geophys. Res., Vol. 101, ND6, p.11465–11481.

Flowerdue R.J., Haigh J.D. (1996). Retrieval of aerosol optical thickness over land using the ATSR–2 dual-look satellite radiometer. Geophys. Res. Lett., Vol. 23, N4, p.351–354.

Foley J.A. (1995). An equilibrium model of the terrestrial carbon budget. Tellus, Vol. 47B, N3, p.310–319.

Folland C.K., Parker D.E. (1995). Correction of instrumental biases in historical sea surface temperature data. Quart. J. Roy. Meteorol. Soc., Part B, Vol. 121, N522, p.319–368.

Folly J.A., Kutzbach J.E., Coe M.T., Levis S. (1994). Feedbacks between climate and boreal forests during the Holocene epoch. Nature, Vol. 371, N6492, p.52–54.

Fortuin J.P.F., Kelder H. (1996). Possible links between ozone and temperature profiles. Geophys. Res. Lett., Vol. 23, N12, p.1517–1520.

Foster J., Liston G., Koster R., Essery R., Behr H., Dümenil L., Verseghy D., Thompson S., Pollard P., Cohen J. (1996). Snow cover and snow mass intercomparisons of general circulation models and remotely sensed datasets. J. Climate, Vol. 9, N2, p.409–426.

Foukal P. (1994). Stellar luminosity variations and global warming. Science, Vol. 264, p.238–239.

Fouquart Y., Isaka H. (1992). Sulfur emission, CCN, clouds and climate: a review. Ann. Geophys., Vol. 10, p.462–471.

Fowler L.D., Randall D.A. (1996a). Liquid and ice cloud microphysics in the CSU general circulation model. Part II: Impact on cloudiness, the earth's radiation budget, and the general circulation of the atmosphere. J. Climate. Vol. 9, N3, p.530–560.

Fowler L.D., Randall D.A. (1996b). Liquid and ice cloud microphysics in the CSU general circulation model. Part III: Sensitivity to modelling assumptions. J. Climate, Vol. 9, N3, p.561–586.

Fowler L.D., Randall D.A., Rutledge S.A. (1996). Liquid and ice cloud microphysics in the CSU general circulation model. Part I: Model description and simulated microphysical processes. J. Climate, Vol. 9, N3, p.489–529.

Fraedrich W. (1995). Steht das Klima voreiner Werde? Neue Erkentnisse aus dem Greenland Icecore Project. Geogr. Heute, Vol. 16, N127, p.30–35.

Fraser P., Cunnold D., Alyea F., Weiss R., Prinn R., Simmonds P., Miller B., Langenfelds R. (1996). Life time and emission estimates of 1, 1, 2-trichlorotrifluorethane (CFC–113) from daily global background observations June 1982—June 1984. J. Geophys. Res., Vol. 101, ND7, p.12585–12599.

Frey R.A., Ackerman S.A., Soden B.J. (1996). Climate parameters from satellite spectral measurements. Part 1: Collocated AVHRR and HIRS/2 observations of spectral greenhouse parameters. J. Climate, Vol. 9, N2, p.327–344.

Friedlingenstein P. (1995). Modelisation du cycle du carbone biospherique et etude du couplage biosphere–atmosphere. Aeron. Acta. A., N392, pp. 1A–XVA, i–vi, 1–278.

Friedlingenstein P., Fung I., Holland E., John J., Brasseur G., Erickson D., Schimel D. (1995). On the contribution of CO_2 fertilization to the missing carbon sink. Global Biogeochemical Cycles, Vol. 9, N4, p.541–556.

Friis-Christensen E., Lassen K. (1991). Length of the solar cycle: an indicator of solar activity closely associated with climate. Science, Vol. 254, p.698–700.

Fröhlich C. (1993). Changes of total solar irradiance. Geophys. Monogr., 75, IUGG Volume 15, Washington, D.C., p.123–129.

Fröhlich C. (1994). Reviews of space observations of total solar irradiance. In: The Sun as a Variable Star: Solar and Stellar Irradiance Variations. Eds.: J.M. Pap, C. Fröhlich, H.S. Hudson, S.K. Solanki. Cambridge University Press.

Fu R., Liu W.T., Dickinson R.E. (1996). Response of tropical clouds to the interannual variation of sea surface temperature. J. Climate, Vol. 9, N3, p.616–634.

Fung I. (1994). Perturbations to the biospheric carbon cycle: uncertainties in the estimates. In: Biotic Feedbacks in the Global Climatic System: Will the Warming Feed the Warming? G. Woodwell and F. MacKenzie (Eds.), Oxford Univ. Press, New York, p.365–374.

Gaffin S.R., O'Neile B.C., Oppenheimer M. (1995). Comment on "The lifetime of excess atmospheric carbon dioxide" by Berrien Moore III and B.H. Braswell. Global Biogeochemical Cycles, Vol. 9, N1, p.167–169.

Gage K.S., McAfee J.R., Williams C.R. (1996). Recent changes in tropospheric circulation over the central equatorial Pacific. Geophys. Res. Lett., Vol. 23, N16, p.2149–2152.

Galbraith J.W., Green Ch. (1992). Inferences about trends in global temperature data. Climatic Change, Vol. 22, N3, p.209–222.

Galindo I. (1992). Extinction of short-wave solar radiation due to El Chichón stratospheric aerosol. Atmósféra, p.259–268.

Gammelsrod T., Foldvik A., Nost O.A., Skagseth O., Anderson L.G., Fogelqvist E., Olsson K.,

Tanhua T., Jones E.P. (1995). Distribution of water masses on the continental shelf in the Southern Weddell Sea. "The Polar Oceans and Their Role in Shaping the Global Environment". Geophys. Monogr. 84. AGU, p.159–176.

Garrat J.R., Prata A.J. (1996). Downwelling longwave fluxes at continental surfaces. A comparison of observations with GCM simulations and implications for the global land-surface radiation budget. J. Climate, Vol. 9, N3, p.646–655.

Garstell M.F., Crisp J., Crisp D. (1995). Radiative forcing of the stratosphere by SO_2 gas silicate ash, and H_2SO_4 aerosols shortly after the 1982 eruption of El Chichón. J. Climate, Vol. 8, N5, p. 1060–1070.

Gassmann F. (1993). Nonlinear dynamical systems: an introduction. Some Physicochemical and Mathematical Tools for Understanding of Living Systems. H.Greppin, M. Gonzon, R. Dogli Agosti (Eds.). University of Geneva, p.45–69.

Gassmann F. (1994). "Abstract Change" scenarios: Global Change scenarios revisited with respect to emerging scientific insights: (Pap.). IPCC Working Group Writing Team II Meet. "Steps towards Decis. Making Framework to Address Clim. Change", Montreaux, March 3–6, 1994. PCI BER, N10, p.3–9.

Gassmann F. (1995). Can biosphere feedback cause homeostasis and abrupt climate changes?. Paul Scherre Institute, Villigen, Switzerland. Preprint. 9 pp.

Gates W.L. (1984) . The use of general circulation models in the analysis of the ecosystem impacts of climate change. Clim. Res. Inst. Rept. Univ. of Oregon, Corvallis. N52. 25 pp.

Gates W.L. (1996). Review of main development since the sixteenth session of the JSC. XV11/doc. 4, Toulouse, France, 11–16 March 1996. 6 pp.

Gayet J.-F., Febvre G., Brogniez G., Chepfer H., Renger W., Wendling P. (1996). Microphysical and optical properties of cirrus and contrails: cloud field study on 13 October 1989. J. Atmos. Sci., Vol. 53, N1, p.126–138.

Gee H. (1996). Changing times, changing teams. Nature, Vol. 382, p.24.

Geleyn J.F. (1994). General circulation models. Final Report. Cloud-Radiation Interactions and Their Parameterization in Climate Models. WCRP–86 (WMO/TD. N6430. Geneva, WMO, p.3–11.

German Advisory Council on Global Change (WBGU), Alfred-Wegener-Inst., Bremerhaven, Germany (Eds.). World in Transition: Ways Towards Global Environmental Solutions. Annual Report 1995. Springer-Verlag, Berlin. 1996. 255 pp.

Gerstell M.F., Crisp J., Crisp D. (1995). Radiative forcing of the stratosphere by SO_2 gas, silicate ash, and H_2SO_4 aerosols shortly after the 1982 eruptions of El Chichon. J. Climate, Vol. 8, N5, p.1060–1070.

Ghan S.J., Leung L.R. (1995). Treatment of cloud-aerosol interactions and orographic subgrid variability in a GCM. In: Workshop on Cloud Microphysics Parameterizations in Global Atmospheric Circulation Models. WCRP–90 (WMO/TD-No.713). WMO, Geneva. p.117–132.

Giambelluca T.W., Henderson-Sellers A. (Eds.) (1996). Climate Change. Developing Southern Hemisphere Perspective. John Wiley & Sons, Chichester. 475 pp.

Gierczak T., Talukdar R., Burkholder J.B., Portmann R.W., Daniel J.S., Solomon S., Ravishankara A.R. (1996). Atmospheric fate and greenhouse warming potentials of HFC 236fa and HFC 236ea. J. Geophys. Res., Vol. 101, ND8, p.12905–12911.

Giorgi F. (1995). Perspectives for regional earth system modelling. Global and Planetary Change, Vol. 10, p.23–42.

Giorgi F., Brodeur C.S., Bates G.T. (1994). Regional climate change scenarios over the United States produced with a nested regional climate model. J. Climate, Vol. 7, N3, p.375–399.

Glantz M.H. (Ed.) (1994). The Role of Regional Organizations in the Context of Climate Change. Springer-Verlag, Berlin. 208 pp.

Gleick P.H. (1989). Climate change and international politics: problems facing developing countries. Ambio, Vol. 18, N6, p.33–339.

Global Climate Observing System Data and Information Management Plan. Version 1.0 (1995). GCOS–13 (WMO/TD-No.677), 30 pp.

Global Climate Observing System (1995a). Plan for the Global Climate Observing System (GCOS). Version 1.0. GCOS–14. WMO/TD-No.681. Geneva. 49 pp.

Global Climate Observing System (1995b). Plan for Space-Based Observations. Version 1.0. GCOS-15. WMO/TD No.684. 79 pp.

Global Climate Observing System (1995c). GCOS/GTOS Plan for Terrestrial Climate-Related Observations. Version 1.0. GCOS–2 (WMO/TD-No.721). UNEP/EAP. TR/95-07. Geneva. 113 pp.

Global Warming Update. Recent Scientific Findings. (1992). George Marshall Institute. Washington, D.C. 30 pp.

Gloersen P., Yu J., Mollo-Christensen E. (1996). Oscillatory behavior in Arctic sea-ice concentrations. J. Geophys. Res., Vol. 101, NC3, p.6641–6650.

Gobbi G.P., Congeduti F., Adriani A. (1992). Early stratospheric effects of the Pinatubo eruption. Geophys. Res. Lett., Vol. 15, p.997–1000.

Goldberg M.D., Fleming H.E. (1995). An algorithm to generate deep-layer temperatures from microwave satellite observations for the purpose of monitoring climate change. J. Climate, Vol. 8, N5, p.993–1004.

Goldman A., Murcray F.J., Rinsland C.P. et al. (1992). Mt. Pinatubo SO_2 column measurements from Mauna Loa. Geophys. Res. Lett., Vol. 19, p.195–196.

Golitsyn G.S., Semenov A.I., Shefov N.N., Fishkova L.M., Lysenko E.V., Perov S.P. (1996). Long-term temperature trends in the middle and upper atmosphere. Geophys. Res. Lett., Vol. 23, N14, p.1741–1744.

Goody R., Haskins R., Abdon W., Chen L. (1995). Detection of climate forcing using emission spectra. Studying the Earth from Space, N5, p.22–33.

Gordon A.H. (1991). Global warming as a random walk. J. Climate, Vol. 4, p.589–597.

Gorka K. (1994). Economic instruments of environmental policy in Polone.

"Industrial and Environmental Policy in Polone". Cracow, p.75–85.
Gorshkov V.G. (1995). Physical and Biological Basis of Life Stability. Man, Biota, Environment. Springer-Verlag, Berlin, Heidelberg. 340 pp.
Gorshkov V.G., Kondratyev K.Ya., Losev K.S. (1992). Global ecological perspectives. Vestnik RAN, N5, p.70–81 (in Russian).
Gorshkov V.G., Kondratyev K.Ya., Losev K.S. (1994). The natural biological regulation of the environment. Chemical Ecology, N3, p.18–29.
Goulden M.L., Munger J.W., Fan S.-M., Daube B.C., Wofsy S.C. (1996). Exchange of carbon dioxide by a decidious forest: response to interannual climate variability. Science, Vol. 271, N5255, p.1576–1578.
Graham N.E. (1995). Simulation of recent global temperature trends. Science, Vol. 267, N5198, p.666–671.
Graham N.E., Barnett T.P., Wilde R. (1994). On the role of tropical and midlatitude SSTs in forcing interannual to interdecadal variability in the winter Northern Hemisphere circulation. J. Climate, Vol. 7, p.1416–1441.
Grassl H. (1975). Albedo reduction and radiative heating of clouds by absorbing aerosol particles. Beitr. Phys. Atm., Vol. 48, p.199–210.
Grassl H. (1988). What are the radiative and climatic consequences of the changing concentration of atmospheric aerosol particles? In: The Changing Atmosphere (Eds.: F.S. Rowland and I.S.A. Isaksen). John Wiley & Sons, Chichester. p.187–199.
Grassl H. (1992). The influence of aerosol particles on radiation parameters of clouds. Idöjàràs, Vol. 86, p.60–75.
Greenwald T.J., Stephens G.L., Christopher S.A., Vonder Haar T.H. (1995). Observations of the global characteristics and regional radiative effects of marine cloud liquid water. J. Climate, Vol. 8, N12, p.2928–2946.
Gregory G.L., Anderson B.E., Browell E.V. (1996). Influence of lower tropospheric ozone on total column ozone as observed over the Pacific Ocean during the 1991 PEM-West A experiment. J. Geophys. Res., Vol. 101, ND1, p.1919–1930.
Grigoryev A.A., Kondratyev K.Ya. (1993). Aerospace Earth-resources and ecological studies in China. Studying the Earth from Space, N2, p.119–123 (in Russian).
Grötzner A., Sausen R., Claussen M. (1996). The impact of sub-grid scale sea-ice inhomogeneities on the performance of the atmospheric general circulation model ECHAM3. Climate Dynamics, Vol. 12, p.477–496.
Gu D., Philander S.G.H. (1995). Secular changes of annual and interannual variability in the tropics during the past century. J. Climate, Vol. 8, N 4, p.864–876.
Gultepe I., Isaac G.A., Leatich W.R., Banic C.M. (1996). Parameterization of marine stratus microphysics based on in situ observations: implications for GCMs. J. Climate, Vol. 9, N2, p.345–357.
Gunarathan M., Young H. (1995). Enhancing global sinks of carbon dioxide. Search, Vol. 26, N8, p.236–238.
Gwiazda R.H., Hemming S.R., Broecker W.S. (1996). Tracking the sources of icebergs with lead isotopes: The provenance of ice-rafted debris in Heinrich layer 2. Paleooceanography, Vol. 11, N1, p.77–93.
Hadj-Sadok T. (1992). The Hague Symposium on Sustainable Development. Network '92, N13, p.15.

Hahmann A.N., Ward D.M., Dickinson R.E. (1995). Land surface temperature and radiative fluxes response of the NCAR CCM2/Biosphere–Atmosphere Transfer Scheme to modifications in the optical properties of clouds, J. Geophys. Res., Vol. 100, ND11, p.23239–23252.

Haigh J.D. (1994). The role of stratospheric ozone in modulating the solar radiative forcing of climate. Nature, Vol. 370, N6490, p.544–546.

Haites E.F. (1994). Portfolio analysis model of climate change decision making (Pap). IPCC Working Group Writing Team II Meet. "Steps Towards Decis. Making Framework to Address Clim. Change", Montreaux, March 3–6, 1994, PSI BER, N10, p.11–114.

Hales J. (1996). Scientific background for AMS Policy Statement on Atmospheric Ozone. Bull. Amer. Meteorol. Soc., Vol. 77, N8, p.1249–1253.

Hales J.M., Cess R.D., Coakley J.A., Hansen J.E., Hofman D.J., Charlson R.J., Schwartz S.E. (1992). Climate forcing by anthropogenic aerosols. Science, Vol. 255, p.423–429.

Hamill P., Tabazadeh A., Kinne S., Toon O.B., Turco R.B. (1996). On the growth of ternary system $HNO_3/H_2SO_4/H_2O$ aerosol particles in the stratosphere. Geophys. Res. Lett., Vol. 23, N7, p.753–756.

Hamilton K. (1995). Comprehensive simulation of the middle atmospheric climate: some recent results. Climate Dynamics, Vol. 11, N4, p.223–241.

Hammit J.K. (1995). Outcome and value uncertainties in global-change policies. Clim. Change, Vol. 30, N2, p.125–145.

Hänninen H. (1995). Assessing ecological implications of climate change: can we rely on our simulation models? Editorial Essay. Climatic Change, Vol. 31, N1, p.1–4.

Hanning-Lee M.A., Brady B.B., Martin L.B., Syage J.A. (1996). Ozone decomposition on alumina: Implications for solid rocket motor exhaust. Geophys. Res. Lett., Vol. 23, N15, p.1961–1964.

Hansen D.V., Bezdek H.F. (1996). On the nature of decadal anomalies in North Atlantic sea surface temperature. J. Geophys. Res., Vol. 101, NC4, p.8749–8758.

Hansen J., Lacis A., Ruedy R., Sato M. (1992). Potential climate impact of Mount Pinatubo aerosols. Geophys. Res. Lett., Vol. 19, p.215–218.

Hansen J., Rossow W., Carlson B., Lacis A., Travis L., Del Genio A., Fung I., Cairns B., Mischchenko M., Sato M. (1995a). Low-cost long-term monitoring of global climate forcings and feedbacks. Clim. Change, Vol. 31, N2–4, p.247–271.

Hansen J., Ruedy R., Sato M., Reynolds R. (1996). Global surface air temperature in 1995: Return to pre-Pinatubo level. Geophys. Res. Lett., Vol. 23, N13, p.1665–1668.

Hansen J., Wilson H., Sato M., Ruedy R., Shah K., Hansen E. (1995b). Satellite and surface temperature data at odds? Reply to John R. Christie and Roy W. Spencer. Climatic Change, Vol. 30, N1, p.103–117.

Hanson H.C., Svenningson B. (1994). Aerosols and clouds. In: Physico-Chemical Behaviour of Atmospheric Pollutants (Eds.: G. Angeletti and G. Restelli). Office for Official Publications of the European Communities (Report EUR 15609/EN), Brussels-Luxemburg, Vol. 2, p.837–846.

Harries J.E. (1996). The greenhouse Earth: A view from space. Quart. J. Roy. Meteorol. Soc., Vol. 122. Part B, N532, p.799–818.

Harries J.E., Russell J.,III, Park J., Tuck A.F., Drayson S.R. (1995). Observations of absorbing layers in the Antarctic stratosphere in October 1991. Quart. J. Roy. Meteorol. Soc., Vol. 121 , N523, p.655–668.

Harshvardhan (1993). Aerosol–climate interactions. In: Aerosol–Cloud–Climate Interactions (Ed.: V.Hobbs). Academic Press, San Diego. p.76–96.

Hartmann D.L., Barkstrom B.R., Crommelynck D., Foukal P., Hansen J.E., Lean J., Lee R.B., III, Schoeberl M.R., Willson R.C. (1993). The solar total irradiance monitoring. The Earth Observer, Vol. 5, N6, p.23–27.

Hartmann D.L., Michelson M.L. (1993). Large scale effects on the regulation of tropical sea surface temperature. J. Climate, Vol. 6, p.2049–2062.

Harvey L.D.D. (1995). Solar-hydrogen electricity generation in the context of global CO_2 emission reduction. Climatic Change, Vol. 29, N1, p.53–89.

Harzallah A., Sadourny R. (1995). Internal versus SST-forced atmospheric variability as simulated by an Atmospheric General Circulation Model. J. Climate, Vol. 8, N3, p.474–495.

Hasselmann K., Bengtsson L., Cubasch U., Hegerl G.C., Rodhe H., Roeckner E., von Storch H., Voss R., Waszkewitz J. (1995). Detection of anthropogenic climate change using a fingerprint method. Max Planck Institut für Meteorologie Report No. 168. 20 pp.

Hasselmann K., Hasselmann S., Giering R., Ocana V., von Storch H. (1996). Optimization of CO_2 emissions using coupled integral climate response and simplified cost models. A sensitivity study. Max Planck Institut für Meteorologie Report No.192. 39 pp.

Hayasaka T., Nakajima T., Fujiyoshi Y., Ishizaka Y., Takeda T., Tanaka M. (1995a). Geometrical thickness, liquid water content, and radiative properties stratocumulus clouds over the Western North Pacific. J. Appl. Climatol., Vol. 34, N2, p.460–470.

Hayasaka T., Kikuchi N., Tanaka M. (1995b). Absorption of solar radiation by stratocumulus clouds: aircraft measurements and theoretical calculations. J. Appl. Meteorol., Vol. 34, N5, p.1047–1055.

Haywood J.M., Shine K.P. (1995). The effect of anthropogenic sulphate and soot aerosol on the clear sky planetary radiation budget. Geophys. Res. Lett., Vol. 22, N5, p.603–606.

He D., Huang M. (1996). A numerical study of tropospheric ozone over China. Atmos. Environ., Vol. 30, N13, p.2449–2451.

Heal O.W., Menaut J.C., Steffen W.L. (Eds.) (1993). Towards a Global Terrestrial Observing System (GTOS): detecting and monitoring change in terrestrial ecosystems. MAB Digest 14 and IGBP Global Change Report 26, Unesco, Paris and IGBP, Stockholm. 71 pp.

Hecht A.D., Tirpak D. (1995). Framework agreement on climate change: a scientific and policy history. Climatic Change, Vol. 29, N4, p.371–402.

Hegerl G.C., von Storch H., Hasselman K., Santer B.D., Cubasch U., Jones P.D. (1996). Detecting greenhouse-gas-induced climate change with an optimal fingerprint method. J. Climate, Vol. 9, N

Hegg D.A., Hobbs P.V., Ferek R.J., Waggoner A.P. (1995). Measurements of some

aerosol properties relevant to radiative forcing on the East Coast of the United States. J. Appl. Meteorol., Vol. 34, N10, p.2306–2315.
Heimann M. (1995). Dynamics of the carbon cycle. Nature (Gr. Brit.), Vol. 375, N6533, p.629–630.
Heimann M., Maier-Reimer E. (1996). On the relation between the oceanic uptake of CO_2 and its carbon isotopes. Global Biogeochemical Cycles, Vol. 10, N1, p.89–110.
Hempel G. (Ed.) (1994). Antarctic Science. Global Concerns. Springer, Heidelberg. XII. 287 pp.
Henderson-Sellers A. (1996a). Soil moisture: A critical focus for global change studies. Global and Planetary Change, Vol. 13, N1–4, p.3–9.
Henderson-Sellers A. (1996b). Soil moisture simulation: Achievements of the RICE and PILPS intercomparison workshop and future directions. Global and Planetary Change, Vol. 13, N1-4, p.99–115.
Henderson-Sellers A., McGuffie K. (1994). Land-surface characterization in greenhouse climate simulations. Int. J. Climatol., Vol. 14, N10, p.1065–1094.
Henderson-Sellers A., McGuffie K., Gross C. (1995a). Sensitivity of global climate model simulations to increased stomatal resistance and CO_2 increases. J. Climate, Vol. 8, N7, p.1738–1756.
Henderson-Sellers A., Pitman A.J., Love P.K., Irannejad P., Chen T.H. (1995b). The Project for Intercomparison of Land-Surface Parameterization Schemes (PILPS): Phases 2 and 3. Bull. Amer. Meteorol. Soc., Vol. 76, N4, p.489–503.
Hendricks J.R., Leben R.R., Born G.H., Koblinsky C.J. (1996). Empirical orthogonal function analysis of global TOPEX/POSEIDON altimeter data and implications for detection of global sea level rise. J. Geophys. Res., Vol. 101, NC6, p.14131–14145.
Hense A., Romer U. (1995). Statistical analysis of tropical climate anomaly simulations. Climate Dynamics, Vol. 11, N3, p.178–192.
Herber A., Thomason L.W., Dethloff K., Viterbo P., Radionov V.F., Leiterer U. (1996). Volcanic perturbation of the atmosphere in both polar regions: 1991–1994. J. Geophys. Res.., Vol. 101, ND2, p.3921–3928.
Herman J.R., Bhartia P.K., Ziemke J., Ahmad Z., Larko D. (1996). UV-B increases (1979–1992) from decreases of total ozone. Geophys. Res. Lett., Vol. 23, N16, p.2117–2120.
Hewitt C.N., Sturges W.T. (Eds.) (1993). Global Atmospheric Chemical Change. Elsevier Science Publ., London and New York. 470 pp.
Hidy G.M., Wolf M. (1995). Projection of trends in anthropogenic aerosol cources. In: Proc. Dahlem Workshop on Aerosol Forcing of Climate, Berlin, April 1994.
Hinsman D. (1995). WMO major global scientific programmes. EARSel Newsletter, N21, p.14–21.
Hobbs P.V. (1993). Aerosol–cloud interactions. In: Aerosol–Cloud–Climate Interactions (Ed.: P.V. Hobbs). Academic Press, San Diego. p.33–75.
Hobbs P.V. (1995). Basic Physical Chemistry for the Atmospheric Sciences. Cambridge Univ. Press. 224 pp.
Hofmann D.J., Oltmans S.J., Koenig G.L., Bodhaine B.A., Harris J.M., Lathrop J.A., Schnell R.C., Barnes J., Chin J., Kuniyuki D., Ryan S., Uchida R., Yoshinaga A.,

Neale P.J., Hayes D.S., Jr., Goodrich V.R., Komhyr W.D., Evans R.D., Johnson B.J., Quincy D.M., Clark M. (1996). Record low ozone at Mauna Loa Observatory during winter 1994–1995: A consequence of chemical and dynamical synergism? Geophys. Res. Lett., Vol. 23, N12, p.1533–1536.

Hoinka K.P., Clause H., Köhler U. (1996). On the correlation between tropopause pressure and ozone above Central Europe. Geophys. Res. Lett., Vol. 23, N14, p.1753–1756.

Hood L.L., Jirikowic J.L., McCormack J.P. (1993). Quasi-decadal variability of the stratosphere: influence of long-term solar ultraviolet radiation. J. Atmos. Sci., Vol. 50, N24, p.3941–3958.

Houghton J.T. (1986). The Physics of Atmospheres. Cambridge Univ. Press. 271 pp.

Houghton J.T. (1994). Global Warming. The Complete Briefing. Lion Publ., Oxford, England.192 pp.

Houghton J.T., Meira Filho L.G., Callander B.A., Harris N., Kattenberg A., Maskell K. (Eds.) (1996). Climate Change 1995. The Science of Climate Change. Cambridge Univ. Press. 572 pp.

Hoyt D., Kyle L. (1990). An alternative derivation of the Nimbus–7 total solar irradiance variations. In: Climate Impact of Solar Variability. Eds.: K.H. Schatten, A. Arking, NASA CP–3086, p. 293–300.

Hoyt D.V., Schatten K.H. (1993). A discussion of plausible solar irradiance variations, 1700–1992. J. Geophys. Res., Vol. 98, NA11, p.18895–18906.

Hsu N.C., Herman J.R., Bhartia P.K., Seftor C.J., Torres O., Thomson A.M., Gleason G.F., Eck T.F., Holben B.N. (1996). Detection of biomass burning smoke from TOMS measurements. Geophys. Res. Lett., Vol. 23, N7, p.745–748.

Hu H. (1996). Water vapour and temperature lapse rate feedbacks in the mid-latitude seasonal cycle. Geophys. Res. Lett., Vol. 23, N14, p.1761–1764.

Huang T.Y.W., Brasseur G.P. (1993). Effect of long-term solar variability in a two-dimensional interactive model of the middle atmosphere. J. Geophys. Res., Vol. 98, ND11, p.20413–20427.

Hudson R.D., Kim J.-H., Thompson A.M. (1995). On the derivation of tropospheric column ozone from radiances measured by the total ozone mapping spectrometer. J. Geophys. Res., Vol. 100, ND6, p.11137–11145.

Hudson R.J.M., Gherini S.A., Goldstein R.A. (1996). Modeling the global carbon cycle: Nitrogen fertilization of the terrestrial biosphere and the "missing" CO_2 sink. Global Biogeochemical Cycles, Vol. 8, N3, p.307–333.

Huebert B.J., Bates T.S., Tindale N.W. (1994). Marine aerosol and gas exchange and global atmospheric effects. In: Global Atmospheric-Biospheric Chemistry. Plenum Press, New York and London. p.39–55.

Hughen K.A., Overpeck J.T., Peterson L.C., Trumbore S. (1996). Rapid climate changes in the tropical Atlantic region during the last deglaciation. Narure, Vol. 380, N6569, p.51–54.

Hughes M.K., Diaz H.F. (Eds.) (1994). The Medieval Warm Period. Kluwer Acad. Publ., Dordrecht. 342 pp.

Hulme M. (1995). Estimating global changes in precipitation// Weather. Vol. 50, N2, p.34–42.

Hulme M. (1996). Recent climatic change in the world's drylands. Geophys. Res. Lett., Vol. 23, N1, p.61–64.

Hunter D.E., Schwartz S.E., Waggoner R., Berkovitz C.M. (1993). Seasonal, latitudinal, and secular variations in temperature trend: Evidence for influence of anthropogenic sulphate. Geophys. Res. Lett., Vol. 20, p.2455–2458.

Hurrell J.W. (1996). Influence of variations in extratropical winter time teleconnection on Northern Hemisphere temperature. Geophys. Res. Lett., Vol. 23, N6, p.665–668.

Hurrell J.W., Trenbert K.E. (1996). Satellite versus surface estimates of air temperature since 1979. J. Climate, Vol. 9, p.901–917.

Hutchinson K.D., Hardy K.R., Cao B.-C. (1995). Improved detection of optically thin cirrus clouds in nighttime multispectral meteorological satellite imagery using total integrated water vapour information. J. Appl. Meteorol., Vol. 34, N5, p.1161–1168.

IPCC 1992: Supplement. Intergovernmental Panel on Climate Change. WMO/UNEP. Geneva. Febr. 1992. 70 pp.

IPCC Special Report–1994. Summaries for Policymakers and Other Summaries. WMO. Geneva. 1994. 54 pp.

IPCC 1994: Climate Change 1994: Radiative Forcing of Climate Change and an Evaluation of the IPCC IS92 Emission Scenarios. J.H. Houghton, L.G. Meira Filho, J. Bruce, Hoesung Lee, N.A. Callander, E.F. Haites, N. Harris, K. Maskell (Eds.). Cambridge University Press, Cambridge, U.K. 339 pp.

Ivlev L.S. (1992). Chemical Composition and Size Distribution of Atmospheric Aerosols. Leningrad Univ. Press, Leningrad, 466 pp. (in Russian).

Ivlev L.S., Kondratyev K.Ya., Korostina O.M. (1992). Numerical simulations of the optical properties of cloud particles of a complex composition. Optics Atmos. Ocean, N5, p.1194–1207.

Jackman C.H., Fleming E.L., Chandra S., Considine D.B., Rosenfield J.E. (1996). Past, present and future modeled ozone trends with comparisons to observed trends. J. Geophys. Res., Vol.101, ND22, p.28,753–758,768.

Jackson T. (1994). Joint Implementation and the Climate Convention. Renewable Energy Dev. Vol. 7, N3, p.5–6.

Jacob D., Leovy C.(1995). Atmospheric trace gas measurements for the year 2000 and beyond. The Earth Observer, Vol. 7, N1, p.31–38.

Jacoby G.C., DArrigo R.D., Davaajamts J. (1996). Mongolian tree rings and 20th century warming. Science, Vol. , 273, N5276, p.771–773.

Jäger H. (1992). The Pinatubo eruption cloud observed by lidar in Garmisch-Partenkirchen . Geophys. Res. Lett., Vol. 19, p.191–194.

Jain A.K., Bach W. (1992). A climate-policy model: the effectiveness of counter-measures to reduce the man-made greenhouse effect. Proc. Third Symp. on Global Change Studies. Amer. Meteorol. Soc., Boston, Mass., p.102–109.

Janssen P.A.E.M., Viterbo P. (1996). Ocean waves and the atmospheric climate. J. Climate, Vol. 9, N6, p.1269–1287.

Jastrow R., Nierenberg W., Seitz F. (1990). Scientific Perspectives on the Greenhouse Problem. The Marshall Press. Jameson Books, Inc. Ottawa, III. 254 pp.

Jaworowski Z. (1994). Ancient atmosphere-validity of ice records. Environ. Sci. & Pollut. Res., Vol. 1 (3), p.161–171.

Jaworowski Z., Segalstad T.V., Ono N. (1992). Do glaciers tell a true atmospheric CO_2 story? The Sci. Tot. Environ., Vol. 114, p.227–284.
Jenkins G.S. (1993). A general circulation model study of the effects of faster rotation rate, enhanced CO_2 concentration, reduced solar forcing: implications for Faint Young Sun paradox. J. Geophys. Res., Vol. 98, ND11, p.20803–20811.
Jennings S.G. (Ed.) (1993). Aerosol Effects on Climate. The University of Arizona Press. Tucson. 305 pp.
Jepma C.J. (Ed.) (1995). The Feasibility of Joint Implementation. Parts I–IV. Kluwer Acad. Publ., Dordrecht.
Jiang Y., Yung Y.L. (1996). Concentrations of tropospheric ozone from 1979 to 1992 over tropical Pacific South America from TOMS data. Science, Vol. 272, N5262, p.714–716.
Jiang Y., Yung Y.L., Zurek R.W. (1996). Decadal evolution of the Antarctic ozone hole. J. Geophys. Res., Vol. 101, ND4, p.8985–8999.
Johannessen O.M., Muench R.D., Overland J.E. (Eds.) (1994). The Polar Oceans and Their Role in Shaping the Global Environment. The Nansen Centennial Volume. Amer. Geophys. Union, Washington, D.C. 524 pp.
Johannessen O.M., Miles M., Bjorgo E. (1995). The Arctic's shrinking sea-ice. Nature, Vol. 376, N6538, p.297–298.
Jones A. (1995). Estimating indirect aerosol forcing. GEWEX News, Vol. 5, N1, p.5–6.
Jones A., Roberts D.L., Slingo A. (1994). A climate model study of indirect radiative forcing by anthropogenic sulphate aerosols. Nature (Gr. Brit.), Vol. 370, N6489, p.450–453.
Jones I.S.F., Sugimori Y., Stewart R.W. (Eds.) (1993). Satellite Remote Sensing of the Ocean Environment. Seibutsu Kenkyusto Co. 528 pp.
Jones P.D. (1994). Recent warming in global temperature series. Geophys. Res. Lett., Vol. 21, p.1149–1152.
Jones P.D. (1995). Recent variations in mean temperature and the diurnal temperature range in the Antarctic. Geophys. Res. Lett., Vol. 22, N11, p.1345–1348.
Jones P.D. (1996). A la recherche du temps perdu. Nature, Vol. 381, N6581, p.375–376.
Jones P.D., Bradley R.S., Jouzel J. (1996). Climate Variations and Forcing Mechanisms of the Last 2000 years. Springer-Verlag. Berlin. X + 649 pp.
Jones P.D., Briffa K.R. (1992). Global surface and temperature variations during the 20th century. Part I: Spatial, temporal and seasonal details. The Holocene, Vol. 2, N2, p.165–179.
Jones P.D., Briffa K.R., Schweingruber F.H. (1995). Tree-ring evidence of the widespread effects of explosive volcanic eruption. Geophys. Res. Lett., Vol. 22, N11, p.1333–1336.
Jonsson H.H., Wilson J.C., Brock C.A., Dye J.E., Ferry G.V., Chan K.R. (1996). Evolution of the stratospheric aerosol in the northern hemisphere following the June 1991 volcanic eruption of Mount Pinatubo: Role of tropospheric–stratospheric exchange and transport. J. Geophys. Res., Vol. 101, ND1, p.1553–1570.
Joos F., Siegenthaler U. (1995). Anthropogenic carbon emissions and future levels of

atmospheric carbon dioxide. In: "CO_2—Eine Herausforderung für die Menscheit. Forum für Allgemeine Ökologie". Interlaken, Switzerland, p.32.

Joos F., Bruno M., Fink R., Siegenthaler U., Stocker T.F., Le Quéré C., Sarmiento J.L. (1996). An efficient and accurate representation of complex oceanic and biospheric models of anthropogenic carbon uptake. Tellus, Vol. 48B, N3, p.397–417.

Joussame S. (1995). Klima: Gestern, Heute, Morgen. Springer-Verlag, Berlin. 150 pp.

Jouzel J., Vaikmae R., Petit J.R., Martin M., Duelos Y., Stievenard M., Lorius C., Toots M., Melieres M.A., Burckle L.H., Barkov N.L., Kotlyakov V.M. (1995). The two-step shape and timing of the last deglaciation in Antarctica. Climate Dynamics, Vol. II, N3, p.151–161.

Kang I.-S. (1996). Association of interannual and interdecadal variations of global-mean SST appearing in a model and observations temperature with tropical Pacific. J. Climate, Vol. 9, N2, p.455–464.

Kapitsa A.P., Ridley J.K., de Q. Robin G., Siegert M.J., Zotikov I.A. (1996). A large deep freshwater lake beneath the ice of central East Antarctica. Nature, Vol. 381, N6584, p.684–686.

Kärcher B. (1996). Aircraft-generated aerosols and visible contrails. Geophys. Res. Lett., Vol. 23, N15, p.1933–1936.

Kärcher B., Hirschberg M.M., Fabian P. (1996). Small-scale chemical evolution of aircraft exhaust species at cruising altitudes. J. Geophys. Res., Vol. 101, ND10, p.15169–15190.

Karl T.R. (1994). Smudging the fingerprints. Nature (Gr. Brit.), Vol. 371, N6496, p.380–381.

Karl T.R., Knight R.W., Christy J.R. (1994). Global and hemispheric temperature trends: uncertainties related to inadequate spatial sampling. J. Climate, Vol. 7, N7, p.1144–1163.

Karl T., Bretherton F., Easterling W., Trenberth K. (1995a). Long-term climate monitoring of the Global Climate Observing System (GCOS). Clim. Change, Vol. 31, N2–4, p.135–147.

Karl T., Knight R.W., Kukla G., Gavin J. (1995b). Evidence for radiative effects of anthropogenic sulphate aerosols in the observed climate record. Earth System Monitor, Vol. 5, N4, p.10.

Karl T., Bretherton F., Easterling W., Trenberth K., Quayle R.G. (1995c). Long-term climate monitoring of the GCOS. GCOS Newsletter, N3, p.4–6.

Karl T.R., Knight R.W., Plummer N. (1995d). Trends in high frequency climate variability in the twentieth century. Nature, Vol. 377, N6546, p.217–220.

Karl T.R., Derr V.E., Easterling D.E., Folland C.K., Hofmann D.J., Levitus S., Nicholls N., Parker D.E., Withee G.W. (1995e). Critical issues for long-term climate monitoring. Clim. Change, Vol. 31, N2–4, p.185–221.

Karl T.R. (Ed.) (1996). Long-Term Climate Monitoring by the Global Climate Observing System. Kluwer Acad. Publ., Dordrecht. 648 pp.

Karl T.R., Knight R.W., Easterling D.R., Quayle R.G. (1996). Indices of climate change for the United States. Bull. Amer. Meteorol. Soc., Vol. 77, N2, p.279–292.

Kärner O. (1996). Global temperature deviations as a random walk. J. Climate, Vol. 9, N3, p.656–658.

Karoly D.J., Cohen J.A., Meehl G.A., Mitchell J.F.B., Oort A.H., Stouffer R.J., Wetherald R.T(1994). An example of fingerprint detection of greenhouse climate change. Climate Dynamics, Vol. 10, p.97–105.

Kaufman Y.J., Chou M.-D. (1993). Model simulations of the competing climatic effects of SO_2 and CO_2 . J. Climate, Vol. 6, p.1241–1252.

Kawamura R., Sugi M., Sato N. (1995). Interdecadal and interannual variability in the northern extratropical circulation simulated with the JMA global model. Part 1: Wintertime leading mode. J. Climate, Vol. 8, p.3006–3019.

Keeling C.D., Whorf T.P., Wahlen M., van der Plicht (1995). Interannual extremes in the rate of rise of atmospheric carbon dioxide since 1980. Nature (Gr. Brit.), Vol. 375, N6533, p.666–670.

Keeling C.D., Chin J.F.S., Whorf T.P. (1996). Increased activity of northern vegetation inferred from atmospheric CO_2 measurements. Nature, Vol. 382, N6587, p.146–149.

Keller A.A., Goldstein R.A. (1995). Oceanic transport and storage of carbon emissions. Clim. Change, Vol. 30, N4, p.367–395.

Keller A.A. (1996). The CO_2 fertilization factor and the "missing" carbon sink. Climatic Change, Vol. 33, N1, p.63–68.

Kellogg W.W. (1995). Contaminants affecting the Arctic climate and the role of the oceans. The Science of the Total Environ., Vol. 160/161, p.769–775.

Kemp D. (1994). Global Environmental Issues. Routledge, London. 224 pp.

Kent G.S., Williams E.R., Wang P.H., Mc Cormick M.P., Skeens K.M. (1995). Surface temperature related variations in tropical cirrus clouds measured by SAGE. J. Climate, Vol. 8, N11, p.2577–2594.

Kergoat L., Ficher A., Moulin S., Dedieu G. (1995). Satellite measurements as a constraint on estimates of vegetation carbon budget. Tellus, Vol. 47B, N1/2, p.251–263.

Kerr R.A. (1995). Did the tropical Pacific drive the world's warming? Science, Vol. 266, N 5185, p.544–545.

Kerr R.A. (1996). Sky-high findings drop new hints of greenhouse warming. Nature, Vol. 273, N5271, p.34.

Kheshgi H.S., Jain A.K., Wuebbles D.J. (1996). Accounting for the missing carbon-sink with the CO_2-fertilization effect. Climatic Change, Vol. 33, N1, p.31–62.

Kiehl J.T., Briegleb B.P. (1993). The relative roles of sulphate aerosols and greenhouse gases in climate forcing. Science, Vol. 260, p.311–314.

Kiehl J.T., Rodhe H. (1995). Modeling geographical and seasonal forecing due to aerosols. In : Proc. Dahlem Workshop on Aerosol Forcing of Climate, Berlin, April 1994.

Kiehl J.T., Hack J.J., Zhang N.H., Cess R.D. (1995). Sensitivity of a GCM climate to enhanced shortwave cloud absorption. J. Climate, Vol. 8, N9,p.2200–2212.

Kim K.Y., Crowley T.J. (1994). Modeling the climate effect of unrestricted greenhouse emissions over the next 10,000 years. Geophys. Res. Lett., Vol. 21, N8, p.681–684.

Kim K.Y., North G.R. (1993). EOF analysis of surface temperature field in a stochastic climate model. J. Climate, Vol. 6, N9, p.1681–1690.

Kim Y., Cess R.D. (1993). Effect of anthropogenic sulphate aerosols on low-level cloud albedo over oceans. J. Geophys. Res., p.14883–14886.

Kim K.Y., North G.R., Hegerl G.C. (1996). Comparison of the second-moment statistics of climate models. J. Climate, Vol. 9, p.921–936.

King A.W., Emanneul W.R., Wullschleger S.D., Post W.M. (1995a). In search of the missing carbon sink: model of terrestrial response to land-use change and atmospheric CO_2 . Tellus, Vol. 478, N5, p.501–519.

King M.D. (1993). Radiative properties of clouds. In: Aerosol–Cloud–Climate Interactions. Ed.: P.V. Hobbs. Academic Press. p.123–149.

King M.D., Radke L.F., Hobbs P.V. (1993). Optical properties of marine stratocumulus clouds modified by ships. J. Geophys. Res., Vol. 98, p.2729–2739.

King M.D., Tsay S.-C., Platnic S.E. (1995b). In situ observations of the indirect effects of aerosol on clouds. In: Aerosol Forcing on Climate. Eds.: R.J. Charlson and J.E. Heintzenberg. John Wiley & Sons. p.227–248.

Kitoh A., Koide H., Kodera K., Yukimoto S., Noda A. (1996). Interannual variability in the stratospheric–tropospheric circulation in a coupled ocean–atmosphere GCM. Geophys. Res. Lett., Vol. 23, p.543–546.

Ko M.K.W., Schneider H.R., Shia R.-L., Weinstein D.K., Sze N.-D. (1993). A two-dimensional model with coupled dynamics, radiation, and photochemistry. 1. Simulation of the middle atmosphere. J. Geophys. Res., Vol. 98, N211, p.20429–20440.

Kodera K. (1995). On the origin and nature of the interannual variability of the winter stratospheric circulation in the Northern Hemisphere. J. Geophys. Res., Vol. 100, ND7, p.14077–14087.

Kodera K. (1996). Spatial and seasonal characteristics of recent trends in the Northern-Hemispheric troposphere and stratosphere. Draft Manuscript, 3 July 1996. 26 pp., ill.

Kodera K., Chiba M., Koide H., Kitoh A., Nikaidou Y. (1996). Interannual variability of the winter stratosphere and troposphere in the Northern Hemisphere. J. Meteorol. Soc. Japan, Vol. 74, N3, p.365–382.

Kogan Z.N., Kogan Y.L., Lilly D.K. (1996). Evaluation of sulphate aerosols indirect effect in marine stratocumulus clouds using observation-derived cloud climatology. Geophys. Res. Lett., Vol. 23, N15, p.1937–1940.

Kolomeev M.P., Nikonov S.A., Sorokovikova O.S., Khmelevtsov S.S. (1993). Modelling of the Northern Hemisphere climate response to the Pinatubo volcanic eruption. Meteorology and Hydrology, N4, p.15–19 (in Russian).

Kondratyev K.Ya. (1954). Radiative Energy of the Sun. Leningrad: Gidrometeoizdat. 386 pp. (in Russian).

Kondratyev K.Ya. (1969). Radiation in the Atmosphere. New York: Academic Press. 786 pp.

Kondratyev K.Ya. (1972). Complex Atmospheric Energetics Experiment. GARP Publ. Ser. No.12, WMO, Geneva. 38 pp.

Kondratyev K.Ya. (1982). The World Climate Research Program: present state and perspectives. Progress in Sci. & Technol., Meteorology and Climatology, Vol. 8. Moscow, VINITI. 275 pp. (in Russian).

Kondratyev K.Ya. (1983a). Satellite Climatology. Leningrad: Gidrometeoizdat. 264 pp. (in Russian)

Kondratyev K.Ya. (1983b). Earth Radiation Budget, Aerosol and Clouds. Progress in Sci. and Technol., Vol. 10. Moscow, VINITI. 315 pp. (in Russian).

Kondratyev K.Ya. (1984). Volcanoes and Climate. World Climate Paper No.54, WMO, Geneva. 103 pp.

Kondratyev K.Ya. (1985). Volcanoes and Climate. Progress in Sci. and Technol., Meteorology and Climatology. Vol. 13. Moscow, VINITI. 284 pp. (in Russian).

Kondratyev K.Ya. (1988). Climate Shocks: Natural and Anthropogenic. John Wiley & Sons, New York. 296 pp

Kondratyev K.Ya. (1989a). Global Dynamics of Ozone. Progress in Sci. & Technol., Geomagnetism and Higher Atmosphere. Vol. 19. Moscow, VINITI. 212 pp. (in Russian).

Kondratyev K.Ya. (1989b). The International Geosphere–Biosphere Programme: The role and place of sun–atmospheric interrelationships. Geofis. Int., Vol. 28, N3, p.453–466.

Kondratyev K.Ya. (1990a). Key Problems of Global Ecology. Progress in Sci. & Technol.,Theoretical and General Problems of Geography. Vol. 9. Moscow, VINITI. 454 pp. (in Russian).

Kondratyev K.Ya. (1990b). New assessments of global climate change. Izv. All-Union Geogr. Soc., Vol. 122, issue 6, p.615–621 (in Russian).

Kondratyev K.Ya. (Ed.) (1991a). Aerosol and Climate. Leningrad: Gidrometeoizdat. 542 pp. (in Russian).

Kondratyev K.Ya. (1991b). The Second World Climate Conference. General characteristic and conception of climate studies. Izv. All-Union Geogr. Soc.. Vol. 123, issue 5, p.401–408 (in Russian).

Kondratyev K.Ya. (1991c). New assessments of global climate change. Atmosfera, Vol. 4, N3, p.177–188.

Kondratyev K.Ya. (1992a). The "aerosol–clouds–climate" interaction. Part I. Aerosol. Optics of the Atmosphere and Ocean, Vol. 5, N3, p.317–323 (in Russian).

Kondratyev K.Ya. (1992b). The "aerosol–clouds–climate" interaction. Part II. Clouds. Optics of the Atmosphere and Ocean, Vol. 5, N3, p.324–335 (in Russian).

Kondratyev K.Ya. (1992c). The Second World Climate Conference. Assessments of global climate change and actions to prevent global changes. Izv. All-Union Geogr. Soc, Vol. 123, issue 6, p.473–480 (in Russian).

Kondratyev K.Ya. (1992d). Global Climate. St. Petersburg: Nauka Publ. 359 pp. (in Russian).

Kondratyev K.Ya. (1992e). Global climate monitoring. Studying the Earth from Space, N1, p.96–106 (in Russian).

Kondratyev K.Ya. (1992f). Global water and energy cycle: the role of observations from space. Studying the Earth from Space, N5, p.116–124 (in Russian).

Kondratyev K.Ya. (1992g). Global climate change: some results and perspectives. Izv. RAN. Ser. Geography, N4, p.123–129 (in Russian).

Kondratyev K.Ya. (1992h). Correction of the scientific program for the EOS measuring complex. Studying the Earth from Space, N6, p.120–123 (in Russian).

Kondratyev K.Ya. (1992i). The Framework Convention on Climate Change: problems and perspectives. Izv. USSR Acad. Sci., Ser. Geography, N2, p.52–63 (in Russian).

Kondratyev K.Ya. (1992j). On the International Framework Convention on Climate Change: global climate change in the context of global change. Il Nuovo Cimento, Vol. 15C, Ser. 1, N1, p.87–98.

Kondratyev K.Ya. (1993a). The Second UN Conference on Environment and Development: some results and perspectives. Izv. All-Union Geogr. Soc., Vol. 125, issue 3, p.1–8 (in Russian).

Kondratyev K.Ya. (1993b). Complex monitoring of the consequences of Pinatubo eruption. Studying the Earth from Space, N1, p.111–122 (in Russian).

Kondratyev K.Ya. (1993c). Ecology and policy. Part I. Izv. Russian Geogr. Soc., Vol. 125, issue 1, p.3–11 (in Russian).

Kondratyev K. Ya. (1993d). Ecology and policy. Part II. Izv. Russian Geogr. Soc., Vol. 125, issue 2, p.1–10 (in Russian).

Kondratyev K.Ya. (1993e). On the priorities in the field of international cooperation to solve the global environmental problems. Studying the Earth from Space, N4, p.94–108 (in Russian).

Kondratyev K.Ya. (1993f). Letter to the editor. Climatic Change, Vol. 24, p.383–384.

Kondratyev K.Ya. (1993g). Global Climate Observing System. Studying the Earth from Space, N6, p.104–115.

Kondratyev K.Ya. (1994a). The Global Terrestrial Observing System (GTOS): detection and monitoring of global ecosystems. Studying the Earth from Space, N3, p.122–125 (in Russian).

Kondratyev K.Ya. (1994b). Encyclopaedia in the field of numerical modelling of the climate system. Izv. RAN. Ser. Geogr. , N2, p.139–142 (in Russian).

Kondratyev K.Ya. (1994c). The biosphere and climate. Proc. Int. Symp. on State of the Art of Remote Sens. in Biosphere Studies. UNEP, Nairobi, p.4.1–4.18.

Kondratyev K.Ya. (1994d). Satellite observations in global ecology. In: "Remote Sensing and Global Climate Change (Eds. R.A. Vaughan and A.P. Cracknell). Springer-Verlag, Berlin, p.67–94.

Kondratyev K.Ya. (1995a). Prospects for satellite monitoring of atmospheric trace gases and aerosol. Optics of the Atmosphere and Ocean. (in press) (in Russian).

Kondratyev K.Ya. (1995b) . Key aspects of ecological policy. I. Ecodynamics. Izv. Russian Geogr. Soc., Vol. 127, issue 3, p.36–43 (in Russian).

Kondratyev K.Ya. (1995c). Key aspects of ecological policy. II. Socio-economic aspects. Izv. Russian Geogr. Soc., Vol. 127, issue 4, p.1–10 (in Russian).

Kondratyev K.Ya. (1995d). Prospects for satellite monitoring of trace gases and aerosol. Studying the Earth from Space, N5, p.111–116 (in Russian).

Kondratyev K.Ya. (1995e). State of the art in studies of global changes: the USA program. Studying the Earth from Space, N2, p.98–105 (in Russian).

Kondratyev K.Ya. (1996a). Volcanic eruptions and climate change. In: The Mount Pinatubo Eruption. Effects on the Atmosphere and Climate (G. Fiocco, D. Fuà, and G. Visconti, Eds.). John Wiley & Sons, New York, p.273–288.

Kondratyev K.Ya. (1996b). New book on climate. Izv. RAN. Ser. Geogr. (in press) (in Russian).

Kondratyev K.Ya. (1996c). Greenhouse warming versus aerosol cooling in the context of global climate change. Idöjàrás. (in press).

Kondratyev K.Ya., Barteneva O.D., Chapursky L.I., Chernenko A.P., Grishechkin V.S., Ivanov V.A., Korzov V.I., Lipatov V.B., Prokofyev M.A., Tolkachev V.K., Vasilyev O.B., Zhvalev V.F. (1976). Aerosol in the GATE area and its radiative properties. Atmospheric Science Paper N247. Colorado State University, Fort Collins, CO. 46 pp.

Kondratyev K.Ya., Hunt G.E. (1982). Weather and Climate on Planets. Pergamon Press. Oxford. 432 pp.

Kondratyev K.Ya., Moskalenko N.I. (1983). Greenhouse Effect of the Planetary Atmospheres. Progress in Sci. & Technol. Studies in Space, Vol. 10. Moscow, VINITI. 156 pp. (in Russian).

Kondratyev K.Ya., Moskalenko N.I., Pozdniakov D.V. (1983). Atmospheric Aerosol. Leningrad: Gidrometeoizdat. 224 pp. (in Russian).

Kondratyev K.Ya., Binenko V.I. (1984). Impact of Clouds on Radiation and Climate. Leningrad: Gidrometeoizdat, 240 pp. (in Russian).

Kondratyev K.Ya., Moskalenko N.I. (1984). Atmospheric Greenhouse Effect and Climate. Progress in Sci. & Technol., Meteorology and Climatology. Vol. 12. Moscow: VINITI. 262 pp. (in Russian).

Kondratyev K.Ya., Kozoderov V.V., Fedchenko P.P. (1990). Biosphere: Techniques and Results of Remote Sounding. Moscow: Nauka Publ. 224 pp. (in Russian).

Kondratyev K.Ya., Buznikov A.A., Pokrovsky O.M. (1992a). Global Ecology: Remote Sensing. Progress in Sci. & Technol. Atmosphere–ocean–space. Program "Sections". Vol. 14. Moscow, VINITI. 310 pp. (in Russian).

Kondratyev K.Ya., Ortner J., Preining O. (1992b). Priorities for global ecology now and in the next century. Space Policy, Vol. 8, N1, p.39–48.

Kondratyev K.Ya., Belchansky G.I., Petrosian V.G. (1993). Conceptual aspects of the satellite monitoring of biodiversity. Doklady RAN, Vol. 98, N1, p.38–42 (in Russian).

Kondratyev K.Ya., Galindo I. (1994). Global climate change in the context of global ecodynamics. Geofis. Intern., Vol. 33, N3, p.487–496.

Kondratyev K.Ya., Grassl H. (1994) . Global climate dynamics as an interactive component of global change. Environ. Conserv., Vol. 21, N3, p.256–257.

Kondratyev K.Ya., Ivlev L.S. (1995). On the climatic implications of anthropogenic aerosol. Doklady RAN, Vol. 350, N1, p.98–99 (in Russian).

Kondratyev K.Ya., Nikolsky G.A. (1995a). Solar activity and climate. 1. Observational data, condensation and ozone hypotheses. Studying the Earth from Space, N5, p.3–17 (in Russian).

Kondratyev K.Ya., Nikolsky G.A. (1995b). Solar activity and climate. 2. Direct impact of changes in the extra-atmospheric spectral distribution of solar radiation. Studying the Earth from Space, N6, p.4–18. (in Russian).

Kondratyev K.Ya., Varotsos C.A. (1995). Atmospheric greenhouse effect in the context of global climate change. Il Nuovo Cimento, Vol. 18C, N2, p.123–151.

Kondratyev K.Ya., Varotsos C.A., Fedchenko P.P. (1995). Global total ozone dynamics, its impact on the variability of surface UV radiation and on ecosystems. Studying the Earth from Space, N4, p.104–116. (in Russian).

Kondratyev K.Ya., Cracknell A.P. (1998a). Observing Global Climate Change. Taylor & Francis, London (in press).

Kondratyev K.Ya., Cracknell A.P. (1998b). Key Issues of Global Change. World Sci. Publ., London (in press)
Kondratyev K.Ya., Galindo I. (1997). Volcanic Eruptions and Climate. Williamsburg: Deepak Publ. (in press).
Kondratyev K.Ya., Peña F.M., Galindo I. (1997). Sustainable Development and Population Dynamics. University of Colima Press, Colima, Mexico — 128 pp.
Kondratyev K.Ya., Binenko V.I., Melnikova I.N. (1996) On "excessive" absorption of solar radiation by clouds in the visible. Doklady RAN. (in press) (in Russian).
Kondratyev K.Ya., Johannessen O.M., Melentyev V.V. (1996a). High Latitude Climate and Remote Sensing. Chichester: Wiley–Praxis. 252 pp.
Kondratyev K.Ya., Buznikov A.A., Pokrovsky O.M. (1996b). Global Change and Remote Sensing. Chichester: Wiley–Praxis. 396 pp.
Kondratyev K.Ya., Tanaka T. (1997). Advanced Earth Observing Satellite-II (ADEOS-II): new perspectives of global environmental monitoring. Studying the Earth from Space, N1, p.105–121.
Koop T., Carslaw K.S. (1996). Melting of $H_2SO_44H_2O$ particles upon cooling: implications for polar stratospheric clouds. Science, Vol. 272, p.1638–1641.
Koppany G. (1994). Temperature variation in Europe and North America since the beginning of instrumental observations. Idöjàrás, Vol. 98, N4, p.227–236.
Kouimtzis T., Samara C. (Eds.) (1995). Airborne Particulate Matter. The Handbook of Environmental Chemistry. Vol. 4, Part D. Springer-Verlag, Berlin. 339 pp.
Krabill W., Thomas R., Jazek K., Kuivinen K., Manizade S. (1995). Greenland ice sheet thickness changes measured by laser altimetry. Geophys. Res. Lett., Vol. 22, N17, p.2341–2344.
Krzyscin J.W. (1996). UV controlling factors and trends derived from the ground-based measurements taken at Belsk, Poland, 1976–1994. J. Geophys. Res., Vol. 101, ND11, p.16797–16805.
Kubota M., Shikauchi A. (1995). Air temperature at ocean surface derived from surface-level humidity. J. Oceanogr., Vol. 51, p.619–634.
Kuik O.J., Peters P., Schrijver N. (Eds.) (1994). Joint Implementation to Curb Climate Change: Legal and Economic Aspects. Kluwer Acad. Publ., Dordrecht. 228pp.
Kukla G., Karl T.R. (1992). Nighttime temperatures in the Northern Hemisphere. CDIAC Communication, N17, p.10.
Kukla G.J., Went E. (Eds.) (1992). Start of a Glacial. Springer-Verlag, Berlin. 353 pp.
Kukla G.T., Karl M. Riches (1994). Minimax Workshop Report.
Kulmala M., Korhonen P., Laaksonen A., Vesala T. (1995). Changes in cloud properties due to NO_x emissions. Geophys. Res. Lett., Vol. 22, N3, p.239–242.
Kumar A., Hoerling M.P.(1995). Prospects and limitations of seasonal atmospheric GCM predictions. Bull. Amer. Meteorol. Soc., Vol. 76, N3, p.335–345.
Kunina I.M. (1993). Direct effects of CO_2 on the growth and productivity of higher plants. Problems of Ecological Monitoring and Modeling of Ecosystems, Vol. 15, p.138–150.
Kurz W.A., Apps M.J., Beukema S.J., Lekstrum T. (1995). 20th century carbon budget of Canadian forests. Tellus, Vol. 47B, N1/2, p.170–177.

Kushnir Y. (1994). Interdecadal variations in North Atlantic sea surface temperature and associated atmospheric conditions. J. Climate, Vol. 7, p.141–157.

Kyle H.L., Hoyt D.V., Hickey J.R., Maschoff R.H., Valette B.J. (1993). Nimbus–7 earth radiation budget calibration history. Part 1: The solar channels. NASA Reference Publication 1316.

Kyle H.L., Weiss M., Ardanuy P. (1995). Cloud, surface temperature, and outgoing radiation for the period from 1979 to 1990. J. Climate, Vol. 8, N11, p.2644–2658.

Labitzke K. (1987). Sunspots, the QBO, and the stratospheric temperature in the North Polar region. Geophys. Res. Lett., Vol. 14, p.535–537.

Labitzke K., McCormick M.P. (1992). Stratospheric temperature increases due to Pinatubo aerosols. Geophys. Res. Lett., Vol. 19, p.215–218.

Labitzke K., van Loon H. (1992). Association between the 11-year solar cycle and the atmosphere. Part V: Summer. J. Climate, N3, p.240–251.

Labitzke K., van Loon H. (1993). Some recent studies of probable connections between solar and atmospheric variability. Annales Geophysicae, Vol. 11, p.1084–1094.

Labitzke K., van Loon H. (1995). Connection between the troposphere and stratosphere on a decadal scale. Tellus, Vol. 47A, p.275–286.

Lanchbery J. (1995). Hard targets. The case for. Options. Spring'95, p.7–9.

Land-Use and Land-Cover Change. Science/Research Plan (1995). IGBP Report No.35/HDP Report No.7. Stockholm. 132 pp.

Landsea C.W., Nicholls N., Gray W.M., Avila L.A. (1996). Downward trends in the frequency of intense Atlantic hurricanes during the past five decades. Geophys. Res. Lett., Vol. 23, N13, p.1697–1700.

Langner J., Rodhe H., Crutzen P.J., Zimmerman P. (1992). Anthropogenic influence on the distribution of tropospheric sulphate. Nature, Vol. 359, p.712–716.

Larsen N., Knudsen B.M., Rosen J.M., Kjome N.T., Kyrö E. (1996). Balloon-borne backscatter observations of type 1 PSC formation: Inference about physical state from trajectory analysis. Geophys. Res. Lett., Vol. 23, N10, p.1091–1094.

Latif M., Barnett T.P. (1996). Decadal climate variability over the North Pacific and North America: dynamics and predictability. J. Climate, Vol. 9, p.981–985.

Lau K.-M., Sui C.-H., Chou M.-D., Tao W.-K. (1994). An inquiry into the cirrus-cloud thermostat effect for tropical sea surface temperature. Geophys. Res. Lett., Vol. 21, N12. p.1157–1160.

Lau K.-M., Weng H. (1995). Climate signal detection using wavelet transform: How to make a lime series sing. Bull. Amer. Meteorol. Soc., Vol. 76, N12, p.2391–2402.

Law K.S., Nisbet E.G. (1996). Sensitivity of the CH_4 growth rate to changes in CH_4 emissions from natural gas and coal. J. Geophys. Res., Vol. 101, ND9, p.14387–14397.

Lawrence M.G. (1993). An empirical analysis of the strength of the phytoplankton–dimethylsulfide–cloud–climate feedback cycle. J. Geophys. Res., Vol. 98, p.20663–20674.

Lean J. (1991). Variations in the sun's radiative output. Rev. Geophys., Vol. 29, p.505–535.

Leatich W.R., Isaac G.A. (1994). On the relationship between sulphate and cloud droplet number concentration. J. Climate, Vol. 7, p.206–212.

Leavsley G.H. (1994). Modeling the effects of climate change on water resources—a review. Clim. Change, Vol. 28, N1–2, p.159–177.

Leck C., Persson C. (1996). The central Arctic Ocean as a source of dimethyl sulfide. Seasonal variability in relation to biological activity. Tellus, Vol. 48, N2, p.156–177.

Leck C., Bigg E.K., Covert D.S., Heintzenberg J., Maenhaut W., Nilsson E.D., Wiedensohler A. (1996). Overview of the atmospheric research program during the International Arctic Ocean Expedition of 1991 (IAOE–91) and its scientific results. Tellus, Vol. 48B, N2, p.136–155.

Lee W.H., North G.R. (1994). Small ice cap instability in the presence of fluctuations. Climate Dynamics, Vol. 11, N4, p.242–246.

Lefohn A.S. (Ed.) (1992). Surface Level Ozone Exposures and Their Effects on Vegetation. Lewis Publ., Chelsea, MI. 366 pp.

Lelieveld J., Heintzenberg J. (1992). Sulfate cooling effect through in-cloud oxidation of anthropogenic SO_2 . Science, Vol. 258, p.117–120.

Lemke P. (1995). Sea ice modelling. Proc. Workshop on Global Coupled General Circulation Models. Scripps Inst. Oceanogr., La Jolla, USA, Oct. 1994. WCRP–87 (WMO/TD-N655). p.24–25.

Le Treut H. (Ed.) (1996). Climate Sensitivity to Radiative Perturbations. Physical Mechanisms and Their Validation. Springer-Verlag. Berlin. IX + 331 pp.

Li X., Maring H., Savoie D., Voss K., Prospero J.M. (1996). Dominance of mineral dust in aerosol light scattering in the North Atlantic trade winds. Nature, Vol. 380, p.416–419.

Li Z. (1996). Influence of absorbing aerosols on the inference of surface solar radiation budget. J. Climate (submitted).

Li Z., Barker H.W. (1994). Solar energy disposition: intercomparison between satellite estimation, GCM simulation and surface observation. Cloud–Radiation Interactions and Their Parameterization in Climate Models. WCRP–86 (WMO/TD-No.648). Geneva, WMO, p.110–163.

Li Z., Moreau L. (1996). Alteration of atmospheric solar absorption by clouds: simulation and observation. J. Appl. Meteorol., Vol. 35, N5, p.653–670.

Li Z., Barker H.W., Moreau L. (1995). The variable effects of clouds on atmospheric absorption of solar radiation. Nature, Vol. 376, N6540, p.486–490.

Li Z., Leighton H.G., Masuda K., Takashima T. (1993). Estimation of SW flux absorbed at the surface from TOA reflected flux. J. Climate, Vol. 6, p.317–330.

Li Z., Moreau L., Arking A. (1996). On solar energy disposition: a perspective from observation and modelling. Bull. Amer. Meteorol. Soc. (submitted).

Liang X.-Z., Wang W.-C., Dudek M.P. (1995). Interannual variability of regional climate and its change due to the greenhouse effect. Global and Planetary Change, Vol. 10, p.217–238.

Liang X.-Z., Wang W.-C., Dudel M.P. (1996). Northern Hemispheric interannual teleconnection patterns and their changes due to the greenhouse effect. J. Climate, Vol. 9, N12, p.465–479.

Liao X., Street-Perrott A., Mitchell J.F. (1995). GCM experiments with different cloud parameterization: Comparisons with palaeoclimatic reconstructions for 6000 years BC. Paleoclimates. Data and Modeling, Vol. 1, N1, p.99–1212.

Lindzen R.S. (1994). Climate dynamics and global change. Ann. Rev. Fluid Mech., Vol. 26, p.353–378.
Lindzen R.S., Pan W. (1994). A note on orbital control of equator-pole heat fluxes. Climate Dynamics, Vol. 10, p.49–57.
Liou K.-N. (1992). Radiation and Cloud Processes in the Atmosphere. Theory, Observations and Modelling. Oxford Univ. Press, New York. 487 pp.
Liou K.-N. (1994). Parameterization of the radiative properties of cirrus clouds: problems and some solutions. Cloud–Radiation Interactions and Their Parameterization in Climate Models. WCRP–86 (WMO/TD-No.648). WMO, Geneva, p.136.
Liu Y.F., Georgi F., Washington W.M. (1994). Simulation of summer monsoon climate over east Asia with an NCAR regional climate model. Mon. Weather Rev., Vol. 122, N10, p.2331–2348.
Lockwood G.W., Skiff B.A., Baliunas S.L., Radick R.R. (1992). Long-term solar brightness changes estimated from a survey of sun-like stars. Nature, Vol. 360, p.653–655.
Lohmann U., Roeckner E. (1996). Design and performance of new microphysics scheme developed for the ECRAM general circulation model Climate Dynamics, Vol. 12, N8, p.557–572.
Long C.S., Miller A.J., Lee H.-T., Wild J.D., Przywarty R.C., Hufford D. (1996). Ultraviolet index forecasts issued by the National Weather Service. Bull. Amer. Meteorol. Soc., Vol. 77, N4, p.729–748.
Lovejoy T.E. (1992). Some thoughts on biological diversity. Network'92. Special Supplement. Review of Agenda 21. N IX. 2 pp.
Lovelock J. (1995). The Ages of Gaia. A Biography of Our Living Earth. W.W. Norton and Co., New York. London, 255 pp.
Lubchenko J., Olson A.M., Brubaker L.B. et al. (1991). The sustainable biosphere initiative: An ecological research agenda. Ecology, Vol. 72, N2, p.371–412.
Lubin D. (1994). The role of the tropical super greenhouse effect in heating the ocean surface. Science. Vol. 265, N5169, p.224–227.
Luo Y., Sims D.A., Thomas R.B., Tissue D.T., Ball J.T. (1996). Sensitivity of leaf photosynthesis to CO_2 concentration is an invariant function for C3 plants: A test with experimental data and global implications. Global Biogeochemical Cycles, Vol. 10, N2, p.209–222.
Ma C.C., Mechoso C.R., Robertson A.W., Arakawa A. (1996). Peruvian stratus clouds and the tropical Pacific circulation: a coupled ocean–atmosphere GCM study. J. Climate, Vol. 9, N7, p.1635–1645.
MacCracken M.C. (1995). Climate change: The evidence mounts up. Nature, Vol. 376, N6542, p.645–646.
Macdonald A.M., Wunsch C. (1996). An estimate of global ocean circulation and heat fluxes. Nature, Vol. 382, p.436–439.
Macdonald R. (1996). Awakenings in the Arctic. Nature, Vol. 380, p.286–287.
MacLeod K.G., Huber B.T. (1996). Reorganization of deep ocean circulation accompanying a Late Cretaceous extinction event. Nature, Vol. 380, p.422–426.
Mahfouf J.-F., Manzi A.O., Noilhan J., Giordani H., Déqué M. (1995). The land

surface scheme ISBA within the Météo-France climate model ARPEGE. Part I: Implementation and preliminary results. J. Climate, Vol. 8, N8, p.2039–2057.
Mahlman J.D. (1995). Toward a scientific centered climate monitoring system. Clim. Change, Vol. 31, N2–4,p. 223–230.
Malanotte-Rizzoli P., Robinson A.R. (Eds.) (1994). Ocean Processes in Climate Dynamics: Global and Mediterranean Examples. Kluwer Acad. Publ. Dordrecht. 437 pp.
Malin G., Turner S., Liss P., Holligan P., Harbour D. (1993). Dimethylsulfide and dimethylsulphoniopropionate in the northeast Atlantic during summer coccolithosphere bloom. Deep-Sea Res., Vol. 40, p.1487–1508.
Manabe S., Stouffer R.J. (1993). Century-scale effects of increased atmospheric CO_2 on the ocean–atmosphere system. Nature, Vol. 364, p.215–218.
Manabe S., Stouffer R.J. (1995). Simulation of abrupt climate change induced by freshwater input to the North Atlantic Ocean. Nature, Vol. 378, p.165–167.
Manabe S., Stouffer R.J. (1996). Low-frequency variability of surface air temperature in a 1000-year integration of a coupled atmosphere—ocean–land surface model. J. Climate, Vol. 9, N2, p.376–393.
Mann M.E., Lees J.M. (1996). Robust estimation of background noise and signal detection in climate time series. Climatic Change, Vol. 33, N3, p.409–445.
Mann M.E., Park J. (1996a). Greenhouse warming and changes in the seasonal cycle of temperature: Model versus observations. Geophys. Res. Lett., Vol. 23, N10, p.1111–1114.
Mann M.E., Park J. (1996b). Joint spatiotemporal modes of surface temperature and sea level pressure variability in the Northern Hemisphere during the last century. J. Climate, Vol. 9, N7, p.874–886.
Manney G.L., Froidevaux L., Waters J.W., Santee M.L., Read W.G., Flower D.A., Jarnot R.F., Zurek R.W. (1996). Arctic ozone depletion observed by UARS MILS during the 1994–95 winter. Geophys. Res. Lett., Vol. 23, N1, p.85–88.
Manzini E., Bengtsson L. (1996). Stratospheric climate and variability from a general circulation model and observations. Climate Dynamics, Vol. 12, p.615–639.
Marchetti C. (1991). The future. IIASA Res. Rept. 91-5. Laxenburg, p.399–416.
Marchuk G.I. (1975). Modeling climate change and the problem of long-range weather forecast. Meteorology and Hydrology, N7, p.25–36. (in Russian).
Marchuk G.I., Kondratyev K.Ya., Kozoderov V.V., Khvorostyanov V.I. (1986). Clouds and Climate. Leningrad: Gidrometeoizdat. 512 pp. (in Russian).
Marchuk G.I., Kondratyev K.Ya., Kozoderov V.V. (1992). Earth's Radiation Budget: Key Issues. Nauka Publ., Moscow, 231 pp.
Marley N.A., Gaffney J.S., Cunningham M.M. (1993). Aqueous greenhouse species in clouds, fogs, and aerosols. Environ. Science and Technol., Vol. 27, p.2864–2869.
Marton-Lefevre J. (1992). Science prepares for UNCED. Network'92. N13, p.4.
Marzocchi W., Mulargia F. (1996). Scale analysis to sort the different causes of mean sea level changes: an application to the northern Adriatic Sea. Geophys. Res. Lett., Vol. 23, N10, p.1119–1122.
Maslanik J.A., Serreze M.C., Barry R.G. (1996). Recent decreases in Arctic summer ice

cover and linkages to atmospheric circulation anomaly. Geophys. Res. Lett., Vol. 93, N13, p.1677–1680.
Masood E. (1996). Industrial world is unlikely to meet promised carbon cuts. Nature, Vol. 382, N6587, p.103.
Matthews E. (1994). Nitrogen fertilizers: Global distribution of consumption and associated emissions of nitrous oxide and ammonia. Global Biogeochemical Cycles, Vol. 8, N4, p.411–439.
Matthews J.T. (1989). Redefining security. Foreign Policy, N74, p.23–41.
Mayewski P.A., Twickler M.S., Whitlow S.I., Meeker L.D., Yang Q., Thomas J., Kreutz K., Grootes P.M., Morse D.L., Steig E.J., Waddington E.D., Saltzman E.S., Whung P.-X., Taylor K.C. (1996). Climate change during the last deglaciation in Antarctica. Science, Vol. 272, p.1636–1641.
McBean G.A., Liss P.S., Schneider S.H. (1996). Advancing our understanding. Preprints Seventh Symp. Global Change Studies. Amer. Meteorol. Soc., Boston, MA. p.41–48.
McCormick M.P. (1992). Initial assessment of the stratospheric and climatic impact of the 1991 Mount Pinatubo eruption: Prologue. Geophys. Res. Lett., Vol. 19, p.149–150.
McCormick M.P., Thomason L.W., Trepte Ch.P. (1995). Atmospheric effects of the Pinatubo eruption. Nature, Vol. 373, p.399–404.
McGuffie K., Henderson-Sellers A., Zhang H., Durbidge T.B., Pitman A.J. (1995). Global climate sensitivity to tropical deforestation. Global and Planetary Change, Vol. 10, p.97–128.
McKee D.J. (Ed.) (1994). Tropospheric Ozone. CRC Press, Roca Baton, Fl. 333 pp.
Mearns L.O., Giorgi F., McDaniel L., Shields C. (1995). Analysis of variability and diurnal range of daily temperature in a nested regional climate model: comparison with observations and doubled CO_2 results. Climate Dynamics, Vol. 11, N4, p.193–210.
Meehl G.A. (1995). Global coupled general circulation models. Bull. Amer. Meteorol. Soc., Vol. 76, p.951–957.
Meehl G.A., Washington W.M. (1995). Cloud albedo feedback and supergreenhouse effect in a global coupled GCM. Climate Dynamics, Vol. 11, N7, p.399–411.
Meehl G.A., Washington W.M. (1996). El Niño-like climate change in a model with increased atmospheric CO_2 concentrations. Nature, Vol. 382, p.56–60.
Meehl G.A., Wheeler M., Washington W.M. (1994). Low-frequency variability and CO_2 transient climate change. Part 3. Intermonthly and interannual variability. Climate Dynamics, Vol. 10, p.277–303.
Mendoza B., Pérez-Enriquez R., Alvarez-Madrigal M. (1991). Analysis of solar activity conditions during periods of El Niño events. Ann. Geophys., Vol. 9, p.54–59.
Meszaros E. (1981). Atmospheric Chemistry. Fundamental Aspects. Akad. Kiado, Budapest. 210 pp.
Meszaros E. (1982). On the atmospheric input of sulphur into the ocean. Tellus, Vol. 34, p.277–282.
Meszaros E. (1985). On climatic effects of radiatively active trace gases and aerosol particles. Idöjárás, Vol. 89, p.302–308.

Meyers S.D., O'Brien J.J. (1995). Pacific Ocean influences atmospheric carbon dioxide. EOS, Vol. 76, N52, p.533–537.

Michaels P.J., Singer S.F., Knapperberger O.P.C., Kerr J.B., McElroy C.T. (1994). Analyzing ultraviolet-B radiation: is there a trend? Science, Vol. 264, p.1341–1343.

Miller R.L., Jiang X. (1996). Surface energy fluxes and coupled variability in the Tropics of a coupled general circulation model. J. Climate, Vol. 9, N7, p.1599–1620.

Miller A.J., Hollandsworth S.M., Flynn L.E., Tiao G.C., Reinsel G.C., Bishop L., McPeters R.D., Planet W.G., DeLuisi J.J., Mateer C.L., Wuebbles D., Kerr J., Nagatani R.M. (1996). Comparisons of observed ozone trends and solar effects in the stratosphere through examination of ground-based Umkehr and combined solar backscattered ultraviolet (SBUV) and SBUV2 satellite data. J. Geophys. Res., Vol. 101, NC4, p.9017–9021.

Miller Ch., Bretherton F. (1995). Long-term climate monitoring by the Global Climate Observing System. Climate forcings and feedbacks. Climatic Change, Vol. 31, N2–4, p.149–162.

Minnis P., Harrison E.F., Stowe L.L., Gibson G.G., Denn F.M., Doeling D.R., Smith W.L. (1993). Radiative climate forcing by the Mount Pinatubo eruption. Science, Vol. 259, p.1411–1415.

Mintzer I., Von Hippel D., Kolar S. (1994). Greenhouse Gas Assessment Methodology. The World Bank, Washington, D.C. 17 pp., Appendices.

Mirin A.A., Ambrosiano J.J., Rolstad J.H., Bourgeois A.J., Brown J.C., Chan B., Dannevik W.P., Duffy P.B., Eltgroth P.G., Matarazzo C., Wehner M.F. (1994). Climate system modelling using a domain and task decomposition message-passing approach. Comput. Phys.Commun., Vol. 84, N1–3, p.278–296.

Mitchell J.F.B., Johns T.C., Gregory J.M., Tett S.F.B. (1995a). Climate response to increasing levels of greenhouse gases and sulphate aerosols. Nature, Vol. 376, N6540, p.501–504.

Mitchell J.F.B., Davis R.A., Ingram W.J., Senior C.A. (1995b). On surface temperature, greenhouse gases and aerosols: models and observations. J. Climate, Vol. 8, N10, p.2364–2386.

Mitosek H.T. (1995). Climate variability and change within the discharge time series: a statistical approach. Clim. Change, Vol. 29, N1, p.101–116.

Modelling climate change, 1860–2050. (1995) Hadley Centre for Climate Prediction and Research. The U.K. Met. Office, Bracknell, p.1–13.

Modelling, validation and assimilation of clouds. Proc. of a Workshop held at ECMWF 31 October–4 November 1994. (1995). ECMWF, Reading, U.K. 464 pp.

Monastersky R. (1995). World climate panel charts path for action. Sci. News, Vol. 148, N19, p.293.

Montzka S.A., Butler J.H., Myers R.C., Thompson T.M., Swanson T.H., Clarke A.D., Lock L.T., Elkins J.W. (1996). Decline in the tropospheric abundance of halogen from halocarbons: implications for stratospheric ozone depletion. Science, Vol. 272, N5266, p.1318–1322.

Moore B., III, Braswell B.H. (1994). The lifetime of excess atmospheric carbon dioxide. Global Biogeochemical Cycles, Vol. 8, N1, p.23–38.

Morrissey M.L., Graham N.E. (1996). Recent trends in rain gauge precipitation measurements from the Tropical Pacific: evidence for an enhanced hydrological cycle. Bull. Amer. Meteorol. Soc., Vol. 77, N8, p.1207–1219.

Mote P.W., Holton J.R., Russel J.M.,III. Boville B.A. (1993). A comparison of observed (HALOE) and modeled (GCM2) methane and stratospheric water vapor. Geophys. Res. Lett., Vol. 20, N14, p.1419–1422.

Mototaka N. (1996). Effects of the ice albedo and runoff feedbacks on the thermohaline circulation. J. Climate, Vol. 9, N7, p.919–929.

MTPE/EOS Reference Handbook–1995. NASA Goddard Space Flight Centre, Greenbelt, MD. (1995). 277 pp.

Munk W., Worcester P., Wunsch C. (1995). Ocean Acoustic Tomography. Cambridge University Press. 433 pp.

Murphy J.M. (1995a). Transient response of the Hadley Centre Coupled Ocean–Atmosphere model to increasing carbon dioxide. Part I: Control climate and flux adjustment. J. Climate, Vol. 8, N1, p.36–56.

Murphy J.M. (1995b). Transient response of the Hadley Centre Coupled Ocean–Atmosphere model to increasing carbon dioxide. Part III: Analysis of global-mean response using simple models. J. Climate, Vol. 8, N3, p.496–514.

Murphy J.M., Mitchell J.F. (1995a). Transient response of the Hadley Centre Coupled Ocean–Atmosphere model to increasing carbon dioxide. Part II: Spatial and temporal structure of response. J. Climate, Vol. 8, N1, p.57–80.

Murphy J.M., Mitchell J.F. (1995b). Interhemispheric and land/sea contrasts in the transient response of a coupled model. Proc. Workshop on Global Coupled General Circulation Models. Scripps Inst. Oceanogr., La Jolla, USA, Oct. 1994, WCRP–87 (WMO/TD-No.655), Geneva, p.43.

Myneni R.B., Los S.O., Asrar G. (1995). Potential gross primary productivity of terrestrial vegetation from 1982–1990. Geophys. Res. Lett., Vol. 22, N19, p.2617–2620.

Nagurny A.P. (1995). Multi-year trends in changes of the sea-ice thickness in the Arctic. Meteorology and Hydrology, N6, p.80–83 (in Russian).

Naja M., Lal Sh. (1996). Changes in surface ozone amount and its diurnal and seasonal patterns: from 1954–55 to 1991–93, measured at Ahmedabad (23°N), India. Geophys. Res. Lett., Vol. 23, N1, p.81–84.

Nakajima T. (1993). Study of climate effect of cloud–aerosol interaction. In: Satellite Remote Sensing of the Oceanic Environment. Eds.: I.S.F. Jones, Y. Sugimori and R.W. Stuart. Seibutsu Kenkyusha. p.235–239.

Nakajima T.Y., Nakajima T. (1995). Wide area determination of cloud microphysical properties from NOAA AVHRR measurements for FIRE and ASTEX region. J. Atmos. Sci., Vol. 52, N23, p.4043–4059.

Nakajima T., Tsukamoto M., Tsusima Y., Numaguti A. (1995). Modelling of the radiative processes in AGCM. Reports of a new programme for creative basic research studies. Studies of global environment change with special reference to Asia and Pacific regions. 1–3. p.104–123.

Nakajima T., Tonna G., Rao R., Boi P., Kaufman Y., Holben B. (1996). Use of sky brightness measurements from ground for remote sensing of particulate polydispersions. Appl. Optics, Vol. 35, N15, p.2672–2686.

Nakamura M. (1996). Effects of the ice albedo and runoff feedbacks on the thermohaline circulation. J. Climate, Vol. 9, p.941–955.

Nasrallah H.A., Balling R.C., Jr. (1993). Analysis of recent climatic changes in the Arabian Gulf region. Environm. Conservation, Vol. 20, p.18–30.

Natural air pollution sulphur released by the ocean. Geophys. Inst. Quarterly (Univ. of Alaska. Fairbanks). 1996. N1, p.4–5.

Neelin J.D., Dijkstra H.A. (1995). Ocean-atmosphere interaction and the tropical climatology. Part I: The dangers of flux correction. J. Climate, Vol. 8, p.1325.

Nemani R.R., Running S.W., Pielke R.A., Chase T.N. (1996). Global vegetation cover changes from coarse resolution satellite data. J. Geophys. Res., Vol. 101, ND3, p.7157–7162.

Nerem R.S. (1995). Measuring global mean sea level variations using TOPEX/POSEIDON altimeter data. J. Geophys. Res., Vol. 100, NC12, p.25135–25151.

Nesme-Ribes E. (Ed.) (1994). The Solar Engine and Its Influence on Terrestrial Atmosphere and Climate. NATO ASI Series. Series I: Global Environmental Change. Vol. 25. Springer-Verlag, Berlin. 561 pp.

Nesme-Ribes E., Ferreira E.N., Sadourny R., Le Treut H., Li Z.-X. (1993). Solar dynamics and its impact on solar irradiance and terrestrial climate. J. Geophys. Res., Vol. 98, NA11, p.18923–18936.

Newell R.E., Weare B.C. (1976a). Ocean temperature and large-scale atmospheric variations. Nature, Vol. 262, p.40–41.

Newell R.E., Weare B.C. (1976b). Factors governing tropospheric mean temperature. Science, Vol. 194, p.1413–1414.

Newell R.E., Newell N.E., Zhu Y., Scott C. (1994). Tropospheric rivers? A Pilot Study, Vol. 12, N24, p.2401–2404.

Newton P. (1995). Climate system. Perspectives past and present. Nature, Vol. 378, N6552, p.6552.

Nilsson S., Schopfhauser W. (1995). The carbon-sequestration potential of a global afforestation program. Clim. Change, Vol. 30, N3, p.267–293.

Nitta T., Kachi M. (1994). Interdecadal variations of precipitation over the Tropical Pacific and Indian Oceans. J. Meteorol. Soc. Japan, Vol. 72, N6, p.823–831.

Noda A. (1995). Transient response study to the gradual increase of atmospheric CO_2 concentration with a coupled atmosphere–ocean model developed at the Meteorological Research Institute. Proc. Workshop on Global Coupled General Circulation Models. Scripps Inst. Oceanogr., La Jolla, USA, Oct. 1994. WCRP–87 9WMO/TD-No.655).

Nogai T., Uchino O., Fujimoto T., Sai Y., Tamashiro K., Nomura R., Sunagawa T. (1993). Lidar observation of the stratospheric aerosol layer over Okinawa, Japan, after the Mt. Pinatubo volcanic eruption. J. Meteorol. Soc. Japan, Vol. 71, N6, p.749–755.

Nomura A. (1995). Global sea-ice concentration data set for use with the ESMWF re-analysis system. ECMWF Re-Analysis Project Techn. Rept. No.76. Reading, UK. p.1–25.

Normile D. (1995). Polar regions give cold shoulder to theories. Science, Vol. 270, N5242, p.1566.

Norris J.R., Leovy C.B. (1995). Comments on "Trends in global marine cloudiness and anthropogenic sulphur". J. Climate, Vol. 8, N8, p.2109–2110.

North G.R., Kim K.-Y. (1995). Detection of forced climate signals. Part II: Simulation results. J. Climate, Vol. 8, N3, p.409–417.

North G.R., Kim K.-Y., Shen S.P., Hardin J.W. (1995). Detection of forced climate signals. Part I: Filter theory. J. Climate, Vol. 8, N3, p.401–408.

Novakov T., Penner J.E. (1993). Large concentration of organic aerosol to cloud condensation nuclei concentrations: Evidence from a marine site. Nature, Vol. 365, p.823–826.

Novakov T., Rivera-Carpio C., Penner J.E., Rogers C.F. (1994). The effect of anthropogenic sulphate aerosols on marine cloud droplet concentrations. Tellus, Vol. 46B, p.132–141.

Ohlmann J.C., Siegel D.A., Gautier C. (1996). Ocean mixed layer radiant heating and solar penetration. J. Climate, Vol. 9, p.619–628.

Ohmura A., Wild M., Bengtsson L. (1996). A possible change in the mass balance of Greenland and Antarctic ice sheets in the coming century. J. Climate, Vol. 9, N7.

Okamoto K., Tanimoto S., Okano K. (1995). Statistical Analysis of the increase in atmospheric CO_2 concentrations and its relation to the possible existence of CO_2 fertilization on a global scale. Tellus, Vol. 47B, N1/2, p.206–211.

Oort A.H., Pan Y.H., Reynolds R.W., Ropelewski C.F. (1987). Historical trends in the surface temperature of the ocean based on the COADS. Climate Dynamics, Vol. 2, N1, p.29–38.

Oreskes N., Shrader-Frechette K., Belitz K. (1994). Verification, validation and confirmation of numerical models in the earth sciences. Science, Vol. 263, p.641–646.

Ou S.C., Liou K.N., Takano Y., Rao N.X., Fu Q., Heymsfield A.J., Miloshevich L.M., Baum B., Kinne S.A. (1995). Remote sounding of cirrus optical depths and ice crystal sizes from AVHRR data: Verification using FIRE II IFO measurements. J. Atmos. Sci., Vol. 52, N23, p.4143–4158.

Ou S.C., Liou K.N., Baum B.A. (1996). Detection of multilayer cirrus cloud system using AVHRR data: Verification based on FIRE II IFO composite measurements. J. Appl. Meteorol., Vol. 35, N2, p.178–191.

Overpeck J.T. (1996). Warm climate surprises. Science, Vol. 271, p.1820–1821.

1994 Ozone Assessment: ozone shown to be in continuing decline. World Climate News. (1995). N6, p.10.

Pachauri R.K. (1992). The Climate Change Convention. What it may mean for the poor. Network'92. N19, p.14–15.

Paillard D. (1995). The hierarchical structure of glacial climatic oscillations: interactions between ice-sheet dynamics and climate. Climate Dynamics, Vol. 11, N3, p.162–177.

Pap J.M. (Ed.) (1994). The Sun as a Variable Star: Solar and Stellar Irradiance Variations. Kluwer Acad. Publ., Dordrecht. 328 pp.

Parikh J., Babu P.G., Parikh K. (1994). Decision making framework to address climate change (Pap.). IPCC Working Group Writing Team II Meet. "Steps Towards Decis. Making Framework to Address Clim. Change", Montreaux, March 3–6, 1994. PCI BER, N10, p.77–97.

Parker D.E., Folland C.K. (1995). Comments on the global climate of 1994. Weather, Vol. 50, N8, p.283–287.

Parker D.E., Jones P.D., Folland C.K.,Bevan A. (1994). Interdecadal changes of surface temperature since the late nineteenth century. J. Geophys. Res., Vol. 99, p.14373–14399.

Parker D.E., Folland C.K., Jackson M. (1995). Marine surface temperature: observed variations and data requirements. Climatic Change, Vol. 31, p.559–600.

Parmesan C. (1996). Climate and species' range. Nature, Vol. 382, N6594, p.765–766.

Parungo F. (1995). Reply. J. Climate, Vol. 8, N8, p.2111–2112.

Parungo F., Boatman J.F., Sievering H., Wikinson S.W., Hicks B.B. (1994). Trends in global marine cloudiness and anthropogenic sulphur. J. Climate, Vol. 7, p.434–440.

Pawson S., Meyer A., Leder S. (1995). Internal variability in a perpetual January intergration of a troposphere–stratosphere–mesosphere GCM. Quart. J. Roy. Meteorol. Soc. Part B, Vol. 121, N522, p.369–398.

Peck S.C., Teisberg T.J. (1996). Optimal CO2 control policy with stochastic losses from temperature rise. Climatic Change, Vol. 31, N1, p.19–34.

Peixoto J.P., Oort A.H.(1992). Physics of Climate. New York: Amer. Inst. of Physics. XXXIX + 520 pp.

Peltier W.R. (1996). Global sea level rise and glacial isostatic adjustment: An analysis of data from the east coast of North America. Geophys. Res. Lett., Vol. 23, N7, p.717–720.

Penner J. (1995). Organic aerosols. In: Proc. Dahlem Workshop on Aerosol Forcing of Climate, Berlin, April 1994.

Penner J.E., Atherton C.S., Graedel T.E. (1993). Global emissions and models of photochemically active compounds. Lawrence Livermore Nat. Lab. Preprint UCRL-JC–112273. 28 pp.

Penner J.E., Atherton C.S., Graedel T.E. (1994a). Global emissions and models of photochemically active compounds. In: Global Atmospheric–Biospheric Chemistry. Plenum Press, New York. p.223–247.

Penner J.E., Charlson R.J., Hales J.M., Laulainen N.S., Leifer R., Novakov T., Ogren J., Radke L.F., Schwartz S.E., Travis L. (1994b). Quantifying and minimizing uncertainty of climate forcing by anthropogenic aerosols. Bull. Amer. Meteorol. Soc., Vol. 75, p.375–400.

Pierrehumbert R.T. (1995). Thermostat, radiator fins, and local runaway greenhouse. J. Atmos. Sci., Vol. 52, N10, p.1784–1806.

Pierrehumbert R.T. (1996). Anomalous scaling of high cloud variability in tropical Pacific. Geophys. Res. Lett., Vol. 23, N10, p.1095–1098.

Pilewskie P., Valero F.P.J. (1995). Direct observations of excess solar absorption by clouds. Science, Vol. 267, p.1626–1629.

Pilinis Ch., King D.B., Saltzman E.S. (1996). The oceans: A source or a sink of methyl bromide? Geophys. Res. Lett., Vol. 23, N8, p.817–820.

Pillet G. (1994). Betting on climate states: the impact of climate change on decision making under "strong" uncertainty and collective risks. (Pap.): IPCC Working Group Writing Team II Meet. "Steps Towards Decision Making Framework to Address Climate Change", Montreux, March 3–6, 1994. PSI BER, N10, p.112–137.

Pincus R., Baker M.B. (1994). Effect of precipitation on the albedo susceptibility of clouds in the marine boundary layer. Nature (Gr. Brit.), Vol. 372, N6503, p.250–252.

Pincus R., Szczodrak M., Gu J., Austin P. (1995). Uncertainty in cloud optical depth estimated made from satellite radiance measurements. J. Climate, Vol. 8, N5, p.1453–1462.

Pinnock S., Hurley M.D., Shine K.P., Wallington T.J., Smyth T.J. (1995). Radiative forcing of climate by hydrochlorofluorocarbons and hydrofluorocarbons. J. Geophys. Res., Vol. 100, ND11, p.23227–23238.

Platnick S.E., Twomey S. (1994). Determining the susceptibility of cloud albedo to changes in droplet concentration with the Advanced Very High Resolution Radiometer. J. Appl. Meteorol., Vol. 33, p.334–347.

Platt C.M.R. (1994). Current activities in cloud-radiation research in Australia. Cloud–Radiation Interactions and Their Parameterization in Climate Models. WCRP–86 (WMO/TD-No.648), Geneva, WMO, p.120–122.

Poetzsch-Heffter C., Liu Q., Ruprecht E., Summer C. (1995). Effect of cloud types on the earth radiation budget calculated with the ISCCP C1 dataset: methodology and initial results. J. Climate, Vol. 8, N4, p.829–843.

Ponting C. (1991). Historical perspectives on sustainable development. Environment, Vol. 32, N9, p.4–9, 31–33.

Portman D.A., Gutzler D.S. (1996). Explosive volcanic eruptions, the El Niño—Southern Oscillation and US climate variability. J. Climate, Vol. 9, N1, p.17–33.

Post M.J., Grund C.J., Langford A.O., Profitt M.H. (1992). Observations of Pinatubo ejecta over Boulder, Colorado, by lidars of three different wavelengths. Geophys. Res. Lett., Vol. 19, p.195–198.

Prabhakara C., Nucciarore J.J., Yoo J.-M. (1995). Examination of global atmospheric temperature monitoring with satellite microwave measurements: 1. Theoretical considerations. Climatic Change, Vol. 30, N3, p.349–366.

Prather M., Midgley P., Rowland S.R., Stolarski R. (1996). The ozone layer: the road not taken. Nature, Vol. 381, N6583, p.551–554.

Prening O. (1991). Aerosol and climate—an overview. Atmos. Environ., Vol. 25A, p.2443–2444.

Prinn R.G. (Ed.) (1994a). Global Atmospheric–Biospheric Chemistry. Plenum Press. New York. 261 pp.

Prinn R.G. (1994b). Global atmospheric-biospheric chemistry. In: Global Atmospheric–Biospheric Chemistry. Plenum Press, New York and London. p.1–18.

Prinn R.G., Hartley D. (1992). Atmosphere, ocean, and land: critical gaps in earth system models. In: "Modeling the Earth System", Ed. D. Ojima. UCAR-OIES. Boulder, CO, p.9–38.

Prinn R., Weiss R., Miller B., Huang J., Alyea F., Cunnold D., Fraser P., Hartley D., Simmonds P. (1995). Atmospheric trends and lifetime of trichloroethane and global average hydroxyl radical concentrations based on 1978–1994 ALE/GAGE measurements. Science, Vol. 269, p.187–192.

Proceedings of Workshop on Global Coupled General Circulation Models. Scripps Inst. Oceanogr., La Jolla, USA, Oct. 1994. WCRP–87 (WMO/TD-No,655). Geneva. 1995, p.1–10.

Proceedings of the First International AMIP Scientific Conference (Monterey, California, USA, 15–19 May, 1995). Ed. by W.L. Gates. WCRP–92 (WMO-TD No. 732). WMO, Geneva (1995). 532 pp.

Proceedings of the Fourth Atmospheric Radiation Measurement (ARM) Science Team Meeting. February 28—March 3, 1994. Charlston, South Carolina, USA. US Dept. of Energy, Washington, D.C., Conf. 940277. 396 pp.

Program of Actions. Agenda–21 and other documents of the Rio-de-Janeiro Conference in popular presentation (Comp. by M. Keating). The Centre "For Our Common Future". Moscow. (1993). 70 pp. (in Russian).

Pudykiewicz J.A., Dastoor A.P. (1995). On numerical simulation of the global distribution of sulphate aerosol produced by a large volcanic eruption. J. Climate, Vol. 8, N3, p.464–473.

Putaud J.-P., Belviso S., Nguyen B.C., Mihalopoulos N. (1993). Dimethylsulfide, aerosols and condensation nuclei over the tropical northeastern Atlantic Ocean. J. Geophys. Res., Vol. 98, p.14863–14871.

Quinn P.K., Marshall S.F., Bates T.S., Covert D.S., Kapustin V.V. (1995). Comparison of measured and calculated aerosol properties relevant to the direct forcing of tropospheric sulphate aerosol on climate. J. Geophys. Res. D, Vol. 100, N5, p.8977–8992.

Radiative Forcing of Climate Change. The 1994 Report of the Scientific Working Group of IPCC Summary for Policymakers. IPCC. WMO/UNEP. (1995). 28 pp.

Raga G.B., Jonas P.R. (1993). On the link between cloud-top radiative properties and subcloud aerosol concentration. Quart. J. Roy. Meteorol. Soc., Vol. 119, p.1419–1425.

Radionov V.F., Marshunova M.S., Rusina E.N., Lubo-Lesnichenko K.E., Pimanova Yu.E.(1995). Aerosol turbidity of the atmosphere in polar regions. Izv. RAN. FAO, Vol. 30, N6, p.797–801.

Rahmstorf S. (1995). Bifurcations of the Atlantic thermohaline circulation in response to changes in the hydrological cycle. Nature, Vol. 378, N6557, p.145–149.

Ramanathaan V., Collins W. (1991). Thermodynamic regulation of ocean warming by cirrus clouds deduced from the 1987 El Niño. Nature, Vol. 351, N6321, p.27–32.

Ramanathan V., Collins W. (1993). A thermostat in the tropics. Nature, Vol. 361, p.410–411.

Ramanathan V., Subasilar B., Zhang G.J., Conant W., Cess R.D., Kiehl J.T., Grassl H., Shi L.(1995). Warm pool heat budget and shortwave cloud forcing: a missing physics? Science, Vol. 267, p.499–503.

Ramanathan V., Valero F.P., Cess R.D. (1996). Excess solar absorption in cloudy atmospheres. GEWEX News, Vol. 6, N1, p.6–7.

Ramaswamy V., Schwarzkopf M.D., Randel W.J. (1996). Fingerprint of ozone depletion in the spatial and temporal pattern of recent lower-stratospheric cooling. Nature, Vol. 382, p.616–619.

Randall D.A., Dazlich D.A., Zhang C., Denning A.S., Sellers P.J., Tucker C.S., Bourona L., Berry J.A., Collatz C.J., Field C.B., Los S.O., Justice C.O., Fung I., (1995). A revised land surface parameterization (SiB2) for atmospheric GCMs. Part 3: The greening of the CSU GCM.J.Climate, Vol. 9, N4, p.738–762.

Randall D.A., Xu K.-M., Sommerville R.C.J., Iacobellis S. (1996). Single-column models and cloud ensemble models as links between observations and climate models. J. Climate, Vol. 9, p.828–890.

Randeniya L.K., Vohralik P.F., Plumb I.C., Rayan K.R., Baughcum S.L. (1996a). Impact of heterogeneous hydrolysis of $BrONO_2$ on calculated ozone changes due to the HSCT aircraft and increased sulphate aerosol level. Geophys. Res. Lett., Vol. 23, N4, p.343–346.

Randeniya L.K., Vohralik P.F., Plumb I.C., Rayan K.R., Baughcum S.L. (1996b). Impact of heterogeneous $BrONO_2$ on ozone trends and transient ozone loss during volcanic periods. Geophys. Res. Lett., Vol. 23, N13, p.1633–1636.

Rango A. (1994). Application of remote sensing methods to hydrology and water resources. J. Sci. Hydrol., Vol. 39, N4, p.309–320.

Rao N.X. , Ou S.C., Liou K.N. (1995). Removal of the solar component in the AVHRR 3.7-m radiances for the retrieval of cirrus cloud parameters. J. Appl. Meteorol., Vol. 34, N2, p.482–499.

Rasch P.J., Boville B.A., Brasseur G.P. (1995). A three-dimensional general circulation model with coupled chemistry for the middle atmosphere. J. Geophys. Res., Vol. 100, ND5, p.9041–9071.

Ratcliffe R.A.S. (1995). Back to basics: is our climate changing? Weather, Vol. 50, N2, p.54–57.

Raven P.H. (1987). The global ecosystem in crisis. A MacArthur Foundation Occasional Paper. Chicago. 24 pp.

Raynaud D., Barnola J.-M., Chapellaz J., Martinerie P. (1996). Changes in trace gas concentrations during the last 2000 years and more generally the Holocene. In: NATO ASI Series. Vol. 141. Climatic Variations and Forcing Mechanisms of the Last 2000 Years. Eds.: P.D. Jones, R.S. Bradley, J. Jouzel. Springer-Verlag, Berlin. p.547–561.

Reichardt J., Ansmann A., Serwazi M., Weitkamp C., Michaelis W. (1996). Unexpectedly low ozone concentration in midlatitude tropospheric ice clouds: A case study. Geophys. Res. Lett., Vol. 23, N15, p.1229–1232.

Reid G.C. (1991). Solar total irradiance variations and the global sea surface temperature record. J. Geophys. Res., Vol. 96, p.2835–2844.

Report of the Thirteenth Session of the Joint Scientific Committee. Victoria, B.C., Canada, March 1992. WMO/TD-No.487. Geneva. (1992). 68 pp.

Report of the GCOS Data and Information Management Panel. First Session. Washington, D.C., February 7–10, 1995. GCOS–11 (WMO/TD-No.673). 16 pp.

Restelli G., Angeletti G. (Eds.) (1992). Dimethylsulfide: Oceans, Atmosphere and Climate. Kluwer Acad. Publ., Dordrecht. 412 pp.

Revel M., Sinko J., Grousset F.E., Biscaye P.E. (1996). Sr and Nd isotopes as tracers of North Atlantic lithic particles: palaeoclimatic implications. Paleooceanography, Vol. 11, N1, p.95–113.

Review of Main Development Since the Sixteenth Session of the JSC (Submitted by the Chairman of JSC). WCRP Joint Scientific Committee—XVII/ Doc. 4. Item 2. Toulouse, France (1996). 6 pp.

Riches M.R. (1993). Climate variability: a basis of uncertainty for global change research. In: "Climate Variability" (Eds. D. Ye, Q. Zeng, R. Zhang). China Meteorol. Press, Beijing. p.42–50.

Ridout J.A., Rosmond T.E. (1996). Global modelling of cloud radiative effects using ISCCP cloud data. J. Climate, Vol. 9, N7, p.1479–1496.

Rind D. (1995). Drying out the tropics. News Sci., Vol. 146, N1976, p.36–40.

Rind D. (1996). The potential for modelling the effects of different forcing factors on climate during the past 2000 years. In: NATO ASI Series. Vol. 141. Climate Variations and Forcing Mechanisms of the Last 2000 Years. Eds.: P.D. Jones, R.S. Bradley, J. Jouzel. Springer-Verlag. Berlin. p.563–586.

Rind D., Overpeck J. (1994). Hypothesized causes of modelled decade to century scale variability. Quart. Sci. Rev., Vol. 12, p.357–374.

Rind D., Healy R., Parkinson C., Martinson D. (1995). The role of sea-ice in CO_2 climate model sensitivity. Part I: The total influence of sea-ice thickness and extent. J. Climate, Vol. 8, N3, p.449–463.

Risbey J.S., Stone P.H. (1996). A case study of the adequacy of GCM simulaions for input to region climate assessment. J. Climate, Vol. 9, N7, p.1441–1467.

Roberts W.O. (1975). Relationship between solar activity and climate change. In: W.R. Bandeen and S.O. Moran (Eds.) Goddard Space Flight Center Special Report, NASA SP–366.

Robock A., Mao J. (1995). The volcanic signal in surface temperature observations. J. Climate, Vol. 8, N5, p.1086–1103.

Robock A., Taylor K.E., Stenchikov G.L., Liu Y. (1995). GCM evaluation of a mechanism or El Niño triggering by the El Chichón ash cloud. Geophys. Res. Lett., Vol. 22, N17, p.2369–2372.

Rodhe H., Langner J. (1993). Atmospheric concentration of DMS and its oxidation products estimated in a global 3-D model. In: G. Restelli and G. Angeletti (Eds.) Dimethylsulfide: Oceans, Atmosphere, and Climate. Springer, Berlin. p.333–343.

Roeckner E., Siebert T., Feichter J. (1995). Climatic response to anthropogenic sulphate forcing simulated with a general circulation model. In: Proc. Dahlem Workshop on Aerosol Forcing of Climate, Berlin, April 1994.

Roederer J.G. (1996). Solar variability effects on climate. Paleoclimate Research, (in press).

Roehl C.M., Boglu D., Brühl C., Moortgat G.K. (1995). Infrared band intensities and global warming potentials of CH_4, C_2F_6 , C_3F_8, C_4F_{10} , C_5F_{12}, and C_6F_{14}. Geophys. Res. Lett., Vol. 22, p.815–818.

Rogers D.C. (1994). Comments on "Quantifying and minimizing uncertainty of climate forcing by anthropogenic aerosols". Bull. Amer. Meteorol. Soc., Vol. 75, N12, p.2312–2314.

Romero J., Fröhlich C., Wehrli Ch. (1995). Solar total irradiance variability measured by SOVA–2 on board EURECA. Adv. Space Res., Vol. 16, N8, p.29–32.

Ropelewski C.F. (1995). Long-term observations of land surface characteristics. Clim. Change, Vol. 31, N2–4, p.415–425.

Rosema J., Lambers H., van de Geijn S.C., Cambridge M.L. (Eds.) (1993). CO_2 and Biosphere. Kluwer Acad. Publ., Dordrecht. 484 pp.

Rossow W.B., Cairns B. (1995). Monitoring changes of clouds. Clim. Change, Vol. 31, N2–4, p.305–347.
Rossow W.B., Zhang Y.-C. (1995). Calculation of surface and top of atmosphere radiative fluxes from physical quantities based on ISCCP data sets. 2. Validation and first results. J. Geophys. Res., Vol. 100, p.1167–1197.
Rott H., Skvarca P., Nagler T. (1996). Rapid collapse of Northern Larsen ice shelf, Antarctica. Science, Vol. 271, p.788–792.
Rudman S.D., Saunders R.W., Kilsby C.G., Minnett P.J. (1994). Water vapour continuum absorption in mid-latitudes: Aircraft measurements and model comparisons. Quart. J. Roy. Meteorol. Soc.,Vol. 120, N518, p.795–808.
Ruimi A., Dedien G., Saugier B. (1996). TURC: A diagnostic model of continental gross primary productivity. Global Biogeochemical Cycles, Vol. 10, N2, p.269–285.
Russel J.M., III, Luo M., Cicerone R.J., Deaver L.E. (1996). Satellite confirmation of the dominance of chlorofluorocarbons in the stratospheric chlorine budget. Nature, Vol. 379, N6565, p.526–529.
Santer B. (1995). Climate change detection studies. Proc. Workshop on Global Coupled General Circulation Models. Scripps Inst. Oceanogr., La Jolla, USA, Oct. 1994. WCRP–87 (WMO/TD-No.655). Geneva, p.50–51.
Santer B.D., Taylor K.E., Wigley T.M.L., Penner J.E., Jones Ph.D., Cubasch U. (1995a). Have sulphate aerosols masked regional-scale features of a greenhouse warming signal? Third Int. Conf. on Modeling of Global Climate Change and Variability. 4–8 Sept., 1995. Abstracts. Max-Planck Inst. für Meteorol., Hamburg, p.158.
Santer B.D., Taylor K.E., Wigley T.M.L., Penner J.E., Jones P.D., Cubash V. (1995b). Towards the detection and attribution of anthropogenic effect on climate. Climate Dynamics, Vol. 12, p.77–100.
Santer B.D., Taylor K.E., Wigley T.M.L., Jones T.C., Karoly D.J., Mitchell J.F.B., Oort A.H., Penner J.E., Ramaswamy V., Schwartzkopf M.D., Stouffer R.J., Tett S. (1996). A search for human influences on the thermal structure of the atmosphere. Nature, Vol. 382, p.39–46.
Saravanan R., McWilliams J.C. (1995). Multiple equilibria, natural variability, and climate transitions in an idealized ocean-atmosphere model. J. Climate, Vol. 8, p.2296–2322.
Sarmiento J.L., Le Quéré C. Pacala S.W. (1995). Limiting future atmospheric carbon dioxide. Global Biogeochemical Cycles, Vol. 9, N1, p.121–137.
Sato M., Hansen J.E., McCormick M.P., Pollack J.B. (1993). Stratospheric aerosol optical depths, 1850–1990. J. Geophys. Res., Vol. 98, p.22987–22994.
Schatten K., Myers D.J., Sofia S. (1996). Solar activity forecast for solar cycle 23. Geophys. Res. Lett., Vol. 23, N6, p.605–608.
Schlesinger M.E. (Ed.) (1988). Physically-Based Modelling and Simulation of Climate and Climate Change. Parts 1 and 2. Kluwer Acad. Publ., Dordrecht. 1084 pp.
Schlesinger M.E. (Ed.) (1990). Climate–Ocean Interaction. Kluwer Acad. Publ., Dordrecht. 385pp.
Schlesinger M.E., Jiang X., Charlson R.J. (1992). Implication on anthropogenic atmospheric sulphate for sensitivity of the climate system. Climate Change and Energy

Policy (Eds.: J. Rosen and R. Glasier). Amer. Inst. of Physics, New York, p.75–108.
Schlesinger M.E., Ramankutty N. (1992). Implications for global warming of intercycle solar irradiance variations. Nature, Vol. 360, p.330–333.
Schlesinger M.E., Ramankutty N. (1994). An oscillation in the global climate system of period 65–70 years. Nature, Vol. 367, p.723–726.
Schlesinger M.E., Ramankutty N. (1995). Is the recently reported 65-to–70-year surface-temperature oscillation the result of climatic noise? J. Geophys. Res., Vol. 100, ND7, p.13,767–13,774.
Schlesinger W.H. (1991). Biogeochemistry. An Analysis of Global Change. Academic Presss, San Diego. 443 pp.
Schmidt E.O., Arduini R.F., Wielicki R.S., Tsay S.-C. (1995). Consideration of modelling thin cirrus effects via brightness temperature differences. J. Appl. Meteorol., Vol. 34, N2, p.447–459.
Schneider E.K. (1995) . Global coupled modelling at COLA. Proc. Workshop on Global Coupled General Circulation Models. Scripps Inst. Oceanogr., La Jolla, USA, Oct. 1994, WCRP–87 (WMO/TD-No.655).
Schneider N., Barnett T., Latif M., Stockdale T. (1996). Warm pool in a coupled GCM. J. Climate, Vol. 9, N1, p.219–239.
Schönwiese Ch.-D., Ulrich R., Beck F., Rapp J. (1994). Solar signals in global climate change. Climatic Change, Vol. 27, N3, p.259–282.
Schramm G., Warford J.J. (Eds.) (1991). Environmental Managements and Economic Development. A World Bank Publ. The Johns Hopkins Univ. Press. Baltimore, MD. 208 pp.
Schulze E.D., Caldwell M.M. (Eds.) (1994). Ecophysiology and Photosynthesis. Springer. Heidelberg. XXII, 576 pp.
Schwartz S.E. (1994). Climate forcing by gases and aerosols. Pure and Appl. Chem., Vol. 66, p.178–187.
Schwartz S.E., Andreae M.O. (1996). Uncertainty in climate change caused by aerosols. Science, Vol. 272, N5265, p.1121–1122.
Schweiger A.J., Key J.R. (1994). Arctic ocean radiative fluxes and cloud forcing estimated from the ISCCP C2 cloud data set, 1983–1990. J. Appl. Meteorol., Vol. 33, N8, p.948–963.
Scientific Design for the Common Module of the Global Ocean Observing System and the Global Climate Observing System: An Ocean Observing System for Climate. OOSD Panel Report. (1995). Texas A&M Univ., College Station. 265 pp.
Scientific Plan for the World Climate Research Program. WCRP Publ. Ser., WMO, Geneva. (1984). 235 pp.
Sedunov Yu.S., Uspensky A.B., Voyakin S.N., Evdokimov V.P., Vasiukhina T.M. (1995). Main directions of development and application of operative space-borne observing systems and environmental monitoring. Meteorology and Hydrology, N4, p.7–21 (in Russian).
Sellers P., Hall F.G. (Eds.) (1987). First ISLSCP Experiment (FIFE). Konza Prairie 1987–1989. Experiment Plan. 141 pp. Appendices.
Sellers P., Hall F., Margolis H., Kelly B., Baldocchi D., den Hartog G., Cihlar J., Ryan M.G., Goodison B., Crill P., Ranson K.J., Lettenmaier D., Wickland D.E. (1995).

The Boreal Ecosystem–Atmosphere Study (BOREAS): An overview and early results from the 1994 field year. Bull. Amer. Meteorol. Soc., Vol. 76, N9, p.1549–1577.

Sellers P.J., Randall D.A., Colatz C.J., Berry J.A., Field C.B., Dazlich D.A., Zhang C., Collelo G.D. (1996a). A revised land surface parameterization (SiB2) for atmospheric GCMs. Part 1: Model formulation. J. Climate, Vol. 9, N4, p.676–705.

Sellers P.J., Los S.O., Tucker C.J., Justice C.O., Dazlich D.A., Collatz C.J., Randall D.A. (1996b). A revised land surface parameterization (SiB2) for atmospheric GCMs. Part 2: The generation of global fields of terrestrial geophysical parameters from satellite data. J. Climate, Vol. 9, N4, p.706–737.

Sellers P.J., Bounoua L., Collatz G.J., Randall D.A., Dazlich D.A., Los S.O., Berry J.A., Fung I., Tucker C.J., Field C.B., Jensen T.G. (1996c). Comparison of radiative and physiological effects of doubled atmospheric CO2 on climate. Science, Vol. 271, p.1402–1406.

Senft D. (1995). FACE-ing the Future. Agricultural Research, Vol. 43, N4, p.4–6.

Senior C.A., Mitchell J.F.B. (1993). Carbon dioxide and climate: The impact of cloud parameterization. J. Climate, Vol. 6, p.393–418.

Shabalova M.V., Konnen G.P. (1995). Climate change: comparisons of palaeoreconstructions with recent temperature changes. Climate Change, Vol. 29, N4, p.409–428.

Shackley S., Wynne B. (1995). Integrating knowledge for climate change. Pyramids, nets and uncertainties. Global Environmental Change, Vol. 5, N2, p.113–126.

Shao Y., Henderson-Sellers A. (1996). Validation of soil moisture simulation in land surface parameterization schemes with HAPEX data. Global and Planetary Change, Vol. 13, N1–4, p.11–46.

Shapiro L.T. (1991). Snow on sea-ice: competing effects in shaping climate. J. Geophys. Res., Vol. 98, ND9, p.17,195–17,208.

Shaw G.E. (1987). Aerosol as climate regulators: a climate-biosphere linkage? Atmos.-Environ., Vol. 21, p.985–986.

Shen S.S.P., North G.R., Kim K.-Y. (1994). Spectral approach to optimal estimation of the global average temperature. J. Climate, Vol. 7, N12, p.1999–2007.

Shimel D.S., Sulzman F. (1995). Variability in the earth climate system: Decadal and longer timescales. Review of Geophysics, Supplement. July 1995. p.873–882.

Simmonds I., Jacka T.H. (1995). Relationship between the interannual variability of Antarctic sea-ice and the Southern Oscillation. J. Climate, Vol. 8, N2, p.637–647.

Simpson J., Adler R.F., Morth G.R. (1988). A proposed Tropical Rainfall Measuring Mission (TRMM) satellite. Bull. Amer. Meteorol. Soc., Vol. 69, p.278–295.

Sinha A., Harries J. (1995). Water vapour and greenhouse trapping: The role of far infrared absorption. Geophys. Res. Lett., Vol. 22, N16, p.2147–2150.

Sinha A., Shine K.P. (1995). Simulated sensitivity of the earth's radiation budget to changes in cloud properties. Quart. J. Roy. Meteorol. Soc., Vol. 121, N524, p.797–820.

Sinha A., Toumi R. (1996). A comparison of climate forcings due to chlorofluorocarbons and carbon monoxide. Geophys. Res. Lett., Vol. 23, N1, p.65–68.

Sirotenko O.D., Pavlova V.N. (1994). Greenhouse effect of food problems. Meteorology and Hydrology, N7, p.5–16 (in Russian).

Sirotenko O.D., Abashina E.V., Pavlova V.N. (1995). Sensitivity of agriculture in Russia to changes in climate, atmospheric chemistry, and soil fertility. Meteorology and Hydrology, N4, p.107–114 (in Russian).

Sixth Symposium on Global Change Studies. Preprints. January 15–20, 1995. Dallas, Texas, USA. Amer. Meteorol. Soc., Boston, MA. 238 pp.

Skea J. (1995). Assessing uncertainties in the climate change debate. In: Climate Change and Energy Use. Status and Options for European Policy. Contractor Reports. Part B. Luxembourg, The European Parliament. XX + 60 pp.

Skilas J.W. (1995). Modelling climate change in the absence of climate change data. Clim. Change, Vol. 30, N1, p.1–6.

Sklyarov Yu.A. (1994). Active cavity radiometer on board Meteor–3 satellite. Studying the Earth from Space, N3, p.20–29 (in Russian).

Sleszynski J. (1994). Carbon tax as an environmental policy instrument. In: "Industrial Environment Policy in Poland". Crakow, p.125–147.

Slingo A., Nicholls S., Schmetz J. (1982). Aircraft observations of marine stratocumulus during JASIN. Quart. J. Roy. Meteorol. Soc., Vol. 108, p.833–856.

Slingo J.M., Sperber K.R., Boyle J.S., Ceron J.-P., Dix M., Dugas B., Ebisuzaki W., Fyfe J., Gregory D., Guermy J.-F., Hack S., et al. (1995). Atmospheric Model Intercomparison Project (AMIP).Intraseasonal oscillations in 15 atmospheric general circulation models (results from an AMIP diagnostic subproject). WCRP-88 (WMO/TD-No.661), Geneva, p.1–32.

Sloan L.C., Rea D.K. (1995). Atmospheric carbon dioxide and early Eocene climate: A general circulation modelling sensitivity study. Paleogeography, Paleoclimatology, Paleoecology, Vol. 119, p.275–292.

Smith I.N. (1995). A GCM simulation of global climate interannual variability: 1950–1988. J.Climate, Vol. 8, N4, p.709–718.

Smith N.R., Needler G.T. and the Ocean Observing System Development Panel (1995). An Ocean Observing System for Climate. The Concept Design. Clim. Change, Vol. 31, N2–4, p.475–494.

Smith T.M., Reynolds R.W., Livezey R.E., Stokes D.E. (1996). Reconstruction of historical sea surface temperatures using empirical orthogonal functions. J. Climate, Vol. 9, p.847–859.

Soden B.J., Fu R. (1995). A satellite analysis of deep convection, upper tropospheric humidity, and the greenhouse effect. J. Climate, Vol. 8, N10, p.2333–2351.

Soden B.J., Lanzante J.R. (1996). Satellite and radiosonde climatologies of upper-tropospheric water vapor. J. Climate, Vol. 9, p.860–872.

Sofia S., Fox P. (1994). Solar variability and climate. An editorial essay. Climatic Change, Vol. 27, N3, p.249–258.

Solomon S., Portmann R.W., Garcia R.R., Thomason L.W., Poole L.R., McCormick M.P. (1996). The role of aerosol variations in anthropogenic ozone depletion at northern midlatitudes. J. Geophys. Res., Vol. 101, ND3, p.6713–6727.

Southern Hemisphere Marine Aerosol Characterization Experiment (ACE–1). Radiative Effects of Aerosols in the Remote Marine Atmosphere. Final Science and Implementation Plan. NOAA/PMEL, Seattle, WA. (1995). 99 pp.

Spinhirne J.D., Hart W.D., Hlavka D.L. (1996). Cirrus infrared parameters and shortwave reflectance relations from observations. J. Atmos. Sci., Vol. 53, N10, p.1438–1458.

Spivy-Weber F. (1992). The missing forest principles// Network'92, N12, p.3.

Sreenath N. (1993). Systems Representation of Global Climate Change Models. Foundation for a Systems Science Approach. Springer-Verlag. London. 238 pp.

Srinivasan G., Hulme M., Jones C.G. (1995). An evaluation of the spatial and interannual variability of tropical precipitation as simulated by GCMs. Geophys. Res. Lettt., Vol. 22, N16, p.2139–2142.

Starr D.O.C. (1994). Physical processes in cirrus clouds: fundamental scales and implications. Cloud–Radiation Interactions and Their Parameterization in Climate Models. WCRP–86 (WMO/TD-No.648). Geneva, WMO, p.24–33.

Stendel M., Bengtsson L. (1996). Monitoring the temperature of the troposphere by means of a general circulation model. Max Planck Institut für Meteorologie Report No. 186. 26 pp.

Stensrud D.J. (1996). Importance of low-level jets to climate: a review. J. Climate, Vol. 9.

Stephens G.L. (1994a). Dirty clouds and global cooling. Nature (Gr. Brit.), Vol. 370, N6489, p.420.

Stephens G.L. (1994b). Satellite Observations Final Report. Cloud–Radiation Interactions and Their Parameterization in Climate Models. WCRP–86 (WMO/TD-No.648), Geneva, WMO p.12–23.

Stephens G.L., Paltridge G.W., Platt C.M.R. (1978). Radiation profiles in extended water clouds. III: Observations. J. Atmos. Sci., Vol. 35, N11, p.2133–2141.

Stephens G.L., Tsay S.C. (1990). On the cloud absorption anomaly. Quart. J. Roy. Meteorol. Soc., Vol. 116, p.671–704.

Stephens G.L., Cess R.D., Zhang M.H., Pilewskie P., Valero F.P.J. (1996). How much solar radiation do clouds absorb? Science, Vol. , 271, N5252, p.1131–1136.

Stigliani W.M., Brower F.M., Munn R.E., Shaw R.W., Antonovsky M. (1989). Future environment for Europe: some implications of alternative development paths. IIASA, Laxenburg, Austria. RR–89–5. 102 pp.

Stoll H.M., Schrag D.P. (1996). Evidence for glacial control of rapid sea level changes in the early Cretaceous. Science, Vol. 272, p.1771–1774.

von Storch H., Navarra A. (Eds.) (1995). Analysis of Climate Variability. Applications of Statistical Techniques. Springer-Verlag, Berlin. 334 pp.

Stothers R.B. (1996). Major optical depth perturbations to the stratosphere from volcanic eruptions: Pyrheliometric period, 1881–1960. J. Geophys. Res., Vol. 101, ND2, p.3901–3920.

Stouffer R.J. (1995). An assessment of the observed warming trend using a 1000-year integration of a coupled ocean–atmosphere model. Proc. Workshop on Global Coupled General Circulation Models. Scripps Inst. Oceanogr., La Jolla, USA, Oct. 1994. WCRP–87 (WMO/TD-No.655). Geneva, p.11.

Strzepek K.M., Smith J.B. (Eds.) (1995). As Climate Changes. Cambridge Univ. Press. 231 pp.

Sud Y.C., Yang R., Walker G.K. (1996a). Impact of in situ deforestation in Amazonia on the regional climate: General circulation model simulation study. J. Geophys. Res., Vol. 101, ND3, p.7095–7109.

Sud Y.C.. Walker G.K., Kim J.-H., Liston G.E., Sellers P.J., Lau K.-M. (1996b). Biogeophysical effects of a tropical deforestation scenario: A GCM simulation study. J. Climate (in press).
Sugden D.E., Marchant D.R., Potter N.,Sr., Souchez R.A., Denton G.H., Swisher C.C.,III, Tison J.-L. (1995). Preservation of Miocene glacier ice in East Antarctica. Nature, Vol. 376, N6539, p.412–414.
Sumi A. (1996). Pattern formation of convective activity over the aqua-planet with globally uniform sea surface temperature (SST). Selected Papers in Climate Research at CCSR. Five Yearly Report, 1991–1995. Center for Climate System Research, University of Tokyo. p.68–89.
Sun Z., Shine K.P. (1995). Parameterization of ice cloud radiative properties and its application to the potential climatic importance of mixed clouds. J. Climate, Vol. 8, N7, p.1874–1888.
Sun D.-Z.,Liu Z. (1996). Dynamic ocean–atmosphere coupling: A thermostat for the Tropics. Science, Vol. 272, N5265, p.1148–1150.
Sundqvist A., Starr V. (1994). Process Studies. Final Report. Cloud–Radiation Interactions and Their Parameterization in Climate Models. WCRP–86 (WMO/TD-No.648), Geneva. WMO, p.24–33.
Sundqvist E.T. (1993). The global carbon dioxide budget (1996). Science, Vol. 259, p.934–941.
Suzuki Y., Ueta K., Mori S. (Eds.) (1996) . Global Environmental Security. From Protection to Prevention. Springer-Verlag. Berlin. XII + 262 pp.
Swap R., Ulanski S., Cobbett M., Garstang M. (1996). Temporal and spatial characteristics of Saharan dust outbreaks. J. Geophys. Res., Vol. 101, ND2, p.4205–4220.
Szilder K., Lozowski E.P. (1995). Cloud feedback examined using a two-component time-dependent climate model. Beitr. Phys. Atmosph., Vol. 68, N1, p.43–57.
Taguchi Sh. (1996). A three-dimensional model of atmospheric CO_2 transport based on analysed winds: Model description and simulation results for TRANSCOM. J. Geophys. Res., Vol. 101, ND10, p.15,099–15,109.
Takano Y., Liou K.-N. (1995). Radiative transfer in cirrus clouds. Part III: Light scattering by irregular ice crystals. J. Atmos. Sci., Vol. 52, N7, p.818–837.
Tang B., Weaver A.J. (1985). Climate stability as deduced from an idealized coupled atmosphere–ocean model. Climate Dynamics, Vol. 1, N3, p.141–150.
Tanre D., Geleyn J.-F., Slingo J. (1984). First results of the introduction of an advanced aerosol radiation interaction in the ECMWF low resolution global model. In: Aerosol and Their Climatic Effects (Eds.: H.E. Gerber and A. Deepak). A Deepak Publ., Hampton, VA, p.133–177.
Tans P.P., Bakwin P.S. (1995). Climate change and carbon dioxide forever. Ambio, Vol. 24, N6, p.376–378.
Tao X., Walsh J.E., Chapman W.L. (1996). An assessment of global climate model simulations of arctic air temperature. J. Climate, Vol. 9, N5, p.1060–1076.
Taubman S.J., Kasting J.F. (1995). Carbonyl sulfide: No remedy for global warming. Geophys. Res. Lett.,Vol. 22, N7, p.803–805.
Taylor J.P., Edwards J.M., Glew M.D., Hignett P., Slingo A. (1996). Studies with a

flexible new radiation code. II. Comparisons with aircraft short-wave observations. Quart. J. Roy. Meteorol. Soc., Vol. 122, Part B, N532, p.839–861.
Taylor K.E., Penner J.E. (1995). Response of the climate system to atmospheric aerosols and greenhouse gases. Nature, Vol. 369, p.734–737.
Tegen I., Lacis A.A., Fung I. (1996). The influence on climate forcing of mineral aerosols from disturbed soils. Nature, Vol. 380, p.419–422.
Ten interesting facts about climate change. The Global Issue 24. 1995, p.7.
Thaker W.C. (1996). Climate fingerprints, patterns and indices. J. Climate. Vol. 9, p.875–887.
Thaker W.C., Lewandowicz R. (1996). Climatic indices, principal components, and the Gauss-Markov theorem. J. Climate, Vol. 9, N7, p.888–1000.
The all new and improved Agenda 21. Network'92. 1992, N4, p.1, 10.
The Bellagio Declaration on the Environment. Bull. Amer. Acad. Arts & Sciences. (1991). Vol. 45, N2, p.7–9.
The Berlin Mandate: Decision 1/CP.1. Clim. Change Bull., Issue 7, 1995, p.7–8.
The Climate Agenda. (1995) International Climate-Related Programmes. A proposal for an Integrating Framework by FAO, ICSU, Unesco and its IOC, WMO (1995). WMO. Geneva. 124 pp.
The GCCIP Newsletter. 1995, Vol. 2, issue 3. 8 pp.
The Global Climate. Ed. John T. Houghton. (1984). Cambridge Univ. Press. Cambridge. 235 pp.
The Global Warming Experiment. George C. Marshall Institute. Washington, D.C., 1995, p.1–3.
The Greenhouse Debate: Part 2. (1992). EOS, Vol. 73, N 30, p.321, 324–327.
Theodorsen A., Henriksen K., Bersås S., Sirota V., Vasiliev A. (1996). Surface ozone at high latitudes. Geophys. Res. Lett., Vol. 23, N1, p.77–80.
The Spirit of Rio: Earth Summit 1992. (1992). Brundtland Bull., N16, p.3–5.
The World Bank and the Environment. A Progress Report. Fiscal Y1991.The World Bank. (1991). Washington, D.C. 131 pp.
Third International Conference on Modeling of Global Climate Change and Variability. 4–8 September, 1995. Abstracts. Max-Planck Inst. für Meteorologie. Hamburg (1995). 199 pp.
Thompson S.L., Pollard D. (1995). A global climate model (GENESIS) with a land-surface transfer scheme (LSX). Part II: CO_2 sensitivity. J. Climate, Vol. 8, N5, p.1104–1121.
Thornton D.C., Bandy A.R., Blomqvist B.W., Davis D.D., Talbot R.W. (1996). Sulfur dioxide as a source of condensation nuclei in the upper troposphere of the Pacific Ocean. J. Geophys. Res., Vol. 101, ND1, p.1883–1890.
Those elusive conventions. Network'92. (1992), N15, p.3.
Tie X.X., Brasseur G. (1995). The response of stratospheric ozone to volcanic eruptions: Sensitivity to atmospheric chlorine loading. Geophys. Res. Lett., Vol. 22, N22, p.3035–3038.
Tie X., Brasseur G.P., Granier C., de Rudder A., Larsen N. (1995). Model study of polar stratospheric clouds and their effect on stratospheric ozone. 2. Model results. J. Geophys. Res., Vol. 101, ND7, p.12,570–12,584.

Tiedtke M. (1993). Representation of clouds in large-scale models. Mon. Weather Rev., Vol. 121, p.3040–3061.

Tiedtke M. (1996). An extention of cloud-radiation paramweterization in the ECMWF model: The representation of subgrid-scale variations of optical depth. Mon. Weather Rev., Vol. 124, N4, p.745–750.

Tinsley B.A., Heelis R.A. (1993). Correlations of atmospheric dynamics with solar activity: evidence for a connection via the solar wind, atmospheric electricity, and cloud microphysics. J. Geophys. Res., ND6, p.10,375–10,384.

Tolbert N.E., Preiss J. (1994). Regulation of Atmospheric CO_2 and O_2 by Photosynthetic Carbon Metabolism. Oxford Univ. Press. 262 pp.

Tolbert M.A. (1996). Polar clouds and sulphate aerosols. Science, Vol. 272, p.1597.

Trenberth K.E. (Ed.) (1992). Climate System Modelling. Cambridge University Press. 788 pp.

Trenberth K.E. (1995). Long-term climate monitoring by the Global Climate Observing System (GCOS). Report of Breakout Group B—Climate responses and feedbacks. Climatic Change, Vol. 31, N2–4, p.163–180.

Toon O.B. (1995). Modelling the relationships between aerosol properties: the direct and indirect effects of aerosols on climate. In: Proc. Dahlem Workshop on Aerosol Forcing of Climate, Berlin, April 1994.

Toon O.E., Ackerman A., Jensen J. (1994). Interaction between aerosols and clouds. Proc. Third RM Science Team Meeting, March 1993. U.S. Department of Energy, CONF–9303112, UC–402.

Tsonis A.A. (1996). Widespread increases in low-frequebcy variability of precipitation over the past century. Nature, Vol. 382, N6593, p.700–702.

Turner D.P., Koerper G.J., Harmon M.E., Lee J.S. (1995). Carbon sequestration by forests of the United States. Current status and projections to the year 2040. Tellus, Vol. 47B, N1/2, p.232–239.

Twohy C.H., Durkee Ph.A., Huebert B.J., Charlson R.J. (1995). Effects of aerosol particles on the microphysics of coastal stratiform clouds. J. Climate, Vol. 8, N4, p.773–783.

Twomey S. (1991). Aerosols, clouds and radiation. Atmos. Environ., Vol. 25A, p.2435–2442.

Uchino O. (1995). Lidar observations of the Pinatubo aerosol layers. The Review of Laser Engineering, Vol. 23, N2, p.161–165.

Uchino O., Nagai T., Fujimoto T., Fujimara M., Akiyoshi H., Yasumatsu S., Hayashida S., Sasano Y., Nakane H., Iwasaka Y., Hase U., Shibata T., Itabe T.,Asai K., Nomura A., Saito Y., Kano T., Sai Y., Tamaki K., Nomura A.,Nomura R., Sunagawa T., Nagasaw C., Abo M., Idesako Y., Kai K. (1993). Observation of Pinatubo volcanic cloud by lidar network in Japan. J. Meteorol. Soc. Japan, Vol. 71, N2, p.285–295.

Unal Y.S., Ghil M. (1995). Interannual and interdecadal oscillation patterns in sea level. Climate Dynamics, Vol. 11, N7, p.255–278.

U.S. Country Studies Program. Guidance for Vulnerability and Adaptation Assessments. Version 1.0. Washington, D.C., October 1994. XXII + 215 pp.

Valero P.P.J., Pilewskie P. (1992). Latitudinal survey of spectral optical depth of the

Pinatubo volcanic cloud-derived particle sizes, columnar mass loadings, and effects on planetary albedo. Geophys. Res. Lett., Vol. 19, p.163–166.

Van Loon H., Labitzke K. (1993). Review of the decadal oscillations in the stratosphere of the Northern Hemisphere. J. Geophys. Res., Vol. 98, NA11, p.18,919–18,922.

Varotsos K., Kondratyev K.Ya. (1995a). Interaction between UV solar radiation and total ozone content: results from observations in Greece. Optics of the Atmosphere and Ocean, Vol. 8, N4, p.608–613 (in Russian).

Varotsos K., Kondratyev K.Ya. (1995b). Tropospheric pollution and solar UV radiation. Optics of the Atmosphere and Ocean, Vol. 8, N4, p.614–618 (in Russian).

Varotsos C., Kondratyev K.Ya., Katsikis S. (1995). On the relationship between total ozone and solar ultraviolet radiation at St. Petersburg, Russia. Geophys. Res. Lett., Vol. 22, N24, p.3481–3484.

Vaughan D.G., Doake C.S.M. (1996). Recent atmospheric warming and retreat of ice shelves on the Antarctic Peninsula. Nature, Vol. 379, N6563, p.328–331.

Vavrousek J. (1992). A UN system that is efficient and more environment-oriented (some ideas for UNCED). Network'92, N13, p.6,7.

Vavrus S.J. (1995). Sensitivity of the Arctic climate to leads in a coupled atmosphere-mixed layer ocean model. J. Climate, Vol. 8, N2, p.158–171.

Verbitsky M.Y., Oglesby R.J. (1995). The CO_2—induced thickening/thinning of the Greenland and Antarctic ice sheets as simulated by a GCM (CCM1) and an ice-sheets model. Climate Dynamics, Vol. 11, N4, p.247–253.

Verwey W.D. (Ed.) (1989). Nature Management and Sustainable Development. IOS. Amsterdam. 531 pp.

Victor D.G. (1995). On writing good histories of climate change and testing social science theories. An editorial comment. Clim. Change, Vol. 29, N4, p.363–369.

Victor D., Salt J. (1995). Hard Targets: the case against. Options. Spring'95, p.5–6.

Viner D., Hulme M. (1994). The Climate Impacts Link Project. U.K. Department of Environment. London. 24 pp.

Vinnikov K.Ya., Robock A., Stouffer R.J., Manabe S. (1996). Vertical patterns of free and forced climate variations. Geophys. Res. Lett., Vol. 23, N14, p.1801–1804.

Visser H., Molenaar J. (1995). Trend estimation and regression analysis in climatological time series: an application of structural time series models and the Kalman filter. J. Climate, Vol. 8, N5, p.969–979.

Viterbo P., Beejaars A.C.M. (1995). An improved land surface parameterization scheme in the ECMWF model and its validation J. Climate. (in press).

Volk C.M., Elkins J.W., Fahey D.W., Salawitch R.J., Dutton G.S., Gilligan J.M., Proffitt M.H., Loewebstein M., Podolske J.R., Minschwaner K., Margitan J.J., Chan K.R. (1996). Quantifying transport between the tropical and mid-latitude lower stratosphere. Science, Vol. 272, p.1763–1768.

Waliser D.E. (1996a). Formation and limiting mechanisms for very high sea surface temperature: linking the dynamics and thermodynamics. J. Climate, Vol. 9, N1, p.161–181.

Waliser D.E. (1996b). Some considerations on the thermostat hypothesis. Bull. Amer. Meteorol. Soc., Vol. 77, N2, p.357–360.

Waliser D.E., Collins W.D., Anderson S.P. (1996). An estimate of surface shortwave cloud forcing over the western Pacific during TOGA COARE. Geophys. Res. Lett., Vol. 23, N5, p.519–522.

Walker G.K., Sud Y.C., Atlas R. (1995). Impact of the ongoing Amazon deforestation study. Bull. Amer. Meteorol. Soc., Vol. 76, N3, p.346–361.

Wallace J.M., Zhang Y., Renwick J.A. (1995). Dynamic contribution to hemispheric mean temperature trends. Science, Vol. 270, p.780–783.

Wallace J.M., Zhang Y., Bajuk L. (1996). Interpretation of interdecadal trends in Northern Hemisphere surface air temperature. J. Climate, Vol. 9, N2, p.249–259.

Walsh J.E., Chapman W.L., Shy T.L. (1996a). Recent decrease of sea level pressure in the Central Arctic. J. Climate, Vol. 19, p.480–486.

Walsh J.E., Tanaka H.L., Weller G. (1996b). Wadate Conference on global change and polar climate, 7–10 November 1995, Tsukuba, Japan. Bull. Amer. Meteorol. Soc., Vol. 77, N6, p.1268–1273.

Walsh K., McGregor J.L. (1995) . January and July climate simulations over the Australian region using a limited-area model. J. Climate, Vol. 8, N10, p.2387–2403.

Wang H.J., Cunnold D.M., Bao X. (1996). A critical analysis of Stratospheric Aerosol and Gas Experiment ozone trends. J. Geophys. Res., Vol. 101, ND7, p.12,495–12,514.

Wang S., Ye D. (1993). An analysis of global warming during the last one hundred years. In: "Climate Variability" (Eds.: D. Ye, Q. Zeng, R. Zhang). China Meteorol. Press, Beijing, p.23–32.

Wang W.-C., Isaksen I.S.A. (Eds.) (1995). Atmospheric Ozone as a Climate Gas: General Circulation Model Simulations. Springer-Verlag. Berlin. 459 pp.

Ward D.M. (1995). Comparison of the surface radiation budget derived from satellite data with that simulated by the NCAR CCM2. J. Climate, Vol. 8, N11, p.2824–2842.

Ward M.N., Hoskins R.J. (1996). Near-surface wind over the global ocean, 1949–1988. J. Climate, Vol. 9, N7, p.871–883.

Washington W.M., Meehl G.A. (1996). High-altitude climate change in a global coupled ocean–atmosphere–sea-ice model with increased atmospheric CO_2. J. Geophys. Res., Vol. 101, ND8, p.12,795–12,801

Weare B.C., Mokhov I.I., and Project Members (1995). Evaluation of total cloudiness and its variability in the Atmospheric Model Intercomparison Project. J. Climate, Vol. 8, N9, p.2224–2238.

Weaver A.J. (1995). Driving the ocean conveyor. Nature, Vol. 378, p.135–136.

Weaver A.J., Hughes T.M.C. (1994). Rapid interglacial climate fluctuations driven by the North Atlantic circulations. Nature, Vol. 367, p.447–450.

Weaver C.P., Ramanathan V. (1996). The link between summertime cloud radiative forcing and extratropical cyclones in the North Pacific. J. Climate, Vol. 9, N7.

Weber G.R. (1995). Global Warming. The Rest of the Story. Dr. Bettinger Verlags GmbH, Wiesbaden. 188 pp.

Webster P.J., Clayson C.A., Curry J.A. (1996). Clouds, radiation and the diurnal cycle of sea surface temperature in the tropical West Pacific. J. Climate, Vol. 9, N7.

Wehrli Ch., Fröhlich C., Romero J. (1995). Results of solar spectral irradiance measurements by SOVA–2 on EURECA. Adv. Space Res., Vol. 16, N8, p.25–28.

Weisenstein D.K., Ko M.K.W., Sze N.-D., Rodriguez J.M. (1996). Potential impact of SO_2 emissions from stratospheric aircraft on ozone. Geophys. Res. Lett., Vol. 23, N2, p.161–164.

Weller G. (1995). Global pollution and its effect on the climate of the Arctic. The Science of the Total Environm., Vol. 160/161, p.19–24.

Wetherald R.T., Manabe S. (1995). The mechanisms of summer dryness induced by greenhouse warming. J. Climate, Vol. 8, N12, p.3096–3108.

White W.B., Peterson R.G. (1996). An Antarctic Circumpolar Wave in surface pressure, wind, temperature and sea-ice extent. Nature, Vol. 380, N6576, p.699–702.

Wielicki B.A., Cess R.D., King M.D., Randall D.A., Harrison E.F. (1995). Mission to Planet Earth: role of clouds and radiation in climate, Bull. Amer. Meteorol. Soc., Vol. 76, N11, p.2125–2153.

Wielicki B.A., Barkstrom B.R., Harrison E.F., Lee R.B. III, Smith G.L., Cooper J.E. (1996). Clouds and the Earth's Radiant Energy System (CERES): An Earth Observing System Experiment. Bull. Amer. Meteorol. Soc., Vol. 77, N5, p.853–868.

Wigley T.M.L. (1994). Outlook becoming more hazier. Nature (Gr. Brit.), Vol. 369, N6483, p.709–710.

Wigley T.M.L. (1995a). Global-mean temperature and sea-level consequences of greenhouse gas concentration stabilization. Geophys. Res. Lett., Vol. 22, N1, p.45–48.

Wigley T.M.L. (1995b). A successful prediction? Nature, (Gr. Brit.), Vol. 376, N6540, p.463–464.

Wigley T.M.L., Richels R., Edmonds J.A. (1996). Economic and environmental choices in the stabilization of atmospheric CO_2 concentrations. Nature, Vol. 379, N6562, p.240–243.

Wild M., Ohmura A., Gilden H., Roeckner E. (1995). Validation of general circulation model radiative fluxes using surface observations. J. Climate, Vol. 8, N5, p.1309–1324.

Williamson D.L., Bromwich D.H., Tzeng R.-Y. (1996). Further descussion on simulation of the modern Arctic climate by the NCAR CCM1. J. Climate, Vol. 9, N7, p.1669–1672.

Willson R.C. (1984). Measurements of solar total irradiance and its variability. Space Sci. Rev., Vol. 38, p.203–242.

Willson R.C. (1993). Solar irradiance. In: Atlas of Satellite Observations Related to Global Change. Eds.: R.J. Gurney, J.L. Foster, C.L. Parkinson. Cambridge University Press. p.5–18.

Willson R.C., Hudson H.S. (1988). Solar luminosity variations in solar cycle 21. Nature, Vol. 332, p.810–814.

Willson R.C., Hudson H.S. (1991). The sun's luminosity over a complete solar cycle. Nature, Vol. 351, p.42–44.

Winguth A.M.E., Heimann E., Kurz K.D., Meier-Reiman E., Mikolajewicz M., Segschneider J. (1994). El Niño—Southern Oscillation related fluctuations of the marine carbon cycle. Global Biogeochemical Cycles, Vol. 8, N1, p.39–63.

Winker D.M., Osborn M.T. (1992). Airborne lidar observations of the Pinatubo volcanic plume. Geophys. Res. Lett., Vol. 19, p.167–170.

Wirth M., Renger W. (1996). Evidence of large scale ozone depletion within the arctic polar vortex 94/95 based on airborne LIDAR measurements. Geophys. Res. Lett., Vol. 23, N8, p.813–816.

Wiscombe W.J. (1995). An absorbing mystery. Nature (Gr. Brit.), Vol. 376, N6540, p.466–467.

WMO Statement on the Status of the Global Climate in 1994. WMO-No.826, Geneva. 1995. 20 pp.

Woodward F.I., Smith T.M., Emanuel W.R. (1995). A global land primary productivity and photogeography model. Global Biogeochemical Cycles, Vol. 9, N4, p.471–490.

Woodward W.A., Gray H.L. (1995). Selecting a model for detecting the presence of a trend. J. Climate, Vol. 8, N8, p.1929–1937.

Woodwell G.M. (1995). Biotic feedbacks from the warming of the Earth. In: Biotic Feedbacks in the Global Climatic System. Eds.: G.M. Woodwell and F.T. MacKenzie. Oxford Univ Press, New York, Vol. 3, p.21–29.

Woodwell G.M., Hobbie J.E., Houghton R.A., Melilo J.M., Moore B., Peterson B.J., Shaver G.R. (1983). Global deforestation: Contribution to atmospheric carbon dioxide. Science, Vol. 222, p.1081–1086.

Woodwell G.M., MacKenzie F.T., Houghton R.A., Apps M.J., Gorham E., Davidson E.A. (1995). Will the warming speed the warming? In: Biotic Feedbacks in the Global Climatic System. Ed. by G.M. Woodwell and F.T. MacKenzie. Oxford Univ. Press, New York, Oxford, p.393–411.

Workshop on Cloud Microphysics Parameterizations in Global Atmospheric Circulation Models (Kananaskis, Alberta, Canada, 23–25 May 1995) (1995). WCRP–90 (WMO/TD-No.713). WMO, Geneva. 353 pp.

World Bank Country Assistance Strategies and the Framework Convention on Climate Change. Case study results in Hungary, Poland and the Czech Republic. Facing the Global Challenge: A Progress Report on World Bank Global Environmental Operations. March—May 1995. The World Bank, Washington, D.C. 1995, p.6–7.

World Commission on Environment and Development. Our Common Future. Oxford Univ. Press. Oxford, UK, 1987. 383 pp.

Wuebbles D.J., Jain A.K., Patten K.O., Grant K.E. (1995). Sensitivity of direct climate warming potentials to key uncertainties. Climatic Change, Vol. 29, N3, p.265–297.

Xi X.X., Brasseur G. (1995). The response of stratospheric ozone to volcanic eruptions: Sensitivity to atmospheric chlorine loading. Geophys. Res. Lett., Vol. 22, N22, p.3035–3038.

Yamanouchi T., Charlock T.P. (1995). Comparison of radiation budget at the TOA and surface in the Antarctic from ERBE and ground surface measurements. J. Climate, Vol. 8, N12, p.3109–3120.

Yang Z.-L., Pitman A.S., McAveney B., Henderson-Sellers A. (1995). The impact of implementing the bare essentials of surface transfer land surface scheme into the BMRC GCM. Climate Dynamics, Vol. 11, N7, p.279–298.

Yang Z.-L., Dickinson R.E. (1996). Description of the Biosphere–Atmosphere

Transfer Scheme (BATS) for the Soil Moisture Workshop and evaluation of its performance. Global and Planetary Change, Vol. 10, N2, p.117–134.

Yatagai A., Yasunary T. (1994). Trends and decadal-scale fluctuations of surface air temperature and precipitation over China and Mongolia during the recent 40-year period (1951–1990). J. Meteorol. Soc. Japan, Vol. 72, p.937–957.

Ye D. (1993). Climatic jumps in the history. In: "Climate Variability" (Eds.: D. Ye, Q. Zeng, R. Zhang). China Meteorol. Press, Beijing. p.3–14.

Yu W., Doutriaux M., Sèze G., Le Treut H., Desbois M. (1996). A methodology study of the validation of clouds in GCMs using ISCCP satellite observations.Climate Dynamics, Vol. 12, p.389–401.

Zecca A., Brusa R.S. (1992). SO_2 cooling and detection of the greenhouse effect. Nuovo Cimento, Vol. 15C, N4, p.481–483.

Zhang G.J., McFarlane N.A. (1995). Sensitivity of climate simulations to the parameterization of cumulus convection inthe Canadian Climate Centre General Circulation Model. Atmosphere–Ocean, Vol. 33, N3, p.407–446.

Zhang G.J., McPhaden M.J. (1995). The relationship between sea surface temperature and latent heat flux in the equatorial Pacific. J. Climate, Vol. 8, N3, p.589–605.

Zhang H., Henderson-Sellers A, McGuffie K. (1996). Impacts of tropical deforestation. Part I: Process analysis of local climatic change. J. Climate, Vol. 9, N7, p.1497–1517.

Zhang M., Fletcher J.O. (1993). Variation of global ocean cloudiness during 1990–1990 and its correlation with sea surface temperature. In: "Climate Variability" (Eds.: D. Ye, Q. Zeng, R.Zhang). China Meteorol. Press, Beijing. p.36–41.

Zhang M.H., Cess R.D., Xie S.C. (1996). Relationship between cloud radiative forcing and sea surface temperatures over the entire Tropical Ocean. J. Climate, Vol. 9, N6, p.1374–1384.

Zhang R., Leu M.-T., Molina M.J. (1996). Formation of polar stratospheric clouds on preactivated background aerosols. Geophys. Res. Lett., Vol. 23, N13, p.1669–1672.

Zhang Y., McGuffie K., Henderson-Sellers A. (1996a). Impact of tropical deforestation. Part II: The role of large-scale dynamics. J. Climate, Vol. 9, p.611–627.

Zhang Y., Wallace J.M., Iwasaka N. (1996b). Is climate variability over the North Pacific a linear response to ENSO? J. Climate, Vol. 9, N7, p.1468–1478.

Zhao X., Turco R.P., Kao C.-Y.J., Elliott S. (1996). Numerical simulation of the dynamical response of the Arctic vortex to aerosol-associated chemical perturbaions in the lower stratosphere. Geophys. Res. Lett., Vol. 23, N12, p.1525–1528.

Zhao Y., Kondo Y. (1996). Wintertime stratospheric ozone changes over Japan since 1991. Geophys. Res. Lett., Vol. 23, N15, p.1969–1972.

Zhu Y., Newell R.E. (1994). Atmospheric rivers and bombs. Geophys. Res. Lett., Vol. 12, N18, p.1999–2002.

Ziel der CO_2 —Minderrung durch weltweit steigenden Energieverbrauch gefardet. Wochenber. Deutsch. Inst. Wirtschaftsforsch. 1995, Vol. 10, p.228–235.

Zielinski G.A., Mayewski P.A., Meeker L.D., Whitlow S., Twickler M.S. (1996). Potential atmospheric impact of the Toba mega-eruption 71,000 yers ago. Geophys. Res. Lett., Vol. 23, N8, p.837–840.

Ziemke J.R., Chandra S., Thompson A.M., McNamara D.P. (1996). Zonal asymmetries in Southern Hemisphere column ozone: Implications of biomass burning. J. Geophys. Res., Vol. 101, ND9, p.14,421–14,427.

Zubakov V. (1995). 21st Century. Scenarios of the Future: Analysis of the Consequences of a Global Ecological Crisis. St. Petersburg: The Marine Technical University Centre. 87 pp.

Zuidema P., Hartmann D.L. (1995). Satellite determination of stratus cloud microphysical properties. J. Climate, Vol. 8, N6, p.1638–1657.

Zwick A. (1995). Climate Change Research and Policy: Update. Inst. for Prospective Technol. Studies. Seville. EC Joint Res. Centre. N7. 58 pp.

3

Ecological problems of Europe in the context of global change

3.1 INTRODUCTION

It has been mentioned earlier that there are sources of present-day ecological problems which result from specific features of socio-economic development, such as the utilization of ecologically harmful technologies causing environmental pollution, or the unpleasant fact that the developing world is often forced to solve its economic problems at the expense of excessive exploitation of its own natural resources (see References). The continuing population growth on the planet produces an enormous ecological loading due to the continuing increase of energy and resources consumption. As has justly been noted (see Stigliani et al., 1991; Surprising Futures . . . , 1987), it is important that the twenty-first century will be very different from the current one, at least, for the following three reasons:

(1) The working out of new technologies must greatly change the interrelationship between man and nature, providing a broad use of ecologically justified technologies, e.g., a gradual transition to biotechnologies (instead of chemical fertilizers and means of protection of agricultural crops) in agriculture. At the same time, some negative consequences of this transformation are highly probable, e.g. those due to the development of bioengineering (especially, in the absence of proper control over organisms created on the basis of bioengineering).
(2) Considerable climatic changes are possible due to "greenhouse" gases (GHG) emissions to the atmosphere and other factors.
(3) The increase of spatial scales and long-term consequences of various demonstrations of interactivity and synergism in the impacts on the environment which can result in chemical (or, generally speaking, ecological) time bombs.

In the context of global ecological problems, an important place belongs to the European continent which has the longest history of industrial development, and therefore is characteristic of a considerable transformation of the environment. This transformation is felt on a global scale; on the other hand, the European continent is subject to the influence of other regions of the planet. The latter circumstance implies that the problems of European ecology should be examined and regarded as an integral (and interactive) part of the range of problems of global ecology. This conclusion becomes more obvious if one takes account of the global interdependence of socio-economic development. There are, however, regional ecological problems that are specific to Europe, such as transboundary pollution transport, acid rains, the acidification of water bodies and soils, degradation of forests, emissions of toxic substances, etc.

Because of the ever-growing urgency of ecological problems on the European continent, the International Institute of Applied Systems Analysis (IIASA) initiated and started to develop in 1985 a project, "Environmental Future in Europe", whose main objective was substantiation of the long-term strategy of management of the European environment (for a period up to the year 2030) under the abovementioned specific conditions which must be a characteristic feature of the next century. Here it is important to mention that the notion of the "environment" in our case involves not only air, water, soil, forests, etc., but also the quality of life in a general sense (the latter notion requires a separate clear definition), whose basis is the concept of sustainable development (another term which should be defined more clearly).

It is assumed that a substantiation of the long-term strategy of control over the environment is possible, in spite of considerable uncertainties with regard to forecasts of its evolution in the future. Since such forecasts are non-reliable, the task is set of controlling the present-day environment in such a way that technological and socio-economic development in the future will not cause any unexpected ("shock") situations for the society, but, on the contrary, will bring about positive consequences. To be more specific, here are some objectives of the project discussed:

(1) Description and analysis of large-scale transformations of the environment which may be relevant to the perspectives of socio-economic development of Europe during the next 40 years. Here account should be taken of alternative development scenarios with provision of sudden changes (e.g., in the field of technologies).
(2) Assessment of the efficiency of various technological and organizational measures aimed at the solution of problems of long-term management on the basis of accounting for large-scale interaction between socio-economic development and the dynamics of the natural environment.
(3) Substantiation of priorities of scientific efforts to ensure the most efficient fulfilment of the objectives of the project.

In accordance with the above-described circumstances, in the assessment of the ecological future of Europe, one should proceed from an analysis of political and socio-economic factors in order to substantiate possible scenarios of the development of the society, on the basis of which forecasts of eco-development must be based. The main difficulty of such an analysis is the large number of factors involved and the necessity to restrict their number while taking account of priority issues. One of the versions of the solution of a problem of this kind, based, first of all, on the

Table 3.1. Priority aspects of socio-economic development, environmental conditions and relevant indicators

Aspects of socio-economic development	Indicators
Population	Total number and age composition from year to year
Energy	Energy production using different kinds of fuel, emissions of sulphur and carbon dioxides
Industry and transport	Labour resources, volume of production, emissions of pollutants
Agriculture	Annual grows rates of supplies and the use of wood
Intersectoral problems	Need in water
Environmental conditions	**Indicators**
Climate	Interannual variability of air temperature and precipitation
Hydrology	Ocean level rise, water resources, soil moisture, levels of lakes and underground water, river runoff, and their interannual variations
Air quality	SO_2, O_3, NOx and CO_2 concentrations; sulphur deposition
Soil quality	Oxidation, salinisation, erosion, fertilization level, absorptivity, toxication
Water quality	Oxidation, eutrophication, toxication
Biota	Initial productivity, biodiversity, threatened species
Land use	Portion of land occupied with forests, agricultural areas and settlements

conclusions drawn by the Brundtland Commission (World Commission..., 1987), is presented in Table 3.1. Another problem consists in the necessity of taking account of time shifts of various impacts on the environment reaching decades and sometimes taking the form of "ecological time bombs", as well as shifts in space (Wilman, 1991). An important facet of the problem is connected with the interactivity of various aspects of socio-economic development and environmental conditions. An essential feature of the scenarios discussed consists in their "smoothness". One must, however, bear in mind the rare enough, but potentially possible, abrupt changes which may be called "turning points" (TP). In principle, the possibility of suddenly changed situations requires very serious attention in ecological forecasting.

The interdependence of processes in the environment, which can be characterized by different origin and spatial-temporal scales, determines the necessity of a system's approach to the solution of the problems of ecodynamics. This system's approach can take different forms. In particular, it is certainly reasonable to apply a system's approach to studying ecodynamics over different spatial scales—from an individual environmental object (such as an agricultural field, a pond, a forest region, etc.) and large natural systems (especially, river and lake systems and their water catchment areas: the Rhine, the Danube, the Volga, the Dnieper, lakes Ladoga and Onega, etc.) up to the entire globe. The final task is to analyse how an integration of different-scale phenomena leads to the formation of global changes. One of the important aspects of the system's approach is a study of the ecodynamics of an entire continent. The aim of this chapter is a discussion of a set of problems relevant to the European continent. The discussion of the subject is especially timely in connection with a recent creation of the European Environment Agency (EEA), and prior to it—the organization of the European Space Agency (ESA).

The growing urgency of environmental alarms on the European continent (among which a special place belongs to the Chernobyl catastrophe of 1986), the numerous but poorly coordinated efforts aimed at the solution of such problems have stimulated the adoption in 1990 of the Regulation (EEC) 1210/90, which came into force on 30 October 1993. The Regulation established the European Environment Agency (EEA) and aimed also at setting up a European Environment Information and Observation Network (EIONET) (Annual Workprogramme 1994–1995, 1994; European Environment Agency, 1994; European Environment Agency Newsletter, 1994; Europe's Environment, 1994; Kondratyev, 1995 a, b; Kondratyev et al., 1995; Multiannual Workprogramme, 1994). Of special interest is the strategy of Environmental Monitoring (Achieving..., 1993) and relevant experience which has been achieved in the Netherlands (Enthhoven, 1993; Kondratyev, 1995d).

The EEA Headquarters are located in Copenhagen, Denmark. In the beginning, in addition to the Executive Director (Dr D. Jimenez-Beltran), the Agency had twelve persons working for it, six of whom had a scientific background, and the other six a technical or administrative background. Four of the six scientific members of the workforce were national experts from the United Kingdom, the Netherlands, Denmark and Spain.

Although the very first collective EEA members have become Member States of the European Union, the Regulation states that "the Agency is open to countries which are not members of the European Communities but which share the concern of the Communities and the Member States for the objectives of the Agency" (European Environment Agency, 1994). The note is certainly relevant to Russia and to other states of Central and Eastern Europe.

An important aspect of the EEA activity is its readiness to include in the Agency's potential user or client constituency (above all, from the viewpoint of the propagation of information already available) a broad range of users—from such official institutions as the European Commission Directorates, the Council of Ministers, the

European Parliament, other European Union (EU) bodies, national environmental authorities, international organizations, both governmental and non-governmental, to representatives of major economic sectors, the media and the general public.

The decisions mentioned above and concerning the establishment of the EEA had a lengthy pre-history which started in 1972, after the Stockholm Conference on the Human Environment. The Conference called on the European Community to develop an environmental action programme for Europe (paradoxical as it may seem, Africa and Latin America were far ahead in their understanding that ecological programmes on a continental scale were a must). It is very important in this context that in the course of the 1980s, the "emergency-type" nature of response to threats and crises in the environment gave way to the strategy of preventive response and risk management. (It should be emphasized, however, that practical implementation of this strategy has not always been a success; the tragic consequences of floods in Greece and Northern Italy which resulted in the autumn of 1994 from abnormally intensive and durable precipitation provide sufficient evidence of the above.) Positive examples of this preventive strategy are recommendations concerning the prohibition of freon emissions (diminishing the danger of the ozone layer depletion) and the reduction of greenhouse gas emissions to the atmosphere (aimed at the prevention of the dangerous global climate warming, although, in this particular case, the problem has proved to be much more complex than anticipated) (see Kondratyev, 1992a,b; Kondratyev et al., 1993).

The basic objective of setting up a European Environment Information and Observation Network (EIONET) consists in providing the European Union and member countries with objective, reliable and comparable information on the environment that can be used directly and efficiently enabling them to:

- take the requisite measures to protect the environment;
- assess the results of such measures;
- ensure that the public is properly informed about the state of the environment and the pressures upon it.

EIONET will comprise National Focal Points, the main elements of national networks and European Topic Centres. The latter will serve as expertise centres working on contracts with the Agencies to coordinate collection and processing of data concerning specific issues (air, water, soil, flora, fauna, etc.) and to substantiate assessments of various situations on a European scale.

3.2 EUROPEAN ENVIRONMENT AGENCY (EEA)

Here are the basic functions of the European Environment Agency (European Environment Agency.., 1994; European Environment Agency Newsletters, 1994–1996):

- coordination of the EIONET effort, and in this connection, taking responsibility for collection, processing and analysis of observational data;

- providing the European Union (EU) and member countries with objective information required for framing efficient ecological policy (including necessary legislative measures);
- assessment of environmental data to produce specialized reports characterizing environmental quality and anthropogenic impacts on it over the EU territory;
- taking measures to ensure the homogeneity and comparability ("harmonization") of observational data;
- efforts aimed at the incorporation of Europe-relevant environmental information into international ecological monitoring programmes supervised by the United Nations Organization and its specialized institutions;
- ensuring dissemination of reliable information available on the state of the environment, including publication of summary reports every three years;
- stimulation of developments and application of techniques for forecasting the environmental state, aimed at the obtainment information for taking timely and adequate preventive measures;
- support of broad information exchange (especially with regard to technologies) which is important for preventing or at least diminishing the damage to the environment;
- stimulation of further development of methods for assessing environmental costs (to compensate the usual impacts on the environment, as well as to protect and restore the environment);
- ensuring cooperation with other similar organizations.

In accordance with the Regulation, the main tasks of the Agency consist in provision of the currently existing and predictable future states of the environment from the viewpoint of the following aspects: environmental quality, human-induced pressures on the environment and its sensitivity to external impacts. The following problems have been classified as first-priority ones: air quality and pollution emissions to the atmosphere; water quality, its pollution and water resources; the state of soils, fauna, flora and biotopes; the use of land and natural resources; waste utilization; noise; chemical compounds that are harmful for the environment; protection of the environment of coastal zones. Of highest priority are problems of transboundary transport, phenomena on inter-state and global scales. The "human dimension" of the range of ecological problems is doubtless very important. An essential task of the Agency is to avoid a duplication of its efforts in relation to other similar international institutions.

Since in many cases a complex (systems) analysis of the role of anthropogenic impact on the environment is required, various interdisciplinary studies have become especially urgent; among these there are investigations in: unfavourable effects of environmental changes on human health; the quality of the environment in urban conditions, in mountain and coastal zones; ecological aspects of the development of various branches of the economy (power production, transport, agriculture, tourism, etc.).

Here are the main strategic guidelines of the EEA:

- clarifying, through gathered experience, the application of the "precautionary

principle" of risk prevention in industry and natural resource development;
- guidelines for integrating international environment laws and agreements into the character of regional, national or local approaches to environmental care and protection;
- the use of "no regret" strategies in economic development;
- practical means of assessing environmental liability, such as the proper and extended use of the "polluter pays" and "resource user pays" principles;
- more structured information exchange on the definition of "best available technologies".

One of the primary objectives of the Agency is broad distribution of environmental information presented in such a form that free access to this information can be ensured.

Assessing priorities formulated by the EEA, one can see that, although all the environmental problems described there are really very important, there are two consideration lacking, and these two are integrating and fundamental factors, especially in the context of the interrelationship of regional- and global-scale problems (see Danilov-Danilyan et al., 1994; Gorshkov, Kondratyev and Losev, 1992, 1994; Kondratyev, Asimakopoulos, Varotsos, 1995; Losev et al., 1993):

1. The key role of the biosphere as a regulator of environmental quality, and consequently, the basis of ecologically sustainable development; its quantitative criteria may be characteristics of the closedness of biogeochemical cycles (first of all, carbon, sulphur, nitrogen and phosphorus, which has been justly noted in Europe's Environment... (1994)), which are determined by the main criterion: the level of consumption of biospheric resources.

   Special attention has been paid to the problem of indicators in the documents of the Second UN Conference on Environment and Development (Alexander, 1994; Kondratyev et al., 1993). The working out of indicators of sustainable development is necessary to provide a reliable basis for decision-making at all levels, as well as for the understanding of self-regulation potentialities of the sustainable development of the interactive system "society-environment".

2. Man (mankind) as an interactive component of nature, whose extent of impact on the environment has in many cases surpassed the permissible limits (Gorshkov, Kondratyev and Losev, 1994; Kondratyev et al., 1994; Losev et al., 1993). This implies that the "human dimension" (socio-economic aspects) should be a focal point of ecological studies (Kondratyev et al., 1993). An objective of primary importance should be an assessment of the prospects of development of European civilization, taking account of the resources available. This is a problem that can be adequately solved only in the context of the subject of global change (first of all, because European countries use a considerable part of global natural resources).

   The presence in Europe of numerous unsolved problems relevant to ecology is an obvious fact. Mr Domingo Jimenez-Beltran, Executive Director of the EEA, states (European Environment Agency..., 1994): "Europe still has important environmental problems which go virtually unnoticed, real time-bombs ticking away... such as pollution of underground aquifers and industrial lands, degradation of forest lands by fires, the decline of urban environments and the loss of biotopes and biodiversity". Planning the perspectives of its efforts, the Agency must take proper care of all kinds of environmental problems in Europe, as well as of socio-economic factors determining the possibility of sustainable development on a global scale. While the first Multiannual Work Programme (1994–1999) of the Agency addresses this large and demanding agenda, the basic environmental concerns can be tackled in the first Annual Work Programme (1994/95) where priority will be given to air and water quality, soil conditions, nature conservation, and "horizontal" tasks like integrated environmental assessments of entire economic sectors.

By acting as an interface between scientists and environmental policy makers and the public, the EEA can fulfil the following aims (European Environment Agency..., 1994):

- provide added value to scientific research projects and results, by translating them into objective information for policy development and implementation;
- translate identified policy developments and implementation needs into research priorities;
- integrate the results of research and development into public information on the state and trends of the environment.

In this connection, special attention should be paid to interdisciplinary problems which lie at the interface between policy-making and ecology, such as:

- improved environmental forecasting and modelling methodologies;
- developing substantive indicators and indices for risk management, including socio-economic indicators;
- determining reliable criteria for judging and measuring sustainable and environmental quality;
- clarifying the connections between technological development and environment and quality of life.

In the longer term, the solution of the abovementioned problems will be of vital importance for the process of decision-making and for assuring public participation in achieving ecologically sustainable development. The role of these issues will grow as Europe seeks a living standard and a development ethic compatible with environmental security and the sustainable use of natural resources worldwide. A Europe preoccupied with an illusory, self-contained environmental dynamic would be a Europe deprived of a perspective of sustainable development on a continental and a global scale.

We shall now discuss particular environmental problems relevant to the European continent.

3.3 EUROPEAN ENVIRONMENT: THE STATE AND PROBLEMS

In addition to information offered by Danilov-Danilyan et al., 1994, and by Losev et al., 1993, we shall consider some evaluations presented in a comprehensive report entitled: "Europe's Environment: The Dobris Assessment" (Europe's Environment.., 1994). The report contains estimates of the ecological situation in over 40 European countries, and is based on materials prepared for the first "Pan-European" ministerial conference which took place in Dobris Castle (in the former Czechoslovakia) in June 1991. The second conference was held in Lucerne (1993), and the third took place in 1995 in Sofia, with the aim of adopting an Environmental Programme for Europe (EPE). The Dobris Report is the basis for the development of the EPE, and consists of six sections: (1) Introduction; (2) Assessments of the state of the environment in eight different fields (air, water, soils, etc.); (3) Anthropogenic impact on the environment; (4) Man's economic activity as a source of this impact; (5) Twelve major environmental problems in Europe; (6) Conclusion.

We shall now briefly discuss each of the sections.

3.3.1 Introduction

Unlike the abovementioned document (European Environment Agency, 1994), the Dobris Report (Europe's Environment, 1994) distinctly puts special emphasis on the necessity of considering European problems in the context of global change issues resulting from the human pressure on the global environment on an unprecedented scale. Moreover, the Report reveals that it is absolutely necessary to analyse the interaction between man and environment in the context of the state of biochemical cycles of carbon, nitrogen, phosphorus and sulphur. (We would like to refer the reader to earlier detailed substantiations of the concept (Danilov-Danilyan et al., 1994; Gorshkov, Kondratyev and Losev, 1992, 1994; Kondratyev et al., 1993; Losev et al., 1993)). Demographic and economic growth are key to understanding human-induced pressures on the environment. (The thesis acquires special significance in connection with the UN Conference on Population and Environment which took place in Cairo, in September 1994.)

It is a known fact that during the recent century the global population increased by a factor of 3, economic development by a factor of 20, fossil fuel consumption by a factor of 30, industrial development by a factor of approximately 50. About 6 million square kilometres of forests have been cut down since the beginning of the eighteenth century (which exceeds the area of Europe). The effect of man's economic activity on the global cycles of sulphur and nitrogen has become comparable to (if it does not surpass) the contribution of naturally induced factors. It should be noted here that, regrettably, the problem of the carbon cycle has not been mentioned in the context, and no reference has been given to the results of relevant research in Russia.

Europe is the second smallest continent, comprising little more than 7 per cent of the Earth's land area. The continent is surrounded by nine major seas, including the world's two largest land-locked seas: the Caspian and the Black Sea. The area included 46 states, of which 19 are in the European Union (EU) area, 21 in Central, Eastern and Southeastern Europe, 4 are small continental states and 2 are small island territories. (It should be noted that the dynamics of European political geography is so intensive that the figures must be renewed on a regular basis.) The continent spans three climatic zones: the circumpolar, the temperate and the subtropical. From the viewpoint of environmental problems, data on land use in Europe are very important. At present, the proportion of European land devoted to agriculture varies from less than 10 per cent in Finland to more than 70 per cent in Hungary, Ireland, Ukraine and the United Kingdom. Forest cover varies from 6 per cent in Ireland to 66 per cent in Finland. Each decade 2 per cent of agricultural land is lost to urbanization. Europe is two to three times more densely populated than the United States and Africa, but only half as densely as Asia.

3.3.2 State of the environment

We shall consider environmental assessments for different natural media, and for more integrated functional units (landscapes, nature, wildlife, cities, etc.).

3.3.2.1 Air

The dynamics of air quality in Europe during recent years has been rather controversial. On the one hand, air quality has been improving due to decreased emissions of SO_2, but on the other hand, the pollution of the atmosphere by other gases increased. Short-term air pollution exceeds WHO (World Health Organization) standards for air quality at least once a year in most large European cities. Short-term summer peak levels of ozone affect more than 100 million Europeans (a typical example of such a situation is the city of Athens, Greece, as revealed by Varotsos and Kondratyev, 1994). Stratospheric ozone depletion and increased concentrations of greenhouse gases are looming problems for Europe and the entire world (Kondratyev, 1989; Kondratyev, Asimakopoulos and Varotsos, 1995; Mohnen, Goldstein and Wang, 1994). Critical loads for acidification are exceeded in more than 60 per cent of Europe's area.

3.3.2.2 Inland waters (land hydrology)

Analysis of the state of rivers, lakes and groundwater relates both the state and the trends to natural processes and human activities. It shows that, on the average, 15 per cent of Europe's renewable water resources is abstracted every year (but there are large regional variations). The water use may be characterized as follows: industry withdraws about 53 per cent, agriculture 26 per cent, and the domestic sector 19 per cent of the total. 65 per cent of the population is supplied from groundwater; in many areas, groundwater is overexploited and its quality under threat. Nitrate and pesticide levels in soil water exceed EU drinking water standards over much of the continent. Eutrophication of rivers and lakes is widespread. Acidification is a severe problem in a large part of the Nordic countries.

3.3.2.3 The seas

Analysis of the state of Europe's nine major seas—the Mediterranean, Black Sea, Caspian Sea, White Sea, Barents Sea, Norwegian Sea, Baltic Sea, North Sea, North Atlantic Ocean—has revealed quite a number of severe problems, such as: the lack of effective catchment management; coastal zone pollution; conflict of uses in the coastal zones; introduction of non-indigenous species; lack of control of offshore activities; over-exploitation of resources; sea level rise as a result of global warming (it should, however, be noted that the latter statement is hypothetical, and in the case of the Caspian Sea, even wrong because the marked increase of its level has completely different causes).

Observational data indicate that all the seas, except for sub-Arctic ones, face eutrophication problems. Nitrate has increased two to threefold in some coastal areas on the Black Sea and the Sea of Azov. Insufficient control of offshore activities is creating pollution problems in the Black, North and Caspian Seas. Introduction of non-indigenous species had severe ecological impacts on the Black Sea. North Sea pollution has had a catastrophic effect on part of its fauna. In the Mediterranean, endemic species are in real danger.

3.3.2.4 Soil

The key role of the soil in the functioning of ecosystems determines their significance and the need for soil protection for the maintenance of a healthy environment. The problem has become most severe due to intensive processes of soil degradation, especially the loss of fertility (a decrease in the concentration of humus). Soil erosion in Europe affects 115 million hectares, causing loss of fertility and water pollution. Critical loads of acidification are exceeded in 75 million hectares of Europe's forest soils. Widespread over-application of fertilizers results in leaching and runoff, and leads to eutrophication and contamination of drinking water with nitrates.

3.3.2.5 Landscapes

A classification of landscapes has been used which implies the presence of 33 types of landscape in Europe. Negative anthropogenic impact has been discovered in practically all types of landscape. European landscapes are undergoing changes or disappearing because of agricultural intensification and abandonment, urban expansion and the development of transport systems. Only 6 per cent of Europe's land area is under landscape protection, but with generally a weak legal status.

3.3.2.6 Protection of nature and wildlife

To analyse the state of ecosystems, fauna and flora, on the one hand, and measures for nature protection, on the other hand, eight natural (or quasi-natural) major ecosystem types have been examined, including the specific features of their distribution (geographical), management qualities and prevailing ecological stresses. The assessment of Europe's fauna and flora deals with seven groups of species. Special attention is given to threatened species according to Red Data lists. The most important facts concern forests that once covered 80–90 per cent of Europe, and are now reduced to 33 per cent; a considerable part of wetlands (in Western and Southern Europe, up to 60 per cent) has been lost; 30 to 50 per cent of all fish, reptiles, mammals and amphibians in Europe are under threat. The total area of

European protected sites has trebled since 1972, but most are small and fragmented, and lack funds and staff for effective protection.

3.3.2.7 Cities

The fact of urban development in Europe is general knowledge. During the 1980s and 1990s, it has been particularly intensive in some cities (e.g., Istanbul and Madrid), but has slowed down or stabilized in others (London, Paris, Milan, Moscow). Two-thirds of Europeans live in urban areas covering 1 per cent of the total land area.

Urban air quality has improved in many cities, but is still frequently unsatisfactory in large cities. An average city of 1 million inhabitants consumes daily about 11 500 tons of fossil fuels, 320 000 tons of water and 2000 tons of food and emits to the atmosphere 2500 tons of carbon dioxide. Urban water supply is neither allocated nor managed efficiently, as a rule. There is no doubt that a complex approach is required to the solution of ecological problems of large cities.

3.3.2.8 Human health

It is only natural that the deterioration of environmental quality which is mostly observed in contemporary Europe affects human health. If one uses the simplest indicator of health, which is self-assessment, then, according to the data of WHO, we can see that, out of 14 European countries examined, the residents of Norway and Sweden are most satisfied with their health. Other (more objective) indicators that have been examined are life expectancy and infant mortality, and the major causes of death in Europe: circulatory and respiratory diseases, cancers, communicable diseases (including AIDS), as well as injury and poisoning. According to the data for 1987–91, circulatory diseases and cancer are "leading" issues (almost 70 per cent of death causes throughout the entire continent). The situation is especially severe in Russia (Losev et al., 1993).

With regard to air pollution effects, of all pollutants, suspended particulate matter (aerosols) poses the greatest problem to health, causing asthma and obstructive airway disease. Water contamination results in more than 2 million cases of gastro-intestinal diseases annually in Europe. Life expectancy at birth is several years lower, and the infant mortality rate higher, in Central and Eastern Europe than in the rest of the continent.

3.3.3 Factors responsible for pressure on the environment

It is obvious that the main causes of the increasing anthropogenic pressure on the environment are population growth, the increased scale of consumption of natural resources, and of industrial and agricultural production. However, it should be emphasized that the interrelationship between population, consumption of natural resources and production has not been studied adequately enough, particularly from the viewpoint of long-term impact of socio-economic development on the environment.

3.3.3.1 Population, production and consumption

12.8 per cent of the world population are Europeans, but this proportion is declining due to the dropping birth rate in Europe and the growing population in African and Asian countries. Table 3.2 characterizes the dynamics of energy consumption levels from 1970 to 1990 (in tons of oil equivalent, per cent) for the entire world, Western and Eastern Europe and the former USSR. These data illustrate the changing ratios of different energy carriers, such as coal, oil, natural gas and some others, where the first three have been predominant, with the distinct increase of the gas share.

Table 3.2. World and European primary energy consumption, 1970–90 (tons of oil equivalent, per cent)

	Coal	Oil	Gas	Nuclear	Hydro	Others
World 1970	31.7	46.8	18.8	0.9	1.7	—
(total) 1990	28.5	40.5	22.2	6.3	2.5	—
Western 1970	27.6	60.3	6.3	1.2	3.0	1.6
Europe 1990	19.3	44.0	16.6	14.4	3.0	2.7
Eastern 1970	60.1	23.3	13.1	—	1.0	2.4
Europe 1990	48.4	24.2	20.1	4.5	1.1	2.4
Former 1970	36.8	34.8	22.3	0.2	1.6	4.3
USSR 1990	21.0	30.0	42.0	4.0	1.6	1.3

A specific feature of economic development in Central and Eastern European countries is that they mostly rely on food imports (especially, meat) from Western European countries.

3.3.3.2 Exploitation of natural resources

Europe has 8 per cent of worldwide renewable freshwater resources, but loses about 15 per cent of these every year. Europe had 1.9 million hectares more forest and wooded land in 1990 than in 1981 (mainly, due to forestation). European countries (except for Russia) have depleted most of their high-grade mineral reserves and now rely mostly on imports, mainly from Africa. Thus, the problem of the limited number of non-renewable natural resources has become a global-scale concern; its socio-economic aspects will become more and more acute.

3.3.3.3 Pollutant emissions to the environment

Europe accounts for about 25 per cent of global human-induced emissions of CO_2, SO_2 and nitrogen oxides. Sulphur dioxide emissions due to transport have considerably increased in recent years in France (to 150 thousand tonnes/year) and the United Kingdom, but simultaneously their emissions from stationary sources have markedly decreased (to about 1 million tonnes/year). Hence the need for an integrated means of collecting data on emissions and waste to all media, as well as for the adoption of a uniform methodology of assessing the emission.

3.3.3.4 *Waste*

The problem of waste disposal is very well known and is a most acute one (particularly, from the viewpoint of potential threats to human health). The problem of transboundary transport is especially critical. On the average, Europeans produce 350 kg of municipal waste per capita per year. This level is, however, different in different countries, particularly with regard to the composition of waste. For instance, in Belgium and Greece the major part of waste is food and garden waste, whereas in the United Kingdom and the Ukraine the shares of such waste is 19 per cent and 27 per cent (against 47 per cent and 50 per cent for the former two countries), and the prevailing amounts of waste are formed by paper and paper products (39 per cent and 28 per cent, respectively). An increasing amount of industrial waste is considered hazardous. Of crucial importance, of course, is the task of waste minimization and recycling.

3.3.3.5 *Noise and radiation*

In the highly industrialized countries of Europe more than 50 per cent of the population is exposed to noise levels from traffic which exceed the acceptable level. About 113 million Europeans (17 per cent of the population) are exposed to levels of noise that have serious negative impacts.

Overexposure to ultraviolet radiation is so far due mostly to lifestyle; there are indications of a 5 per cent increase in UV-B radiation in winter in the northern hemisphere (it should be mentioned here that there exists a multi-year trend of the intensification of UV-B radiation due to the total ozone content depletion (Kondratyev, 1989; Kondratyev, Varotsos and Cracknell (in press)). According to the WHO data, radon exposure is estimated to cause 10 000 cancer deaths a year in Europe.

3.3.3.6 *Chemicals and genetically modified organisms*

The manufacture, marketing and use of chemicals result in the release of many compounds to the environment, often with undesirable effects on human health, welfare and ecosystems. About 100 000 chemicals are marketed in the EU, and between 200 and 300 new ones appear each year. An issue of special concern is the increasing number of genetically modified organisms. Between 1991 and 1994, nearly 300 modifications of field releases of genetically modified organisms were made in the EU and emitted to the environment. The greatest contributions of this kind have been made by France (93 organisms), Belgium (56), the United Kingdom (47) and the Netherlands (42). More evidence is urgently needed about their impact on the environment.

3.3.3.7 *Natural and technological hazards*

Apart from volcanic eruptions responsible for global-scale after-effects, and earthquakes, naturally induced hazards and catastrophes are relevant to such phenomena as storms and floods, heatwaves and dust storms, forest fires and droughts, and their impact on the environment increases as civilization continues its development. The main problem is the lack of sufficiently reliable methods for forecasting them.

Technogenic hazards and catastrophes which are due to industrial accidents are no less dangerous. It appears that the majority of accidents occur in the petroleum industry. Precise knowledge of the full consequences of the Chernobyl accident is unlikely to be obtained, but some unexpected effects of this unprecedented catastrophe have already emerged. Tanker accidents are harmful enough, although they contribute only about 10–15 per cent of all the oil that reaches the sea as a result of human activity.

3.3.4 Anthropogenic impacts

The authors of (Europe's Environment, 1994) single out eight key types of anthropogenic impact on the environment.

3.3.4.1 Enenrgy production

The increasing role of energy production (in the ecological context) is determined by the needs of the growing total energy consumption, although, after the petroleum crisis of the 1970s, notable progress has been achieved in the development of energy- and resource-saving technologies (however, this progress has slowed down markedly in Western Europe during recent years). On the average for Europe, energy consumption is distributed over various sectors as follows: 41 per cent for industry, 22 per cent for transport, and 37 per cent for the commercial and domestic sectors. Here are the main energy production sources (according to data for 1990): solid fuel (37 per cent) and nuclear power (35 per cent) in the EU countries; solid fuel (67 per cent) and nuclear power (12 per cent) in Central Europe; solid fuel (25 per cent), natural gas (34 per cent), petroleum (14 per cent) and nuclear power (12 per cent) in Russia. The contribution of renewable energy sources is 9 per cent, 8 per cent, and 15 per cent, respectively. Among the most harmful ecological consequences of energy production, there are: the pollution of the atmosphere, natural soils and waters; the acidification of the environment; increases in the tropospheric ozone and greenhouse gas concentrations.

3.3.4.2 Industry

The impact of industry on the environment is characteristic of considerable differences between various parts of Europe. Its concrete effects are environmental pollution and exploitation of natural resources. The chemical, pulp and paper, cement, steel and non-ferrous metal industries lead to the most severe environmental impacts. In 1990, according to data for 20 countries, the contribution of SO_2 and nitrogen oxides emissions by industry was 25 per cent and 14 per cent, respectively. However, there is a trend towards reduction of all emissions from industrial plants in most recent times. An important result of an advancement of chemical technologies has been a 30 per cent reduction in energy consumption per unit output in the Western European chemical industry between 1980 and 1989.

3.3.4.3 Transport

Development of various kinds of transportation (particularly of road transport) determines its special importance from the viewpoint of its impact on the environment. Private car ownership doubled in Western Europe between 1970 and 1990. Transport accounts for 25 per cent of energy-related CO_2 emissions, with road transport making up 80 per cent of the total. Although emissions per vehicle will be reduced in the next few years, the total emission of pollutants will increase due to the constantly increasing number of vehicles. Land fragmentation due to the transport infrastructure will impact increasingly on land use. Road traffic intensification has brought about a considerable increase in the number of traffic accidents during recent years: over the past 20 years in the EU countries alone, more than one million people have died in road traffic accidents. The perspectives of rapid development of air transport require a serious analysis, especially with regard to the impacts of subsonic and supersonic aviation on the ozone content in the troposphere and stratosphere (Kondratyev, 1989; Mohnen, Goldstein and Wang, 1994).

3.3.4.4 Agriculture

In spite of the fact that agricultural production affects the environment considerably, the actual extent of this impact has not been studied adequately so far. Today agriculture accounts for 42 per cent of Europe's total land area; (the figure is possibly overestimated because it has been obtained without taking account of the situation in Russia). Farms tend to increase in size in Western Europe, but the total contribution of agriculture to the economy is declining. There have been consistent gains in agricultural production (alongside constant reduction of ploughed land) and labour productivity. However, the intensification of agricultural production leads to water pollution, decline of soil quality, loss of biodiversity and landscape changes. On the other hand, agriculture is also a victim of environmental degradation.

3.3.4.5 Forestry

Forests cover 33 per cent of Europe's land area (a 10 per cent increase over the last 30 years), but this is mostly characteristic of Southern and Western Europe, whereas in many Eastern European countries the forest cover has remained stable or even declined. The introduction of non-indigenous species has altered the composition of forests. Since 1965, European wood production has increased by 18 per cent, and consumption by 28 per cent, which reveals that a considerable amount of wood is imported into Europe. The dynamics of European forests has ecological aspects of critical importance; the subject has been discussed in detail in the paper by Gorshkov, Kondratyev and Losev (1994).

3.3.4.6 Fishing and aquaculture

In this context, the issue of special concern is the unacceptable level of fishing intensification in basic fishing regions of Europe: the north-east Atlantic, the Mediterranean, Black and Caspian Seas and the Sea of Azov. Fish stocks are certainly overexploited there. The utilization of ecologically harmful fixed and drift nets have a negative impact on marine animals. The development of aquaculture causes water pollution and suppression of native species which are often replaced by new exotic species cultivated artificially.

3.3.4.7 Tourism and recreation

Since in recent decades tourism has become one of the most important social and economic activities, it certainly affects the environment. Its impact shows itself distinctly in such places as protected areas, rural and mountain zones (the Alps region alone receives 100 million tourists per year), coastal areas (the coastal Mediterranean alone received 157 million tourists in 1990), cities and heritage sites (urban-based tourism is gaining popularity), theme and leisure parks. Reliable evaluation of the ecological consequences of developing tourism is certainly required, especially because this aspect has not been adequately studied so far.

3.3.4.8 Households

The environmental impacts of households should be investigated in terms of resource use and emissions. The importance of this impact on the environment is determined by the high level of consumption of resources and production (about 70 per cent of industrial production in Europe), and the enormous size of emissions and waste. Western European households now own at least one car each, thus causing significant environmental disruption. In spite of realistic potentialities of utilizing waste, less than 10 per cent is actually collected for recycling. An important factor is water consumption (European households account for approximately 19 per cent of total water supplied for all purposes, and this fraction is increasing). As for tourism, ecological consequences of contemporary household activity has not been sufficiently studied.

3.3.5 Key problems of the environment

3.3.5.1 Climate change

This range of problems has been studied in great detail in numerous publications (see, for example, Kondratyev, 1992; Kondratyev et al., 1994; Kondratyev, Buznikov and Pokrovsky, 1996), hence, no special comments are necessary. However, it should be emphasized that, in this context, one must not reduce the entire complex problem to estimating the possible "greenhouse" climate warming and its consequences for the environment and the biosphere, as has been done in Europe's Environment, 1994. The actual problem is far more complex.

3.3.5.2 Total ozone content (TOC) depletion

The problem of anthropogenic impact on the global ozone layer, which is no less important, has been discussed in detail earlier (see, for example, Kondratyev, 1989). We shall draw attention only to some specific circumstances which are characteristic of Europe: during the past decade, a 6–7 per cent TOC decline has been observed at mid-latitudes over Europe (Europe contributed about one-third of global annual emissions of ozone-depleting substances); a consequence of the TOC drop must be increased UV-B radiation, and hence, skin cancer deaths are expected to reach two per million inhabitants by the year 2030. Even if the earlier adopted agreements on banning ozone-depleting emissions to the atmosphere (from the London Protocol to the Vienna Convention) are fully implemented, it will be at least 70 years before ozone depletion stops. An addition to information that may be found in Europe's Environment, 1994, we would like to note that: (1) the ozone content decrease in the lower stratosphere may cause a certain cooling of the global climate (to counterbalance the "greenhouse" warming); (2) an anthropogenically induced growth of the ozone content takes place in the troposphere (see Section 3.3.5.6), which may be important both from the viewpoint of climate change (tropospheric ozone is one of the greenhouse gases (Kondratyev, 1992; Mohnen, Goldstein and Wang, 1994)), and in terms of its impact on human health (especially due to the formation of photochemical smog in large southern cities).

3.3.5.3 The loss of biodiversity

European ecosystems include more than 2500 habitat types and some 215 000 species, of which 90 per cent are invertebrates. Almost every European country has endemic species that can be found nowhere else. The European centres of biodiversity include the Mediterranean Basin and the Caucasus Mountains on the south-eastern margin of Europe. Since the ecological role of biodiversity is largely unknown, the wisest course is to adopt the precautionary principle of avoiding any actions that needlessly reduce biodiversity. We should like to refer the reader to a paper by Gorshkov, Kondratyev and Losev (1994) where the key importance of biodiversity is substantiated for maintaining the sustainability of the global environment. Measures for maintaining biodiversity should be based on the recommendations of the Convention on Biodiversity (see Kondratyev et al., 1993).

3.3.5.4 Major accidents

In the context of the problem of major accidents, that has already been discussed earlier, it is most important to work out criteria for acceptable risk levels for both human health and the environment. In the Netherlands, for example, processes that have a probability of causing ten deaths more frequently than once every 100 000 years are considered unacceptable. Risk management should be based on estimates of possible levels of consequences of accidents and probabilities of their happening (special attention should certainly be given to possibilities of accidents at atomic power stations). It is important to discuss emergency response or contingency plans in both national and transboundary situations.

3.3.5.5 Acidification of the environment

Combustion of fossil fuels emits sulphur and nitrogen dioxides into the atmosphere, where the gases are converted into acids (gas-to-particle conversion), which, after deposition, lead to a series of undesired changes in vegetation (especially forests), soils and waterbodies (aquatic ecosystems). Severe acidification of freshwater is occurring, for example, over large areas of southern Scandinavia, causing widespread fish kills. Coniferous forests are being damaged in the Czech and Slovak Republics, Germany and Poland as a result of acidification and high concentrations of ozone and sulphur dioxide in the air. Although acid deposition is expected to decrease in Europe following emission reductions, in more than half the area critical acidification levels will still be exceeded.

3.3.5.6 Tropospheric ozone and other photochemical oxidants

Observational data on the surface concentration of tropospheric ozone reveal that in many cases (in cities) this concentration exceeds the maximum allowable concentration (MAC) established by the WHO regulations (its growth rate for Europe has been estimated at 1 per cent a year). This severely endangers human health by causing premature ageing of the lungs, eye, nose and throat irritation, chest discomfort, coughs and headaches. This situation makes it necessary to carry out monitoring of not only the surface ozone concentration, but also its precursors, such as nitrogen oxides, volatile organic compounds, methane and carbon monoxide, whose concentration levels are also increasing. These minor gaseous components (first of all, ozone) may also influence building constructions, crops and, possibly, forests. Regrettably, no definite goals have yet been set to prevent these adverse impacts, and the actions already undertaken are not thought to be sufficient in Europe.

3.3.5.7 The management of freshwater resources

Key problems in this case are: the imbalance of water availability and demand (especially, the deficit of drinking water), the pollution of rivers and waterbodies, damage to (and sometimes destruction of) aquatic ecosystems. These problems can only be solved through complex studies of the "waterbody-catchment" systems. Much water is lost in distribution systems. The losses have been estimated at 25–30 per cent for France, the United Kingdom and Spain, but the figure may reach 50 per cent for other countries. Economic measures are absolutely necessary to stimulate water saving. Urgently needed is international cooperation for the management of transboundary rivers.

3.3.5.8 Forest degradation

Two very important causes of forest degradation in Europe are air pollution and forest fires, although it is not always the case that concrete causes of forest degradation are established definitely (a much more comprehensive monitoring of the state of European forests is required). A 1992 survey of 113 tree species in 34 European countries showed that 24 per cent of trees were damaged, in that defoliation exceeded 25 per cent, and a further 10 per cent were suffering from discoloration. As much as 54 per cent of the forests of the Czech Republic may have suffered irreversible damage. An average of 700 000 hectares of wooded land are burnt each year by a total of 60 000 fires in Europe (fires are more frequent in southern European countries, particularly in the south of Italy, in Portugal and in north-western Spain). Since forest fires are usually of strictly anthropogenic origin, it is important to analyse the socio-economic aspects of the problem. Still more urgent is investigation of the role of forests in ensuring the sustainability of the global environment (Gorshkov, Kondratyev and Losev, 1994).

3.3.5.9 Coastal zones

Ecological problems of coastal zones are exceptionally important because of the extended European coastline (which is at least 148 000 km long) and the high concentration of the population in coastal areas (about 200 million people living within a 50-km-wide band). Hence the ever-increasing anthropogenic pressure on the coastal environment. Marine pollution of the coastal zone is a serious problem in all of Europe's seas. In spite of the observed marked deterioration of the quality of coastal waters and damage to coastal ecosystems, no comprehensive coastal zone management scheme yet exists for Europe.

3.3.5.10 Waste problems

Europe produces more than 250 million tons of municipal waste and more than 850 million tons of industrial waste annually. Only in the OECD countries of Europe there are 10 000 annual transboundary movements of hazardous waste, totalling 2 million tonnes. More than 55 000 contaminated sites have been registered in just six European countries, and the total contaminated area in Europe is estimated to be between 47 000 and 95 000 km^2. The problem of toxic waste is especially acute. The scale of minimization and processing of waste is utterly insufficient, and, as a rule, waste (including toxic materials) remains outside proper control. That is why the problem of waste disposal is so urgent.

3.3.5.11 Urban stress

Urban areas in Europe show increasing signs of environmental stress, notably in the form of poor air quality, excessive noise and traffic (mostly automobile) congestion. (It is also important that urban transport accounts for about 30 per cent of total energy use.) One of the goals and means to achieve sustainable urban patterns in Europe and a satisfactory quality of urban environment is improved city planning which might ensure efficient city management, including its many infrastructures.

3.3.5.12 Chemical risk

Intensive production of more than 10 million chemical compounds has given rise to the problem of preventing a chemical risk following from the emission to the environment of a huge amount of chemical compounds and their consequent impact on human beings and the biosphere. An urgent task is that of the reduction of the content of anthropogenic chemical compounds in various natural environments to a permissible level. Here the question of priorities is of primary importance. This is why during the years 1991–1994, within the framework of the EU programme, an ecological assessment has been completed of 1700 chemicals produced or imported in amounts of more than 1000 tons a year.

3.3.6 Conclusion

Materials prepared by the European Environment Agency (EEA) (European Environment Agency, 1994; European Environment Agency Newsletters, 1994–1996; Europe's Environment, 1994) contain ample information concerning the most essential problems existing today and possible means of their solution. It is only natural that the multiplicity, diversity and interdependence of European environment problems have not made it possible for the authors of these documents to exhaust the entire range of relevant problems. Perhaps, the main drawback of the report "Europe's Environment, 1994" is the lack of a proper analysis of the ecological situation in high-latitude European regions, especially in the Arctic, in spite of the increasing role of this part of the continent in economy (Kondratyev, 1995c). It will suffice to remind the reader about present-day gas and petroleum production in the European north, and the soon-to-be forthcoming exploitation of gas and oil deposits on the shelves of Arctic seas, whose ecological aspects are most acute. The fishing industry in the Arctic contributes very considerably to the economy of Europe. One should take into account the consequences of Arctic nuclear tests and major accidents with nuclear power submarines. Although the formation of a gigantic "ozone hole", which is a regular springtime phenomenon of Antarctica, is impossible around the North Pole, there are clear results of observations revealing the appearance of "mini-holes", not only in the Arctic, but also in the St. Petersburg region. All this indicates a necessity to work out a special Section devoted to the high-latitude environment problems, all the more so since its high sensitivity to external impacts is very well known (it will suffice to mention, for example, the possible global consequences of permafrost melting which may accompany climate warming). The ecological problems of mountain regions and small islands certainly deserve special attention. Of peculiar importance also is the problem of the substantiation and implementation of an optimized pan-European ecomonitoring system which should be based on a combined use of conventional and satellite observation means (Kondratyev, Buznikov and Pokrovsky, 1996).

The main conclusions of the accomplishment of the Fifth Environment Action Programme include the following (European Environment Newsletter, 1995):

- the European Union is making progress in reducing certain pressures on the environment;
- this progress is not enough to improve the general quality of the environment and even less to progress towards sustainability;
- without accelerated policies, pressures on the environment will continue to lead to the transgression of human health standards and the often-limited carrying capacity of the environment;
- actions taken to date, will not lead to full integration of environment considerations into economic sectors or to sustainable development;
- transport appears a key sector on which to focus future policy.

The following key issues require further attention at the European level: climate change (i.e. CO_2 emissions), acidification, (urban) air quality, water abstraction and the quality of ground water, habitat destruction and fragmentation, and waste management. Other issues, which have not been comprehensively tackled at European level, are: the degradation of soil quality, which is an important natural resource, coastal management and chemicals in the environment.

In general, population and economic growth show upward trends, translating into more energy use, tourism and, in particular, transport. If these cannot be combined with sufficient (and cost-effective) abatement measures, (further) decoupling of economic growth from these trends is essential in order to be able to secure sustainable development.

3.4 BASIC FEATURES OF THE SOCIO-ECONOMIC DEVELOPMENT OF EUROPE

Typical general features of socio-economic development are the following: (1) relative stability of certain hierarchical structures (related, for instance, to demographic dynamics or educational systems); (2) need in information prior to decision-making (this is also relevant to data on the environment); (3) the market syndrome of industrial development and commercial operations; (4) tendency towards growth. It should be noted in this connection that damage to the environment may take place both in the industrial countries (e.g., due to a high level of power consumption), and in the developing countries (due to excessive exploitation of natural resources).

The following features of socio-economic development in Europe may be considered to be most probable (until recently, this was valid for Western Europe; today this holds for the whole Europe):

1. The growing tendency towards achieving a higher quality of the environment, and hence, the internationalization of expenses for the solution of relevant problems. One of the examples is the planned or already implemented projects aimed at the reduction of the emissions of ozone-destroying and greenhouse gases (Kondratyev, 1992).
2. The growth of house-building industry and branches of industry providing the bettering of living conditions and the increase of the standard of living.
3. Gradual displacement of industry away from Western Europe.
4. The growth of unemployment, the reduction of the working week and the increase of free time (consequence: the development of tourism and the entertainment industry).
5. The weakening of the role of national borders, and respectively, the intensification of migration processes (the problem of borders in Eastern Europe is certainly an anomaly).

Among possible turning points (TP) there are: war (events in many regions demonstrate this serious danger); a sudden and great economic decline; migrations caused by socio-economic or ecological factors; mass riots originating in social dissatisfaction and provoking serious actions on the part of the state.

Naturally, the ecodynamics of the European continent is determined by interactive contributions of the development of the ecological situation in individual regions, among which the Baltic Sea region occupies a specific place. This territory belongs to several states with developed industry and agriculture. The urgency of the problem of the ecological safety of the Baltic countries stimulated the organization of three international conferences during the recent years, the first of which took place in Gdansk (1988), the second in Leningrad (1989), and the third in Copenhagen (1991).

There can be no doubt that the socio-economic and political aspects (and the "human dimensions" in general) of changes taking place in the environment become more and more acute. Population growth, the economic level and other factors in socio-economic development are finite causes of ecological ill-being observed on local, regional, or global scales (see Danilov-Danilyan et al. (1994); Gorshkov et al. (1994); Kondratyev (1991, 1992, 1993a); Kondratyev et al. (1993); Kondratyev and Cracknell (1998); Kondratyev and Grassl (1993); Kondratyev, Moreno-Pena and Galindo (1994); Losev et al. (1993)).

A decisive aspect of future ecodynamics is environmental politics and economics. This subject has been discussed in detail in an excellent monograph, *Earth Politics* by Professor Ernst Ulrich von Weizsacker (1994). Some of the conceptual results contained in this book have been discussed in detail in Chapter 1.

3.5 ECODYNAMICS OF THE EUROPEAN CONTINENT

From the point of view of the continent's ecodynamics, it is important to consider interactive processes of two types: (1) having a global scale (e.g., emissions of greenhouse gases); (2) restricted, as a rule, to the scale of a continent (e.g., emissions of sulphur dioxide or nitrogen oxides). The processes of the first type can cause, in particular, climatic changes in Europe accompanied by relevant socio-economic consequences (the shortening of the heating season and reduction of power consumption in winter, but an increase in energy consumption in summer due to air conditioning needs; the enhancement of evapotranspiration in summer, a drop in river run-off and its effect on the extent of energy production by hydro-power stations; the deterioration of the quality of natural water; the lengthening of the vegetation season at high latitudes; the enhancement of the eutrophication of water basins; the drying up of wetlands in Southern Europe, etc.). Examples of processes of the second type are: the growth in area of regions where the admissible limit of pollution concentration will be exceeded; an increase in the acidification of lakes and rivers; a decrease in biodiversity in forests and lakes which are subject to anthropogenic impact.

As has been mentioned earlier, a problem of primary importance is the substantiation of an observational system which would be adequate for obtaining the ecological information required for political decision-making. Naturally, this kind of a system should be based on the employment of already-existing observation means. The most difficult task in the analysis of observational data is discrimination of an anthropogenic "signal" against the background of "noise" (e.g., the natural variability of environmental parameters). Since the original source of ecological problems are peculiar features of socio-economic development, of crucial importance become: (1) the working out of models describing the interaction between socio-economic factors and changes in the environment; (2) estimations of the impact of new technologies, legislative and political measures, the influence of cultural traditions on the environment.

Substantiation of observational systems is certainly dependent on specific features of problems under consideration and (to a great extent) on scenarios for forecasting ecodynamics. However, it is possible to formulate some basic principles which take into account the following considerations (see Kondratyev, 1990; Kondratyev and Pokrovsky, 1989; *Surprising Futures*, 1987):

1. The increasing probability of rapid changes in the environment under conditions of an intensification of socio-economic development.
2. The presence of ecodynamics parameters (ecodynamics is defined as interrelated changes in the environment and socio-economic development) which are most sensitive to external impacts (the finding of such parameters is particularly significant for an early recognition of anthropogenic "signals").
3. The difference between ecological consequences of various socio-economic changes.
4. The necessity of solving the problem of disc "signals" from early warning data for integrated natural systems (e.g., the water catchment regions for rivers and lakes).
5. Data series for ecodynamics of the past.

Of principal importance is the choice of ecodynamics indicators and the determination of their priorities. Among the first priority indicators there are: (1) energy consumption (its efficiency, the role of various kinds, ecological consequences, etc.); (2) water resources (their types, users' distribution, consumption per unit GNP (gross national product), etc.); (3) natural renewable resources (production per unit resources, the buffer capacity of soils and lakes, the decrease of the oxygen concentration in seas and lakes, the nutrient balance, the dynamics of forests and agricultural areas, the accumulation of toxicants, etc.; (4) adjustable renewable resources (the type and volume of crops; the consumption of fertilizers, herbicides and pesticides; import and export of primary and processed food products, etc.); (5) the level of pollution and aids used for its monitoring; (6) characteristics of land use.

Substantiation of a limited number of possible scenarios of ecodynamics requires the choosing of certain key variants which comprise, in cases under examination, a set of scenarios based upon the following assumptions: (1) preservation of contemporary trends in Europe and other regions of the globe (a slow economic growth and moderate progress in the slowing down of negative ecological tendencies, in particular, moderate climate warming by the year 2030); (2) rapid development of economy accompanied by serious negative ecological consequences, including considerable climate warming and the 1-m world ocean level rise by the year 2030; (3) ecologically beneficial socio-economic development in Europe and other regions (effective energy and resources saving, control over GHG emissions, etc.) which is accompanied by slight climate warming; (4) ecologically beneficial socio-economic development in Europe but not over the remaining part of the world. The purpose of choosing the latter two versions of development has been an estimation of the effectiveness of the ecologically beneficial development in Europe in the absence of a similar situation in the remaining part of the globe. The answer to the question has naturally been negative: global efforts alone are capable of ensuring ecological well-being over the entire planet, and consequently, in its individual regions.

The data of Table 3.3 illustrate qualitative prognostic estimates of the European ecodynamics for the case of the first version of development. Preventability implies the capability of the society to eliminate or weaken in time negative ecological consequences. The table presents estimates for various aspects of preventability (it is clear, for instance, that in all cases there are considerable expenses involved, and that almost always there are serious socio-political difficulties in the ensuring of ecologically beneficial development). The notion of the "time bomb" implies catastrophic consequences of the accumulated pollution of the environment.

Table 3.4 presents integral (aggregated) estimates of ecodynamics for different versions of development. Examination of the data of Table 3.4 shows that, in comparison with the conditions for 1980, the preservation of present-day trends of development (version 1) will lead almost in all cases (with partial exception of urbanization and summer episodes of pollution by toxicants) to a grave complication of the ecological situation. It is clear that version 2 is fraught with still graver consequences. Comparison of estimates for versions 3 and 4 testifies in favour of the above conclusion on the necessity of ecologically safe development over the entire globe.

Most sensitive to changing environmental conditions are: control over water resources, acidification of soils, wood production and the sea level; relatively non-sensitive are the development of transport and cities. On the whole, the results considered show that *only ecologically safe development both in Europe, and in the entire world can ensure the ecological well-being of mankind.*

Among measures that are required to ensure ecologically sustainable development, there are: (1) saving of energy and resources in industry and agriculture; (2) reduction of emissions to the environment through utilization of low-waste technologies; (3) improvement of the quality of instruments intended for control of pollution; (4) stimulation of development of ecologically safe technologies and production; (5) reduction of the use of chemical fertilizers and pesticides in agriculture, etc.

From the viewpoint of perspectives for further research and development, the following issues must be given priority: (1) numerical simulation modelling of land use under conditions of increasing anthropogenic loadings; (2) study of the "industrial metabolism" (anthropogenic cycles of matter and energy) for integrated natural systems and variations in the chemical composition of natural media (including pollution, oxidation, salinization, eutrophication, etc.), an example of which may serve ecological research for the basin of the Rhine; (3) assessments of the limits of stability of ecosystems under anthropogenic loadings; (4) substantiation of optimized observation systems, taking account of specific features of socio-economic development (Kondratyev, 1990, 1992; Kondratyev and Pokrovsky, 1989; Litovka and Novikov, 1991; Matrosov, Golovchenko and Noskov, 1991).

The basic conclusion is that ecologically beneficial socio-economic development on the European continent is only possible if the same situation holds for the entire globe. This conclusion is in perfect harmony with the concept of an international programme devoted to the socio-economic aspects of global change.

Table 3.3. Europe's ecodynamics by the year 2030 (first version). Figures 1,2,3 define the level of an anthropogenic impact on relevant phenomena and processes:1, not serious; 2, moderately serious; 3, very serious

Processes and phenomena	Intensity	Coverage	Preventability						Response time of the environment	Societal adaptability
			Early warning	Uncertainty	Existing technologies	Expenses	Social acceptability	Average		
Water resources management	2	2	(2)	(2)	(2)	(2)	(2)	2	2	2
Acidification of soil	3	2	(3)	(1)	(1)	(3)	(2)	2	2	3
Wood production	2	1	(2)	(3)	(1)	(2)	(1)	2	2	2
Threatened land areas	2	2	(2)	(1)	(1)	(2)	(2)	2	—	2
Coastal waters:										
Sea level	2	2	(2)	(1)	(2)	(2)	(2)	2	1	2
Pollution	2	2	(2)	(1)	(2)	(2)	(2)	2	1	2
Chemical "time-bomb"	2	1	(3)	(3)	(3)	(2)	(2)	3	3	2
Spatial distribution of toxicants	2	1	(3)	(3)	(3)	(2)	(2)	3	3	2
Transport development	3	3	(2)	(3)	(3)	(3)	(2)	3	3	3
Urbanization	1	1	(1)	(1)	(1)	(2)	(2)	1	1	1
Summer episodes of pollution by toxicants	3	2	(1)	(1)	(2)	(2)	(2)	2	1	2

Table 3.4. Europe's ecodynamics during the 1980s and by 2030

Processes and phenomena	1980	1	2	3	4
Water resources management	—	2	3	1	3
Acidification of soils	2	3	3	1	1
Wood production	1	2	3	1	3
Threatened land areas	1	2	2	1	2
Coastal waters:					
Sea level	1	2	3	1	3
Pollution	1	2	2	1	1
Chemical "time-bomb"	1	2	3	1	2
Spatial distribution of toxicants	1	2	3	1	1
Transport development	2	3	3	1	2
Urbanization	1	1	2	1	2
Summer episodes of pollution by toxicants	2	2	3	1	3
Total	14	23	30	12	22

3.6 CONCLUSION

The authors of *Surprising Futures* (1987) were quite right to emphasize the fact that mankind is going beyond acute, but local and relatively simple, environmental problems. It is entering the era of chronic, large-scale and exceptionally complex syndromes of interdependence between the global economy and the environment of the entire planet. In comparison with the preceding generations of problems, the syndromes appearing at present are characteristic of widely spread scientific ignorance, the huge costs of decision making and realization, and temporal-spatial scales which exceed by far the scales of most social institutions. Difficulties due to such changes will increase in the next century as the Earth's population grows, and the volume of industrial production goes up, to say nothing of agricultural production which is likely to be doubled (in response to growing demands).

The major challenge of the forthcoming decades is to learn how to control long-term and large-scale interactions between the environment and development to achieve the ecologically stable well-being of people. What long-term trends in environmental changes will restrict the development of society most of all? How can one obtain pragmatic and reliable information about relevant limits and about ways of overcoming or avoiding relevant difficulties? What are the necessary scientific efforts and institutional changes or technological innovations in the next decade that will make it possible to increase freedom of development and at the same time ensure the sustainable development of the biosphere? All these questions are on the agenda and require a solution. The authors of *Surprising Futures* (1987) tried to give answers to some of the questions—an unrewarding task. As an illustration of an unusual prediction that shows only too well how difficult the above-discussed tasks are, take the following statement: "By the year 2020, Russia removed its socialist name and shifted its orientation westwards".

As has been discussed earlier, sufficiently reliable assessments reveal that the most acute problem of our days is population growth, and already the existing overpopulation gives rise to an unacceptable anthropogenic pressure on the biosphere, that is fraught with catastrophic consequences (Gorshkov, Kondratyev, Losev, 1992; Kondratyev, 1989, 1990, 1992). However, some researchers are of the opinion that, even if the population continues to grow, the world can avoid a global ecological catastrophe by using on a broader scale ecologically justified technologies which are at once low-waste and reserve-saving. That is why it is essential to understand the present demographic situation and its possible future trends. Interesting relevant information can be found in the writings of Keyfitz (Keyfitz and Flieger, 1990; Keyfitz, 1991, 1992). Keyfitz proceeds from the concept of "carrying capacity" of the planet which is characterized by the maximum permissible population level that cannot damage natural resources and the environment (in this context, Keyfitz agrees with the conclusion of Forester that the developing world has no chance of reaching the standard of living characteristic of developed industrial countries).

The basic feature of the contemporary demographic situation consists in the fact that the demographic situation was better in 1970 than today, in both more and less developed countries (MDC and LDC). According to the UN data, a drop in the annual population growth rate took place on a global scale from over 2 per cent in 1950 to 1.7 per cent in 1975–80, which gave a good reason for predicting future decrease (although slow) of the global population growth rate. In fact, the growth rate has stabilized at the level of 1.7 per cent and, in all probability, will remain there at least till 1995. If, during the 1970–75 period, the birth-rate dropped by 3 units per 1000, after 1975 the drop decreased to 1 unit per 1000. In the meanwhile, the absolute population growth rate continued increasing and surpassed the figure of 400 million during 1985–90. Hence the necessity for a considerable correction of the United Nations forecasts concerning population growth. Population growth was accompanied by the increase of the consumption share of the vegetative primary production on the land of up to 40 per cent (Holdgate, 1991).

In the period 1970–75, in LDCs a drop of birth-rate from 4.5 to 1.5 units per 1000 occurred. Beginning with 1980, the birth-rate decrease began slowing down and coincided with aggravating economic difficulties. It is important that there was a shift in people's mentality in LDCs where more and more individuals began considering that further population growth would be inadmissible. By the year 1988, the population of 68 developing countries (63 per cent of the world population number) recognized the fact. Population growth slowed down, particularly in those countries of south-eastern Asia where noticeable economic progress has been achieved.

The deceleration of the birth-rate in LDCs was accompanied by the same process in MDCs where this turn of events could not be anticipated. According to forecasts of 1980, the very low birth-rate level that has been registered at the time was supposed to be temporary. However, in actual fact, the birth-rate decline there continued and has not stabilized yet. (In this connection, some countries, e.g. Canada, plan to compensate the birth-rate decline by the inflow of immigrants.)

It follows from the UN data that in the least developed countries with the population number of about 425 million (for 1990), the population growth rate reached the figure of 2.68 per cent per year, which markedly exceeded the average level for LDCs (2.1 per cent in 1985–90). It is highly probable that this growth rate will still increase during the 1990s to 2.77 per cent. Of great importance is the considerable change in the age structure of the population of LDCs where the proportion represented by the younger generation grows. All the above-presented facts show that social contrasts between the South and the North are enhanced, against a background of worsening economic inequity. The situation is paradoxical because, after the cold war was over, the enormous resources that thus became available and could have been used to help LDCs (in the form of loans and grants), were not used for the purpose; in fact the scale of this help diminished. (Burton (1987), states that, by the year 1990, global military expenditure reached 1 trillion US dollars per year, which is equivalent to about $200 per capita, whereas in many countries the per capita income is less than $200; the situation is still very similar, although global military confrontation has lessened).

A set of nine new scenarios of world population growth has been substantiated by Lutz and Prinz (1994). In accordance with these scenarios:

1. World population will continue to grow. By 2030, world population will increase by at least 50 per cent and may even double in size.
2. Developing countries will account for a greater share of the world population. By 2030, today's developing countries will represent between 85 and 87 per cent of the world population (Africa's share of world population will increase most rapidly).
3. All populations will become older.

Table 3.5 data characterize demographic dynamics in various regions of the globe in accordance with the central scenario suggested by Lutz and Prinz (1994).

According to the data discussed during the UN Conference on Population and Development in 1994 (Cairo Conference..., 1994; Rousch, 1994; Rowley, 1994) by 2050 the world's population will reach 7.8–12.5 billion.

After a reference to a persuasive statement by Zarsky (1990):

> The need for new, sustainable development paths is particularly urgent in light of expected population growth. Currently at about 5 billion, the world's population could climb to between 8–14 billion before stabilising sometime before the end of the next century. More that 90 per cent of the increase will occur in the world's poorest countries. Significant new economic resources will be needed to satisfy even the simplest needs of these members of the human family. Yet if future development paths mirror those of the past the resulting ecological degradation will jeopardise the survival of the entire planet. It is clear that rich country consumption patterns are not sustainable in the long term, particularly if emulated by developing countries. Developed countries . . . generate 74 per cent of carbon dioxide emissions and some 65 per cent of all

Table 3.5. Total population size in 12 world regions under the central scenario (in millions)

	1990	2010	2030	2050	2070	2100
Northern Africa	140	226	332	440	529	595
Sub-Saharan Africa	502	924	1499	2097	2561	2700
North America	277	325	376	420	475	577
Central America and the Caribbean	147	219	289	342	370	371
South America	294	407	516	604	667	727
Western and Central Asia	197	312	442	553	632	682
South Asia	1191	1806	2428	2874	3065	2855
China and Hong Kong	1159	1469	172	1873	1945	1968
South East Asia	518	735	937	1076	1129	1082
Japan, Australia, and New Zealand	144	158	160	158	154	151
Eastern Europe	345	368	380	385	392	427
Western Europe	377	404	416	416	415	426
Developing regions	4149	6097	8167	9859	10897	10980
Industrialized regions	1142	1255	1333	1378	1437	1582
World total	5291	7352	9499	11238	12334	12562

greenhouse gases. The developing countries together produce only 16 per cent of carbon dioxide and some 35 five per cent of total greenhouse gases.

Hunt (1994a,b) has justly concluded that:

to achieve sustainability, it is imperative to radically change both economics (in that it is the fundamental tool of allocating scarce resources and the new scarcity is the global environment) and world institutions (in that they must be the instruments of global sustainable development). However, the argument brings out contradictions and paradoxes. While the case for a new economics can be relatively easily made, its practical application is another matter. In the case of traded resources, prices serve to allocate them between competing users. With scarcity, prices rise and less of the traded resource is used, or substitutes are found. In contrast, the critical and increasingly scarce world resources that we have cited as being used up are, in most cases, public goods and as such do not have prices of markets. This problem can be looked upon another way: the external disbenefits—"externalities"—of global commons pollution and degradation may not be amenable to valuation or internalisation.

The finite nature of global resources means that Adam Smith liberalism and self-interest will not globally achieve "the common good" where the global commons are concerned. Instead the "prisoners dilemma" will apply. A promotion of self-interest by a country means that all countries are worse off. To avoid a mutually destructive outcome, allocation of scarce global environmental resources must be made through planning and agreement, co-opting market forces where appropriate. While such planning will be international, its effects will inevitably trickle down to regions and communities.

There is great inequality between nations in their use of scarce global resources. The fundamental question is whether increasing shares of resources are to be allocated to developing nations so that they can meet their legitimate material aspirations. There is a moral commitment and sacrifice required by the industrialised countries in re-cutting the cake. The level of international agreement and planning required to administer the shares is surely beyound any historical precedent except world war. But there is a paradox except in the call for an increase in planning when the world markets are being opened up and few countries would want to adopt central economic control. Nevertheless, it would appear that global agreements are not necessarily contradictory to free market economies.

In the less industrialised countries, where Malthusian scarcity is evident, sustainable development will require a different approach. More emphasis will need to be placed on the maintenance of the resource base and the consequent reduction in exports of non-renewables or of products that lead to environmental damage or inequity. A policy of self-sufficiency, which directed more production to home consumption, would be preferable. The emphasis would switch to providing "livelihoods", rather than GNP and/or exports, thus necessitating more smaller scale "grass roots" development programs. However, such policies are contrary to the present free trade policies of most industrialised countries and, more importantly, of the powerful world institutions which they control.

The first task of the industrialised countries is not how to get agreement with the developing countries about resource shares. It is to reach consensus among themselves on a radical plan to reduce their collective impacts on the global commons—they are the debtors in terms of the equitable use of the biosphere. When this has been done they will be in a moral and technical position to help the developing countries to adopt the most efficient and pollution-free industrial strategies.

Formulating ecological priorities from the viewpoint of the most dangerous trends determining the unstable global ecodynamics, Holdgate (1991) identified the following six trends: population growth, deforestation, desertification, loss of ecological diversity, environmental pollution, and climatic changes, emphasizing the true fact that the means for overcoming these trends and a transition to a stable ecodynamics are in the hands of politicians, and not the scientists.

Summarizing the discussion of the contemporary global demographic situation and relevant environmental problems, Keyfitz (1991) emphasized, along the same lines as earlier Gorshkov, Kondratyev and Losev (1992) and Kondratyev and Pokrovsky (1989), and also Kondratyev (1989, 1990, 1992), that no reliable recipes have been developed so far to ensure the transition to the stage of ecologically beneficial socio-economic development in the world. However, some fragmentary considerations may be offered for at least partial solution of the problem. Here are some measures that are likely to yield certain changes for the better: (1) socially-oriented pricing policy taking account of replacement (prevention) of ecological losses; (2) placing the basic financial responsibility upon the users of natural resources; (3) economic stimulation of material- and energy-saving and low-waste technologies; (4) broad dissemination of knowledge about ecologically justified technologies; (5) stimulation of a rise in the cultural level of the agricultural zone and its role in the development of industrial production; (6) enhancement of measures to restore destroyed nature; (7) family planning aimed at, at least, the limitation of population growth.

Ghimire (1993) has justly emphasized that demographic dynamics, environmental changes and development processes have to be treated in their "totality", looking at how they interact and are interdependent.

It can hardly be doubted that all the abovementioned measures (and many others, not mentioned here) are important. However, the key problem consists in the substantiation of priorities. Doubtless, the problem of first priority is reduction of the absolute population number level, and not only its growth rate (Gorshkov, Kondratyev and Losev, 1992; Kondratyev, 1990; Holdgate, 1991). Colourfully expressed by Keyfitz (1991), the main fact is that approximately 1.1 billion people who drive automobiles and consume meat just for pleasure (without any physiologically justified necessity) are in combination with about 4 billion people who live under much more severe conditions—this is a bit more than can be acceptable for our planet. It should be added here that the example given above is a convincing illustration of how dangerous and actually explosive the present global socio-economic situation is.

The cold war period is over; the danger of a nuclear war has not ceased to exist but is much less nowadays. This implies that all resources available now should be utilized to reorganize the world on a fairer basis, which might ensure the existence of all humans on this planet under conditions of equal (similar) standards of living, and a sustainable environment and biosphere. The solution of this problem of unprecedented complexity may be achieved only on the basis of cooperation between all the states within the United Nations system, and that system must get stronger. The conceptual basis of this task should be reflected in the documents of a Third Conference on Environment and Development, thorough preparations to which should be started immediately and should take into account the experience of the recently held Second UN Conference on Environment and Development.

REFERENCES

Achieving long term environmental objectives. Proposal for greater international cooperation. 1993. Preprint Conference on Environmental Technology for the Mediterranean Region. Athens, Greece, 1–4 April, 1993.—4 pp.

Annual Workprogramme Mid-1994–1995. 1994. European Environment Agency. Copenhagen,17 November 1994, 78 pp.

Banuri. 1993. The landscape of diplomatic conflicts. In: W. Sachs (Ed.): Global Ecology. London, Zed Books Ltd.
Blowers A. (Ed.). 1993. Planning for a Sustainable Environment. A Report by the Town and Country Planning Association. Earthscan, London. 256 pp.
Burton I. 1987. Our Common Future. The World Commission on Environment and Development. Vol. 29, No. 5, p.2529.
Caccia Ch. 1991. Reducing CO_2 emissions. Environment. Vol. 33, No. 6, p.2–3.
Cairo Conference maps out "path to a better reality". 1994. Newsletter of the ICPD, No. 19, p.1, 4–6, 8.
Changing the Global Environment. Perspectives of Human Involvement. 1989. Eds.: D. B. Botkin, M. F. Caswell, J. E. Estes, A. A. Orio. Academic Press, London, 459 pp.
Chiras D. D. 1993. Lessons from Nature. Learning to Live Sustainably on the Earth. Island Press. Earthscan Publ. 310 pp.
Cothern C. R., Ross N. P. (Eds.). 1994. Environmental Statistics, Assessment, and Forecasting. Boca Raton, FL., CRC Press.—432 pp.
Cropper M. L. and Griffiths Ch. 1994. Interaction of population growth and environmental quality. Amer. Econom. Rev., Pap. and Proc. Vol. 84, No. 2. p.250–265.
Croze H. 1994. Overall new programme. Earth Views.—Vol. 1, No. 1.—2 pp.
Danilov-Danilyan V. I., Gorshkov V. G., Arsky Yu. M. and Losev K. S. 1994. Environment between Past and Future: The World and Russia. Moscow, VINITI Publ., 134 pp.
Dixon J. A., Scura L. F., Carpenter R. A., Sherman P. B. 1994. Economic Analysis of Environmental Impacts. Earthscan Publ., Ltd., London.—160 pp.
El Ashri M. T. 1992. The GEF and its future. Network'92. No. 14, p.5.
El-Badry M. A. 1992. World population change—a long-range perspective. Tellus, Vol. XXI, No. 1, p.18–23.
Enthhoven M. E. 1993. Strategic Environmental Initiative. Preprint Conference on Environmental Technology for Mediterranean Region. Athens, Greece, 1–4 April, 1993.—11 pp.
Environmental Quality. Twentieth Annual Report. 1990. The Executive Office of the President. Washington, DC, 494 pp.
European Environment Agency. Putting information to work. Earthscan Publ., 1994, 14 pp.
European Environment Agency Newsletter, 1994–1996.
Europe's Environment. The Dobriš Assessment. An overview. Earthscan Publ., 1994, 122 pp.
Europe through a Glass Darkly, 1994, Proc. of the American Academy of Arts and Sciences. Vol. 123, No. 2, p.1–223.
Fischer K., Schott J. 1993. Environmental Strategies for Industry. International Perspectives on Research Needs and Policy Implications. Island Press.—Earthscan Publ.—300 pp.
Gallopin R. and Herman D. 1992. The Limits to Poverty: The Bariloche Report (manuscript).
Ghimire K. B. 1993. Linkages between Population, Environment and Development.

Case Studies from Costa Rica, Pakistan and Uganda. UNRISD Report 93.1, Geneva.—118 pp.
Goodland R. and Herman D. 1992. Ten Reasons Why Northern Income Growth Is Not the Solution to Southern Poverty. Washington, DC, World Bank, Environment Dept.
Gore A. 1993. Earth in the Balance. Ecology and the Human Spirit. PLUME, New York, 408 pp.
Gorshkov V. G., Kondratyev K. Ya. and Losev K. S. 1992. Global ecological perspectives. Vestnik of RAS (Russian Academy of Sciences), No. 5, p.70–81.
Gorshkov V.G., Kondratyev K.Ya., Danilov-Danilyan V. I. and Losev K. S. 1994. The Environment: from New Technologies to New Thinking. Moscow, VINITI Publ., 134 pp.
Gorshkov V. G., Kondratyev K. Ya., Losev K. S. 1994. Global ecodynamics: now and in the future// Int. J. of Global Energy Issues. Vol. 6, No. 3/4/5.—p.254–262.
Haas R. 1994. Eastern Europe: A subsidy strategy for ecological recovery. Int. J. of Global Energy Issues. Vol. 6, No. 3/4/5.—p.133–138.
Harou P., Daly H., Goodland R. 1994. Environmental sustainability through project approisel. Sustainable Development.—Vol. 2, No. 3.—p.13–21.
Holdgate M. W. 1991. The environment of tomorrow. Environment, Vol. 33, No. 3, p.14–20, 40–44.
Hollander J. M. (Ed.). 1993. The Energy-Environment Connection. Island Press. Earthscan Publ. 450 pp.
Homer-Dixon T. F. 1990. Environmental change and violent conflict. Emerging Issues. Amer. Acad. of Arts and Sciences. Cambridge, MA. Occasional Paper No. 4, 46 pp.
Homer-Dixon T. F. 1991. On the threshold: environmental changes as causes of acute conflict. Intern. Security, Vol. 16, No. 2, p.76–116.
Hunt C. 1994a. The sustainable development debate revisited. Current Affairs Bull.—Vol. 70, No. 10.—p.4–11.
Hunt C. 1994b. Ecologically sustainable development: Spatial, Economic and Institutional Issues In Its Implementation. PhD Thesis. Flinders Univ. of South Australia. March 1994.—484 pp.
Johnson S. P. and Corcelle G. (Eds.). 1989. The Environment Policy of the European Communities. Kluwer Acad. Publ., 336 pp.
Keyfitz N. 1991. From Malthus to sustainable growth. IIASA Working Paper-91-23. Laxenburg, 33 pp.
Keyfitz N. 1992. Completing the worldwide demographic transition: the relevance of past experience. Tellus, Vol. 448, No. 1, p.26–30.
Keyfitz N. and Flieger W. 1990. World Population Growth and Aging. Demographic Trends in the Late Twentieth Century. The Univ. of Chicago Press, 608 pp.
Kondratyev K. Ya. 1989. Global Ozone Dynamics. Itogi Nauki i Tekhniki. Geomagnetism and Atmospheric Upper Layers. Vol. 11, VINITI Press, Moscow, 212 pp.
Kondratyev K. Ya. and Pokrovsky O. M. 1989. The International Geosphere-Biosphere Programme. Key aspects of requirements to Earth's observational data from space. Izvestiya of USSR Acad. Sci., Ser. Geography, No. 1, p.20–28.

Kondratyev K. Ya. 1990. Key Aspects of Global Ecology and Requirements to Observational Data. Itogi Nauki i Tekhniki. Theoretical and General Problems of Geography. Vol. 9, VINITI Press, Moscow, 454 pp.

Kondratyev K. Ya. 1991. On the international Framework Convention on Climate Change. Izvestiya of Rus. Acad. Sci., Ser. Geography. Issue 5, p.117–122.

Kondratyev K. Ya. 1992. Global Climate. Nauka Press, St. Petersburg, 359 pp.

Kondratyev K. Ya. 1993a. Monitoring of atmospheric pollution in Europe: EUROTRAC Programme. Optics of the Atmos. and Ocean. Vol. 6, No. 9.—p. 1159–1168.

Kondratyev K. Ya. 1993b. The Second UN Conference on Environment and Development: Some results and perspectives. Izvestiya of Rus. Geograph. Soc., Vol. 125, issue 3, p.1–8.

Kondratyev K. Ya. and Grassl H. 1993. Global Climate Dynamics as an Interactive Component of Global Change. St. Petersburg, Acad. Sci. Publ. 195 pp.

Kondratyev K. Ya., Danilov-Danilyan V. I., Donchenko V. K. and Losev K. S. 1993. Ecology and Politics. St. Petersburg, Res. Centre for Ecolog. Safety of Rus. Acad. Sci. Publ., 286 pp.

Kondratyev K. Ya., Moreno-Pena F. and Galindo I. 1994. Global Change: Environment and Society. Univ. de Colima, Colima, 47 pp.

Kondratyev K. Ya. 1995a. Key aspects of ecological politics. 1. Ecodynamics. Proc. Russian Geographical Soc.—Vol. 127, No. 3.

Kondratyev K. Ya. 1995b. Key aspects of ecological politics. 2. Socio-economic problems. Proc. Russian Geographical Soc.—Vol. 127, No. 4.

Kondratyev K. Ya. 1995c. The European Programme of Polar Regions and Ocean Studies. Proc. of the Russian Geographical Soc., Vol. 127, No. 1.—p.38–42.

Kondratyev K. Ya. 1995d. Strategic environment management: Dutch experience. Proc. Russian Geographical Soc.—Vol. 127, No. 1.

Kondratyev K. Ya., Asimakopoulus D. N., Varotsos C. A. 1995. European ecodynamics in the context of global change. Studying the Earth from Space, No. 2.

Kondratyev K. Ya., Buznikov, A.A. and Pokrovsky, O.M. 1996. Global Change and Remote Sensing. Wiley–Praxis, Chichester—370 pp.

Kondratyev K. Ya. and Cracknell A. P. 1998. Observing Global Climate Change. London, Taylor & Francis (in press).

Lang T., Hines C. 1993. The New Protectionism. Protecting the Future against Free Trade. Earthscan Publ. Ltd., London.—192 pp.

Litovka O. P. and Novikov E. A. 1991. Natural Resources Potential of Regional Socio-Economic Complexes. Nauka Press, Leningrad, 128 pp.

Losev K. S., Gorshkov V. G., Kondratyev K. Ya., Kotlyakov V. M., Zalikhanov M. I., Danilov-Danilyan V. I., Gavrilov I. T., Golubev G. N., Revyakin V. S. and Grakovich V. F. 1993. Problems of Russia's Ecology. Moscow, VINITI Publ., 348 pp.

Lough J. M., Wigley T. M. L., Palutikof J. P. 1983. Climate and climate impact scenarios for Europe in a warmer world. J. Climate and Appl. Meteorol.—Vol. 22, No. 10.—p.1673–1684.

Lutz W. and Prinz Ch. 1994. New world population scenarios. Options.—Autumn'94.—p.4–7.
Making Development Sustainable. 1994. The World Bank, Washington, DC, 270 pp.
Matrosov V. M., Golovchenko V. D. and Noskov S. I. 1991. Modeling and Forecasting of Regional Socio-Economic Development Indices. Nauka Press, Novosibirsk, 144 pp.
Meadows D. H., Meadows D. L., Rangers J. and Behrens W. W.,III. 1972. The Limits to Growth. Report to the Club of Rome. New York, A Signet Book, 207 pp.
Mohnen V.A., Goldstein V., Wang W.-C. 1994. Assessing tropospheric ozone as a climate gas. Global Change Newsletter, No.15, p.1–3.
Multiannual Workprogramme 1994–1999. European Environment Agency. Copenhagen, 10 November 1994, 50 pp.
O'Riordan T. and Rayner S. 1991. Risk management for global environmental change. Global Environm. Change, Vol. 1, No. 2, p.91–108.
Rousch W. 1994. Population: the view from Cairo. Science, Vol. 265, No. 5176.—p.1164–1167.
Rowley J. 1994. A victory for women. People and the Planet, Vol. 3, No. 4.—p.34–36.
Stigliani W. M., Doelman P., Salomons U., Schulin R., Schmidt G. R. B. and Van der Zee S. 1991. Chemical time bombs: predicting the unpredictable. Environment, Vol. 33, No. 4, p.4–9, 26–30.
Surprising Futures. 1987. Notes from an International Workshop on Long-Term World Development. Friiberg Manor, Sweden, Jan. 1986. Swedish Council for Planning and Coordination of Research. Ord and Form AB, Uppsala, 128 p.
Varotsos C., Kondratyev K.Ya. 1994. Environmental dynamics in Athens: from rural to urban region. Optics of the Atmos. and Ocean, 7, No.3, p.346–361.
Von Weizsacker E. U. 1994. Earth Politics. Zed Books, London and New Jersey.—243 pp.
Water Resources: Good to the last drop? 1996. IIASA Options, Summer 1996, 6–11.
Wilman L. B. 1991. Implications of environmental complexity for science and policy: contributions from systems theory. Global Environm. Change, Vol. 1, No. 3, p.235–247.
World Commission on Environment and Development. Our Common Future. 1987. Oxford Univ. Press, Oxford, UK, 383 pp.
Zarsky L. 1990. Our Common Future: the Brundtland Report revisited. Canberra Bull. of Public Administration.—No. 62, 128 pp.

4

Socio-economic and environmental problems in China

The Second UN Conference on Environment and Development (SUNCED) and consequent approval of the documents adopted by the SUNCED at the United Nations General Assembly have been a powerful stimulus to taking up the closely interrelated problems of the environment and socio-economic development. Of course, many problems lack a distinct formulation, which implies considerable uncertainties with regard to priorities of global and regional ecology and ways of solution of the most urgent problems of the environment and socio-economic development (see Kondratyev (1993); Kondratyev et al. (1992); Kulpin (1990); Marchuk and Kondratyev (1992); Suy Disin (1990)). In this connection, it is interesting to analyse approaches to the solution of problems of this kind in different countries. China, a state with the highest population in the world, attracts special attention. The Report submitted to SUNCED by Chinese experts (National Report..., 1992) and other materials (Grigoryev and Kondratyev, 1993; Kennedy, 1994; Kondratyev, 1991; Kulpin, 1990; Suy Disin, 1990; Sun, 1992; Zhong, 1994) make it easier to form the idea of this country's experience in the solution of the many problems of socio-economic development and ecology.

4.1 NATURAL CONDITIONS. POPULATION DYNAMICS AND SOCIO-ECONOMIC DEVELOPMENT

The territory of China, which is characteristic of a great diversity of natural conditions, encompasses about 6.5% of the land surface of the globe (9.6 million, square kilometres). About two-thirds of the territory is occupied by mountainous and other areas unsuitable for agriculture. China's territory is composed of: mountainous regions (33%), plateaux (26%), river basins (19%), plains (12%), hilly areas (10%). Out of 960 million hectares of the entire territory of the country, about 62% (592 Mha) have been economically developed; there are prospects of economic utilization (by improving the quality of land) for another 10% (98 Mha), whereas almost 270 Mha belong to the category of land unsuitable for any use in the economy (these are deserts and mountain areas). Agricultural land occupies about

96 Mha of the territory, which is equivalent to 0.085 ha per capita (40 years ago the figure was 0.18 ha). The land areas are distributed most inhomogeneously, being mainly concentrated in the south-east and south-west of the country, with 80% of land areas with a vegetation cover—in semi-arid and arid regions of the northwestern part of China.

The water resources of the country constitute about 2800 km^3 (2700 km^3 of river runoff and 870 km^3 of subsurface water), which is equivalent to 2400 m^3 of water per capita (which is only 25% of the global average). The urgency of the whole problem is determined by the inhomogeneous distribution of water resources and strong fluctuations of precipitation and river runoff.

The mineral resources of China are very diverse. China possesses all 160 basic types of mineral resources utilized at present in the world; 148 types out of this number are in sufficient amounts and occupy places not lower than no. 20 in the world list. Among mineral resources prospected there are rich amounts of coal, tungsten, tin, molybdenum, mercury, lead, zinc, vanadium, titanium and rare minerals. The needs of the country's economy are sufficiently satisfied with resources of iron, manganese, copper, aluminium, gold, silver, oil and natural gas. Large prospected resources of fossil fuels as energy carriers (901.5 billion tons of coal; 78.75 billion tons of oil; 33.3 trillion cubic metres of natural gas) complement water energy resources occupying the first place in the world (380 million kilowatts of energy, with the present level of utilization of 9.1%). In 1990, 1.08 billion tons of coal and 138 million tons of oil were extracted in China; 66.04 tons of steel and 618 billion kilowatts of electric power were produced.

China is the largest coal producer and consumer in the world. At the end of 1987, China's proven coal reserves of 859.4 billion tons were among the largest in the world, representing about 50% of the world total (Zhong, 1994). Table 4.1 data characterize reserves and utilization rates of fossil fuels.

According to the assessment made in 1987, the total oil reserves in China were 78.8 billion tons (Zhong, 1994). The proven recoverable reserves of oil were estimated at 3.2 billion tons, or 5.6% of the proven recoverable reserves of coal. The proven recoverable reserves of gas were estimated at only 1.4 trillion cubic metres in 1992, or 1.0% of the world total.

Table 4.1. Reserves and utilization rates of fossil fuels

Resources	Reserves	Proven recoverable reserves[a]	R/P ratio[b] (years)		Per capita proven recoverable reserves[c]	
	China	China	China	World	China	World
Coal (10^9 t)	859.4	114.5 (11.0 %)	103.2	232.0	96.4	189.7
Oil (10^9 t)	78.8	3.2 (2.4 %)	43.1	43.1	2.7	24.9
Natural gas (10^{12} t)	33.3	1.4 (1.0 %)	64.8	64.8	1178	25242

[a] Figures in parentheses denote percentage of world total.
[b] R/P ratio stands for the lifetime of proven recoverable reserves at 1992 rates of production.
[c] Measured in tons for coal and oil, and in cubic metres for natural gas.

Table 4.2. Regional distribution of energy resources, economy and energy use (%)

	North China	Northeast China	East China	Central South	Southwest China	Northwest China
Energy resources, of which:	43.9	3.8	6.9	5.6	28.6	12.1
Coal	64.0	3.1	6.5	3.7	10.7	12.0
Hydropower	1.8	1.8	4.4	9.5	70.0	12.5
Oil, and natural gas	14.4	48.3	18.2	2.5	2.5	14.0
GNP in 1990	13.5	11.7	33.2	25.3	10.6	5.7
Energy use in 1990, of which:	18.9	17.5	25.2	19.4	10.9	8.1
Coal	22.4	17.8	24.4	17.1	10.9	7.4
Electricity	16.51	5.3	29.0	21.6	9.3	8.3

Table 4.3. Final energy consumption by sector (%)

	China			USA	FRG	UK
	1980	1985	1990	1987	1987	1987
Agriculture	6.0	5.5	5.1			
Industry	66.6	65.2	67.1	34.2	34.8	31.6
Construction	1.7	1.8	1.3			
Transport and communication	1.0	5.0	4.7	35.0	23.6	27.9
Services	4.1	4.4	5.0			
Residential	16.6	18.1	16.8			
Others[a]	—	—	—	30.8	41.6	40.5

[a] Including residential, commercial, public and agricultural sectors.

Regional distribution of energy resources in China is characterized by high Table inhomogeneity (Table 4.2). For the present-day situation low per capita energy consumption, but high energy use per GNP, is typical (Zhong, 1994).

Table 4.3 data show final energy consumption by sectors of economy. Zhong (1994) has summarized the situation with China's energy resources and their development, energy consumption patterns and relevant environmental consequences as follows:

1. Reserves and utilization rates of China's fossil fuels are given in Table 4.1. Because of the low level of exploration for oil and natural gas, the proven recoverable coal reserves in China are currently 12.7 times her proven recoverable oil and natural gas reserves combined. For hydropower, the economically exploitable capacity in China is estimated at as much as 378 GW, the largest in the world. Because of the lack of investment and unfavourable exploitation conditions, however, hydropower has so far been underdeveloped. As for China's uranium resources, the development of nuclear power in China

has only just begun, and her domestic uranium supply can meet the needs of short-term nuclear power development, but not those of a long-term large-scale development programme if future nuclear power stations are to be based on the currently used PWRs. With regard to renewables, their role in the energy balance is currently negligible, and there is little prospect for a dramatic increase because renewable systems are not yet competitive with conventional energy supply. China is thus bound to rely mainly on coal as a fuel in the foreseeable future.

2. The characteristics of the Chinese energy consumption pattern are: main reliance on domestic energy resources; coal-dominant structure of energy consumption; uneven geographical distribution of energy resources and economy; low per capita energy consumption but high energy use per GNP; heavy reliance on biomass energy by rural households; and industry-dominated composition of final energy consumption.
3. Total CO_2 emissions from fossil in China amounted to 586.87 MtC in 1990. This means that China ranks second in global CO_2 emissions when Soviet emissions are split among the newly independent republics. But on a per capita basis, China's CO_2 emissions of 0.5 tC in 1990 were very low in comparison with the world average. With regard to the breakdown of CO_2 emissions by fuel, not surprisingly, coal predominates, accounting for 83.4% of the total emissions in 1990. Analysis of the contributions to CO_2 emissions growth over the period 1980–90 shows that China's economic growth measured in per capita GDP was an overwhelming factor, alone resulting in an increase of 314.67 MtC. The population expansion was responsible for an increase of 68.27 MtC. The change in fossil fuel mix also contributed an increase of 5.33 MtC. By contrast, a reduction of 154.67 MtC was achieved because of great success in energy conservation. The penetration of carbon-free fuels also contributed to a small fall in emissions (5.33 MtC).
4. Driven by the threat of further degradation of the environment and the harmful economic effects of energy shortages, China is already determined to make great efforts towards energy conservation and enhanced energy efficiency in general, and towards using coal much more efficiently in particular. Moreover, with little prospect of significant change in the coal-dominated energy structure, it is conceivable that the implementation of these measures will be accelerated when curbing global CO_2 emissions requires special action on China's part.

Very important is forestation (5.7 Mha) to obtain wood and use it as fuel; in combination with natural forests and bushes, this yields 88.60 Mt of biomass as fuel. China is rich in renewable energy resources (solar, wind, tidal). Forests occupy 12.98% (124.65 Mha) of the country's territory and provide 10.572 billion cubic metres of wood. Grass stands suitable for utilization occupy an area of 224.34 Mha (23.4% of the land territory). China's biological resources involving over 30 thousand kinds of higher plants are very rich (there is an excellent genobank of agricultural plants in Beijing); there are also about 104 000 kinds of animals. China is one of the group of countries possessing main global resources in biodiversity.

An exceptionally rich zone of natural resources are the coastal areas, with four seas washing the continental coastline 18 thousand kilometres long and the coastlines of over 6500 islands (about 14 thousand kilometres). The areas are inhabited with over 1500 kinds of fish, 300 species of which are industrial fish, and more than 100 types of algae (also of industrial value).

According to the census data for 1 July 1990, the population of China constituted 116 002 billion people. The first stage of development of the Chinese People's Republic (1949–1972) was characteristic of a rapid population growth (from 54 167 to 87 177 million people); then the annual mean growth rate reached 2.09%. During the 1970s and later, the policy of birth-rate control was implemented (family planning), which lowered the population growth rate to 1.56% during the period 1972–1990. The population growth rate has been reduced, during recent years, to 11.0-14.4 per one thousand people. It should be noted that this is the lowest population growth rate in the entire developing world. The recent overall population growth rate stands at 1.19% (The rapid... 1996). In 1995, the growth rate of the 60-and-older population was nearly three times higher at 3.37%. According to Koptyug (1996), the size of population was 1.21% billion and, by the year 2030, will increase to 1.6 billion.

Naturally, the reduction of the population growth rate took place unevenly in different regions of the country, getting weaker in the direction east–west (75% of China's population is concentrated in the northern, northeastern, eastern and central southern parts of the country which comprise only 44% of the land territory, whereas 25% of the population is scattered in the southwestern and northwestern parts, occupying 56% of the territory). This trend is due to both natural, and socio-economic features. Although during recent decades (especially in the 1980s when rapid industrial development began) a reduction of the share of the peasant population took place and, consequently, the share of the city population increased (from 19.4% in 1980 to 26.23% in 1990), the peasant proportion of the population remains very large. With a great ethnic diversity (56 nationalities), the representatives of the Khan nationality prevail (92%); however, the growth rate is higher among the small nationalities (minorities). Therefore, their relative population number increases.

During the four decades since the formation of the Chinese People's Republic, great success has been achieved in the development of national economy, education, science, culture, medicine and sports, and the standard of living of the population has become much higher. The decade of the 1980s is marked by great progress. As a result of reforms implemented during the decade, and thanks to the policy of open doors to the outer world (this policy has been proclaimed again at the 14th Congress of China's Communist Party in October of 1992 as the basic policy), the gross national product (GNP) was doubled, and China now holds the eighth place in the world with regard to GNP. In 1990 the absolute GNP value constituted 1940 trillion yuans (which is equivalent to approximately 300 trillion US dollars), and the national income was 1430 trillion yuans. During the period 1980–1990, the annual average GNP growth rate (in comparable prices) reached 9%, and that of national income 8.7%. An intensive development of transport, communications and the building industry took place. The production of the latter reached the level of 283.4 billion yuans by 1989, which exceeds by a factor of 3.7 the level of 1970. This ensured, in particular, a considerable improvement of living conditions (during the 40 years of the existence of the Chinese People's Republic, 1.77 billion square metres of living space has been built and occupied, with a total volume of capital investment of 255.6 billion yuans). The three-stage strategy of socialist modernization of China provides for achieving a per capita GNP by the middle of the next century at the level of countries of medium development. Due to the World Bank Report (Numbers... 1996), the Chinese GNP per person measured on a purchasing-power-parity basis was $1800 in 1994. Koptyug (1996) has discussed impressive data which characterize the successful development of the Chinese economy and the people's well-being. For instance, real individual income was growing at a rate of 5–7% annually (with $460 and $200 per person for urban and rural population, respectively).

Foreign trade developed very rapidly, and its volume (the sum of imports and exports) increased from 4.59 billion US dollars in 1970 to 115.41 billion in 1990, i. e. by a factor of 24. The volume of exports constituted 62.06 billion dollars, and that of imports 53.35 billion dollars. China keeps first place in the world in coal-mining and cement production, and fourth place in the oil output, and production of steel and energy, With regard to agriculture, China is the world leader in the production of grain, cotton and meat. In 1990 the production of these issues reached 435, 4.47 and 25.04 million tons, respectively, and that of sugar 71.8 million tons. Possessing less than 7% of the world's cultivated land, China has successfully solved the problem of providing food and clothing for 22% of the world's population, which certainly is an outstanding achievement. Due to economic predictions (Numbers... 1996) by the year 2001 the Chinese economy will be the largest in the world.

4.2 DYNAMICS OF SOCIO-ECONOMIC DEVELOPMENT AND ENVIRONMENTAL CONSERVATION

The development of the Chinese economy takes place under conditions of the pressure of the ever-growing population number and still insufficiently advanced economy, science and technology. Naturally, since China's population (and consequently, consumption) grows (the average per capita consumption in 1989 constituted 700 yuans), the pressure on the environment also increases. The influence of agricultural sphere in China is higher than in any other country due to the sharp contrast between population growth and cultivated land area reduction, and also to the deficit of water, soil erosion, pasture desertification, and such natural calamities as floods and droughts. If the annual population growth is 17 million people, the cultivated land area reduction reaches the figure of 300 kha per year. One third of the land area is subject to soil erosion and water deficit.

The need in forest resources and their potential are still at variance. Important functions of forests are insufficient (the per capita share of forest area in China is 0.11 ha, which is only 11.3% of the global mean value). In the northern part of the country there is a sharp deficit of water resources, especially in cities (300 cities exist under conditions of a permanent water deficit constituting 10 million cubic metres), in the upper and middle parts of the Liskhe and Yellow rivers (where industry and mining are concentrated), in the Khaikhe river valley, and in the coastal belt of the Shandong province, as well as in arid regions and some plateau areas in the north-east of China, where both in urban and agricultural areas a chronic water deficit, for industrial and household use, is the pattern of existence (81% of China's water resources are concentrated in the Yangtze river valley and southwards from it).

Today the main source of natural water pollution is formed by industrial wastes containing heavy metals and toxicants. In 1989 the total volume of wastes reached the figure of 35.5 billion tons (25.2 billion tons was the contribution of industrial wastes, 80% of which penetrated rivers, lakes and seas without purification). Out of 532 rivers investigated in 1988, 436 were polluted to a certain extent. Industrial development is obviously ahead of water treatment complexes operation.

As coastal regions are more and more developed and the resources of the coastal sea areas exhausted, their pollution is intensified. In 1990 along China's coastline 34 relatively heavy "red tides" have been registered, which is 70 times greater than during the 1961–1980 period. Due to excessive fishing, the volume of coastal zone fishery is reduced, and some symptoms of commercial fishery degradation are observed.

Since coal will remain the main source of energy in China for a long time in the future, the lack of radical progress in the advancement of the technology of combustion and chemical conversion of coal will bring about heavier atmospheric pollution and acid rains.

The acuteness of ecological problems in China is due to rapid industrial development under conditions of outdated technologies and management methods. Due to insufficient financial resources, measures that are taken to control and diminish pollution are rather backward, and their scale is smaller than is needed, a situation that will persist for some time yet. There is no doubt that the process of urbanisation will accelerate (during the last 40 years, the number of cities increased from 132 to 450, and by the end of the century will go up to 600), but the rate of creation of the necessary ecological infrastructures will not be sufficient. Although environmental conditions in cities may be partly improved, radical positive moves in the quality of the urban environment can hardly be expected. In spite of the difficulties and negative aspects mentioned above, there are significant positive tendencies.

The implementation of the economic reform was accompanied by paying more attention to problems of environmental protection. In spite of the increase of the GNP that took place during the 1980s, the state of the environment in the country was mostly stable owing to relative reduction of non-purified liquid and solid emissions (the share of the emissions per unit production costs was reduced considerably), better purification of gaseous emissions, utilization of waste and rigid administrative measures (e.g., according to the data of the National Environmental Protection Agency, 1500 enterprises violating ecological norms were closed in 1991).

The quality of the urban environment has been somewhat improved. Although in some industrial centres the consumption of coal increased by 2–3 times, nature-protection measures yielded a noticeable decrease (or stabilization) of the dust content in the atmosphere and the fall-out of dust from the atmosphere. The water quality in the largest rivers and in the coastal zone mostly remained good enough.

The programme of reforestation was developed rapidly: by the end of 1988, trees were planted on 9.44 Mha and formed the "Great Green Wall" in China. Active participation of the population in the forestation programme made it possible to plant about 10 billion trees over the entire country; as a result, the afforested area reached 38.30 Mha, i.e. became one of the largest in the world. By the year 2000, the forest area will be enlarged by 164 Mha (17% of China's territory). Due to measures in providing 800 million peasants with fuel, the damage to natural vegetation has become much less. By 1989, 606 preserves of various ecological categories had been created in China over a territory of more than 40 Mha, which has improved considerably the protection of rare species, including those that were threatened with complete elimination.

The above-described achievements in environmental protection are the result of a policy which is based on the principle of ecological problems having first priority in the course of socialist modernization, taking account of specific features of socio-economic development of the country. Here are basic components of this policy:

- Support of the strategy of synchronous planning and implementation of socio-economic development and nature-protection measures, aimed at ensuring a completely balanced, sustainable development. The principle is based on the statement in the Constitution: "The country protects and improves human environment, controls and prevents pollution and other kinds of danger to the people". "The country ensures rational use of natural resources and protects animals and plants". Of great importance has been the adoption of the Law on Environmental Protection which provisions harmonization of socio-economic development and nature protection measures.
- Incorporation of nature protection measures in long-term programmes of socio-economic development and government plans. Beginning with the Sixth Five-Year Plan of national economic and social development (1981–1985), the solution of problems of environmental protection has become an inherent part of all Five-Year Plans.
- Steadily following the line of family planning as the state policy eventually aimed at the possibility to control the quantity and improve the quality of the country's population (the latter problem is being solved by taking some special measures, such as prohibition of early marriages, or marriages between close relatives and medically unfit individuals). It has already been noted above that China is the only country where the task of reduction of the population growth rate (and, consequently, weakening the pressure on the environment) has become state policy.
- Working out of a strategy in industry, agriculture and other sectors of the economy, whose kernel is the solution of the problem of rational use of natural resources and environmental protection by making use of technologies that are not ecologically harmful, and also a relevant policy.

- Creation and support of administrative structures involved in environmental protection at different governmental levels (presently, the system is a five-level one) to provide an interactive system of coordination and cooperation of ecoeconomic development, with a clear distribution of functions between organizations responsible for economic management and environmental protection.
- Following three basic principles: policy of management improvement, preventive measures to eliminate unfavourable ecological consequences and following the principle "polluter pays"; the policy of maximum encouragement of local initiative and creative activity at all levels: governmental bodies, state and private enterprises, social organizations.
- Working out and realization of laws, legislative acts, instructions and standards which form a legal basis for environment protection; continuous improvement, harmonization and practical employment of various management systems and environmental protection measures aimed at eventual creation of a national environmental management system upon an adequately institutionalized, standardized and scientific basis.
- Intensification of efforts in the field of environmental protection; organization of scientific and technological teams (including industrial) for the implementation of core projects, which must ensure the fulfilment of environmental protection tasks on the contemporary level. During the last two decades, about 300 research institutions have been organized which are involved in environmental problems and relevant technological developments.
- Organization and advancement of ecological monitoring and a relevant information service for operational control of the dynamics of environmental quality (particularly, environmental pollution), regular publication of reports on the state of the environment on the national and local levels (one of the purposes is to inform the public).
- Broad propaganda of relevant knowledge and intensification of special education in the field of environmental protection in order to expand the number of experts in the area and to raise the level of public knowledge and understanding of related problems.
- Development of international cooperation, encouragement of the solution of global ecological problems and promotion of the progress of humankind. China joined many of international conventions, agreements and protocols.

As a developing country, China should make an issue of first priority the progress of national economy to ensure a rapid rise in its people's living conditions in the course of socialist modernization. The economy must be the kernel around which all other fields are developed. In this context, environmental protection is a major constituent of national socio-economic development whose main objective is to improve the standard of living of the people. In the course of economic development, control must be taken over ecodynamics, and ecological degradation should be prevented; at the present stage of the country's development, the problems of environmental protection should only be solved taking account of specific conditions of rational economic development and realistic possibilities. This is the only way to reach an agreement between the rate of the solution of ecological problems and the demands of the developing national economy.

Due to Hu Angang (1996) the following nine areas of environmental concern are of prime importance:

- serious water loss and soil erosion;
- rapid desertification;
- fast degeneration of grasslands;
- sharp reduction of forest resources;
- lowering groundwater levels and dwindling lakes;
- worsening water pollution;
- serious air pollution;
- vast amounts of solid waste;
- spread of environmental pollution to rural areas.

4.3 AN EXPERIENCE OF SATELLITE ENVIRONMENTAL MONITORING

The huge territory of China, and its diverse natural conditions dictate the necessity of a complex monitoring of the country's ecodynamics involving the use of satellite observation means. Environmental investigations in China are closely connected with a broad utilization of remote sensing data, especially of satellite information. We shall examine the experience of remote sensing (RS) in China following the reviews by Grigoryev and Kondratyev (1993) and Kondratyev (1991).

Satellite information is passed over to ground-based centres of data processing from two Chinese meteorological satellites (Atlas..., 1990). The data of American meteorological and natural resources satellites are also widely used. The data of American satellites of the NOAA type are received at a station built by American experts on the Chinese territory.

In China there are a number of centres and laboratories involved in environmental remote sensing. The largest of them are: the National Centre of Satellite Meteorology, the National Remote Sensing Centre of the State Commission on Science and Technology, and the Unified Remote Sensing Centre of the Academy of Sciences. A substantial contribution to the development of RS problems is made by the Laboratory for Information System on Natural Resources and Environment of the Institute of Geography of the Chinese Academy of Sciences.

The abovementioned institutions are involved in important studies which analyse the information content of the materials of aerospace surveys of natural formations, using the advantages of image types obtained in different wavelength intervals. In practical remote sensing, multispectral images in the visible and IR spectral regions, and radar images of the Earth's surface have been broadly used. Experiments in synchronous three-level ecological monitoring are performed: from satellites, from aircraft and on the surface. Methods of computer processing of RS data on the basis of information about the spectral properties of natural and anthropogenic formations are also widely used.

Huge territories in China are located in desert areas. At desert borders, in regions of settling of so-called marginal land areas (steppes and semi-deserts), which are ecologically "fragile", the desertification of land takes place. Anthropogenic desertification develops most intensively in Northern China, bringing a lot of trouble to the economic activity of 35 million dwellers. To control desertification, which causes degradation of living space, RS techniques can be very helpful. Aerospace photography data have revealed that the desertification area in China expanded by 39 000 km^2, having reached the figure of 176 000 km^2 during the period from the late 1950s to the middle 1970s. During these 25 years, 1560 km^2 of land was subject to desertification every year (Cao Shuhu, 1991).

To evaluate the dynamics of desertification during the period 1975–1987, apart from aerospace surveys, satellite images were also used (Zhu Zhenda and Wang Tao, 1991). Analysis of observational data has shown that the land degradation during this period encompassed 25 200 km^2, and the rate of desertification increased to 2100 km^2 per year.

Serious attention is given to the use of RS data for monitoring forests, The total forest area in China is 124.65 Mha, which constitutes only 7.8% of the entire territory. The anthropogenic impact on forests is very strong, Therefore, during the last 30 years, forest degradation took place annually over the area of 1560 km^2 (Xu Guanhua et al., 1991). In this connection, a project has been worked out to utilize RS data to compile a series of thematic maps suitable for the solution of the problem of forest restoration. In the implementation of this project, the data of Landsat and the Chinese satellite FY-1 were broadly used. By making use of this information, a series of thematic maps has been compiled, with the following scales: 1:100 000; 1:200 000 and 1:500 000, including forest areas and land appropriate for afforestation.

The RS data are also widely employed for forest taxation; techniques for automatic (computer) deciphering of various tree stand indices have been developed. These data are also used for control over damage to forests from insects infestations. One of widespread kinds of forest degradation is damage to basic tree stands from caterpillars, in some years, up to 2.7 Mha of forest area can be invaded by caterpillars. Such phenomena are called smokeless fires or forest plague. The experience of using a thematic mapper (Landsat) data for the detection of forest areas infected with the "forest plague" on a test site "Gushan forest" in the Engui province is of interest (Dai Changda et al., 1991), On the basis of spectral differences, forest areas degraded to a larger or smaller extent as a result of caterpillar invasion have been detected. Comparison of the data of automatic (computer) deciphering of images with those of surface observations revealed a high reliability of detection of harmed forests. Errors in determining strongly and slightly harmed tree stands were about 1% and about 5%, respectively.

Investigations of other types of vegetation cover were also conducted. In particular, the propagation of reed thickets, an important resource broadly used in economic activity, has been detected from multispectral data of the Landsat satellite. A test site for the investigations became the coastal zone of Dongting lake, one of the largest freshwater lakes in China. From the materials of field and aircraft observations, the spectral features of reeds and other natural formations have been first studied (Liu Jiykan and Li Shuang, 1989; Lu Guonian, 1991), and then deciphering of satellite data has been performed taking account of the ground-based and aircraft information. Computer processing of RS data made it possible to evaluate the reed mass over the entire coastal zone of the lake in 1980; the figure obtained was 601 000 tons.

RS data are also broadly used in the solution of land-use problems, Thus, for instance, the regions of the Jiangkhan plain and Dongting lake (34 thousand square kilometres) were investigated. From the Landsat images, a map of land use for the region on a scale 1:250 000 was compiled. Ten types of land use and a number of natural formations are presented on the map. The map was machine-made on the basis of accounting for symptoms of various formations (Zeng Zhiyuan and Pan Xianzhang, 1991).

Similar maps were compiled for other regions, both densely and sparsely populated. One of the latter regions is Tibet. The investigation of Tibet is based on the analysis of aircraft survey materials and even more on the Landsat data (Liu Xia et al., 1991).

Priority directions in the investigations of natural resources were assessments of land-use conditions, of soil and of vegetative resources. For optimization of information use, preliminary classification of various natural formations was performed, and deciphering signs were established. This kind of information was used, for instance, in the compilation of land-use maps for the region of probable inundation on the Yangtze river in connection with the designed building of a powerful hydroelectric station (Zhou Wancun, 1990).

One of the areas of employing RS data is mapping of the country. Satellite information has often been employed in the inventory of forest cover, the compilation of geological maps, and maps of floods, etc. A marked achievement is the compilation of the first Chinese atlas of satellite images of the country (Cheng Shupeng et al., 1990). The atlas contains images obtained from the Landsat satellites which characterize various natural and anthropogenic phenomena, as well as map-schemes of their interpretation.

The atlas is divided into four sections which characterize: (1) types of the Earth's surface and land use; (2) hydrological phenomena (in dynamics); (3) regional features of the relief; (4) linear peculiar features of geological structure. Of course, the publication of an atlas offering this rich geographical information can be useful not only for China. The atlas is certainly of interest for experts in other countries. The deciphering of images which is very precise because account is taken of ground-based data, made it possible to obtain unique information on the geography of China, including such difficult to reach and inadequately studied regions as, for instance, the Taklamakan desert, the Kuen-Lun mountains, and regions of intensive economic development (of interest, in particular, is information on the land-use dynamics in the Beijing region).

A most important direction of research is the analysis of the possibilities of using remote sensing for studying natural catastrophic phenomena. The objective of remote sensing here is obtaining the operational information required for the prediction of such phenomena, as well as of data on their propagation and dynamics, and, finally, on their consequences.

Frequent and numerous floods are a calamity for the Chinese people, especially with regard to big rivers. Annually about 8 Mha of land in densely populated areas are inundated. This circumstance provoked the development of a system of controlling floods on the basis of RS data. Such developments have played an important role in the control of some particular floods during recent years, and thus have been helpful in providing solutions concerning the creation of a national system of flood control (Jing-shan Jiang and Shu-hu Cao, 1991).

During inundations remote sensing was carried out from aircraft and satellites. Simultaneously, at some points direct surface measurements were made. The entire body of information was passed over, via a communications satellite, to a data centre in Beijing for further processing, analysis and decision-making. Close attention in the system of flood monitoring is given to information obtained with synthetic aperture radar (SAR) installed on aircraft, which ensures all-weather information. The system of monitoring has, in particular, been employed successfully in 1991. During that year, rich with precipitation, several large-scale inundations took place on the Yangtze and other rivers. The analysis of the catastrophic situation caused by the floods was performed from radar aircraft photographs and images from the NOAA and Landsat satellites. Similar efforts have been made for the Dongting lake and Khuankhe river (Cao Shuhu, 1991), and also during a flood in the Pankhin region at the overflow of the Layao river (Shu Shiwei, 1991). Here both multispectral satellite images and aircraft radar photographs have been used. During the inundation of the Layao river, 22 hours of aircraft surveying yielded data for a territory of 11 000 km^2. Mosaic pictures compiled from radar images (scales 1:100 000 and 1:500 000) played an important role in taking the necessary measures to overcome the consequences of the calamities.

Great damage to the population and national economy of China is done by forest fires. Beginning with 1987, the annual number of forest fires is about 4000 (before that date, there were more fires about—15 000 per year). Forest fires have their harmful effect over the area of about 70 kha (before 1987 the damaged area was about 1 Mha). The utilization of operational satellite information, encompassing large territories, opened up new possibilities for control over this kind of natural calamity.

Beginning with 1985, for monitoring forest fires images received from meteorological satellites of the NOAA type (the AVHRR data) were used. The data were successfully utilized, for instance, to control the development of catastrophic forest fires which occurred from 6 May till 2 June, 1987 in the mountainous region Xin An. Analysis of a combination of images of the fire region obtained from the NOAA, Landsat and SPOT satellites, made it possible to detect new sources of burning, and follow the expansion of the fire zone (Xu Guanhua et al., 1991). It has also appeared possible to detect different degrees of affected forests, and to differentiate four classes of forest territories. On this basis, appraisal of losses of wood has been made.

One of the important directions of remote sensing of natural calamities is relevant to studies of seismic phenomena. It should be emphasized that every year the number of people in China who perish during earthquakes is greater than the total number of victims of all other natural catastrophes (54%). The figure is somewhat higher than the world average (51%), which is due to the seismic activity in the region and its dense population. Satellite images are used for studying active ruptures, many of which have proved to be in close correlation with the orientation of epicentre groups of earthquakes reflected on the map compiled in 1991 (Ma Zongjin and Cao Xianglin, 1991).

Of interest are investigations of thermal anomalies in earthquake zones, detected in infrared images. In particular, thermal anomalies have been detected in zones of the development of large-scale earthquakes: one which took place in November 1986 in the Yun-nan province (the Lenshend region) and two which occurred in October 1989 in the Shensi province (the Datong region). These anomalies formed just before subground shocks began, and were registered both from satellite, and ground-based data.

RS data are employed in the assessment of damage brought about by subground shocks. Such investigations have been made for regions damaged by tectonic phenomena in Tangshan (1976) and Yun-nan (1988). Data of satellite RS and materials of detailed aerial surveying (with scales up to 1:1000–1:200 in key regions) have been used. The first made it possible to relate damaged regions with zones of activated fractures. The detailed surveys were helpful in estimating the damage to settlements.

Just as in the rest of the world, in China more and more attention is paid to efforts in the field of collection, analysis and storage of information about the environment and its changes. At present, projects are developed to create a national geoinformation system (GIS) concerning the environment (Jin Zhang and Qi Tuan, 1990; Lan Yunchao and Yan Ken, 1990). One of the significant sources of information is the data from the NOAA satellites.

Geoinformation systems are developed also to solve various individual problems. There, as a rule, RS data are a component part of the block of the GIS database. For instance, they are employed in the analysis of the interaction between cities and the environment, and city planning (Zhou Wancun, 1990). Various RS data, including satellite, are an information source for the formation of the database of the GIS intended for the evaluation of the flood risk (Chen Shupeng, 1989). This kind of an information system has been created for the Yellow River region, in its lower part. The choice of this particular region is due to the fact that during the last 2500 years, the Yellow River (Khuankhe) has overflowed and flooded the surrounding territory bringing a great damage to the population. Besides, in 26 cases (in 1955 the latest case was registered) considerable displacements of its bed took place. As a result, a GIS atlas has been compiled to assess flood risk in one of the regions exceptionally important for China's national economy.

An experience is known of creating a GIS based on the use of RS data and aimed at the solution of water problems (Su Yabin and Jin Hongtao, 1991). It was tested in the Li Shui region (the Khunan province), in the basin of the Yangtze river in southern China. The basic information contribution was made by analysis of the data of the Landsat satellites. The interpretation of satellite images has made it possible to obtain information about some peculiar features of the geological, geomorphological and hydrological structure of the region. In turn, these data have served as a basis for the estimation of water resources.

4.4 PERSPECTIVES OF THE SOLUTION OF ECOLOGICAL PROBLEMS

The decade of the 1990s is of critical importance for the solution of problems of the socio-economic development of China and environmental protection; the task of optimizing efforts in the fields of economy and ecology remains extremely difficult. To achieve this optimization, the Chinese government formulated the following major objectives (National Report..., 1992):

- Achievement of an annual GNP growth rate of 6% due to advanced economic efficiency, and thus providing the people with food and clothing, and with relatively comfortable living conditions (the most recent forecasts show a possibility of a 9% GNP growth rate). In this context, some indices of economic development to be reached by the year 2000 may be mentioned: grain yield, 500 million tons; coal and oil outputs, 1.4 billion tons and 145 million tons, respectively; electric power production, 1100 billion kW/h; total transportation volume, 1.9 billion tons (the capacity of ports, 1 billion tons); production of steel, fertilizers and ethylene, up to 80, 120 and 3 million tons, respectively.
- Organization by the year 2000, in accordance with further economic development, of control over environmental pollution; improvement of the environmental quality of cities of key importance; overcoming naturally induced negative ecological consequences. This implies the establishment of the upper limit of industrial emissions of sewage waters of 32 billion tons with an 84% purification of waters; ensuring a 70% purification of water in cities; a reduction of sulphur dioxide emissions to 20 million tons; purification of 82% of industrial gaseous emissions; access to gas utilization for 60% of city dwellers; central heating for 470 million m^2; utilization of 37% of solid industrial emission products (320 million tons); reduction of street noise in cities to below 70 dB.

To be able to fulfil the above-described plans, it is necessary not only maintain and advance the already existing approaches, but also to enhance special priority measures along the following lines:

- To formulate and implement additional strategies with regard to policy in the fields of economic, industrial and technological development to ensure coordinated socio-economic development and environmental protection.
- To continue the employment and improvement of those strategies and measures in environmental protection which have proved to be efficient, and concentrate attention upon the realization of measures of nature protection, taking account of modern scientific and technological achievements.
- To ensure gradual increase of financing environmental protection measures, as the national economic potential grows.

- To obtain the necessary financial and technological support through international cooperation on the basis of the principles of equality and mutual concern.

In our modern world, global and regional ecological problems are becoming more and more acute, namely climate warming, ozone layer depletion, acid rain, desertification, deforestation, rapid reduction in biodiversity, enhancement of water loss and soil erosion and freshwater deficit, sea pollution, etc. (with regard to the primary significance of global ecological problems, the matter has been discussed in detail by Kondratyev (1993), Kondratyev et al., (1992), Kulpin (1990), Marchuk and Kondratyev (1992) and Suy Disin (1990) who have analysed the special role of biosphere global dynamics in the context of the problem of global ecological security). For many developing countries, which struggle for their survival in the literal sense of the word, the issues of priority are saving their people from poverty and starvation, and getting rid of their foreign debts, because otherwise the only opportunity they have is excessive exploitation of natural resources and selling them at low prices. This results, of course, in the destruction of the environment.

The destruction of the biosphere on a global scale and the poverty of a substantial part of the population on this planet are the two main interrelated problems mankind is facing. Therefore, such tasks as changing the present unfair economic order to reduce the gap between the levels of development of LDC and MDC, and to preserve the global environment under conditions when man's dignity is in danger, and promoting adequate socio-economic development for many of the third world countries, are urgent objectives, both for the developing, and the industrial states. These problems were the focus of the SUNCED (Kondratyev, 1993).

The solution of all the abovementioned problems requires coordinated efforts on the part of all countries and regions on the basis of effective cooperation between them. Economic development and environmental protection are two important constituents of the new economic order. To implement the efforts, the following is necessary (National Report . . ., 1992):

- All countries (large and small, rich and poor) should be involved in international cooperation, for the purpose of achieving harmony between socio-economic development and ecodynamics (naturally, the contributions and responsibilities of large and rich countries must be much greater).
- Every country should have the right to choose its own strategy of economic development and environment protection, taking account of its realistic possibilities. Each country must possess a permanent sovereignty over its own natural resources. At the same time, international cooperation should be supported in the interests of sustainable socio-economic development.
- The interests of various countries with different natural conditions and levels of development should be taken into consideration, and the needs of the developing states, where the majority of the world population lives, should be adequately reflected in relevant international agreements.
- A clear understanding must be found of the main responsibility (both in the past and in the present) for the contemporary degradation of the global environment,

proceeding from the principle of impartiality and differentiated responsibility.

- The developing countries must take more efficient measures for environmental protection, with the help of relevant technical and financial support from the international community; this will enlarge the contribution of the developing countries to the solution of global problems of ecology and socio-economic development.

History and today's reality teach us that economic progress and the solution of environmental problems in every country require a peaceful and stable international situation. Now, when the international situation has radically changed for the better, the global situation is still full of uncertainties and instability. Since our historical experience has proved that large-scale wars and regional military conflicts are not only a threat to humankind's survival, but also endanger the environment, it is of exceptional importance that the solution of all international problems be based on such principles as national sovereignty and territorial integrity, non-interference in internal affairs, the elimination of aggression, equality and mutual benefit, and peaceful coexistence.

As a developing country, China is facing not only threats originated by the presence of global ecological problems, but also the necessity of solving such environmental protection problems that appear in the process of economic development. In this connection, China declares its readiness to make all the necessary efforts for the solution of this kind of problem, and also to contribute to the progress of common achievements in environmental protection on a global scale (National Report..., 1992). In this respect, the whole problem is focused on the harmonization of socio-economic development and ecodynamics. Separation or counterposition of these two aspects is the same thing as, according to a Chinese proverb, "climbing a tree to catch a fish".

The inseparability of the problems of environment and development should always be taken into consideration in the implementation of international collaboration; one should proceed from the fact that initial causes (roots) of the deterioration of the environmental quality are, so characteristic for many developing countries, a low level of socio-economic development and actual impoverishment. The only way to break the vicious circle of poverty, uncontrolled population growth and environmental degradation is economic development, and liquidation of poverty on its basis. This is why it is so vital to develop international cooperation for the sake of the solution of environmental problems on the basis of equality, within the framework of enhanced collaboration between the North and the South, and for the establishment of a new economic order, capable of providing conditions for sustainable development of all countries, and particularly, the developing states.

For many developing countries, such ecological problems as desertification, floods, droughts and deficit of water resources, which have long been waiting their solution, are a real hindrance to socio-economic development; they create difficulties and threats that are more dangerous for those countries than global ecological problems, such as climate change or the ozone-layer depletion.

The interrelationship of ecological problems means that the solution of global and regional problems is inseparable. Here it is important to realize that, with regard to the present state of the environment, the developing countries are, to a great extent, victims of the current situation. Up to now, the developed countries are the main consumers of the Earth's limited natural resources and the main sources of environmental pollution. Hence follows the principle of common, but differentiated responsibility and, consequently, the necessity of a greater contribution from industrial countries to the solution of ecological problems on the basis of international cooperation by way of serious (not symbolic) financial and technical support from developing countries in their solution of problems of both ecology and development.

Technologies must be regarded as mankind's common property, and their passing over to the developing countries as a contribution to the solution of problems common to the whole humankind (however, the right to intellectual property should be protected). Of course, international ecological cooperation must rely on the principle of sovereignty for all countries, and must keep away from the tendency to interfere in the internal affairs of other countries. It is also important for the whole world to listen better to the voice of the developing countries, since, without effective participation of this part of the world in international ecological collaboration, one can hardly expect any productive results in the field of management and protection of the environment.

The high quality of the National Report of the Chinese People's Republic prepared for the Second UN Conference on Environment and Development is due to the multi-aspect approach, the comprehensiveness and the good organization of the processing of numerous relevant materials under the guidance of an inter-ministerial Coordination Group which integrated the efforts of the Ministry of Foreign Affairs, the State Planning Commission, the State Commission on Science and Technology, and the National Environmental Agency. The materials have been preliminarily discussed by the scientific community (the Chinese Academy of Sciences), experts in industry and agriculture, and various social organizations.

REFERENCES

Atlas of Fengyun (FY-I) Meteorological Satellite. 1990. Beijing. China Meteorol. Press, 62 pp.

Cao Shuhu. 1991. Real-time and quasi-real-time monitor system of remote sensing information on the flood risk and losses. Bull. of Flood Damage Evaluation Inform. System and Wetland Use. Beijing. p.38–46.

Chen Shupeng (Ed.). 1989. Index atlas of geographical information system of China. Vol. I. Flood risk forecasting and disaster countermeasures information system in the lower Yellow river. Science Press, Beijing, 106 pp.

Cheng Shupeng, Huang Xuan and Li Huiguo. 1990. Geoscience analysis of Landsat imagery—an image atlas of China. Bull. Resources & Environm. Inform. System. Beijing, p.29–35.

Dai Changda, Lei Liping, Hu Deyong and Wang Jiesheng. 1991. Detection of pine caterpillar pest by LANDSAT Thematic Mapper. Preprint. Remote Sensing Satellite Ground Station of Chinese Acad. Sci. 4 pp.

Grigoryev A. A. and Kondratyev K. Ya. 1993. Aerospace natural resources and ecological investigations in China. Issled. Zemli iz Kosmosa (Studying the Earth from Space). No. 2, p.119–123.

Hu Angang. 1996. "The Daily Yomiuri", November 1, 1996, p. 16A.

Jiang charts China's path to consolidate power of base. "The Daily Yomiuri", 150 October 1996, p.5.

Jin Zhang and Qi Tuan. 1990. Research and establishment of the nation-wide natural environment information system. Proc. Second Intern. Workshop on Geograph. Inform. System. Beijing. p.198–208.

Jing-shan Jiang and Shu-hu Cao. 1991. All weather flood monitoring and loss estimation by using space technique. Preprint. Space Sci. & Application Res. Centre. Acad. Sinica. Beijing, 10 pp.

Kennedy P. 1994. Preparing for the Twenty-First Century. Ch. 9: India and China. Vintage Books, New York, p.163–192.

Kondratyev K. Ya. 1991. On the development of environmental sciences in the Chinese People's Republic. Izvestiya of USSR Geograph. Soc., Vol. 123, No. 4, p.349–354.

Kondratyev K. Ya. 1993. The Second UN Conference on Environment and Development: some results and perspectives. Izvestiya of Russian Geograph. Soc., Vol. 125, No. 3, p.1–8.

Kondratyev K. Ya., Buznikov A. A. and Pokrovsky O. M. 1992. Global Ecology: Remote Sensing. Itogi Nauki i Tekhniki. "Atmosphere-Ocean, Programme Sections". Vol. 14, 310 pp.

Koptyug V. A. 1996. Shock discoveries behind the Chinese wall. "Soviet Russia", May 12, 1996, p.3.

Kulpin E. S 1990. Man and Nature in China. Moscow, Nauka Press, 246 pp.

Lan Yunchao and Yan Ken 1990. An application of aerial remote sensing and GIS for urban system analysis and planning. Proc. Second Intern. Workshop on Geograph. Inform. System. Beijing, p.400–406.

Liu Jiykan and Li Shuang. 1989. A study on the application of remote sensing technology to land resources investigation in Tibet Plateau. Proc. 3rd Pacific Basin Intern. Symp. on Advances in Space Sci. Technol. and its Applications. Los Angeles, USA. p.1–7.

Liu Xia, Zhang Shulin and Su Wensheng. 1991. Application of LANDSAT to the investigation of reed resources in the Dongting lake. Bull. of Flood Damage Evaluation Inform. System and Wetland Use. Beijing, p.124–128.

Lu Guonian. 1991. Planning for development and utilization of the beach land in Dongting lake. Bull, of Flood Damage Evaluation Inform. System and Wetland Use. Beijing, p.129–134.

Marchuk G. I. and Kondratyev K. Ya. 1992. Priorities of Global Ecology. Moscow, Nauka Press, 305 pp.

Ma Zongjin and Cao Xianglin. 1991. Applications of space techniques to earthquake hazards reduction in China. Preprint. Inst. of Geology, Beijing, p.1–15.

National Report of the People's Republic of China on Environment and Development. "China Environmental Science Press". 1992. Beijing, 119 pp.
Numbers game blurs state of China. "The Daily Yomiuri", 15 October 1996, p. 5.
Shu Shiwei. 1991. All-weather flood surveillance by SLR. Bull, of Flood Damage Evaluation Inform. System and Wetland Use. Beijing, p.4753.
Su Yabin and Jin Hongtao. 1991. The use of geographic information system Toolkit (GIST) for water resources study based on remote sensing technique. Bull, of Flood Damage Evaluation Inform. System and Wetland Use. Beijing, p.140–145.
Sun F. H. 1992. Man's impact on the environment in historical Fujian (Southeast China). In: Global 2000 Revisited Mankind's Impact on Spaceship Earth (Ed. by Hygh W. Ellsaesser). Paragon House, New York. —p.65–85.
Suy Disin. 1990. Ecological Problems of China. Moscow, Progress Publ., 230 pp.
The rapid graying of China. "Los Angeles Times World Report", 22 July 1996, p.9.
Xu Guanhua, Sun Siheng and Zhang Yugui. 1991. The afforestation in China and its monitoring by remote sensing data. Preprint. Natl. Lab. of Resources and Environm. Inform. System and Chinese Acad. of Forestry. p.16.
Zeng Zhiyuan and Pan Xianzhang. 1991. Applying LBV transformed image to the study of land use in the area of Jianghan plain and Dongting lake. Bull. of Flood Evaluation Inform. System and Wetland Use. Beijing, p.140–145.
Zhong Z. X. 1994. Analysis of the Chinese energy system: implications for future CO_2 emissions. Int. J. of Environ. and Pollution. Vol. 4, No. 3/4. p.181–198.
Zhou Wancun. 1990. Remote sensing investigation and appraisal of land use and capacity in the region of the Yangtze river. Gorge's Project. Remote Sensing Inform., No. 2, Beijing, p.43–44.
Zhu Zhenda and Wang Tao.1991. Desertification in China. Preprint. Inst. of Desert Research. Lanzhou, p.1–5.

5

The ecological situation in Russia

(K. S. Losev)

5.1 INTRODUCTION

At present, the national and regional ecological programmes, as well as the ecological legislation, are being worked out in Russia. This work raises a number of complications which are due to the practical development of the situation in a country where the process of transition from a centralized system of economic management to new market mechanisms of economic regulation is under way. Therefore, in the development of the ecological legislation, and various ecological programmes, it is necessary to take into account the transition process itself, and the realities of the future market mechanism, which is not perfect for the solution of ecological problems.

Ecological legislative acts, programmes and monitoring require an analysis of the realistic ecological situation in the country and its individual regions. Such assessments as, for instance, "severe ecological crisis", "ecological crisis" or "a possibility of a crisis", etc., require different approaches in the elaboration of ecological regulations and programmes, the organization of monitoring systems and modelling methods, the choice of strategies and techniques for the solution of ecological problems. That is why it is so important to estimate the state of the environment in Russia of today, and the potential for changing the present-day situation.

5.2 NATURAL RESOURCES

Russia occupies a territory of 17.1 million square kilometres. There is a great variety of landscapes and ecosystems in the country. Territories are organized in accordance with the laws of geographical zonality and altitudinal belt structure; the continental features range from sea shores to central regions of Siberia, However, natural systems are preserved mainly in the Arctic, in Eastern Siberia, in the northern part of the far east and in high-mountain areas.

The republic of Russia may be subdivided into three large regions: the European part with an area of 3470 thousand square kilometres, the Urals region with an area of 1865 thousand square kilometres, and Siberia and Far East with an area of 11 740 thousand square kilometres. These regions differ from each other with regard to their geographical conditions, the history of their settlement, the extent of their economic development, the structure of their population and the ecological situation.

The population of Russia, according to the data of a census taken in 1989, is 147 386 000, of which 108 419 000 (73.6%) are city dwellers and only 38 967 000 (26.4%) are peasants. Out of the total number of cities (1037), there are 13 cities with a population of 1 million people or more. The distribution of the population over the three abovementioned large regions is non-uniform: 95 022 million people (64.6%) live in the European part (69 745 000 in cities and 25 277 000 in villages); in the Urals, 20 279 000 people (13.7%) live in the Urals, (15 155 000 in cities and 5 124 000 in villages); 32 098 000 live in Siberia and the Far East, 23 856 000 (21.7%) in cities and 8 572 000 in villages.

The density of population in the European part is, on the average, 27 persons per square kilometre, in the Urals 10.7, and in Siberia and the Far East less than 3 persons per square kilometres, with an average density of 8.6 people per square kilometre. The major part of the population is concentrated along the southern border of the Russian republic, southwards from 60°N. To the north of this latitude there lies a territory which bears the name of "North". This is a large territory (11 million square kilometres). Its population is only 9.3 million people (out of which 6 million people populate its smaller part—the European North—with an area of 1.5 million square kilometers), and the population density is only 0.8 persons per square kilometre.

A territory is a resource, and not only a spatial resource; first of all, it is a resource in the form of a space occupied by biota, its natural communities whose functions provide a stable environment.

There are different estimates with regard to an amount of territory necessary to maintain the living conditions of a single person. According to some estimates for the temperate zone, for the normal existence of a single individual, about 100 m^2 of living and working space is required, the same amount for the infrastructure (roads, power lines, etc.), 12 thousand m^2 of pastures and hayfields, 4.6 thousand m^2 of agricultural fields and 700 m^2 of forest area necessary for the absorption of the released carbon dioxide and production of oxygen. The sum totals to 17.5 thousand m^2. Other estimates yield the figures of 1 to 2 ha. Proceeding from such estimates, the conclusion is drawn that the Earth is able to support about 5.7 billion people, taking account of the fact that 149 million square kilometres are unsuitable for living (glaciers, deserts, high-mountain areas, etc. (see Problems..., 1993).

Such estimates for Russia's territory make it possible to suppose the presence of certain territorial reserves that may be populated, thus assuming the possibility of population growth. However, the notion of a large territorial reserve in Russia is illusionary and only makes the basis for carelessness and irresponsibility. In actual fact, the territorial resources are restricted, not by the above-given figures, but by the conditions providing the possibility to maintain favourable living conditions as a result of the activity of natural communities of living organisms.

The amount of primary biological production provided by various communities of living organisms over the territory of Russia characterizes the biotic production ensuring environmental stability. The main contribution to this production is made by forests. Primary biological production over the whole country varies from 0.25 (tundra and deserts) to 3.75 kg/m^2 (deciduous forests). Recalculated per hectare, the average primary biological production over the Russia's territory is 20 t/ha annually; the figure for the European region is 25 t/ha, for the Urals 20 t/ha, and for Siberia and the Far East 15 t/ha.

The population is concentrated mostly in zones of the largest amounts of primary biological production; its density decreases sharply in areas of lower indices of the above. There are 230 tons of photosynthesis production per year per person over the Russian territory. The total photosynthetic mass production is 34 109 tons per year. Since for the creation of unit of photosynthetic production, the amount of moisture required is two orders of magnitude greater (usually 100–500 times larger) than that transpired by plants, it follows that every year, over Russia's territory, the biota transpires much more than 3400×10^9 m^3 of water or 3400 km^3, i.e. it regulates a major part of the continental water cycle. So, one can see that natural communities are a powerful regulator of the environment over the territory of Russia.

Water resources are one of the most important components of the environment; it is consumed in the largest quantities. The long-period annual mean run-off of Russian rivers is 4300 km^3 which constitutes almost 10% of the world river run-off. 22.4 thousand m^3 of water per year is the share of a single person in Russia. The amount of water provided for 1 km^2 of the territory is a little more than 250 thousand m^3 (on the average). 1000 km^3 of the annual river run-off falls on the European part of Russia, whereas a similar figure for Siberia and the Far East is about 3300 km^3; this shows that a dweller in Siberia or the Far East is much better provided with water than an individual in the European part of the country.

About 84% of the run-off of Russian rivers goes into the Arctic Ocean, another 8% goes into the Pacific; the remaining 8% is distributed between the Atlantic Ocean (3%) and the Caspian Sea (5%). The realistic exploitation power of all Russian rivers is about 7×10^5 MW.

Substantial amounts of fresh water are concentrated in Russian lakes. The largest and unique storage of the cleanest fresh water is Lake Baikal which encompasses 23 thousand km^3 of water, which constitutes over a half of the total world river run-off. The remaining part of fresh lake water is concentrated mainly in the Great European Lakes of Russia—Ladoga and Onega.

Potential resources of fresh subsurface water in the territory of Russia have been estimated at 229.3 km^3/year, out of which 60% falls on the European part, 25% on Western Siberia, and 15% on Eastern Siberia and the Far East. This kind of distribution shows that the underground waters of Siberia and the Far East have not been sufficiently explored, The exploitation resources of subsurface water constitute only 22 km^3/year, or 10% of the potential resources.

One more source of fresh water is mountain glaciers, whose area in Russia is about 3500 km^2. Major mountain glaciers are concentrated on the northern slopes of the Caucasus, the Altai, the Koryak ridge, the Suntar-Hayata mountains and Chersky ridge. The glaciation zone of the islands of the Russian Arctic is considerable, having an area of about 56 125 km^2.

The huge permafrost zone in Siberia and the Far East contain abundant reserves of subsurface ice—up to 19 thousand km^3. Ample amounts of swamp water are located in the north of the European part and in the taiga regions of Siberia and the Far East.

The major part of the 50 000-kilometre-long borderline of the former USSR belongs to Russia; this is mainly the coastlines of the Arctic and Pacific Oceans. The Atlantic coastal zone of Russia is limited to the belt of Black Sea resorts in the Krasnodar region, and the area along the shores of the Gulf of Finland in the estuary of the Neva river. The exit to the Caspian Sea is in the region stretching from the Volga river estuary to the northern foothills of the Eastern Caucasus.

The sea is a complex resource (including a resource of water). The internal and territorial marine waters of Russia have an area of about 0.8 million km^2. The shelf out to the 200-m isobath occupies about 1.2 million km^2, and the economic zone to 4 million km^2.

The soil resources of Russia are a zonal system outside mountains, and a vertical belt system in the mountains. Soil is a layer where the decomposition of primary photosynthetic production takes place. Therefore, its role in maintaining stable living conditions is extremely important. The basic soils over the plain area of the territory of Russia are situated in latitudinal zones and are of the following types:

- Tundra soils with a slight peaty humus horizon. Bog gley soils are formed in areas of excessive wetness.
- Forest podzolic and bog soils which are replaced by soddy podzolic soils in areas with a thick grass cover, and by bog peaty gley soils in moist areas. The forest soils occupy the major part of Russia's territory—about 1 billion hectares.
- Grey wooded soils and leached orthic black soils are characteristic of the wooded steppe. Today woods in this zone have practically all been felled.
- Chernozems and chestnut soils are characteristic of steppes and semi-deserts. Sierozems (grey soils) in Russia can be found only in the Caspian lowlands.

Considering the fact that the tundra occupies from 200 to 300 thousand ha, and the wooded area about 1 billion ha, the share of all other types of soils is about 0.4–0.5 million ha, i.e. the land suitable for efficient agricultural development constitutes one-third of the entire territory, at most. Taking account of regions of deserts and semi-deserts, the amount of land appropriate for effective agricultural use is still less. Here climate remains an important factor, and climate in the depth of Russia's territory is sharply continental. The climatic factors restrict the propagation of many agricultural crops to regions of the southern part of the country.

Over the territory of Russia, the basic mass of photosynthetic production is yielded by forests which constitute about one-fifth of the forest reserve of the world, and whose wood reserve, 75 billion m^3, is even more than one-fifth of the global

Table 5.1. Prospected resources of fossil fuel in Russia (rounded-off data for 1987)

Oil (million tons)	Gas (billion tons)	Coal (million tons)	Brown coal (million tons)	Uranium
6000	31 000	100 000	800 000	Substantial amount

abundance. Over 25% of wooded territory is located in the European part and in the Urals; about 75% in Siberia and the Far East. Three-quarters of Russian forests are coniferous woods; the remaining part is formed by deciduous forests and other kinds of tree and bush.

Russian forests yield annually up to 2 million tons of cedar nuts, 2 million tons of red bilberries, 1.5 million tons of blueberries, 0.3 million tons of cranberries and about 0.8 million tons of mushrooms (Ecological..., 1990).

Here are some estimates for other renewable resources: solar radiation power (absorption by the earth's surface over Russia's territory), less than 2×10^2 TW; wind power, no more than 4 TW; geothermal power, 100 million tons of equivalent fuel. The utilization of solar energy is limited by the territory to the south of 52°N.

Russia possesses considerable reserves of non-renewable resources. This is primarily fossil fuel (Table 5.1). The main reserves of oil and gas in Russia are concentrated in Western Siberia (more than two-thirds), and coal reserves there constitute almost three-quarters of the reserves of the whole country, as well as the reserves of hydropower resources. On the continental shelf, the reserves of oil were estimated at 0.5 billion tons, and those of gas as 4.5 trillion m^3 (for the year 1988).

During recent years, the growth rate of prospected reserves of oil and coal has been low; the growth rate was sufficient only for gas.

Iron ore is mainly deposited in the following regions: in the European part of Russia, the Kursk magnetic anomaly with a 42.2 billion-ton-reserve of ore, and the Kola-Karelia reserve of 3 billion tons; in the Urals, 15 billion tons; in Siberia and the Far East, 10.1 billion tons. On the whole, the iron ore reserves of Russia constitute 30% of the world abundance.

Russia possesses considerable reserves of the ores of heavy and light non-ferrous metals and of such chemical raw materials as, for instance, potassium and common salts, and phosphorus ores. Various building materials are very abundant in Russia, and their output production reaches 1 billion tons per year.

5.3 CONSUMPTION OF NATURAL RESOURCES

11.5 ha is an individual's share of territory in Russia, which is much more than the global average (3 ha). Out of these 11.5 ha about 60% are wooded areas, 20%—tundra, and the remaining 20% partially wooded steppes, steppes and semi-deserts. The territorial distribution per capita is most non-uniform with regard to large

Table 5.2. Land areas undeveloped economically

	Africa	Asia	North America	South America	Europe	Australia	Russia
ha	918	373	900	422	19	229	752
%	31	14	42	24	4	33	34

Table 5.3. Utilization of Russia's territory for different kinds of economic activity and as a result of anthropogenic pressure

Kind of activity	Agriculture	Settlements and infrastructure	Test sites with waste and disposal areas	Water resources	Anthropogenic pressure	Total
Total (million ha)	224.7	3.5	16.5	5	32	280.7
Per capita	1.5	—	—	—	—	1.89

regions: in the European part it is 3.65 ha (which is close to the global average), in the Urals 9.3 ha; in Siberia and the Far East almost 37 ha.

During the 1980s, with the help of 65 maps compiled on the basis of aerospace data for scales 1 : 1 000 000 and 1 : 2 000 000, an evaluation of the land area which has not been economically developed was made (Table 5.2). On the whole, the world territory undeveloped economically is 5089 million ha, or 39%. In Russia, this index is within 50–60% of the total territory. This is characteristic of a small number of countries; at temperate latitudes, a similar index is characteristic for a single country only, viz. Canada.

The most significant portion of Russia's territory which is economically developed, is occupied by agricultural lands. Their share is 224.7 million ha, or 13.1%. Among these, arable land occupies 132.8 million ha (7.8% of Russia's territory), hayfields 25.4 million ha (1.5%) and pastures 64.8 million ha (3.8%). Table 5.3 presents information about the utilization of Russia's territory for different kinds of economic activity.

If one considers that 11.5 ha of territory falls to a single person in Russia, and that the mean primary biological production is 20 t/ha per year, one can calculate that 230 tons of the production is the share of an individual. Since every person consumes completely the biological production obtained from a territory of 1.89 ha, the value of per capita consumption is 37.8 tons, or 17.7% of the total per capita primary biological production.

Proceeding from the value of per capita consumption of primary production, the shares of consumed primary production for the three large regions were recalculated: approximately 40% of primary biological production is consumed in the European part, about 20% in the Urals, and 10% in Siberia and the Far East. This shows that natural communities have already lost their sustainability, and the disturbed biota uses its power not for the restoration of its sustainability, but for the acceleration of the process of decay and imbalance. The decaying biota emits at least the following pollutants: carbon dioxide, methane, and soluble compounds of nitrogen and phosphorus. Therefore, the growing amount of nitrates and phosphates in water bodies reflects not only the broad employment of fertilizers and detergents, but also the decay of the biota.

From the viewpoint of the index of consumption of primary biological production, Russia is in the state of a severe ecological crisis, which is especially profound in its European part. Over huge territories, ecosystems have lost the ability of self-purification; the production of primary biological products has sharply dropped; a rapid decay of ecological niches is taking place, the genomes of many kinds of living organisms are being destroyed, and quite a number of these have already disappeared or are disappearing; technogenic desertification is developing, and the decaying biota becomes itself a source of environmental pollution. The level of consumption of other renewable resources is very high.

Water is a product which is consumed most of all from the viewpoint of its mass, in Russia, and over the whole world. In 1989 the total amount of 117.3 km^3 of water (117.3×10^9 tons) has been extracted from Russian waterbodies, which corresponds to approximately 800 m^3 (800 tons of water per capita). However, this is not the whole amount of water consumed for economic activity and for drinking, since a considerable portion of this water (168.5 km^3) is used repeatedly after recycling. Therefore, the actual figure of per capita water consumption per year is nearly 2000 m^3 (2000 tons). This corresponds to about 3 m^3 (tons) per day, or 1000 litres per hour.

In 1989, from surface land waters, 96.6 km^3 of water has been extracted, from underground sources 14.5 km^3, and from marine waterbodies 6.1 km^3. The basic water mass (water intake) was used for industrial purposes (59.4 km^3); for, agricultural needs 39.2 km^3 was used, for transport 4.2 km^3, and for municipal needs 14.0 km^3. On the whole, water intake from water sources grows permanently, and its growth rate during the last 10 years, on the average, was 0.7 km^3/year.

Water intake from rivers is small in relation to the average annual long-period river run-off: it constitutes only 2.3% of the run-off. However, water consumption is not limited by water intake alone. Since ancient times, water arteries have been used as transport pathways. The mean distance of passenger and freight traffic over Russian rivers is 375 km, and freight turnover there reaches 200 billion t/km. Small rivers, especially in the North, in Siberia and in the Far East, are broadly used for timber rafting, including loose floating. Surface waters are particularly broadly used for hydropower purposes. 11 hydroelectric power stations have been built on the Volga and Kama rivers only; their total power is 13.5 MW. As a result of creating a cascade of large water reservoirs on the Volga, many irreversible changes have taken place on the river. Finally, surface land waters, as well as underground and sea water, are broadly utilized for the dilution and purification of polluted water.

Soil resources are employed in Russia to a greater extent than in many other countries: in Russia the per capita amount of arable land is 1 ha, in the USA 0.7 ha, in Hungary 0.5 ha, in the Czech Republic or France 0.3 ha.

Table 5.4. Growth rate of crops per ton of fertilizer in various countries for 1986

Country	USA	China	India	USSR
Crops in tons per ton of fertilizer	18	18	16	8

Russia possesses a considerable amount of irrigated land (6 million ha, mainly in the Northern Caucasus and the Volga region), and ample drained land (6.3 million ha, chiefly in the non-chernozem zone of the country). However, this land is utilized inefficiently owing to the low quality of reclamation systems: hundreds of thousands of hectares of improved land are not used any longer; at least, 20–25% of this land needs complex reconstruction measures.

The entire volume of ploughland is used in Russia very ineffectively. The Czech Republic or France, where the per capita share of arable land is three times lower, obtain 1.3–1.6 times more agricultural production than Russia. The efficiency of using mineral fertilizers in the former USSR and in Russia was and remains very low (Table 5.4).

The quality of agricultural land, especially arable land, deteriorates continuously. During the recent 15 years, the land area which is subject to water and wind erosion, has increased by 7.5 million ha (250 thousand ha/year) over the Russian territory and reached 60 million ha, out of which 40 million ha are subject to water erosion.

In the non-chernozem zone, soils are severely lacking in phosphorus and potassium. The content of humus in the ploughland decreases: it becomes 0.64 ton/year less per hectare. Most of examined arable lands are characterized by a medium, low or very low concentration of humus: such are 87.8% of the examined lands. This shows that agricultural lands are rapidly degrading.

Soils are the habitation medium for heterotrophs. The latter are represented by microorganisms and fungi; consuming organic matter, they restore its constituents to primary non-organic forms and thus close the biogeochemical cycle of nutrients. That is why one may state that over huge territories in Russia, the balanced process of mineralization of organic matter is disturbed.

The soil bacteria and fungi play the role of filters which regulate the flow of gases coming from the bowels of the Earth. Over thousands of years, a certain balance has been established between their "filtering capability" and the gas flux. At present, this balance has been disturbed because of the destruction of the soil cover over large territories.

One of the most important renewable resources in Russia is forest. Wood cutting is carried out over an area of 2 million hectares. Forest planting in 1989 constituted 0.6 million ha, and protection of forest to facilitate natural restoration embraced another 1.3 million ha. On the whole, in 1989 the gap between wood cutting and restoration was 1 million ha.

During the last 10 years, forest planting in Russia has been reduced by 22 thousand ha annually; however, forest protection to facilitate its natural restoration was enhanced. At the same time, the volume of wood cutting during the last five years increased annually by 7 million m^3, or by 2%. It should also be noted that the number and extension of forest fires in Russian forests have also increased: during the last 4 years, the area of burnt forests increased by a factor of 3 and amounted to 1.6 million ha, out of which 1 million ha of burnt forest is the result of anthropogenic impact.

The consumption of timber leads to a reduction of forest areas, and to a substantial consumption of both the primary production, and the basic biomass producing the above and thus maintaining the biogeochemical balance. The consumption of forest resources is also most ineffective. When wood is felled, branches, bark, and pine-needles are wasted; tree stumps and the entire root system are left behind, and practically all young growth perishes. In wood cutting and timber transportation, and in the process of rafting, about 35 million m^3 of forest is wasted or simply lost. The processing of timber is accompanied by other losses. Here are some data for Russia: out of 1 thousand m^3 of wood, 6.1 tons of plywood, 18.9 tons of chip-board, and 28 tons of paper and cardboard are produced. For Finland these figures are, respectively: 13.5, 13.5 and 190 tons. In the European–Urals part of the country, where about 25% of all forests are situated, approximately 200 million m^3 of wood is produced, i.e. nearly two-thirds of all logging (State of Natural..., 1990; RSFSR..., 1990).

Larger and larger forest areas become subject to pollutants emitted into the air by industrial enterprises; among these are sulphur dioxide, nitrogen oxides, hydrogen fluoride, etc.

Every 2 million ha of wood cutting implies the destruction of about 1.4 billion tons of biological mass.

Thus, one can see that low-efficiency consumption of basic renewable resources increases the consumption of primary biological production over the territory of the country and aggravates the ecological crisis; the worst deformation takes place in the European part and the Urals region.

The mining industry of Russia produces up to 18% of the world production of mineral raw materials; previously about 40% of all capital investment was expended in this area. Russia's annual mining output is 15 billion tons of rock mass. Raw materials constitute a significant portion of Russia's exports—up to 76%. Within the borders of the former USSR, Russia was also the supplier of raw materials. 53 tons of raw materials per year is the per capita output; these raw materials are partially exported, but mainly are processed, using 2000 tons of water per capita per year and 1.5×10^3 W of power. The final product is only 2–5 tons; the rest is waste. Thus mineral raw materials are used inefficiently, with great losses of the basic component, and not in a complex set (Ecological..., 1990; Economic..., 1990; State of Natural..., 1989).

5.4 NECESSITY FOR CHANGING STEREOTYPES IN THE SPHERE OF CONSUMPTION OF NATURAL RESOURCES

The pattern of consumption and use of the territory, primary biological production, renewable and non-renewable resources makes it possible to conclude that consumption stereotypes, as they exist today and are advertised by mass media and even scientific literature, should be changed.

The statement that, since Russia is a huge country, there still remain large reserves for population and economic activity, is wrong: the major part of non-developed and non-fully-developed territories (60–65%) are located in regions of severe climate, where permafrost is common and the photosynthesis output is very low. The zone of relatively favourable climatic conditions, with the highest level of photosynthetic production is already populated, occupied with tens of thousands of economic objects and fully developed. So, one can see that, in spite of the presence of a large physical space, there are ecological limitations to the territory. Therefore, proposals to move industry eastwards, since the European regions of Russia are industrially overcrowded and comparable to Germany or Japan, are actually very dangerous in the present conditions of severe ecological crisis. It should be realized that, in spite of the physical space available in Russia, the ecological space has been completely exhausted.

Another stable stereotype, existing perhaps for as long as mankind itself, is the idea of the necessity of population growth. Many scientific papers and especially mass media automatically repeat the old-fashioned demographic concept that population must increase; such notions are the result of ignorance where the ecological and economic facets of the matter are concerned.

The demographic concept of encouraging population growth was always supported by the state itself in the form of additional taxes for childlessness or awards to mothers of large families. Meanwhile, it is a known fact that in order to maintain the same standard of living with a 1% increment of population, the gross national product (GNP) must be increased by 2%. Without such an increase in GNP, the new-born babies will grow to be poorer, less healthy and less happy. If a society does try to attain this or a greater population increment, ecological changes that will increase in the same proportion will make these children less healthy and will cause genetic violations, and in the foreseeable future biosphere destruction and resource exhaustion will bring about poverty, starvation and downfall for humankind.

The current situation in Russia shows only too well that the regions of tension and instability are those where the birth rate and population density are high, and that the greatest problems arise in the larger families.

In many cases, the well-known notion is supported that it is necessary to expand the utilization of renewable resources. Although the grandiose plans of turning the rivers of the north of the European part of Russia and Siberia have been abandoned, there are plans for more intense consumption of water resources, especially over the European territory of Russia, under the pretext of a growing population and industrial and agricultural potential. Since the population increases, one may think that this is the only possible way out. However, the deficit of water resources is, and will for a long time be, caused by their inefficient utilization, the lack of advanced technologies, excessive watering and the low quality of amelioration systems, and, finally, the pollution of water sources.

One can find mention of arable land reduction as a negative event both in scientific literature, and in the mass media. However, the area of ploughland and other agricultural land objects is indeed too large. From the point of view of the amount of arable land per capita, Russia occupies one of the foremost places. But the effectiveness of using this arable land is so low that Russia has become one of the biggest importers of grain in the world. The enormous area of arable land requires a great amount of machinery for its treatment, and a great number of work-hands for harvesting. In fact, the area of ploughland should be sharply reduced, by a factor of 2 or even 3, and simultaneously the efficiency of land-use should be increased. It is these huge arable land areas that form one of the factors of land-use inefficiency, although the decisive factor there is the form of ownership.

Forest, as a most important renewable resource, should also be mentioned here. There is a notion that, since Russia suffers from having insufficient quantities of wood, wood cutting should be expanded (the deficit in wood exceeded 50 million m^3 by 1989). However, the loss of wood due to the lack of proper selection of trees to be cut down accounts for this deficit. As to the total loss of wood, it exceeds the deficit by a factor of 2 or 3. Wood consumption can be efficiently reduced by using advanced technologies which make it possible to obtain a good deal more final production per unit of raw wood. However, this is only part of the whole problem. The thing is that wood cutting is followed by an abrupt disturbance of the nutrient cycle: as the natural vegetation cover is restored, the openness of the cycle decreases. To bring a forest back to its undisturbed state, as much as 300 years is required. However, repeated wood cutting takes place, as a rule, every 50 years, after ripe wood is formed, which breaks the process of restoration of primary forest with a closed matter cycle. Therefore, it would be reasonable to increase the intervals between cuttings by a factor of 6, i.e. to reduce the number of cuttings as much as necessary, taking account of the expansion of the area of used forests.

With regard to non-renewable resources, the stereotype is very similar to that relevant to the problem of Russia's territory: the notion of the abundant mineral resources in the country and strong hopes concerning geological prospecting and finding new deposits. However, a number of minerals are decreasing in quantity; a more serious factor is that both the prospecting for, and the mining of, mineral reserves become more and more expensive. For instance, in 1988 the mining process became twice as expensive in comparison with 1975. The reduction of reserves of non-renewable resources, which we take away from future generations, is not taken into account in economic calculations: hence the illusion of their being free. Of special concern are the resources of oil and gas which are exported in huge amounts and at extremely low prices.

One more stereotype should be mentioned: one relevant to the disappearance or great reduction in number of numerous species of living organisms over Russia's territory. There are attempts to save these organisms and list them in the "red book". But this cannot stop their disappearance, because it is first of all due to the fact that man consumes a lot more biological primary production than he really needs, and thus deprives other organisms of food and thereby condemns them to death. Another cause of the disappearance of living organisms is that man, by destroying ecological systems and changing the environment, destroys also the ecological niches of organisms, which leads to the decomposition of the genotype of species and stops the operation of powerful mechanisms in their preservation.

Giving up the old stereotypes would be the best way to reinforce ecologically oriented thinking in every field of human activity—from the propagation of education and scientific knowledge to the elaboration of laws, regulations and programmes to solve the problems of the environment and of using the riches of nature.

5.5 ENVIRONMENTAL POLLUTION

Man has created a system of synthesis and decomposition of substances which is parallel to the biota. Up to 10 million different substances are synthesized all over the world, 50 000 of which are produced on a large scale, and 5000 on a very large scale. This is approximately the number of substances produced in Russia. Pollutants enter the environment in accordance with the technological processes of the producer—the pollutant supplier—except in emergency cases. After entering the environment, the presence and dynamics of pollutants are determined by geographical conditions and geophysical laws. Therefore, the extent of pollution is largely determined by local and regional features of the territory (Losev, 1989).

Pollutants usually are gases, aerosols, dust, liquids, matter dissolved or suspended in water, and solid substances. Gases, aerosols and dust are usually emitted into the atmosphere; liquids and substances dissolved in water enter waterbodies; and solid (and sometimes liquid) pollutants are accumulated on the Earth's surface or in its natural or artificial voids.

Atmospheric pollutants are carried and transported by air fluxes. For the major part of Russia, west–east transport is characteristic, i.e. western winds prevail, and therefore, the flux of atmospheric pollutants moves mainly from west to east.

Water pollutants from the basic industrial regions of the European part of Russia are transported, chiefly by rivers, to southern regions—to the Caspian and Black Seas, and partially westwards—to the Baltic Sea—and northwards—to the Arctic Ocean. From the eastern part of the Urals region, from Western and Eastern Siberia the flux of pollutants is transported by rivers northwards, to the Arctic Ocean. From a considerable part of the Far East, pollutants follow water courses eastwards, to the Pacific Ocean.

Solid waste is concentrated around mining and industrial centres.

Tables 5.5–5.10 present data on atmospheric air pollution over Russia's territory, including a comparison with some other countries; they also show the dynamics and loadings as calculated per capita (Ecological..., 1990; State of Natural..., 1989, 1990).

The geographical distribution of the pollutant loading per capita (for city dwellers) shows the structure of the location and concentration of industrial enterprises in industrial zones. It becomes obvious that, instead of plans of transferring industrial enterprises from west to east, plans should be made concerning restructuring and relocating industry in the eastern part of Russia: in

Table 5.5. Carbon emissions by various countries for 1960 and 1987

Country	Carbon emissions (million tons)		Carbon emissions per dollar GNP (tons)		Carbon emissions per capita (tons)	
	1960	1987	1960	1987	1960	1987
USA	791	1224	420	276	4.38	5.03
USSR	396	1035	416	436	1.85	3.68
China	215	594	—	2024	0.33	0.56
Germany[a]	149	182	410	223	2.68	2.98
Great Britain	161	156	430	224	3.05	2.73

[a] Data for West Germany only.

Table 5.6. World production of clorofluorocarbons in 1988

	EEC countries	USA	Japan	USSR	Other countries	Total
Thousand tons	520	455	130	123	71.5	11 995
Percentage	40	35	10	9.5	5.5	100

Table 5.7. Emissions of individual pollutants to the atmosphere in Russia (million tons)

Year	1981	1985	1986	1987	1988	1989
Solid substances	9.7	9.1	8.8	8.7	8.3	7.7
Anhydride sulphide	12.1	11.8	11.2	11.1	10.5	9.8
Carbon monoxide	9.1	8.5	8.7	8.3	8.9	8.3
Nitrogen oxides	2.4	2.5	2.7	2.9	2.8	2.8
Hydrocarbons	6.8	6.7	7.0	6.7	6.6	6.5
Total emission	41	42.4	40.9	39.9	38.5	36.1

the Urals region and beyond. The same data indicate possible ways of protection against atmospheric pollution.

There are 99 cities over the Russian territory, whose ecological situation has been recognized as unfavourable. Among these, 56 are located in the European part, 28 in Siberia and the Far East, and 15 in the Urals region. In the list of 56 Russian cities having the highest level of pollution, there are 25 cities in Siberia and the Far East, 8 in the Urals, and 23 in the European part of Russia.

The main pollution suppliers over Russia's territory are enterprises of the following industrial complexes: metallurgy, 27.3% of pollution; energy, 21.1%; oil and gas industry and petroleum chemistry, 19.8%; all together these industrial

Table 5.8. Per capita pollutant emissions to the atmosphere in Russia, in 1989 (kg/year)

Solid substances	Anhydride sulphide	Nitrogen oxides	Hydrocarbons	Carbon monoxide	Total
52	67	19	44	56	238

Table 5.9. Pollutant emissions by large regions per capita

Regions	European part	Urals region	Siberia and Far East
Total emissions (million tons)	14.4	8.3	13.0
Per capita emissions (kg)	150	415	400

Table 5.10. Per capita loading of atmospheric pollutants in cities of large regions for 1989 (kg/year)

Regions	Russia	European part	Urals region	Siberia and Far East
Loading	324	195	550	560
Solid substances	713	91	40	119
Anhydride sulphide	90	55	140	165
Nitrogen oxides	26	19	46	34
Hydrocarbons	60	36	60	123
Carbon monoxide	77	46	159	119

complexes are responsible for two-thirds of the total atmospheric pollution (Ecological..., 1990).

Pollutants presented in Tables 5.7 and 5.10 are only a part of contaminants emitted to the atmosphere. The general-form index "solid substances" involves, apart from soot, also lead, cadmium, mercury and other toxic metals, and hydrocarbons of these compounds are specific for each city or industrial area.

Regional features of the background pollution of Russia's territory correspond to the distribution of population and industry. The heaviest background pollution is observed in the European part of the country, where, with respect to some indices (e.g., sulphur dioxide), it is close to the Western European level and is approaching the maximum permissible concentrations; in relation to lead, its concentration in Russia is even somewhat higher. With regard to the same indices, the background atmospheric pollution in Siberia and the Far East is, as a rule, an order of magnitude lower.

Acid rains are not very significant for the major part of Russia's territory; pH of melted waters is usually 5.5–6.0. Acid rains occur mostly in the north-west of Russia, in Karelia and the Kola peninsula.

An essential role in the formation of the background pollution belongs to transboundary transport. The western transport brings atmospheric pollutants to the Russia's territory from Western and Eastern Europe, the Baltic region, the Ukraine and Belarus. Thus, for instance, out of every 1.04 tons of sulphur which falls annually upon an area of 1 km^2 of the European territory of Russia, 38% is sulphur brought from the west, across the state border, by transboundary transport. Then, 61% of sulphur emitted to the atmosphere in the European part of Russia is displaced eastwards, to the Urals and Siberia, where the mean value of sulphur emission per year is 0.59 t/km^2 (State of Natural..., 1990).

Characterizing atmospheric pollution in Russia in general terms, one should emphasize that the level of the background atmospheric pollution in the most densely populated regions of the country is approximately the same as in Western Europe and the USA. About one half of Russians—75 million people—live in conditions of a strong atmospheric pollution when one or another pollutant concentration is above the maximum permissible level, by a factor of 1–10, 100 or even a few hundred.

The amount of sewage entering Russian waterbodies reached in 1989 the figure of 76.3 km^3/year. The growth rate of sewage constituted, on the average, 1 km^3 annually during the recent decade. According to official data, 35.5%, or 27.15 km^3 of this sewage is untreated sewage.

Information about water pollution is not at all precise. The fact that this information has been distorted where sewage amounts are concerned has become recognized. The estimates of sewage should be revised.

Table 5.11 presents a list of branches of economy which are responsible for 95% of sewage; three of them (energy production, agriculture and municipal economy) are responsible for 80% of the entire sewage (Ecological..., 1990).

Within this list of main water polluters, perhaps, only part of the sewage created by power engineering may be regarded as non-polluted, since the basic amount of water there is used for cooling purposes. This is a water volume of about 32 km^3. The rest of sewage is more or less contaminated; this means that in actual fact, the amount of polluted sewage waters over Russia's territory constitutes not 27, but 55 km^2.

In 1989 water treatment systems were supplied with 22.1 km^3 of water. If one considers the fact that a quarter of such systems are still low-efficiency mechanical treatment systems, and that out of the inspected systems, over one-half operated

Table 5.11. Sewage by discharge by various branches of economy in Russia (km^3)

Branch of economy:	Agriculture	Municipal services	Power engineering	Timber and paper	Ferrous/ non-ferrous metallurgy		Oil-extracting industry	Petroleum chemistry	Chemical industry
Sewage amount:	14.5	12.4	33.9	3.1	1.5	2.2	0.2	1.6	3.1

Table 5.12. Discharge of pollutants with sewage waters in 1989

Substance	Sulphates (million tons)	Chlorides (million tons)	Phosphorus total (thousand tons)	Nitrogen total (thousand tons)	Ammonia nitrogen (thousand tons)	Oil products (thousand tons)
1989	11.5	8.5	52	142	176	29

Substance	Nitrates (thousand tons)	Surface-active substances (thousand	Fat, oil (thousand tons)	Iron (thousand tons)	Heavy metals thousand tons)
1989	76	12.6	45	34	2.5

ineffectively, one may conclude that, out of the whole water volume, less than one half was adequately treated (the usual extent of purification does not exceed 80%). Thus, no less than 45 km^2 of polluted water, which requires a 10–100-fold dilution, were discharged into natural waterbodies, i. e. from 450 to 4500 km^3/year. The latter value exceeds the long-period annual mean run-off of Russian rivers (4300 km^3) (Ecological..., 1990).

Dry and wet deposition of pollutants from the atmosphere becomes a more and more significant factor of waterbody pollution. For example, the density of sulphur deposition over Russia's territory varies from 2 to 0.25 t/km^2/year, and that of ammonia nitrogen is 0.3 t/km^2, or 5.5 million tonnes. A large amount of pollutants is washed away from agricultural fields: the wash-out of pesticides and fertilizers reaches 30–50%. The total number of 20 million tons of organic matter is discharged to Russian waterbodies with sewage waters (see Table 5.12) (Ecological..., 1990).

The annual per capita share of sewage waters in Russia is 520 m^3, out of which 370 m^3 are polluted. This volume of water contains approximately 170 kg of pollutants, mainly, chemical compounds, dangerous organic compounds and heavy metals. The processes of decomposition of some compounds, and the formation of others, sometimes more toxic, take place in water.

Only one-quarter of the water resources of Russia is concentrated in its European part and in the Urals; however, three-quarters of the entire water intake for industry and agriculture is taken there, and it is there that 75% of sewage is discharged into water reservoirs. The Volga basin rivers are especially heavily polluted. The water intake from the Volga is 36 km^3/year, i.e. 30% of the total amount of water intake in Russia, and the volume of sewage is 23 km^3/year, i.e. a little less than one-third of the total volume of sewage.

On the whole, in the European part of Russia, all large and medium-size rivers are polluted throughout the year. The rivers of the Urals region are strongly polluted too. In Siberia and the Far East, the Ob' river basin is heavily polluted throughout the year; the Amur river is also heavily polluted due to large amounts of pollutants washed from the Chinese territory. All these rivers have practically lost the capability of self-purification, and are polluted along their entire course. Lakes are also heavily polluted, among them the largest in Europe, Lakes Ladoga and Onega. The situation with the Azov Sea and the Gulf of Finland is critical.

Along the Pacific coast, permanent pollution of coastal waters, exceeding permissible limits, especially in the vicinity of large ports, is observed. This is due to the fact that most settlements, including large cities as Vladivostok, Magadan, Petropavlovsk-on-Kamchatka, have no water treatment systems.

Also polluted are underground waters over the Russian territory, particularly in its European part. About a thousand polluted regions (224 of which are large ones) have been detected. Large polluted regions are close to industrial centres; other, smaller regions are polluted by agricultural and municipal sewage. The largest polluted regions can be found in Bashkiria, and in the Rostov, Voronezh and Tambov regions.

Many substances, even in insignificant concentrations, are dangerous for man, whose daily water consumption amounts to 2–2.5 litres. Dioxin may serve as an example (it has been found in drinking water in some places, including the cities Ufa and St. Petersburg). In water, pollutants react with each other and with natural substances; as a result, more toxic substances may be produced. For instance, acid rains originate soluble aluminium compounds which are very toxic for fish.

In waterbodies, pollutants, passing through tropic chains, are gradually accumulated in the organisms of animals. Therefore, even though a pollutant may be in an extremely low concentration, its content in the body of fish or aquatic birds can be very large. Finally, pollutants in waterbodies are concentrated in bottom depositions which can become a secondary source of pollution long after the discharge of the contaminants has stopped.

Examination of soil pollution is accompanied by certain difficulties. They are due to different ways of movement of toxicants to man, which is caused by the possibilities of formation of volatile, soluble and insoluble forms, and the extent of assimilation of the toxicants by the biota and by man. A total index of the danger of soil pollution by a complex of metals has been developed, which is defined by a sum of the coefficients of metal concentrations determined in the evaluation of the pollution. The concentration coefficient is the ratio between the content of metal in the soil and its background content. The background content is usually estimated from samples taken at a distance of over 20 km from the source of pollution (*Annual Bulletin*, 1991). Table 5.13 presents an approximate scale of the danger of soil pollution.

During 15 years of investigations, three cities in Russia were singled out which, according to the total pollution index for soils, for the city itself and and area of the 5-km radius around it, were considered the most dangerous. These are Monchegorsk (the Kola peninsula), Revda (the Urals region) and Belovo (the Kemerovo region). If one uses a different criterion in one's assessment—that of maximum stability of pollution throughout the year—the number of such dangerous cities will increase to ten. So, out of the total number of examined Russian cities, 5% are considered extremely dangerous according to the mean value of the sum of soil pollution by heavy metals, and 14% according to stable maximum values of pollution. These are

Table 5.13. Approximate assessment scale of the danger of soil pollution

Degree of pollution	Total index	Indices of human health in polluted settlements
Admissible	less than 16	The lowest level of morbidity among children, minimum frequency of functional pathologies
Moderately dangerous	16–32	Increasing number of general sick rate
Dangerous	32–128	Increasing number of general sick rate, the number of children who are often subject to sickness, children with chronic diseases, disturbances of the functional state of the cardiovascular system
Extremely dangerous	more than 128	Increasing number of child sick rate, disturbances of reproductive function in women (increasing number of toxicoses in pregnant women, premature birth cases, stillborn babies, hypotrophy in new-born babies, etc.)

mainly centres of non-ferrous metallurgy. The cities that belong to the group of "extremely dangerous" according to maximum pollution indices, can be referred to as "dangerous" according to averaged values. Another seven cities are considered moderately dangerous from the viewpoint of the level of soil contamination. Thus, over one in five of Russian cities examined may be considered dangerously polluted, mainly with regard to the level of concentration of heavy metals, without taking account of other harmful toxicants (Table 5.14).

The following Russian cities where non-ferrous metallurgy enterprises are located and where maximum values of toxic substances of the first class of dangerousness reached, by 1989, five maximum permissible concentrations (MPC), or exceeded the background values by a factor of 10 where concentrations of lead, cadmium, arsenic and zinc are concerned: Belovo, Vladikavkaz, Dalnegorsk, Zyryanovsk, Kirovograd, Konstantinovka, Leninogorsk, Mednogorsk, Monchegorsk, Novokuznetsk, Novosibirsk, Revda, Rudnava Pristan' and smaller settlements Glubokoye and Yaroslavsky.

Toxicants of the second class of dangerousness contaminate mostly cities with ferrous metallurgy enterprises and power engineering plants (coal and heat power stations), where, as a rule, MPC and background levels of pollution are exceeded by a factor of several units (sometimes by an order of magnitude). This mainly concerns such pollutants as manganese and vanadium.

Sources of the first-class toxicant, fluorine, are aluminium plants. For example, in

Volgograd the source of fluorine pollution is an aluminium plant which emits 710 tons of fluorine compounds every year. The average concentration of fluorine in this city is at the level of 3 MPC. Soil contamination with fluorine has been detected at a distance of 20 km from the source. The concentration of water-soluble fluorine at a distance of 0.5–20 km varies from 120 to 20 ppm, i.e. from 12 to 2 MPC; the concentration of its total content varies from 280 to 85 ppm. In the radius of 5 km from the pollution source, the content of water-soluble fluorine is, on the average, 8 MPC. A similar situation obtains with aluminium plants in Bratsk and Irkutsk.

Apart from toxic industrial pollution caused by heavy metals and other toxicants, an agriculture-induced contamination of soils takes place. Especially harmful is the contamination with pesticides which have been more and more broadly used, beginning in 1950, to combat weeds, insects, etc.

125 different pesticides and 16 mixed preparations are being at present used in Russia, out of which 7% are highly toxic, 46% are of medium toxicity (22% the second class of dangerousness, and 24% the third class), 37% are low-toxic, and another 10% have not been examined for toxicity. Some of the estimates cannot be considered reliable enough. The volume of consumption of pesticides in 1989 was 150 thousand tons (with the maximum consumption, 197 thousand tons, in 1987), and the supplies of pesticides constituted, in 1989, 292 thousand tons (the largest supplies, 544 thousand tons, were in 1986) (*Annual Bulletin* (Content), 1991).

With regard to 34 examined pesticides, Russia's territory can be divided into medium or slightly polluted areas. Among the medium-polluted areas, there are the Moscow region and the Central-Chernozem zone, where contaminated areas during the spring-to-autumn period constituted, respectively, 10–20% and 17–12 %, with the average levels of 0.8–1 MPC and 1.3–1.1 MPC. The middle part of the Volga region may also be considered a zone of medium pollution. In Siberia, the Omsk region can be characterized by approximately the same level of pollution. However, pesticides causing this pollution differ from region to region, depending on the type of agricultural activity.

Regions of slight soil contamination are the Irkutsk region, the Kurgan region, the Primorsky region, and Northern Caucasus, where polluted areas constitute from 2% to 7%. Among zones with pure soils there are the upper Volga region and Western Siberia, with the exception of mass recreation areas.

Table 5.15 presents data on the pollution of vegetative production by pesticides (*Annual Bulletin*..., 1991).

The testing of vegetative production for the content of nitrates has revealed that in Russia the pollution level is very high: one-sixth part of the whole amount contained nitrates in excess of the established norms. Fodder appears to be the most polluted kind of vegetation production.

It may be stated that the control of industrial and agricultural toxicants in Russia is insufficient, especially with regard to vegetative and cattle-breeding areas. The control is absent altogether where the natural biota is concerned. That is why no final conclusions can be drawn concerning the extent of pollution by toxicants of soils, agricultural production or natural biota. However, there is no doubt that the level of pollution by toxicants is considerable and further accumulation of toxic

Table 5.14. Metal concentration in the soil depending of the type of prevailing industry (ppm)*

Prevailing industry	Lead		Manganese		Vanadium		Copper		Nickel		Zinc	
	m	a	m	a	m	a	m	a	m	a	m	a
Non-ferrous metallurgy	270	870	1260	3050	130	190	210	1100	100	200	560	3350
Ferrous metallurgy	40	90	2850	7200	110	200	60	155	280	280	140	300
Chemical industry	35	110	950	2000	90	290	50	95	85	85	90	200
Power engineering	40	90	1350	110	110	170	80	165	50	50	160	210
Percentage abundance in soil	10		850	50	50		20		40	40	50	
MPC, background	10		1500	150	150							210

* m, mean values; a, means of three stable maximum values.

Table 5.15. Dynamics of the number of probes and the results of finding residual amounts of pesticides in the vegetative production of 1987–1990

Year	Number of probes	Share of probes with detected pesticides (%)	Share of probes with excess of norms
1987	6484	21.7	10.0
1988	5567	19.7	5.4
1989	5429	17.1	5.1
1990	9699	11.9	3.1

substances in soils and in living organisms takes place. Therefore, over large territories natural communities are suppressed and destroyed, and their productivity decreases. Many toxicants remain outside the system of observation; in many cases, their maximum permissible concentrations (MPC) have not been determined, and their effects on individual organisms or genotypes of plants and animals are not known.

A huge amount of solid waste is formed in the Russian territory (Table 5.16) (*Problems* ..., 1993). Solid industrial waste is accumulated over large areas and are sources of pollution for the atmosphere, water and soil. Disposal areas in dry weather mostly produce a lot of dust; as a result, solid particles penetrate into the atmosphere and then fall back to the ground, in the vicinity of waste dumps, over an area of about 5-10 km in radius, thus polluting soil and vegetation. Refuse dumps around coal mines often catch fire, since they contain coal particles and dust, and

Table 5.16. Production and accumulation of some kinds of solid waste per capita in the former USSR (per year/total accumulation, tons)

Rock	Ash dump	Halites	Phosphogypsums	Metallurgy waste	
				Ferrous	Non-ferrous
53/176	0.27/3.5	0.25/2.2	0.06/0.47	0.4/2.8	0.05/1.4

may smoulder for long periods. Refuse dumps pollute water bodies with their runoff. Some of the disposal areas contain highly toxic substances, such as zinc, cadmium, mercury. Thus, dangerously contaminated zones are formed around such refuse dumps.

Solid waste is complemented by sludge and mud which are accumulated in water-treatment systems and in city sewage. The volume of dried-up sludge amounts to 100 million tons.

Large amounts of solid household waste are formed in cities, towns and smaller settlements. The volume of such waste reaches 130 million m^3 per year, or 0.9 m^3 per capita. Since the density of household waste is about 0.2 t/m^3, their per capita share per year is about 0.2 tons. About one-third of this amount is food garbage; another third, paper and cardboard; the rest comprises waste of wood, textiles, metals, polymers, glass.

There are seven incinerators across Russia's territory which recycle about 3% of solid household garbage; another 9% is removed from cities to disposal areas. There are over a thousand such areas over the country, and most of them do not meet sanitary norms.

The share of polymer materials, metals and glass in domestic refuse gradually increases. This is due to the fact that disposal areas for domestic garbage are more and more often used for industrial waste; among these, dangerous waste can often be found, such as mercury lamps and conventional radioactive waste. The growing share of polymers in the garbage leads to an additional danger of burning domestic refuse—the formation of dioxin. Apart from special disposal areas, there are numerous unofficial or simply illegal dumping grounds around cities and settlements. These are complemented by a large number of dumps around summer-cottage communities. It is extremely difficult to assess the volume of all these waste dumps, but one may be sure that the above-presented figure (130 million m^3, or 26 million tons, or 0.2 tons per capita) is too low. Here are some figures concerning amounts of domestic garbage per capita in other countries, for the sake of comparison: USA, 0.744 tons; Australia, 0.681 tons; Canada 0.635 tons; the Netherlands, 0.449 tons.

Domestic garbage dumps have a certain lifetime. From time to time they are made more compact, and then closed. Such closed dumping grounds contain no less than 0.5 billion tons of refuse. Such areas are sources not only of dangerous toxicants, but also of various infections.

There is a special kind of solid waste originated both in households, and in industry and economy. This is metallic waste—scrap. According to data available, about 1 billion tons of scrap have been accumulated over the former USSR territory. However, only 30% of steel and 25% of non-ferrous metals have been produced from recycling scrap.

Dumping grounds occupy about 15 Mha of Russia's territory. Over this area, biological primary production is consumed completely; over adjoining areas in a radius up to 10 km, it is substantially reduced, owing to permanent pollution of the soil and natural communities of living organisms (*Problems*..., 1993).

Every year over the whole country, over 20 million tons of non-utilized high-toxicity industrial waste is accumulated; these should be stored in a specific way within the premises of corresponding enterprises. Part of high-toxicity waste is utilized. However, very often, instead of storing this waste, it is thrown away into sewage systems, into gorges and ravines, into domestic garbage dumps. Nobody knows how much and what precisely is being discharged, although in 1990 a special decision was taken concerning taking account of toxic waste. It requires a complete inventory of toxic waste. Apart from the acting sources—of toxic waste, there also are latent sources their burial places. Therefore, search and inventory of toxic waste burial places are urgently required, first of all on new building sites or rebuilding areas in cities.

For a long time, some extremely dangerous toxic products have not been discussed openly. For example, the problem of dioxins had not been mentioned in public until very recently. Even so, already in 1987 a network of special centres for studying the problem of dioxins had been established in several Russian cities: Moscow, St. Petersburg (then Leningrad), Volgograd. The data obtained at the centres are inaccessible. In the meantime, dioxins are contained in many chlorine products. For instance, the chemical plants of the city of Ufa alone discharge about 0.3 kg of dioxins per day to the Ufa river. It is also accumulated in the soil and sludge.

Dioxin waste has been also buried in the ground on the premises of water treatment systems; thence dioxins are emitted to the atmosphere. Many people suffered from dioxin poisoning back in the 1960s and 1970s because they took part in starting the production of pesticides on the chlorine-phenol basis at a chemical plant in Ufa. Over a hundred people picked up a specific illness as a result of regular contact with high doses of dioxin. Table 5.17 presents the general pollutant loading over Russia's territory, and Table 5.18 the loading of particularly dangerous pollutants (*Problems*..., 1993).

Table 5.17. Pollutant mass created in the process of economic activity per capita and per square kilometre of territory (1989)

	Pollutants				
	Atmospheric		Sewage waters	Solid waste	
	Total	Harmful		Total	Harmful
Per capita (tons/year)	13	0.24	500	52	17.7
Per km^2 (tons/year)	66	1.6	4500	700	230

Table 5.18. Per capita amount of atmospheric and water dangerous pollutants, for Russia, 1989

Carbon monoxide	Atmosphere					Water	
	Hydrocarbons	Fluorides	Phenol	Heavy metals	Lead	Mercury	Benzopyrene
60 kg	40 kg	90 g	30 g	14 g	23 g	170 mg	100 mg

There hardly exists a region over the entire territory of Russia where one cannot detect anthropogenic pollutants, at least in small concentrations, to say nothing of such global pollutants as carbon dioxide, methane or chlorofluorocarbons.

5.6 RADIOACTIVE POLLUTION

Up to 1 January 1989, 1747 nuclear tests had been performed in the world: USA, 932; USSR, 649; France, 188; Great Britain, 43 (presently, Great Britain has transferred its nuclear tests to a test site in the state of Nevada, USA); China, 34; India, 1. Up to 1 January 1990, the total number of nuclear explosions increased to 1806. In 1963 nuclear tests in the atmosphere and in water were prohibited. However, by that time, substantial amounts of radioactive substances had been emitted to the atmosphere, and the dates of nuclear tests can be very well detected from traces of pollution produced by the tests and deposited on ice layers on mountain cover and on glaciers throughout the world, including the Antarctic. Particularly well pronounced were layers formed in 1963 and 1964 (Izrael, 1996).

Dozens of states possess atomic reactors used for power production. In the process of "burning" of the uranium fuel in the reactor, plutonium is produced which can be utilized to produce nuclear arms. The legal trade in nuclear fuel creates favourable conditions for attempts at making nuclear arms, which is very well demonstrated by such examples as Israel, Iraq or Pakistan. Now that comparatively small groups of terrorists have been able to seize hostages, a situation is imaginable when entire states may become hostages of terrorists who have taken possession of nuclear arms or a nuclear reactor. If an event such as nuclear disarmament ever takes place, there will remain people involved in nuclear weapons production and information sources that will make it possible to restore the production of nuclear arms. This kind of knowledge, as a part of the cultural heritage, has been fixed forever in the accumulated extra-genetic information of mankind. Since nuclear reactors are actually nuclear arms factories, the danger of the re-creation of this kind of weapons will always be realistic in the presence of nuclear power engineering in the country, and the time necessary for the re-creation will be reduced to a small time interval. Hence the question: is nuclear power engineering compatible with the long-term objectives of humankind and the strategy for its survival?

Industrial production of plutonium and other fissionable materials began as far back as 1949. This year may be considered the starting point for the development of the potential of dangerous atomic industry, although experimental atomic reactors in Russia were created prior to that date.

Over the territory of the country, at least four enterprises producing fission materials are in operation; they produce nuclear weapons as well. These complexes are located in the European part, in Siberia and in the Urals. The enterprise entitled "Mayak" ("Beacon") in the vicinity of Kyshtym in the Urals, but better known as "Chelyabinsk-40", has been functioning for over 40 years. Apart from the problem of the stability of the operational regime of its reactors and auxiliary systems, there is an important problem of disposing of its liquid and solid radioactive waste.

Russia possesses 11 atomic power stations (Table 5.19) which yield about 12% of electric power in the country.

The total number of operating reactors over Russia's territory is 31, and 6 are being built. The main loading with energy reactors falls on the European part of Russia where there are 22 reactors. In their vicinity, in an area with the radius of 250 km, 9 325 000 people are exposed to their impact.

After the Chernobyl accident, the designing and building of dozens of atomic reactors, mainly in the European part of Russia, was stopped. Most of these atomic stations were planned to be built in the upper and middle Volga regions.

Many power reactors (of the order of several hundred units) have been installed on atomic submarines, which are also provided, as a rule, with rockets with nuclear warheads. There are at least four bases for such submarines two in the Arctic Ocean, on the Kola peninsula, and two in the Pacific Ocean, in the Primorsky

Table 5.19. Russian atomic power stations and the distance between these and the closest settlements

Name of atomic power station	Number of reactors: operating/ being built	Distance from settlements (km)			Population (thousand people)
		less than 50 km	50–100	100-150	
Kolskaya	4/2	—	Kirovsk, Apatity	—	69
Leningradskaya	4/0	—	St. Petersburg	—	5000
Smolenskaya	3/0	—	—	Smolensk	316
Kurskaya	4/1	Kursk	—	—	397
Kalininskaya	2/2	Vyshny Volopchek	—	Tver'	71 — 428
Dmitrovgradskaya	2/0	Dmitrovgrad	Ulyanovsk	—	165 — 381 — 170
Balakovskaya	3/1	Balakovo	—	Engel's	
Beloyarskaya	1/0	Ekaterinburg	Kamensk-Uralsky Pervouralsk	—	1316 — 193 — 132
Sibirskaya	4/0	Troitsk	—	—	89
Bilibinskaya	4/0	—	—	—	—

region and Kamchatka. Two plants building such submarines should be added here.

Apart from the nuclear submarine fleet, there is also a surface fleet with power reactors which is mostly represented by ice-breakers, having their base in Murmansk and engaged in transportation along the course of the Arctic Sea Route.

The majority of silos for intercontinental rockets of the former USSR, as well as bases of strategic bombers capable of carrying nuclear bombs, are concentrated in Russia. These silos and bases are mostly located in the most developed and populated area, and therefore present a considerable potential danger.

An individual nuclear reactor contains the same amount of fission material as 1000 bombs similar to that dropped on Hiroshima. The Chernobyl accident has shown that, from the viewpoint of the contamination of a territory and health threat for the population, the explosion of a reactor, its burning and the emission of a huge amount of fission material, creates a radioactive threat over an area exceeding by far that affected by the Hiroshima atomic explosion.

However, in the absence of accidents, about 250 radioactive isotopes enter the environment when uranium fuel is used. These invisible radionuclides fall out, are absorbed and assimilated by the environment around us and the biological community.

There are 15 disposal areas in Russia intended for the burial of radioactive waste, and these are also sources of potential radiative danger.

The fission material is widely used in scientific research, in numerous instruments, and in the production sphere. Over Russia's territory, approximately 17 000 various institutions use this kind of material.

Smaller-scale accidents and emergency situations take place from time to time in atomic power reactors. There are data revealing that such cases are frequent enough. Little is known about large-scale accidents on atomic objects. For instance, it was only in 1989 that information was published about a serious accident at the military atomic enterprise "Mayak", near the town of Kyshtym, which happened as far back as in 1957.

Then the explosion caused an emission of radioactive substances which propagated across three large regions—the Chelyabinsk, Sverdlovsk (now Ekaterinburg) and Tiumen' regions. The length of the main radioactive trail extended northeastwards as far as the town of Kamensk-Uralsky, over 100 km, and its width reached 10 km. The territory contaminated with strontium-90 was over 15 km^2. 270 000 people were then living in the contaminated zone. During the whole period since that accident, about 1 million people were exposed to radiation, and about half that number took a large dose of radiation. 65 000 people, as a result, need medical treatment. Some categories of people involved have not been taken into account at all (liquidators of the consequences of the accident, the military, prisoners). The total emission of radioactive substances amounted to 20 million curies, but only 2 million curies penetrated outside the premises of the enterprise, thus having formed the "Urals trail". The major part of the harmful matter remained within the territory of the enterprise.

The mass media treated the subject of a big fire at the Beloyarskaya nuclear power station in 1978. It seems that there was a series of repeated fires. However, there are no data available concerning radioactive matter emissions during the fires. In 1974–75 and in 1992 accidents accompanied by emissions of radioactive substances took place at the Leningradskaya atomic power station. But the most dreadful catastrophe in the entire history of nuclear power engineering occurred outside Russia, in the Ukraine—the Chernobyl power station accident. Its scale was such that a considerable area of Russia's territory was also involved.

The accident took place on 26 April 1986. It has been described in detail in documents, scientific literature, mass media productions, and even in fiction. The main cause of the accident apparently were defects of the reactor. One of them is the so-called "positive coefficient of vapour reactivity". Due to the breaking of some regulations and other erroneous actions of the reactor operators, the effect of vapour formation, i.e. the release of heat in the reactor, was enhanced. This started the "runaway" of the reactor. American experts are of the opinion that the runaway was sharply enhanced by other defects in the construction, especially safety rods. As a result, two explosions in a row followed. The intensive emission of radioactive matter from the burning reactor continued till 6 May. 50 million curies of various radionuclides and 50 million curies of chemically inert radioactive gases were emitted to the atmosphere. Of these, according to different estimates, from 7 to 14 million curies of iodine-131 and from 1 to 2.4 million curies of caesium-137 were emitted (Soviet and American estimates). As a result, 31 people were killed and over 200 people took doses of radiation which led to the development of radiation sickness.

As a consequence of the Chernobyl catastrophe, a territory with a radius of over 2 thousand kilometres, encompassing 20 states, has become subject to radioactive contamination to some extent. Over the former USSR territory, 11 large regions with a population of 17 million people, including 2.5 million children younger than 5 years old, have been exposed to radiation. The contamination was non-uniform, appearing as spots and bands, and was dependent on the direction of air fluxes, and also possibly on gravity fields. Table 5.20 presents areas contaminated by caesium-137, for December 1989.

In Russia the bands and spots of contamination as a result of the Chernobyl accident affected chiefly the Bryansk region and the southern parts of the Kaluga and Tula regions, and also the Oryol region. Smaller areas of contamination are also located in the north of the Belgorod regions, the west of the Ryazan' region and the south-east of the Smolensk region. A small spot of contamination by caesium-137

Table 5.20. Areas contaminated by caesium-137 for December 1989 (in km^2)

States	Extent of contamination (curies/km^2)			
	5–15	15–40	over 40	Total
Russia	4 540	2350	360	7250
Ukraine	2 355	740	680	3775
Belorussia	7 202	4060	2210	13472
Total	14 097	7150	3250	24495

Table 5.21. Areas contaminated by caesium-137 for December, 1989 (in km^2)

	Surface contamination by caesium-137 (curies/km^2)		
	5–15	15–40	over 40
Population	1130100	800900	4600
Settlements	413	167	26

has also been noted in the south-west of the Leningrad region (Grigoryev, 1991). The total area of pollution by caesium-137 and strontium-90 with values from 1 curie/km^2 and above in Russia is about 1.5 Mha.

Table 5.21 presents data on the population which inhabited the polluted zones of different degrees of contamination for Russia's territory, without taking account of those who migrated to other areas (Expert..., 1991; International..., 1991).

There are 11 reactors of the Chernobyl type in Russia: 4 each at the Leningradskaya and Kurskaya atomic power stations and 3 at the Smolenskaya station; these constitute half of the reactors of the European part of Russia. This means that about 1 million square kilometres of territory and about a quarter of the population of the European Russia remain hostages of nuclear power engineering and potential victims of possibly incompetent authors and executives of the programme of development of nuclear power engineering.

Information on radioactive waste disposal remains, as before, one of the least accessible spheres in the field of the atomic industry and nuclear power engineering. At many of their plants the waste is accumulated directly within their premises. For example, at the oldest of such complexes, the "Mayak" enterprise ("Chelyabinsk-40"), during the 40 years of its operation, the amount of just liquid waste that has been collected exceeds by an order of magnitude the quantity emitted during the Chernobyl accident, i.e. over 1 billion curies. The waste is concentrated in several artificial waterbodies, and in two lakes in the basin of the Techa river. Some specialists suspect that, apart from water filtration and underground horizon pollution, in this limestone area, other accidental events could occur, such as dam breaks and even falls-through of underground cavities. If such a thing were to happen, millions of cubic metres of radioactive water would be carried by a chain of small rivers to the big rivers Tobol, Irtysh and, finally, the Ob' river which would bring it to the Arctic Ocean. Considering the complicated pattern of sea currents, radioactive water could reach as far as the shores of Alaska. There is hardly ever any mention of radioactive waste of other centres of atomic industry and nuclear power engineering in the specialist literature or the mass media.

According to the data available relevant to the "Mayak" enterprise, one may calculate that it produced, on the average, about 25 million curies of liquid radioactive waste per year. This constitutes about 1.2 curies per capita in the Urals region; during the whole period of "Mayak's" operation, the per capita amount of accumulated liquid radioactive waste is 50 curies. Within the premises of the enterprise, a certain portion of fission material is stored; there also are some solid radioactive waste quantities, whose potential is comparable with that of liquid waste (*Problems*..., 1993).

Another problem is the storage of radioactive waste from passenger ships, freighters and warships, mostly atomic icebreakers and submarines. There are documents available now that reveal the facts of submersion of containers with radioactive material in the region of Novaya Zemlya and in the Kara Sea. Liquid radioactive waste was discharged to the Barents Sea, in a square with the coordinates: 74 and 78°N and 48 and 52°E. It may be supposed that similar practices took place in the Pacific Ocean.

Quite a problem remains from the system of burial of radioactive waste from those 17 000 Russian enterprises that use fissionable materials in the form of small point sources of different intensity. The fact is that no accurate account has been taken of the storage, or liquidation, or burial of such sources. Therefore, in the regions of industrial or scientific centres and their localities, powerful radioactive sources are sometimes discovered.

Still another problem is the so-called conventional-radioactive waste of low power (30–300 mrh (milliroentgens per hour)). For instance the worked-out elements of X-ray instruments can be classified as such waste. Up to very recently, such waste could often be found in city dumping grounds.

Since the curtain of secrecy has been lifted a little recently, it has become obvious that the situation with radioactive wastes and their disposal is very similar to the status quo of the atomic industry and nuclear power engineering. Hence the enhanced risk for the population of taking a dangerous dose of radiation anywhere, including in people's own dwellings. Finally, it should be remembered that, as a result of utilizing uranium fuel, 250 radionuclides penetrate to the environment, thus increasing the background radiation, especially in the vicinity of relevant plants. This is also a form of waste that cannot be removed. The same applies to pollution from nuclear explosions. This non-removable waste has been rapidly growing during recent decades. This growth rate is hardly compatible with the strategy of survival of humankind.

5.7 HEALTH PROBLEMS

In Russia, just as in many other countries, first of all industrially developed, civilization has been conducive to a discrepancy between man's genetic programme and the ecological niche he occupies. This did not ensure the functioning of the mechanisms of preservation of the human genome. As a result, such genetically fixed functions as love and kindness were not always realized. They were replaced by cruelty, rudeness, the need to use foul language, smoke, drink alcohol and finally, by drug addiction and violence (Tables 5.22 and 5.23), (Health Care..., 1990). Data presented in Tables 5.22 and 5.23 are actually only the tip of an iceberg, since such diseases as alcoholism and drug addiction are to a great extent "underground" ones. Thus, for instance, there are indications that the number of drug addicts is actually an order of magnitude larger than officially registered, and the number of alcoholics is greater by a factor of 3–4.

Table 5.22. The number of cases of alcoholism and psychosis in Russia (in thousands of people)

	Years			
	1980	1985	1988	1989
Cases registered for the first time	338.2	380.7	291.4	280.7
Registered chronic cases	2027	2812	2946	2786

Table 5.23. The number of drug addicts in Russia (in thousands of people)

	Years			
	1980	1985	1988	1989
Cases registered or the first time	1.8	3.7	8.2	8.0
Registered chronic cases	17.3	16.4	30.3	32.5

Table 5.24. The number of psychiatric cases in the former USSR)

	Years					
	1980	1985	1986	1987	1988	1989
Total number (thousand people)	485	570.7	600.8	585.0	407.7	343.2
Number of cases per 100 000 people	181.6	206.4	215.2	207.5	143.3	119.8

The increase of violence is characterized by especially high growth rates, particularly with regard to grave crime (premeditated murder, grievous bodily harm, rape, attempted rape, etc.). The growth rates are from 25% to 35% per year, depending on the type of crime. Teenage crime grows fast: e.g., during the period from 1985 to 1989, the number of teenagers (14–17 years of age) convicted increased from 8.7% to 13.1% of the total number of convicted persons.

Psychiatric diseases are one of the indicators of the decay of the genome. Table 5.24 presents data on the number and dynamics of psychiatric cases in Russia. One may erroneously deduce from the data of Table 5.24 that, beginning with 1987, the number of psychiatric cases decreased. This, however, is the result of changed (in 1987) criteria for registering such cases. The same applies to registering chronic psychiatric cases. In 1989 approximately 3100 thousand chronic psychiatric cases were registered in Russia, and over 2 800 000 alcoholics and drug addicts.

Among psychiatric cases registered for the first time in Russia in 1989, children (0–14 years old) constituted almost 70%. About one-third of the children were diagnosed as mentally retarded. Among chronic psychiatric cases, children constitute 40%.

The number of children-invalids who are subject to registration and medical control and treatment on a permanent basis also grows rapidly; the most frequent diagnoses of such children are: psychiatric disorder, diseases of the nervous system,

Table 5.25. Number of cases relevant to genetic disorders and diseases registered for the first time

	Psychiatric disorders	Diseases of the nervous systems and sense organs	Congenital anomalies
Children			
(thousand people)	215.3	2254.7	89.2
per 100 000	633	6628	262
Other cases			
(thousand people)	1043.2	4638	14.3
per 100 000	921	4094	13
Total			
(thousand people)	1258.5	6893.4	103.5

deafness, blindness, congenital defects of development, consequences of traumas. From 1985 to 1989 the number of such children increased by 54%, i.e. it increased annually by over 13%. The total number of children-invalids in Russia exceeds 0.5 million individuals. The number of oligophrenic cases among children grows at a rate of 5.5% per year (Lemeshev, 1991).

Table 5.25 presents the distribution of cases of genetic diseases registered for the first time (Health Care, 1990). The total number of genetic disease cases registered for the first time in Russia constituted 8255.4 thousand people, i.e. approximately one case for every 18 persons. The data cannot be comprehensive for obvious reasons. Very alarming are data concerning children. A serious signal indicating the increasing rate of genetic disorders is that, out of 182.8 thousand people registered in Russia in 1989 in connection with psychiatric disorders, 83.8 thousand (over 40%) were children.

Here the data published in the book, *Health Care in the USSR* (1990), are presented. This is the first collection of statistical papers on the subject presenting information that previously was not open to the public. However, the book cannot be comprehensive enough because of the imperfections of medical registration in the country. However, the "Introduction" to the book states that in reality half of the children born in Russia nowadays have disorders or deficiencies of some kind.

Here are some more data concerning children and teenagers. Over the former USSR territory, about 1.1 million young children and 350 000 teenagers (15–18 years of age) suffer from severe neuroses and psychiatric disorders. Only 8.5% of schoolchildren have the right body build, weight and height; only 25–30% of schoolchildren may be considered healthy; 40–45 % have disorders at the level of functional disturbance that may be conducive to serious illnesses; 25–35% already have chronic diseases. Out of the total number of young men called up for military service, only 12–15% are really healthy people.

There is medical evidence that during the last 25 years, the frequency of congenital inherited deficiencies in children has doubled. Because of the poor health of mothers,

Table 5.26. Distribution of cases in Russia, according to the type of illness (registered for the first time)

Type of illness	Children up to 15 years of age		Teenagers and adults	
	Total (thousand persons)	Per 100 000 persons	Total (thousand persons)	Per 100 000 persons
Total number	41 186	121 069	58 063	51 507
Malignant tumours	35.6	106	794.3	701
Vascular system diseases	135.3	398	48.5	43
Psychiatric disorders	215.3	633	1 043.2	921
Diseases of the nervous system and sense organs	22.54.7	6 628	4 638.7	4 094
Diseases of respiratory organs	2 894.3	85 079	23 366	20 621
Diseases of urinary and genital systems	250.4	736	2 693.3	2 377
Diseases of skin and subcutaneous fat	1 645.2	4 836	3 573.9	3 154
Congenital anomalies	89.2	262	262	13

10% of new-born babies arrive too early; their central nervous system is under-developed, their protective reactions are weak.

Thus, the gradual background growth of genetic diseases, first of all psychiatric ones, takes place, and the growth rate increases. The birth rate in Russia drops from year to year, but the number of sick people grows. The number of premature birth and stillbirth cases increases rapidly. This is first of all due to the disturbance of mechanisms in the human population that stabilize at a normal level the decomposition of the human genome. Apart from this, the decay of the genome was accelerated after 1950 by the rapid growth of the loading upon the human organism that resulted from the increase in the volume of toxic waste and the low quality of life.

During the last 40 years, the most frequent causes of death have become diseases of the vascular system and malignant tumours: from 1970 to 1988, the number of deaths caused by the former increased from 381.2 to 551.9, and that caused by the latter from 130.6 to 162.5 per 1000 deaths. During the last 10 years, the number of deaths from malignant tumours has increased by over 13% per 100 000. These are indicators of the deteriorating living conditions and health of Russia's population (Dvoirin et al., 1991).

Table 5.26 presents data on the range of diseases that are due, to some extent, to the deformation of the environment. Table 5.26 shows only too well that the danger of a more rapid growth rate of most grave diseases in subsequent generations in Russia is very realistic. The number of cases of malignant tumours in Russia increased by about 25% during the nine years from 1980 to 1989. There are 2 656 000

Table 5.27. Statistics of medical cases in Russian cities with high air pollution (number of cases per 100 000 persons)

City	Type of disease	1987	1988	1989
Arkhangelsk	Malignant tumours	244	278	414
	Skin diseases	398	434	448
	Diseases of respiratory organs	56 291	71 725	60 714
Nizhny Novgorod	Malignant tumours	258	360	405
	Skin diseases	2 270	2 248	2 081
	Diseases of respiratory organs	48 847	63 063	54 955
Berezniki	Malignant tumours	258	360	327
	Skin diseases	2 270	2 116	2 261
	Diseases of respiratory organs	56 555	75 746	74 006
Krasnoyarsk	Malignant tumours	—	260	421
	Skin diseases	—	797	757
	Diseases of respiratory organs	—	60 356	58 135
Ekaterinburg	Malignant tumours	311	234	502
	Skin diseases	1 595	2 023	1 924
	Diseases of respiratory organs	24 862	29 796	35 418
Usolye-Sibirsk	Malignant tumours	226	157	246
	Skin diseases	2 662	2 801	2 361
	Diseases of respiratory organs	29 920	49 583	44 099

psychiatric disorder cases in Russia (the data for 1989). Thus, at least 1 out of every 33 individuals in the country is registered in connection of this or that kind of disease closely relevant to the ecological situation.

Very pronounced is the result of chemical impact on the population when one analyses the medical case situation in cities with a high level of pollution (Table 5.27). The growth rates of these diseases are very high. For instance, for cancer the growth rate reaches 20–70% per 3 years. According to the data for oncological cases in the cities of Siberia and the Far East, the number of cases is much greater in more polluted cities, both for men (by 25%) and for women (by 38%).

A sharp increase of medical cases among children can be seen from the data of Table 5.28.

In agricultural regions where the level of employment of mineral fertilizers and pesticides is high (above 10 kg/ha), changes in the immunity status of the population are registered, to say nothing of an increased number of respiratory system diseases, and a series of grave illnesses in children: blood circulation anomalies, infantile cerebral paralysis, psychiatric disorders, etc.

On the whole, the statistical data concerning the health status of Russia's population can be assessed as alarming.

For October 1990, the number of people infected with the AIDS virus reached in Russia the figure of 429, of which 256 are children. The children were mostly infected in medical institutions in such cities as Elista (75 persons),

Table 5.28. Number of medical cases among children for the Yaroslavl region (per 1000 persons)

Diseases	1988	1989	1990
Congenital heart defects	6.76	21.3	25.0
Malignant tumours	1.24	2.68	2.36
Kidney diseases	1.24	45.9	30.9
Diseases of the digestive system	27.4	45.9	30.9

Volgograd, Rostov-on-Don, Stavropol and Grozny. The forecast is about 1 million virus-carriers by the year 2000, and possibly some tens of thousands of AIDS cases.

Thus, the high level of chemical loading and the low quality of life in Russia, supplemented by the decay of the genome, which is due to leaving the human ecological niche, taken together, determine the catastrophic situation with the population health and the tendency towards rapid deterioration of the status quo.

5.8 ECOLOGICAL LOSSES AND MEASURES TO RESTORE ENVIRONMENTAL QUALITY

There have been numerous attempts in various industrially developed countries to assess the "ecological" losses as a percentage of the gross national product (GNP). Thus, for a number of countries of the European Economic Union (EEU), direct losses have been estimated as 5%, on condition that the expenses for nature protection measures constitute 1.25% of the GNP. At the same time, strict control is provided of technologies, and also of cases of violating the established quotas for resources, emissions, etc. In Germany the losses are evaluated as 7%, at 1.5% expenditure for nature protection measures. In the USA the total direct expenses relevant to technological causes, including accidents and catastrophes, chronic diseases partly due to environmental pollution, etc., constitute 4–6% of the GNP and ensure a 15–25% reduction of untimely death cases. According to other estimates, the imperfection of the technology of industrial management systems and the resultant technological risk in enterprises in the USA bring about a detriment evaluated at 10–15% of the GNP. The USA spends up to 2% of the GNP on nature protection measures (Porfiryev, 1990).

There exist different assessments of ecological losses for the former USSR. According to some data, such issues as soil erosion and water and air pollution bring about direct losses estimated at 5% of the GNP per year. However, many experts are of the opinion that the losses should be evaluated taking account of indirect damage caused by depredatory exploitation of natural resources; if so, the losses should be doubled, thus amounting to 10% of the GNP. This must be the lower limit of ecological losses; their upper limit is somewhere about 20%. There are firm grounds to suppose that these figures can be appropriate for Russia which occupies about 80% of the territory of the former USSR.

Both in industrially developed countries and in Russia the assessments of ecological losses were made on the basis of a stereotype, according to which the main indices of human impact on the environment were its pollution and deforestation.

The total expenditure for nature protection measures in Russia was a little above 1% of the GNP in 1981 and about 0.8% of the GNP in 1989. This shows that, instead of getting higher, the funding of nature protection measures decreased (Collected Papers, 1991; Social Development..., 1991; RSFSR..., 1990).

The capital investment in environmental protection constitutes about 0.2% of the GNP; the remaining money goes to the maintenance and exploitation of the existing objects of environmental protection and the financing of the managerial staff.

The USA spends from 1.5% to 2% of their GNP on environmental protection. The value varies from year to year within the range 60–80 billion US dollars and is dependent on the planned programmes and the legislative and public response to these plans. The structure of the expenditure also varies: for instance, if previously special emphasis was put on the protection of water and air, approximately in equal shares of financing, later more and more attention has been given to the problem of dangerous toxic waste, and the major portion of money is spent on its solution. The structures of expenditure on environmental protection in the USA and Russia are very different and can hardly be compared, since in the USA the major part of the money goes for the improvement of drinking water quality, whereas in Russia the money is mostly spent on the building of water treatment systems for processing sewage waters (*Problems...*, 1993).

The insignificant share of budgetary funds that has been allotted in Russia in 1989 for capital investment has not been coped with: part of the work has not been fulfilled. Three large-scale industrial complexes (petroleum chemistry, metallurgy and wood processing), which are the main polluters of the environment, have not fulfilled their capital investment plans, respectively, by 63%, 78% and 72%. The building of nature protection systems is never completed in time, in spite of the fact that their number is absolutely insufficient for the stabilization of the pollution level in the air basin and water reservoirs. In 1992 the centralized funding of nature protection measures has been sharply reduced, and the budget for 1993 had no mention of such needs.

The smallest portion of nature protection financing is allocated for specially protected areas, i.e. the territories intended for the conservation of primary biological production which may become starting points for the restoration of the biosphere destroyed by man. Preserves, national parks and reserved hunting (zoological) grounds are such territories.

In 1989 the total sum spent on such territories constituted less than 0.7% of the whole sum intended for nature protection measures. The total number of reserves and national parks over Russia's territory in 1989 was 80, 69 of which were state reserves occupying an area of 18.7 million km^2. Three new ones were organized in 1989, whose total area was 485 000 ha: one in the Orenburg region (steppe), one in the Penza region (partially wooded steppe) and one more in the Kemerovo region (foothills) (Ecological..., 1990).

There are 11 national parks over Russia's territory, with a total area of 1.3 Mha.

However, they are not properly managed: in some of them industrial activity, forest cutting or hunting take place, sometimes on a large scale.

In 1989 the total area of reserves and natural national parks in Russia was 20 Mha, and that of reserved hunting grounds 50 Mha. Taken together, the area is still three times smaller than the agricultural land of Russia.

Over Russia's territory, the excessive territorial consumption which destroys ecosystems amounts to about 260 Mha. These calculations do not take account of other expenditures relevant to the decay of the human genome and environmental pollution. If all losses are properly taken into account, the entire GNP will not be sufficient to cover them. This shows only too well how expensive for mankind is the invisible part of the "ecological" iceberg—the consumption of primary biological production exceeding all reasonable limits and thus leading to the destabilization of the environment, the accelerating decay of the present-day biosphere and the human genome.

5.9 PERSPECTIVES OF SURVIVAL

The two decades that have passed since the Stockholm Conference in 1972 have revealed that the basic tendencies of rapid deterioration of global and regional ecological conditions have not changed for the better, in spite of a huge sum of money (over 1 trillion US dollars) invested in nature protection measures. The ecological situation in industrially developed countries has not undergone any cardinal changes. Deformations of the environment increase even in comparatively "safe" regions: for instance, in Western Europe the area of damaged forests has expanded, the area of soils which are subject to acidification, as well as the number of waterbodies that are, are increasing, and mushrooms disappear. Within the framework of the existing ecological stereotypes, the cause of these mishaps is attributed to insufficient nature protection activity.

Now, after the Rio de Janeiro Conference in 1992, a revision and re-evaluation of ecological problems has taken place (see Chapter 1). It becomes clear at last that nature is not the capital of humankind, but is its natural environment, man cannot live without his natural environment where he is only one of many elements. The biosphere maintains stable environmental conditions which are favourable for life in general, and for the life of man in particular. Hence it follows that the limits of human development are determined by the degree of ecological disturbance, and not by simple consumption of the resources. It has become obvious that man's interference in natural processes has gone so far that the environmental changes it induced may appear to be irreversible, and its destructive activity and its consequences cannot be compensated by nature protection measures and capital investment.

Numerous programmes and projects of environmental protection which were adopted and implemented on a global, regional or local level after the Stockholm Conference (1972) proved to be insufficient and inefficient because they had no influence on the rate of development of negative global environmental changes

(Table 5.29). The factor that makes it impossible to increase the efficiency of nature protection measures, becomes more and more clear: the growth of population and its demands that can only be satisfied on the basis of expanding consumption and destruction of the biota involving an advance in the development of power engineering.

There have been some successful programmes of environmental "purification", but these were individual events, whereas global tendencies were the same and even deteriorating.

Absolutely all global tendencies towards a deteriorating ecological situation are relevant to the territory of Russia. Moreover, Russia is one of the world regions which substantially contribute to the preservation, development and enhancement of negative global ecological tendencies. Russia's contributions are even greater because the resource and energy consumption per unit GNP in this country is 2–3 times higher than in Western Europe or the USA, and 5–6 times higher than in Japan. Besides, some natural peculiar features of Russia's territory, such as the large areas of swamp lands and permafrost, apparently provoke an enhancement of some anthropogenic impacts, thus increasing pollutant emissions as a result of the destruction of natural communities of organisms and environmental degradation in these regions.

Expenditure on nature protection measures in Russia are low-efficiency, just like the nature protection systems themselves, since they employ old-fashioned and ineffective technologies. These programmes lack a proper accounting for possible expansion of industry, and hence a potential increase of harmful emissions, and of the number of pollutants which must be subject to removal. Therefore, in areas of ecological catastrophes nature protection measures do not fulfil their function.

An analysis of a whole spectrum of disturbances and deformations of the environment over Russia and the former USSR territory performed in the late 1980s made it possible to detect regions with the greatest changes, and to compile a map of their distribution.

It can be seen that, against the background of a global ecological crisis, a grave regional ecological crisis takes place over Russia's territory which can be characterized by frequent local deformations of the environment with a characteristic scale of about 10–15 000 km^2—on the average, one every 70 000 km^2 of the territory. Often individual areas with strong local disturbances of the environment amalgamate, forming large regions with environmental disturbances. The severe ecological crisis in Russia is aggravated by social and economic deformations.

Over 70 years of centralized management put Russia in a position of functioning in its own way and being surrounded by free market systems with which it had to compete. The "command" system of economic management was organized in absolute contrast with the basic principle of functioning of biological systems—a free market. The overhead charges in the command system have proved to be much greater than in the free market system. The growth of a more and more complicated structure of the command system in conditions of scientific and technological progress, in spite of the expansion of management structures (and the increasing overhead costs), did not make it possible to pursue a strictly balanced economic activity.

Table 5.29. Environmental trends

Characterization	Tendency 1972–1990	Scenario 2030
Consumption of primary biological production	Increase of consumption: 40% on land, 25% global (estimate of 1985)	Increase of consumption: 80 - 85% on land, 50- 60% global (the year 2030)
Change of greenhouse gas concentration in the atmosphere	Increase of greenhouse gas concentration from tenths of a percentage point to a few percent annually	Concentration growth, concentration growth acceleration of CO_2 and CH_4 due to accelerated destruction of the biota
Ozone layer depletion, an increase of the ozone hole over Antarctica	Ozone layer depletion by 1–2% annually, annual growth of ozone hole area	Preservation of the tendency even if chlorofluorocarbon emissions are cancelled
Forest, particularly tropical forest, area reduction	Reduction rate from 117 (1989) to 180 ± 20 thousand km^2 (1989) per year; ratio between forest restoration and forest cutting of 1:10	Preservation of the tendency; reduction of tropical forest area from 18 (1990) to 9–11 million km^2 (2030); reduction of forest area in the temperate belt where forest diseases will increase
Desertification	Expansion of desert area (60 000 km^2 per year); increase of technogenic desertification	Preservation of the tendency, its possible growth due to decreased water cycle on land
Degradation of land	Increase of erosion (24 billion tons annually); fertility decrease; pollutant accumulation, acidification, salinization	Preservation of the tendency, its possible growth due to decreased water cycle on land
Ocean level rise	A 1–2 mm/year ocean level rise	Preservation of the tendency, possible increase and acceleration of ocean level rise up to 7 mm/year
Disappearance of organism species	Rapid disappearance of organism species	The tendency can be enhanced as the biosphere is destroyed; a 25% reduction by the year 2050
Qualitative exhaustion of land waters	Increase of sewage volumes, point and areal pollution sources, the number of pollutants	Preservation and increase of the tendencies

Accumulation of pollutants in media and organisms, migration	Increased accumulation of the mass and number of pollutants in media and organisms; growth of the radioactive medium	Preservation and possible enhancement of the tendencies
Deterioration of living conditions of humans, increase of diseases and diseases relevant to ecological disturbances, appearance of new diseases	Increase of poverty; food shortage, high in-fantile death rate; high level of illness, lack of pure drinking water in developing count-ries; living in badly polluted zones, increase of genetic diseases, increased number of still-birth and premature birth cases; high number of accidents, increased medicines consumption, increase of allergic sicknesses in developed world and of AIDS throughout the world	Preservation of the tendencies; increase of food shortage; growth of genetic diseases and those due to ecological disturbances, expansion of infectious disease areas, appearance of new diseases

A huge resource- and power-consuming economy was formed in Russia: the group A hyperstructures and the military-and-industrial complex required enormous amounts of metal, energy carriers and other resources. The one-sided structure of the command system economy was one of the basic factors of serious disturbances and violations of the natural environment. Russia, which has in the field of extraction and processing of raw materials, and in the military industry, a structure close to those of developed Western countries, contributes to environmental pollution on a similar scale: huge emissions of greenhouse and toxic gases, a broad variety of pollutants entering all types of media, an abundant mass of specifically toxic pollutants, the accumulation of pollution in waterbodies, soil and vegetation, and frequent leaks of radioactive matter. At the same time, the grave deformation of the environment follows the pattern of the developing countries: rapid land degradation, forest area reduction, exhaustion of natural minerals. As a result, the nature of the country is subject to double loading—that of the developed and of the developing country.

The ecological catastrophe was aggravated by the economic growth of the 1980s which was developing (at least at 70%) due to extensive factors. It is particularly characteristic of the resource and military branches of the economy which require their own fuel/power and raw material bases and enormous territorial resources.

Such branches of the economy as power engineering, raw material extraction, artificial material production and the military industry appear to be most ecologically dangerous, and their concentration in certain areas creates a dangerous territorial and economic combination. Such combinations are very stable. Their transformation requires huge expense. Such ecologically dangerous combinations occur in the Urals and Volga regions, at the Kola peninsula, and in the south of Siberia. In the meanwhile, up to very recently, the expansion and complication of the structures of such territorial complexes in ecologically dangerous directions was continued, by the building of new and the broadening of already existing ecologically dangerous centres.

Inefficient or old-fashioned technologies and equipment and the low level of technological culture have also greatly contributed to this tense ecological situation.

Having this dead weight of deformations and disproportions in its economy and technology, Russia entered a new road, having turned from the old command system of economic management to a free market type of economy. The period of this transition which will probably take a lot of time, perhaps the decade of the 1990s, is a turning point in Russia's development. Complete re-building of the entire economic structure is necessary to form a normal economy. The period will be a very difficult one for the solution of ecological problems.

At the present stage of the country's development, one can hardly expect any significant funding of ecological programmes, or an increased level and quality of life for the population. Industrial production on the whole will decrease because of the dismounting of the group A hyperstructures in industry and the military-industrial complex. Non-profitable enterprises will be forced out, and, as a consequence, unemployment will grow. A most important issue will be control of the inflation growth rate. To ensure the possibility of population growth, a 2% increase of the

GNP is required per 1% increment of population. (This is the so-called demographic compensation.) Since the GNP continues to fall, it is clear enough that for the time being, and for a certain period in the future, the level and quality of life of the Russian population will go down.

Under these difficult economic conditions, a reasonable choice of the ecological strategy for the transition period should be made. It is now that the basis of ecological conditions for future Russian generations is being formed.

In conditions of a structural reformation of the economy, a justifiable ecological strategy implies, above all, a reasonable ecologically admissible technological and technical policy. Such a policy may be briefly formulated as follows: to produce more with lower costs, i.e. save resources, use them with the greatest effect, improve and often change the technologies, implement and expand recycling. One may take Japan as a good example: at present the production output there per unit of resource expenditure is 81% greater than in 1973. In other words, a strategy of preventive ecological measures should be worked out involving the use of advanced technologies accompanied by a structural reconstruction of the economy to guarantee energy- and resource-saving and thus open up opportunities for a frequent change of technologies, the implementation of recycling and a minimization of waste.

Efforts should be concentrated on the development of the production of consumer goods and the increase of the consumption share. On the whole, Russian economy must reduce as much as possible the consumption of energy and resources in the GNP production, and per capita of its population. The free market system itself and competence must facilitate the realization of this strategy, if such obstacles as inefficient economic management and national conflicts within the country can be overcome.

The same preventive ecological strategy should be employed with regard to the problem of the export of resources. The share of resources in the total volume of sales should be continuously reduced. Very important is the problem of the right of foreign investors to extract resources in Russia. At present, the tendency on the part of foreign investors to invest capital, first of all, in resource-extraction branches of economy, especially in the petroleum and gas industry, is very obvious. If this tendency is let alone and becomes predominant, the country will be further shifted towards the Third World.

Many problems should be solved now in the course of preventive ecological strategy: possible expansion of ecologically dangerous branches of the economy (metallurgy, chemical and power engineering), especially in places where such complexes are concentrated; the implementation of projects that can seriously affect nature; the burial of harmful waste brought from abroad, including the Commonwealth of Independent States (CIS) members.

One of the main objectives of the ecological strategy for the transition period must be the task of maximum reduction of the consumption of primary biological production in Russia, first of all, due to a decrease in territorial consumption and wood cutting, as well as by way of intensive forest restoration. Within the framework of the same task, possible expansion of protected territories should be encouraged, above all, of rreserves and national parks.

During the transition period, due to the temporary decrease of the level and quality of life in Russia, a birth-rate reduction should be expected. A parallel flow of people immigrating from former non-Russian republics of the USSR (which are presently unstable independent states where the restriction of the rights of the Russian-language population has become everyday practice) can be anticipated, and in actual fact takes place already. Under these conditions, a promising and ecologically justified demographic policy should be the strategy of stabilization of the population. And the first step in this direction could be giving up the support of raising large families. The strategy is dictated also by purely economic reasons, since the country is at present unable to ensure a demographic compensation. Therefore, the growth of population, apart from negative ecological consequences, may have negative economic consequences as well, such as an additional drop in the level and quality of life, first of all, for large families.

The struggle against pollution should be continued. This problem should be in the hands of regional and local authorities, the "greens", and the public.

An important task in the transition period must be the increase of the scientific potential in the sphere of ecology, especially promising fundamental research which will enable to formulate a strategy capable of ensuring the ecological sustainability of Russia in the twenty-first century.

This is a specific period, when any kind of ecological programme will be difficult to implement due to fundamental ecological disturbances that have inertia and cannot be removed quickly enough.

The basic difficulty in the realization of ecological programmes consists in the absence of the infrastructure required for the solution of ecological problems. A system of special ecological legislative acts and the "ecologization" of the whole Russian law system are necessary. The problem of the legal basis of ecological activity has not been solved so far, although it is a priority problem in the realization of any ecological programmes.

The problems of nature utilization and protection can only be solved with the help of specific legislative acts. The infrastructure of financing of ecological tasks can only be formed via a system of legislative acts.

Legislative acts must form the basis for the system of collection, storage, processing and distribution of ecological information. Structures responsible for observation and control over the state of the environment should be legally determined; public organizations should have a free access to ecological information. The observational infrastructure should put emphasis on the regional and local monitoring systems, since ecological problems are specific and have a strictly geographical localization. Global monitoring is needed for observing and control of only a part of components.

On the basis of legislative acts, a special infrastructure should be formed for detailed and timely information of the population about the state of natural environment, possible ecological accidents and self-protection, and also for ecological education and taking account of the public opinion.

Ecological legislative acts should be formulated in such a way that they could form the basis for future environmental sustainability. Therefore, all laws and legal acts concerning forms of property, resources, science, education, etc., should be formulated taking account of the ecological approach.

World experience shows that industrially developed countries with the free market systems of economy solve ecological problems via systems of legislative acts and executive structures, which, along with the judicial bodies, are responsible for meeting legal regulations, as well as for funding large-scale ecological projects and research, and for control over following the law and for the expenditures involved. In such countries legal acts are effective. The experience accumulated there should be used in the elaboration of Russian ecological legislative acts, and also directly, through the organization of an expertise of legal projects with the help of international institutions, such as, for example, UNEP.

So, it can be seen that the first priority task for Russia should be the formulation of a programme of creating an infrastructure, above all a legal one, for the solution of problems of the protection and restoration of the violated natural environment. Such a programme will not require large capital investment, since there are special bodies which, according to their status, must implement such a programme, viz. the Parliament's Committee on Ecology and Use of Nature and the Ministry of Environment and Use of Natural Resources, as well as the staff of the Counsellor on Ecology and Medicine to the President of Russia.

Substantial financing of ecological programmes on the part of the government in Russia is hardly possible in the near future; besides, it is known from the earlier experience that such funding is not very effective: the ecological crisis is being aggravated. It is more reasonable to finance ecological programmes, proceeding from priority directions, or from considering individual regions or river basins where the solution of ecological problems is most urgent. The packet of such programmes must include projects involved in international agreements accepted and signed in the times of the USSR and later inherited by Russia, its legal successor. These programmes can be complemented by administrative projects on a local scale, financed from local budgets.

The first priority programme should be that of fundamental research, where the most important component must be the study of the functioning of the biosphere in its interaction with the human society, the mechanisms of this interaction, the existing limitations of the consumption of the environmental resources, and the limits of development both on a global and on regional scales. An analysis of the available concepts of ecological safety is necessary, involving an examination of the possibilities of implementing these concepts. By way of comparison, the most appropriate basic concept should be chosen, so that a long-term strategy can be formulated to ensure the ecological sustainability of Russia in the next century as a part of the global ecological space.

REFERENCES

Annual Bulletin: 1991. Content of residual amounts of pesticides in natural environment objects of the USSR. Books 1 and 2. Obninsk, 298 pp.

Annual Bulletin: 1991. Soil pollution in the USSR by toxicants of industrial origin in 1990. Obninsk, 204 pp.

Collected Papers of Statistical Materials, 1989. 1990. Moscow, Finances and Statistics Publ., 237 pp.

Collected Papers of Statistical Materials, 1990. 1991. Moscow, Finances and Statistics Publ., 352 pp.

Dvoirin V. V., Tserkovny G. F. and Akel E. M. 1991. Cancer cases in the population of the USSR, 1985–1989. Problems of Oncology, Vol. 37, No. 4, p.401–436.

Ecological situation in the RSFSR in 1989. 1990. Problems of Environment and Natural Resources. VINITI, issue 11–12, 172 pp.

Economic and social geography in the USSR. 1990. Moscow, Prosveshcheniye Publ., 272 pp.

Expert estimates of programmes and resolutions concerning liquidation of consequences of the Chernobyl Atomic Power Station accident. 1991. The World Day of Environment. Moscow, VINITI, 128 pp.

Grigoryev A. A. 1991. Ecological Lessons of the Past and Present. Leningrad, Nauka Publ., 252 pp.

Health Care in the USSR. 1990. Statistical Coll. Pap., Moscow, Finances and Statistics Publ., 240 pp.

International Chernobyl Project. 1991. An assessment of radiological consequences and protective measures. Moscow, AT Press, 96 pp.

Izrael Yu. A. 1996. Radioactive Fallout after Nuclear Explosions and Accidents. St. Petersburg, "Progress-Pogoda", 356 pp.

Lemeshev M. Ya. 1991. Before It Is Not Too Late. Moscow, Molodaya Gvardya Publ., 239 pp.

Losev K. S 1989. Water. Leningrad, Gidrometeoizdat Publ., 272 p.

Northern Test Site: Evidence and Documents. 1990. Arkhangelsk, "Infoclub" Agency, 107 pp.

Porfiryev B. N. 1990. Ecological expertise and technological risk. Itogi Nauki i Tekhniki. Series: Nature Protection and Reproduction of Natural Resources, Vol. 27. Moscow, VINITI, 204 pp.

Problems of Russia's Ecology. 1993. Losev K. S., Gorshkov V. G., Kondratyev K. Ya., Kotlyakov V. M., Zalikhanov M. Ch., Danilov-Danilyan V. I., Gavrilov I. T., Golubev G. N., Revyakin V. S., Grakovich V. F. Moscow, 348 pp.

RSFSR in Figures in 1989. 1990. Moscow, Finances and Statistics Publ., 254 pp.

Social Development in the USSR, 1989. 1991. Statistical Coll. Pap. Moscow, Finances and Statistics Publ., 416 pp.

State of Natural Environment in the USSR in 1988. 1989. Problems of Environment and Natural Resources, VINITI, issue 10–11, 204 pp.

State of Natural Environment and Nature Protection Activity in the USSR in 1989. 1990. Moscow, State Committee of the USSR on Nature Protection, 359 pp.

State of the World, 1990. 1990. New York, London, W. W. Norton and Co., XII, 253 p.

World Resources, 1988–89. 1988. New York, Basic Books, Inc., XII, 372 pp.

Conclusion: Key issues of global change at the end of the second millennium

After the Second UN Conference on Environment and Development (UNCED) which was held in Rio de Janeiro in June 1991 the recognition of the importance of global environment problems has greatly increased (Kondratyev et al., 1996a,b; Kondratyev and Cracknell, 1997). The approval of the UNCED documents such as Agenda 21 and Framework Convention for Climate Change (FCCC) by the UN General Assembly has opened certain perspectives of international efforts to combat growing environmental dangers of global scale (climate change, ozone depletion, desertification etc.) A special session of the UN General Assembly took place in June 1997 to discuss the results achieved during the 5-year period after the UNCED. Under such circumstances it is very important to assess the global ecological situation at the end of the second millennium and potential perspectives of future development as a result of interaction between society and environment.

In view of the great number of problems, the critically important task is the establishment of priorities. Unfortunately, it has so happened that an attempt to work out "The Earth Charter" as a general document defining priorities, which was undertaken before the UNCED, failed and, as a substitution, a fairly vague document of pure declarative nature was suggested, although in 1990 a monograph was published with clear enough establishment of basic global ecological priorities (Kondratyev, 1990). At the same time the principle of the biotic regulation of the environment has been suggested as the cornerstone of global ecological safety (Gorshkov, 1995; Gorshkov et al., 1996). The totality of ecological problems of different scale—from global to local—has been discussed by Kondratyev et al. (1996b).

To avoid repetitions we shall first dwell on a few important illustrations which characterize the present-day global ecological situation and then discuss global change priorities as well as potential ways of overcoming growing ecological dangers in the next century.

Undoubtedly, the most annoying feature of the present time is the continuing growth of global population which may be characterized by Table 1 data (see Kondratyev, 1996). The specific features of this growth have been: (1) its

Table 1. World population 1950–95

Year	Total (billion)	Annual addition (million)	Year	Total (billion)	Annual addition (million)
1950	2.555	37	1980	4.457	77
1955	2.779	53	1981	4.533	81
1960	3.038	41	1982	4.614	81
1965	3.345	70	1983	4.695	80
1966	3.414	69	1984	4.775	81
1967	3.484	71	1985	4.856	83
1968	3.555	74	1986	4.941	87
1969	3.629	75	1987	5.029	88
1970	3.704	78	1988	5.117	88
1971	3.782	77	1989	5.205	90
1972	3.859	77	1990	5.295	86
1973	3.936	76	1991	5.381	88
1974	4.012	74	1992	5.469	88
1975	4.086	73	1993	5.556	88
1976	4.159	73	1994	5.644	88
1977	4.231	73	1995 (prel.)	5.732	87
1978	4.304	76			
1979	4.380	77			

concentration in developing countries; (2) the growth of urban population (Table 2). It is obvious that such a situation leads to strengthening socio-economic conrasts (and controversies) between developed ("golden billion") and developing countries under conditions of worsening economic situation in many developing countries. Since the most broadly discussed environmental problem is so-called global warming (it would be adequate to call it global climate change) and in this context—the growth of greenhouse gases (GHGs) emissions to the atmosphere—it is justified to consider what is going on with the CO_2 (as a principal GHG, apart from water vapour) in light of the FCCC recommendation to reduce GHGs emissions to the level of 1990 by the year 2000. Tables 3 and 4 contain relevant information (see Kondratyev, 1996). As can be seen the world carbon emissions from fossil fuel burning are still growing although some countries have undertaken certain measures to reduce emissions. The situation in different countries may be characterized by Table 5 data (Matsarski, 1996). The projections shown in Table 5 demonstrate that quite a number of developed countries (Australia, Canada, Italy etc.) are far from being able to satisfy the FCCC requirements. The low level of carbon emissions in the former USSR countries is a result of declining economics (it is highly probable that this trend will reverse at a later stage of economic development). An important feature of Table 4 data are strong contrasts between emissions per person in developed and developing countries. Even China, in spite of a high level of total

Table 2. World urban population 1950–95

Year	Population (billion)	Share of world population that is urban (%)
1950	0.738	29.3
1955	0.864	31.4
1960	1.033	34.2
1965	1.186	35.5
1970	1.353	36.6
1975	1.538	37.7
1980	1.752	39.4
1985	1.994	41.2
1990	2.277	43.1
1995 (prel.)	2.584	45.2

Table 3. World carbon emissions from fossil fuel burning, 1950–95, and atmospheric concentrations of carbon dioxide, 1960–95

Year	Emissions (million tons of carbon)	Carbon dioxide (parts per million)	Year	Emissions (million tons of carbon)	Carbon dioxide (parts per million)
1950	1.620	**n. a.**	1979	5.239	336.7
1955	2.020	**n. a.**	1980	5.172	338.5
1960	2.543	316.8	1981	5.000	339.8
1965	3.095	319.9	1982	4.960	341.0
1966	3.251	321.2	1983	4.947	342.6
1967	3.355	322.0	1984	5.109	344.3
1968	3.526	322.8	1985	5.282	345.7
1969	3.735	323.9	1986	5.464	347.0
1970	4.006	325.3	1987	5.584	348.8
1971	4.151	326.2	1988	5.801	351.3
1972	4.314	327.3	1989	5.912	352.7
1973	4.546	329.5	1990	5.943	354.0
1974	4.553	330.1	1991	6.010	355.5
1975	4.527	331.0	1992	5.926	356.2
1976	4.786	332.0	1993 (est.)	5.940	357.0
1977	4.920	333.7	1994 (est.)	5.990	358.8
1978	4.960	335.3	1995 (prel.)	6.056	360.7

Table 4. Carbon emissions from fossil fuel burning: top 20 emitters, 1994

Country	Total emissions (million tons)	Emissions per person (tons)	Emissions per dollar GNP[a] (tons per million dollars)	Emissions growth 1990–94 (%)
United States	1371	5.26	210	4.4
China	835	0.71	330	13.0
Russia	45	3.08	590	−24.1
Japan	299	2.39	110	0.1
Germany	234	2.89	140	−9.9
India	22	0.24	160	23.5
United Kingdom	153	2.62	150	−0.3
Ukraine	125	2.43	600	−43.5
Canada	116	3.97	200	5.3
Italy	104	1.81	110	0.8
France	90	1.56	80	−3.2
Poland	89	2.31	460	−4.5
South Korea	88	1.98	200	43.7
Mexico	88	0.96	140	7.1
South Africa	85	2.07	680	9.1
Kazakstan[b]	81	4.71	1250	n.a.
Australia	75	4.19	230	4.2
North Korea[b]	67	2.90	960	n.a.
Iran[b]	62	1.09	270	n.a.
Brazil	60	0.39	70	15.8

[a]GNP data are adjusted for purchasing power parity and are for 1993.
[b]Latest data available are for 1992

Table 5. Anthropogenic emissions of CO_2, excluding land-use change and forestry: relative inventory figures for 1991–94 and projections for 2000 (percentage relative to 1990, 1990 = 100)

	Data from inventories					Data from projections
	1990 (Gg) 1	1991 (%) 2	1992 (%) 3	1993 (%) 4	1994 (%) 5	2000 (%) 6
Australia	288 965					115
Austria	59 200					110
Bulgaria (1990)	82 990					84
Bulgaria (1988)	96 878					72
Canada	462 643	98	101	102	105	113
Czech Republic	165 792	94	86	84		83
Denmark	520 025	121	110	114	121	103

Table 5. *(continued)*

	Data from inventories					Data from projections
	1990 (Gg) 1	1991 (%) 2	1992 (%) 3	1993 (%) 4	1994 (%) 5	2000 (%) 6
Denmark (elect. trade adjusted)	58 278	105	104	103	101	92
Estonia	37 797	97	74	55	57	46–61
Finland	53 900	100	96	97	108	130
France (temp. adjusted)	366 536	106	102	100		104
Germany	1 014 155	96	91	90		90
Greece	82 100					115
Hungary (1990)	71 673					99
Hungary (1985–1987)	83 676					84
Iceland	2 172	96	101	106		105
Ireland	30 719					< 120
Italy	428 941					114
Japan	1 155 000	102	103	101	107	102
Latvia	22 976					< 74
Liechtenstein	208					118
Luxembourg	11 343					67
Monaco	71					
Netherlands	167 600	104	103	104	105	> 100
Netherlands (temp. adjusted)	174 000	100	101	100	102	> 96
New Zealand	25 476	102	110	107	108	> (114–117)
Norway	35 514	95	96	101	106	114
Poland (1990)	414 930	96	90			
Poland (1988)	478 880	83	78			74-99
Portugal	42 148					140
Romania (1990)	171 103	83	72	70		
Romania (1989)	198 479	71	62	61		
Russian Federation	2 388 720					83-87
Slovakia	58 278					84
Spain	227 322					116
Sweden	61 256	89	92	90	95	104
Switzerland (temp. adjusted)	45 070	103	101	98	96	96
United Kingdom	577 012	102	99	97	96	92-96
United States	4 957 022	99	100	103	103	> 103

Note: Gg = 1,000 tonnes

emissions, is at the very end of the list of countries with low per person emissions (India is the last on this list). In view of the necessity of further industrial development in such countries as China, India and others it is obvious that their total emissions will increase in the future.

An important question in this context is the adequacy of the FCCC recommendation to reduce GHGs emissions. This question has been discussed in detail by Kondratyev et al. (1996a,b) from the viewpoint of the analysis of the UNCED and the IPCC (Intergovernmental Panel on Climate Change) documents. On the one hand it is obvious that, generally speaking, the reduction of GHGs emission is a very useful measure. But, on the other hand, it is equally clear that such a measure is not a panacea against global change dangers. It is very well known truth that global climate change results from the interaction between all components of the climatic system "atmosphere–ocean–land" surface (including biosphere)—cryosphere" under the still unknown impact of solar activity. Present-day numerical modelling (even in case of 3-D coupled global models) is far from being able to reliably simulate real climate change and, consequently, to identify contributions of various climate-forming factors (including the enhancing greenhouse effect of the atmosphere). A curious illustration of this circumstance is the recent emphasis on the consideration of "aerosol cooling" as counteracting "greenhouse warming" (one has to point out that the necessity of studying the atmosphere as a colloidal medium was discussed by Schmauss and Wiegand about 70 years ago and subsequently this problem has been considered in numerous publications, including a number of recent monographs on aerosol and climate problems which have not been referenced in the most recent documents mentioned).

It is in view of the joint consideration of "greenhouse warming" and "aerosol cooling" that the statement has been made in the Second IPCC Assessment Report (Bolin, 1996; Houghton et al., 1996) that "the balance of evidence now suggest that there is a discernible human influence on climate". Nobody doubts that at the present time human influence on climate is a reality. The problem is, however, that an anthropogenic contribution to climate change has not been assessed quantitatively as yet. This is why the authors of the statement could go no further than to mention "a discernible human influence" (this is, unfortunately, fairly vague definition) under non-specified "the balance of evidence". Apart from the choice of better semantics (by definition, "the balance" can indicate only an equality of different factors), it would be worthwhile to accomplish a concrete analysis of "the balance of evidence". It is well known, for instance (although from the publications which have not been referenced in the IPCC Report—see Kondratyev et al., 1996b), that an assumption of pure scattering aerosol (hence "aerosol cooling") is not acceptable, because real atmospheric aerosols always contain components absorbing in both the short wave and the long wave regions. Aircraft observations of spectral shortwave radiation fluxes under different condition have shown, for example, that, on the average, solar radiation absorption by aerosols and water vapour is of the same order of magnitude.

Thus, as far as climate is concerned, the task is to study climate in all its complexity without overemphasis on certain individual factors of climate change such as the atmospheric greenhouse effect. An important aim is to identify the place and role of climate change within a more general domain of global change. As has been shown by Gorshkov (1995), the basic processes which regulate environmental dynamics are based on the principle of the biotic regulation of the environment. If we accept such a concept then the following succession of global change priorities may be suggested as a basis for further discussion:

1. Biotic regulation of environmental dynamics.
   Hence priorities of:

   1.1 Biosphere dynamics: biochemical cycles (C, N, S, P, etc.); use of biospheric resources (thresholds); global monitoring of both terrestrial and marine biosphere dynamics (relevant indicators).
   1.2 Consumption of biospheric resources. Use of renewable and non-renewable resources. Developed and developing countries; countries with transitional economy.
   1.3 Life standards: "golden billion" and the rest of the world. Sustainability: ecological, socio-economic, political, ethical.
   1.4 Carrying capacities on national, regional and global levels for various components of nature: land, forests, seas, inland waters, etc.
   1.5 Sustainable development and population dynamics.

2. Water. Drinking water deficit.
3. Energy. Fossil fuels versus renewable sources (wind, tides, etc.). Future of nuclear and hydrogen energy. Coupled energy production development and environmental dynamics.
4. Food. Impacts of environmental dynamics.
5. Environment. Key issues for life support:

   5.1 Climate change (internal variability, external impacts: anthropogenic contribution (GHGs and aerosols), volcanic eruptions, solar activity).
   5.2 Stratospheric ozone depletion.
   5.3 Tropospheric ozone increase (UV-B enhancement and subsequent impacts on man and ecosystems).
   5.4 Polluted urban atmospheres and health.
   5.5 Environmental economics (ecological taxes, etc.).
   5.6 Environmental ethics.

This preliminary scheme of priorities demonstrates the subordinate role of the climate change problem within a much more general global change domain. In view of a certain overemphasis of the UNCED on global climate change as well as an overemphasis of the FCCC on the atmospheric greenhouse effect it would be extremely important to undertake a re-analysis of relevant documents in the course of preparations to the Special UN Assembly devoted to the UNCED process. A part of the effort could be accomplished in connection with the preparations to the Third Conference of the Parties (COP-3) of the FCCC held in Kyoto in December 1997.

In the context of such a suggestion the following draft of a declaration based on conceptual views discussed by Gorshkov et al. (1996) may be put forward for further consideration as the equivalent of the Earth Charter:

(1) The Earth is the unique planet in the Solar System because of the existence of life on this planet in the form of biota—a totality of all living organisms, including human beings. Important properties of life include: biological stability of species and their communities as well as very rigid distribution of energy fluxes absorbed by biota over organisms of various sizes. The biota itself is responsible for the formation of its environment and for the stability of optimal environmental properties in accordance with its needs. Only because of this has such a long-term existence of the Earth's biota has been possible.

Supporting a stable environment is the principal purpose of all living organisms.

(2) Like other species *Homo sapiens* is one of the species of the biota and therefore its principal aim is to also support the global biosphere. Otherwise it would be impossible to develop sustainably. Long ago human beings left their natural ecological niche and started to consume much more biospheric resources than ecological equilibrium allows. But, after the industrial revolution started, this process of the violation of the natural equilibrium accelerated continuously under conditions of rapid growth of population.

(3) Simple approximate assessments demonstrate that to guarantee a stable biosphere it is possible to use not more than 1% of biospheric resources. Now it is close to 10%. Similar figures characterize the evolution of global biogeochemical cycles. For instance, the natural global carbon cycle was closed with an accuracy of the order of 0.01% (this is where biodiversity plays a decisive role). Now this accuracy is of the order of 0.1% (the threat of global ecological catastrophe has become visible).

Hence: we have to consider the biosphere not just as a resource, but the fundamental condition of the continuation of life on Earth.

Now, the basic aim is to restore the already disturbed biosphere and keep it under conditions which will guarantee sustainable development.

There are, however, many unknown or poorly quantified aspects of assessing biosphere dynamics. In this context a decisive role belongs to simulation modelling and adequate observing systems. The latter problem is of special importance since there is still no adequate life support (global change) observing systems (even at a planning stage).

(4) A critically important feature of the development of contemporary civilization is inequality between developed and developing countries expressed as an unequal use of existing resources. Per capita consumption of the "golden billion" is incomparably higher than for rest of the world. A new social order is therefore necessary to be agreed upon the basis of international cooperation and partnership.

One can have healthy life support systems only through relevant international agreements.

(5) Undoubtedly, more efforts are necessary to develop clean and efficient technologies, minimizing the use of energy and raw materials.

The decisive measure is, however, only a reduction of the size of global population. This is, of course, a prohibitively difficult problem which involves serious socio-economic, religious, ethical, and other aspects.

Only global partnership, under conditions of democracy and respect for all people and nations, can lead to the solution of the problem.

REFERENCES

Bolin, B. (1996). A brief overview of the IPCC Second Assessment Report, UN Climate Change Bulletin, Issue 10, 1st quarter 1996, pp.1–2.

Gorshkov, V. G. (1995). Physical and Biological Bases of Life Stability—Man, Biota, Environment, Springer-Verlag, Berlin, 340 pp.

Gorshkov, V. G., Kondratyev, K. Ya., Losev K. S. (1996). Ecological Problems of Contemporary Civilisation, Vestnik of the Russian Acad. Sci., No 2. (in Russian).

Houghton, J. T., Meira Filho, L. G., Callander, B. A., Harris, N., Kattenberg, A., Maskell, K. (Eds.) (1996). Climate Change 1995, The Science of Climate Change, Cambridge University Press 572 pp.

Kondratyev, K. Ya. (1990). Key Problems of Global Ecology, VINITI Press, Moscow (in Russian).

Kondratyev, K. Ya. (1996). Global Change and Demographic Dynamics, Izv. of the Russian Geogr. Soc. Vol. 128, No. 3 (in Russian).

Kondratyev, K. Ya. and Varotsos C. (1996). Environmental Physico-Chemistry, Vol. 1, Radiation, Greenhouse effect, Climate Change, Edited by P. Travlos and E. Kostarkis (in Greek).

Kondratyev, K. Ya., Buznikov, A. A., Pokrovsky, O. M., 1996a: Global Change and Remote Sensing, Wiley–Praxis, Chichester, 284 pp.

Kondratyev, K. Ya., Donchenko, V. K., Losev, K. S., Frolov, A. K. (1996b). Ecology – Economics – Policy, Russian Acad. Sci./Nansen Int. Environ. & Remote Sens. Centre, St. Petersburg, 828 pp.

Kondratyev K.Ya., Peña F.M., Galindo I. (1997). Sustaining Development and Population Dynamics. Univ. of Colima Press, Mexico, 128 pp.

Kondratyev, K. Ya., Cracknell, A. P. (1998). Key Aspects of Global Change, World Sci. Publ., London (in press).

Matsarski, V. (1996). Implementing the Convention: A second review, UN Climate Change Bulletin, Issue 11, 2nd quarter 1996, p.4-7.

Additional references

INTRODUCTION

Abe T., Levin S.A., Higashi M. (1996). Biodiversity. An Ecological Perspective. Springer, Berlin. xii + 294 pp.

Anderson J., Brook C., Cochrane A. (1995). A Global World? Re-Ordering Political Space. The Open University Press, Oxford, UK. 287 pp.

Ayres R. (1997). Turning Point. The End of the Growth Paradigm. Earthscan. London. 224 pp.

Bak P. (1996). How Nature Works. The Science of Self-organised Criticality. Springer, Berlin. ix + 212 pp.

Bakkes J., Van Woerden J. (eds.). (1997). The Future of the Global Environment: A Model-based Analysis Supporting UNEP's First Global Environment Outlook. RIVM 402001007 and UNEP/DEIA/TR. 97–1. BA Bilthoven, The Netherlands. 154 pp.

Bate R (ed.) (1997). What Risk? Butterworth, Heinemann, Oxford. 328 pp.

Boehmer-Christiansen S. (1995). Reflections on the politics linking science, environment and innovation. Innovation, Vol.8, N.3, p.275–287.

Boehmer-Christiansen S. (1996). Political pressure in the formation of scientific consensus. Energy & Environment, Vol.7, Iss.4, p.365–375.

Brune D., Chapman D.V., Gwynne M., Paryna J.M. (eds.). (1997). The Global Environment. Vol.1: Basic Sciences, Exploitation and Degradation. Vol.2: Human Interactions, Technology, Restriction, Management and Preventive Strategy. Wiley-VCH, UK.

Carew-Reid J., Prescott-Allen R., Bass S., Dalal-Clayton B. (1994). Strategies for National Sustainable Development. A Handbook for their Planning and Implementation. Earthscan Publ. Ltd., London, 203 pp.

Clymer M.G., Mechoso C.R. (eds.). (1997). Mission Earth: Modeling and Simulation for a Sustainable Global System. Simulation Councils, Inc., San Diego, 96 pp.

Cohen M.J. (1997). Risk society and ecological modernisation. Alternative visions for post-industrial nations. Futures, Vol.29, N.2, p.105–120.

Dahl A.L. (1996). The Eco Principle. Ecology and Economics in Symbiosis. George Ronald, Oxford, Zed Books Ltd, London & New Jersey, 180 pp.

Daily G.C., Alexander S., Enlich P.R., Goulder L., Lubchenko J., Maton P.A., Mooney H.A., Postel S., Schneider S.H., Tilman D., Woodwell G.M. (1997). Ecosystem services: benefits supplied to human societies by natural ecosystems. Issues in Ecology, N.2, p.1–15.

Döös R. (1997). Can large-scale environmental migrations be predicted? Global Environmental Change, Vol.7, N.1, p.41–61.

Döös R. (1997). Predicting environmental migration. Tiempo, Issue 24, p.1–10.

den Elzen M. (1996). Global Environmental Change: An Integrated Modelling Approach. Jon Carpenter Publ., The Netherlands, 263 pp.

Global Environmental Outlook. (1997). UNEP/Oxford University Press. New York, 264 pp.

Grigoryev A.A., Kondratyev K.Ya., (1997). Global population dynamics. Proc. Russian Geographical Soc., Vol.129, N.4, p.1–10 (in Russian).

Hardi P., Zdan T. (1997). Assessing Sustainable Development: Principles in Practice. IISD, Winnipeg, Manitoba, Canada, 166 pp.

Holland H.D., Petersen U. (1996). Living Dangerously: The Earth, Its Resources, and the Environment. Princeton University Press. 490 pp.

Huntley B., Cramer W., Moran A.V., Prentice H.C., Allen J.R.M. (eds.). (1997). Past and Future Rapid Environmental Changes. The Spatial and Evolutionary Responses of Terrestrial Biota. Springer, Berlin, NATO ASI Ser. 1, Vol.47, xiv + 523 pp.

Jänicke M., Weidner H. (eds.). (1997). National Environmental Policies. A Comparative Study of Capacity-Building. Springer, Berlin, 320 pp.

Kates R.W. (1996). Population, technology, and the human environment: a thread through time. Daedalus, Vol.125, N.3, p.43–71.

Kondratyev K.Ya. (1997). Key issues of global change at the end of the second millennium. Newsletter European Geographical Society, N.63, p.4–8.

Kondratyev K.Ya. (1997). The Present-Day Global Ecodynamics. Studying the Earth from Space, N.5, p.105–126.

Lutz W. (ed.). (1996). The Future Population of the Earth. What Can We Assume Today? Revised 1996 Edition. IASA/Earthscan Publ. Co. Ltd., London, 500 pp.

May G. (1996). The Future is Ours: Foreseeing, Managing and Creating the Future. Adamantine Press, London, 253 pp.

Munn R.E., La Rivière J.W.M., van Lokeren-Campagne N. (1996). Policy Making Global Change: Policy Making in an Era of Global Environmental Change. Kluwer Acad. Publ., Norwell, MA., 225 pp.

O'Riordan T. (ed.). (1996). Ecotaxation. Earthscan, London, 192 pp.

Our Changing Planet. The FY 1998 US Global Change Research Program. (1997). Washington, DC, 118 pp.

Population Dynamics and Management of Urban Environment (1996). Special Plenary Session. IAF 47th Congress. Beijing. 36 pp.

Porteous J.D. (1996). Environmental Aesthetics. Ideals, Politics, and Planning. Routledge, London, 290 pp.

Sadgrove K. (1997). A–Z Corporate Environmental Management. Earthscan, London, 356 pp.

Shulz-Baldes M. (1995). Wege zur Lösung globaler Umweltprobleme. Springer-Verlag. Heidelberg, xii + 275s.

van der Sluijs J.P. (1997). Anchoring amid Uncertainties. Ludy Feyen, Leiden. 266 pp.

Spirin A.S. (1997). Modern biology and biological security. Herald of the Russian Acad. Sci., Vol.67, N.7, p.579–588.

Starr C. (1996). Sustaining the human environment: the next two hundred years. Daedalus, Vol.125, N.3, p.235–253.

Stepin V.S. (1997). Philosophy at the turn of the centuries. Herald of the Russian Acad., Sci., Vol.67, N.5, p.387–395.

The Potential Benefits of Integration of Environmental and Economic Policies. An Incentive Based Approach to Policy Integration by European Commission. (1996). Kluwer Acad. Publ., Dordrecht. 448 pp.

Vitousek P.M., Mooney H.A., Lubchenko J., Melillo J.M. (1997). Human domination of earth's ecosystems. Science, Vol.277, p.494–499.

Vogler J., Imber M.F. (eds.). (1996). The Environmental and International Relations. Routledge. London, 236 pp.

Walker B., Will S. (1996). Global Change and Terrestrial Ecosystems. Cambridge Univ. Press, 619 pp.

von Weizsäcker E., Lovins A.B., Lovins L.H. (1997). Factor Four. Doubling Wealth, Halving Resource Use. Earthscan, London, 352 pp.

Welford R. (1996). Corporate Environmental Management. Systems and Strategies. Earthscan, London, 288 pp.

Welford R. (1997). Hijacking Environmentalism. Corporate Responses to Sustainable Development. Earthscan Publ. Ltd. 251 pp.

Welt im Wandel. Wege zur Lösung globaler Umweltprobleme. Jahresgutachten 1995 (1996). Springer, Berlin, 247 pp.

World in Transition: Ways Towards Global Environmental Solutions. (1997). German Advisory Council on Global Change (WRGU). Annual Report 1996. Springer, Berlin, ix + 234 pp.

Yablokov A., Ostroumov S. (1991). Conservation of Living Nature and Resources Problems, Trends and Prospects. Springer-Verlag, Berlin, 271 pp.

Yablokov A. (1993). Nature Conservation and Biodiversity. The European Common Garden. Towards the Building of a Common Environmental Policy. Strasbourg. p.117–135.

Zillman J.W. (1997). International co-operation in Integrated Global Observing systems. Keynote Lecture. First Symposium on Integrated Observing Systems. Amer. Meteorol. Soc. 77th Annual Meeting, Long Beach, CA, 21 pp.

CHAPTER 1

Advancing Sustainable Development. The World Bank and Agenda 21. (1997).

Environmentally Sustainable Development Studies and Monographs Series N.19. The World Bank. Washington, DC. 80 pp.
Agarwal B. (1997). Gender, environment, and poverty interlinks: regional variations and temporal shifts in rural India, 1971–91. World Development, Vol.25, N.1, p.23–52.
Andrews J.E., Brimblecombe P., Jickells T.D., Liss P.S. (1996). An Introduction to Environmental Chemistry. Blackwell Science Ltd., Oxford, 209 pp.
Arrow K.J., Cropper M.L., Eads G.C., Hahn R.W., Lave L.B., Noll R.G., Portney P. R., Russel M., Schmalennsee R., Smith V.K., Stavins R.N. (1996). Is there a role for benefit–cost analysis in environmental, health, and safety regulation. Science, Vol.272, N.5259, p.221–222.
Ausubel J.H. (1996). The liberation of the environment. Daedalus, Vol.125, N.3, p.1–18.
Bowman K.S. (1997). Should the Kuznets effects be relied on to induce equalizing growth: evidence from post-1950 development. World Development, Vol.25, N.1, p.127–143.
Briceño S. (1996). Seeking synergies among the Rio Conventions. Climate Change Bull., Iss.13, p.2–3.
Brophy M., Netherwood A., Starkey R., Welford R., Young W. (1995). International Sustainable Development Research Network: Conference Report. Sustainable Development, Vol.3, N.2, p.55–62.
Bugmann H.K.M., Yan X., Sykes M.T., Martin P., Linder M., Desanker P.V., Cumming S.G., (1996). A comparison of forest gap models: model structure and behaviour. Clim. Change, Vol.34, N.2, p.289–313.
Campbell J.B. (1996). Introduction to Remote Sensing. Second Edition. Taylor & Francis, London, 620 pp.
Chelyshev E.P. (1997). Stable development of fishery: is it possible? Herald of the Russian Acad. Sci., Vol.67, N.7, p.589–595.
Chepurnykh N.V., Novoselov A.A. (1996). Economics and Ecology: Development, Catastrophes. "NAUKA", Moscow. 272 pp. (in Russian).
Coenen R., Klein-Vielhauer S. (1997). The significance of environmental technology for economically and ecologically sustainable development. The IPTS Report, N.14, p.5–11.
Cohen J.E. (1995). How many People can the Earth Support? Norton, New York, 532 pp.
Dobrovol'skii G.V. (1997). The silent crisis of the planet. Herald Russian Acad. Sci., Vol.67, N.4, p.313–320.
Driver T.S., Chapman G.D. (eds.). (1996). Time-Scales and Environmental Change. Routledge, London, 275 pp.
Eddy J.A., Oeschger H. (1993). Global Changes in the Perspective of the Past. John Wiley & Sons, Chichester, 383 pp.
Edmann K.-H., Kastenholz H.G. (eds.). (1995). Umwelt-und Naturschultz am Erde des 20. Jahrunderts. Probleme, Aufgaben und Lösungen. Springer, Berlin, 265 pp.
Environmental matters at the World Bank. Winter/Spring 1997. Washington DC, 32 pp.

Filatov N.N. (1997). Geographic Information Systems. GIS Applications for Environmental Studies. Petrozavodsk, KGPU, 102 pp. (in Russian).

Five Years after Rio. Innovations in Environmental Policy. (1997). Environmentally Sustainable Development Studies and Monograph Series N.18, The World Bank, Washington DC, 53 pp.

GCOS/GTOS Plan for Terrestrial Climate-Related Observations. Version 2.0 1997. GCOS-32. WMO/TD N.796, UNEP/DEIA/TR.97-7. Geneva.

Gitel'zon I.I., Bartsev S.I., Okhonin V.A., Sukhovol'skii V.G., Khlebopros R.G. (1997). The strategy of development: what kind. Herald of the Russian Acad. Sci., Vol.67, N.5, p.415–420 (in Russian)

Global Ocean Ecosystem Dynamics (GLOBEC) Science Plan. (1997) SCOR/IGBP Report 40. CLOBEC Report 9. Stockholm, 82 pp.

Goody R. (1997). Observing the Earth: a Frankenstein dilemma. Studying the Earth from Space. N.4, p.93–99 (in Russian).

Gumilev L.N. (199). Ethnogenesis and the Earth's Biosphere. Leningrad: Gidrometeoizdat, 528 pp. (in Russian).

Häberle H. (1995). Environment reflecting human development. Annals of the European Academy of Sciences, Vol.10, Evolutio Humana Europea 1995, Salzburg, p.43–51.

Hammitt J.K. (1995). Outcome and value uncertainties in global change policy. Clim. Change, Vol.30, N.2, p.125-145.

Human Diversity. (1997). J. Amer. Acad. Arts and Sci., Vol.126, N.2, 232 pp.

Janetos A.C. (1997). Do we still need nature? The importance of biological diversity. Consequences, Vol.3, N.1, p.17–26.

Kondratyev K.Ya., Sumi A., Pokrovsky O.M. (1997). Global Change and Climate Dynamics: Optimisation of Observing Systems. University of Tokyo for Climate Systems Study CCSR Report N.3, 213 pp.

Kondratyev K.Ya. (1997). Global ecodynamics and sustainable development: nature aspects and human dimensions Izv. Russian Geograph. Soc., Vol.129, N.6 (in Russian).

Kramer H.J. (1996). Observation of the Earth and Its Environment. Springer-Verlag, Berlin, 960 pp.

Liss P.S., Duce R.A. (eds.). (1996). The Sea Surface and Global Change. Cambridge University Press, 496 pp.

Lubchenco J., Olson A.M., Brubaker L.B., Carpenter S.R., Holland M.M., Hubbell S.P., Levin S.A., MacMahon J.A., Matson P.A., Melilo J.M., Mooney H.A., Regal Ph.J., Risser P.G. (1991). The sustainable biosphere initiative: an ecological research agenda. Ecology, Vol.72, N.2, p.371–412.

Marchuk G.I. (1997). Conjugated equations and sensitivity of functionals. Studying the Earth from Space, N.4, p.100–125 (in Russian).

Nentwig W. (1995). Humanökologie.Fakten-Argumente-Ausblicke. Springer-Verlag, Berlin, xiv + 5885.

Population and Global Security (Environmental Challenges II). (1994). Ed. by N. Polunin and M. Nazim. Limited Geneva Edition for UN Population Fund et al. Geneva, 285 pp.

Programme for the Further Implementation of Agenda 21. Adopted at the Nineteenth Special Session of the UN General Assembly (23–27 June 1997). Advance unedited text. 7 July 1997. E/1997/MESC.3. 6E.97 62490, 52 pp.
Rao U.R. (1996). Space Technology for Sustainable Development. Tata McGraw-Hill Publ. Co. Ltd., New Delhi, 564 pp.
Ravalion M. (1997). Good and bad growth: the human development reports. World Development, Vol.25, N.5, p.631–638.
Schlesinger W.H. (1991). Biogeochemistry. An Analysis of Global Change. Academic Press, San Diego, 443 pp.
Schott J.R. (1997). Remote Sensing. The Image Chain Approach. Oxford Univ. Press, New York, 394 pp.
Smith K. (1996). Environmental Hazards: Assessing Risk and Reducing Disaster. Second edition. Routledge, London, 389 pp.
Stigliam W.M. (1993). Chemical Time Bombs: Definition, Concepts and Examples. Ecoscript 50. Foundation for Ecodevelopment, Amsterdam, 22 pp.
Sustainable Development. The UK Strategy. (1994). London: HMSO, 267 pp.
The Impact of Environmental Assessment. The World Bank's Experience. Second Environmental Assessment Review 1996. Experimental Department. The World Bank. Washington DC, 146 pp.
The Istanbul Declaration and the Habitat Agenda. (1997). UN Centre for Human Settlements (Habitat). Nairobi, 172 pp.
Trofimov A.M., Kotlyakov V.B., Yu P. (1997). Theoretical aspects and main conditions in modelling of complex ecological-economic systems. Izv. Russian Geograph. Soc., Vol.129, N.3, p.6–14 (in Russian).
Vasilenko I.A. (1997). Universal values: can they be protected? Herald Russian Acad. Sci., Vol.67, N.6, p.514–517.
Waggoner P.E. (1996). How much land can ten billion people spare for nature? Daedalus, Vol.125, N.3, p.73–93.

CHAPTER 2

An Introduction to Simple Climate Models used in the IPCC Assessment Report. (1997). Ed. by J.T. Houghton, L.G. Meira Filho, D.J. Griggs, K. Maskell. IPCC Tech. Paper II, WMO, Geneva, 50 pp.
Arrhenius S. (1896). Über den Einfluss des atmosphärishen Kohlensäueregehalts auf die Temperatur der Erdoberfläche. Bih. Kunge. Sr. Vet. Akad. Handl. 22 (Sect.1, N.1), p.1–102.
Arrhenius S. (1896). On the influence of carbonic acid in the air upon the temperature of the ground. Phil. Mag., Vol.41, p.237–276.
Baliunas S., Frick P., Sokoloff D., Soon W. (1997). Time scales and trends in the Central England temperature data (1659–1990): A wavelet analysis. Geophys. Res. Lett., Vol.24, N.11, p.1351–1354.
Baliunas S.L., Soon W.H. (1997). An assessment of the sun–climate relation on time scales of decades to centuries: the possibility of total irradiance variations. Harvard-

Smithsonian Center for Astrophysics. Preprint Series N.4565, Cambridge, MA, 10 pp.

Barnett T.P. (1995). Monte Carlo climate forecasting. J. Climate, Vol.8, N.5, p.1005–1022.

Barrrett J. (1996). Do CO_2 emissions pose a global threat? In: Emsley J. (ed.), The Global Warming Debate, Bourne Press Ltd, Bournemouth, p.60–70.

Basist A.N., Ropelewski C.F., Grody N.C. (1995). Comparison of tropospheric temperature derived from the Microwave Sounding Unit and National Meteorological Center analysis. J. Climate, Vol.8, N.4, p.668–681.

Bengtsson L. (1997). A numerical simulation of anthropogenic climate change. Ambio, Vol.26, N.1, p. 58–65.

Beran M.A. (1995). Carbon Sequestration in the Biosphere Processes and Prospects. NATO ASI Series I: Global Environmental Change, Vol.33, Springer-Verlag, Berlin, 305 pp.

Berger A., Loutre M.-F. (1997). Paleoclimate sensitivity to CO_2 and insolation. Ambio, Vol.26, N.L, p.32–37.

Berner R.A. (1997). The rise of plants and their effort on weathering and atmospheric CO_2. Science, Vol.276, N.5312, p.544–546.

Berntsen T., Isaksen I.S.A., Wang W.-C., Liang X.-Z. (1996). Impacts of increased anthropogenic emissions in Asia on tropospheric ozone and climate. A global 3-D model study. Tellus, Vol.48B, N.1, p.13–32.

Biggs R.H., Joyner M.E.B. (eds.) (1994). Stratospheric Ozone Depletion /UV-B Radiation in the Biosphere. NATO ASI Series. Series I: Global Environmental Change, Vol.18, Springer-Verlag, Berlin, 358 pp.

Bintanja R., Fortuin J.P.F., Kelder H. (1997). Simulation of the meridionally and seasonally varying climate response caused by changes in ozone concentration. J. Climate, Vol.10, N.6, p.1288–1311.

Blersh D. (1997). The National Polar-orbiting Operational Environmental Satellite System. (NPOESS). Earth System Monitor, Vol.7, N.3, p. 6–8, 16.

Bodeker G.E., McKenzie R.L. (1996). An algorithm for inferring surface UV irradiance including cloud effects. J. Appl. Meteorol., Vol.35, N.10, p.1860–1877.

Boehmer-Christiansen S. (1996). Political pressure in the formation of scientific consensus. In: Emsley J. (ed.), The Global Warming Debate. Bourne Press Ltd, Bournemouth, p.234–248.

Bojkov R.D. (1995). The Changing Ozone Layer. WMO/UNEP. Geneva, 25 pp.

Bojkov R.D., Fioletov V.E. (1995). Estimating the global ozone characteristics during the last 30 years. J. Geophys. Res., Vol.100, N.D8, p. 16537–16553.

Böttcher F. (1996). Climate change: forcing a treaty. In: Emsley J. (ed.) The Global Warming Debate, Bourne Press Ltd, Bournemouth, p.267–285.

Boyd Ph., Newton Ph., Rivkin R.B., et al. (1997). Measuring biogenic carbon flux in the ocean. Science, Vol.275, p.554–555.

Braatz B.V., Jallow B.P., Molnár S., Murdiyarso D., Perdomo M., Fitzgerald J.F. (eds.) (1996). Greenhouse Gas Emission Inventories. Kluwer Acad. Publ., Dordrecht, 404 pp.

Brauch H.G. (ed.) (1996), Klimapolitik. Springer, Berlin, 450 pp.

Brekke A. (1997). Physics of the Upper Polar Atmosphere. Wiley-Praxis, Chichester, England, 491 pp.

Bryan F.O., Gent P.R. (1997). Progress in coupled climate system modeling. U.S. WOCE Report 1997. WOCE Office, College Station, TX. p.34–36.

Bryson R.A. (1997). The paradigm of climatology. Bull. Amer. Meteorol. Soc., Vol.78, N.3, p.449–456.

Bush A.B.C. (1997). Numerical simulation of the Cretaceous téthys circumglobal current. Science, Vol.275, N.5301, p.807–810.

Cai W. (1996). Surface thermohaline forcing conditions and the response of the present-day global ocean climate to global warming. J. Geophys. Res., Vol.101, N.C1, p.1079–1094.

Cai W.J., Syktus J., Gordon H.B., Ofarrell S. (1997). Response of a global coupled ocean–atmosphere–sea ice climate model to an imposed North Atlantic high-latitude freshening. J. Climate, Vol.10, N.5, p.929–948.

Cane M.A., Clement A.C., Kaplan A., Kushnir Y., Pozdnyakov D.V., Sager R., Zebiak S.E., Murtugudde R. (1997). Twentieth-century seasurface temperature trends. Science, Vol.275, p.957–960.

Carton J.A. (1997). See-saw sea. Nature (Gr. Brit.), Vol.385, N.6616 p.487–489.

Chahine M.T. (1997). GEWEX strategies and goals are affirmed by Joint Scientific Committee. GEWEX News, Vol.7, N.3, p.2.

Chandra S., Jackman Ch.H., Fleming E.L. (1995). Recent trends in ozone in the upper stratosphere: Implications for chlorine chemistry. Geophys. Res. Lett., Vol.22, N.7, p. 843–846.

Chandra S., Lean J.I., White O.R., Prinz D.K., Rottman G.L., Bruechner G.E. (1995). Solar UV irradiance variability during the declining phase of the solar cycle 22. Geophys. Res. Lett., Vol.22, N.18, p.2418–2484.

Chandra S., Varotsos C., Flynn L.E. (1996). The midlatitude total ozone trends in the northern hemisphere. Geophys. Res. Lett., Vol.23, N.5, p.555–558.

Chapin F.S., III, Starfield A.M. (1997). Time lags and novel ecosystems in response to transient climate change in Arctic Alaska. Clim Change, Vol.35, N.4, p.449–461.

Chatterjee K. (1997). AIJ: Developing countries' perspectives. Tiempo, Issue 23, p.11–14.

Chen W.Y., Van den Dool H.M. (1997). Atmospheric predictability of seasonal, annual, and decadal climate means and the role of the ENSO cycle: A model study. J. Climate, Vol.10, N.6, p.1236–1254.

Chen T.H., Henderson-Sellers A., Milly P.C.D. et al. (1997). CABAW Experimental results from the project for intercomparison of land-surface parameterization schemes. J. Climate, Vol.10, N.6, p.1194–1215.

Christy J.R., Spencer R.W., McNider R.T. (1995). Reducing noise in the MSU daily lower-tropospheric global temperature dataset. J. Climate, Vol.8, N.4, p.888–896.

CLIVAR International DecCen Workshop on atmosphere–ocean interactions and their influence on decadal-scale climate variability (Vancouver, Canada, September 4–6, 1996). Summary. ICSU/IOC/WMO, 22 pp.

Collier U., Löfstedt R.E. (1997). Cases in Climate Change Policy. Earthscan Publ. Ltd., London, 256 pp.

Cook E. (1996). Milestones in Ozone Protection. Earthscan Publ. Ltd, London, 100pp.
Cook E.R., Meko D.M., Stockton C.W. (1997). A new assessment of possible solar and lunar forcing of the bidecadal drought rhythm in the Western United States. J. Climate, Vol.10, N.6, p.1343–1356.
Corbyn P. (1996). Does CO_2 respond to global temperature changes rather than cause them? Counter evidence to greenhouse assumptions from ice core data and volcanoes. In: Emsley J. (ed.), The Global Warming Debate, Bourne Press Ltd, Bournemouth, p.71–77.
Corbyn P., Golipour M. (1996). What is global temperature? The inadequacy of estimation methods. In: Emsley J. (Ed.) The Global Warming Debate. Bourne Press Ltd, Bournemouth, p.80–86.
Covey C., Sloan L.C., Hoffert M.I. (1996). Paleoclimatic data constraints on climate sensitivity: the paleocalibration method. Clim. Change, Vol.32, N.2, p.165–184.
Crommelynck D., Fichot A., Domingo V., Lee III, R. (1996). SOLCON solar constant observations from the ATLAS missions. Geophys. Res. Lett, Vol.23, N.17, p.2293–2296.
Crowley T.J. (1993). Use and misuse of the geologic "analogs" concept. In: Global Changes in the Perspective of the Past, Eddy J.A. and Oeschger H. (eds.). John Wiley & Sons Ltd, Chichester, UK, p.17–27.
Crowley T.J. (1997). The problem of paleo-analogs. Climate Change, Vol.35, N.1, p.119–121.
Crutzen P.J., Grooss J.-U., Brühl C., Müller R., Russell III, J.M. (1995). A reevaluation of the ozone budget with HALOE UARS data: No evidence for the ozone deficit. Science, Vol.268, N.5211, p.705–707.
Currie R.G. (1996). Variance contribution of luni-solar (Mn) and solar (Sc) signals to climate data. Int. J. Climatol., Vol.16, N.12, p.1343–1364.
Dalfes H.N., Kukla G., Weiss H., (eds.), (1997). Third Millennium BC Climate Change and Old World Collapse. NATO ASI Series, Vol.49, Springer, Berlin, xiv + 728 pp.
Daniel J., Solomon S., Albrition D. (1995). On the evaluation of halocarbon radiative forcing and global warming potentials. J. Geophys. Res., Vol.100, N.D1, p.1271–1285.
Davis C., Godin S., Megie G., Mery Y., Flesia C. (1997). Physical state and composition of polar stratospheric clouds. J. Atmos. Chem., Vol.e27, N.1, p.1–16.
Delobbe L., Tricot C., Van Ypersele J.P. (1997). Tropopause-height feedback and radiative forcing in 2D climate simulations. Atmosfera, Vol.10, N.2, p.75–100.
Delworth T.L., Manabe S., Stouffer R.J. (1997). Multidecadal climate variability in the Greenland Sea and surrounding regions. A coupled model simulation. Geophys. Res. Lett., Vol.24, p.257–260.
Demirchian K.S., Demirchian K.K (1995). Calculated global carbon cycle model. Izv. Russian Acad. Sci. Ser. Energetics, N.5, p.3–15 (in Russian).
Dijkstra H.A., Neelin J.D. (1995). Ocean–atmosphere interaction and the tropical climatology. Part II: Why the Pacific cold tongue is in the East? J. Climate, Vol.8, N.5, p.1343–1359.
Doerrel P. (1996). Die CO_2 — Steuer wäre eine Katastrophe. Brennstoftspiegel, N.12, p.14–15.

Dong B., Valdes P.J. (1995). Sensitivity studies of Northern Hemisphere glaciation using an atmospheric general circulation model. J. Climate, Vol.8, N.10, p.2471–2496.

Douville H., Royer J.-F. (1997). Influence of the temperate and boreal forecasts on the Northern Hemisphere climate in the Météo-France climate model. Climate Dynamics, Vol.13, N.1, p.57–74.

Downing T.E. (1996). Climate Change and World Food Security. Springer, Heidelberg, 662 pp.

Dutton J.A. (1995). An analytical model of atmospheric feedback and global temperature change. J. Climate, Vol.8, N.5, p.1122–1139.

Eckman R.S., Grose W.L., Turner R.E., Blackshear W.T. (1996). Polar ozone depletion: A three-dimensional chemical modeling study of its long-term global impact. J. Geophys. Res., Vol.101, N.D17, p.22977–22990.

Eckman R.S., Grose W.L., Turner R.E., Blackshear W.T., Russel J.M.III, Froidevaux L., Waters J.W., Kumer J.B., Roche A.E. (1995). Stratospheric trace constituents simulated by a three-dimensional general circulation model: Comparison with UARS data. J. Geophys. Res., Vol.100, N.D7, p.13951–13966.

Egger J. (1997). Flux correction: tests with a simple ocean–atmosphere model. Climate Dynamics, Vol.13, N.4, p. 285-292.

Ehrmann M. (1997) Der Genfer Klimagipfel. Natur und Recht, Vol.19, N.5, p.229–233.

Eltahir E.A.B. (1996). Role of vegetation in sustaining large-scale atmospheric circulations in the tropics. J. Geophys. Res., Vol.101, N.D2, p.4255–4268.

Emsley J. (1996). Global warming: The Report of the European Science and Environment Forum. Introduction. In: Emsley J. (ed.), The Global Warming Debate. The Report of the European Science and Environment Forum. Bourne Press Ltd, Bournemouth, p.22–27.

Epperson D.L., Davis J.M., Bloomfield P., Karl T.R., McNab A.L., Gallo K.P. (1995). Estimating the urban bias of surface shelter temperatures using upper-air and satellite data. Part II: Estimation of the urban bias. J. Appl. Meteorol., Vol.34, N.2, p.358–370.

Fearnside P.M. (1997). Greenhouse gases from deforestation in Brazilian Amazonia: Net committed emissions. Climatic Change, Vol.35, N.3, p.321–360.

Felix R.W. (1997). Not by Fire but by Ice. Discover What Killed the Dinosaurs . . . and Why it Could Soon Kill Us. Sugarhouse Publ., Bellevue, WA, 255 pp.

Feng X., Epstein S. (1996). Climatic trends from isotropic records or tree rings: The past 100–200 years. Climatic Change, Vol.33, N.4, p.551–562.

Flatøy F., Hov Ø. (1996) Three-dimensional model studies of the effect of NO_x emissions from aircraft on ozone in the upper troposphere over Europe and the North Atlantic. J. Geophys. Res., Vol.101, N.D1, p.1401–1422.

Fleagle R.G. (1996). The course of US responses to global change: Can't we do better? An Editorial Essay. Climatic Change, Vol.33, N.4, p.447–452.

Francis P.N., Taylor J.P., Hignett P., Slingo A. (1997). On the question of enhanced absorption of solar radiation by clouds. Quart. J. Roy. Met. Soc., Vol.123, Part B., N.538, p.419–434.

Fronval T., Jansen E. (1996). Rapid changes in ocean circulation and heat flux in the

Nordic Seas during the last interglacial period. Nature (Gr. Brit.), Vol.383, N.6603, p.806–610.
Fuglestvedt J.S., Isaksen I.S.A., Wang W.-C. (1996). Estimates of indirect global warming potentials for CH_4, CO and NO_x. Clim. Change, Vol.34, N.3-4, p.405–437.
Gautier C., Landsfeld M. (1997). Surface solar radiation flux and cloud radiative forcing for the Atmospheric Radiation Measurement (ARM) Southern Great Plains (SGP): A satellite, surface observations, and radiative transfer model study. J. Atmos. Sci., Vol.54, N.10, p.1289–1307.
Glantz M.H. (1996). Currents of Change: El Niño's Impact on Climate and Society. Cambridge Univ. Press, xiii + 194 pp.
Global Climate Observing System (1995). Plan for the Global Climate Observing System (GCOS). Version 1.0 GCOS-14. WMO/TD, N.681. Geneva, 49 pp.
Global Climate Observing System. (1996). Plan for Space-Based Observations. Version 1.0. GCOS-15. WMO/TD, N.684, 79 pp.
Goldberg M.D., Fleming H.E. (1995). An algorithm to generate deep-layer temperatures from microwave satellite observations for the purpose of monitoring climate change. Climate Change, Vol.8, N.5, p.993–1004.
Goody R. Abdon W. (1996). Reversible and irreversible sources of radiation entropy. Quart. J. Meteorol. Soc., Vol.122, Part B, N.530, p.483–494.
Gray W.N., Sheaffer J.D. (1996). Global scale climate trends associated with variable Atlantic thermohaline transport as inferred from changes in intense hurricane activity. Preprints. Seventh Symp. Global Change Studies. Amer. Meteorol. Soc., Boston, MA, p.92–97.
Greenhouse forecasting still cloudy. Model gets it right—without fudge factors (1997). Science, Vol.276, N.5315, p.1040–1042.
Gu D., Philander S.G.H. (1997). Interdecadal climate fluctuations that depend on changes between the tropics and extratropics. Science, Vol.275, p.805–809.
Gualdi S., Navarra A., von Storch H. (1997). Tropical intraseasonal oscillation appearing in operational analysis and in a family of general circulation models. J. Atmos. Sci., Vol.54, N.9, p.485–1202.
Guilyardi E., Madec G. (1997). Performance of the OPA/ARPEGE-T21 global ocean–atmosphere coupled model. Climate Dynamics, Vol.13, N.2, p.149–165.
Gunst R.F., Sabyasachi B., Brunell R. (1992). Defining and estimating global mean temperature anomalies. J. Climate, Vol.6, p.1368–1374.
Guyot G. (1997). Physics of the Environment and Climate. John Wiley & Sons Ltd, Chichester, UK, 256 pp.
Haigh J.D. (1996). The impact of solar variability on climate. Science, Vol.272, p.981–984.
Haigh J.D. (1996). Stratospheric ozone and the impact of solar variability on climate. Proc. of the First SPARC Gen. Assembly, Melbourne, 4 pp.
Hall A., Manabe S. (1997). Can local linear stochastic theory explain sea surface temperature and salinity variability? Climate Dynamics, Vol.13, p.167–186.
Hanna E. (1996). The role of Antarctic sea ice in global climate change. Progress in Physical Geography, Vol.20, N.4, p.371–401.
Hannachi A. (1997). Low-frequency variability in a GCM-three-dimensional flow regimes and their dynamics. J. Climate, Vol.10, N.6, p.1357–1379.

Hansen J., Sato M., Ruedy R. (1997). Radiative forcing and climate response. J. Geophys. Res., Vol.102, N.D6, p.6831–6864.
Harris R.N., Chapman D.S. (1997). Borehole temperature and a baseline for 20th-century global warming estimates. Science, Vol.275, p.1618–1620.
Harvey L.D.D. (1996a). Polar boundary layer plumes and bottom water formation: A missing element in ocean general circulation models. J. Geophys. Res., Vol.101, N.C9, p.20799–20808.
Harvey L.D.D. (1996b). Development of a risk-hedging CO_2 -emission policy. Part I: Risks of unrestrained emissions. Climatic Change, Vol.34, N.1, p.1–40.
Harvey L.D.D. (1996c). Development of a risk-hedging CO_2 -emission policy. Part II: Risks associated with measures to limit emissions, synthesis, and conclusions. Climatic Change, Vol.34, N.1, p.41–71.
Harvey M.W., Fraser G.J. (1996). Dynamical influence on the Antarctic ozone hole. Weather, Vol.51, N.8, p.266–273.
Hasselmann K. (1997). Are we seeing global warming? Science, Vol.276, N.5315, p.914–916.
Hasselmann K. (1997). Multi-pattern fingerprint method for detection and attribution of climate change. Climate Dynamics (in press).
Hayashi Y. Golder D.G. (1997). United mechanisms for the generation of low- and high-frequency tropical waves. Part I. Control experiments with moist convective adjustment. J. Atmos. Sci., Vol.54, N.9, p.1262–1276.
Haywood J.M., Stouffer R.J., Wetherald R.T., Manabe S., Ramaswamy V. (1997). Transient response of a coupled model to estimated changes in greenhouse gas and sulfate concentrations. Geophys. Res. Lett., Vol.24, N.11, p.1335–1338.
Hegerl G.C., North G.R. (1997). Comparison of statistically optimal approaches to detecting anthropogenic climate change. J. Climate, Vol.10, N.5, p.1125–1133.
Hegerl G.C., Hasselmann K., Cubasch U., Mitchell J.F.B., Roeckner E., Vass R., Waszkewitz J. (1997). Multi-fingerprint detection and attribution analysis of greenhouse gas, greenhouse gas-plus-aerosol and solar forced climate change. Climate Dynamics (in press).
Heimann M. (1996). Closing the atmospheric CO_2 budget: influences from new measurements of $^{13}C/^{12}C$ and O_2/N_2 radius. Global Change Newsletter, N.28, p.9–11.
Henderson-Sellers A. (1997). Can we integrate climatic modelling and assessment? Environmental Modelling and Assessment, Vol.1, N.1,2, p.59–70.
Hobbs P. (1997). Comments on "An Open letter to Ben Santer" and the IPCC Report. Bull. Amer. Meteorol. Soc., Vol.78, N.1, p.83.
Hobbs P.V., Reid J.S., Kotchenruther R.A., Ferek R.J., Weiss R. (1997). Direct radiative forcing by smoke from biomass burning. Science, Vol.275, p. 1776-1779.
Horton E.B., Parker D.E. (1997). Global and regional climate in 1996. Weather, Vol.52, N.6, p.174–181.
Hoyle F. (1996). The great greenhouse controversy. In: Emsley J. (ed.). The Global Warming Debate, Bourne Press Ltd, Bournemouth, p.179–189.
Hoyt D.V., Schatten K.H. (1997). The Role of the Sun in Climate Change. Oxford University Press, 279 pp.
Huffman G.J., Adler R.F., Rudolf B., Schneider U., Keehn P.R. (1995). Global

precipitation estimates based on a technique for combining satellite-based estimates, rain gauge analysis, and NWP model precipitation information. J. Climate, Vol.8, N.5, p.1284–1295.

Huffman G.J., Adler R.F., Arkin P., Chang A., Ferraro R., Gruber A., Janowich J., McNab A., Rudolf B., Schneider U. (1997). The Global Precipitation Climatology Project (GPCP) combined precipitation data set. Bull. Amer. Meteorol. Soc., Vol.78, N.1, p.5–20.

Hughen K.A., Overpeck J.T., Peterson L.C., Trumbore S. (1996). Rapid climate changes in the topical Atlantic region during the last deglaciation. Nature (Gr. Brit.), Vol.380, N.6569, p.51–54.

Hughes T. (1996). Can ice sheets trigger abrupt climate change? Arct. and Alp. Res. Vol.28, N.4, p.448–465.

Hughes M.G., Robinson D.A. (1996). Historical snow cover variability in the Great Plains region of the United States: 1910 through 1993. Int. J. Climatol., Vol.16, N.9, p.1005–1018.

Human effect on global climate? (1996). Comments on the paper by B.D. Sunter et al. Michaels P.J., Knappenberg P.C., Weber G.R. Authors reply. Santer B.D., Boyle J.S., Parker D.E. Nature (Gr. Brit.), Vol.384, N.6609, p.522–524.

Hunt B.G., Davies H.L. (1997). Mechanisms of multidecadal climatic variability in a global climatic model Int. J. Climate, Vol.17, N.6, p.565–580.

Iacono M.J., Clough S.A. (1997). Application of infrared interferometer spectrometer clear sky spectral radiance to investigations of climatic variability. J. Geophys. Res., Vol.101, N.D23, p.29439–29460.

Jackman C.H., Fleming E.L., Chandra S., Considine D.B., Rosenfield J.E. (1996). Past, present, and future modeled ozone trends with comparisons to observed trends. J. Geophys. Res., Vol.101, N.D22, p.28,753–28,768.

Jamieson D. (1996). Ethics and International Climate Change. Climatic Change, Vol.33, N.3, p.323–336.

Jaworowski Z. (1996). Reliability of ice core records for climatic projections. In: Emsley J. (ed.), The Global Warming Debate. Bourne Press Ltd, Bournemouth, p.95–105.

Jaworowski Z. (1997). Ice core data show no carbon dioxide increase. 21st Century Science and Technology, Vol.10, N.1, p.42-52.

Johannessen O.M., Biørgo E., Miles M.W. (1996). Global Warming and the Arctic. Science, Vol.271, N.5245, p.129.

Johns T.C., Carnell R.E., Crossley J.E., Gregory J.M., Mitchell J.F.B., Senior C.A., Tett S.F.B., Wood R.A. (1997). The Second Hadley Centre coupled ocean–atmosphere GCM: model description, spinup and validation. Climate Dynamics, Vol.13, p.103–134.

Kacholia K., Reck R.A. (1997). Comparison of global climate change simulations for $2CO_2$ induced warming. An intercomparison of 108 temperature change projections published between 1980 and 1995. Climate Change, Vol.35, N.1, p.53–69.

Karl T.R. (1996). The IPCC Scientific Assessment of Climate Change: observed climate variability and change. Preprints. Seventh Symp. Global Change Studies. Amer. Meteorol. Soc., Boston, MA, p.7–13.

Kasting J.F. (1997). Warming early Earth and Mars. Science, Vol.276, N.5316, p.1213–1215.

Kaufmann R.K. (1997). Assessing the dice model: uncertainty associated with the emission and retention of greenhouse gases. Clim. Change, Vol.35, N.4, p.435–448.

Keeling R.F., Piper S.C., Heinmann M. (1996). Global and hemispheric CO_2 sinks deduced from changes in atmospheric O_2 concentration. Nature, Vol.381, p.218–221.

Kelly P.M., Jones P.D. (1996). Removal of the El Niño-Southern Oscillation signal from the gridded surface air temperature data set. J. Geophys. Res., Vol.101, N.D14, p.19013–19022.

Kheshgi H.S., Schlesinger M.E., Lapenis A.G. (1997). Comparison of paleotemperature reconstructions as evidence for the paleo-analog hypothesis. Clim. Change, Vol.35, N.1, p.123–129.

Kiehl J.T., Trenberth K.E. (1997). Earth's annual global mean energy budget. Bull. Amer. Meteorol. Soc., Vol.78, N.2, p.197–208.

Knutson T.R., Manabe S., Gu D. (1997). Simulated ENSO in a global coupled ocean–atmosphere model: multidecadal amplitude modulation and CO_2 sensitivity. J. Climate, Vol.10, N.1, p.138-161.

Kogan F.N. (1997). Global drought watch from space. Bull. Amer. Soc., Vol.78, N.4, p.621–636.

Kondratyev K.Ya. (1996). Greenhouse warming versus aerosol cooling in the context of global climate change: contribution of satellite observations. World Resource Review, Vol.8, N.4, p.456–466.

Kondratyev K.Ya. (1997). Comments on "Open Letter to Ben Santer". Bull. Amer. Meteorol. Soc., Vol.78, N.4, p.689–691.

Kondratyev K.Ya., Varotsos C. (1997). A review on greenhouse effect and ozone dynamics over Greece. In: Atmospheric Ozone Dynamics. Observations in Mediterranean Region. (ed.), by C. Varotsos. Springer, Berlin, p.175–228.

Kukla G., McManus J.F., Rousseau D.D., Chuine I. (1997). How long and stable was the last interglacial? Quaternary Sci. Revs, Vol.16, N.6, p.605–612.

Kurganski M.V., Dethloff K., Pisnichenko I.A., Gernardt H., Chmielewski F.-M., Jansen W. (1996). Long-term climate variability in a simple, non-linear atmospheric model. J. Geophys. Res., Vol.101, N.D2, p.4299–4314.

Lambert S.S. (1997). The effect of enhanced greenhouse warming on winter cyclone frequencies and strengths. J. Climate, Vol.10, N.5 (Part 2), p.1447–1452.

Lanchbery J., Victor D. (1995). The role of science in the global climate negotiations. In: Bergesen H.E., Partmann (eds.), Green Global Yearbook of International Co-operation and Development. Oxford University Press, p.29–40.

Lanzante J.R. (1996). Resistant, robust and non-parametric techniques for the analysis of climate data: Theory and examples, including applications to historical radiosonde station data. Int. J. Climatol., Vol.16, N.11, p.1197–1226.

Latif M. (1997). Dynamics of interdecadal variability in coupled ocean–atmosphere models. Int. WOCE Newsletter, N.26, p.17–24.

Lean J., Rowntree P.R. (1997). Understanding the sensitivity of a GCM simulation of Amazonian deforestation to the specification of vegetation and soil characteristics. J. Climate, Vol.10, N.6, p.1216–1235.

Lefohn A.S. (ed.) (1992). Surface Level Ozone Exposures and Their Effects on Vegetation. Lewis Publ., Chelsea, MI, 366 pp.

Leslie L.M., Fraedrich K. (1997). A new general circulation model: formulation and preliminary results in a single- and multi-processor environment. Climate Dynamics, Vol.13, N.1, p.35–43.

Lindzen R.S., Kirtman B., Kirkdavidoff D., Schneider E.K. (1997). Seasonal surrogate for climate. J. Climate, Vol.10, N.6, p.1681–1684.

Lupo A.R., Oglesby R.J., Mokhov I.I. (1997). Climatological features of blocking anticyclones: a study of Northern Hemisphere CCM 1 model blocking events in present-day and double CO_2 concentration atmosphere. Climate Dynamics, Vol.13, N.3, p.181–195.

Mabey N., Hall S., Smith C., Gupta S. (1997). Argument in the Greenhouse. The International Economics of Controlling Global Warming. Routledge, London, 464 pp.

MacKay R.M., Ko M.K.W., Shia R.-L., Yang Y., Zhan Sh., Molnar G. (1997). An estimation of the climatic effects of stratospheric ozone losses during the 1980s. J. Climate, Vol.10, N.4, p.774–788.

Mahasenan N., Watts R.G., Downlatabi H. (1997). Low-frequency oscillations in temperature-proxy records and implications for recent climate change. Geophys. Res. Lett., Vol.24, p.563–566.

Mahlman J.D. (1997). Dynamics of transport processes in the upper troposphere. Science, Vol.276, N.5315, p.1079–1084.

Makhijani A. Gurney K. (1995). Mending the Ozone Hole. Science, Technology and Policy. MIT Press, 360 pp.

Manabe S. (1997). Early development in the study of greenhouse warming: the emergence of climate models. Ambio, Vol.26, N.1, p.47–51.

Manabe S., Stouffer R.J. (1997). Climate variability of a coupled ocean–atmosphere–land surface model: implication for the detection of global warming. the Walter Orr Roberts Lecture. Bull. Amer. Meteorol. Soc. Vol.78, N.6, p.1177–1185.

Masera O.R., Ordóñez M.J., Dirzo R. (1997). Current emissions from Mexican forests: current situation and long-term scenarios. Clim. Change, Vol.35, N.3, p.265–295. Mayewski P.A., Dibb J.E., Mecker C.D. et al. (1995). Highlights of the GISP2 glaciochemical records: Climate change and atmospheric chemistry. Ant. Res. US, Vol.9, Fall–Winter, p.32–36.

McBean G.A., Liss P.S., Schneider S.H. (1996). Advancing our understanding. Preprints Seventh Symp. Global Change Studies. Amer. Meteorol. Soc., Boston, MA, p.41–48.

Mearns L.O., Rosenzweig C., Goldberg R. (1997). Mean and variance change in climate scenarios: methods, agricultural applications, and measures of uncertainty. Clim. Change, Vol.35, N.4, p.367–396.

Meehl G.A. (1996). Vulnerability of freshwater resources to climate change in the tropical Pacific region. Water, Air and Soil Pollution, Vol.92, p.203–213.

Meehl G.A., Kiladis G.N., Weickmann K.M., Wheeler M., Gutzler D.S., Compo G.P. (1996). Modulation of equatorial subseasonal convective episodes by tropical–extratropical interaction in the Indian and Pacific Ocean regions. J. Geophys. Res., Vol.101, N.D10, p.15033–15049.

Meehl G.A., Washington W.M. (1996). El Niño-like climate change in a model with increased atmospheric CO_2 concentrations. Nature, Vol.382, p.56–60.

Michaels P.J., Knappenberger p. C. (1996). The United Nations Intergovernmental Panel on Climate Change and the scientific "consensus" on global warming. In: Emsley J. (ed.), The Global Warming Debate. Bourne Press Ltd, Bournemouth, p.158–178.

Mierzejewska A.W., Wu Z., Newell R.E., Miyashita T. (1997). Japanese whaling ships' sea surface temperatures 1946–84. Bull Amer. Meteorol. Soc., Vol.78, N3, p.442–447.

Miller R.L. (1997). Tropical thermostats and low cloud cover. J. Climate, Vol.10, N.3, p.409–440.

Miller G.H., Magee J.W., Jull A.J.T (1997). Low-latitude glacial cooling in the Southern Hemisphere from amino-acid racemization in emu eggshells. Nature (Gr. Brit.), Vol.185, N.6613, p.241–244.

Mitchell J.G.B., Johns T.C. (1997). On modification of global warming by sulphate aerosols. J. Climate, Vol.10, N.2, p.245–267.

Moene A. (1996). Random natural global temperature changes superior to anthropogenic signals. In: Emsley J. (ed.), The Global Warming Debate. Bourne Press Ltd, Bournemouth, p.139–143.

Mudge F.B. (1997). The development of the "greenhouse" theory of global climate change from Victorian times. Weather, Vol.52, N.1, p.13–17.

Munhoven G., François L.M. (1996). Glacial-interglacial variability of atmospheric CO_2 due to changing continental silicate risk weathering: A model study. J. Geophys. Res., Vol.101, N.D16, p.21423–21437.

Müller R., Crutzen p. J., Grooss J.-U., Brühl C., Russell J.M., Tuck A.F. (1996). Chlorine activation and ozone depletion in the Arctic vortex: Observations by the Halogen Occultation Experiment on the Upper Atmosphere Research Satellite. J. Geophys. Res., Vol.101, N.D7, p.12531–12554.

Munk W., Worcester P., Wunch C. (1995). Ocean Tomography. Cambridge University Press, 433 pp.

Murray D.R. (1997). Carbon Dioxide and Plant Responses. Research Study Press, UK, 250 pp.

Nakićenović N., Nordhaus W.D., Richels R., Toth F.L. (eds.). (1994). Integrative Assessment of Mitigation, Impacts, and Adaptation to Climate Change. IIASA, Laxenburg, 669 pp.

Nilsson A. (1996). Ultraviolet Reflections: Life Under a Thinning Ozone Layer. John Wiley & Sons Ltd, Chichester, UK, 164pp.

O'Riordan T., Jäger J. (eds.). (1996). Politics of Climate Change. A European Perspective. Routledge, London, 396 pp.

Orsolini Y.J., Hansen G., Hoppe U.-P., Manney G.L., Fricke K.H. (1997). Dynamical modelling of wintertime lidar observations in the Arctic: Ozone laminae and ozone depletion. Q.J. Roy. Meteorol. Soc., Vol.123, Part A, N.539, p.785–800.

Palmer T.N. (1996) Predictability of the atmosphere and oceans: from days to decades. NATO Adv. Study Inst. "Decadal Climate Variability: Dynamics and Predictability". Springer, Berlin, p.82–141.

Parker D.E., Folland C. (1997). The "30 years war": More on climate reference period. Weather, Vol.52, N.1, p.26–27.

Parry M., Carter T. (1997). Climate Impact and Adaptation Assessment. The IPCC Method. Earthscan Publ. Ltd, London, 192 pp.

Parson E.A. (1996). Three dilemmas in the integrated assessment of climatic change. An Editorial Comment. Clim. Change, Vol.34, Nos.3–4, p.315–326.

Parsons M. (1995). Global Warming: The Truth Behind the Myth. Plenum Press, New York, 271 pp.

Paté-Cornell E. (1996). Uncertainties in global climate change estimates. An Editorial Essay. Climatic Change, Vol.33, N.2, p.145–149.

Pavlich M.J. (1996). Marine–terrestrial correlation as a means of testing global climate change. US Geol. Surv. Circ., N.1119, p.61–66.

Pelletier J.D. (1997). Analysis and modeling of the natural variability of climate. J. Climate, Vol.10, N.6, p.1331–1342.

Peñaloza M.A. (1996). A determination of the greenhouse parameter for dry and unpolluted air. Atmosfera, Vol.9, N.2, p.99–118.

Perry A. (1996). IPCC second assessment and beyond. Progress in Physical Geography, Vol.20, N.4, p.477–481.

Peters D., Egger J., Entzian G. (1995). Dynamical aspects of ozone mini-hole formation. Meteorol. and Atmos. Phys., Vol.55, N.3–4, p.205–214.

Portmann R.W., Solomon S., Garcia R.R., Thomasson L.W., Poole L.R., McCormick M.P. (1996). Role of aerosol variations in anthropogenic ozone depletion in the polar regions. J. Geophys. Res., Vol.101, N.D17, p.22991–23006.

Prabhakara C., Yoo J.-M., Maloney S.P., Nucciarone J.J., Arking A. (1996). Examination of Global Atmospheric Temperature Monitoring with Satellite Microwave Measurements: 2. Analysis of satellite data. Climatic Change, Vol.33, N.4, p.459–476.

Preparing for the Future Through Science and Technology. An Agenda for Environmental and Natural Resource Research (1995). National Science & Technology Council. Washington DC.

Preparing for an Uncertain Climate Summary. (1997). Office of Technology Assessment. US Congress. Washington, DC, 63 pp.

Proceedings of the Sixth Atmospheric Radiation Measurement (ARM) Science Team Meeting (1997). US Department of Energy, Washington DC, 392 pp.

Przybylak R. (1997). Spatial and temporal changes in extreme air temperatures in the Arctic over the period 1951–1990. Int. J. Climatol., Vol.17, N.6, p.615–634.

Pszenny A., Brasseur G. (1997). Tropospheric ozone: an emphasis of IGAC research. Global Change Newsletter, N.30, p.2-10.

Putaud J.-Ph., Ngnyen B.C. (1996). Assessment of dimethylsulphide sea–air exchange rate. J. Geophys. Res., Vol.101, N.D2, p.4403–4411.

Putnam S.N., Chapman D.S. (1996). A geothermal climate change observatory: First year results from Emigrant Pass in northern Utah. J. Geophys. Res., Vol.101, N.B10, p.21877–21890.

Räisänen J. (1997). Objective comparison of patterns of CO_2 induced climate change in coupled GCM experiments. Climate Dynamics, Vol.13, N.3, p.197–212.

Ramaswamy V., Chen C.-T. (1997). Linear additivity of climate response for combined albedo and greenhouse perturbations. Geophys. Res. Lett., Vol.24, p.567–560.

Randall D.A., Wielicki B.A. (1997). Measurements, models, and hypothesis in the atmospheric science. Bull. Amer. Meteorol. Soc., Vol.78, N.3, p.399–406.

Ravishankra A.R. (1997). Heterogeneous and multiphase chemistry in the troposphere. Science, Vol.276, N.5315, p.1058–1065.

Report of Twelfth Session of the CAS/JSC Working Group on Numerical Experimentation (Japan Meteorol. Agency, Tokyo, Japan, 28–31 October 1996). (1997).CAS/JSC WGNE report N.12, Geneva, 31 pp.

Research Activities in Atmospheric and Oceanic Modelling (1997). Ed. by A. Staniforth. Report N.25. WMO/TD N.792, Geneva, WMO.

Rind D., Lonergen P., Shah K. (1996). Climate effect of water vapor release in the upper troposphere. J. Geophys. Res., Vol.101, N.D23, p.29395–29405.

Ringer M.A. (1997). Interannual variability of the earth's radiation budget: some regional studies. Int. J. Climatol., Vol.17, N.9, p.929–952.

Russell J.M. III, Luo M., Cicerone R.J., Deaver L.E. (1996). Satellite confirmation of the dominance of chlorofluorocarbons in the stratospheric chlorine budget. Nature, Vol.379, N.6565, p.526–529.

Rybak O.O. (1997). Statistical structure of air surface temperature time series. Atmosfera, Vol.10, N.2, p.59–74.

Sagan C., Chyba C. (1997). The early faint Sun paradox: organic shielding of ultraviolet-labile greenhouse gases. Science, Vol.276, N.5316, p.1217–1221.

Sakai K., Peltier W.R. (1997). Dansgaard-Oeschger-oscillations in a coupled atmosphere–ocean climate model. J. Climate, Vol.10, N.5, p.949–970.

Schafer J.S., Saxena V.K., Wenny B.N., Barnard W., DeLuisi J.J. (1996). Observed influence of clouds on ultraviolet-B radiation. Geophys. Res. Lett., Vol.23, N.19, p.2625–2628.

Schiller A., Nikolajewicz U., Voss R. (1997). The stability of the North Atlantic thermohaline circulation in a coupled ocean–atmosphere general circulation model. Climate Dynamics, Vol.13, N.5, p.325–347.

Schneider E.K., Lindzen R.S., Kirtman B.P. (1997). A tropical influence on global climate. J. Atmos. Sci., Vol.54, N.10, p.1349–1358.

Schoeberl M.R., Douglass A.R., Kawa R., Dessler A.E., Newman P.A., Stolarski R.S., Roche A.E., Waters J.W., Russell J.M., III. (1996). Development of the Antarctic ozone hole. J. Geophys. Res., Vol.101, N.D15, p.20909-20924.

Schröder W., Treder H.-J. (1997). Doubts over the sunspot–climate link. Weather, Vol.52, N.2, p.60–61.

Segalstad T.V. (1996). The distribution of CO_2 between atmosphere, hydrosphere, and lithosphere; minimal influence from anthropogenic CO_2 on the global greenhouse effect. In: Emsley J. (ed.), The Global Warming Debate. Bourne Press Ltd, Bournemouth, p.41–50.

Sellers P.S., Dickinson R.E., Randall D.A., Betts A.K., Hall F.G., Rerry J.A., Collatz C.J., Denning A.S., Mooney H.A., Nobre C.A., Sato N., Field C.B., Henderson-Sellers A. (1997). Modeling the exchanges of energy, water, and carbon between continents and the atmosphere. Science, Vol.275, p.502–509.

Semenov M.A., Barrow E.M. (1997). Use of a stochastic weather generator in the development of climate change scenarios. Clim. Change, Vol.35, N.4, p.397–414.

Singer S.F. (1996). Eine vorläufige Kritik am "Second Assessment Report on Climate Change" des IPCC. Fusion (BRD), Vol.17, N.3, p.11–15.

Singer S.F., Boe B.A., Decker F.W., et al. (1997). Comments on "Open letter to Ben Santer". Bull. Amer. Meteorol. Soc., Vol.78, N.1, p.81–82.

Singer S.F. (1996). A preliminary critique of IPCC's Second Assessment of Climate Change. In: Emsley J. (Ed.), The Global Warming Debate, Bourne Press Ltd, Bournemouth, p.146–157.

Singer S.F. (1997). Global Warming: Unfinished Business (in press).

Slingo J.M., Sperber K.R., Boyle J.S., Ceron J.P., Dix M., Dudas B., Ebisuzuki W., Fyle J., Gregory D., Gueremy J.-F., Hack J., Harzallah A., Innes P., Kitoh A., Lan W.K.-M., McAverney B., Madden R., Matthews A., Palmer T.N., Park C.-K., Randall D., Renno N. (1996). Intraseasonal oscillation in 15 atmospheric general circulation models: results from an AMIP diagnostic subproject. Climate Dynamics, Vol.12, N.5, p.325–257.

Slingo A., Webb M.J. (1997). The spectral signature of global warming. Q.J. Roy. Meteorol. Soc., Vol.123, Part B, N.538, p.293–307.

van der Sluijs J.P., De Bruyn G.J., Westbroek P. (1996). Biogenic feedbacks in the carbonate–silicate geochemical cycle and the global climate. Amer. J. Sci., Vol.296, N.8, p.932–953.

van der Sluijs J.P. (1997). Anchoring Amid Uncertainty. On the management of uncertainties in risk assessment of anthropogenic climate change. Leiden, Ludy Feyen, 260 pp.

Smith I.N., Dix M., Allan R.J. (1997). The effect of greenhouse SSTs on ENSO simulations with an AGCM. J. Climate, Vol.10, N.2, p.342–352.

Soden B.J. (1997). Variations in the tropical greenhouse effect during El Niño. J. Climate, Vol.10, N.5, p.1050–1055.

Sonechkin D.M., Dacenko N.M., Ivaschenko N.N. (1997). Estimation of global warming trend with the use of wavelet analysis. Proc. Russian Acad. Sci. Physics of the Atmos. and Ocean, Vol.33, N.2, p.184–194 (in Russian).

Sonnenflecken, nicht CO_2, strenern das Erdklima (1996). Fusion (BRD), Vol.17, N.3, p.31.

Soon W.H., Posmentier E.S., Baliunas S.L. (1996). Inference of solar irradiance variability temperature changes., 1880-1993: an astrophysical application of the sun–climate connection. The Astrophys. J., Vol.472, p.891–902.

Spencer R.W., Christy J.R., Grody N.C. (1996). A technical comment: Analysis of "Examination of 'Global Atmospheric Temperature Monitoring with Satellite Microwave Measurements' ". Climatic Change, Vol.33, N.4, p.477–489.

Spencer R.W., Braswell W.D. (1997). How dry is the tropical free troposphere? Implications for global warming theory. Bull. Amer. Meteorol. Soc., Vol.78, N.6, p.1097–1106.

Stevens M.J., North G.P. (1996). Detection of the climate response to the solar cycle. J. Atmos. Sci., Vol.53, N.18, p.2594–2608.

von Storch H., Navarra A. (eds.) (1995). Analysis of Climate Variability. Springer,

Berlin, 334 pp.
Straus D.M., Shukla J. (1997). Variations of midlatitude transient dynamics associated with ENSO. J. Atmos. Sci, Vol.54, N.7, p.777–790.
Syktus J., Chappell J., Oglesby R., Larson J., Marshall S., Saltzman B. (1997). Latitudinal dependence of signal-to-noise patterns from two general circulation models with CO_2 forcing. Climate Dynamics, Vol.13, N.5, p.293–302.
Taylor F.W. (1991). The greenhouse effect and climatic change. Rep. Prog. Phys., Vol.54, p.881–918.
Tett S., Johns T., Mitchell J., Gregory J. (1996). Global warming: natural change or human induced change? Preprints. Seventh Symp. Global Change Studies. Amer. Meteorol. Soc., Boston, MA, p.49–52.
Tett S.F.B., Johns T.C., Mitchell J.F.B. (1997). Global and regional variability in a coupled AOGCM. Climate Dynamics, Vol.13, N.5, p.303–325.
Thompson A.M., Tao W.-K., Pickering K.E., Scala J.R., Simpson J. (1997). Tropical deep convection and ozone formation. Bull. Amer. Meteorol. Soc., Vol.78, N.6, p.1043–1054.
Tie X., Brasseur G.P., Granier C., De Rudder A., Larsen N. (1996). Model study of polar stratospheric clouds and their effect on stratospheric ozone. 2. Model results. J. Geophys. Res., Vol.101, N.D7, p.12575–12584.
Timbal B., Mahfouf J.E., Royer J.F., Cubasch U., Murphys J.M. (1997). Comparison between doubled CO_2 time-slice and coupled experiments. J. Climate, Vol.10, N.6, p.1463–1469.
Tinsley B.A. (1997). Do effects of global atmospheric electricity on clouds cause climate changes? EOS, Vol.78, N.33, p.341–342,349.
Tinsley B.A., Beard K.V. (1997). Links between variations in solar activity, atmospheric conductivity, and clouds: an informal workshop. Bull. Amer. Meteorol. Soc., Vol.78, N.4, p.685–687.
Trenberth K.E. (1997). Short-term climate variations: recent accomplishments and issues for future progress. Bull Amer. Meteorol. Soc., Vol.78, N.6, p.1081–1096.
Twickler M. (1997). GISP2 ice core is long, continuous and detailed record. Witness the Arctic, Vol.5, N.1, p.1–2.
Valero F.P.J., Collins W.D., Pilewskie P., Bucholtz A., Flatau P.S. (1997). Direct radiometric observations of the water vapor greenhouse effect over the equatorial Pacific Ocean. Science, Vol.275, p.1773–1775.
Varotsos C., Kondratyev K.Ya., Katsikis S. (1995). On the relationship between total ozone and solar ultraviolet radiation at St. Petersburg, Russia. Geophys. Res. Lett., Vol.22, N.24, p.3481–3484.
Villalba R., Cook F.R., D'Arrigo R.D., Jacoby G.C., Jones P.D., Salinger M.J., Palmer J. (1997). Sea-level pressure variability around Antarctica since A.D. 1790 inferred from subantarctic tree-ring records. Climate Dyn., Vol.13, N.6, p.375–390.
Vitart F., Anderson J.L., Stern W.F. (1997). Simulation of interannual variability of tropical storm frequency in an ensemble of GCM integrations. J. Climate, Vol.10, N.4, p.745–760.
Wagner D. (1996). Scenarios of extreme temperature events. Climatic Change, Vol.33, N.3, p.385–407.

Weber G.-R. (1996). European temperature variations since 1525. In: Emsley J. (ed.), The Global Warming Debate, Bourne Press Ltd, Bournemouth, p.113–138.
Weber G.-R. (1997). Spatial and temporal variations of 300hPa temperatures in the northern hemisphere between 1966 and 1993. Int. J. of Climatology, Vol.17, p.1–15.
Weber G.-R. (1997). Smudged fingerprint: The elusive search for a human impact on the climate system. Manuscript Essen, Germany, 21 pp.
Wiin-Nielsen A. (1997). A note on hemispheric and global temperature changes. Atmosfera, Vol.10, N.3, p.125–136.
Williams M. (ed.) (1997). Climate Change Information Kit. UNEP's Information Unit for Conventions (UC). Châtelaine, Switzerland, 60 pp.
WMO Statement on the Status of the Global Climate in 1996. (1997). WMO, N,858, Geneva, 11 pp.
Wu D.L., Waters J.W. (1996). Satellite observations of atmospheric variances: a possible indication of gravity waves. Geophys. Res. Lett., Vol.23, N.24, p.3631–3634.
Xie Sh.-P. (1997). Stability of equatorial symmetric and asymmetric climates under annual solar forcing. Quart. J. Roy. Meteorol. Soc., Vol.123, Part A, N.541, p.1359–1376.
Xue Y. (1997). Biosphere feedback on regional climate in tropical north Africa. Quart. J. Roy. Meteorol. Soc., Vol.123, Part B, N.542, p.1428–1516.
Ye H., Mather J.R. (1997). Polar snow cover changes and global warming. Int. J. of Climate, Vol.17, N.2, p.155–162.
Ye D.-Z., Wu G.X. (eds.) (1996). Variability of the Climate System. Theoretical and Appl. Climatol., Vol.55, N.1–4, 228 pp.
Zhang R.-H., Levitus S. (1997). Interannual variability of the coupled tropical ocean–atmosphere system associated with the El Niño-Southern oscillation. J. Climate, Vol.10, N.6, p.1312–1330.
Zhang Y., Wallace J.M., Battisti D.C. (1997). ENSO like interdecadal variability. J. Climate, Vol.10, N.5, p.1004–1020.
Zillman J.W. (1997). Atmospheric science and public policy. Science, Vol.276, N.5315, p.1084–1092.
Zwick A. (1997). Climate Change Research and Policy: Updates. A periodic survey for the Commission of the European Communities. EC Joint Res. Centre Inst. for Prospective Technological Studies. Seville, N.9 (EUR 17303 EN), 76 pp.

CHAPTER 3

Billharz S., Moldan B. (eds.) (1996). Scientific Workshop on Indicators of Sustainable Development. Wuppertal, Germany, November 15–17, 1995. SCOPE Publ., Wuppertal, 69 pp.
Boehmer-Christiansen S. (1995). The politics of the European environmental policy. Energy & Environment, Vol.6, N.3, p.169–195.
Bonazzi M., Gomez Y., Paloma S. (1997). EU and the Mediterranean Region: a Future Dilemma. The IPTS Report, Vol.14, p.36–43.
Cheddadi R., Yu G., Guiot J., Harrison S.P., Prentice I.C. (1997). The climate of

Europe 6000 years ago. Climate Dynamics, Vol.13, N.1, p.1–9.
Eronen M., Zetterberg P. (1996). Climatic changes in Northern Europe since late glacial times, with special reference to dendroclimatological studies in northern Finnish Lapland. Geophysica, Vol.32, Vol.1–2, p.35–60.
Expanding the Measure of Wealth, Indicators of Environmentally Sustainable Development (1997). Environmentally Sustainable Development Studies and Monographs Series. N.17. The World Bank. Washington, DC, 109 pp.
Grigoryev A.A., Kondratyev K.Ya. (1996). Global change: indicators of sustainable development. Proc. Russian Geogr. Soc., Vol.128, N.4, p.26–37.
Jol A., Wieringa K. (eds.) (1996). Climate change in the European Union. Environmental Issues Series, N.2. European Environment Agency. Copenhagen, 26 pp.
Jol A., Kielland G. (eds.) (1997). Air Pollution in Europe 1997. EEA Environmental Monograph N4. European Environmental Agency, Copenhagen, 107 pp.
Jones R.G, Murphy J.M., Noguer M., Keen A.B. (1997). Simulation of climate change over Europe using a nested regional model. II: Comparison of driving and regional model responses to a doubling of carbon dioxide. Q.J. Roy. Meteorol. Soc., Vol.123, Part B., N.538, p.265–292.
Maslin M.A., Berger A. (1997). European view of the future of paleoclimatic research. Quarternary Sci. Revs., Vol.16, N.6, p.501–504.
Moldan B., Billharz S., Matravers R. (eds.) (1997). Sustainable Indicators: A Report on the Projection Indicators of Sustainable Development. John Wiley & Sons, Chichester, 415 pp.
Nonhebel S. (1996). Effects of temperature rise and increase in CO_2 concentration on simulated wheat yields in Europe. Climatic Change, Vol.34, N.1, p.79–90.
Piervitali E., Conte M., Colacino M. (1997). Summer air temperature anomalies in Europe during the century 1811–1910. Il Nuovo Cimento, Vol.20C, Ser.2, N.2, p.195–210.
Peterson G. (1996). The current and future direction of numerical models in simulating atmospheric long-range transport of heavy metals over Europe: a review. GKSS Report. Geesthacht, N.E73, 20 pp.
Stanners D., Bourdeau Ph. (eds.) (1995). Europe's Environment. The Dobriš Assessment. European Environment Agency. Copenhagen, xxvi + 712 pp.
Starkel L., Gutry-Korycka M. (eds.) (1996)). Global change, Polish Perspectives. Issue 3. Geographica Polonia, N.67, 139 pp.
Towards the 5th Framework Programme. Scientific and technological objectives. (1997). The European Commission. Brussels, 52 pp.
Wieringa K. (ed.) (1995). Report for the Review of the Fifth Environmental Action Programme. In: Environment in the European Union 1995. European Environment Agency, Copenhagen, 151 pp.

Index